Foundations of Decision Analysis

Ronald A. Howard
Stanford University

Ali E. Abbas
University of Southern California

PEARSON

Boston Columbus Indianapolis New York San Francisco Hoboken
Amsterdam Cape Town Dubai London Madrid Milan Munich Paris Montréal Toronto
Delhi Mexico City São Paulo Sydney Hong Kong Seoul Singapore Taipei Tokyo

Vice President and Editorial Director, ECS: *Marcia J. Horton*
Executive Editor: *Holly Stark*
Editorial Assistant: *Michelle Bayman*
Executive Marketing Manager: *Tim Galligan*
Marketing Assistant: *Jon Bryant*
Senior Managing Editor: *Scott Disanno*
Program Manager: *Erin Ault*
Project Manager: *Rose Kernan*
Operations Specialist: *Maura Zaldivar-Garcia*
Cover Designer: *Black Horse Designs*
Cover Photo: *Bull's Eye/ImageZoo/Prism/Corbis*
Manager, Rights and Permissions: *Rachel Youdelman*
Associate Project Manager, Rights and Permissions: *William Opaluch*
Composition: *Integra Software Services Pvt. Ltd.*
Printer/Binder: *Courier Westford*
Cover Printer: *Courier Westford*
Typeface: *10/12 Times LT Std*

Pearson Education Ltd., *London*
Pearson Education Singapore, Pte. Ltd
Pearson Education Canada, Inc.
Pearson Education—Japan
Pearson Education Australia PTY, Limited
Pearson Education North Asia, Ltd., *Hong Kong*
Pearson Educación de Mexico, S.A. de C.V.
Pearson Education Malaysia, Pte. Ltd.
Pearson Education, Inc., *Upper Saddle River, New Jersey*

Library of Congress Cataloging-in-Publication Data
Howard, Ronald A.
Foundations of decision analysis/Ronald A. Howard, Stanford University, Ali E. Abbas, University of Illinois at Urbana-Champaign.
pages cm
Includes bibliographical references and index.
ISBN-13: 978-0-13-233624-6
ISBN-10: 0-13-233624-3
1. Decision making. I. Abbas, Ali E. (Ali El-Sayed) II. Title.
HD30.23.H683 2014
003′.56—dc23 2013017028

1 2 3 4 5 6 7 8 9 10

www.pearsonhighered.com

ISBN-13: 978-0-13-233624-6
ISBN-10: 0-13-233624-3

BRIEF CONTENTS

PART 8 Characterizing What You Want

PART 9 Some Practical Extensions

PART 10 Computing Decision Problems

PART 11 Professional Decisions

PART 12 Ethical Considerations

CONTENTS

PREFACE

Decisions are the only means you have to change your future life. We make decisions every day. Some decisions are routine, like choosing a television program to watch. Occasionally we make decisions that have profound effects on us and those around us. Gaining competence in decision making is a highly desirable attainment. Although many of the principles of good decision making have been known for centuries, there is little emphasis on this subject throughout our educational lives. The important concepts in this book could be taught in grade school and in high school. Yet when we ask graduate students about having taken previous courses in decision making, few say that they have. Students in professional courses from major companies with global interests have the same response.

The purpose of this book is to provide an opportunity to gain this mastery; to be able to achieve clarity of action in making any decision on which you focus your attention. One of the biggest obstacles in gaining decision competence is that most of us think we are pretty good at making decisions. Yet it is easy to demonstrate that even in relatively simple decision situations people make decisions that they see as unwise when they carefully review them.

Let us preview the major conceptual lessons that we will share. The most challenging phenomenon we face in decision making is uncertainty. Suppose for each alternative we face in making a decision we had a video showing the future course of our lives in as much detail as we wished. Then we could easily make the decision. Uncertainty is inseparable from all significant decisions. To become masters of decision making we must become competent in dealing with uncertainty. We must learn to surf on the sea of uncertainty rather than to drown in it. We must build clear thinking about uncertainty as a precursor to making a decision; we call this achieving clarity of thought.

Learning how to deal with uncertainty does not mean that we do not relish it in our lives. Who would want to live a life with a future calendar that is completely filled out so that, for example, you would know years in advance on each day when, where, and with whom you would be having lunch.

The most important distinction of decision analysis is that between a decision and the outcome that follows it. This distinction, once thoroughly understood, is a powerful aid to achieving clarity of action. Though it is common for people who make a decision followed by an unfortunate outcome to see the decision as bad, this is not clear thinking. Good decisions can have bad outcomes; bad decisions can have good outcomes. The quality of the decision depends only on the quality of the thought and analysis that you have used in making it.

The amount of analysis appropriate to a decision can range from virtually none to extensive computer modeling. Everyone will have extensive conversations about making important decisions–sometimes with others, sometimes with oneself. Mastering the concepts of decision analysis will increase the focus and usefulness of these conversations. While few decisions will warrant the extensive analysis possible using these methods, merely thinking using the concepts in this book can improve many of the choices we make every day. We find that as students become acquainted with decision analysis it changes their conversation with friends and colleagues.

An important distinction about any decision is its degree of revocability. Some decisions are very revocable, like changing the movie you will see once you arrive at the multiplex; other decisions have limited revocability, such as amputating your leg for a medical purpose. The irrevocability of an important decision is a sign to invoke the power of decision analysis.

There is no point in valuing an outcome after the decision is made. You will be living the rest of your life beginning with that outcome or, as we prefer to call it, prospect. Once you commit yourself to making good decisions there is no place in your life for regret or guilt. *Good decisions never become bad; bad decisions never become good.*

Consider how evolution has prepared us for the modern world. What will befall an airplane pilot if he flies into the clouds and has no instruments? Soon he will think he is upside down when he is not, or not turning when he is. Without intervention he is likely to die. This is not a matter of his training or experience as a pilot, but rather that he is human. Humans never developed the ability to operate an aircraft without visual reference. Why not? Because before the invention of aircraft there was no evolutionary advantage to this ability. As long as you are standing on earth or swimming in the water, you know which way is down. Notice that birds that fly have no such problem. Duck hunters watching a flock fly into the clouds do not say, "Drop the guns, we will catch them as they fall." The many pilots of aircraft flying in bad visibility somewhere at this moment also have no difficulty, because the aircraft are equipped with instruments and the pilots are trained to read them. Even if a pilot feels he is upside down, his instruments show him that he is not.

Consider another example. For thousands of years, humans have been able to dive into deep water successfully for food or pearls. They would take a deep breath and hold it till they reached their goal and then return to the surface with their spoils. The development of scuba–self-contained underwater breathing apparatus–equipment has allowed people for many decades to do what only the most athletic of our ancestors could achieve. Suppose you are using scuba equipment and you have dived to a depth of 100 feet, about 3 atm, and then find that your equipment does not function. You are now far below the surface with only a lungful of air, air that is now extremely precious to you: you can't breathe. Your instinct is to head to the surface as quickly as possible and preserve what air you have. Unfortunately, following this instinct will probably kill you, for the lungful of air that you have will expand threefold by the time you reach the surface. This expansion will destroy the alveoli in your lung that allow you to breathe and admit air to your bloodstream. Following your instinct will kill you. Instructors point out that in this situation that as you slowly ascend, no faster than your smallest bubbles, you must blow out the air as you rise to avoid this misfortune. You must give up what is precious to you according to your instinct to save your life. (Of course it is even better to dive with a buddy who can assist you in these circumstances.) Here again before the invention of scuba there was no evolutionary advantage in having this be our natural behavior.

Finally, as you sit reading this, it is possible that under your chair there is a highly radioactive substance whose emissions will kill you by tomorrow. You have no alarm, since the ability to sense radiation was not of evolutionary value to our ancestors. If in our modern world you are concerned about the presence of radiation, there are many instruments that will warn you of its presence.

Now let us consider the evolutionary influence on decision making. While evolution has sensitized us to deal with judging the intentions of those we meet for millions of years, there is not any evolutionary knowledge of dealing with uncertainty. If each of us suddenly heard the roar of a live lion we would immediately react, though the noises of everyday life cause no alarm. There was an evolutionary advantage to being aware of dangers from other predatory forms of life that we have, fortunately, little reason to use frequently today. Yet someone can sign a paper having profound effects on his future welfare without alarm since making marks on a paper does not inspire the natural fear induced by the lion's roar.

Just as knowledge and proper instruments have helped us overcome our evolutionary disadvantages in these areas, so also can they help us in becoming better decision makers. The essential commitment is to use our instruments rather than trusting our intuition.

It is easy to show, and we do so in several occasions in this book, that our intuition on matters of uncertainty is severely flawed. Using our instruments is essential for clear thinking. No matter how long you have studied the subject, solving probability problems intuitively is as likely to be successful as a pilot flying in bad weather without visible reference and without instruments. The list of people who have made reasoning errors about uncertainty looks like the roll call of famous scientists.

Once uncertainty has been mastered, the next step is to use our instruments for making decisions in the face of uncertainty to arrive at clarity of action. The decision procedure will apply to virtually every decision that you face. Once a student in decision analysis said that he could see using the methods we were presenting for financial decisions, but not for medical decisions. We replied that if we had to choose between using it for financial decisions and for decisions about the health of a family member, then we would hire a financial advisor to manage money and use decision analysis for family medical decisions. The reason is that we would want to use the best decision method for the health of family and we know no better method than the one we present in what follows.

This book summarizes what we have learned by teaching decision analysis to thousands of people in the United States and around the world in university classes and special professional educational programs. Dozens of doctoral students and colleagues have contributed to its development. We intend for this book to extend the appreciation and application of this field, with roots in centuries past, to the decision-makers of the future.

HOW TO USE THIS BOOK

Decision making in our daily lives is an essential skill, whose fundamentals should not rely on knowing much more than arithmetic. Often you can make the decision using easily explained concepts without any calculations. We have therefore written the early chapters of this book and certain later chapters to be accessible to a general audience. Readers with more mathematical and computational preparation can benefit from the remainder of the book after understanding the fundamentals.

To be specific, Chapters 1 through 17, Chapters 26, 29, and 33, as well as Chapters 37 through 40 provide the foundations of decision analysis using reasoning. The story is not in the math: a decision maker can, step by step, transform confusion into clarity of thought and action.

Other chapters in this book are intended to expose readers to problems that require a higher level of analysis, such as problems that may appear in organizations. They are covered in Chapters 18 through 25, Chapter 27, Chapter 28, Chapters 30 through 32, and Chapters 34 through 36. While the analyses in these chapters require a higher level of computation, they rely on the basic principles presented in Chapters 1 through 17. No knowledge of calculus is essential to proper understanding of any of these chapters.

The "Decision Analysis Core Concepts Map" at the end of this book is a useful tool to help you understand some of the main concepts presented. You can use this map in several ways. First, it summarizes some of the important concepts, and so it can be used as a checklist for things you need to know. Second, it tells you the chronological order of concepts you need to understand before learning about another concept. An arrow from one concept to another helps you identify what you need to know before understanding a particular concept.

We do not require the reader to use any software for the analyses carried out in this book. Our purpose is to provide the foundations needed to solve the problems from fundamental principles. While software packages and spreadsheets undergo change in versions and upgrades, the concepts needed to solve these problems remain the same. An analyst should understand and know how to analyze problems from the first principles. We have presented much of the sophisticated analysis in tabular form to give the reader exposure to solving these problems numerically. To gain a better understanding of these chapters, we suggest that the reader repeat the tabular analysis on their own instead of just reading the chapters. The replication of these tables in spreadsheets or other current tabular forms can be assigned as homework problems in classes.

Chapter 37 provides an informative case study (The Daylight Alchemy) that has been used in many decision analysis classes as a final take-home exam. It captures many of the tools presented throughout the book.

Below are some suggestions for using this book in a classroom:

When teaching to an audience that has an interest in the foundations of decision making but less emphasis on the math or computations, the following chapters could be covered:

Chapters 1 through 17 introduce the foundations of decision analysis without requiring significant mathematical sophistication. Topics include characterizing a decision, the rules of actional thought, *u*-curves, sensitivity analysis, probability encoding, and framing.

Chapter 26 discusses multi-attribute decision problems with no uncertainty. The presentation prepares the reader to address multi-attribute problems where uncertainty is present.

Chapter 29 presents a fundamental notion about probability: when two people have differences in beliefs, we can construct a deal that both will find attractive, and we can also make money out of constructing those deals.

Chapter 33 analyzes decisions that involve a small probability of death, such as skiing or driving a car.

Chapters 37 through 39 explain how to use the decision analysis approach when there are large groups involved. They also discuss some impediments to quality decision making in organizations.

Chapter 40 discusses ethical considerations in decision making. Like any tool, decision analysis is amoral: you can use it to determine the best way to rob a bank. The ethics must come from the user.

Other chapters in the book are also relevant when teaching to a technical audience that would like to learn about large-scale problems and the computations involved. For example, seniors in an undergraduate engineering curriculum, MS students, or MBA students. For this audience, the instructor may wish to add any of the following chapters to the chapters listed above:

Chapters 18 through 25 discuss advanced information gathering from multiple sources, the concept of creating options in our daily lives, other types of *u*-curves that describe risk aversion, using approximate formulas for valuing deals, and the concept of probabilistic dominance relations that, when present, facilitate the determination of the best alternative.

Chapters 27 and 28 analyze multi-attribute problems where a value function for cash flows is determined and explain how to handle multiattribute decision problems with uncertainty.

Chapter 30 shows how to update probability after observing the results of an experiment.

Chapter 31 examines several auction types and illustrates how to use the basic concepts of decision analysis to determine the best bid and the value of the bidding opportunity.

Chapter 32 presents the concepts of risk scaling and sharing: how a decision maker can determine the best portion of an investment, how a partnership can share an investment, and how to establish the risk tolerance of a partnership.

Chapter 34 analyzes situations where a person is exposed to a large probability of death, such as may be faced in medical decisions.

Chapters 35 and 36 illustrate how to solve decision problems numerically by simulation and discretization.

We hope you enjoy reading the book and then applying this powerful way of thinking about decisions in your daily life.

CHAPTER 1

Introduction to Quality Decision Making

CHAPTER CONCEPTS

After reading this chapter, you will be able to explain the following concepts:

- Normative vs. descriptive pursuits
- Reactive vs. proactive decision making
- Thought vs. action
- Decision vs. outcome
- What constitutes a good decision
- Stakeholders of a decision
- The six elements of decision quality
- The decision basis

1.1 INTRODUCTION

We all make decisions every day, but few of us think about how we do it. Psychological research has shown that people make decisions that after reflection they regard as wrong. Our purpose in this book is to provide a systematic process that enables quality decision making.

1.2 NORMATIVE VS. DESCRIPTIVE

To begin, it is important to distinguish between descriptive and normative pursuits. Descriptive fields do what the name implies; namely, describe reality and actions as they are, while normative fields identify how they should be. For example, it sometimes happens that when we add a column of numbers from the bottom up using pencil and paper, we obtain a different sum than when we add the same column of numbers from the top down. When this occurs, we say that we have made a mistake because we have an arithmetic norm requiring that the sum of numbers be the same regardless of the order in which we add them. If we have no norm for what we are doing, we cannot say descriptively that we have made a mistake. The rules of arithmetic provide norms for arithmetical computations. Similarly, the foundations of decision analysis provide the norms for decision making.

Consider the various fields of study at the University. Is physics a descriptive or a normative field? Although many results in physics have the names of laws, in fact, these findings are models of reality that aim to describe what is so. To confirm their descriptive ability, they must be tested by experiment. Even today, scientists still perform expensive, sophisticated experiments to see whether Einstein's model describes the physical behavior of the universe. While

Newton's model has been used for centuries and continues to be used today, Einstein's model is more descriptive of physical behavior at velocities approaching the speed of light.

In this book, our primary focus will be on normative decision making—how we should make decisions, rather than how we actually make them. Yet for three important reasons, we shall also address descriptive decision making. The first reason is motivational: If we do not learn through demonstration that we are faulty decision makers, we will not see the point of learning a powerful normative process. The second reason is practical: Descriptive models of human behavior may allow us to predict the natural conduct of the people affected by our decisions. Just as the results of a normative process like addition are no better than the numbers entered, the results of our normative decision process will be no better than its inputs. We need to understand that these inputs come from humans displaying various biases and distortions, and we must learn to control for such factors. Finally, our descriptive knowledge of how people receive information will enable us to present our results and have them understood.

You might wonder about the difference between what we naturally do in making decisions and what we would like to do upon reflection. In other words, why is there a difference between descriptive and normative behavior? One possible explanation is that in evolutionary terms, we still have the bodies and brains of our caveman ancestors. Even in the business district of a major city, hearing the roar of a lion will alarm us. For millions of years, this instinctual sense of alarm was critical for survival, but today it is of little value.

One consequence of our origins is that in many cases, our natural capabilities are better suited to the challenges of our ancestors than to the challenges of modern life. Examples abound:

- We cannot sense a highly radioactive environment, even though it could kill us in a matter of hours.
- If we lose visual reference while flying an airplane in bad weather, without instruments, we crash.
- If we are scuba diving and we lose our air supply at depth, our natural instinct to hold our breath and dash for the surface might end up killing us.

Though we do not have these capabilities, we have developed compensations for them all:

- We use Geiger counters to sense radiation.
- We use instruments to fly in bad weather.
- We learn through instruction that the unnatural act of releasing air gradually as you surface is the proper procedure when scuba diving.

Another consequence of our origins is that we have capabilities that once helped us to survive, but now may even lead us to harm.

- Millions of years ago if you had food before you, you ate it before it could spoil or be taken from you by another creature. Now this instinctive behavior at the buffet table can be ultimately harmful if it leads to diabetes or heart disease.
- Millions of years ago if someone challenged you, aggressive behavior toward him could save your life. Now, road rage can lead to injury and death.

Figure 1.1 presents a way to visualize the effects of our nature on decision making. Here we picture our choice of action as determined by the interaction between two decision systems: One **deliberative**, or reasoned; the other **affective**, or emotional. The **affective decision system** is the "hot emotional system." This system existed within the 6+ million year old brain, and was motivated by sex, fear, and hunger stimuli that were directly related to survival. It focused on stimuli that are here and now; proximate and immediate.

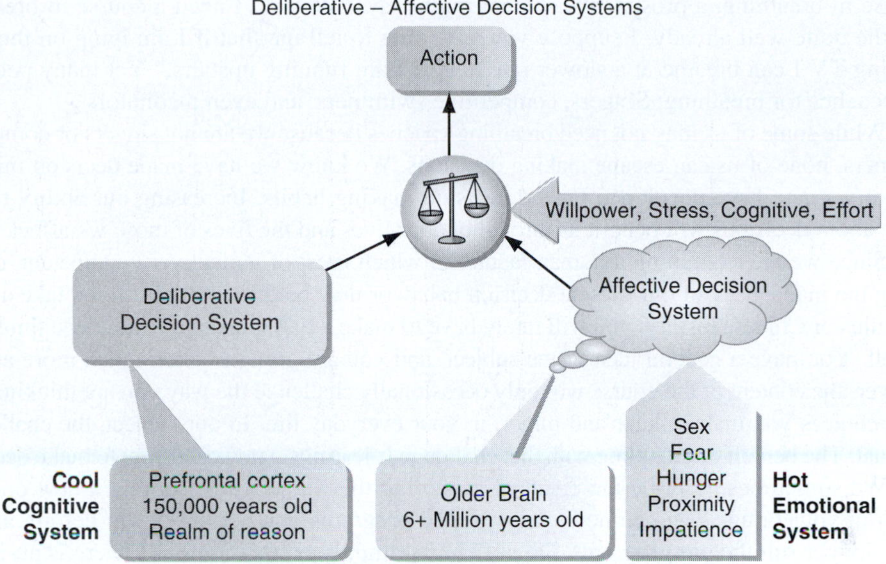

FIGURE 1.1 Deliberative-Affective Systems

In contrast, the **deliberative decision system**, or the realm of reason, is the "cool cognitive system." The final evolution of the human brain some 150,000 years ago resulted in the development of the prefrontal cortex, and along with that, the deliberative decision system. The prefrontal cortex enhanced, but did not replace, our "old" brains: As a result, the two systems coexist, often creating considerable internal conflict.

The pull of each system in determining action is influenced consciously, by willpower, and unconsciously, by factors such as stress and cognitive effort. We usually think of hard work as some kind of physical activity that will leave you exhausted. However, another kind of work, the cognitive effort involved in thinking, can end up exhausting your deliberative decision system, thereby increasing the influence of the affective decision system.

You might use willpower to avoid the tempting high calorie dessert by remembering that eating it will not serve your desire to lose weight. However, the jet lag you experience by flying through several time zones may create stress that will tip the scale toward the "hot" emotional system, resulting in poor reasoning during the next day's business conference.

Perhaps the simplest example of the struggle between the systems is to observe someone at a party eating handful after handful of peanuts and saying "I know I am going to regret this tomorrow."

We were not evolutionarily equipped to make many of the decisions we face in modern life. For example:

- Choosing among medical treatments that have uncertain and long run consequences.
- Making financial decisions, as individuals or companies, that will produce uncertain futures of long duration.

Making such decisions by "gut feel" is to hand them over to the affective decision system. As we proceed, we shall see many examples of affective decision making gone wrong. The purpose of this book is to develop our deliberative decision system and to increase its role in our decision making.

Learning normative decision making poses special challenges. We have all made thousands of decisions in our lives, and most of us think we are good at making them. If we offered a course in breathing, a prospective student might say, "Why do I need a course in breathing? I breathe quite well already. I suppose you are going to tell me that if I am lying on the couch watching TV I can breathe at a slower rate than if I am running upstairs." Yet many people do have coaches for breathing: Singers, competitive swimmers, and even meditators.

While some of us may not need breathing coaches because we are not singers or competitive swimmers, none of us can escape making decisions. We know we have made decision mistakes, and that we may have developed flawed decision making habits. Increasing our ability to think clearly about decisions will benefit us throughout our lives and the lives of those we affect.

Since we are examining a human faculty in which most of us feel very competent, demonstrating the inadequacy of our present decision behavior may be discomfiting. If you take a course in calculus or Chinese history, you will rarely have to make a major change in how you think about yourself. You have a general idea of the subject, and you are going to learn much more about it. However, the content of the course will only occasionally challenge the way you are thinking about all the choices you make, major and minor, in your everyday life. In our subject, the challenge is continual. The benefit of grappling with that challenge is learning a powerful way to make decisions.

We sometimes describe the result of mastering this subject as installing a new operating system in your brain. You can now run powerful programs you could not run before, and you can no longer run the old programs. Do not embark lightly on this journey. There is an Eastern saying, "Better not to begin, but if you begin better to finish."

As Samuel Butler put it, "A little knowledge is a dangerous thing, but a little want of knowledge is also a dangerous thing." This book is not about making decisions only in a specific field, such as business or medicine. The concepts apply everywhere and are useful in all fields, as our examples will show.

1.3 DECLARING A DECISION

Decisions do not arise in nature. No one walks through a forest and says, "I have just spotted a wonderful decision." Decisions are declared by human beings. Sometimes they arise when we have what philosophers call a break in our existence—some change in our circumstances—that impels us to declare a decision. We can consider these decisions as reactive to the change. Whether we experience a change for the worse, like losing a job or falling ill to a disease, or a change for the better, like inheriting money, we face declaring a decision.

We can also declare decisions proactively, without any external stimulus. You can declare a decision about quitting your job or about taking up skydiving just because you want to. Figure 1.2 illustrates the different types of declarations.

Some of the most important decisions you can make are those you declare proactively. When Warren Buffett[1] was asked about the worst decision he ever made, he said, "The things I could have done and didn't do have cost us billions of dollars..." He viewed his worst investment failures as errors of omission, rather than commission—errors arising from a lack of proactivity.

Whether the decision is reactive or proactive, it is yours. The alternatives you have belong to you. You have total power over the alternative you select, but seldom over the consequences of selecting that alternative. We are using the word "alternative" in the American sense, rather than in the European sense. Saying, "We have one alternative" is understandable to an American, but a European might ask, "Alternative to what?" So when we say you have only one alternative,

[1]Warren Edward Buffett (born August 30, 1930) is an American investor, businessman and philanthropist. Buffett has been one of the richest men in the world and has given most of his fortune to charity.

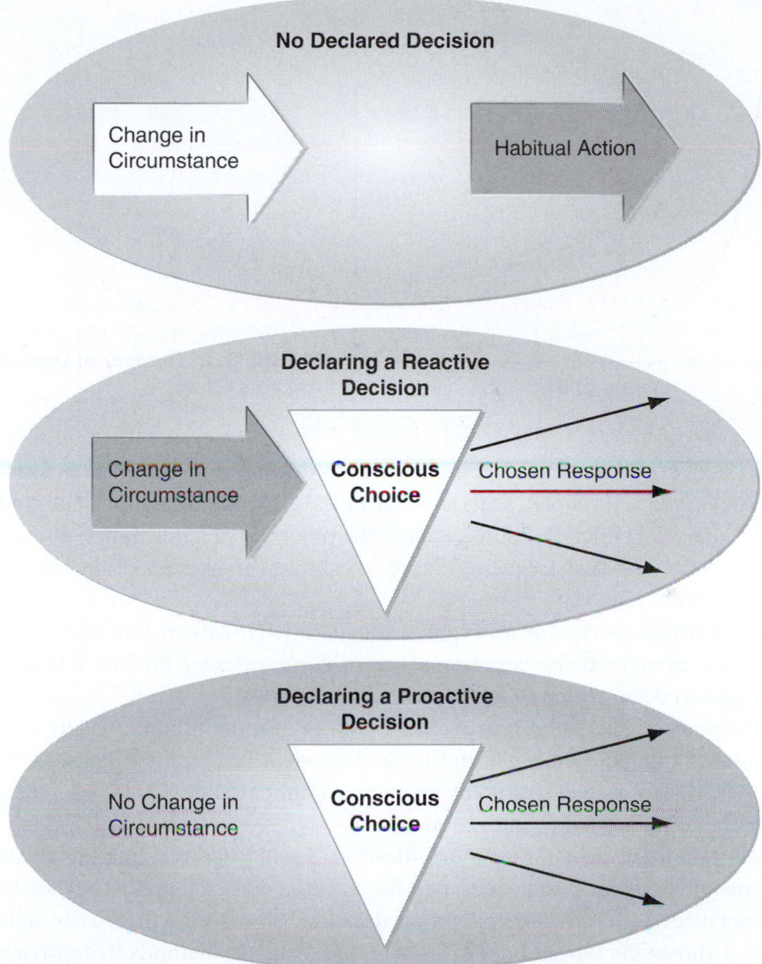

FIGURE 1.2 **Declaring a Decision: Reactive vs. Proactive Decision Making**

we mean that you have no choice. A cartoon once showed a chaplain offering comfort to a convict about to be executed in an electric chair. The caption was, "My advice is to pray to a saint who helps the wrong people by mistake."

Doing nothing is always an alternative. Suppose you go to a restaurant for dinner. The waiter presents the menu and then awaits your order. You say, "I will need a few more minutes." Shortly thereafter, he returns and you again request more time. Whenever the waiter returns, you repeat your request. What happens? The last time you see the waiter, he tells you that the kitchen is closed and that no more food will be served. You have chosen the "do nothing" alternative, and you have suffered the consequences.

To truly have alternatives in making a decision means that they are completely under your control. For example, you may say you have the alternative of getting a job with company ABC, but you do not. You have the alternative of applying for a job with company ABC. You may say you have the alternative of going to graduate school, but your real alternative is to apply to graduate school. Taking care in understanding alternatives is an important step in thinking clearly about decisions.

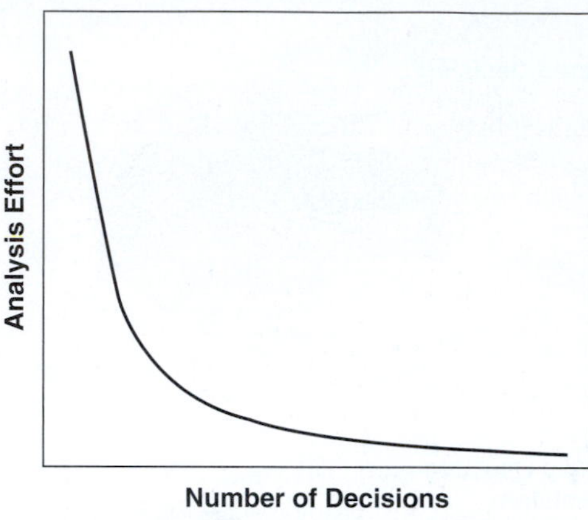

FIGURE 1.3 Number of Decisions vs. Analysis Effort

Once a decision is declared, knowing what to do may require little effort or extensive analysis. Most everyday decisions, such as what to have for breakfast or what clothes to wear, seldom require analysis. Other decisions, such as the purchase of a new home or car, may require more analysis, but are also less frequent. Figure 1.3 shows the number of decisions we face and the analysis effort they require.

Decision analysis works for all types of decisions. However, you can deal with simple decisions in a few minutes using common sense or some rules of thumb. You do not need an extensive analysis to decide what to have for breakfast.

More complicated decisions, however, are worthy of more thought. Using a simple checklist to remind us of things to consider and to help us identify common decision making errors might make the process easier. Examples of more complicated decisions are where to spend a vacation, or whether to buy a new television set.

The most important decisions we face deserve a much more refined analysis. They may involve elements of complexity, dynamics, and far-reaching consequences. They are worthy of, but frequently do not receive, the structured, rigorous decision process we will describe in later chapters.

Figure 1.4 shows the types of decisions we may face and methods to approaching them.

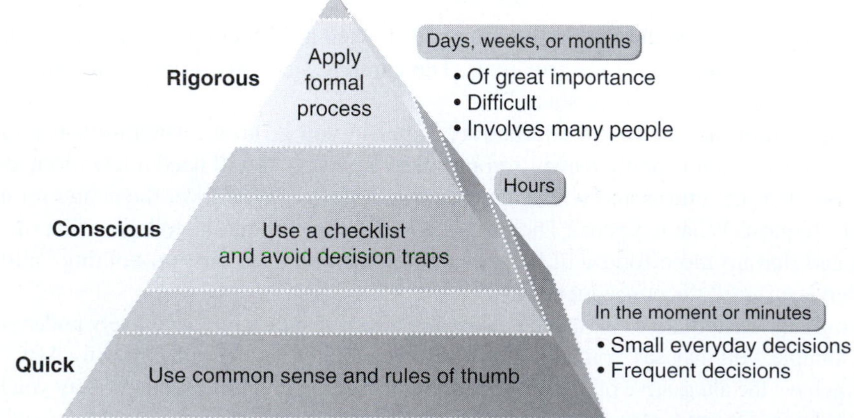

FIGURE 1.4 Hierarchy of Decisions

1.4 THOUGHT VS. ACTION

The mere idea of thinking about something does not mean we have made a decision. To better understand our decisions, we first make a distinction between "thought" and "action." Figure 1.5 illustrates all possible combinations of actions and thoughts. The diagram identifies four different regions.

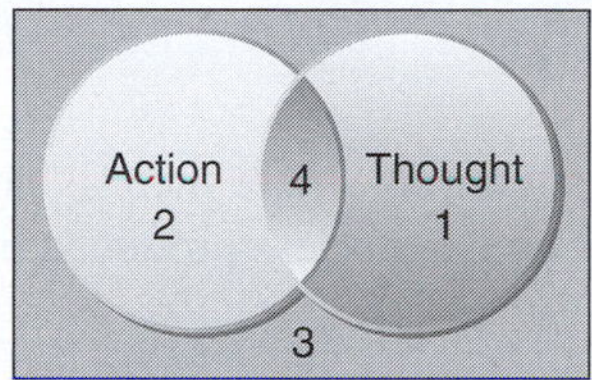

FIGURE 1.5 Regions of Thought and Action

REGION 1: THOUGHT WITHOUT ACTION Consider what you do in your daily life. Are there times when you have thought without action? A little reflection shows that the answer is "Yes." For example, we can think "What a beautiful cloud!," or "I should quit this job!," or "I am bored." This region also includes feelings you may have towards someone or something. Much of our self-talk is thought without action, and perhaps that is a good thing.

Reflection

Think of other situations where you may have thought but not acted.

REGION 2: ACTION WITHOUT THOUGHT Is there action without thought? Once again, the answer is "Yes." A simple example is a reflex response to a stimulus, like crying after cutting an onion. A more thought provoking example is riding a bicycle. Can you imagine trying to use an instruction manual for riding a bicycle? The manual would describe things like steering head angles, tire contact areas, and the center of gravity of the combined human-bicycle system. We learn to ride a bicycle automatically without thinking, so a manual like this would be of little use.

Walking is another great example of an inborn ability or a trained skill. No explanation involving muscle contraction will help.

Reflection

To illustrate the advantage of coaching even when a skill has been learned and performed automatically without thinking, consider the following question:

Suppose you are riding fast on your bicycle and wish to turn to the left around a curve as quickly as possible. In what direction do you turn the handlebar?

The common answer is that you turn the handlebar to the left. Very experienced bicycle and motorcycle riders know that the correct answer is to turn the handlebar to the right. This is clear if you observe motorcycles racing counter clockwise around a circular dirt track. They all have their handlebars turned to the right, away from the center of the circle.

For this to be the correct answer, what must be true of cycle design?

Even without instincts or training, you can take appropriate action in new situations, without thought, by acting on your inner knowledge. Buddha would call it "right action:" Just do it. However, it is often difficult to make spontaneous and correct decisions in our personal or professional lives. Complexity, uncertainty, or conflicting values may confuse us.

Reflection

Think of other situations where there is action but no thought.

REGION 3: NO THOUGHT-NO ACTION Are there situations where there is no thought and no action? One example is being in a coma. The state of consciousness produced by proper meditation might be another.

Reflection

Think of other situations where there is no action and no thought.

REGION 4: THOUGHT AND ACTION—"ACTIONAL THOUGHT" Finally, we may want to think about what to do, which we call **actional thought**. When we think about a decision, we are practicing actional thought. But what constitutes high quality actional thought? One answer is decision analysis, our present endeavor.

1.5 WHAT IS A DECISION?

We now need to ask a fundamental question, "What is a decision?" A frequent answer is that it is a choice, or a choice among alternatives. But we want more precision in our understanding. The following is our definition of a **decision**:

> *A decision is a choice between two or more alternatives that involves an irrevocable allocation of resources.*

Suppose a friend tells you that he has decided to buy a new Rolls-Royce. How will you know when he has actually made the decision? Is it when he has visited the dealer to look at Rolls-Royces, or when he has made an appointment to return to buy the car? You will know he has bought a Rolls-Royce when he gives the dealer his cashier's check for the purchase price and the dealer has given him the registration and the keys. If your friend drives around the block in his new car and decides that he does not like it after all, can he just ask the dealer for his money back? The dealer may well say, "I see, you want to sell us a pre-owned Rolls-Royce in excellent condition. Here is our offer." His offer will typically be less than the number on the cashier's check he recently received. The difference is the monetary resource that your friend has committed in making the purchase.

A resource deserves its name if it is something that is scarce and valuable. Money is a resource; the time in our lives is a resource. Thinking about a decision takes time: The decision to think about a decision is an irrevocable use of that time. The decision to buy the Rolls-Royce by handing over the cashier's check represents an irrevocable loss of resources—the difference between what you pay for it and what you could sell it for after accepting ownership. Every decision, then, is irrevocable in the sense that the resources committed to it will be at least partially lost.

1.5.1 A Mental Commitment or Intention is Not a Decision

You can say that you have decided to diet, but you will not have made a decision until you do not order your customary dessert at a meal. Even if you abandon your diet tomorrow, today's meals are different.

The roots of the word "decision" are consistent with this interpretation. The Latin word corresponding to decide means "to cut off." As long as you are just thinking about the decision, you are not cutting off anything except the time you might have spent doing something else. As soon as you sign the contract, choose not to fasten your seatbelt before driving, or start down the expert-rated ski trail, you have cut off some possible futures and created the possibility of others. As a radio commentator once said, "The past is a canceled check and you have no claim on the future."

As we have seen, the resource allocation of a decision can be irrevocable in whole or in part. For example, if you are merely thinking about where to spend your vacation, you have not yet made a decision. Time spent is irrevocable at the current level of science, so while thinking about your vacation you have indeed decided to spend some time, you have not yet committed monetary resources. You make a decision when you book the tickets, make the hotel reservations, and thereby commit some resources that are at least partially irrevocable due to fees for cancellation or change.

The moment of decision is the moment when changing your mind costs something. If, in anger, you write an email to your boss saying you quit, your moment of decision is when you hit "send." Up until that moment, you can change your mind with little consequence. Once you hit "send," however, you begin a chain of events that will be difficult, if not impossible, to reverse.

Resources are scarce, and we use our methods to allocate them. Love is not a resource because it is not finite. We do not recommend using the methodology of decision analysis when allocating love. This is more a matter of wisdom than engineering.

Reflection

Which of the following represents a decision?

a. I have decided that I do not like vanilla ice cream.
b. I have decided that the stock market will go up.
c. I have decided that the stock market will go up, so I will invest right away. Here is a check for my purchase.
d. I have decided to ace the test.
e. I have decided to diet, and I have thrown away the ice cream in the freezer.

1.5.2 What Makes Decision Making Difficult?

Now let us look back at some of the decisions we have made and think about why they were difficult. Sometimes decisions are difficult because they require making trade-offs among several factors. They may be difficult because of other people who are involved. We call such people stakeholders. We define **a stakeholder** as someone who can affect, or will be affected by, the decision. In personal decisions, stakeholders may be friends or family. For example, suppose you are interested in buying a motorcycle, but you know this will worry your mother. Your mother is a stakeholder in this decision. You will have to balance upsetting her and your personal enjoyment of the motorcycle. Stakeholders in business decisions can be shareholders, employees, and customers. Stakeholders in medical decisions can be the patient, doctors, nurses, and the patient's family.

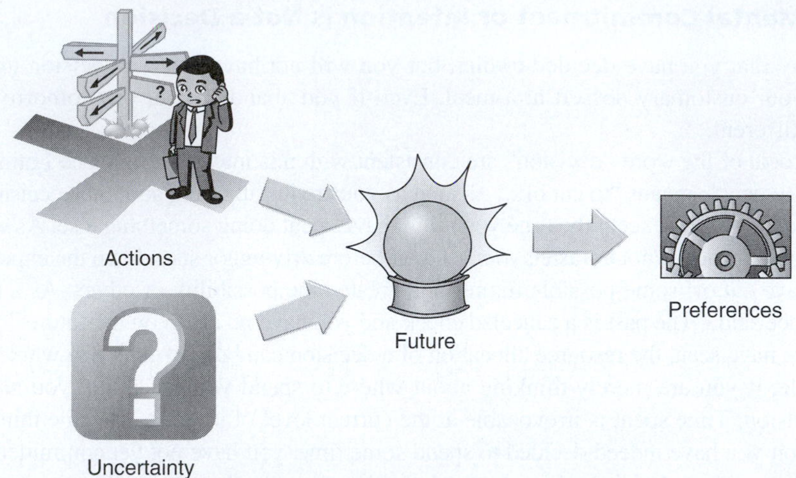

FIGURE 1.6 Decision Making with Uncertainty

Sometimes decisions are difficult because of fear of a bad outcome, or fear of regret, or even fear of blame. In all these cases, the difficulty is our uncertainty about the outcome. Suppose no matter what alternative you chose, there was no resulting uncertainty about the future. Imagine you could magically and instantly play a movie to illustrate your future, based on each potential alternative. After seeing the movie, you choose the best alternative. This opportunity to foresee the future would make decision making easy and free of any regrets or worries. By viewing the movie, you would learn more about your preferences and the types of tradeoffs you would be willing to make to substitute one vision of the future for another. Sometimes people say decision making is difficult because of time pressure or constraints, but even those situations would be simple if you saw the movie of your future lives resulting from each alternative.

Unfortunately, these movies of our future lives do not exist. Consequently, we can only choose the best course of action at a certain moment in time. Our futures are always uncertain, but we do have preferences. What we want to do is choose the best alternative given our preferences by properly considering uncertainty, as depicted in Figure 1.6. Creating the normative process for doing this is the subject of this book.

1.6 DECISION VS. OUTCOME

Suppose you had a choice between two deals (shown in Figure 1.7).

Deal A gives you \$100 if a tossed coin lands on heads and \$0 otherwise;

Deal B gives you \$100 if a rolled die lands on 5 and \$0 otherwise.

Which deal would you choose? Most people would choose deal A.

Suppose you choose deal A and your friend chooses deal B. The coin is tossed and it lands tails, and the die is rolled and it lands on 5. You do not get the \$100 but your friend does. Did you make a bad decision? The answer is no. If you faced this decision situation again, would you still choose deal A? Most people would say yes.

This example, shown in Figure 1.7, illustrates the most fundamental distinction in decision analysis, the difference between the quality of a decision and the quality of its outcome. The distinction implies that we can make good decisions but still get a bad outcome due to uncertainty. Observing the outcome tells us nothing about the quality of the decision—just about the quality of the result.

FIGURE 1.7 Coin vs. Die (Coin photo: Imagedb/Fotolia; dice photo: Piai/Fotolia)

Using the distinction between a decision and its outcome, we can think of four eventualities:

- Making a good decision and getting a good outcome.
- Making a good decision and getting a bad outcome.
- Making a bad decision and getting a good outcome.
- Making a bad decision and getting a bad outcome.

To illustrate using Figure 1.8, imagine you are at a party and you have had a few alcoholic drinks. At the end of the party, you are drunk and must decide whether to drive home in this state. A good decision would be not to drive and to stay at your friend's house until the morning when you can drive home sober. A bad decision would be to drive while drunk.

We consider possible outcomes following each decision. If you decide to stay and drive sober, you could have a car accident on your way home the next morning. A bad outcome has followed a good decision. If you decide to stay and drive sober, and arrive home safely, a good outcome has followed a good decision. On the other hand, if you decide to drive drunk and arrive home safely, a good outcome has followed a bad decision. Finally, if you decide to drive drunk and have an accident, a bad outcome has followed a bad decision.

Ambrose Bierce[2] uses a poem to describe this idea of not judging a decision by its outcome:

"You acted unwisely" I cried, "as you see
By the outcome".... He calmly eyed me:
"When choosing the course of my action," said he,
"I had not the outcome to guide me."

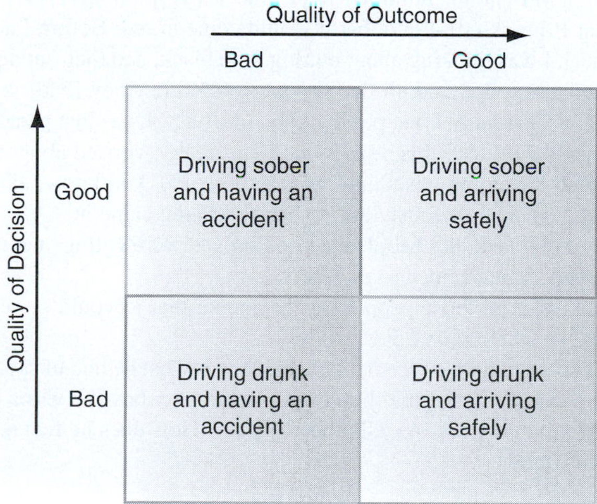

FIGURE 1.8 Decision vs. Outcome

[2]Ambrose Gwinnett Bierce (June 24, 1842–1914) was an American satirist, critic, poet, short story (horror) writer, editor, and journalist.

Ambrose rushes to tell a person that he acted unwisely "made a bad decision" based on the outcome that he received. The response highlights the distinction between a decision and an outcome: the person did not have the outcome available at the time of making the decision to guide his decision making.

Some people may live their lives feeling guilt or regret about something they did when, in fact, they had made a good decision at the time. We often hear statements like, "This did not work last time, so we cannot do it again" or, "This project turned out to be a bad investment, it was a very bad decision." Careful thinking about this statement shows that they are judging the quality of the decision by the past outcome.

On the other hand, people may also make a bad decision, receive a good outcome, and live thinking that they did the right thing. When this happens, we may hear, "We tried this last time and it worked, so it was a good decision and we should just do it again."

Reflection

Reflect on the distinction between a decision and an outcome.

Do you confuse the two in your own decision making? Have you judged the quality of a decision you made based only on the outcome? Have you been judged by the quality of a decision you made based on its outcome?

We see the need for the distinction between a decision and an outcome in understanding the daily news. The following piece aired April 12, 2004 on CNN.

Gambler: Roulette Play "Just a Mad Thing to Do"[3]

(CNN)—Ashley Revell, a 32-year-old man from London, England, sold everything he owned, even his clothes, to try his luck Sunday on one spin of a roulette wheel in Las Vegas, Nevada. He put $135,300 on red, and with friends and family watching, the ball hit the mark, giving Revell $270,600. The event was filmed by Britain's Sky One television as a short reality series called "Double or Nothing." CNN's Anderson Cooper asked Revell what was going through his mind when the wheel was spinning.

REVELL: I was just…pleading that I'd pick[ed] it and that it would come in red. Before I actually walked up to the wheel, I was thinking about putting it on black, and then suddenly the guy was spinning the ball around and all the Sky viewers said…they [had] voted that I should put it on red. So suddenly I just put it all on red. But…I was just pleading that it would come in and I'd get lucky this time. What I was really worried about was that I'd lose and my parents would be upset and my family would, you know, all my friends would be upset. So…I was obviously just so happy when it came in.

COOPER: So you were going to put it on black, but people back in England were voting, and they said you should put it on red? You decided to do that?

REVELL: Yeah, that's right. I mean, with all those people sort of hoping that it would be red, I thought I've got to go red, so that's what I did.

COOPER: Your father was opposed to this whole concept all along. This is what he had to say. He was quoted in an interview as saying: "I told him he was a naughty boy, he was a bad boy, he shouldn't do it. He should work like all other kids do." How does he feel now? I mean, has he changed his mind?

---

[3]Reprinted by permission from "Gambler: Roulette Play 'Just A Mad Thing To Do'" CNN.com, 4/12/04

REVELL: Yeah, I think so. I mean, I obviously went and shook his hand before I did it, and after he was just hugging me and jumping up and down. So, you know, I think most all dads are just concerned, and he's seeing all my friends being married off and having kids and stuff, and he's like any father, he just wants me to settle down and make sure I'm secure.

COOPER: Now, why did you do this? I mean, is it true that you sold all your possessions, even underwear, everything you had, and then put all the money on this? Why? Was it all just to be on TV?

REVELL: Looking back on it now, I mean, at no point before I did the bet did I think about losing. I just felt positive and thought about just going ahead and winning. But now I've actually won, I can think about what would have happened if I'd lost. And to be honest, I was crazy to do this bet. It was the maddest thing. I mean, this is really about all I've got left, the tuxedo, which I'm not allowed to keep. So it was just a mad thing to do. And I'm thinking back now about what would have happened if I lost. I'd have nothing to go back to, nothing to wear. But I'd still have my friends, my family, and they'd always be there for me. So they gave me the security to be able to do this. But you know, never again. I mean, that's – it was mad.

Reflection

Keeping in mind the roulette player's decision, consider the following questions:

- Did Revell make a good decision by playing this game?
- Did he have a good outcome?
- Does the outcome he received change the quality of the decision he made?
- Should his father have a different view of the situation because the outcome was good?
- If you were in Revell's place, would you have made the same decision?
- Do you think Revell would be willing to repeat the same gamble again with his current wealth of $270,600?

1.7 CLARITY OF ACTION

We have discussed the difference between a decision and an outcome, and determined that a good decision does not guarantee a good outcome. What, then, is the role of decision analysis?

The purpose of decision analysis is to enable a decision maker to achieve clarity of action in his decision and, even more fundamentally, to achieve clarity of thought. Furthermore, knowing that we have made the best decision provides peace of mind, since that is all we can do to influence the future.

We can make an analogy between decision making and flying a plane. When the weather is clear, we do not need to rely on all available instruments. However, when the weather is cloudy, we need to rely on our instruments. In a similar manner, when decisions are simple, we can make decisions using our own judgment. When decisions are more complicated, however, our judgment may fail, and we need to rely on rules to help us identify the best course of action.

Continuing the analogy, relying on instruments when flying in bad visibility is necessary not just for beginning pilots, but for the most experienced ones. When deprived of the familiar cues provided by seeing the ground, even an experienced pilot will soon believe, incorrectly, that he is, in fact, upside down, and will make the wrong corrective adjustments to the controls. The same is true for decision makers acting in an uncertain world. Unless they use the instruments we will build, they may also make grievous errors.

A Story by Ron Howard

Many years ago, I had a medical condition that caused occasional debilitating flareups. By taking some medications continually and other medications during flareups, I could control the condition. However, the medications had long run serious negative effects. My doctor recommended a major operation that, if successful, would cure the condition. Yet success was uncertain, and I could die in the operation. I demurred; my doctor thought I was indecisive.

Since this was a major important decision with great uncertainty and with implications for the rest of my life, I did an extensive six-month long analysis with the help of two medical doctors who were in my class. We did dynamic probabilistic modeling of the future. The doctors sent my x-rays to a specialist across the country for advice. We found that my best alternative was to see whether I had another flareup. If I did, then I should have the operation. If I did not, then I should keep postponing it. A flareup occurred and I called my doctor to schedule the operation. He asked when, and I replied "right away." He had difficulty understanding why his indecisive patient was now so decisive. On the day of the operation, just before going under anesthesia, a kind nurse assured me that everything would be fine. I thanked her, told her I had a 2% chance of dying on the operating table, and that I was ready to go.

This story makes three points. First, that I had clarity of action in having the operation in the face of uncertainty about the consequences. Second, that the most irrevocable part of the decision was to allow myself to be anesthetized rather than to get up and leave the hospital. Third, I had a good outcome.

1.8 WHAT IS A GOOD DECISION?

We have probably used this term "good decision" in many of our daily conversations. But what is a good decision? And how do we know that we have made one?

1.8.1 Common Misconceptions about What Makes a Good Decision

There are many common misconceptions about what constitutes a good decision. In graduate classes or in executive seminars, people often answer,

"A good decision is one that produces a desired outcome."

As we have seen, there is a clear distinction between a decision and its outcome, so this cannot be the correct definition.

Another common response is,

"A good decision is one that has the highest chance of getting the best outcome."

Once again, this answer has a problem since this definition takes into account neither the absolute desirability of the best outcome nor the chances of very bad outcomes. Consider a deal with an 80% chance of gaining you $1,000,000 and a 20% chance of costing you $10. Another deal may provide you with a 90% chance of gaining $100 and zero otherwise. Most people would agree that the first deal is more attractive than the second, yet it has a lower chance of the best possible outcome, and a higher chance of the worst outcome. The example also illustrates the problem with the response,

"A good decision is one that has the lowest chance of getting the worst outcome."

Arno Penzias,[4] a Nobel Laureate, was asked how he knew a good project when he saw one. His response was,

> *"Simple, imagine that what you're going to do will be 100% successful; find out how much money it's going to be worth; multiply by the probability of success, divide by the cost, and look at the figure of merit."*

While this approach may sound like a reasonable criterion for project selection, closer examination reveals that it focuses only on the monetary outcome of 100% success and ignores other levels. In some cases, this answer would correspond to the best project, but in others, it may not. Using a ratio does not take into account the actual monetary values involved.

For example, consider two projects that will either succeed or fail. They each have a 90% chance of success. The first project will either cost $10, and, if successful, will yield $100. The Penzias figure of merit is 0.9 times 100 divided by 10, or 9. The second project will either cost $150,000, and, if successful, will yield $1,000,000. Its Penzias figure of merit is 6. Choosing based on the figure of merit would lead us to choose the first project, yet most companies would prefer the second project to the first.

There are still more ways of looking at this same example. Does the company have only two available projects? Can the company do both? Are there other considerations besides monetary reward, such as legal or ethical issues?

1.8.2 The Six Elements of Decision Quality

To answer the question of what constitutes a good decision, we first need to understand the main elements of a decision:

1. The decision-maker;
2. A frame;
3. Alternatives from which to choose;
4. Preferences;
5. Information; and
6. The logic by which the decision is made.

First, every decision requires **a decision-maker**, the person who will act. As we have discussed, decisions are never found in nature: A person speaks them into existence. For example, anyone who says "I am going to decide whether to make this investment... have the operation... set the research budget at $200 million..." must be committed not only to thinking about acting, and but also to deciding. Otherwise, the analysis is useless. Commitment to actional thought is the first element of good decision making.

Next, the person must provide a way of viewing the decision. We call this view **a frame**. For example, a person's frame may be deciding which car to buy from a certain category of cars. The frame could also be whether to buy or lease a car, whether to own a car in the first place or to use public transportation, or even whether to commute to a job or work at home. Each frame presents a different view of the decision problem to be addressed.

The choice of a particular frame will lead to the creation of **alternatives** appropriate to that frame. These alternatives are available courses of action that the person believes would lead to different futures. Making a high quality decision will involve consideration of several

---

[4]Arno Penzias joined Bell Labs in 1961. He conducted research in radio communication and won the Nobel Prize in 1978 for research that enabled a better knowledge of the origins of the universe. He later became Chief Scientist, and continued to search for innovative and new product ideas by visiting small companies around the country.

substantially different alternatives. Note that by an alternative, we mean a choice that is actually available and is under the decision maker's control. You can choose to apply for many different jobs; however, your alternative cannot be to accept a job offer unless you have a job offer available. If you have no alternatives, or, in the American sense, only one alternative, then you have no choice in what to do and you have no decision to make.

Can you have too many alternatives? While a new alternative can sometimes be better than any that you recognize, finding alternatives takes time and effort. For example, suppose you had gone to a carpet store and spent two hours selecting a carpet for your living area. You have found one that is very attractive and reasonably priced, so you are about to buy it. The salesperson then says, "I should mention that we have a warehouse of carpets just behind the store with 10,000 other carpets you could look at." Many of us would say we had made our choice and would rather spend our time in another way.

A decision maker will also have **preferences** on the futures that arise from different alternatives. The preferences describe what the person wants. If the decision maker were indifferent to the possible futures, there would be no need to make a decision, but merely to live in acceptance of what will be. A wise saying from the east is, "The great road is not difficult for he who has no preference." However, many of us prefer pleasure to pain; success to failure; health to illness; wealth to poverty; youth to old age; chocolate to vanilla; and so we have preferences. A high quality decision will have clear, carefully specified preferences.

The linking of what we can do to what we want to do is provided by what we know, also known as our **information**. This information may leave us uncertain about what the future will follow; we must often make the decision in the face of uncertainty. We are always tempted to get more information, but information costs resources. A high quality decision process ensures that information acquisition is neither overdone nor underdone.

Finally, we must use some process to derive the action we should take, from what we can do, what we want, and what we know. If we desire to use a systematic process, such as **logical reasoning**, then we will want to use the best rules we know for this reasoning. We shall soon present such a set of rules for your consideration.

We can depict the six essential elements of decision quality using the metaphor of a three-legged stool, as shown in Figure 1.9. The stool metaphor is useful because it makes clear to anyone working on a decision precisely which aspect of the decision is currently under consideration. The three legs of this stool represent the three essential elements of any decision. One leg is what

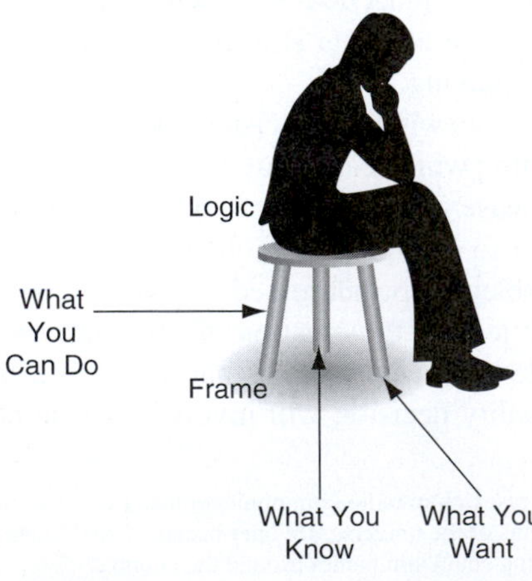

FIGURE 1.9 The Decision Quality Stool

you can do: Your alternatives. The second leg is what you know: The knowledge that relates your alternatives to possible consequences. The third leg is what you want: Your preferences on consequences. The three legs constitute the **decision basis**: The complete description of the decision problem you face. A seat, the logic that will determine your best action for this decision basis, holds the legs together. We shall have much to say about the nature of this logic in what is to come.

One important thing to understand about the stool metaphor is that the stool will collapse if you remove any of its legs. You have no decision to make if you have only one alternative, if you see no connection between any of your alternatives and the future, or if you are indifferent to the possible consequences.

The location of the stool represents your decision frame, which determines the alternatives, information, and preferences that will be germane to your decision. For example, if you need a place to live, you could frame the problem as one of finding a new rental apartment or house. You could also use a larger frame that includes buying a home as an additional alternative. The choice of a frame, then, determines the decision basis and is the most fundamental aspect of making a decision. Later in this book, we will have a much more complete discussion on the subject of framing.

Finally, the most essential element is the person sitting on the stool. There is no decision without the person who constructs the other elements of the stool and is committed to using it for support in making the decision. The person making the decision establishes the frame, seeks and creates alternatives, assembles pertinent information, states preferences, and uses proper reasoning to select the most desirable alternative. That person is, therefore, responsible for placing the stool, fashioning its legs, constructing the seat that connects them, and, finally, sitting on the stool—following the clear course of action.

Another metaphor that contains the six elements of a decision is the chain shown in Figure 1.10. The notion here is that the chain is only as strong as its weakest link. To achieve decision quality, you must assure the quality of each link.

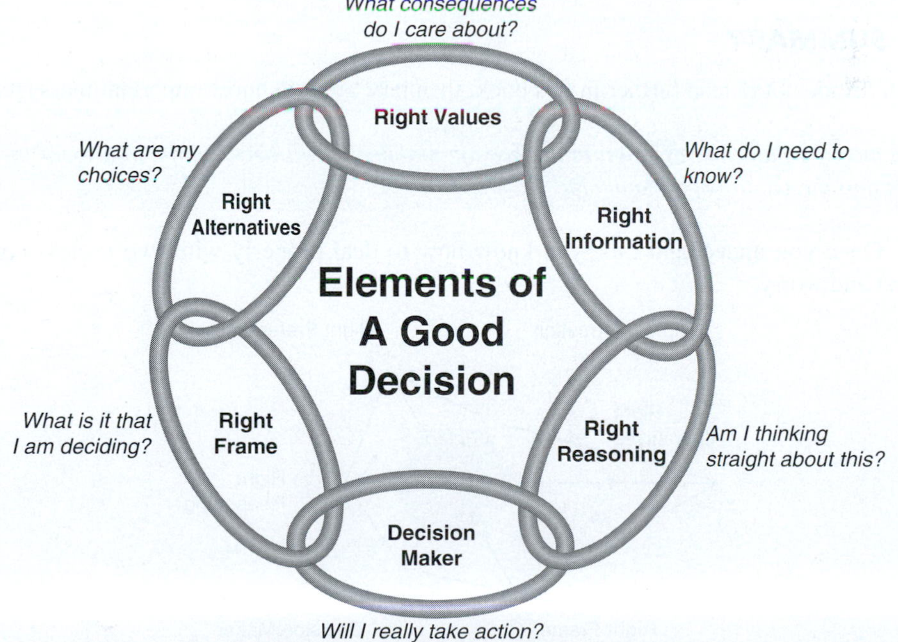

FIGURE 1.10 Decision Quality Chain

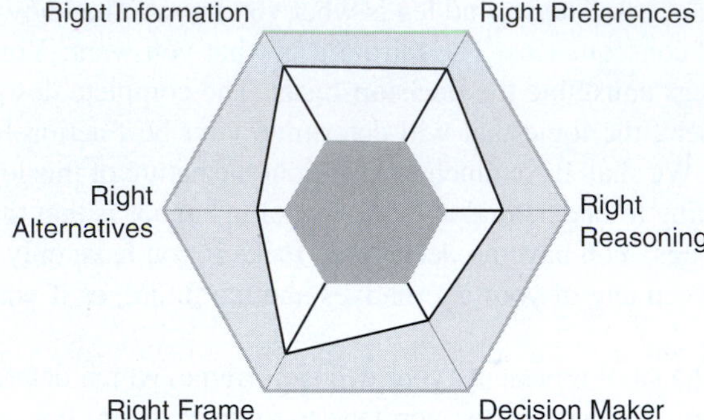

FIGURE 1.11 Decision Quality Spider Web

1.8.3 Graphically Representing Decision Quality

The decision quality spider web shown in Figure 1.11 graphically represents the qualitative attainment of decision quality. It can be a useful tool for individuals and for groups engaged in decision process assessment.

The distance from the inner hexagon to the outer one represents the degree of achievement for each element. The outer hexagon represents the proper balancing of elements for this particular decision. If the line for an element extends beyond the outer hexagon, that element is requiring too much effort. Figure 1.12 depicts an unbalanced analysis because too many alternatives are being considered and too little valuable information of reasonable cost is being gathered. The resulting picture displays the deficiencies or excesses in any of the elements of decision quality.

1.9 SUMMARY

If you decide not to read further in this book, then take away its most important message.

> *The most fundamental distinction in decision making is that between the quality of the decision and the quality of the outcome.*

Once you understand this, you know how to deal properly with two useless concerns: Regret and worry.

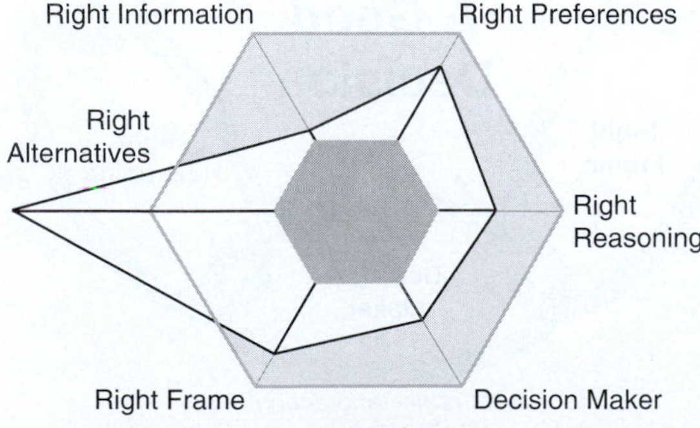

FIGURE 1.12 An Unbalanced Analysis

> If you make good decisions, there is no place for regret in your thinking. Just continue to make good decisions. Why regret if you made a good decision and the outcome was out of your control?
>
> If you find you are worrying about a decision before making it, transfer your energy to making sure it is the best decision you can make.

Annie Duke, a very successful professional poker player, described how she exhibits this wisdom in her playing. She uses all her abilities to make good decisions and pays no attention to whether they actually yield good outcomes or bad outcomes. This behavior often confuses opponents who expect her game to change when she has had a very bad outcome or a very good outcome. In future sections, we will develop the structure of thought necessary for making good decisions.

KEY TERMS

- Normative vs. descriptive
- Recognizing a decision is an essential first step to good decision making
- Declaring a decision
- Thought vs. action
- The importance of actional thought
- The definition of a decision as an irrevocable allocation of resources
- The need to deal with uncertainty in decision making
- The difference between a decision and its outcome
- The role of decision analysis in helping the decision maker achieve clarity of action
- The six elements of decision quality
- The decision basis
- The stool metaphor
- The decision chain
- The spider web diagram

PROBLEMS

Problems marked with an asterisk (*) are considered more challenging.

1. From your readings of this chapter, explain the following terms in decision analysis:
   a. Actional thought
   b. Clarity of action
   c. Decision
   d. Outcome
   e. Normative versus descriptive
   f. Decision basis

2. Which of the following situations represents a decision?
   a. I am thinking of going to Tahoe during the break.
   b. This is a beautiful star.
   c. I need to buy an umbrella.
   d. I have decided to become President.
   e. I have decided that breathing is good from me.

3. What is a decision? What makes decision making difficult? What goal do we pursue in decision analysis? Is decision analysis a normative or descriptive discipline? Explain.

4. Name the six elements of decision quality.

5. Select a newspaper article describing someone facing a decision. Who is (are) the decision makers and what is (are) the decision(s)? What are the uncertainties present? What does (do) the decision maker(s) like or dislike? If you were a consultant hired to help the decision maker(s), what kind of warranty could you give them?

6. Write a brief answer to each of the following.
   a. What is the difference between a decision and its outcome?
   b. Give an example of a good decision followed by a good outcome.
   c. Give an example of a good decision followed by a bad outcome.
   d. Give an example of a bad decision followed by a good outcome.
   e. Give an example of a bad decision followed by a bad outcome.

7. Take some time to think about an important decision situation you are currently facing or will be facing in the near future. Describe your decision situation.
   a. What makes this decision hard? List some of the issues involved in your decision.
   b. Which of these issues describe something that you can control?
   c. Which of these issues describe something over which you have no control?

8. Write a page on a decision that most changed your life. In looking back, how did your decision making fair on each of the six elements of decision quality?

***9.** Write a short paper on what you did today—how you spent your time, what did you eat, etc. Do you feel like today was well spent? Did you get the things done that you wanted to? Did you make good choices about how you spent your time? Why or why not? If you had $10 million dollars in the bank, what would you do differently with your time? What is preventing you from doing this right now—is it really the money or is something else holding you back?

***10.** Consider the following quote from Ghandi talking about the British occupation: "They cannot take away our self-respect if we do not give it to them."

Explain this phrase and show how it relates to Decision Analysis.

***11.** Give other examples of normative and descriptive fields.

*12. Mohammad is considering whether to go to college for a PhD in decision analysis and is figuring out which schools he should consider. Which of the following considerations should be a part of his decision basis?
   a. Mohammad believes that decision analysis will give him the opportunity to find a good job after graduation.
   b. Mohammad has a preference for schools which have historically successful football teams.
   c. Mohammad will choose among the top three US universities that accept him.
   d. All of the above should be considerations for his decision basis.

*13. You are considering buying stock in a Silicon Valley startup. Which of the following statements should not be a part of the decision of whether or not to invest in the company?
   a. You examine the balance sheet for the company, and are encouraged by the slow rate at which they are spending their venture capital.
   b. You attend a presentation by the CEO and CTO, and are greatly impressed by their exciting vision for the future of the company.
   c. You decide that you would rather invest in a conservative mutual fund which pays regular quarterly dividends than take a large risk of losing all of your investment in the company.
   d. All of the above should be parts of the decision basis.

CHAPTER

2

Experiencing a Decision

CHAPTER CONCEPTS

After reading this chapter, you will be able to explain the following concepts:

- Analysis of a simple decision
- What role probability plays in the decision-making process
- Why the thumbtack is a better deal than the medallion in our example
- The sunk cost
- Decision vs. outcome
- The clairvoyant and the value of clairvoyance

2.1 INTRODUCTION

In the last chapter, we presented our definition of a decision and discussed what is meant by having high quality actional thought. In this chapter, we present and analyze a simple decision that has all of the elements of more complex decisions, and will enable us to think about the way we make decisions in our daily lives. To benefit from this exercise, as we go along, imagine yourself facing this decision and think of the choices you would make. As we have discussed, a decision is a choice that involves an *irrevocable allocation of resources*, and is often difficult to make because some degree of *uncertainty* is present. Both of these elements (resources and uncertainty) are part of the demonstration described below.

2.2 ANALYSIS OF A DECISION: THE THUMBTACK AND THE MEDALLION EXAMPLE

In the following situation, an instructor is speaking to his students about decision making.

Note: *"I" refers to instructor and "C" refers to one or more individuals in the class.*

I: To illustrate a decision, we need both resources and uncertainty. For resources, I have here $100 in U.S. currency. Would anyone like that?

C: (Chorus of ***Yes***!)

I: O.K., it is not surprising that you would like this $100 bill. This seems like an easy decision, but now for some uncertainty. I have here something we can toss. (*Instructor withdraws a coin-like object from his pocket.*)

C: Is that a coin?

Head

Tail

FIGURE 2.1 Two Faces of the Medallion
(Courtesy of the authors)

I: No, but it is very similar. It is a medallion with "60th Snap-On Anniversary" and a box wrench on one side and "The Master's Choice" in large script on the other. (*He holds up the object illustrated in Figure 2.1.*)

(*The medallion is being passed around the class.*)

I: Take a look at it, but don't toss it. Notice that it is as heavy as a large coin, very well made, embossed on both faces, and has a milled edge. Is that right?

C: Yes.

I: Fine, but I wouldn't try to spend it. (*Instructor retrieves the medallion.*) I could toss this medallion to introduce uncertainty, and someone could try to call which face would come up. What do we wish to call the faces of the medallion?

C: Heads and tails.

I: How do you know which face is "heads?"

C: A head usually has a person on it; a tail doesn't.

I: Yes, that works in many countries of the world, but not all. You have to check before you wager. In this case, that rule won't help us because there is no person on either face. (*Instructor selects a volunteer.*) How would you like to name the faces?

Volunteer: I name the one with the wrench as "heads," and the one with the script as "tails."

I: O.K., now we all know what calling the toss of the medallion means. Whenever I refer to the possible outcomes of the toss I will now refer to them as either heads or tails. Clear?

C: Clear.

I: I will now construct a deal. If the owner of the deal calls the toss of the medallion correctly, the owner receives the $100; otherwise, the owner receives nothing. This picture shows a certificate that gives the bearer the right to call the flip of the medallion. (*He presents Figure 2.2 as shown below.*) The certificate is worth $100 only if the owner calls the

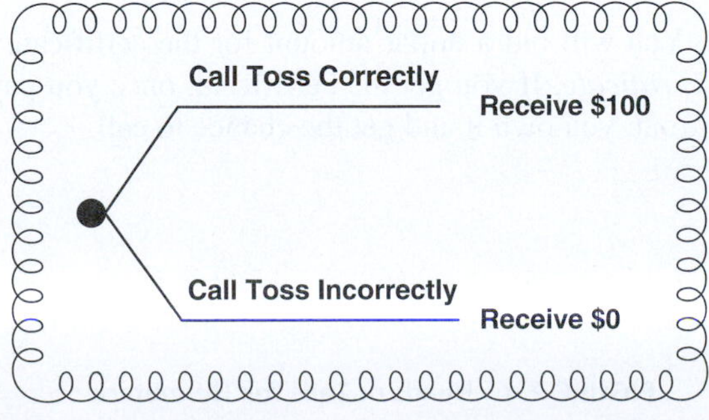

FIGURE 2.2 The Certificate with Scrollwork Around the Edge

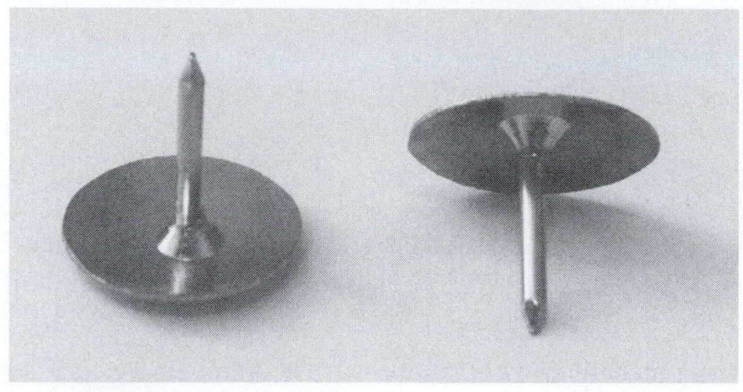

FIGURE 2.3 Two Sides of the Thumbtack: Pin Up and Pin Down (Rachel Youdelman/Pearson Education, Inc.)

toss of the medallion correctly and nothing otherwise. The scrollwork around the edge of the certificate shows that it is a valuable piece of paper. Who would like this certificate?

C: (Chorus of ***I'll take it***!)

I: Of course! Who wouldn't? But first, let's consider another "uncertainty-generating" device. I have here a jar (*opens an opaque, 1-liter plastic jar.*) Into the jar I am going to put a thumbtack. Next, I will screw on the lid, turn it upside down, shake the jar vigorously up and down, and, finally, set it down on the desk on its lid. When I hold the lid and carefully unscrew the jar, we shall see whether the thumbtack has landed "pin down" or "pin up." (*He holds up tack shown in Figure 2.3.*)

To simplify our discussion, we will designate "*pin down*" as *heads* and "*pin up*" as *tails*. (*He presents Figure 2.4 as shown below.*) "Pin down" is the way you would prefer to have it fall if you are just about to sit on it.

To provide the uncertainty in this decision, we might toss either the thumbtack or the medallion. If you call the outcome of the thumbtack toss correctly, you get the $100; if not, you get $0. Who would like the certificate with this uncertainty device?

C: (Chorus of *I'll take it*!)

I: Of course, everyone wants the certificate, but since we have only one we shall have to auction it off and sell it to someone through a bidding process. I get to choose whether the medallion or the thumbtack will be used, and you get to call its outcome—either heads or tails. However, for an additional $3, you can decide which device you want to use. Now these are escrow rules. That means that the bid amount for the highest bidder will be collected upfront before we call the outcome.

2.2.1 Rules for the Bidding Process

I: To summarize the rules of the bidding process:

1. This is a closed bid auction. You will bid a dollar amount for the certificate: *The highest bidder acquires the certificate.* If you get the certificate, once you pay the specified amount, in cash upfront, you own it and get the chance to call.

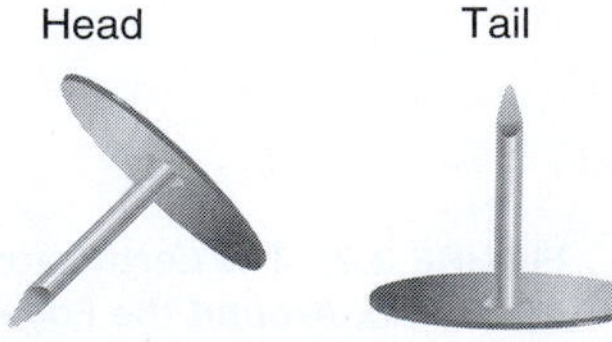

FIGURE 2.4 Heads or Tails for Thumbtack

2. If you are the certificate acquirer, the instructor will choose whether you will call the medallion or the thumbtack. However, at the time of making your bid you can specify which device will be used by agreeing to pay an extra $3 should you become the acquirer. If you acquire the certificate, you will receive either $100 or $0, depending on your call. You will receive the payoff after the demonstration lesson is complete.
3. Any ties between the bids will be resolved by tossing a U.S. 25 cent coin. To avoid ties, you may want to bid including cents.
4. No collusion is allowed. That means you may not collude with a classmate to bid as a group and then divide the possible winnings.
5. The certificate is not transferable (non-negotiable); it cannot be resold.
6. Should you decide to withdraw after making the highest bid, but before making any payment, you must pay a penalty of $10 cash. At that point, the second-highest bidder will be designated the highest bidder, and the process will continue.

I: The rules are summarized here. (*He presents Figure 2.5 as shown below.*)

2.2.2 Starting the Bidding Process

(*At this point, the instructor passes out index cards.*)

I: On your card, please write your name and indicate what you are willing to pay for the certificate. To decrease the chances of having a tie, you may wish to include cents in your bid. It will be in your best interest not to disclose your bid to anyone else. If you want to specify the device for an additional $3, be sure to indicate that as well. The person whose name appears on the card with the largest amount of money will acquire the certificate.

Rules for Bidding on the Certificate:

- Closed bid.
- Bid specifies $ you will pay.
- Device: Medallion or thumbtack.
- We choose the device.
- For an additional $3, you can choose the device.
- Highest $ bidder acquires the certificate.
- Ties resolved by coin tossing. (Suggest bid using cents.)
- Payment is by cash or by check.
- $10 fee for highest bidder to withdraw.
- No Collusion, no syndicates.
- Deal not transferable.
- Acquirer will receive deal payoff after class discussion.

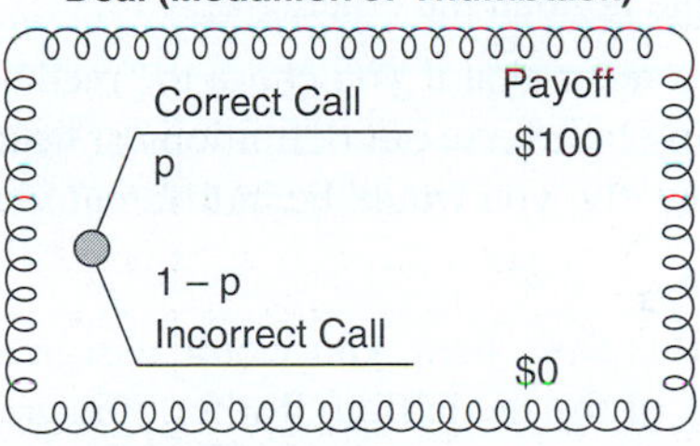

On the Index Card:

Write your
First name, Last name, $bid
Optional: If you would like to pay $3 for the choice of the device, write

"If I am the acquirer, I will pay $3 for choice of device. I choose device ...."

First name: Bill
Last name: Gates
My bid: $xxxx.xx

I will pay $3 to use the ...

FIGURE 2.5 The Rules of the Bidding Process

2.2.3 The "Fair Coin" Fallacy

C: Is this a "fair coin?"

I: First, we need to remember that this is not a coin. It is a medallion. Now, to answer your question, what does the word "fair" mean? I assume you mean it is equally likely to land heads or tails. Correct? (*Student nods.*) Now I am pretty sure that the designer of the medallion did not take any such consideration into account. What chance does the person who owns the certificate have of receiving $100 if he chooses the medallion?

C: 50/50.

I: If this is the case and this is your belief, then perhaps you can rephrase your question as "is the probability of heads or tails 0.5?" You may have a belief that it is equally likely to land either way, but this is just your personal belief. Have any of you tossed a medallion like this before?

C: No.

I: Why, then, do you assign 50/50, or probability one-half, to each face?

C: Because it has only two possible outcomes.

I: The existence of life on Mars has only two possible outcomes: Either there is life or there is not. Do you think there is a 50/50 chance of life on Mars?

C: No.

I: Why, then, do you believe the probability is 50/50?

C: Because the medallion is symmetric.

I: Actually, the medallion is not symmetric at all. As you have seen, both faces of the medallion are, in fact, very different in three dimensions.

C: I believe the probability is 50/50 because I don't think that one face is more likely to come up than the other.

I: You mean that if you chose the medallion and called heads, but then I told you we were going to reverse our definitions of heads and tails before seeing whether you had called it correctly, you would be indifferent to this change.

C: Yes.

I: In this case, then, you would assign a 50/50 chance, but this probability is not a property of the medallion. Rather, it is your belief about how the medallion will land when I toss it. Your indifference to reversal of the definition of heads and tails means that you believe it is equally likely to land either way. Others may have a different view.

Reflection

Take a minute to consider the situation. Think about the three questions below, and the decisions you might make.

1. What device (medallion or thumbtack) would you prefer to call if you had the choice?
2. How much would you bid for the certificate?
3. Would you pay $3 to choose the device you want?

For comparison, we show some statistics on the answers to these questions in Appendix A that is located at the end of this chapter. The data is taken from a graduate class at Stanford University that had an enrollment of 270 students. About 30% of the students chose the medallion and 70% chose the thumbtack. Furthermore, 30% of the students were willing to pay $3 for the choice of the device. Of those, 25% preferred the medallion over the thumbtack.

2.2.4 The Acquirer Revealed

I: Now we have the bids. The acquirer is Sally at $42. Congratulations Sally. Please pay me $42 and the certificate is yours. Thank you Sally. I now write Sally's name on the certificate to acknowledge her ownership. Before we go on, Sally, I see that you have chosen to pay $3 to choose the device. So I will need another $3 from you. (*Sally hands Instructor the $3.*) Thank you. I see that you have chosen the medallion. Why?

Sally: The medallion is like a coin, and I do have some experience in coin tossing, but I know nothing about thumbtack tossing. It is safer to choose the medallion.

I: Your argument may sound appealing and is used frequently by those new to the field, but it is incorrect. In fact, the medallion deal is the worst deal you can get if you believe the probability of heads is one half. No binary device is harder to call than one that is equally likely to come up either way. If you believe that the thumbtack is more likely to fall one way than the other, you are better off using the thumbtack. And I guarantee that it will never be a worse choice than the medallion.

To demonstrate, we can convert the thumbtack deal into a medallion deal at no cost. Does anyone know how to do this?

Sally: Yes. I can simply flip my own coin. If the result is heads, I call heads on the thumbtack. If the result is tails, I call tails on the thumbtack.

I: Correct. To see this, suppose that the thumbtack has already been tossed, but no one had seen how it has fallen. Before calling it, you remove a coin from your pocket and toss it. If it comes up heads, you call heads for the thumbtack; if it comes up tails, you call tails for the thumbtack. Since you believe your coin has a 50/50 chance of coming up heads or tails, you will have a 50/50 chance of winning the $100. Regardless of how the thumbtack has fallen, you have created the medallion deal.

I: You see, Sally, if you were working in a company, it would be difficult to explain to your manager why you would pay an extra $3 to choose the medallion. But this was your decision, you now own the certificate, and you will call the medallion toss. As you know, you cannot sell the certificate. But if you were able to sell it, what it is the least you would accept for it?

Sally: $45.

I: Why do you say $45? (*To the class.*) Should this value depend on what Sally paid for the deal?

C: Sure, she wants to make a profit.

I: Suppose she makes a mistake and bids $95 for the certificate; then what?

C: She would sell it for the most she could get for it.

I: Why doesn't she sell it for the most she can get for it regardless of what she paid for it?

C: Shouldn't what you paid for something affect your selling price?

I: Not really. What you paid for something may tell you a lot about its market value; for example, when buying a souvenir from an artisan in a foreign country. However, what you originally paid for the item becomes more important if you are paying a tax on the difference between your buying price and your selling price. You now care what you paid for it because it will affect your future tax payment. This difference aside, you should generally sell anything you own for as much as you can get for it.

This is called the **sunk cost principle**: How much you spent to get into the situation you are in does not make any difference to your future.

I: Here's an example. Suppose you inherit a house from your grandmother that she paid $5,000 for many years ago. The current market price is about $100,000. Someone approaches you, offers you $10,000, and says, "That is a 100% profit! What more do you want?" You would say, "About $100,000."

I: Sally, let us review your present investment in the certificate. Your bid was $42, plus $3 to call the medallion, for a total of $45. You just said that you would be indifferent to selling it for $45. Is that still true?

Sally: No. Originally, I was thinking that the certificate was worth about $48 to me and that if I could get it for an investment of $45, I would have a good deal. I can now see that I was not thinking straight. If I could sell the certificate now, I would be indifferent to selling it for $44.

I: Sally, here is another way to think about it. With your present understanding, suppose that instead of paying $45 for the certificate, I had given it to you at no charge. However, on the way to class, you lost $45 from your wallet and you had just discovered the loss. Your bank account is down $45 and you own the certificate. Would you be just indifferent to selling it for $44?

Sally: Yes, I would. With the same bank account, how I received the certificate should not affect my selling price. I should not be thinking of profit.

I: Let's continue. Sally, what is the chance you are going to receive the $100?

Sally: 50%.

(*At this point, without warning, the instructor flips the medallion without looking at it and places a piece of paper over it.*)

I: Sally, what is the chance you are going to receive $100 now?

Sally: Still 50/50.

I: Right. Does everyone else agree?

C: Yes.

I: That is correct. It is interesting that some people may see a difference between the probability of an event in the future or the same event in the past, even when they have no new information about it. Knowing the event has occurred does not change your information about calling correctly.

I: (*Instructor peeks under the paper.*) Very interesting. Sally, what is your chance of receiving $100 now?

Sally: Still 50%.

I: And if I were to call the medallion toss, what is my chance of calling it correctly?

Sally: 100%.

I: Right. What is my probability that the medallion has fallen one way or the other?

Sally: For you it's zero for one face and one for the other.

I: So my probability of looking at a head or a tail is not 0.5 on each one, but one for one face and zero for the other.

Sally: Right.

I: Even though the coin has now been tossed for both of us, my probability of heads or tails is definitely different from yours. So, is the probability out in the world or in our heads?

Sally: It's in our heads. It depends on what we know.

2.2.5 The Value of Clairvoyance

I: Right. The probability depends on your state of information. Now, Sally, suppose I offered to tell you what I saw under the paper. Is that information valuable to you?

Sally: Yes. It guarantees that I will call correctly.

I: What is the most you would be willing to pay for it? Class, what do you think she should pay?

C: (*Answers all over the place.*)

I: Is this a matter of logic or opinion? Think of me as a clairvoyant: One who can tell you anything about the past, present, or future as long as the telling requires no judgment. If Sally obtains my services for nothing, what is the certificate worth to her?

C: $100.

I: What if she does not have my services?

C: $44, the lowest price at which she would remain indifferent to selling it.

I: So what is the most she would pay to transform a $44 certificate into a $100 certificate?

C: $56.

I: Right. The most she should pay to know how the medallion has landed, and to call this result, is $56. We call this the **value of clairvoyance** on the result of the medallion toss. Since you would never pay more than that value of clairvoyance for any information source that does not provide clairvoyance, this concept helps you rule out many information-gathering activities.

Reflection: The Clairvoyant

The Clairvoyant can tell anything physically determinable past, present, or future and can compute with infinite resources. However, the clairvoyant cannot exercise judgment. If you ask the clairvoyant, "How many people in the room are happy?" He will respond that he cannot answer. If you ask him the total age of all the people in the room, he can tell you if you specify exactly what you mean by "age."

The *value of clairvoyance* is a key concept in decision analysis because it shows you the most you would pay to know something uncertain. (See Figure 2.6.)

FIGURE 2.6 The Value of Clairvoyance is a Key Concept in Decision Analysis
(Destina/Fotolia)

I: If the clairvoyant wants to be paid less than Sally's value of clairvoyance, she should buy the information, but if he wants more, she should not. Also, note that in this situation, once we know the price at which Sally is indifferent to selling the certificate, the value of clairvoyance is a matter of logic.

I: We say that an information source provides imperfect information if it does not yield clairvoyance. Suppose, for example, a person sitting at the back of the room has binoculars and thinks he saw how the medallion fell. If he asks for more than the value of clairvoyance for his information, then Sally should not care how good his eyesight was, since she should not pay any more than the value of clairvoyance for any type of information gathering activity about how the medallion fell. Companies routinely spend many times more than the value of clairvoyance for information that is not as valuable to them as they believe.

I: To continue, consider the possibility of my offering Sally this deal. To ensure that I carry this out correctly, Sally can appoint an agent. I have a watch that shows seconds digitally. Suppose I glance at the watch and note the number showing the seconds. If the number is between 00 and 49, I will tell her the face of the medallion that is up. If the number is between 50 and 59, I will tell her the face of the medallion that is down. That is, she has a 5/6 chance of correct information about the face that is up and 1/6 chance of incorrect information. Would this imperfect information be worth something to her?

C: Yes.

I: More than $56?

C: No, less.

I: Right, and we have methods to help her figure out what she is willing to pay for such information; we shall present these methods later. Are most of life's information gathering opportunities like clairvoyance, or like the deal with the watch?

C: The watch.

I: Yes. In every field of decision making, and in most information gathering activities—surveys, pilot plants, test wells, medical tests, and controlled experiments—we encounter imperfect information.

2.2.6 The Call

I: Well, Sally, we are almost ready to let you call the toss of the medallion. From this point on, let's be clear about what we are going to learn. Are we going to learn anything about Sally's decision making ability after observing the outcome of the toss?

C: No. At the time, she paid \$42 plus \$3 for something she believed was worth \$48. Now, she believes the medallion is equally likely to land either heads or tails, and none of us has any new information.

I: So really, we are not going to learn anything. All we are going to find out is whether or not she takes home the \$100. In other words, we know the decisions she has made, but we don't know whether she will get the more desirable outcome. Does the world tend to reward people on their decisions or on their outcomes?

C: Outcomes.

I: That is something to consider. Now Sally, what do you call?

Sally: "Heads" ("or Tails")

I: You are right, here's the \$100 (or you are wrong, thanks for being a good sport).

2.3 LESSONS LEARNED FROM THE THUMBTACK AND MEDALLION EXAMPLE

The previous demonstration involved a simple decision, yet presented a challenge to our intuition. There are several lessons we can learn from the thumbtack demonstration and we summarize those below.

2.3.1 Probability is a Degree of Belief

In probability classes, you might be used to seeing terms such as "fair coin" or a "perfectly shuffled deck of cards." These terms often lead us, incorrectly, to believe that probability is a property of the coin since it is fair, or that probability of drawing a card is a property of the deck. In fact, what the instructor really means when he says "fair coin" is that he wants us to assume our belief is equally likely that the coin will land heads or tails.

Reflection: If a Magician Tosses the Coin

We have a magician friend who tosses coins and always makes the coin land heads. If you knew this guy would be tossing a coin for you, would you still believe heads has a probability of 0.5? No. Therefore, probability is nothing more than our degree of belief that a certain event or statement is true.

In many statistics classes, there is a notion of a "long run fraction" of repeated trials that is supposed to represent the probability of an event. However, understanding probability as a degree of belief is much more useful. In real-world decision making, we never encounter infinite repeated trials. For example, consider the probability that it will rain tomorrow in Palo Alto. Suppose you knew the number of times it rained on that day of the month for the last hundred years. Would you use the fraction of times it had rained as your probability of rain tomorrow?

Instead, we would recommend that you consider this data, but also obtain current weather forecasts and then go outdoors to look at the sky. Based on this information, you would assign a probability to represent your belief that it will rain tomorrow. As your state of information changes, your probability assignment may also change. If you look out the window again and see that it is now raining, you would revise your probability based on this new information. Therefore, every probability assignment you make should be conditioned on your *current state of information*.

2.3.2 Probability Comes from a Person

Probability is not a property of the medallion or the thumbtack, but comes from a person. "The medallion has a probability of 0.5 for landing heads" has no meaning. People may feel uncomfortable with the thumbtack demonstration because they are not accustomed to assigning probability. They are used to analyzing data and calculating statistical quantities from data. Probability, however, does not come from data. It represents a person's state of information about an uncertainty. Therefore, before we can talk about any probability assignment, we need a *person*.

There is no such thing as a "correct probability." To illustrate, suppose two people are talking about the probability of rain tomorrow. One says the probability of rain is 0.5; the other says the probability of rain is 0.7. Now suppose it rains. Who was correct? They both were. If it does not rain, they are still both equally correct. There is no such thing as the actual probability of rain; each individual presented a belief about the chance of rain.

In the previous demonstration, at the point of calling the result, Sally believed she had a 50/50 chance of calling correctly, but the instructor knew the result of the toss. His probability was 1 and hers was 0.5. The two probabilities were different, since they represented different states of information. Therefore, every probability assignment should specify the person who is making the assignment.

2.3.3 Thumbtacks and Probability

Over the years, many people have presented their reasons for believing the thumbtack is more likely to land one way or the other. Here is a popular one: We call it the "coin-nail" model. Suppose the thumbtack had its pin cut off, leaving just the round head. It would look more like the medallion, and many people would assign a probability of 0.5 for its landing on either side.

Suppose the pin was very long. Then the thumbtack would almost always fall pin down. The actual length of the pin is somewhere between zero length (no pin at all) and the very long pin. This reasoning would require you to assign a higher probability for it to land pin down than pin up (see Figure 2.7).

Another method of reasoning refers to the principle of minimum potential energy, suggesting that the thumbtack is more likely to land with its pin up because the center of gravity is lower when it is pin up. Most of these arguments, however, do not take into account the actual jar, the

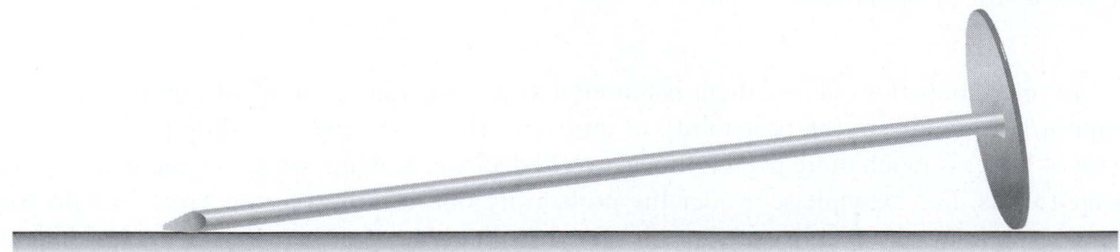

FIGURE 2.7 Thumbtack with Long Pin

shaking mechanism of the jar, or even our existing information about thumbtacks. Once again, we need a person to make the probability assignment.

2.3.4 The Thumbtack Deal is at Least as Good as the Medallion Deal

Even though Sally believed she had a 50/50 chance of calling correctly with the medallion, she should never have paid extra. The demonstration showed how to convert the medallion deal into the thumbtack deal by simply flipping a coin and using the result to call the thumbtack. Paying $3 for the chance to choose the medallion was, therefore, a bad decision. We will refer back to this point in Chapter 6.

2.3.5 The Sunk Cost Principle

The *sunk cost principle* is another fundamental distinction we make in decision analysis. According to the sunk cost principle, a decision is made by considering only the possible futures that it might generate. The historical account of how the situation developed is pertinent only to the extent that it has provided information useful in assessing the likelihood of these futures. Any resources consumed in the past are pertinent to the present decision only through this learning effect. The resources consumed may be things such as money, time, and effort.

Reflection

If you have been working on a "do-it-yourself" plumbing job all day and you still aren't finished, you may need to decide whether to invest more of your time or to hire someone else.

The only question is whether it is a better use of your resources, time, and money, to continue by yourself or to hire a professional. Note that from the time you have already spent, you have learned something about your degree of competence in carrying out this job. However, the time you have wasted is not otherwise pertinent to the decision.

Saying to yourself, "look at all the time I have spent on this, I have got to finish it" is falling prey to a temptation to violate the sunk cost principle.

Often, the sunk cost principle often conflicts with human nature. That is, we find it hard to avoid blaming ourselves for resources wasted in arriving at the present situation. Imagine, for example, you are present at a Board of Directors meeting, helping to decide whether to abandon a foundering project. Before the meeting is over, you will most likely hear something like this:

"Look at all the time and money we have already wasted."

Seldom will anyone laugh, and yet laughter would be appropriate. The lost resources are meaningless to the future of this project. The future value lies in the experience gained from the bad outcome.

For several reasons, the sunk cost principle can create confusion. First, people often believe that experience gained from a bad outcome should never be used in making decisions. In truth, learning from experience should always be a part of good decision making. It is the wasted or expended resources that have no real bearing.

Confusion also arises when people consider making a decision that will limit future action. For example, a person who goes on a diet might decide to eliminate tempting foods from the house to increase their chances for success. Similarly, an alcoholic might check into a rehabilitation facility that will restrict opportunities to drink.

Reflection

In this famous excerpt from Homer's *Odyssey*, Ulysses makes a decision in an attempt to limit future action.

Ulysses had to sail near the coast of the Sirens, who were known to tempt mariners to a watery fate by luring them with their cries. Ulysses had his crew stuff their ears with wax and bind him to the mast, with orders not to release him no matter how much he begged them. On hearing the Sirens' calls, Ulysses struggled to get loose and begged to be released, but his crew refused. As a result, they sailed by in safety. (Of course, Ulysses could also have stuffed his own ears with wax, but being tied to the mast makes a better story.)

Like learning from past experiences, mindfully limiting your future actions can improve your chances of making a good decision. None of the examples above violate the sunk cost principle. In fact, limiting your future actions can be prudent.

Finally, people sometimes are not sure whether keeping records of past purchases will help them to make informed decisions about future spending. Such records are useless as any kind of guide to current market prices, due to their constant fluctuation. However, there is one compelling reason to track your purchases: Taxation. What you have paid for a share of stock, or for your home, may have an important bearing on your future cash flows because you will often be taxed on the difference between what you paid for something and what you receive for it when you sell it.

Common maxims advise us to follow the sunk cost principle. One is "Don't cry over spilt milk." More to the point, another warns "Don't throw good money after bad." A more optimistic way of looking at a decision, consistent with the sunk cost principle, would be "Today is the first day of the rest of your life."

2.3.6 The Value of the Certificate

When making a decision, monetary value is often not the only consideration. When participating in the demonstration described earlier in the chapter, a student once bid $100 for the certificate. He mentioned that the opportunity to get the certificate and be the center of class discussion was worth more to him than the money itself. Furthermore, he explained, the experience would be a nice memory of his learning this material. As a result, he valued the certificate at more than the amount of money he could receive.

Other people may have religious beliefs about not owning such a certificate and would value it at zero; they would refuse it even if they received it at no cost. Still others may be willing to engage in the demonstration and may even have the same beliefs about the probability of heads or tails with the thumbtack, but still place different values on the certificate. In future chapters, we will clarify values for certain and uncertain deals, but for now, understand that, for many reasons, different people may have different values for the same deal.

2.3.7 The Value of Clairvoyance

As we discussed, the value of clairvoyance on uncertainties you face is a key concept in decision analysis. If you knew the future that would follow your choice, the decision simplifies.

> *The value of clairvoyance on any uncertainty is the most you would pay the clairvoyant to know the outcome of that uncertainty.*

The ability to buy clairvoyance is rare in practical decision-making. However, there are many information-gathering activities that will provide imperfect information at a cost: market surveys, medical tests, pilot plants, seismic measurements, wind tunnel experiments, etc. They can be valued using the same principle:

> *The value of clairvoyance on any information gathering activity is the value of clairvoyance on the results of that activity.*

Calculating the value of clairvoyance does not require the existence of the clairvoyant, but only the concept of one. We shall discuss how to exploit the notion of clairvoyance in more detail in future chapters.

2.3.8 Good Decision vs. Good Outcome

This chapter's demonstration provides an excellent example of the difference between a good decision and a good outcome. Once the decision is made, we will not learn anything about the quality of the decision by observing its outcome. Furthermore, once the outcome is revealed, there is no point in seeing it as good or bad: it is simply the outcome, the starting point for future decisions.

2.4 SUMMARY

- The thumbtack deal is better than the medallion deal.
- Probability comes from a person. It is not a physical property. It depends on the person's information.

KEY TERMS

- Sunk cost principle
- Value of clairvoyance

APPENDIX A Results of the Thumbtack Demonstration

Figure A.1 shows a histogram of bids made by 270 graduate students for the opportunity to call either the medallion or the thumbtack deal. In general, the bids are too low for the decision situation they are facing. As shown here, sometimes students misunderstand the demonstration and make a bid of $95 or more. These students often choose to withdraw for $10.

Thirty percent of students were willing to pay $3 for the right to choose the device. 25% of those, who were willing to pay $3, preferred the medallion over the thumbtack. Given our previous discussion showing that you cannot be worse off by receiving the thumbtack deal, it would be difficult to explain why you would pay $3 to get the medallion deal.

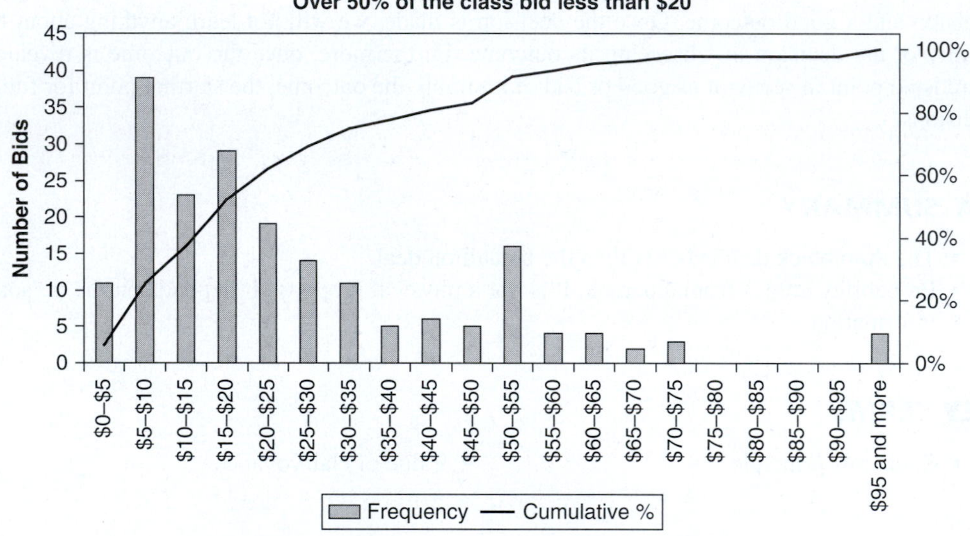

FIGURE A.1 Histogram and Cumulative Histogram of Bids

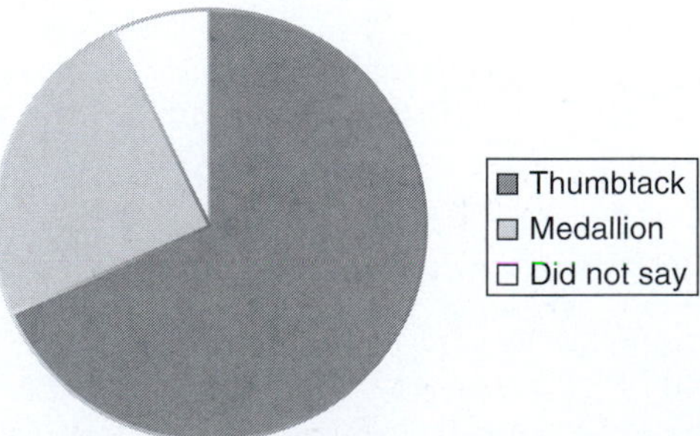

FIGURE A.2 $3 Choice Pie Chart

PROBLEMS

Problems marked with an asterisk (*) are considered more challenging.
Problems marked with a dagger (†) are considered quantatative.

***1.** Suppose that at $45, you were the high bidder for the certificate auctioned off during the chapter demonstration. After winning the certificate, you determine that the least you would be willing to sell the certificate for is $75. The instructor offers to tell you truthfully how the thumbtack has landed before you make your call. What value would you assign to this information?

a. $25
b. $30
c. $55
d. You need more information to answer this question.

***2.** What is the result of a high-quality decision analysis?

a. Knowledge of what will happen as a result of your decision
b. Guarantee of a good outcome
c. Clarity of action
d. None of the above.

***3.** In this chapter, we described a fallacy involving the "fairness" of the coin. What was this fallacy?

a. A "fair coin" implies a 50/50 chance at heads vs. tails, but we can never be sure of the actual probability of the coin landing heads.
b. "Fairness" is not clear, and, therefore, should not be used.
c. "Fairness" is not a property of the coin, but of our beliefs about the coin.
d. A flipped "fair coin" and a spun "fair coin" have different probabilities of landing heads.

***4.** Which of the following is an example of falling into the sunk cost trap?

a. I need to think about how much I paid for the house five years ago, since that affects the taxes I need to pay when I sell it.
b. Let's do some research on the past performance of this company before we invest in it.
c. I bought this for $15; therefore I shouldn't accept to sell it for anything less than $15.
d. All of the above.

***5.** Yosem has already flipped a new Massachusetts quarter three times, and each time it has come up heads. He then says to his friend Pablo, "I'll bet you a dollar that the next flip will come up tails." Pablo agrees to the bet. Yosem flips the coin, it comes up tails, and so he collects a dollar from Pablo. Which of the following best described Yosem's action?

a. The first three coin flips clearly showed that the Massachusetts quarter is not a "fair" coin. Therefore, Yosem's bet on the coin coming up "tails" was a bad decision.
b. Since any ordinary quarter is "fair," tails was bound to come up eventually. Yosem made a good decision to bet on the next coin flip being tails, and had a good outcome.
c. After observing three heads in a row, Yosem should have bet more that tails would come up next. Betting only one dollar was a bad one because it reduced his winnings.
d. It is not unusual to observe three consecutive heads when flipping coins. Yosem was just lucky and got the good outcome.

***6.** Nathalie is an expert in taxes and she prefers more money to less. She bought a ticket to see *Ariadne auf Naxos* at the opera. It cost her $50. Unfortunately, she is sick on the day of the performance and she decides not to go. The opera does not reimburse tickets but considers unused tickets as donations. Her friend Robyn offers to buy her ticket for $15.

I. She rejects that offer feeling that since she paid $50, Robyn should pay her no less than $50.
II. She rejects that offer because she prefers to donate the unused ticket back to the opera. She knows that a donation of $50 will provide her a tax credit of $20.

In which cases is Nathalie violating the sunk cost principle?

a. I only
b. II only
c. Both I and II
d. None

7. How many of the following is an example of a violation of the sunk cost principle?

I. Let's do some research on past performance of this company before we invest in it.
II. I need to think about how much I paid for the house five years ago because I do not want to lose on the sale.
III. I need to think about how much I paid for the house when I sell it if it will reduce my tax bracket and I will have a higher profit.
IV. My car transmission just broke down and I decided to scrap the car because the cost for repairing the car exceeds the amount I paid for the car.

a. Only one is a violation of the sunk cost principle
b. Only two are violations of the sunk cost principle
c. Only three are violations of the sunk cost principle
d. All of the above are violations of the sunk cost principle

***8.** Jasmine believes that the thumbtack has an 80% probability of landing pin down. Winston believes that this probability is closer to 40%. When the instructor flips the tack, however, it lands pin up. What may we conclude from this event?

a. Jasmine's belief about the tack's probability of landing pin down was incorrect. It should have been much less.
b. Winston's belief reflects reality more accurately than Jasmine's.
c. The probability of landing pin down is actually 80%, but Jasmine just got a bad outcome.
d. None of the above.

9. Mary assigns probability 1/7 to a particular die landing on "5" next time it is tossed. Consider the following two statements about her probability assessments:

I. Mary's probability must be wrong. The correct probability of a die landing on "5" is 1/6th.
II. Mary can't assign this probability until she has seen at least seven flips.

Which of the statements are true?

a. I only
b. II only
c. Both I and II
d. Neither I nor II

10. Which of these statements necessarily violates the sunk cost principle?

a. The engine in my car just broke down. I decided to scrap the car because the costs for repairing the car exceed the amount I paid for it.
b. After observing how financial stocks performed last week, I will sell all of my shares of Bank of Amerigo next week.
c. I will buy a $1 book that I would not otherwise want in order to make my online purchase exceed $25. That way, I can save $5 on the shipping fees I would otherwise pay.
d. I would have paid $30 for a ticket if it all went to the venue, but I won't pay it knowing Ticketbuster gets most of the money in fees.

11. Big Game is the annual football rivalry game between Stanford University and UC Berkeley. Jack assigns a probability of 0.6 that Stanford will beat UC Berkeley in Big Game 20XX. Kim believes that the probability is actually 0.4 and challenges Jack to a bet. If UC Berkeley wins, Jack pays Kim $100 and if Stanford wins, Kim pays Jack $100.

Which of the following statements is valid?

a. If Kim wins the bet, we know that she made a good decision.
b. The objective probability that Stanford wins the game is between 0.4 and 0.6.

c. The clairvoyant's probability that Stanford will win Big Game 20XX is greater than 0 and less than 1.
d. None of the preceding statements is valid.

***12.** How many of the following statements violate the sunk cost principle?

I. I bought my tennis racket online at $300, but I found its head was too heavy for me so I choose to put it up for sale. I will not sell it for less than $300 because that would be a loss to me.
II. Despite waiting in line for 3 hours for the Black Friday shopping day, Mary decided to go back home before the store opened since she cannot bear the cold wind.
III. A manager should look at her employee's past performance reviews when deciding whether to grant him a promotion.

a. 0
b. 1
c. 2
d. 3

***13.** Which of the following statements violates the sunk cost principle?

I. I called customer support because my digital camera does not work anymore, and I have been on hold for a few minutes. I am wondering whether I should give up and try again the next day, or whether I should stay on the line and keep waiting. As I make that decision, I consider how much time I have already spent waiting, because it helps me think of how much longer I might need to wait before I can speak to a representative.
II. When selling my small business, I should think about how much I paid to purchase the company 5 years ago because it may change my tax bracket and affect my profit.

a. I only
b. II only
c. Both I and II
d. Neither I nor II

***14.** How many of the following statements violate the sunk cost principle?

I. I want to stay until at least the 7th inning stretch so I can hear Cameron sing *Take me Out to the Ball Game*…
II. I have to finish drinking this beer because I paid $8 for it!
III. I prefer to go to the bar rather than pay $100 for the only baseball tickets left, but since I drove this far to get here, I'll buy the expensive tickets.
IV. Since you are late getting here, the best thing to do is buy outfield tickets!

a. 0
b. 1
c. 2
d. 3

†*15. William is a private equity guru. One year ago, the owners of a troubled retail company accepted William's buyout offer of $15M. At the time, William's PIBP for the company was $30M. William worked very hard to restructure the company over the past year, but, unfortunately, the company went bankrupt. The following statements are thoughts that William had after the bankruptcy.

How many of them **do not** violate the principles of decision analysis?

I. I have the chance to invest an additional $1M in order to earn $3M extra on the liquidation of the company's assets. However, I shouldn't make such an investment because it won't fully recoup my $15M original investment.
II. I am very unhappy with the bankruptcy, but I still feel like the buyout was a good decision.
III. I should remember this outcome and learn from it so it will help make future investment decisions.

a. 0
b. 1
c. 2
d. 3

†***16.** Theater Tickets

You have made plans to attend the theater alone. The tickets cost $20 each. Consider the following two scenarios:

a. You have purchased a ticket for $20. When you arrive at the theatre you discover you have lost your ticket, and find you have another $20 bill in your wallet. Would you buy another ticket and attend?

b. You arrive at the theater only to discover that on the way you have lost one of the two $20 bills in your wallet. Would you spend the other $20 bill to buy a ticket and attend?

†***17.** Project Funding

As the President of XYZ's largest subsidiary, you have two project proposals on your desk.

Note: All monetary values cited are in today's U.S. dollars.

I. Last year, we approved Project I. Upon completion, it will generate $100 M in revenue, and cost $90M. You spent the $90M. The proposal before you now states that an additional $20M is required to complete the project and realize the $100M in revenue. If you do not spend the additional $20M, no revenues will be realized.

II. Project II will generate $80 M in revenue. A trusted advisor tells you that you can secure this deal at a cost of $20 MM.

Now consider the following questions:

a. If you had $100 MM to invest today, which project(s) would you fund?

b. If you had $20 MM to invest today, which project(s) would you fund?

c. If you had known from the start that there would be the additional $20MM required to complete Project I, would it still have deserved funding?

CHAPTER

3

Clarifying Values

CHAPTER CONCEPTS

After reading this chapter, you will be able to explain the following concepts:

- Value in use
- Value in exchange
- Personal Indifferent Buying Price (PIBP)
- Personal Indifferent Selling Price (PISP)
- Cycle of ownership
- Effects of wealth, time and information on the cycle of ownership
- Market buying and selling price
- Value around a cycle of ownership

3.1 INTRODUCTION

In the last chapter, we explored the decision making process through our analysis of the medallion and thumbtack demonstration. We asked a student named Sally to consider the value of a certificate she had acquired. We witnessed her confusion as she considered the difference between her purchase price and the actual certificate value to her. The issue of valuation arises in so many decisions that we shall now present the basic concepts in familiar settings that have no uncertainty. Later we shall address the additional complications in valuation created by uncertainty. The ideas introduced in this chapter will support much of our future development.

3.2 VALUE IN USE AND VALUE IN EXCHANGE

To illustrate the concept of valuation, we begin with another discussion between the same instructor and his class.

Note: *"I" refers to instructor and "C" refers to one or more individuals in the class.*

I: Today, we shall be talking about buying and selling prices. Frank, that is a very nice shirt you are wearing today.

Frank: Thank you.

I: What would you be willing to sell it for?

Frank: $500.

C: (*Laughs*)

I: I know you would, but let me ask you, what is the least you would be willing to sell it for? I assure you we are not really going to buy it, so you can just tell the truth. What is the least amount of money we could give you that would make you indifferent between owning the shirt without that money and not owning the shirt, but having that money?

Frank: $30.

I: You would not take $29, but would definitely take $31?

Frank: Right.

I: We call the $30 Frank specified his Personal Indifferent Selling Price (PISP) for the shirt. PISP is an important concept in decision analysis.

I: Who likes Frank's shirt?

Joe: (*Raises his hand*)

I: Joe, if I take care of the cleaning, how much would you pay for Frank's shirt?

Joe: $5.

I: Joe, I remind you that we are not really going to sell Frank's shirt. So just imagine that we are, and tell me, just between us, what is the most you would pay for it?

Joe: $20.

I: That means that at $19 you would definitely buy it and at $21 you would not.

Joe: Yes.

I: We call the $20 your Personal Indifferent Buying Price (PIBP) for Frank's shirt. PIBP is another important concept in decision analysis.

I: Now, would Frank and Joe have a deal?

C: No.

I: Correct. For two people to have a deal, the Personal Indifferent Buying Price for the one who doesn't own it has to be higher than the Personal Indifferent Selling Price for the one who does own it.

By the way, Joe, what would you pay for a second identical shirt just after you've bought the first?

Joe: $5. I want variety.

I: I see. Indeed, there is no reason why a second identical item would have the same Personal Indifferent Buying Price.

Now, it is important to note that what we have been talking about are ***values in use***. Frank is not in the shirt-selling business and Joe is not in the shirt-buying business.

Note: Personal Indifferent Selling Price (PISP)

A *selling price* arises when you sell something you own. The word *indifferent* signifies that it is the amount at which you are indifferent about whether you continue to own the item, or give it up to receive that sum. The word *personal* indicates that it depends on the person: The amount can change from one person to another.

Note: Personal Indifferent Buying Price (PIBP)

A *buying price* arises when you buy something you do not own. The word *indifferent* signifies that it is the amount at which you are indifferent about whether you buy something, or continue not to own it. The word *personal* indicates that it depends on the person: it can change from one person to another.

Note: Value in Use

The PIBPs and PISPs reflect the *values in use* of the shirt. They do not reflect the actual cost of the shirt or the potential selling price. Out in the world, however, there are indeed people who are in the business of buying and selling things, such as shirts, to make a living. We call this the *market*.

C: And what about brokers, do they have a PIBP and a PISP?

I: Brokers are people who put together deals between potential buyers and sellers. If Barbara is a broker, for example, and she knows someone who will pay $50 for Frank's shirt, she will happily pay Frank his PISP of $30, or even more. Barbara may think this is the ugliest shirt in the world, but since she can sell it for $50, she cares only about the profit she can make in this transaction.

Note: Market Buying and Selling Prices

The *market buying price* is what you would have to pay in the market for a particular good. The *market selling price* is what you would receive in the market for selling something you own.

C: But why doesn't my PISP for an item depend on the market price? For example, suppose I own a house and I know its market price is a lot more than what I paid for it. Should I not sell it for as much as I can get out of it?

I: By all means. Now remember, your PISP is the value you are getting out of the house by living in it and owning it. If the housing market suddenly drops or rises, this value of the house should not change, as long as your use of the house continues to have the same benefit to you. When you are going to sell your house, you should sell it for as much as you can get out of it. *Knowing your PISP does not mean you need to sell it at your PISP*. You would be just indifferent to selling it at that price.

I: You should also understand that the market price is not well defined. For example, if I ask a realtor about the market price of my house, she may respond, "That depends on how long you are willing to wait to sell it. If you need cash by next week, we can get x dollars, but if you are willing to wait six months, it could go for as high as y dollars."

Note: Value in Exchange

The market buying price and the market selling price reflect the *market value, or the value in exchange,* of a given good or service. The market value is specified by actual market buying and selling prices.

C: What about going shopping or buying groceries? In this transaction there is a market, yet you buy something for your own use. Right?

I: Yes. When you go shopping, you buy things for which your personal indifferent buying price is greater than the market selling price. Have you noticed when you buy groceries, you leave saying "Thank you" and the cashier also says "Thank you." You bought things for less than your PIBP, and the storeowner, who is a broker in this case, sold the goods for an amount that is higher than what he paid for it. Both parties are happy with this transaction.

I: We're learning a lot of new terms today, so let's take a moment to recap.

Your **Personal Indifferent Selling Price**, or **PISP**, is the least you would be willing to accept to forgo the use of something that you own. Your PISP does not depend on the market buying price, but on its value in use to you. It also does not depend on what you paid for it. The PISP is the price at which you become indifferent to losing the item.

Your **Personal Indifferent Buying Price**, or **PIBP**, is the most you would be willing to pay to obtain the use of something you do not own. Your PIBP does not depend on the market selling price, but on its value in use to you. The PIBP is the price at which you become indifferent to buying the item.

The degree of value a good or service creates for you is also known as its *value in use*. It has nothing to do with selling or exchanging, and does not depend on the market price, but only on the benefit derived from owning or receiving the item. Value in use is specified by personal indifferent buying and selling prices.

The **market buying price** is what you would have to pay in the market for a particular item or service. The **market selling price** is what you would receive in the market for selling something you own. Note that these prices are fluid because they are observable only within a fluctuating market. Often, you will only learn the market buying and selling price for a unique item— for example, a particular house or piece of art—at the time of the actual transaction.

The **value in exchange** is the market value of any good or service. Buying and selling prices are what determines the value in exchange of a given item.

Why are these concepts important? Once you have established your PIBP for an item, getting it at a lower price is a good deal for you. This does not mean that you should not try to get it for an even lower price, but paying anything less than your PIBP is a good deal for you, since your value in use will be higher than what you paid for it. Conversely, once you have established your PISP, selling something for anything higher than that price will be a good deal for you.

In principle, we can think of having PISPs for everything in our lives that we currently own, and PIBPs for everything that we do not own, but might buy. We would then go through life selling those things with market buying prices higher than our PISPs and buying those things with market selling prices lower than our PIBPs. Of course, after any exchange that significantly affects our wealth, all of our PISPs and PIBPs may need to be reconsidered.

To illustrate the concepts of PIBP and PISP in this demonstration, we used items that had no uncertainty. Frank knows his shirt quite well, and Joe could see the shirt before he gave it a PIBP. We introduced these same concepts in the last chapter, during the thumbtack demonstration. We asked Sally for the least she would be willing to sell the certificate for once she acquired it. She said $45. While the outcome of the certificate was uncertain, the certificate had a definite value for her, which we now refer to as her PISP. We will often refer to the PISP of an uncertain deal, such as the certificate, as the *certain equivalent* of the deal. Sally believed she had made a good deal, since she paid $42 for something that she would not sell for less than $45.

Compared to the complex decisions we often face in our actual lives, the thumbtack deal was relatively easy to evaluate. For this reason, Sally was able to quickly calculate her certain equivalent of this deal. In chapter 11, we shall discuss how to calculate your PIBP and PISP for deals with a larger number of possible monetary outcomes.

Now let us continue our class discussion by defining values around a **cycle of ownership**.

3.3 VALUES AROUND A CYCLE OF OWNERSHIP

I: Let's get back to Frank and his shirt. Suppose that Frank sells the shirt for $30—his PISP. After a few minutes, he changes his mind and considers buying it back. What is his personal indifferent buying price?

C: $30.

I: Correct. To see why this is the case, suppose that at the time of the transaction, the buyer decides not to buy the shirt. If Frank is selling the shirt at his PISP, then he should be indifferent as to whether he sells it or not. He is, then, indifferent between two situations. One where he owns the shirt, and one where he does not own the shirt, but has extra money equal to his PISP. To maintain that state of indifference, when he sells the shirt at his PISP and buys it back, the amount of money he pays must be identical to his PIBP. We call this notion of an instantaneous buying and selling at your PIBP and PISP a cycle of ownership. Within a given cycle of ownership, buying and selling prices must be the same.

Figure 3.1 identifies the two properties most important to a cycle of ownership Wealth and ownership. The grey shaded area represents your wealth when you own the shirt. The striped area

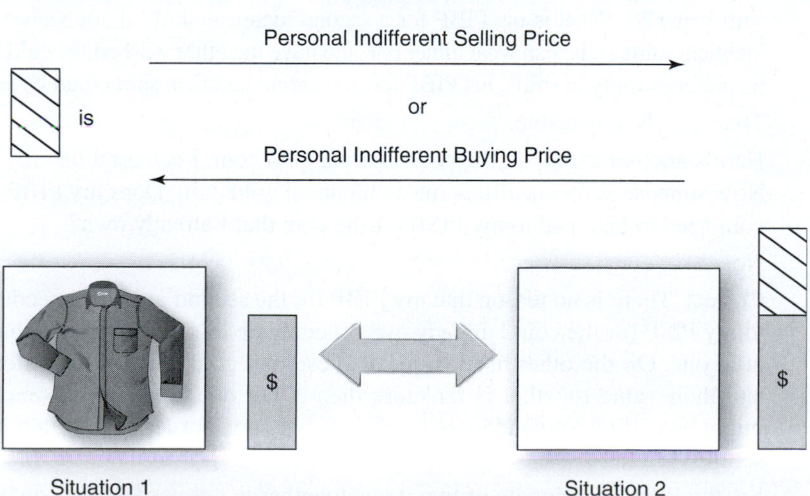

FIGURE 3.1 PIBP and PISP within a Cycle of Ownership

represents either a personal indifferent selling price or personal indifferent buying price at different points in the cycle. Within a cycle of ownership, whether you end up owning the shirt and having less money (equal to the PISP), or passing up the shirt and having more money (PIBP), the personal indifferent selling and buying prices must be the same.

You do not need to actually sell and buy an item at your indifference prices to think about your PIBP & PISP.

The concept of a cycle of ownership simply helps you determine whether you have really indifferent between winning an item and saving less money equal to your PIBP or PISP or not.

Note: Cycle of Ownership

Instantaneous buying and selling at your PIBP and PISP is known as a *cycle of ownership*.

Within a cycle of ownership, the PIBP and PISP must be equal. A cycle of ownership requires no passage of time, no new information, and no change in the state of wealth.

A cycle of ownership only occurs when the price paid for the item is equal to the PIBP for the item. If we pay less or more, there cannot be a cycle of ownership because of our new state of wealth. The cycle of ownership can also change by time or by receiving new information since it can change our values in use.

C: Can my PIBP and PISP include a range of prices?

I: No. The prices must be fixed. By definition, we say the PISP is the least you would be willing to accept to sell an item, while remaining indifferent as to whether you sold it or not. If you had a range, then the PISP is the minimum of that range. The same applies to the PIBP, since it is the maximum you would be willing to pay. If there is a range, the PIBP would be the maximum. There is also another important reason for not having a range. Suppose Frank had a range of PISP from $25 to $35 for his shirt. If he is indifferent to any of these values, then he must be indifferent to receiving $25 instead of $35. Frank's bank account would not survive for long if he had the habit of considering $25 equal to $35.

I: Now, I want you to consider the question of identical items. Joe said his PIBP for Frank's shirt was $20. What is his PIBP for a second identical shirt? If Joe values having two identical shirts so he can wear either one and have the other washed, it could be higher. If he prefers variety in shirts, his PIBP for the second identical shirt could be less.

C: That sounds reasonable.

I: Here's another example: Suppose I own a gold coin. I can establish my PISP for it. Now suppose someone offers me an identical gold coin. Does my PIBP for the new coin need to be equal to my PISP for the coin that I already own?

C: No.

I: Correct. There is no reason that my PIBP for the second gold coin needs to be equal to my PISP for the coin I already own. It could be less, since my keychain can only hold one. On the other hand, suppose these two gold coins were collectors items, and their value together is far more then either one alone. In this case, a broker might be willing to spend more for the second one in order to make a profit later, by selling them as a set.

C: So, the PIBP of a bundle of two items together is not necessarily equal the sum of their individual PIBP's?

I: Correct. The gold coin example illustrates that it can be higher or lower. Consider a matched pair of shoes. If you have two legs, your PIBP for just the left shoe, or just the right, might be zero, since you can't wear either shoe without the other. However, your PIBP for a matching pair of shoes would be much higher.

C: Can the PISP be negative?

I: Think about that. Are there things in our lives that we own and that we would pay money to get rid of?

C: Trash.

I: Yes, that's a good one. Any others?

C: (*Silence*)

I: How about risk? We pay to get rid of risks in our lives in the form of insurance. We sell those risks at a negative PISP. An insurance company may be willing to assume ownership of one of my liabilities for a market buying price, the premium, that is less in magnitude than my PISP.

I: Here's another question for you: Can the concepts of PIBP and PISP apply to services, or just to goods?

C: Yes, they can apply to services as well. For example, you can have PIBPs for things such as haircuts, manicures, and car washes.

I: Correct, and you could be either buying or selling these services. What, then, is the value of clairvoyance?

C: The value of clairvoyance would be the price at which you would be indifferent about paying for the Clairvoyant's services: Your PIBP for clairvoyance.

I: Correct.

3.3.1 Wealth Effects on Values and the Cycle of Ownership

Now we consider changes in our lives that may affect our PIBP, our PISP, or our cycle of ownership.

I: My personal indifferent buying price for a new Rolls-Royce is $30,000. Remember that I have to keep and use the car. This is my value in use so I am not thinking of reselling it. Since the market selling price is much higher than that, I don't own one. My personal indifferent buying price is relatively low because I don't feel comfortable driving it to the airport, I think maintenance would be expensive, and I have not found them to be much fun to drive. If I did buy the Rolls-Royce for $30,000, my personal indifferent selling price for it at that instant, as we have discussed, would also be $30,000.

I: Figure 3.2 represents the cycle of ownership using the Rolls-Royce rather than a shirt. I am currently in *Situation 2* in Figure 3.2. *(He presents Figure 3.2 as shown below.)* I do not have a Rolls-Royce and I have a bank account whose level is indicated on the right hand side of the Figure. To be in *Situation 1* and also own a Rolls-Royce, I would have to pay some money out of my bank account. To be indifferent between the two situations, this amount must be my PIBP of $30,000. Since I am indifferent to the two situations, if I consider returning to Situation 2 from Situation 1, the required increase in my bank account is my PISP, which is the same $30,000. This is the same analysis we did for the shirt.

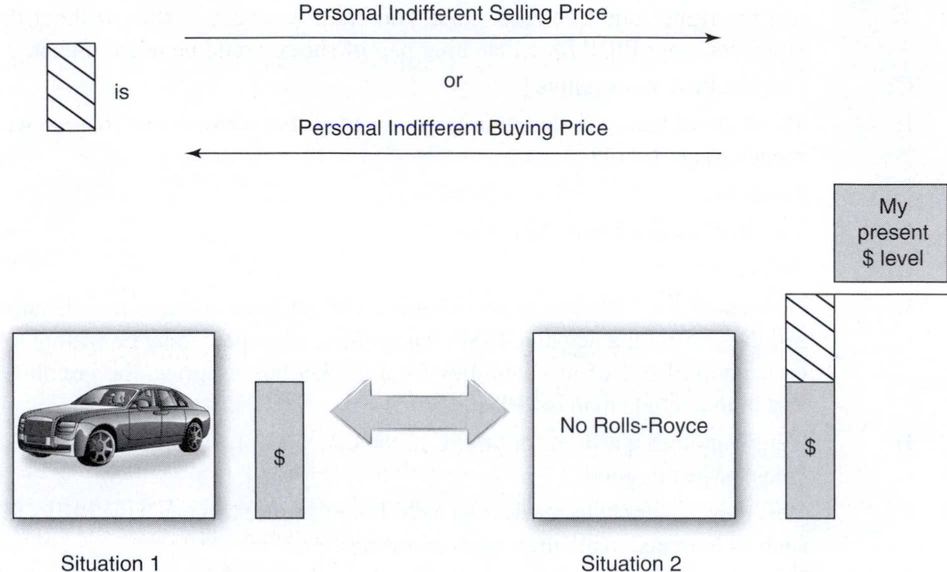

FIGURE 3.2 PIBP and PISP for a Rolls-Royce

I: Suppose that instead of buying a Rolls-Royce, I receive one as a gift. Does this mean that my personal indifferent selling price for it, now that I own it, is still $30,000?

C: Not necessarily.

I: Correct. As you can see in Figure 3.3, I am now in *Situation 4*: I have my original bank account and I now own the Rolls-Royce. Due to this generous gift, I have become a wealthier man, the inconveniences once associated

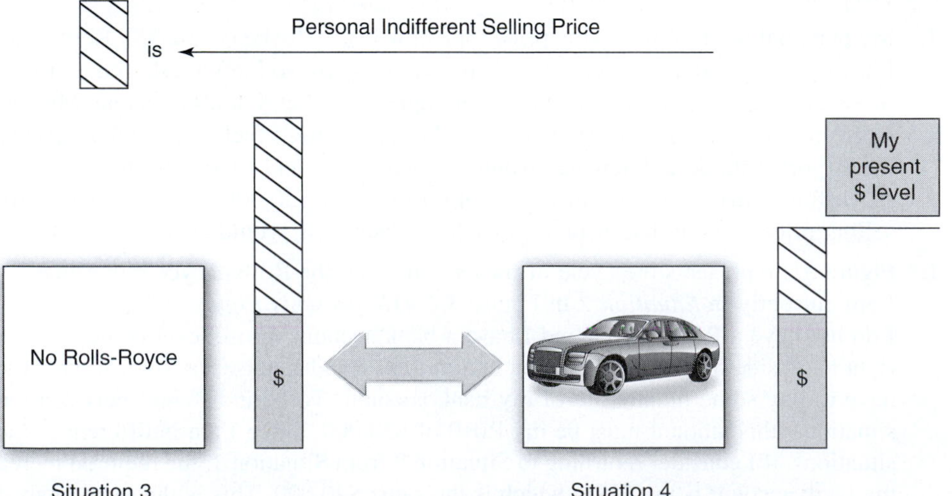

FIGURE 3.3 A New Cycle of Ownership for a Free Rolls-Royce

with the car are now easier to manage, and I might end up using it a lot more frequently. Consequently, my personal indifferent selling price for the Rolls-Royce might now be $40,000 when I contemplate moving to *Situation 3*. If I do move to Situation 3, with no Rolls-Royce and a higher bank account, then my personal indifferent buying price for the Rolls-Royce would be $40,000. There is no reason why my personal indifferent buying price for something in a world where I do not have it should be my personal indifferent selling price for the same thing in a world where everything is the same except that I have it.

I: Within a cycle of ownership, buying and selling prices are only tautologically the same. This is true both in my original world and in the free Rolls-Royce world. When I receive my free Rolls-Royce, I may have a new PISP for it because I am in a new cycle of ownership. Around this new cycle, my PISP is equal to my PIBP if I sold it and bought it at an instant.

C: Does wealth have to increase your PISP (or your PIBP)?

I: Not necessarily. It can go either way. For example, you may own a bicycle that you ride to work every day. At a new wealth level, you may have a driver and a new Rolls- Royce, so you will not use the bicycle as often. Your PISP for the bicycle in this new state may be less than your PISP in the previous state.

3.3.2 New Information Effects on Values and the Cycle of Ownership

I: Returning to original example where I have no free Rolls-Royce, suppose I learn that the price of gasoline will double. Now my personal indifferent buying price will be less than $30,000 in this world of higher operating costs – my value in use will be lower. If I bought it at this price, my personal indifferent selling price at this instant would equal my personal indifferent buying price, since they must be equal in this world.

3.3.3 Time Effects on Values and the Cycle of Ownership

I: As we discussed, the cycle of ownership always occurs at a particular instant in time. The passage of time can change our resources and our preferences. For example, if I get tired of driving the Rolls-Royce, my PISP will fall. Similarly, my PIBP for a sky diving experience may also change as I age. However, within any cycle of ownership, the PIBP and PISP will always remain the same.

3.3.4 Expansion

I: Suppose I have an option to buy a Rolls-Royce that I must drive on a daily basis for the next ten years, but can then opt to sell. In this case, the most I would pay for it would consist of two parts: My PIBP for the ten years of use, and what I believe is today's market value of a used Rolls-Royce ten years from now. Perhaps, if there was a market in used Rolls-Royce futures in which I could sell the decade-later ownership rights today, I could evaluate this second component. This example illustrates that the simple concepts we have been discussing can be extended to clarify many issues of valuation. As we progress in our discussions, our ability to evaluate will continually expand.

C: Can I have a PIBP for an item even if it is impossible to buy it?

I: Yes you can. You can assess a value for something even if it is not available. This kind of thinking is what drives entrepreneurs to create new products and services.

C: Does a corporation have indifferent buying or selling prices?

I: Yes. Suppose, for example, that a company is thinking about obtaining a patent for its exclusive use. They can think about the value the patent provides in improving their own design and the value of preventing some other company from getting the patent and using it. The company would then assess an indifferent buying price for acquiring that patent. This would be the **Corporate Indifferent Buying Price (CIBP)**. If a company owns a patent, it could also assess can also assess the **Corporate Indifferent Selling Price (CISP)** for its sale. Alternatively, an entity might think about buying a patent not for its use but for later sale, in which case it would be a broker concerned with value in exchange.

3.4 SUMMARY

- A *value in use is* different from a *value in exchange*. The value in use considers how much something is worth to us. The value in exchange considers the value we can get by buying an item and selling it to someone else. Value in exchange depends on market prices. Brokers are concerned with value in exchange.
- My *Personal Indifferent Buying Price (PIBP)* is the most I would be willing to pay for the use of something I do not own so that I would be indifferent to the transaction. It does not depend on market price.
- My *Personal Indifferent Selling Price (PISP)* is the least I would be willing to receive for giving up the use of something I own so that I would be indifferent to transaction. It does not depend on the market price.
- My *certain equivalent* is my PISP for an uncertain deal that I own.
- PIBP and PISP cannot have a range of values.
- The *cycle of ownership* is at an epoch in time and can change by changes in wealth, time, and information.
- Around a cycle of ownership, PIBP = PISP.
- PIBP and PISP can change with wealth, information, and time. In a new state, and new cycle of ownership, the new values of PIBP and PISP will be the same.
- If your wealth state changes, your PIBP and PISP for a given item may also change because your usage of the item may change. For example, if you inherit a large sum of money, your lifestyle may change and you may not use certain items that you would have in your previous state, or perhaps have an interest in new items and services.
- Your PIBP for buying a second item does not need to equal your PISP for selling the first item. Your usage of a second item does not need to be the same as the first.
- Corporations also have indifferent buying and selling prices, the *Corporate Indifferent Buying Price (CIBP)* and the *Corporate Indifferent Selling Price (CISP)*.

KEY TERMS

- Personal Indifferent Buying and Selling Prices (PIBP), (PISP)
- Market buying price
- Market selling price
- Cycle of ownership
- Value in use
- Value in exchange
- Market value
- Corporate Indifferent Buying Price and Selling Price (CIBP), (CISP)

PROBLEMS

Problems marked with an asterisk (*) are more challenging.

***1.** Which of the following is always true?

a. PISP is the value in exchange for a deal you own.
b. If you buy groceries for $50, then your PIBP for them is equal to $50.
c. PIBP is your value in use for a deal you own.
d. Your PIBP and PISP do not change around a cycle of ownership.

***2.** Ed bids $46 for the medallion deal of the last chapter. If Ed is following the principles we've taught so far, what must be true at that moment?

a. His PIBP must be equal to $46.
b. His PISP must be equal to $46.
c. His PIBP must be greater than or equal to $46.
d. His PIBP must be less than $46.

***3.** Vicki is shopping for books. She prefers *Great Expectations* to *The Chosen* and prefers *The Chosen* to *Jurassic Park*. She buys *Great Expectations* and *Jurassic Park* for a total of $30. Which of the following MUST be true?

a. Her PIBP for *Jurassic Park* is greater than $15.
b. Her PIBP for *Great Expectations* is greater than $15.
c. Her PIBP for *Jurassic Park* is less than $15.
d. None of the above.

***4.** Mary is selling her car for $ 5,000. Ali thinks about it, decides that his PIBP for the car is $5,000 and buys the car. Immediately after the transaction is completed, what do you know about Ali's PISP for the car?

a. It is equal to Mary's PISP for the car.
b. The exact value of it.
c. Neither (a) nor (b).
d. Both (a) and (b).

***5.** Your PIBP for a new Rolls-Royce is $135,000. Your friend, Ronnie, offers to give you his new Rolls-Royce for free. Three minutes later, you receive a phone call from your friend asking to buy your Rolls Royce. Which of the following must be true?

a. Your PISP must be greater than $135,000.
b. Your PISP must be less than or equal to $135,000.
c. Your PIBP for a second Rolls Royce is $135,000.
d. None of the above.

***6.** Which of the following is always true?

a. Your PIBP must change over time.
b. Your PISP for an object must change with wealth.
c. Value of clairvoyance on the medallion toss, if you owned the deal, is equal to your PIBP for the clairvoyant's services.
d. Your PISP exceeds market value.

***7.** Which one of the following statements about *value in use* is true?

a. John bought groceries for $50, so his PIBP for the groceries was $50.
b. My PIBP for two identical items can be, at most, twice my PIBP for one item.
c. My PIBP for a ticket to the Big Game must remain constant until game day.
d. Xi sold his helmet for $20, so his PISP for the helmet was at most $20.

***8.** On Monday, Jeff bought a ticket to the Big Game for $40. On Wednesday, another friend offered him six tickets, each for the remaining home football games for $80, but Jeff was able to negotiate him down to $60. Which of the following statements must be true?

a. On Monday, Jeff's PIBP for the Big Game ticket was at least $40.
b. On Monday, Jeff's PIBP for all seven home game tickets was at least $100.

c. On Wednesday, Jeff's PISP for the remaining six home game tickets was at most $60.
d. On Wednesday, Jeff's PIBP for each of the remaining six home game tickets was at least $10.

***9.** Two years ago, Alexandros purchased a television from TV Town for $500. Last year, he sold it to his friend Jason for $600. Just yesterday, Jason sold the same television back to Alexandros for $300. In each case, the television was purchased for the purpose of personal use. Given this information and that both follows the principles taught so far, how many of the following statements must be true?

I. Jason's PIBP for the television when he purchased it last year was less than or equal to $600.
II. Alexandros's PIBP for the television when he purchased it from TV Town is less than his PIBP for the television when he purchased it from Jason yesterday.

a. I only
b. II only
c. Both I and II
d. Neither I nor II

***10.** A discount department store has a couple of dresses on sale for $1000. You choose to buy one and take it home. That evening, your friend Mary sees the dress and offers to buy it from you for $2000. After calling the store and having the identical dress put on hold for you, you decide to sell the dress you own to Mary. Given this information, which of the following statements must be true at the time you decide to sell the dress to Mary?

a. Your PISP must be less than or equal to $2000.
b. Your PISP must be greater than $2000.
c. Your PIBP must be greater than $2000.
d. None of the above.

***11.** John sold his used bike for $75, even though he paid over $400 for it three years ago. Assuming John follows the principles taught so far, what can we infer from John's behavior?

a. John's PISP must lie between $75 and $400.
b. John did not violate the "sunk cost" principle.
c. Around a cycle of ownership, the buying price of an item need not equal its selling price.
d. John's PIBP for the bike must have been below $400 three years ago when he bought it.

***12.** Lisa is in the market for a new bass guitar. She sees an Ibanez SR300 Model on sale for $400 at Guitar Grotto, and an identical one for $350 at Axe Mart. She gathers her savings and purchases the guitar from Axe Mart. We denote Lisa's PIBP for this guitar at the time of this purchase as b. A year later a friend offers Lisa $375 for her guitar, but Lisa refuses the offer. If s is Lisa's PISP for this guitar at the time of her friend's offer, which of the following statements must be true?

a. $b \geq \$400$
b. $\$350 \leq b < \$400$
c. $b = s$
d. $s \geq \$375$

***13.** Ben buys an HP computer for $700. If he follows the principles taught so far, how many of the following are true?

I. His PIBP for that computer must be less than $700.
II. If Ben received an offer to buy an additional (identical) HP computer at an 80% discount, he would take it.
III. If Ben were offered a free computer (either the HP or a Dell), he would definitely prefer the Dell because it costs $900.

a. Neither, I, II or III
b. I only
c. II only
d. III only

*14. Gerry is shopping for a new car and sees a gorgeous Porsche 911 sitting outside the dealership. "I would love to buy that!" he thinks, but after seeing the sticker price of $100,000, he realizes that he will not be buying that car. Fortunately, he sees that there is a charity raffle event that day, so he decides to pay $100 for a raffle ticket. As luck would have it, Gerry wins the car! Which of the following statement(s) are necessarily true?

I. Gerry's PIBP for the Porsche was greater than $100,000.
II. Gerry's PIBP for the raffle ticket was greater than or equal to $100.
III. Now that he owns the Porsche, his PISP is the sticker price of $100,000.

a. I
b. II
c. I and II
d. II and III

*15. The Draeger's grocery store in Menlo Park will buy Parmesan Reggiano cheese from any supplier for $2,000 per 100-pound wheel.

a. State your personal indifferent buying price for getting a 100 pound wheel of Parmesan Reggiano cheese, eating what you want (over time), and throwing away the rest. Briefly explain your choice.
b. A friend of yours owns a 100 pound wheel of Parmesan Reggiano cheese. He invites you to bid on the wheel. State the maximum amount you would bid. Briefly explain your choice.
c. Summarize the differences and similarities in your answers to parts a and b.

*16. A friend of yours mentions to you that his PIBP for a shirt is a range of $40 – $50. What argument can you give him to explain it is not wise to have a range for his PIBP?

*17. Give some examples of items you own where you feel your PISP would go up by receiving a substantial amount of wealth. Identify some other items for which you feel your PISP would go down.

FOOD FOR THOUGHT QUESTION

The purpose of this question is to help you practice determining your PIBP.

Using your web browser, refer to one of the Internet auctions (for example, ebay.com, ubid.com, onsale.com, amazon.com). Select an item that interests you. What is it?

- Establish your PIBP for that item. How much is it?
- Could you determine your PIBP? Why or why not?
- Did you identify just one value, or a range of values?
- Would you bid for that item in a closed bid system with your PIBP?

Assuming you did bid with the value of your PIBP and acquired the item. Do you believe you got a good deal? Repeat this step until you narrow it down to where you will feel indifferent to whether or not you acquire the item.

CHAPTER

4

Precise Decision Language

CHAPTER CONCEPTS

After reading this chapter, you will be able to explain the following concepts:

- Considerations for a precise decision language
- Simple speaking
- Familiar speaking
- Fundamental speaking
- Experts and distinctions
- Mastery of a subject
- Creating your own distinctions

4.1 INTRODUCTION

Becoming an expert in a subject requires making powerful distinctions in that subject. To learn decision analysis, you must master its basic distinctions. Collectively, these distinctions form a precise decision language that allows you to think clearly about any decision you or others are facing. We will show that restricting decision language to particular terms that are accurate, familiar, and fundamental contributes to clarity of thought and understanding. You have already begun to learn the language; that learning will expand as we proceed. In the previous chapters, we identified many powerful terms, including *value of clairvoyance*, *sunk cost*, *Personal Indifferent Selling Price (PISP)*, and *Personal Indifferent Buying Price (PIBP)*.

Note the clarity that these terms bring to our discussion: The term "Personal Indifferent Selling Price," for example, is a "selling price" because the seller is going to exchange ownership of an item for money. We use the word "indifferent" because the seller would not accept a lower amount; he does not care whether or not he sells it at that price. We use the word "personal" because other people may have other indifferent prices, and because the seller is forgoing the use of the item rather than thinking as a broker and considering other potential selling opportunities. We note that the term PISP did not involve any notion of a market price, gain, or any other price, except the one that represents our indifference to keeping an item or selling it at that price. This is an example of the type of precision we would like to have in distinctions we create and use in our discussion.

4.2 LEGO-LIKE PRECISION

We shall develop the concepts and language of decision making following the design principles of Legos. Building an easily assembled and durable Lego model requires precisely made components. We shall define our decision components precisely so that they will easily fit together, and yet form a structure of thought that can support the challenges it may face.

A Story by Ron Howard

In the 1960's I made a consulting trip to the Netherlands. One day I spotted in a toy store a set of colorful plastic blocks in a wooden case. They could be snapped together to make many toys. I brought the box home to my children and they became some of the first kids in the United States to play with "Legos." When I was a child, I played with Erector sets and Tinker toys. Why not Legos? Plastics existed back then. I considered the performance requirements: The blocks must be easy for a child to snap together, and yet they must adhere well enough that the wings of "airplanes" do not readily fall off. The answer is the later development of "precision plastic molding," for the performance requirements dictate that the dimensions of the blocks must be precise enough for a structure to hold together (Figure 4.1).

4.3 PRECISE DECISION LANGUAGE

There are several considerations in the design of a precise decision language. The foremost is that it be a language for both learning and practicing decision analysis. To be truly precise, decision language should be plain, familiar, and fundamental.

4.3.1 Simple Speaking

Words should be **simple**—that is, they should say what they mean. As Seneca famously states, "When the words are clear, the mind will be also." The reverse is also true.

Terms that we shall avoid are **expectation** or **expected value**. In probability class, the expectation of an uncertain deal that may produce different amounts of money with different probabilities is computed by multiplying each possible dollar amount received by the probability of receiving it, and then summing over all the possible outcomes. For example, in our demonstration, Sally had a 0.5 probability of calling the medallion toss correctly and thereby receiving \$100. A student in probability class would compute the expectation or expected value of the deal to be $0.5 \times 100 + 0.5 \times 0$ or \$50. However, Mary will receive either \$100 or nothing. One thing she can be sure of is that she will not receive \$50, the expected value. How strange it is to say that something is expected when it cannot happen. Yet that is a commonly used terminology, and one that contributes to confusion rather than to clarity.

Therefore, the term "expected value" has a misunderstanding built into it—a misunderstanding that analysts avoid only by their education. According to a popular teaching maxim, "The expected value is seldom expected." We shall have more to say on this topic in later chapters.

FIGURE 4.1 Precision Plastic Molding
(Adriana Berned/Getty Images)

4.3.2 Familiar Speaking

Not only should words be simple, but they should also be **familiar**. Whether you are in a boardroom or on a construction site, your discussions should involve words that people can easily recognize. Words that are strange or overly technical will interrupt the flow of communication, and should be avoided in the context of a precise decision language.

For example, the word "stochastic" often appears in technical conversation as a substitute for the word "probabilistic"—a word that is much more familiar to most of us. To further complicate matters, the dictionary offers several definitions for stochastic, the first of which is "of, relating to, or characterized by conjecture; conjectural[1]." We doubt anyone would want a client to use this meaning. Whenever possible, try to stick to the words that are most widely recognized and the most easily understood.

4.3.3 Fundamental Speaking

To eliminate any potential confusion, decision language should also be **fundamental**—that is, it should use a limited number of precisely defined words. The language should be rich enough to describe any decision situation, but limited enough that two different people discussing the same situation would have virtually the same description. For example, a doctor was overheard telling one of his patients "there is some kind of a chance of a likelihood of a bad result." The precise decision language translation is clear and to the point: "There is a probability of a bad result."

When describing uncertainty, there is no reason to become fuzzy or verbose. When we conduct the conversation about a decision problem in precise decision language, rather than excessive language, our message becomes much stronger.

4.4 EXPERTS AND DISTINCTIONS

Now we have discussed what we mean by precise language, let us explore the meaning of **expertise**. What distinguishes an expert from a non-expert? In practice, we find that an essential characteristic of an expert is the ability to make and use powerful distinctions about a subject that are beyond the knowledge of non-experts.

The following examples will demonstrate this assertion. Suppose your car was not running right and you took it to an auto mechanic. He opens the hood and exclaims, "There's a lot of metal and wires in there!" How do you evaluate the mechanic's expertise? You would likely say to yourself, "Even *I* knew that! How can this person possibly help me?" Very likely, you would begin searching for a different mechanic to fix your problem.

For another example, imagine you are facing brain surgery, and you are meeting the surgeon to discuss the procedure. You ask, "Doctor, exactly what are you going to do to me?" The surgeon replies, "You know that gray, goopy stuff in your head? I am going to take some of it out." How do you feel? If someone is going to operate on your head, you want that person to have long Latin names for everything that is supposed to be inside your skull, and other long Latin names for everything that is not. Furthermore, you want that person to have the skill to remove the latter while doing as little damage as possible to the former.

To illustrate the nature of expert distinctions, let us examine a few paragraphs from the book *The Silence of the Lambs* by Thomas Harris. This novel and its movie spinoff describe the efforts of an FBI agent, Officer Starling, who is trying to solve a series of murders. The corpse of one of the victims is found with an insect in its mouth. To learn more about the insect, Officer Starling consults two entomologists, Pilcher and Rosen.

Excerpts from their conversation appear below, in italics, interspersed with comments relevant to our discussion on distinctions.[2]

---

[1]http://www.thefreedictionary.com/Stochastics

[2]From SILENCE OF THE LAMBS © 1989 by Thomas Harris. Reprinted by permission of St. Martin's Press. All Rights Reserved.

From: *The Silence of the Lambs*

> *The insect was long and it looked like a mummy. It was sheathed in a emitransparent cover that followed its general outlines like a sarcophagus. The appendages were bound so tightly against the body, they might have been carved in low relief. The little face looked wise.*

I: Up to this point, the discussion uses only terminology that would be familiar to almost everyone.

"In the first place; it's not anything that would normally infest a body outdoors and it wouldn't be in the water except by accident," Pilcher said. "I don't know how familiar you are with insects or how much you want to hear."

I: Now the entomologist is asking just how far Officer Starling wants to get into a technical discussion. She tells him:

"Let's say I don't know diddly. I want you to tell me the whole thing."

"Okay, this is a pupa, an immature insect, in a chrysalis—that's the cocoon that holds it while it transforms itself from a larva into an adult."

I: Now the discussion has reached the level of high school biology.

"Obtect pupa, Pilch?" Roden wrinkled his nose to hold his glasses up.

I: Welcome to the realm of the expert.

"Yeah, I think so. You want to pull down Chu on the immature insects? Okay, this is the pupal stage of a large insect. Most of the more advanced insects have a pupal stage. A lot of them spend the winter this way."

"Book or look, Pilch?" Roden said.

"I'll look." Pilcher moved the specimen to the stage of a microscope and hunched over it with a dental probe in his hand. "Here we go: No distinct respiratory organs on the dorsocephalic region, spiracles on the mesothorax and some abdominals; let's start with that."

"Ummhumm," Roden said, turning pages in a small manual. "Functional mandibles?"

"Nope."

"Paired galeae of maxillae on the ventromeson?"

"Yep, yep."

"Where are the antennae?"

"Adjacent to the mesal margin of the wings. Two pairs of wings, the inside pair are completely covered up. Only the bottom three abdominal segments are free. Little pointy cremaster—I'd say Lepidoptera."

I: Later, the discussion continues:

"What about pilifers?"

"No pilifers," Pilcher said. "Would you turn out the light, Officer Starling?"

What, then, are the characteristics of experts? The first and most important one is that they understand powerful distinctions about the subject of interest. This understanding is not merely knowledge of subject itself, but also knowledge of the most significant distinctions and how those distinctions relate to one another.

A true expert will also have humility based on the awareness of the limits of present knowledge. For example, we once met a combustion expert at the Jet Propulsion Laboratory: An actual rocket scientist. At the end of our conversation, he pulled a book of matches from his pocket and lit one, saying, "It works every time, and yet we don't know why." His knowledge of what we were seeing exceeded that of anyone in the room, or maybe anyone in the city, yet he acknowledged its limitations.

True experts usually have other characteristics as well. One is knowledge of the history of the field, and an understanding of how it arrived at its present state of development. The expert will usually know about any wrong directions the field may have taken, as well as any true progress it has made to date.

Sometimes the expert must have physical skill—for example, if the expert is an auto mechanic or a surgeon. However, physical skill is not always necessary. Even a completely paralyzed internist could be an extremely competent diagnostician.

Now that we know what it means to be an expert, we can refine our knowledge of what it means to be an expert in decision analysis. The expert decision analyst must understand powerful distinctions about decision making that transcend the knowledge of lay people. We have already discussed many of these, such as the distinction between decision and outcome, and the value of clairvoyance. Our task now is to develop a set of powerful distinctions complete enough to achieve clarity of action in any decision problem. We call this process characterization of a decision situation and will discuss it further in the next chapter. First, let's consider another question: What is mastery?

4.5 MASTERY

A master is the ultimate expert. Note that mastery does not reside in the tools of the master. To illustrate, if you require an emergency appendectomy, and faced only these choices, which would you prefer?

- To be operated on in the back of your car by a skilled surgeon using a penknife, needle, and thread.
- To be operated on in the surgeon's state-of-the-art operating room by your favorite actor.

The choice for us is clear, and so we were not surprised to see a recent news story: A woman involved in what appeared to be a minor traffic accident boarded a plane in Hong Kong directly bound for London. After a few hours of flight, she suddenly had great trouble breathing. A surgeon flying as a passenger knew at once that she had a collapsed lung and was in mortal danger. With no hospital nearby, the surgeon decided to operate on the spot. He used a scalpel from the plane's emergency kit, sterilized it with brandy, cut into her chest wall, and inserted a shunt made from a section of emergency oxygen mask tubing stiffened with a piece of coat hanger wire. When the plane arrived in London, doctors said his quick and skillful action had saved her life and that she would have a quick and complete recovery.

So, what is **mastery** of a field? We suggest five characteristics:

1. The master understands powerful distinctions about the field, and understands the importance of each.
2. The master sees and appreciates the relationships among these distinctions.

3. If mastery requires physical action in the world, the master possesses the skills and abilities to carry out that action, whether the master is a surgeon, mechanic, or violinist.
4. The master typically possesses a *deep* and *broad* knowledge of the setting of the field—its history, its relation to other fields, and the roles of major contributors. For example, a master mechanic would tell you where the field of a mechanic ends and the field of a master of auto body repair begins.
5. Most fundamentally, the master exhibits a humble awareness of the limits of present knowledge and an associated commitment to perpetual learning.

4.6 CREATING YOUR OWN DISTINCTIONS

We have presented several distinctions and illustrated how they lead to clarity of thought in a decision. When analyzing a given decision, you will also create your own distinctions. Once you create them, you will also need to make sure that they are clear and that everybody involved in the decision situation is aware of their precise meaning. In the next chapter, we will discuss this process of creating distinctions for a given decision situation in greater detail.

4.7 FOOTNOTE

We learned the importance of distinction from George Spencer Brown in his book, *Laws of Form*. There are now many websites discussing this work, including one by Randall Whitaker[3], who notes: "In his 1979 book *Principles of Biological Autonomy*, Varela intensively explored (and elaborated upon) the British logician George Spencer Brown's 'calculus of indications.' Spencer Brown's *Laws of Form* (1969) outlines a complete and consistent logic based on 'distinctions,' which Maturana and Varela identify as "the elementary cognitive act."

4.8 SUMMARY

A precise decision language is a powerful tool for thinking clearly about decision making. Although some of the terminology you encounter may seem subtle and academic, this terminology was created based on decades of experience in the teaching and practice of decision analysis. It has proven its value over time by helping decision makers avoid common decision mistakes. Throughout the rest of this book, you will have an opportunity to see for yourself the confusion that results from common but misleading decision terminology and how a precise decision language, once mastered, quickly clears up these confusions in thinking.

KEY TERMS

- Expectation
- Expected value
- Simple speaking
- Familiar speaking
- Fundamental speaking
- Expertise
- Mastery
- Distinctions

[3]Whitaker quotation: "In his 1979 book Principles of Biological Autonomy, Varela...as "the elementary cognitive act.", Dr. Randall Whitaker. Reprinted with permission.

PROBLEMS

Problems marked with an asterisk (*) are considered more challenging.

1. Think of some conversations you have had with people you believe are experts in their field. Recall some of the distinctions they used, and the types of conversations that led you to believe they were experts.

2. Think about any misunderstandings you may have had due to imprecise language use.

***3.** List the five characteristics of mastery. Can you think of other characteristics to add to the list?

CHAPTER

5

Possibilities

CHAPTER CONCEPTS

After reading this chapter, you will be able to explain the following concepts:

- Distinctions:
  - Name
  - Kind
  - Degree
  - Clarity
  - Observability
  - Usefulness
- Clarity test
- Mutually exclusive
- Collectively exhaustive
- Elemental possibilities
- Possibility tree
- Measure of an elemental possibility

5.1 OVERVIEW

A decision represents a choice that anticipates possible futures. To think clearly about our decision we must describe these futures for each alternative we face. We call this process **characterization**. In this chapter, we show how to characterize each alternative by creating possibilities depicting the future.

5.2 CREATING DISTINCTIONS

A **distinction** is a division of reality into possibilities. A **simple distinction** is a division of reality into only two possibilities. For example, suppose we are sitting in a lecture hall and considering whether the next person who enters the room will be a Beer Drinker. Once we have defined what we mean by Beer Drinker, then the next person entering will either be a Beer Drinker or will not be a Beer Drinker. There are only two possibilities.

What anyone means by Beer Drinker depends on the background of understanding of that person. Often, the distinction will be sufficiently clear to a listener such that no more specification is necessary. However, the everyday meanings of most terms contain ambiguities significant enough to render them inadequate for careful decision analysis. For example, by Beer Drinker, do you mean a person who has drunk just one beer, a regular Beer Drinker, or someone who drinks beer more than any other beverage? The more we examine the distinction, the more cloudy it becomes. There are several basic elements needed for sharp and meaningful distinctions: **Clarity**, **observability**, and **usefulness**.

5.2.1 Clarity

The first important quality of every good distinction is **clarity**. To ensure that we are dealing only with clear distinctions in our analysis, just like the "precision plastic molding" of Lego pieces we discussed in the last chapter, we must require that every distinction meet the **clarity test**. The clarity test rests on the previously defined notion of the clairvoyant. As you may recall, a clairvoyant is both competent and trustworthy. He knows the value of any physically defined quantity in the past, present, and future. In addition, the clairvoyant can perform any amount of computation or unit conversion required. If the distinction is specific enough that a clairvoyant can tell, without any exercise of judgment, whether that distinction has occurred, then the distinction specification passes the clarity test. Clarity is the first element of good distinctions that we shall define and explain.

For example, let us examine the distinction Beer Drinker. How can we define what we mean by Beer Drinker? In decision analysis classes and in many executive seminars, we often spend a good portion of a class period figuring it out.

Most people think they speak and think clearly. Using an example from everyday speech, the clarity test exercise demonstrates the challenges of achieving spoken clarity. As a first step, a class volunteer creates an initial definition of Beer Drinker, which usually takes a few minutes. They often start with a definition that is something like:

"A Beer Drinker is someone who drinks beer."

After a while, they realize that this definition lacks clarity. Has this person had only one beer in their lifetime, or several beers over the last few weeks? The volunteer then revises their definition:

"Someone who has had a beer at least two times a week in at least 35 weeks over the last year."

We then ask the class if the definition passes the clarity test or if there are elements of the definition that need more clarification.

The volunteer usually confronts many potential interpretations. What about root beer or ginger beer? What about non-alcoholic beer? What about stout, porter, or ale? What about sake, which some call rice beer? What about malt liquor? Does a sip count as having a beer, or do you have to drink a certain quantity? Does it count if the beer is regurgitated? What if someone drinks beer according to your definition, but also drinks three times as much wine?

The distinction of beer itself occupies part of the discussion. Often the volunteer decides to default to a third party for a definition of beer, such as the U.S. government, the California liquor board, or an in-class beer expert. Third parties can certainly be useful in defining terms, but the point is to define what you mean by beer drinker and to be able to communicate your definition to others. You may not define beer in the same way as the U.S. government.

Next, there may be 15 to 30 minutes of nonstop questions and modifications about the definition, until the volunteer accepts the definition as capturing what he meant by the distinction and the class agrees that this definition meets the clarity test.

Taken to the extreme, this exercise would never end. At some point, we need to rely on our common background of understanding. We stop when we believe we have achieved the appropriate level of clarity in communication; namely, that everybody involved in the decision process understands the meaning of the definition. Notice that you can personally disagree with a definition, but it can still be clear. You may personally define a beer drinker as someone who drinks even less beer than the volunteer has specified, but you can still understand the volunteer's use of the term.

Note that if you are the one making the decision without the need to communicate with anyone else, you may need the clarity test only to assure that you are being consistent in your thinking.

The 'right' definition clearly communicates what you mean when you say Beer Drinker. One year, for example, a student named Melanie considered herself a Beer Drinker, but she was pregnant so she was not drinking beer at the moment. To qualify herself, she created the following definition:

"... a person who has drunk a total of 60 ounces or more of beer, defined as a fermented beverage made of hops, barley, and grains with alcoholic content between 3.25% and 12.5% by volume, on at least 3 occasions separated by at least 24 hours in 10 contiguous months over the last 20 months."

Defining the frequency of drinking in the form of 10 contiguous months over the last 20 months was a critical component of Melanie's definition. By most other definitions of Beer Drinker developed in our class, Melanie would not qualify.

The distinction Beer Drinker has been clearly defined when the clairvoyant can determine whether a particular person is a Beer Drinker. In everyday life, thankfully, we do not need to define terms this clearly. Imagine someone asking you "How's it going?" and you trying to apply the clarity test. "What do you mean 'How's it going'? Do you mean my health, wealth, mood, family, job... ?" Forget it. When someone asks you "How's it going?" they really mean "I recognize your existence." While an interchange like this may confuse a Martian, it works well enough for those who share a common background of understanding.

When it comes to any distinction in professional practice, reaching consensus using the clarity test may take up to 45 minutes. For example, 25 lung cancer specialists had the task of defining what it meant for a person to have lung cancer. As experts, they used many medical, biological, and pathological distinctions in the course of their discussions. It took about 40 minutes for the group to settle on a definition satisfactory to all participants.

When drawing up a contract, one of a lawyer's primary concerns is usually to assure clarity of definition in each element of the contract. It will do no good to write in a contract that a product must be "of high quality" unless "high quality" is sufficiently defined, and a measuring procedure adequately specified, so that the event "high quality" can be determined in practice.

Our experience is that any notion of "close enough" in the interests of expediency when meeting the clarity test will ultimately lead to confusion and wasted time. It is better to invest sufficient time up front to achieve real and lasting clarity.

5.2.2 Observability

In addition to clarity, some distinctions also have **observability**. A distinction is observable if it is already known by us, or if it will be known by us in the future. It is necessary that distinctions be observable if a future decision will depend on them. Only observable distinctions can create well–defined bets. In the thumbtack-medallion exercise, the result of either "heads" or "tails" had to be an observable distinction. Distinctions that describe the outcome of any experiment or test must be observable if they are to inform any future decisions.

Observability is a desirable, but not a necessary, quality of a distinction. For example, you might never know the price a competitor is paying for raw material, and yet representing your uncertainty about that distinction could play an important role in your decisions.

Note that whether the next person entering the room is a Beer Drinker may not be observable for many of our clear distinctions of Beer Drinker. If you ask someone using Melanie's definition, the person might reply honestly, "I don't know." Of course, they could also be mistaken, lie, refuse to answer, or provide an offensive retort. Since the distinction is not observable, we would be ill advised to use it as the basis for a bet.

5.2.3 Usefulness

A third distinction about distinctions we create is whether a distinction is **useful** in the sense of improving our sense of clarity of action. We could define many clear, even observable, distinctions that would provide no benefit in achieving decision clarity. For example, if we were thinking about an investment decision in the stock market, the beer drinker distinction may not be a useful distinction to create. However, it may be very useful in deciding the quantities of beer to order for a party.

5.3 THE POSSIBILITY TREE

Once we have created a distinction and made sure it has passed the clarity test, we can depict it in a diagram known as the **possibility tree**. Figure 5.1 shows the possibility tree for the distinction of Beer Drinker.

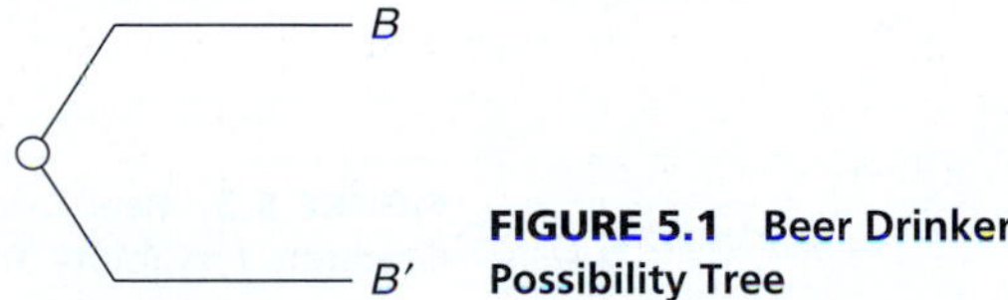

FIGURE 5.1 Beer Drinker Possibility Tree

Here we show a tree with one branch representing the possibility "Beer Drinker," denoted by *B*, and another branch representing the possibility "not a Beer Drinker," represented by *B*' or "*B* prime." This diagram shows that we are considering two possibilities for the next person entering the room: That they will be a Beer Drinker, and that they will not be a Beer Drinker. We can say that this distinction has **two degrees**.

By construction, the degrees for a distinction show all possibilities. The degrees represent every possible response the clairvoyant could provide and are said to be **collectively exhaustive**. Furthermore, exactly one degree must be the answer of the clairvoyant; the degrees are said to be **mutually exclusive**.

Of course, being a Beer Drinker is not the only distinction we might make about the next person entering the room. Another might be whether they are a college graduate, a distinction we denote by *G*. We can ask the same volunteer for a clarity test definition of College Graduate. This discussion usually proceeds more quickly, with the major issue being whether "two-year" or only "four-year" college programs are included. The definition might require that the person be on the list of graduates of accredited institutions providing two or more years of college education or corresponding international levels. If a hiring agency was using this distinction, it would be important to specify the accreditation agencies in the definition.

Figure 5.2 shows the corresponding College Graduate possibility tree.

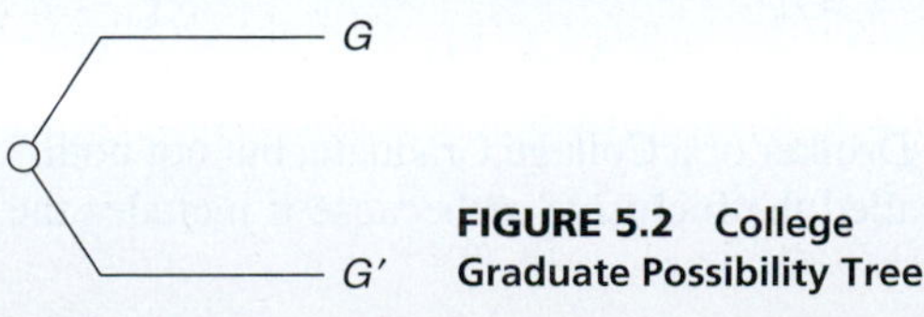

FIGURE 5.2 College Graduate Possibility Tree

5.3.1 Several Kinds of Distinctions

Characterizing most alternatives requires possibilities composed of several kinds of distinctions. For example, you might be interested in whether the next person entering the room is both a Beer Drinker and a College Graduate or a Beer Drinker and not a College Graduate. A possibility tree representing these multiple kinds of distinctions appears in Figure 5.3. Notice that the introduction of the G–G' distinction has increased the number of possibilities to four. The order of this tree indicates that we are first going to think about whether the person is a Beer Drinker, and next, about whether the person is a College Graduate.

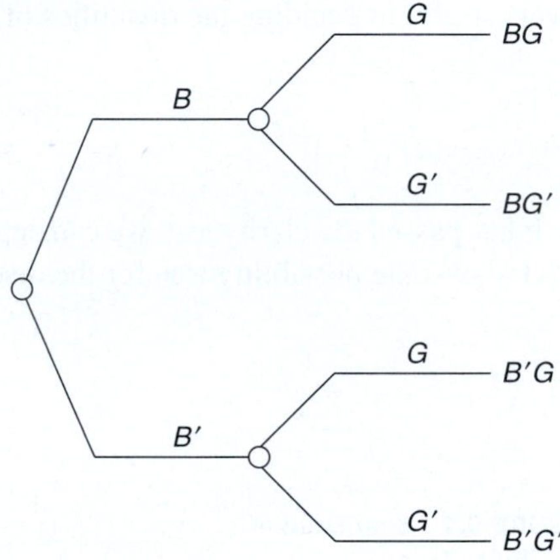

FIGURE 5.3 Beer Drinker/College Graduate Possibility Tree

5.3.2 Elemental Possibilities

An **elemental possibility** is composed of one degree of each kind of distinction. Each path through the possibility tree in Figure 5.3 represents one elemental possibility; this tree has four elemental possibilities. We label the endpoint that characterizes each elemental possibility by writing the sequence of letters that describe the degrees to reach it. Therefore, the elemental possibility that the person is both a Beer Drinker and a College Graduate is represented by the top endpoint BG. The possibility that the person is neither corresponds to the bottom endpoint $B'G'$.

Since the distinctions of Beer Drinker and College Graduate have degrees that are mutually exclusive and collectively exhaustive, the set of elemental possibilities is also mutually exclusive and collectively exhaustive.

5.3.3 Compound Possibilities

We can then use elemental possibilities to construct compound possibilities. A **compound possibility** is a distinction represented by a collection of elemental possibilities. For example, the compound possibility represented by the distinction "either a Beer Drinker or a College Graduate, possibly both" would be represented by the top three endpoints in Figure 5.3. We describe such a compound possibility by joining with plus signs the symbols for the elemental possibilities it contains. Thus, this compound possibility is denoted by $BG + BG' + B'G$.

THE OTHER "OR" We can contrast this compound possibility with the one described by the distinction "either a Beer Drinker or a College Graduate, but not both." The first compound possibility we discussed is called the **inclusive or** because it includes the possibility of both. This one is the **exclusive or** because the possibility of both is excluded. Therefore, the notation for this compound possibility is simply $BG' + B'G$, the two elemental possibilities in the center of the tree. The possibility tree allows us to identify and eliminate the ambiguities created by everyday English phrases such as *either a Beer Drinker or a College Graduate*.

5.3.4 Reversing the Order of Distinctions

Figure 5.3 shows a possibility tree in which have first thought about the distinction Beer Drinker and then the distinction College Graduate. There is no reason why we could not think first about whether the person was a College Graduate and then whether he or she was a Beer Drinker. If we draw the possibility tree in this order, we obtain Figure 5.4.

Note that this tree has the same elemental possibilities as that of Figure 5.3, but in Figure 5.4, the order of the middle two elemental probabilities is switched. Also, note that the order of letters in the notation for each endpoint makes no difference. Being both a College Graduate and a Beer Drinker is the same as being both a Beer Drinker and a College Graduate.

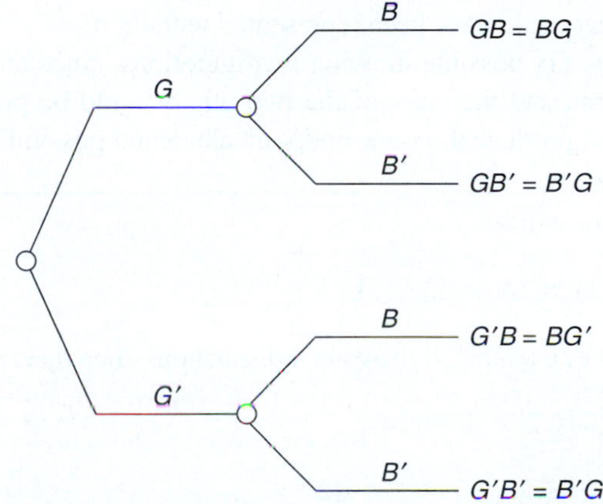

FIGURE 5.4 College Graduate/Beer Drinker Possibility Tree

If we draw Figure 5.4 from right to left, then we can put it in the form of Figure 5.5, which shows clearly that the same endpoints are produced by both trees.

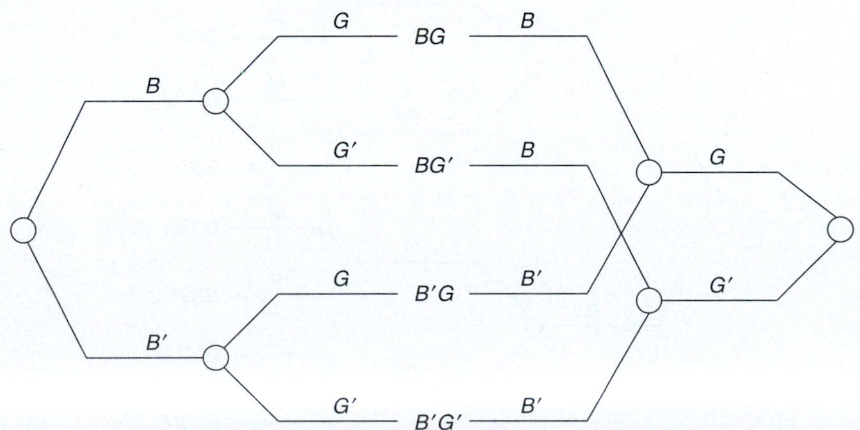

FIGURE 5.5 Beer Drinker-College Graduate Tree Possibilities

5.3.5 Many More Kinds of Distinctions

We can introduce as many kinds of distinctions as we like. For example, in a discussion about whether the next person entering the room drinks beer or has a college education, we might also want to introduce the sex of the person using the distinction Male or Female. Happily, this distinction does not require an extensive clarity test discussion. (However, a colleague who does genetic counseling tells us that such conversations may require 32 degrees of the distinction of gender or sex of a person) We shall use M to designate Male and F to designate Female. While we often use the name of one degree as the name of the distinction, as we did with Beer Drinker and College Graduate, here we have a name for the distinction, Sex, and other names for each degree.

The possibility tree for these three distinctions appears in Figure 5.6.

Now, there are eight elemental possibilities represented by the tree: *BGM*, *BGF*, *BG'M*, and so on. For example, the elemental possibility *B'G'F* represents the situation in which the next person entering the room is a Female who is not a Beer Drinker and not a College Graduate. In this possibility tree, we could represent the compound possibility Male Beer Drinker using a collection of two elemental possibilities: $BGM + BG'M$. Therefore, adding a type of distinction does not prevent us from representing any compound possibility that could have been represented without it.

With three kinds of distinctions there are six possible drawing configurations, since any of the three distinction types could be placed first, and then any of the two others could be placed second. All configurations are equally valid and produce the same endpoint elemental possibilities.

If there are N kinds of distinctions, then there will be

$$(N) \times (N - 1) \times (N - 2)\ldots.1$$

or $N!$ different possibility tree configurations. For example, if there are 4 distinctions, then there are

$$4! = 4 \times 3 \times 2 \times 1 = 24$$

possible configurations.

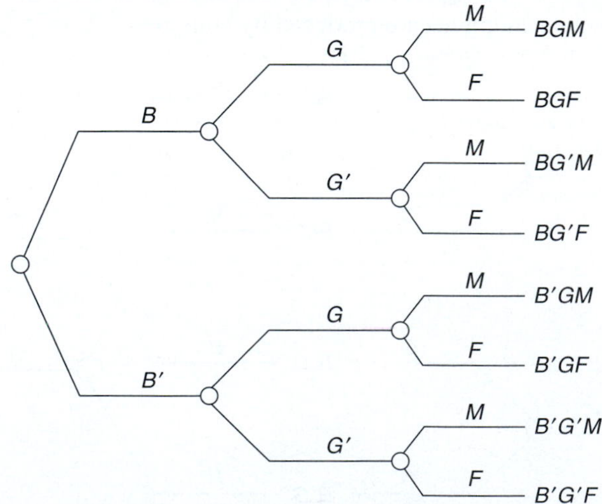

FIGURE 5.6 Result of Adding Sex Distinction

5.3.6 Distinction Degrees

We do not have to limit ourselves to creating simple dichotomies for each kind of distinction. Instead, we can create as many degrees of distinction, within a given kind, as we wish.

MULTIPLE DEGREES OF DISTINCTION FOR BEER DRINKING Suppose you believe that higher education is associated with less beer drinking, and you wish to refine your thinking on the subject. You may want to distinguish three different levels of beer consumption, which we might denote by *B*1, *B*2, and *B*3, as shown in Figure 5.7. You define *B*1 as the event that the person is a light Beer Drinker, drinking no more than 50 quarts of beer per year; *B*2, the event that the person is a medium Beer Drinker, drinking more than 50 quarts of beer per year but not more than 150 quarts per year; and *B*3, the event that the person is a heavy Beer Drinker, drinking more than 150 quarts of beer per year. Therefore, every person who might enter the room can be characterized by one of the three events: *B*1, *B*2, or *B*3.

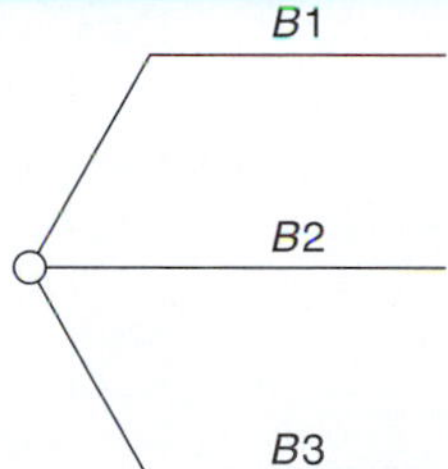

FIGURE 5.7 **Three Degrees of Beer Drinking Distinction**

MULTIPLE DEGREES OF DISTINCTION FOR EDUCATION We might also wish to create additional degrees of distinction for the level of education of the person entering the room. As an example, here we create four degrees of distinction for education: *G*1, *G*2, *G*3, and *G*4. *G*1 is the event that the person has completed no more than grade school; *G*2 is the event that the person has completed more than grade school but no more than high school; *G*3 is the event that the person has completed more than high school, but no more than undergraduate college; and *G*4 is the event that the person has completed more than undergraduate college. Figure 5.8 shows the corresponding possibility tree for the distinction Education with the four degrees of distinction that we have specified.

In creating the different degrees of distinction for both Beer Drinking and Education, we must ensure that all degrees will pass the clarity test. We must also ensure that the possibilities are mutually exclusive and collectively exhaustive. For example, if we define *G*1 as no more than grade school; *G*2 as more than grade school but no more than high school; *G*3 as completing college; and *G*4 as no more than a master's degree, then we would violate both mutual exclusion

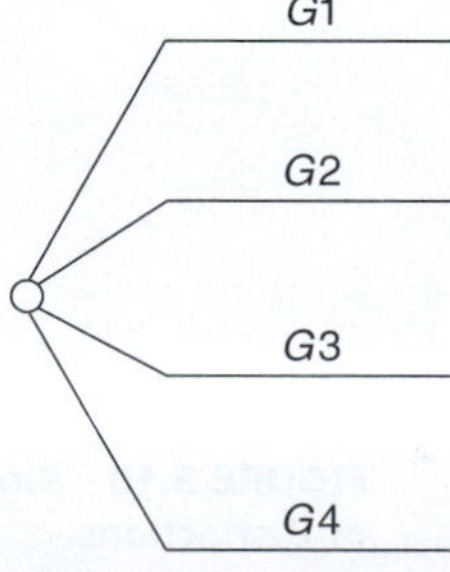

FIGURE 5.8 **Four Degrees of Education Distinction**

and collective exhaustion. We violate mutual exclusion because a person with a college degree can belong to both *G*3 and *G*4. We violate collective exhaustion because a person with PhD degree would not belong anywhere in the tree.

POSSIBILITY TREE FOR TWO DISTINCTIONS WITH SEVERAL DEGREES Consider all the elemental possibilities generated by the two kinds of distinctions we have defined in the possibility tree of Figure 5.9.

This tree shows the twelve elemental possibilities that we can distinguish, ranging from *B*1*G*1, the light Beer Drinker who has completed no more than grade school, to *B*3*G*4, the heavy Beer Drinker who has completed more than undergraduate college. The number of elemental possibilities in a possibility tree is the product of the number of degrees of distinction of each kind—in this case, 3 × 4. We can, therefore, think about the **kind of** distinctions or **number of distinctions** we create. We determine this as we move horizontally across the possibility tree. We can also think about the **number of degrees** for each distinction. This is determined as we move vertically down the possibility tree. For visual clarification, see Figure 5.10.

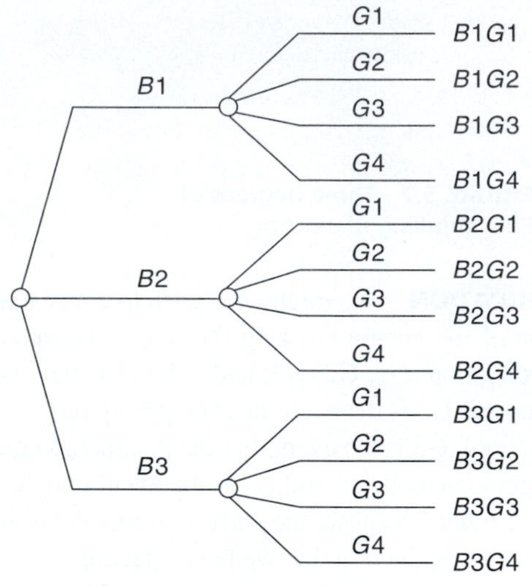

FIGURE 5.9 Possibility Tree with Multiple Degrees of Distinction

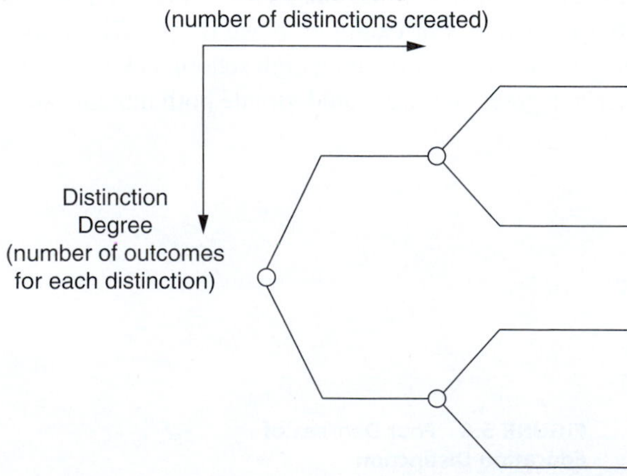

FIGURE 5.10 Kinds and Degrees of Distinctions

5.3.7 Distinction Order

As with possibility trees composed of simple distinctions, we can draw possibility trees with multiple degrees of distinction in any order: They will all generate the same endpoints. Regardless of tree order, the endpoints will be mutually exclusive and collectively exhaustive elemental possibilities. Even when the number of kinds of distinction and degrees of distinction is so large that the tree is too large to be drawn, the concept of the tree of possibilities is still useful as an aid to visualizing the possibilities created.

5.3.8 Representing Several Degrees with Simple Distinctions

The introduction of several degrees of distinction of one kind is a useful, but not a necessary, generalization of the notion of simple distinctions or dichotomies that we have discussed. We can always represent several degrees of distinction using several simple distinctions. For example, in our discussion of the amount of beer drinking, we could define two simple distinctions. The first distinction L represents a light Beer Drinker, a person who drinks no more than 50 quarts of beer per year. The second distinction H represents a heavy Beer Drinker, a person who drinks more than 150 quarts of beer per year. Figure 5.11 shows that the possibility tree generated by these two simple distinctions creates four elemental possibilities *LH*, *LH'*, *L'H*, and *L'H'*.

The figure shows that the elemental possibility *L'H'* is the degree of distinction *B*2. The elemental possibility *L'H* is the degree of distinction *B*3. Finally, the elemental possibilities *LH* and *LH'* or *LH* + *LH'* represent the degree of distinction *B*1. Here we note that *LH* is an elemental possibility that cannot occur because *L* and *H* as defined are mutually exclusive events. Therefore, $B1 = LH'$, $B2 = L'H'$, and $B3 = L'H$. We have interpreted the three degrees of distinction in terms of simple dichotomies. We can repeatedly use this procedure of introducing dichotomies to allow any kind of distinction with multiple degrees to be represented by simple distinctions.

THE IMPORTANCE OF CREATING DISTINCTIONS Complex decisions will require us to face many possibilities. Creating distinctions helps us achieve clarity by breaking thoughts into smaller pieces. Consider, for example, all the possibilities that may occur when we toss a coin and roll a die. By creating two separate distinctions of coin toss and roll of the die, we consider the two degrees of the coin toss distinction (heads or tails) and the six degrees for the roll of a die

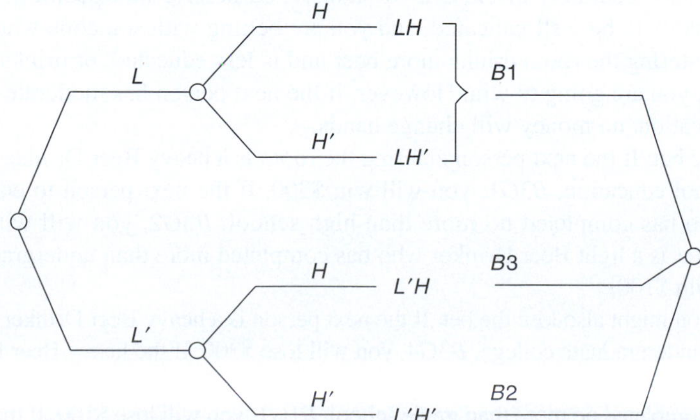

FIGURE 5.11 Two Distinctions and Four Possibilities

(1, 2, 3, 4, 5, 6). By thinking about these two distinctions and the eight degrees associated with them, we can characterize the twelve elemental possibilities for the outcome of the coin toss and the outcome of the die roll.

Consider, for example, a student deciding whether his weekend plans will include skiing or skydiving. To help him decide, he may create any of the following distinctions: (1) whether or not these activities will affect his grades; (2) the trip cost; (3) whether or not he will enjoy the trip; (4) his parents' reaction if they know; (5) whether or not he will meet new people on the trip; and (6) whether or not he will have an accident. Each of these degrees of distinction must pass the clarity test. If the degrees of distinction are all mutually exclusive and collectively exhaustive, then he produces elemental possibilities that are also mutually exclusive and collectively exhaustive. Thus, he can systematically create elemental possibilities that would otherwise be quite difficult to envision.

In general, if we have created n distinctions each with m degrees, then we have m raised to the nth power (m^n) elemental possibilities. For example, if we have created 3 distinctions, each with 4 degrees, then we have $4 \times 4 \times 4 = 64$ elemental possibilities. In this example, however, we thought about only 12 degrees (3 distinctions each with 4 degrees).

Reflection

Think of a decision you are facing. Create the distinctions you feel are most important to the decision situation. Mention the degrees you would construct for each distinction and make sure they pass the clarity test. List all the elemental possibilities you derive.

5.4 MEASURES

In some settings, it is useful to associate a number with each elemental possibility that we have created. We call this number a **measure**. Using the Beer Drinker–College Graduate example, your winnings could be based on a bet concerning someone's education and beer drinking.

Suppose you are hosting a party with a friend and you are betting on the characteristics of the next person to arrive. There is no issue of observability, because the guests are all friends who will readily and accurately disclose their status. You believe that people who drink beer heavily are less likely to be well-educated, and you are betting with someone who disagrees. If the next person entering the room drinks more beer and is less educated, or drinks less beer and is more educated, you are going to win. However, if the next person has moderate levels of beer drinking and education, no money will change hands.

The specific bet: If the next person entering the room is a heavy Beer Drinker with no more than a grade school education, *B*3*G*1, you will win $300. If the next person to enter is a heavy Beer Drinker who has completed no more than high school, *B*3*G*2, you will win $200. If the next person to enter is a light Beer Drinker who has completed more than undergraduate college, *B*1*G*4, you will win $100.

However, you might also lose the bet. If the next person is a heavy Beer Drinker who has completed more than undergraduate college, *B*3*G*4, you will lose $300. If the heavy Beer Drinker has an undergraduate college education only, *B*3*G*3, you will lose $200. Finally, if the person is a light Beer Drinker who has completed no more than grade school, *B*1*G*1, you will lose $100. If the person entering the room is not in any of these categories, neither party to the bet will receive a payoff.

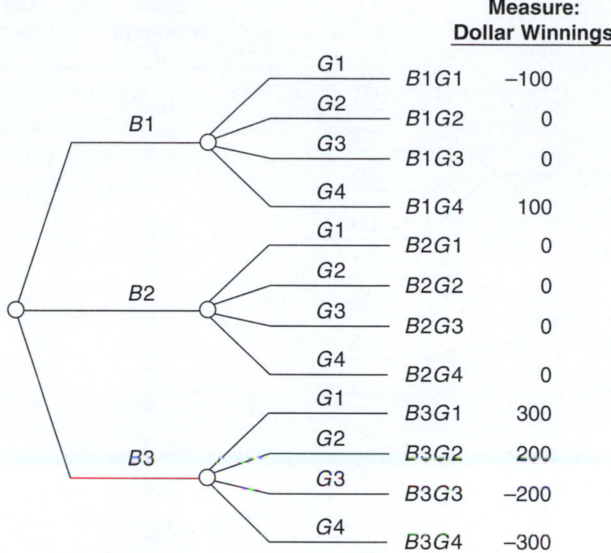

FIGURE 5.12 Possibility Tree with a Measure

The possibility tree with a measure representing your payoff from the bet is shown in Figure 5.12.

We see that six of the elemental possibilities will yield no exchange of money, and that the others will have the payoffs we discussed. In most decision situations, measures are useful. In business, the measure of importance is often profit, which is the difference between the distinctions revenue and costs. In medicine, one measure of interest may be quality of life, which could be a function of days of discomfort and severity of discomfort.

5.4.1 Using Multiple Measures

For each elemental possibility, you could attach several different measures. For example, your bet about beer drinking and education could involve not just money, but washing each other's cars some number of times. In this case, we would attach number of car washes won or lost as a second column of measures at the end of the possibility tree.

Suppose, for example, that when either of you wins money, the other party must also wash your car; otherwise, no car washing will be done. The possibility tree to illustrating such a situation appears in Figure 5.13.

Wouldn't it be wonderful if all contracts were so clearly represented?

Many situations we face involve distinctions with multiple measures. For example, a decision about the purchase of a storage device for your data may include measures such as (1) cost; (2) storage capacity; (3) physical size; (4) weight (for ease of carrying around); and (5) read-write time.

Reflection

Think of a decision situation you are facing. Create the necessary distinctions. Draw the possibility tree. Think of multiple measures that can be associated with the tree.

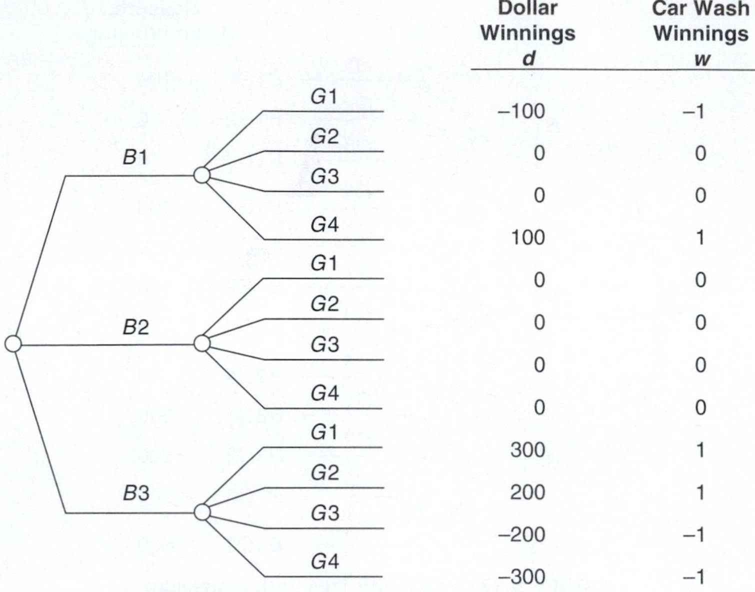

FIGURE 5.13 Possibility Tree with Two Measures

5.5 SUMMARY

- Distinctions divide the world into possibilities. They help us think about the possibilities we face for each alternative we choose. They also help us think clearly about the consequences of our decisions. In many decisions we face, we will have to create our own distinctions.
- We discussed the following distinctions about distinctions:
  - Clear: The distinction should be clear.
  - Useful: The distinction should be useful for thinking about the decision.
  - Observable: Some distinctions are observable but this is not, necessary requirement.
- Clarity Test: Each degree of a distinction should pass the clarity test. The test is satisfied if the clairvoyant, who cannot exercise judgment, can tell whether the degree occurred or did not. In practical settings, we mean every stakeholder should understand the meaning of each degree of a distinction created.
- Distinctions must be:
  - Mutually Exclusive: Only one can occur.
  - Collectively Exhaustive: One must occur.

  If these two conditions are satisfied for every distinction we create (exactly one of its degrees will occur), then the possibilities that occur at the end of the tree are all mutually exclusive and collectively exhaustive.
- Distinctions are often given names such as Age, Sex, Beer Drinker, etc.
- If we have *m* distinctions, then there are *m*! orders for their sequence in drawing the possibility tree.
- If we have *m* distinctions each with *n* degrees, then we have *m*x*n* elemental possibilities.
- Simple distinctions can be used to construct compound distinctions or distinctions with more than two degrees.

KEY TERMS

- Characterization
- Distinction
- Simple distinction
- Clarity
- Observability
- Usefulness
- Possibility tree
- Collectively exhaustive
- Mutually exclusive
- Elemental possibility
- Compound possibility
- Inclusive or
- Exclusive or
- Number of distinctions
- Number of degrees of a distinction
- Measure

PROBLEMS

Problems marked with an asterisk (*) are more challenging.

1. Think of the following two possibilities:
   A. You have never caught a fish.
   B. You have caught a fish longer than 12 inches.
   Which of the following is true?
   i. A and B are mutually exclusive and collectively exhaustive.
   ii. A and B are mutually exclusive, but not collectively exhaustive.
   iii. A and B are not mutually exclusive, but are collectively exhaustive.
   iv. A and B are neither mutually exclusive nor collectively exhaustive.

2. You just heard the statistic "Your chances of being hit by lightning are one in one hundred million." Which of the following statements attempts to add clarity to the distinction "getting hit by lightning?"
   A. "I disagree; I think the chances are one in a billion."
   B. "Is this the chance of being hit by lightning in your lifetime, or on any particular day?"
   C. "How did you know this is the chance of getting hit by lightening?"
   D. "What evidence do you have that leads you to this conclusion?"

3. Decide which of the following possibilities are mutually exclusive, but not collectively exhaustive. The temperature outside is
   A. Under 10°C, over 20°C
   B. Under 20°C, over 10°C
   C. Under 20°C, over 20°C
   D. Under 10°C, under 20°C

4. "Professional athletes are paid too much."
   Which of these is an attempt to make distinctions in the above statement more clear?
   A. "Are you including baseball players?"
   B. "Is $1 million too much?"
   C. Both (A) and (B).
   D. Neither (A) nor (B).

5. Which of the following statements best describes why it is important for a distinction to pass the clarity test?
   A. No value of information can be calculated without clarity on each of the distinctions in a decision problem.
   B. By using the clarity test, a decision analyst will ensure that people discussing the decision problem can communicate information about an uncertainty without any confusion.
   C. Without the clarity test, the clairvoyant cannot determine which degree of a distinction has occurred.
   D. If the distinction has not passed the clarity test, then we will have to assign a probability range, rather than a single number, hindering the accuracy of the analysis.

6. Brad is uncertain about what dessert they will serve at the DA party this year. He constructs three degrees for his distinction: "one ingredient will be sugar," "one ingredient will be milk," and "this dessert will contain neither sugar nor milk."
   A. The degrees of Brad's distinction are mutually exclusive.
   B. The degrees of Brad's distinction are collectively exhaustive.
   Which of the statements are true?
   i. A only ii. B only
   iii. Both A and B iv. Neither A nor B

7. How many of these statements are true?

A. The set of possibilities "I never have drunk cow's milk." and "I have drunk milk from a mammal." is mutually exclusive but not collectively exhaustive.

B. The set of possibilities "My bicycle cost more than $1000," "My bicycle cost more than $500 and less than $900," "My bicycle cost me nothing." is mutually exclusive but not collectively exhaustive.

C. Mutually exclusive means no more than one possibility can be true.

D. A set of possibilities that is collectively exhaustive might not be mutually exclusive.

8. How many of these statements are true?

A. The set of possibilities "My net worth is more than $100,000" and "My net worth is less than $1 million" is mutually exclusive but not collectively exhaustive.

B. The set of possibilities "I have never been to Beijing, China" and "I have been to China" is both mutually exclusive and collectively exhaustive.

C. Mutually exclusive means at **most** one possibility can be true.

D. A set of possibilities that is mutually exclusive might not be collectively exhaustive.

***9.** Consider the following decision situations and for each one, create a set of distinctions. For each situation, list a set of useful mutually exclusive and collectively exhaustive elemental possibilities.

A. Tom is deciding whether or not to quit his job and accept another offer.

B. Anne is deciding where to go for graduate school.

C. Mark is thinking about skipping classes today.

D. Joe is interested in buying hi-tech stocks.

E. Ali tosses a coin and throws a die.

***10.** For each of the following, provide an example:

A. A distinction that does not pass the clarity test.

B. A distinction that is clear, but not observable.

C. A distinction that is clear, observable, but not useful for the decision.

***11.** Think of a situation where an unclear distinction may cause a problem in corporate decision making. If you were asked to help the group achieve clarity of action, what questions would you ask to make the distinction pass the clarity test?

12. Your friend tells you that he is considering cheating on the next exam. Think of the distinctions you might create to help him think about the possible futures if he did go along with this decision. What are your decision alternatives?

FOOD FOR THOUGHT QUESTION

Think about an important decision that you will need to make in the near future.

1. Create two useful distinctions that will help you think about the decision.
2. For each distinction, indicate whether it is observable.
3. Subject one of these distinctions to the clarity test. What are the degrees?

CHAPTER

6

Handling Uncertainty

CHAPTER CONCEPTS

After reading this chapter, you will be able to explain the following concepts:

- Probability
- Background state of information
- Conditional probability
- Elemental probability
- Node probability
- Probability tree
- Adding measures to a probability tree
- Probability distribution
- Cumulative probability distribution
- Excess probability distribution

6.1 INTRODUCTION

In the last chapter, we developed a framework for creating distinctions to describe the possibilities we face in making a decision. The mere construction of these possibilities, however, does not take into account how likely we believe they will occur. For example, you may create a distinction for the possibility of life on Mars with two degrees: life on Mars and no life on Mars. However, you may believe that one of the degrees is much more likely than the other. To specify this belief, we now add probabilities to the possibility tree.

6.2 DESCRIBING DEGREE OF BELIEF BY PROBABILITY

This section explains how to express belief in terms of probability and the notations used when making probability assignments. We shall explore the importance of personal experience when assigning probability, and discuss division of uncertainty; the probability wheel; the use of conditional probabilities, nodal probabilities, and elemental probabilities at the end of the tree.

6.2.1 Using All Information

As discussed in Chapter 5, let us begin with the simple and previously defined distinction of Beer Drinker. Now that we know what we mean by Beer Drinker, how sure are we that the next person entering the room will be a Beer Drinker? The answer depends on our *personal experience*—everything we have learned throughout life. An important part of this experience is the *setting* to which we are applying our knowledge. The question is not what fraction of people are Beer Drinkers, but, rather, how likely it is that the next person entering this particular public room in this building, on this day, after this time, will be a Beer Drinker. For example, in coming to the meeting room you might have noticed a raucous keg party. Conversely, you might have noticed that the building was host to a religious convention. You might know that you have arranged to meet a beer-loving friend in this room in just a few minutes. All such information is included in your total experience, your current state of knowledge. We refer to your current state of knowledge at the time of making the probability assignment as the symbol, &.

6.2.2 Dividing Certainty

You can describe the likelihood that the next person entering the room will be a Beer Drinker by dividing your certainty between the two degrees Beer Drinker, B, and not a Beer Drinker, B'. If, all things considered, you think it is four times as likely that the person will not be a Beer Drinker, then you must divide your certainty between the two degrees, in the ratio 1 to 4. Since you only have one unit of certainty, this means that you must place a fraction 0.2 on the degree Beer Drinker, and 0.8 on the degree not a Beer Drinker to maintain this 1 to 4 ratio. Figure 6.1 shows this division. The number 0.2 has been written adjacent to the upper branch and the number 0.8 adjacent to the lower branch. We call these two numbers **probabilities**. Given the state of information represented by your total experience, we say that the probability you assign to the next person entering the room will be a Beer Drinker is 0.2.

We can think of this **division of certainty** as water flowing through pipes. The pipes represent the branches of the tree and split at its nodes. The total flow at the end of the pipes (end of the tree) must be equal to the total flow entering the pipes (our certainty at the beginning of the tree). Twenty percent of the water flows into the Beer Drinker, B, degree and eighty percent flows into the other degree.

6.2.3 Notation

Our **notation** for probability assignment to the possibility Beer Drinker is $\{B \mid \&\} = 0.2$. The braces, { }, mean that you are making a probability assignment. The possibility for which you are making this assignment appears to the left of the vertical bar; in this case it is the degree Beer Drinker, with symbol B. The state of knowledge on which the probability assignment is made appears to the right of the vertical bar. In this case, it is simply your total experience, which we denote by &. Similarly, your probability that the next person entering the room will not be a Beer Drinker is $\{B' \mid \&\} = 0.8$. Using this notation asserts that probability is a degree of belief at a current state of information. See Figure 6.2.

As shown in Figure 6.3, you might also wish to divide your certainty between the two degrees of College Graduate, G and G', to describe your belief about whether the next person entering the room will be a College Graduate.

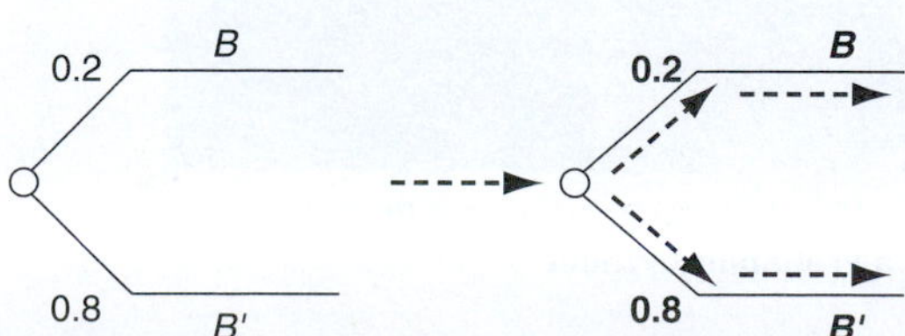

FIGURE 6.1 Division of certainty is analogous to water flowing through a pipe

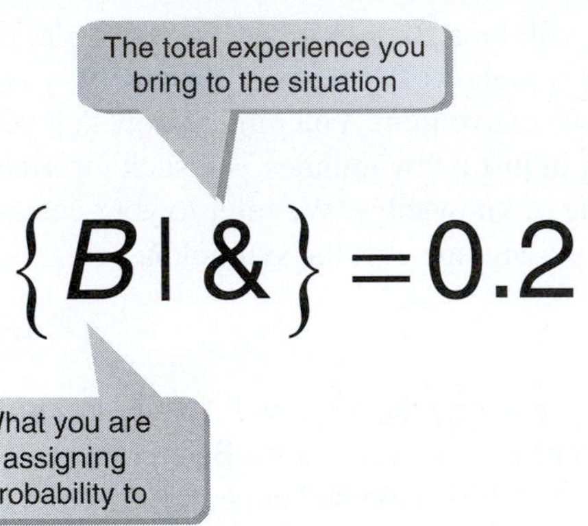

FIGURE 6.2 Notation for Probability Assignment

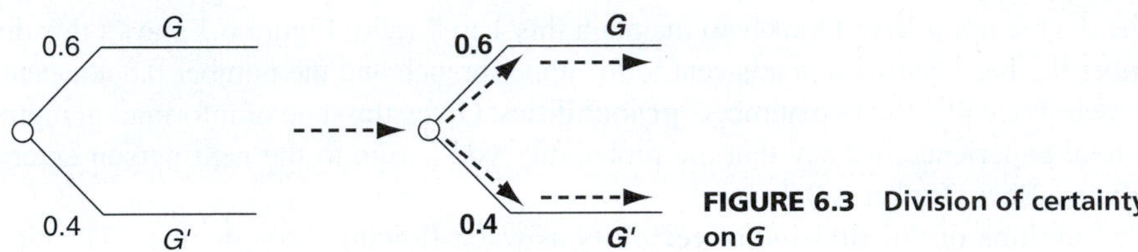

FIGURE 6.3 Division of certainty on G

Here, you have assigned a probability 0.6 to the degree College Graduate: $\{G|\&\} = 0.6$, and you have assigned $\{G'|\&\} = 0.4$. Therefore, you have divided your certainty between the two degrees, G and G', in the ratio of 3 to 2.

6.2.4 The Probability Wheel

The **probability wheel** (Figure 6.4) is a very useful tool for probability assignment. The probability wheel has two circular parts. One is blue, and the other is orange. Using a radial slit in each part, we can adjust the ratio of orange to blue by sliding one color over the other. There is also a pointer on this same side. The wheel can spin and, at the end of the spin, the pointer can land on either orange or blue. The wheel is graduated on the other side, so we can read the probability that corresponds to any given area of orange and blue.

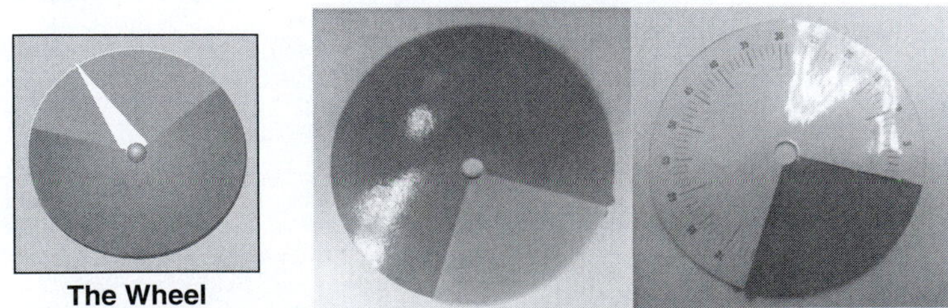

FIGURE 6.4 The Probability Wheel. (Left) Pointer on a Probability Wheel. (Right) Both Sides of a Probability Wheel

For example, to assign a probability to person X being the next U.S. President (defined using the clarity test), we ask the person making the probability assignment, the *assessor*, which of the following two deals he or she prefers on the day of presidential inauguration.

Deal 1 pays \$100,000 if person X is inaugurated and pays zero otherwise.

Deal 2 pays \$100,000 if we spin the wheel and the pointer lands on orange, and pays zero if it lands on blue.

If the assessor prefers Deal 1, then we increase the area of orange on the wheel. By changing the proportions of orange and blue we will reach a point where the assessor is just indifferent between the two deals. This probability can then be read directly off the graduated side of the probability wheel.

The probability we assign to the degree of a distinction can change with time or with receiving information. For example, we may learn that presidential candidate X has withdrawn from the race. This information would change the probability assignment to virtually zero. To account for this, we need to have a time stamp associated with every probability assignment.

We once asked all members of a class to assign the probability that a particular presidential aspirant would be elected president of the United States. This person was not yet an official nominee. One woman in the class assigned a probability of 0.95—a probability much higher than that assigned by anyone else. When asked why it was so high, she said it had to be high because she was working on the campaign of this aspirant.

She had confused two legs of the stool: The preference leg and the information leg. She had let her strong desire for his election lead her to the belief that it was very likely. Wanting something to happen does not make it more likely, nor does wanting something not to happen make it less likely.

6.2.5 The Odds of a Simple Distinction

We can express our uncertainty about a **simple distinction** in terms of a ratio. We refer to this as the odds. The odds is the ratio of the probability of obtaining one degree of the distinction to the probability of obtaining all other degrees. For the Beer Drinker probability assignment, the probability $\{B|\&\} = 0.2$ and therefore $\{B'|\&\} = 0.8$ shows that the odds on B is the ratio,

$$\frac{\{B|\&\}}{\{B'|\&\}} = 0.25.$$

The same information is conveyed by saying that the odds on B' is 4. Using the probability wheel, the ratio of orange to blue where the assessor is just indifferent between the two deals of betting on Beer Drinker or betting on the wheel is 1:4. This shows that the assessor believes that the person entering the room is four times as likely not to be a Beer Drinker as to be a Beer Drinker.

Given the odds on one degree of a simple distinction, we determine the probabilities of each degree using the relations:

$$\{B \mid \&\} = \frac{\text{odds on } B}{1 + \text{odds on } B} = \frac{0.25}{1 + 0.25} = 0.2 \text{ and } \{B'|\&\} = \frac{1}{1 + \text{odds on } B} = \frac{1}{1 + 0.25} = 0.8.$$

6.2.6 Conditional Probabilities

We have elicited your probability that a person is a College Graduate as 0.6, $\{G|\&\} = 0.6$. We can also elicit your probability that a person is a College Graduate, G, given that you know he or

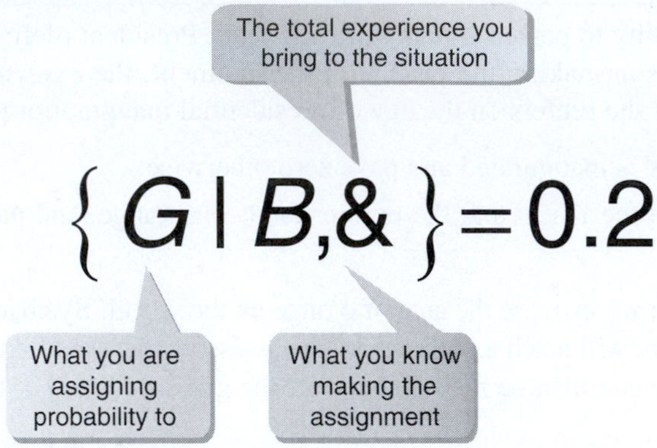

$$\{G \mid B,\&\} = 0.2$$

FIGURE 6.5 Notation for Conditional Probability Assignment

she is either a beer drinker, B, or not a beer drinker, B'. We call such probabilities **conditional probabilities**, for they are conditioned on the knowledge of the first distinction. Any conditional probability is conditioned on all distinctions that precede it in the possibility tree.

The notation for conditional probabilities follows the same rule for notation that we defined previously; the conditioning event is included to the right of the vertical bar. Therefore, the probability that the person will be a College Graduate given that the person is a Beer Drinker is written as $\{G \mid B,\&\}$. If you assess this probability as 0.2, you are saying that if you know the person is a Beer Drinker, then you believe the person is four times as likely not to be a College Graduate as to be one. Figure 6.5 illustrates this assessment. This probability is considerably different from your 0.6 probability assignment to the person being a College Graduate when you did not know whether the person is a Beer Drinker or not.

In a final assessment, you assign the conditional probability of College Graduate given not a Beer Drinker, written $\{G \mid B',\&\}$, as 0.7.

You can also use the probability wheel to assess conditional probabilities. In this case, the question becomes

> *"If you knew the person was a Beer Drinker, would you prefer to receive the big payoff if the person is a College Graduate or if the wheel spin results in orange at its current setting?"*

The point of indifference achieved by varying the fraction of orange is the conditional probability assessed.

As we discussed, we can also express the conditional probability of one degree of a distinction in terms of its odds for the given state of information. Figure 6.5 shows that the odds that a person is a College Graduate given we know he is a Beer Drinker and given our current state of information is the ratio $0.2/0.8 = 0.25$.

6.3 THE PROBABILITY TREE

The Beer Drinker and College Graduate distinctions enable us to discuss the connection between being a Beer Drinker and being a College Graduate. To represent your beliefs about the relationship of these distinctions, you now attach the probabilities and conditional probabilities you have assigned to the possibility tree of Figure 4.3 to produce Figure 6.6. We call the result a **probability tree**.

The beginning of the tree shows the division of certainty according to the probabilities of 0.2 and 0.8 on the events B and B', as we saw in Figure 6.1. Now, however, we must consider further division of certainty. The B branch of the tree represents the possibility that you know

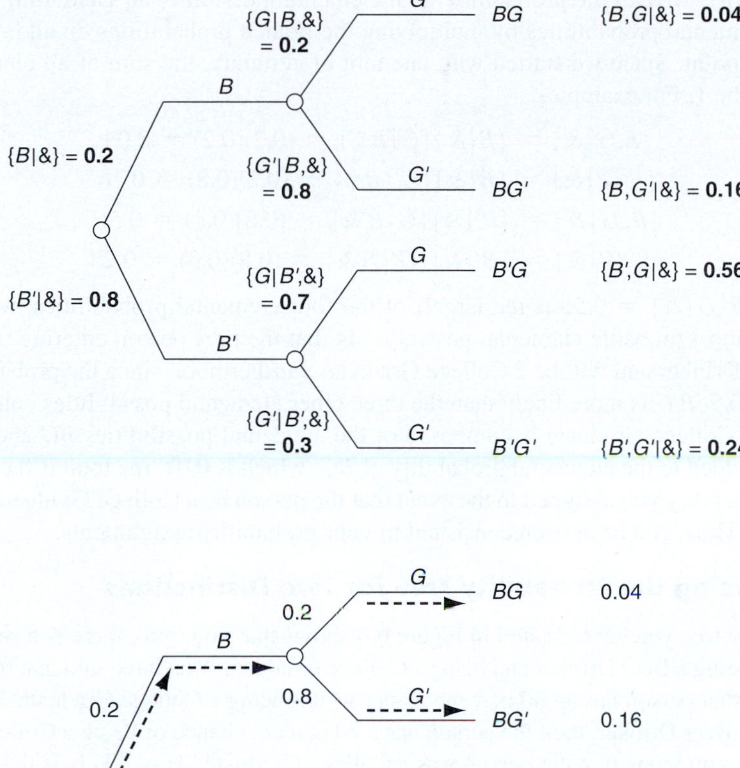

FIGURE 6.6 The Probability Tree. (Top) Notation (Bottom) Division of Certainty

that the person is a Beer Drinker. In this case, you have said you would split your certainty in the ratio 1 to 4 and hence assign a probability 0.2 to the person's being a College Graduate, and 0.8 to the person's not being a College Graduate.

Similarly, the split following the branch B' reflects your assessment that if you knew that the person was not a Beer Drinker, you would assign a probability of 0.7 to the person's being a College Graduate and 0.3 to the person's not being a College Graduate.

6.3.1 Elemental Probabilities

Figure 6.6 shows that your certainty has been split twice: Once by the distinction Beer Drinker, and once by the distinction College Graduate. If 20 percent of your certainty flows through the branch B and 20 percent of that flows through the branch G, then the fraction of the original certainty represented by the elemental possibility BG is 0.04. This is the probability that the person is both a Beer Drinker and a College Graduate, written $\{B,G|\&\}$.

For brevity, we call the probability of an elemental possibility an **elemental probability**. We obtain elemental probabilities by multiplying the branch probabilities on all branches leading to the endpoint. Since we started with one unit of certainty, the sum of all elemental probabilities must be 1. For example,

$$\{B,G|\&\} = \{B|\&\}\{G|B\&\} = (0.2)(0.2) = 0.04$$
$$\{B,G'|\&\} = \{B|\&\}\{G'|B\&\} = (0.2)(0.8) = 0.16$$
$$\{B',G|\&\} = \{B'|\&\}\{G|B'\&\} = (0.8)(0.7) = 0.56$$
$$\{B',G'|\&\} = \{B'|\&\}\{G'|B'\&\} = (0.8)(0.3) = 0.24$$

Since $\{B',G|\&\} = 0.56$ is the largest of the four elemental probabilities, we see that in this case, the most probable elemental possibility is that the next person entering the room will not be a Beer Drinker and will be a College Graduate. Furthermore, since the probability of $B'G$ is larger than 0.5, $B'G$ is more likely than the three other elemental possibilities combined. Note that the event College Graduate is composed of the elemental possibilities BG and $B'G$. When $\{B',G|\&\}$ is added to the elemental probability of BG, which is 0.04, the total is 0.6. This is precisely the probability you assigned to the event that the person be a College Graduate as recorded in Figure 6.3. Thus, you have been consistent in your probability assignments.

6.3.2 Reversing the Probability Tree for Two Distinctions

The probability tree you have created in Figure 6.6 shows that you think there is a strong connection between being a Beer Drinker and being a College Graduate. You have said that if a person is a Beer Drinker, that person has an 80 percent chance of not being a College Graduate, but that if the person is not a Beer Drinker, then the person has a 70 percent chance of being a College Graduate.

Suppose you knew that the person was a College Graduate? How likely is the person to be a Beer Drinker? In other words, what is the conditional probability of $\{B|G,\&\}$?

STEP 1: REVERSE THE ORDER OF DISTINCTIONS We can answer this question by reversing the order of the distinctions, as we did with the possibility tree of Figure 5.4. Remember that reversing the order of distinctions creates the same elemental possibilities. The reversed possibility tree has the order College Graduate—Beer Drinker and is shown in Figure 6.7. As we demonstrated in Figure 5.4, this will require reversing the two middle elemental possibilities.

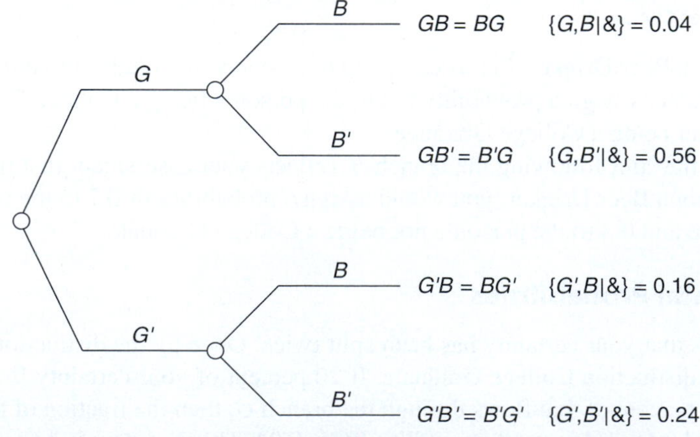

FIGURE 6.7 Reversing the Order of Distinctions. First Step is to Match the Elemental Probabilities

STEP 2: COPYING THE ELEMENTAL PROBABILITIES Since Figure 6.7 has the same elemental *possibilities* as the probability tree in Figure 6.6, it must also have the same elemental *probabilities*. Reversing the order in which you describe an elemental possibility, such as *BG* or *GB*, should not change the probability you assign to this possibility. If you believe that $\{B,G|\&\} = 0.04$, there should be no reason to assign any other probability but 0.04 to the possibility *GB*. Inconsistencies are not allowed.

Logically, the elemental probabilities must be the same for both probability trees. To insist that they must be different is the equivalent of insisting that $BG \neq GB$. In other words, this insistence means that the probability the next person who walks in the room is a Beer Drinker and a College Graduate is different from the probability person is a College Graduate and a Beer Drinker. This is incorrect.

We can, therefore, assign the following elemental probabilities to the reversed possibility tree, $\{B,G|\&\} = \{G,B|\&\} = 0.04$, $\{B,G'|\&\} = \{G',B|\&\} = 0.16$, $\{B',G|\&\} = \{G,B'|\&\} = 0.56$, and $\{B',G'|\&\} = \{G',B'|\&\} = 0.24$. This also means that we can simply copy the elemental probabilities associated with the elemental possibilities of Figure 6.6 directly to their corresponding positions for those possibilities in Figure 6.7.

STEP 3: RECONSTRUCT THE DIVISIONS OF CERTAINTY FOR THE NEW TREE Once you know how much of your certainty has arrived at each endpoint of the tree, it is a simple matter to construct all the probabilities in the tree. For example, we know that whatever certainty flowed through branch *G* must end up at either the endpoint *GB* or *GB'*, as shown in Figure 6.8. The amount of certainty that went down that branch must be the sum of 0.04 and 0.56 or 0.6. (Using the water flow analogy, the amount of water flowing in *G* must be the total amount flowing through both *GB* and *GB'*). We write this number as the probability of *G* associated with the initial division of certainty.

Recall that we previously commented on your consistency in having assigned a 60 percent chance to the person being a College Graduate. Similarly, the 0.4 on the branch *G'* is the sum of the probabilities 0.16 and 0.24 associated with the elemental possibilities *G'B* and *G'B'*.

We now proceed to obtain the conditional probabilities *B* and *B'* given *G* and *G'*.

STEP 4: FIND THE CONDITIONAL PROBABILITIES IN THE REVERSED TREE Let us begin with the conditional probabilities following the branch *G*. Here, we know that your certainty was split in such a way that the upper endpoint received a probability 0.04 and the lower one 0.56. This means that it must have been split in the ratio 1 to 14, and the sum of the conditional probabilities

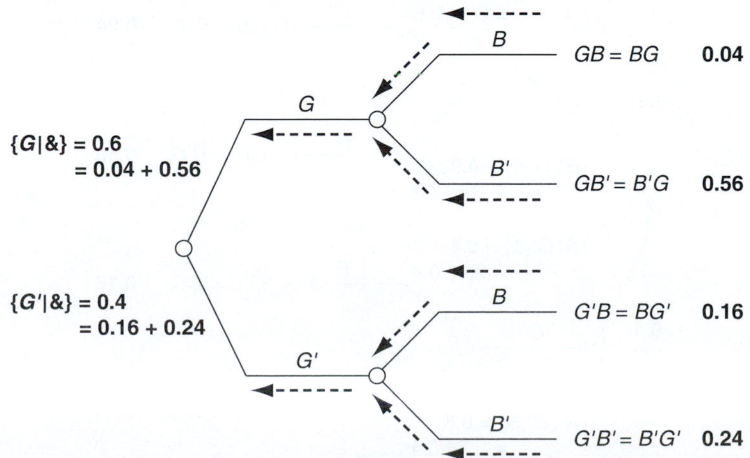

FIGURE 6.8 Reconstructing Divisions of Certainty

must be 1. Given that the person is a College Graduate, you must assign a conditional probability $1/15$, or 0.067, to the person also being a Beer Drinker. In our notation, we write $\{B|G,\&\} = 1/15 = 0.067$. Of course, this means the conditional probability that the College Graduate will not be a Beer Drinker, $\{B'|G,\&\}$, is $14/15 = 0.933$.

Continuing the conditional probability assignment to the branches that follow G', we see that the conditional probabilities on the two branches must be in the ratio 0.16 to 0.24, which means that they must be 0.4 and 0.6. Thus, the probability that the person will not be a Beer Drinker if the person is not a College Graduate, $\{B'|G',\&\}$, is 0.6.

Another way of determining a conditional probability is to think of what it must be if we wish to have the same elemental probability at the end of the tree. Since $\{G|\&\}\{B|G,\&\} = \{G,B|\&\} = 0.04$, and we have already calculated $\{G|\&\} = 0.6$, then we must have $(0.6)\{B|G,\&\} = 0.04$, and $\{B|G,\&\} = .04/0.6 = 0.067$. Therefore, we can calculate a conditional probability by dividing the elemental probability (0.04) by the probability of the node preceding it ($\{G|\&\} = 0.6$) to get 0.067. Similarly, we can calculate all conditional probabilities in the tree by dividing the elemental probabilities by the probabilities of the nodes preceding them. Figure 6.9 shows the complete probability tree in reversed order.

6.3.3 Interpreting the Results

This reversed probability tree shows the probabilities that you must assign to the distinction Beer Drinker if you know the results of the distinction College Graduate. For example, we see that if you know that the person is a College Graduate, you assign a .933 chance that the person will not be a Beer Drinker. And if you know that the person is not a College Graduate, then the person has a 60 percent chance of not being a Beer Drinker. Figure 6.9 contains no information that was not presented in Figure 6.7. The information is merely presented in a new way to show more completely the relationship you believe exists between the two distinctions.

If you do not agree with the conditional probabilities of the reversed probability tree, it means your original probability assignments must have been ill-considered, or you have learned something new. This happens occasionally and should be no cause for alarm. Make adjustments on one of the probability trees, original or reversed, and recalculate the probabilities until you have a set of probability assignments that accurately reflect your beliefs.

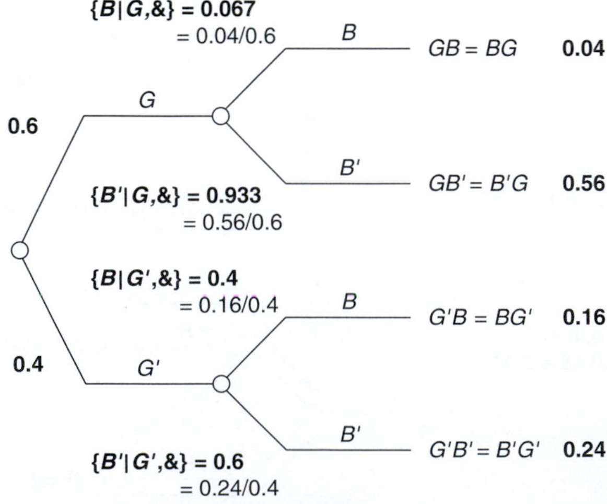

FIGURE 6.9 Complete Probability Tree in Reversed Order

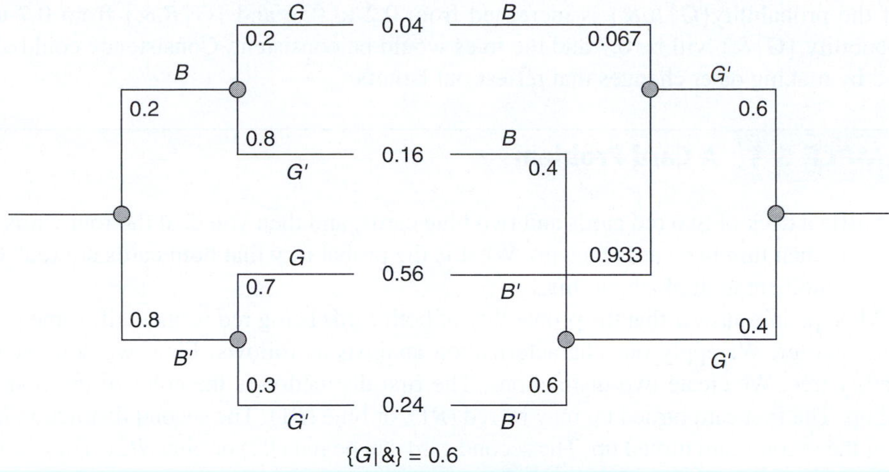

FIGURE 6.10 Tree Reversal

Another way to view what we have been discussing appears in Figure 6.10. Here we connect the trees in each direction by a set of pipes that carry the certainty (or water). The flow enters on the left divides twice (*B*-*B′* and then *G*-*G′*), then it coalesces twice (*B*-*B′* and then *G*-*G′*). The 0.2 probability of $\{B|\&\}$ results in a 0.6 probability of $\{G|\&\}$ on the right hand side of the figure. Note that if we imagine the flow going from right to left, we obtain the same results. We see immediately that a two-distinction probability tree drawn in one order can be interpreted immediately in the reverse order. The elemental probabilities are identical when matched to their corresponding possibilities.

We have observed that up to this point the probabilities assessed have been consistent. The person assigns $\{B|\&\} = 0.2$, $\{G|\&\} = 0.6$, and reversing the tree on the left results in a tree with $\{G|\&\} = 0.6$. We mentioned that if the probabilities were not consistent, then we must make adjustments in probability assignments until they match our beliefs in both trees. To illustrate, suppose that initially the probability $\{G|\&\}$ were assigned as 0.7. Figure 6.10 shows that this is not consistent with the probabilities assessed in the left tree, since those require a probability of 0.6 for $\{G|\&\}$. Figure 6.11 shows one way to resolve the discrepancy by changing the conditional probabilities $\{G|B,\&\}$ and $\{G|B',\&\}$.

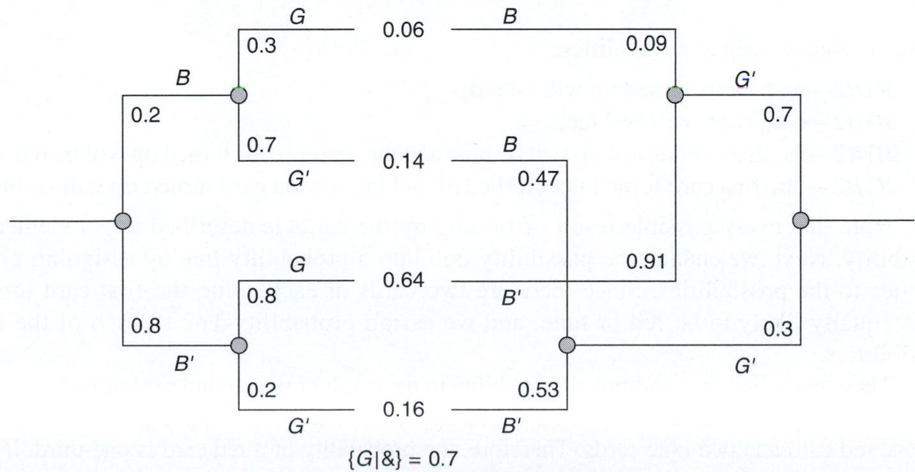

FIGURE 6.11 Resolving Inconsistencies

If the probability$\{G \mid B,\&\}$ is increased from 0.2 to 0.3. and $\{G \mid B,\&\}$ from 0.7 to 0.8, the probability $\{G \mid \&\}$ will be 0.7 and the trees would be consistent. Consistency could also be obtained by making other changes that reflect our beliefs.

EXAMPLE 6.1 A Card Problem

You shuffle a deck of two red cards and two blue cards, and then you deal the four cards face down. You then turn two cards face up. What is the probability that both cards are red? Give yourself a moment to think about this.

Most people answer that the probability of both cards being red is one-half, some people say one-quarter. We apply our characterization analysis as follows: First, we construct the possibility tree. We create two distinctions. The first distinction is the color of the first card turned up. The first card turned up may be red (*R*1), or blue (*B*1). The second distinction is the color of the second card turned up. The second card can be red (*R*2) or blue (*B*2). The events in Figure 6.12 constitute the possibility tree for this situation.

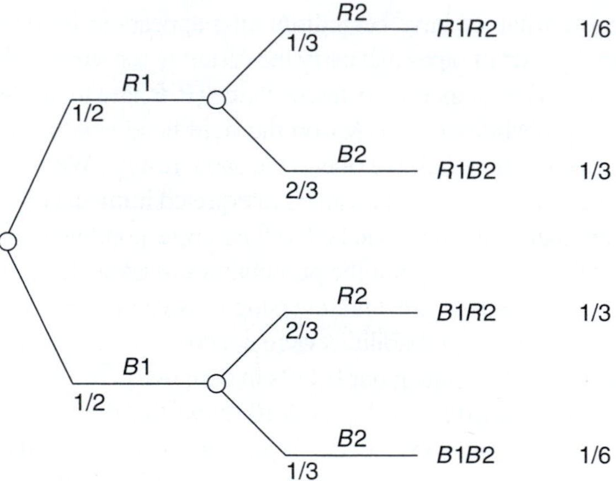

FIGURE 6.12 Possibility Tree for Card Problem

There are four elemental possibilities:

*R*1*R*2—both cards turned up will be red;
*B*1*B*2—both cards will be blue;
*B*1*R*2—the first card turned up will be blue and the second card turned up will be red, and
*R*1*B*2— the first card turned up will be red and the second card turned up will be blue.

Note that every possible result of turning up the cards is described by an elemental possibility. Next, we change the possibility tree into a probability tree by assigning probabilities to the possibilities. Since there are two cards of each color, the first card turned up is equally likely to be red or blue, and we assign probability $1/2$ to each of the two possibilities.

Then, we assign the conditional probability to the result of the second card turned up given the result of the first card. If the first card turned up was red, then the remaining face down cards are one red card and two blue cards. Therefore, the probability of a red card is one-third. If the first card turned up is a blue card, there are two red cards and one blue card remaining face down, so the probability that the second card is red is now two-thirds. These probabilities are recorded in the tree. We now multiply the branch probabilities to obtain the four elemental probabilities. We see that the probability of the outcome we are interested in, $R1R2$, is $1/2 \times 1/3$ or $1/6$.

EXAMPLE 6.2 The Thumbtack Revisited

Let us now refer back to the thumbtack demonstration and show that you cannot be worse off with the thumbtack deal than with the medallion deal if you believed the medallion has a 0.5 probability of landing heads. To see this, suppose you believed the thumbtack had a probability, p, of landing heads or tails. Now suppose you tossed any other coin you believed had a 0.5 probability of landing heads or tails. The probability tree for this situation is shown in Figure 6.13.

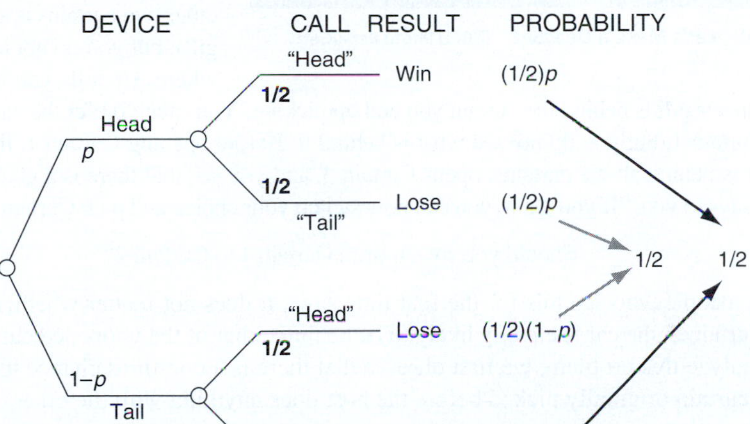

FIGURE 6.13 Probability for the Thumbtack Deal

Supposing you observed the outcome of the coin toss and called that same outcome for the thumbtack toss. What is the probability of calling the thumbtack toss correctly? From the tree of Figure 6.13, this probability is equal to the probability of the coin landing "Head" and the thumbtack landing "Head", $\frac{1}{2}p$, plus the probability of the coin landing "Tail" and the thumbtack landing "Tail", $\frac{1}{2}(1-p)$,

$$\{\text{Win} \mid \&\} = \frac{1}{2}p + \frac{1}{2}(1-p) = \frac{1}{2}$$

Regardless of the probability you assign to the thumbtack being heads, if you use this method, you will always have a probability equal to 0.5 for calling correctly with the thumbtack. This is true even if the thumbtack always lands heads or always lands tails.

EXAMPLE 6.3 Let's Make a Deal

FIGURE 6.14 Let's Make a Deal Set (NBCU Photo Bank/Getty Image)

The following problem appeared as a popular television game show *Let's Make A Deal* (Figure 6.14). It was also discussed and debated by several distinguished academics and practitioners. Imagine you are a contestant on the show. Your experience might go like this:

You walk up to the stage and the TV host asks you to choose from one of three curtains. He says that behind one curtain is a car, and behind each of the other two curtains is a goat (a joke gift), but you do not know what is where. He tells you that you will receive whatever gift is behind the curtain you end up picking. You much prefer the car to the goat. You pick Curtain 1, but you do not see what is behind it. Before opening Curtain 1, the host, who knows what is behind all the curtains, opens Curtain 3, and you see that there is a goat behind that curtain. He says to you, "If you want, you can now switch your choice and pick Curtain 2."

Should you switch from Curtain 1 to Curtain 2?

Many people who see this for the first time say, "It does not matter whether or not you switch to Curtain 2, the car is equally likely to be behind either of the unopened curtains."

To analyze this problem, we first observe that there is a one-third chance that the car is behind the curtain originally picked before the host does anything with the other curtains; the car is never moved.

Imagine you decide to stick with your original choice of curtain. By following this advice, you have a one-third chance of picking the door behind which is a car, and a two-thirds chance of picking the door behind which is a goat. Figure 6.15(a) shows this situation. Since you get what is behind the door you picked, you therefore have a one-third chance of getting a car and a two-thirds chance of getting the goat.

The key to understanding why you should switch is to observe that by switching, you get the opposite of what you had picked originally. To see why this is the case, suppose there was a goat behind Curtain 1, which you picked. If the host reveals another goat behind Curtain 3, then if you switch to Curtain 2, you will get the car for sure. On the other hand,

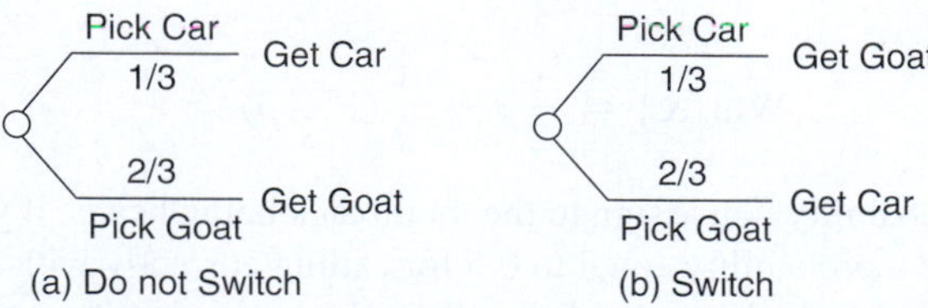

FIGURE 6.15 (a) Do Not Switch. (b) Switch

if there was a car behind Curtain 1, which you chose, and the host reveals a goat behind Curtain 3, then if you switch, you will get the goat for sure. Therefore, if you pick a car and you switch, you will end up with a goat, and if you pick a goat and you switch you will end up with a car. Figure 6.15(b) shows this situation. By switching, you will get the opposite of what you had picked originally and actually double your chance of receiving a car.

A major controversy arose surrounding this problem. In the appendix of this chapter, we show examples of the correspondence received and published in Marilyn Vos Savant's book *The Power of Logical Thinking: Easy Lessons in the Art of Reasoning…and Hard Facts About Its Absence in Our Lives.*

6.4 SEVERAL DEGREES OF DISTINCTION

The main rule of the probability game is that there is one unit of certainty to be divided among all degrees of distinction. If we had several degrees of distinction for drinking beer, such as the three degrees shown in Figure 5.7, then the numbers on all branches must add up to 1. Figure 6.16 shows an example of the division of certainty you might have among the three distinctions $B1$, $B2$, $B3$, that we defined in Chapter 5.

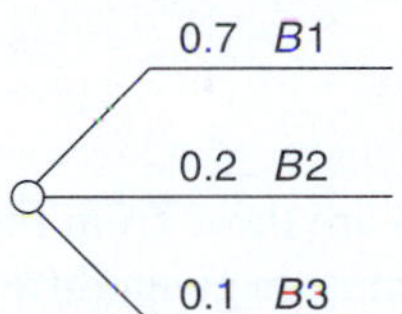

FIGURE 6.16 Division of Certainty Among Three Degrees of Distinction

Here, we have assigned probability 0.7 to $B1$, 0.2 to $B2$, and 0.1 to $B3$. This assignment says that since $\{B1 | \&\} = 0.7$ and $\{B3 | \&\} = 0.1$, then the next person entering the room is, in your opinion, seven times as likely to be someone who drinks less than 50 quarts of beer per year as to be a person who drinks more than 150 quarts of beer per year.

If you wished to use the more refined levels of education we have previously defined, then you might divide your certainty as shown in Figure 6.17.

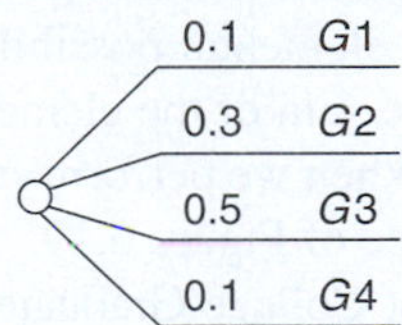

FIGURE 6.17 Division of Certainty Among Four Degrees of Distinction

The possibility $G = G3 + G4$. Therefore, the two branches corresponding to $G3$ and $G4$ in the figure must receive a total of 60 percent certainty to be consistent with your earlier assessment of the probability of G, and they do. Since $\{G3 | \&\} = 0.5$, you see it as likely as not that the person has an undergraduate degree, but nothing higher.

6.5 MULTIPLE DEGREES OF DISTINCTION

Now, we will examine probability trees with multiple degrees of distinction. Recall the possibility tree of Figure 5.9, in which we had three levels of distinction for drinking beer and four levels of distinction for education. We show probability assignments for this possibility tree in Figure 6.18.

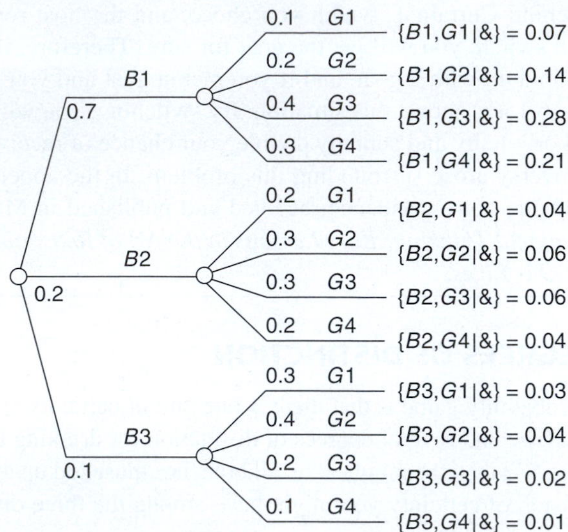

FIGURE 6.18 Probability Tree with Multiple Degrees of Distinction

The probabilities on the first distinction $B1$, $B2$, $B3$ are those from Figure 6.16. You now have to assign probabilities to different educational attainments given different degrees of beer drinking. We record these in the second layer of the tree. Note that as the level of beer drinking increases, this person believes the probabilities assigned to different levels of education shift to be greater for lower levels. The figure also shows the elemental probabilities obtained by multiplying the branch probabilities.

6.5.1 Deducing Probabilities

From this tree of Figure 6.18, we can deduce probabilities we have assigned to different levels of education alone by considering them as collections of elemental possibilities. Therefore, the probability of a grade school education at most, $G1$, is the sum of the elemental probabilities of $B1G1$, $B2G1$, and $B3G1$, or $0.07 + 0.04 + 0.03 = 0.14$. When we perform such a calculation for all four degrees of education, we obtain the probability tree of Figure 6.19.

Note that this tree implies a probability of the event College Graduate, $G = G3 + G4$, of 0.62. This is slightly different from the probability of the event that you assigned earlier, without the detailed thought that went into Figure 6.19. Such discrepancies are to be expected. However, consistency requires that you modify this assessment, as well as either or both of your earlier assessments, until there is no discrepancy.

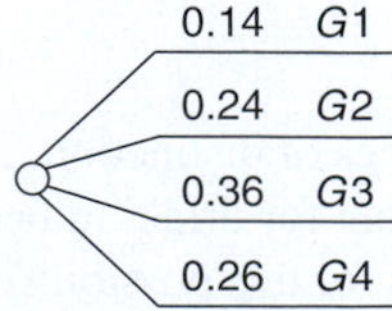

FIGURE 6.19 Probability Tree with Four Degrees of Distinction

EXAMPLE 6.4 A Coin Problem

Three coins are placed in a bag. One has two heads, one has two tails, and the third has both a head and a tail. We take a coin from the bag and toss. The result is heads. What is the probability that this is the two-headed coin? Think about it for a minute.

We create two kinds of distinctions. The first distinction is the *type* of coin drawn from the bag. It can be a two-headed coin, $2H$, a two-tailed coin, $2T$, or a normal coin, with a head and a tail, HT. The probability tree appears on the left side of Figure 6.20. The second distinction is the *result* of the toss. It can be a head, H, or a tail, T. The probability tree appears on the left side of Figure 6.20, showing the six elemental possibilities.

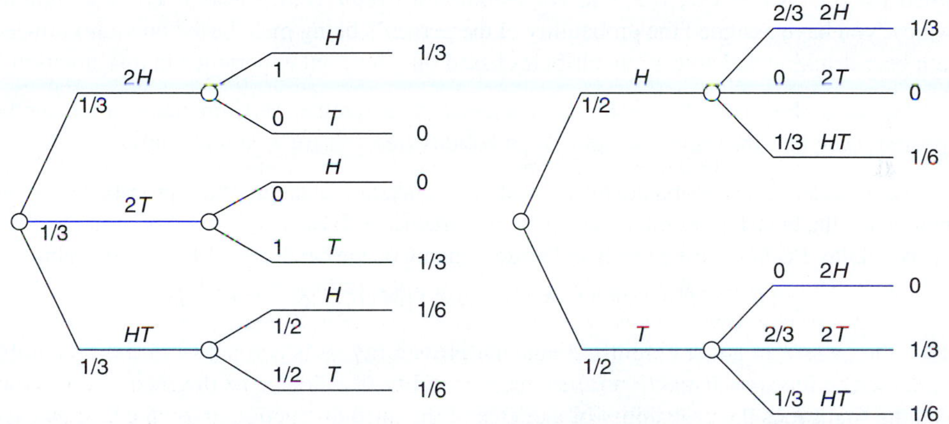

FIGURE 6.20 Probability Tree and Reversed Tree for a Coin Problem

We are now ready to assign probabilities. By the statement of the problem, we are equally likely to draw each type of coin, so we assign probability $1/3$ to each possibility for the first distinction. If you should have different beliefs about the likelihood of a coin selected, you can incorporate those as well.

Next, we assign the probability of tossing a head with each type of coin. If the coin is two-headed, then the probability of throwing a head is 1; if the coin is two-tailed, the probability of tossing a head is 0; if the coin is a normal coin, then the probability of throwing a head is $1/2$. We then multiply the branch probabilities to obtain the elemental probabilities. For example, the probability that the coin was two-headed and that it produced a head is $1/3$.

Finally, we want to compute the probability that the coin was the two-headed coin, given that it produced a head. This requires that we reverse the order of distinctions in the tree so that this conditional probability will appear on the reversed tree. The reversed tree appears on the right of Figure 6.20. As usual, we begin by copying the elemental probabilities into their appropriate places. We see that there is a total probability of $1/2$ following each of the branches H and T, and that the probability that the coin was two-headed given that it produced a head is $(1/3)/(1/3 + 0 + 1/6)$, or $2/3$. Given that a head was produced, there is no chance of a two-tailed coin and a one-third chance of a normal coin. Note that our analysis provides the answer not only to the question at hand, but also to any other potential related questions.

6.6 PROBABILITY TREES USING MULTIPLE DISTINCTIONS

We can construct probability trees for as many kinds of distinctions as we like. For example, if we add the sex of the person entering the room to our existing distinctions, as we did in the possibility tree of Figure 5.6, we can construct a corresponding probability tree, as shown in Figure 6.21.

The probability assignments for the first two distinctions Beer Drinker and College Graduate are identical to those shown in Figure 6.6. We should expect this, since they are conditioned on the same state of information. All that you must add are the conditional probability assignments on the sex of the person entering the room conditional on the path through the distinctions Beer Drinker and College Graduate. The tree shows that you have assigned the probability that the person entering the room will be Male given that the person is a both Beer Drinker and a College Graduate as $\{M|B,G,\&\} = 0.8$, and therefore that $\{F|B,G,\&\} = 0.2$, and you have also assigned $\{M,B,G',\&\} = 0.7$, $\{M|B',G,\&\} = 0.6$ and $\{M|B',G',\&\} = 0.3$. These assignments show how you have assigned the probability of the person's being male based on your knowledge of both beer drinking and education attributes based on your total experience in this situation, &.

No matter how complex the problem in terms of the number of distinctions, or the number of degrees each has, the same principles of probability tree construction will apply.

The eight elemental possibilities represented by the endpoints of this tree have probabilities obtained by multiplying the branch probabilities along the path that leads to each endpoint. For example, the elemental possibility *BG'M* referring to a Beer Drinker—non-College Graduate—Male has probability

$$\{B,G',M|\&\} = \{B|\&\}\{G'|B,\&\}\{M|B,G',\&\} = 0.112$$

Notice the pattern in generating this elemental probability. It is composed of the probability of a degree of the first distinction times the probability of a degree of the second distinction given the first times the probability of a degree of the third distinction given the first and second. In Appendix B, we place this result in general form.

Adding distinctions to a probability tree, however, makes probability assessment increasingly difficult. For example, suppose we add yet another distinction, Over 21 years old. Now, we have to assess the probability that the next person entering the room is Over 21 years old given

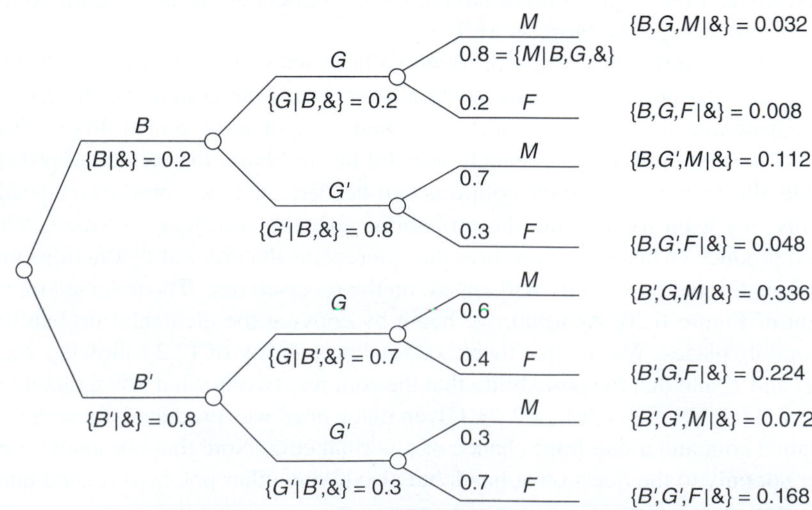

FIGURE 6.21 Probability Tree for Three Distinctions

that the person is a Beer Drinker, a College Graduate, and Male. Notice how challenging it is to keep in mind all the information to the right of the vertical bar.

Even if we could make the conditional probability assessments required, the sheer number grows exponentially with the number of distinctions. In general, a decision problem with only ten binary distinctions would elicit 1023 probability assignments. In the next chapter, we discuss the concept of irrelevance, and show by creating distinctions that are irrelevant, we can minimize the number of probability assignments needed.

6.6.1 Probabilities of Compound Events

> We can obtain the probability of any compound event by adding the probabilities of the elemental possibilities upon which it is composed.

Therefore, if we wish to find the probability that the next person entering the room will be a Male using the probability tree in Figure 6.21, we note that the event Male is composed of four elemental possibilities. M appears in the elemental possibilities of BGM, $BG'M$, $B'GM$, $B'G'M$; therefore, the probability $\{M|\&\}$ is the sum of the elemental probabilities associated with these possibilities, such as,

$$\begin{aligned}\{M|\&\} &= \{B,G,M|\&\} + \{B,G',M|\&\} + \{B',G,M|\&\} + \{B',G',M|\&\} \\ &= 0.032 + 0.112 + 0.336 + 0.072 \\ &= 0.552\end{aligned}$$

Thus implicit in your probability assignments of Figure 6.21 is the assignment of probability 0.552 to the event that the next person entering the room in this situation will be a Male.

Suppose that your real purpose all along had been to assign this probability. Further, suppose you found it easier to assign the probability conditional on whether the person was a Beer Drinker and also conditional on whether the person was a College Graduate. Then, you would have constructed precisely the probability tree of Figure 6.22 as an aid to making the probability assignment on the distinction of Male. We would describe your actions as introducing the distinctions Beer Drinker and College Graduate into the discussion to aid your thought about the distinction Male–Female.

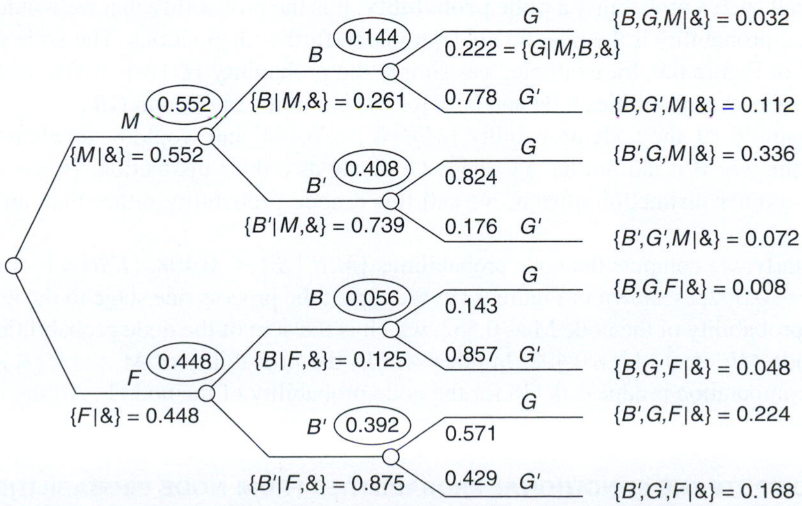

FIGURE 6.22 Reordered Probability Tree with Three Distinctions

Once you have constructed the probability tree, you can answer any question related to the probabilities of the elemental possibilities. For example, suppose we were interested in the probability that the person is either a Beer Drinker, a Male, or both. We identify the elemental possibilities that include this statement, namely,

$$BGM, BGF, BG'M, BG'F, B'GM, B'G'M$$

and then simply sum up their probabilities

$$\{B,G,M|\&\} + \{B,G,F|\&\} + \{B,G',M|\&\} + \{B,G',F|\&\} + \{B',G,M|\&\} + \{B',G',M|\&\} =$$
$$0.032 + 0.008 + 0.112 + 0.048 + 0.336 + 0.072 = 0.608$$

6.6.2 Probability Tree for a Specified Assessment Order

Steps in creating the probability tree in Figure 6.22 corresponding to an assessment order of Sex–Beer Drinker–College Graduate from the tree of Figure 6.21.

STEP 1: DRAW THE POSSIBILITY TREE IN THE SPECIFIED ORDER AND COPY THE ELEMENTAL PROBABILITIES. Our first step in creating the probability assignments is to record at the appropriate endpoints the elemental probabilities from the probability tree of Figure 6.21. Therefore, the path *MB'G* in Figure 6.22 has a probability 0.336 since it is the probability of the endpoint *B'GM* in Figure 6.21. With all elemental probabilities properly recorded, we can begin to compute the conditional probabilities in the tree.

STEP 2: CALCULATE THE NODE PROBABILITIES. In Figure 6.22, look at the node following branch *MB* where the division into the two distinctions *G* and *G'* occurs. The probability of arriving at this point must be the sum of the probabilities of the two elemental possibilities MBG and *MBG'* or $0.032 + 0.112 = 0.144$. In other words, the probability of $MB = \{M,B|\&\} = 0.144$. Therefore, the amount of certainty at the point *MB* to be divided between the distinctions *G* and *G'* is 0.144. We record this in an oval adjacent to the node to show that this is the probability of arriving at that point in the tree.

We call such a probability a **node probability**; it is the probability that we would record as an elemental probability if the tree did not continue to further distinctions. The node probability for node *G* in Figure 6.9, for example, was simply the probability $\{G|\&\} = 0.6$, or the sum of the two elemental probabilities 0.04 and 0.56 for the possibilities *GB* and *GB'*.

In Figure 6.19, the node probability $\{M,B|\&\}$ is 0.144, and would be an elemental probability for the tree if it did not have College Graduate as a third distinction. However, since it does have another distinction after it, we call this a node probability rather than an elemental probability.

Similarly, we compute the node probabilities $\{M,B'|\&\} = 0.408$, $\{F,B|\&\} = 0.056$, and $\{F,B'|\&\} = 0.392$, as shown in Figure 6.19. We repeat the process one stage to the left by computing the probability of the node M as 0.552, which is the sum of the node probabilities for node *MB* and node *MB'* or $0.144 + 0.408$. In other words, the probability of $M = \{M|\&\} = 0.552$. A similar computation produces 0.448 for the node probability of the node F. All the node probabilities are shown in ovals adjacent to the nodes in the tree of Figure 6.22.

STEP 3. COMPUTE THE CONDITIONAL PROBABILITIES FROM NODE PROBABILITIES. Once we have the node probabilities at all nodes, we can compute the conditional probability at any branch by dividing the node probability at the end of the branch by the node probability just before it. We can begin in any part of the tree.

For example, to find the conditional probability of B given M, we write $\{B|M,\&\} = \{B,M|\&\}/\{M|\&\} = 0.144/0.552 = 0.261$, as shown in the tree. To compute the probability of G given F and B', we write $\{G|F,B',\&\} = \{G,F,B'|\&\}/\{F,B',\&\} = 0.224/0.392 = 0.571$. Of course, for the M-F distinction the conditional probabilities are the node probabilities. We see that it is a simple mechanical procedure to change the order of any probability tree. In professional settings, computers automatically perform the computations.

Now that we have the probability assignments for the tree in Figure 6.22, let's examine them in greater detail. Note, for example, that $\{G|M,B,\&\} = 0.222$ and $\{G|F,B,\&\} = 0.143$. From either Figure 6.6 or Figure 6.17, we know that $\{G|B,\&\} = 0.2$. This means that you have assigned the probability of College Graduate given Beer Drinker as 0.2 in the absence of any additional information. However, if you are told additionally that the person is a Male, you would increase that probability to 0.222. If you were told that the person is a Female, you would decrease it to 0.143. Therefore, we can easily determine the inferential value of any piece of information.

6.6.3 Using the Chain Rule and Selected Node Probabilities

We have said that as long as all distinctions are represented in a probability tree, the tree is sufficient to compute the probability of any set of distinctions given any other set of distinctions. To compute a conditional probability that is not initially represented in the tree, we can always reorder the tree, as previously shown, to create a tree that contains the conditional probability we seek. For example, if we want to know the probability of being Male–College Graduate, we can reorder the tree of Figure 6.21 in the order College Graduate–Sex–Beer Drinker and simply read the probability of $\{M|G,\&\}$ at the second layer.

If we do not want to reorder the tree, we can still determine conditional probabilities, and we have a choice of methods. For example, suppose we are interested in the conditional probability that the person is a College Graduate given that he is a Male, $\{G|M,\&\}$, which cannot be read directly from the tree. First, we recall from our chain rule discussion that

$$\{G,M|\&\} = \{M|\&\}\{G|M,\&\}$$

and that the probability we seek is given by

$$\{G|M,\&\} = \{G,M|\&\}/\{M|\&\}$$

From the tree, we can compute the probability $\{G,M|\&\}$ as the sum of $\{M,B,G|\&\}$ and $\{M,B',G|\&\}$; or $0.032 + 0.336 = 0.368$; the probability of M, $\{M|\&\}$, is 0.552 as shown. Hence $\{G|M,\&\} = 0.368/0.552 = 0.667$. You have assigned a probability of about two-thirds that if a Male enters the room, he will also be a College Graduate.

Another way to compute this same probability is to consider the portion of the probability tree in Figure 6.22 that follows the branch M to be a probability tree that is already conditioned on the person being a Male. Within this tree, we can then compute the probability that the person is also a College Graduate. This is just the probability of traversing the remaining branches BG or $B'G$, or $(0.261)(0.222) + (0.739)(0.824) = 0.667$, and gives the same result.

We may also be interested in the probability that if a Female enters the room she will be a College Graduate, $\{G|F,\&\}$. Using any of the methods we have discussed will result in about 0.518, as you may wish to check. Thus, the probability assignments you have made indicate that in this situation, the probability that the next person entering the room will be a College Graduate is higher if the person is Male than if she is Female.

6.6.4 Computing the Probability of Any Set of Distinctions Given Any Information Represented by Distinctions

Sometimes, it is important to calculate probabilities conditioned on information that is represented by more than one degree of a distinction. In these situations, it is useful to create a new distinction that represents the information in question. To illustrate, consider the following problem that appeared in *Parade* Magazine.

> A woman and a man (unrelated) each have two children. At least one of the woman's children is a boy, and the man's older child is a boy. Assuming each birth is equally likely to be a boy or a girl, do you believe the chances that the woman has two boys are equal to the chances that the man has two boys?

SOLUTION

Most people who listen to this story for the first time would say yes, the chances are equal. Using the tools we have developed, let's explore further. We will use the notation B for boy, G for girl, and use numbers 1 and 2 to represent birth order.

The Woman: What we know about the woman is that she has two children, and that at least one of those children is a boy. Refer to the probability tree in Figure 6.23.

Now, let I be the event that at least one child is a boy. We can create a distinction, I and I', at the end of the tree, as in Figure 6.24. By looking at the beginning of the tree, it is easy to determine whether I has occurred. For example, if we have $B1$ for first child and $B2$ for second child, then we have at least one child and the conditional probability $\{I \mid B1, B2, \&\} = 1$. This must imply that $\{I' \mid B1, B2, \&\} = 0$.

We are interested in the conditional probability of two boys, given that at least one is a boy. This can be written using our notation as $\{B1, B2 \mid I, \&\}$. To determine this conditional probability,

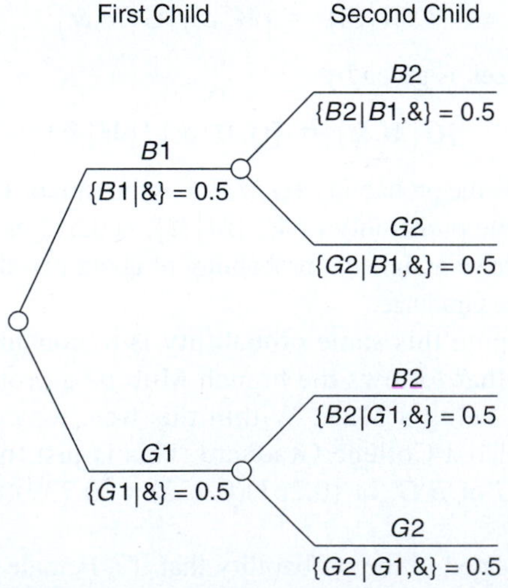

FIGURE 6.23 Probability Tree for Two Children

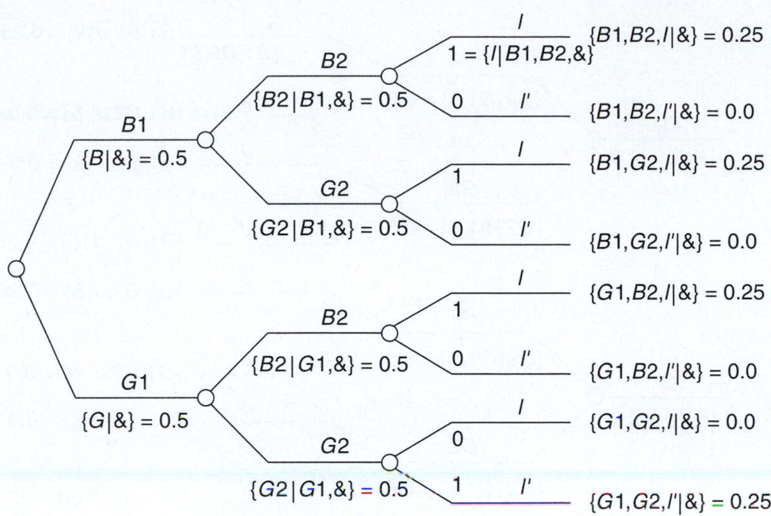

FIGURE 6.24 Distinction *I* Added to Tree

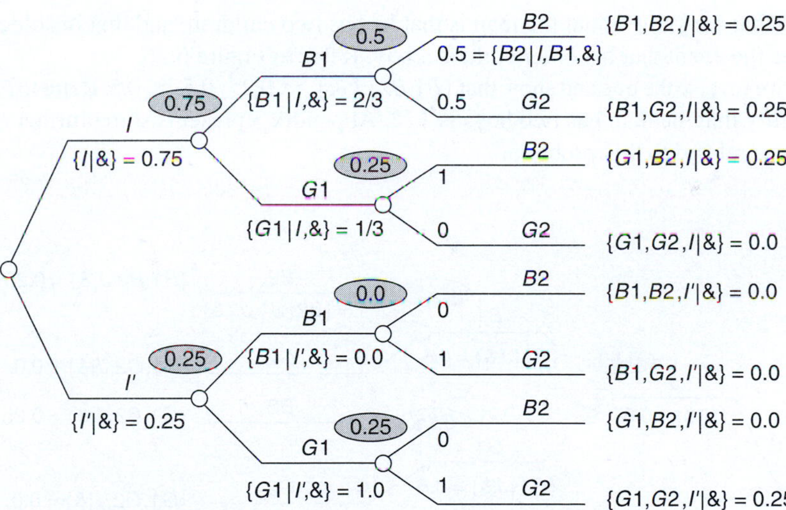

FIGURE 6.25 Probability Tree Starting with Distinction *I*

therefore, we need to reverse the order of the tree of Figure 6.24 and start with the distinction I, as shown in Figure 6.25. As we discussed, the first step is to copy the elemental probabilities and calculate the node probabilities. The second step calculates the conditional probabilities in the tree by dividing each node probability by the one that precedes it.

To determine the probability $\{B1,B2 \mid I,\&\}$, we now divide the elemental probability $\{B1,B2,I \mid \&\}$ by the node probability $\{I \mid \&\}$ to get $0.25/0.75 = 1/3$. The probability that the woman has two boys is $1/3$.

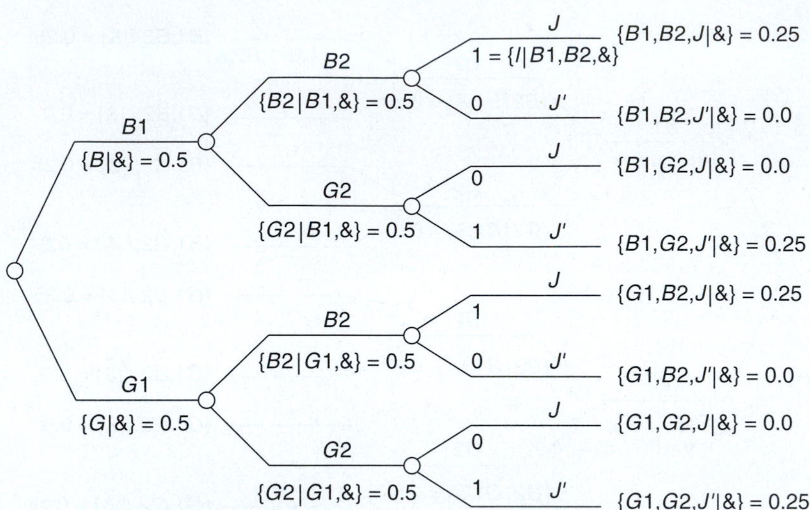

FIGURE 6.26 Distinction *J* Added to Tree

The Man: What we know about the man is that he has two children, and that his older child is a boy. Let *J* be the event that his older child is a boy. Refer to Figure 6.26.

Now, we reverse the tree and show that $\{B1,B2 \mid J,\&\} = 0.25/0.5 = 0.5$. Refer to Figure 6.27. The probability that the man has two boys is $1/2$. Appendix 3 presents some further discussions and exchanges related to this problem.

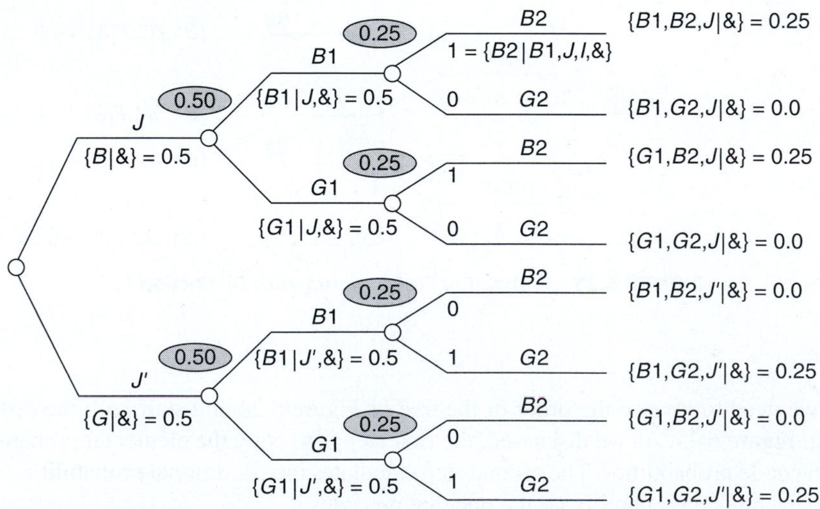

FIGURE 6.27 Probability Tree Starting with Distinction *J*

Reflection

Using the same ideas applied to the tree of Figure 6.22, calculate the probability that a person is a College Graduate given he is either a Male or a Beer Drinker. Use both inclusive and exclusive meanings of the OR statement.

6.7 ADDING MEASURES TO THE PROBABILITY TREE

We can use the probability tree of Figure 6.18 to further explore the subject of measures.

At each endpoint in Figure 6.28, we have shown the dollar winnings you will receive from the bets, as originally specified in Figure 5.9. For each elemental possibility, we now know both the elemental probability and the associated measure value. For example, the probability of losing \$100 is 0.07 and the probability of winning \$100 is 0.21.

The same measure value may be associated with several different endpoints. For example, you will receive a payoff 0 for 6 different elemental possibilities. The probability of receiving 0 is the sum of the probabilities of these six elemental possibilities, or $0.14 + 0.28 + 0.04 + 0.06 + 0.06 + 0.04 = 0.62$.

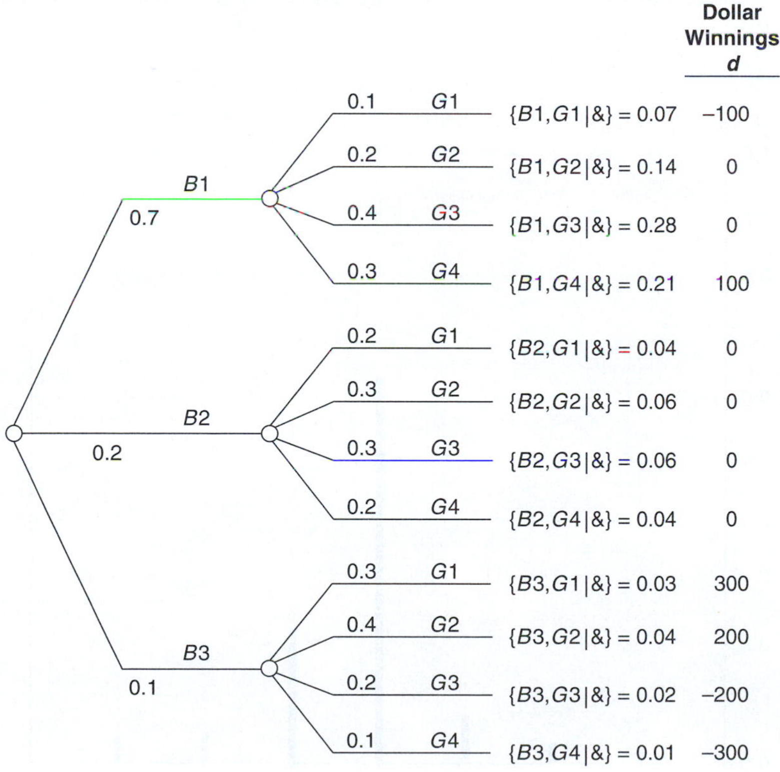

FIGURE 6.28 Probability Tree with a Measure

6.7.1 Probability Distributions

Although the probability tree contains all information we have on the probabilities to be assigned to different measure values, there is a more convenient way to describe the information. For each possible value of the measure, we can construct a graph that shows the probability of achieving that value. We call this graph a **probability distribution**. For every possible value of the measure *m*, we construct a bar whose height is the probability that that value will be attained. We give this distribution the notation $\{m \mid \&\}$ and call it the probability distribution on m given your background information, &. The probability distribution on dollar winnings, *d*, for this problem, appears as Figure 6.29.

Compared to the tree, the probability distribution is a much more compact and informative representation of the probabilities assigned to *d*. Just by looking at it, we can see that more of your certainty lies to the right of 0 than to the left. This means that you are more likely to win the bet than you are to lose it. However, the probability of 0.62 at 0 shows that the most likely result is that no money will change hands.

CUMULATIVE PROBABILITY DISTRIBUTIONS An alternate representation of the probabilities assigned to a measure is the ***cumulative probability distribution,*** which shows the probability that the measure will take on any value less than or equal to a specified number. For any specified number ***c***, the cumulative probability that *m* will be less than or equal to ***c*** is, in our notation, $\{m \leq c \mid c, \&\}$.

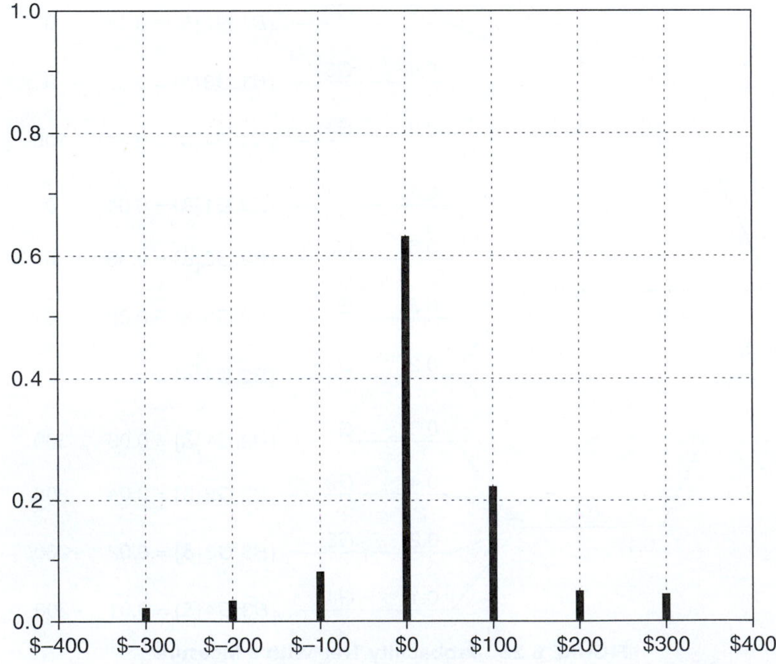

FIGURE 6.29 Probability Distribution of Dollar Winnings {*d*|&}

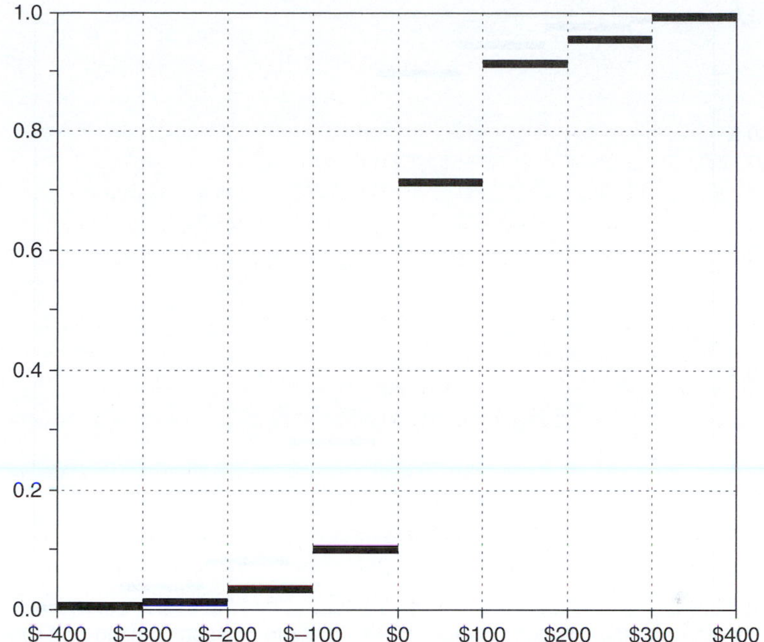

FIGURE 6.30 Cumulative Probability Distribution of Dollar Winnings $\{d \le c|c,\&\}$

The cumulative probability distribution $\{d \le c \mid c,\&\}$ on dollar winnings, *d*, is shown in Figure 6.30.

To construct the cumulative distribution, we begin at the left side. For numbers less than –300, there is no chance that the dollar winnings will be less than or equal to such a number; therefore the height of the cumulative distribution is 0. For numbers between –300 and –200, the probability that the winnings will be less than or equal to such a number will be 0.01, just the chance of losing \$300. For numbers between –200 and –100, the probability of winnings less than or equal to such a number is 0.03, the sum of the probabilities of losing \$200 and of losing \$300.

We see that we are generating a staircase. Steps occur where there are bars in the probability distribution; the height of each step is the height of the bar. Since the sum of all the bar heights in the probability distribution is 1, eventually the cumulative distribution rises to a height of one; in this case, at the value of + \$300. With probability 1, the dollar winnings will be less than or equal to any amount greater than \$300. Note that the probability of no exchange of money is represented by the jump from 0.1 to 0.72 in the value of the cumulative probability distribution at 0.

EXCESS PROBABILITY DISTRIBUTIONS Sometimes it is more convenient to talk about the probability that the value of a measure will exceed any specified number, rather than be less than or equal to it. We call the probability distribution defined in this way an **excess probability distribution**. For any number, *c*, the excess probability distribution on a measure, *m*, shows the probability that *m* will be greater than *c*; in our notation, $\{m > c \mid c,\&\}$.

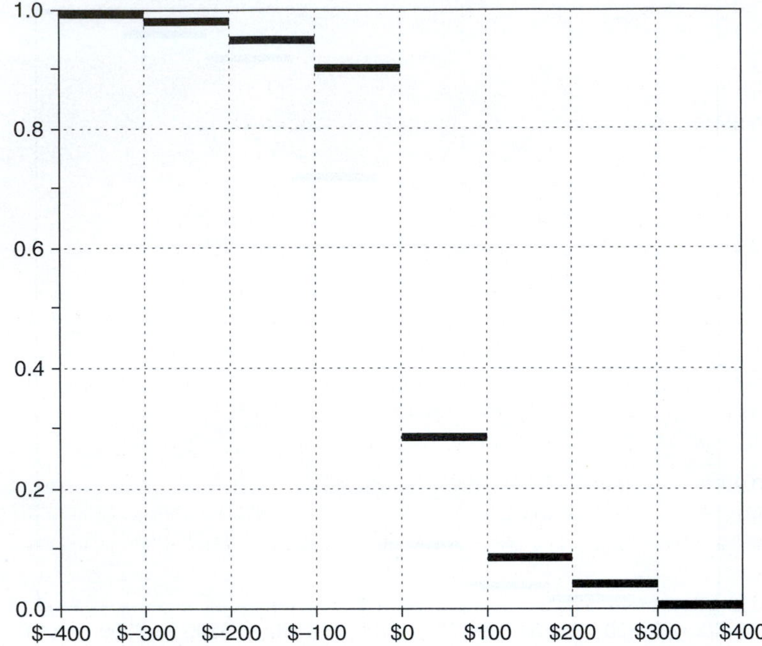

FIGURE 6.31 Excess Distribution of Dollar Winnings $\{d > c|c,\&\}$

Figure 6.31 shows the excess probability distribution $\{d > c \mid c,\&\}$ on dollar winnings d.

The probability that you will win more than any amount less than –\$300 is 1. The probability that you will win more than –\$300 is 0.99. The probability that you will win more than –\$200 is 0.97. Note that we constructed a staircase in the opposite direction. The excess distribution starts at the left at 1 and falls anywhere there is a bar in the probability distribution by an amount equal to the height of that bar.

The excess distribution may be a better communication tool than the cumulative in some situations. For example, the cumulative distribution on profit may appear to an executive to imply that as the level of profit increases it becomes more probable. The excess distribution correctly shows that as the amount of profit increases the less like is it to be exceeded.

COMPLEMENTARITY WITH CUMULATIVE DISTRIBUTIONS Since $d \leq c$ and $d > c$ represent two mutually exclusive and collectively exhaustive possibilities, the sum of their probabilities must be one. This means that if we add together the values of the cumulative probability distribution and the excess probability distribution at any point, the sum must be 1. If you cut out the excess distribution with a pair of scissors and invert it on top of the cumulative probability distribution, the fit would be perfect.

The question of which of the three representations of the probabilistic assignments on a measure is most appropriate will depend on the application.

EXAMPLE 6.5 The Game of Craps

Craps is a very common game played by rolling two cubical dice with faces numbered 1 through 6 by dots. The craps table can accommodate up to about 20 players, who each get a chance to be the "shooter" who will roll the dice. The shooter and all other players can bet on the results of the throws.

Rules of the game: A new game begins with a new shooter who rolls two dice. The result of any roll of the dice is the sum of the numbers facing up. If the sum is 7 or 11, the shooter wins. If the sum is a 2, 3, or 12, it is "craps" and he loses. Finally, if the sum is any of the numbers 4, 5, 6, 8, 9, or 10, then this number establishes his "point." For example, if the sum is 4, we say his point is 4. He now continues rolling the dice and will win if his point comes up again before he rolls a 7. If 7 comes up before he rolls his point, he loses.

Now, let's analyze the game of craps using the tools we have developed to find the chance that the shooter will win the game. If he does, he will receive a gain of what he has bet; otherwise, he will lose his bet. See Figure 6.32.

It is easiest to imagine that we are dealing with a red die and a green die. The number on the red die has degrees from one through six, as does the number on the green die. If we think of the probability tree in the order red die, then green die, Table 6.1 shows in a convenient

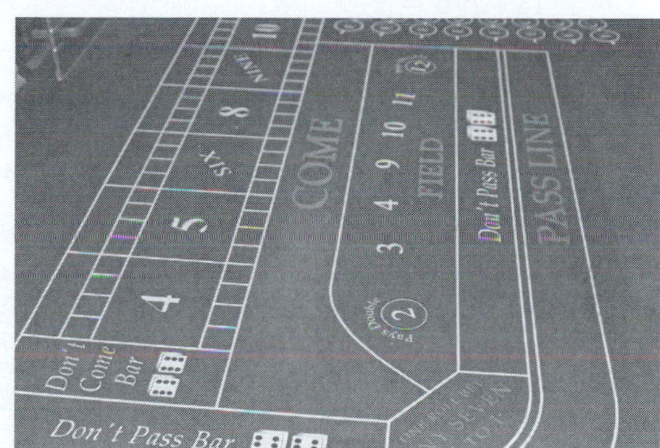

FIGURE 6.32 **Craps Board–Rolling Seven** (Dbvirago/Fotolia)

TABLE 6.1 Possibilities for the Sum of Two Dice

		Red Die Outcome					
		1	**2**	**3**	**4**	**5**	**6**
Green Die Outcome	**1**	2	3	4	5	6	7
	2	3	4	5	6	7	8
	3	4	5	6	7	8	9
	4	5	6	7	8	9	10
	5	6	7	8	9	10	11
	6	7	8	9	10	11	12

way the resulting 36 elemental possibilities. Each elemental possibility has a measure equal to the sum of the numbers on the two dice; this number is recorded in the table, and ranges from 2 to 12. If we believe that each degree on each die has probability 1/6, and that there is no interaction between the dice, then each of these elemental possibilities would have elemental probability 1/36.

We are now ready to compute the probability distribution on the measure. For example, the table shows that there are five elemental possibilities that produce the sum 6, and therefore, the probability of rolling a six is 5 times 1/36, or 5/36. The chance of rolling a seven is the sum of 6 elemental probabilities of 1/36, or 1/6.

Table 6.2. shows the number of possible ways of rolling each sum and the probability of rolling it.

Figure 6.33 shows the probability distribution for the sum of the two dice {Sum | &}.

What is the probability of craps on the first roll?

This is the probability of getting the sums 2, 3, or 12. From Table 6.2, this has a probability of $1/36 + 1/18 + 1/36 = 1/9$. So with the first throw of the dice, you have a 1 /9 chance of losing.

What is the probability of winning on the first roll?

This is the probability of getting the sums 7 or 11. From Table 6.2, this has a probability of $1/18 + 1/6 = 2/9$.

What is the probability of rolling a second time (establishing a point)?

This is the probability of not getting craps and not winning on the first roll. This probability is equal to $1–(1/9)–(2/9) = 2/3$. This is also the probability of getting the sums 4, 5, 6, 8, 9, or 10 on the first roll. From Table 6.2, this probability is equal to $1/12 + 1/9 + 5/36 + 5/36 + 1/9 + 1/12 = 2/3$.

Now suppose you did establish a point. Suppose you rolled a sum of 4 on the first roll, what is the probability of winning?

TABLE 6.2 Ways of Rolling a Sum

Sum of Dice	Number of Possible Ways	Probability of Rolling
2	1	1/36
3	2	1/18
4	3	1/12
5	4	1/9
6	5	5/36
7	6	1/6
8	5	5/36
9	4	1/9
10	3	1/12
11	2	1/18
12	1	1/36
	36	1

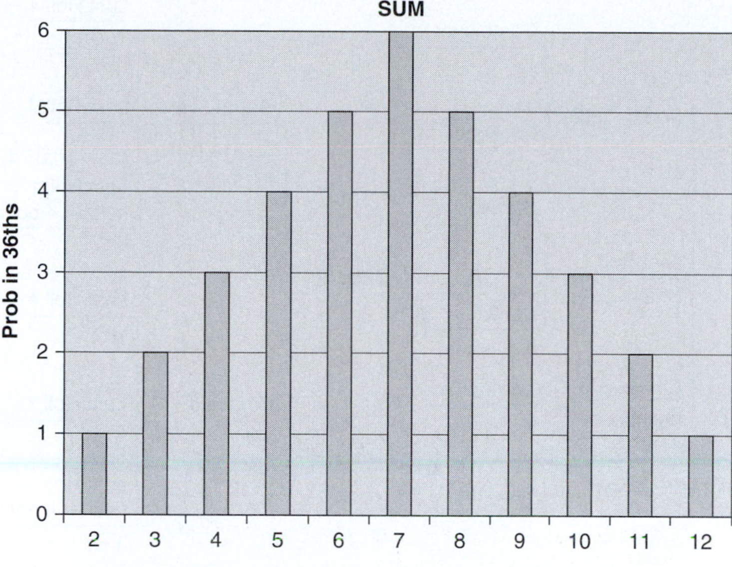

FIGURE 6.33 Probability Distribution for Sum of Two Die

This is the probability of not getting 7 before getting another sum of 4. There are many ways to calculate this probability, but creating the right distinctions can simplify it significantly. With any throw, we can either win (throw a sum of 4 with probability $1/12$), or lose (throw a 7 with probability $1/6$). Throw anything else and we will need to throw again.

When the game stops, however, we have either won or lost. This means that if we know the game has stopped, then we must have thrown either a sum of 4 or 7 on the last roll of the dice. Knowing the game has stopped does not update our probability of whether we won or lost when the game has stopped. Figure 6.34 shows the probability for the distinction of Last Roll and Game Status.

$$\begin{aligned}\{\text{win} \mid \&\} &= \{\text{Sum 4 before Sum 7} \mid \&\} \\ &= \{\text{Sum 4 before Sum 7} \mid \text{Game has stopped}, \&\} \\ &= \{\text{Sum 4 on the last throw} \mid \text{Sum 4 Or Sum 7 were rolled on the last throw}, \&\}\end{aligned}$$

So, if we know the game has stopped, what is the probability that we rolled a sum of 4 on the last throw of the dice?

This is equal to the ratio

$$\frac{\{\text{Sum} = 4 \mid \&\}}{\{\text{Sum} = 4 \mid \&\} + \{\text{Sum} = 7 \mid \&\}} = \frac{\left(\frac{1}{12}\right)}{\left(\frac{1}{12} + \frac{1}{6}\right)} = \frac{1}{3}$$

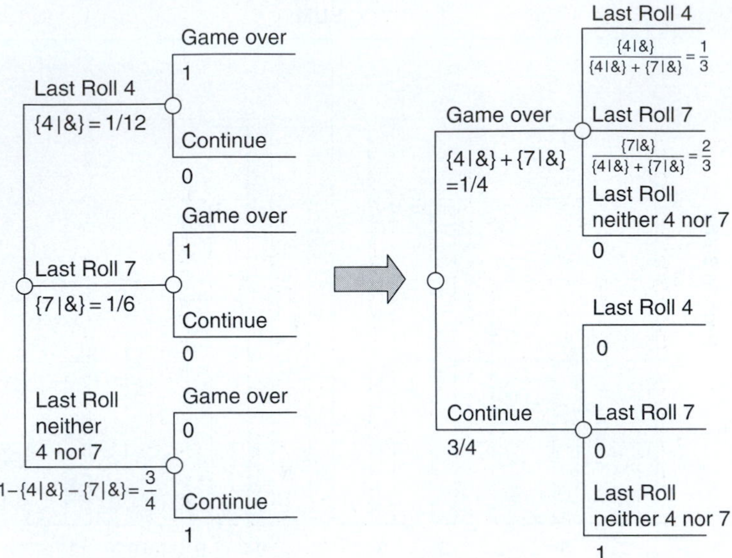

FIGURE 6.34 Tree Reversal to Calculate Probability

TABLE 6.3 Probability of Winning for a Given Sum

Sum on First Roll	Probability of Winning Given Sum on First Roll
4	1/3
5	2/5
6	5/11
8	5/11
9	2/5
10	1/3

Using similar analysis, Table 6.3 shows the probability of winning given the sums of 4, 5, 6, 8, 9, and 10 in the first roll of the dice.

We can now draw our probability tree for the game of craps, and, for convenience, put a measure to each possibility that represents a unit monetary reward for winning and a unit loss for losing. See Figure 6.35.

We can also sum up all the elemental probabilities for winning (244/495) and losing (251/495) and draw the following simplified probability tree. See Figure 6.36.

Given you have a lower chance of winning than losing, would you still play craps? People play to be entertained.

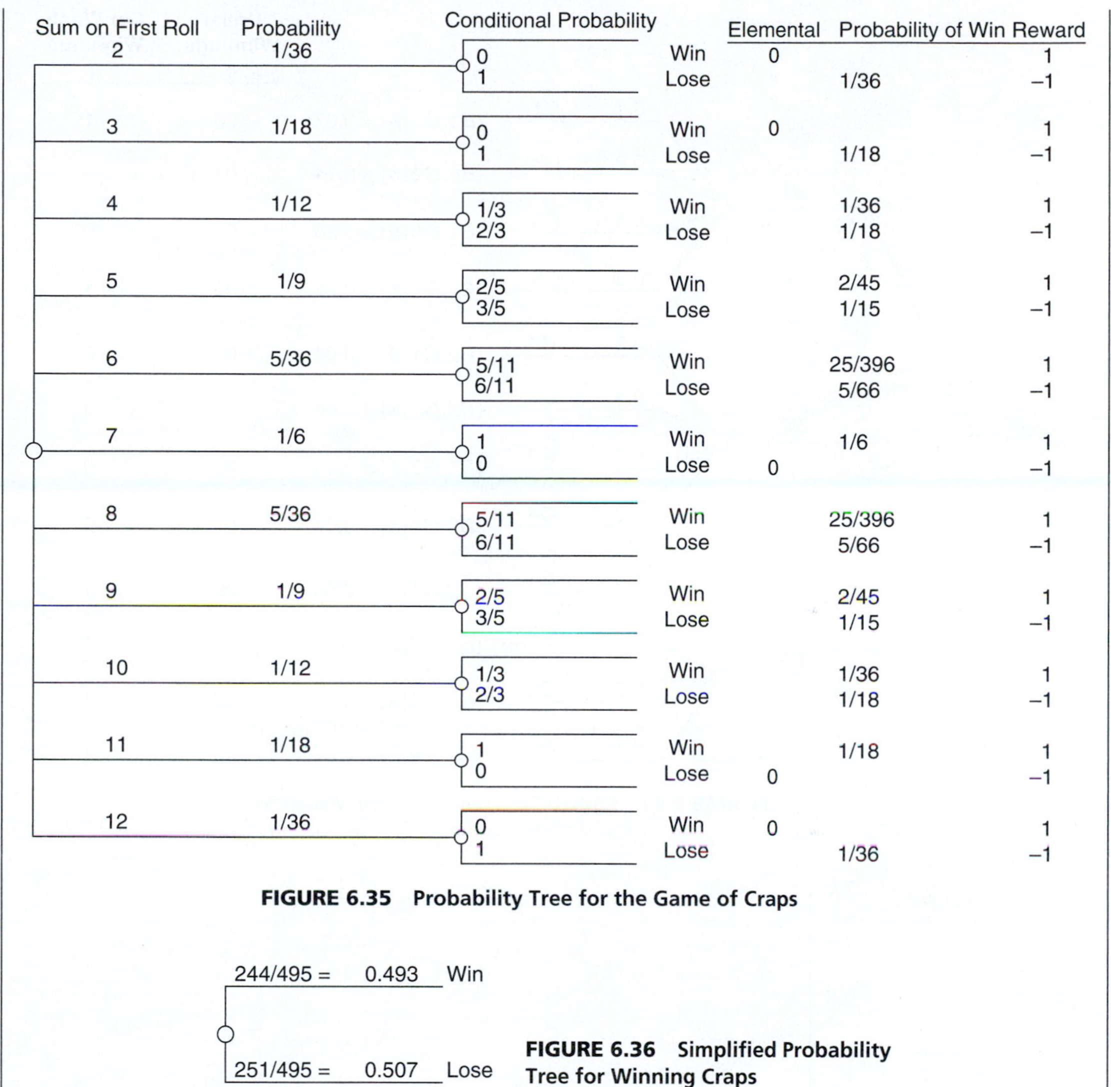

FIGURE 6.35 Probability Tree for the Game of Craps

FIGURE 6.36 Simplified Probability Tree for Winning Craps

6.8 MULTIPLE MEASURES

Figure 6.37 adds a second measure called car wash winnings, w, to Figure 6.30.

Whenever you lose money you must wash the other person's car; whenever you win money you get your car washed. The probability distribution of the measure car wash winnings appears in Figure 6.38.

Since you win money whenever you get your car washed, the probability of 0.28 for the value $w = 1$ is your probability of winning money shown by the height of the excess probability distribution of Figure 6.31 at the value zero.

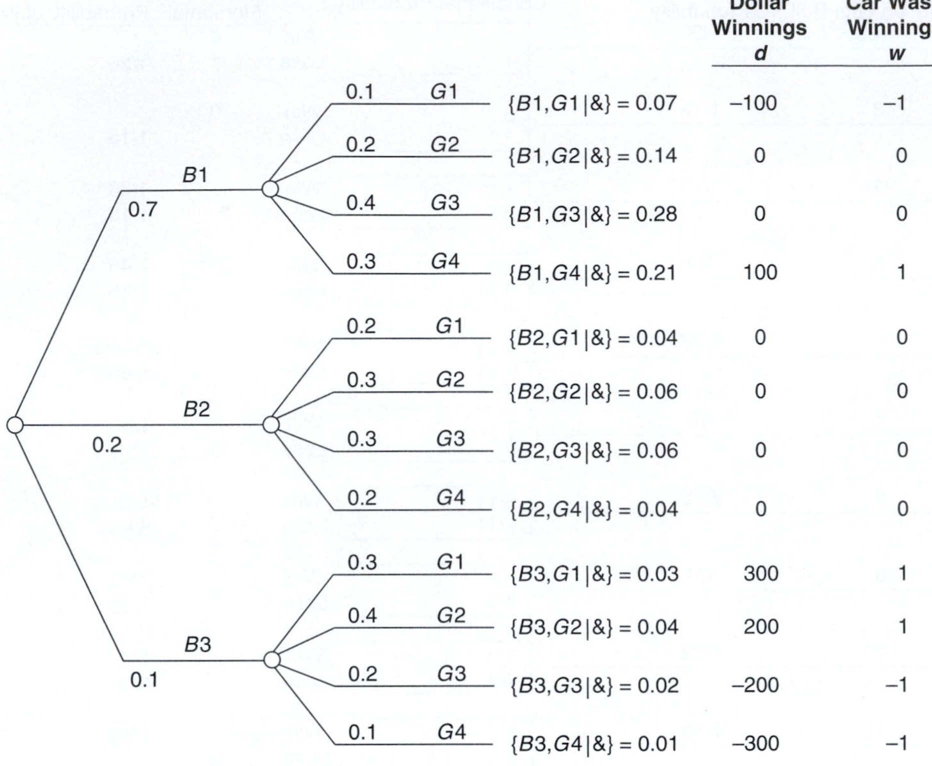

FIGURE 6.37 Probability Tree with Two Measures

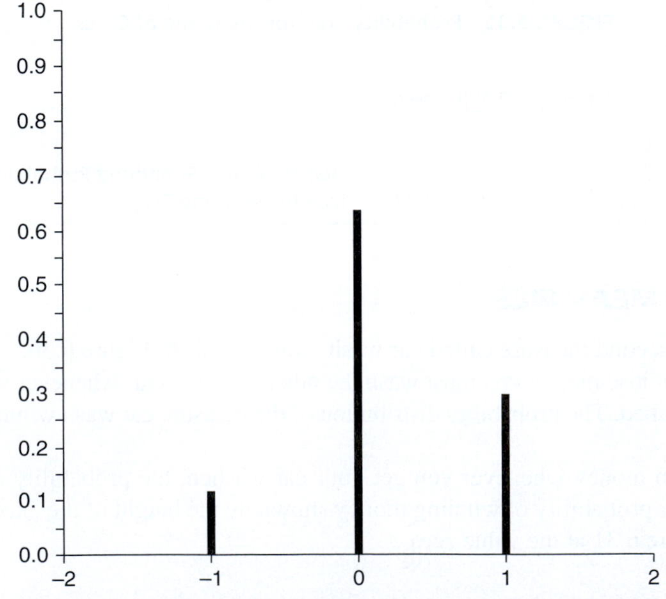

FIGURE 6.38 Probability Distribution of Car Wash Winnings {w|&}

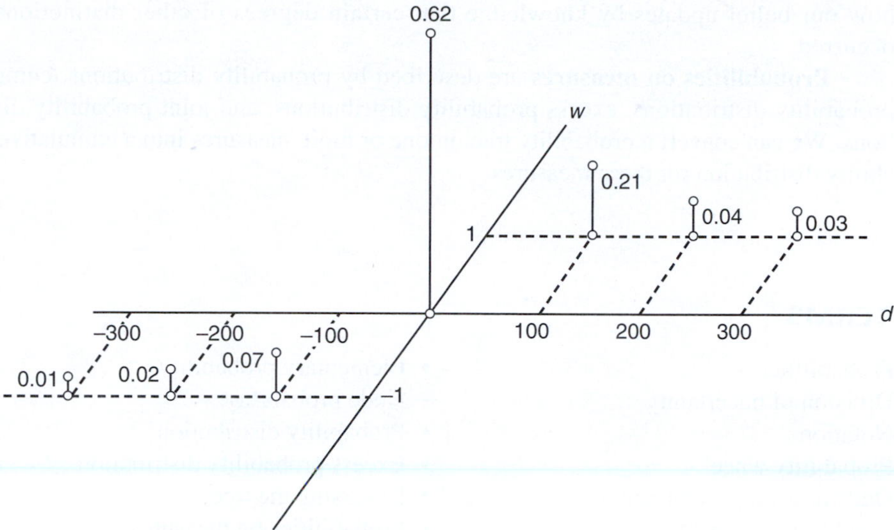

FIGURE 6.39 Joint Probability Distribution

6.8.1 Joint Probability Distributions

We can also construct a joint probability distribution on both dollar winnings and car wash winnings as shown in the three-dimensional plot of Figure 6.39.

Here, for every point in the *d*,*w* plane described by dollar winnings and car wash winnings, we construct a bar whose height is the probability that this combination of values will be attained. Our notation for this joint probability distribution is $\{d, w \mid \&\}$. We can also define joint cumulative and joint excess distributions when necessary.

6.9 SUMMARY

Probability is a measure of belief. It depends on the person's state of information. ***Notation:*** We use the notation $\{A \mid \&\}$ to express the probability of *A* occurring given our state of information, &.

We use the notation $\{B \mid A,\&\}$ to express the conditional probability of *B* occurring given we know that A has occurred and given our state of information, &.
When a distinction has two degrees, *A* and *A'*, we define the odds on *A* occurring by the ratio $\dfrac{\{A \mid \&\}}{\{A' \mid \&\}}$.

Probability trees represent the division of certainty graphically through many distinctions, each having multiple degrees. Probability trees can be reordered. To change the order of the tree, we determine Node Probabilities and Elemental Probabilities. We then match the elemental probabilities of the trees in both orders, and calculate the probability of each node in the reverse tree by reconstructing the division of uncertainty. **Reversing the tree** is the process of changing the order of a tree to determine how our belief updates by knowledge that certain degrees of other distinctions have occurred.

Probabilities on measures are described by probability distributions, cumulative probability distributions, excess probability distributions, and joint probability distributions. We can convert a probability tree on one or more measures into a cumulative probability distribution for these measures.

KEY TERMS

- Probabilities
- Division of uncertainty
- Notation
- Probability wheel
- Odds of a simple distinction
- Conditional probabilities
- Probability tree
- Elementary probability
- Node probability
- Probability distribution
- Excess probability distribution
- Reversing the tree
- Probabilities on measures

APPENDIX A The Chain Rule for Distinctions: Calculating Elemental Probabilities

To generalize our calculation of the elemental probabilities, suppose that we have distinctions A, B, C, each with several degrees of distinction, any one of which we designate by A_i, B_j, and C_k. Then, in accordance with our visualization of the probability tree, we can write for the probability of the elemental possibility A_i, B_j, C_k, as

$$\{A_i, B_j, C_k | \&\} = \{A_i | \&\}\{B_j | A_i, \&\}\{C_k | A_i, B_j, \&\}$$

We call this formula the ***chain rule for distinctions***. The chain rule shows how to compute the probability of any elemental possibility, $\{A_i, B_j, C_k \mid \&\}$, as the product of conditional probabilities. Note that we have six orders in which we can draw a tree with the three distinctions, A, B, C. They are A-B-C, A-C-B, B-A-C, B-C-A, C-A-B, and C-B-A. Since the tree that generates the elemental possibility $A_iB_jC_k$ could be drawn in any of these six possible orders, the chain rule for three events can be written in six possible ways.

For example, this elemental probability of A_i, B_j, C_k is also given by,

$$\{A_i, B_j, C_k | \&\} = \{C_k | \&\}\{A_i | C_k, \&\}\{B_j | A_i, C_k, \&\}$$

obtained from the order C-A-B.

Equating these two expressions enables us to relate conditional probabilities. For example, since

$$\{A_i, B_j, C_k | \&\} = \{A_i | \&\}\{B_j | A_i, \&\}\{C_k | A_i, B_j, \&\} = \{C_k | \&\}\{A_i | C_k, \&\}\{B_j | A_i, C_k, \&\}$$

we can deduce that

$$\{B_j | A_i, \&\} = \frac{\{C_k | \&\}\{A_i | C_k, \&\}\{B_j | A_i, C_k, \&\}}{\{A_i | \&\}\{C_k | A_i, B_j, \&\}}$$

Expanding the Conversation:

Suppose we have the distinction A, with typical degree A_i, and we wish to assign the probability $\{A_i | \&\}$. If we find it easier to think about A_i by introducing another distinction B with typical degree B_j, then we could draw a probability tree in the order B-A and obtain the probability of A_i by adding the probabilities of the elemental possibilities that constitute the compound event A_i in this tree,

$$\{A_i | \&\} = \sum_j \{A_i, B_j | \&\} = \sum_j \{B_j | \&\}\{A_i | B_j, \&\}$$

The symbol $\sum_j$ is a summation. This means that we add up all expressions that have j.

For example, if we have two simple distinctions A and B, with corresponding degrees A_1, A_2, B_1, B_2. This expression would imply that

$$\{A_1 | \&\} = \{A_1, B_1 | \&\} + \{A_1, B_2 | \&\} = \sum_j \{A_1, B_j | \&\}$$

This result is equivalent to applying the chain rule to compute each of the elemental probabilities and then summing them over the possible degrees of B.

If we wish to introduce the further distinction C with typical degree C_k, then we would write,

$$\{A_i | \&\} = \sum_k \sum_j \{A_i, B_j, C_k | \&\}$$

Here, we have introduced two distinctions B and C, computed the elemental probabilities using the chain rule, and then summed over all elemental possibilities contained in the compound possibility A_i to obtain the probability we desire. For example, if B and C were simple distinctions with degrees B_1, B_2, C_1, C_2, the expression implies

$$\{A_1 | \&\} = \{A_1, B_1, C_1 | \&\} + \{A_1, B_2, C_1 | \&\} + \{A_1, B_1, C_2 | \&\} + \{A_1, B_2, C_2 | \&\}$$

We use this general method of introducing new distinctions into the conversation in virtually every application of decision analysis, as it is often very helpful to condition our belief on additional distinctions.

APPENDIX B *Let's Make a Deal*[1] Commentary

As discussed in this chapter, an article concerning the television show *Let's Make a Deal* appeared in a 1990 issue of *Parade* magazine. Following the publication of the article, author Marilyn vos Savant received numerous (incorrect) responses criticizing her analysis. These responses came from distinguished people around the country. Here, we present some of that correspondence to illustrate that probabilistic reasoning can be misleading, even for people quite distinguished in the field. For further reading, see the link below.

http://www.marilynvossavant.com/articles/gameshow.html

Marilyn's Problem Statement:

Suppose you're on a game show, and you're given the choice of three doors. Behind one door is a car; behind the others, goats. You pick a door, say #1, and the host, who knows what's behind the doors, opens another door, say #3, which has a goat. He says to you, "Do you want to pick door #2?" Is it to your advantage to switch your choice of doors?

Marilyn's Answer:

Yes; you should switch. The first door has a one-third chance of winning, but the second door has a two-thirds chance. Here's a good way to visualize what happened. Suppose there are a million doors, and you pick door #1. Then the host, who knows what's behind the doors and will always avoid the one with the prize, opens them all except door #777,777. You'd switch to that door pretty fast, wouldn't you?

Readers' Responses:

Since you seem to enjoy coming straight to the point, I'll do the same. You blew it! Let me explain. If one door is shown to be a loser, that information changes the probability of either remaining choice, neither of which has any reason to be more likely, to $1/2$. As a professional mathematician, I'm very concerned with the general public's lack of mathematical skills. Please help by confessing your error and in the future being more careful.

xxxxx, Ph.D.
George Mason University

You blew it, and you blew it big! Since you seem to have difficulty grasping the basic principle at work here, I'll explain. After the host reveals a goat, you now have a one-in-two chance of being correct. Whether you change your selection or not, the odds are the same. There is enough mathematical illiteracy in this country, and we don't need the world's highest IQ propagating more. Shame!

xxxxx, Ph.D.
University of Florida

Your answer to the question is in error. But if it is any consolation, many of my academic colleagues have also been stumped by this problem.

xxx, Ph.D.
California Faculty Association

Marilyn's Response:

Good heavens! With so much learned opposition, I'll bet this one is going to keep math classes all over the country busy on Monday.

My original answer is correct. But first, let me explain why your answer is wrong. The winning odds of 1/3 on the first choice can't go up to 1/2 just because the host opens a losing door. To illustrate this, let's say we play a shell game. You look away, and I put a pea under one of three shells. Then I ask you to put your finger on a shell. The odds that your choice contains a pea are 1/3, agreed? Then I simply lift up an empty shell from the remaining other two. As I can (and will) do this regardless of what you've chosen, we've learned nothing to allow us to revise the odds on the shell under your finger.

The benefits of switching are readily proven by playing through the six games that exhaust all the possibilities. For the first three games, you choose #1 and "switch" each time, for the second three games, you choose #1 and "stay" each time, and the host always opens a loser. Here are the results.

	Door 1	Door 2	Door 3	Result
GAME 1	AUTO	GOAT	GOAT	Switch and you lose.
GAME 2	GOAT	AUTO	GOAT	Switch and you win.
GAME 3	GOAT	GOAT	AUTO	Switch and you win.
GAME 4	AUTO	GOAT	GOAT	Stay and you win.
GAME 5	GOAT	AUTO	GOAT	Stay and you lose.
GAME 6	GOAT	GOAT	AUTO	Stay and you lose.

When you switch, you win 2/3 of the time and lose 1/3, but when you don't switch, you only win 1/3 of the time and lose 2/3. You can try it yourself and see.

Alternatively, you can actually play the game with another person acting as the host with three playing cards—two jokers for the goat and an ace for the prize. However, doing this a few hundred times to get statistically valid results can get a little tedious, so perhaps you can assign it as extra credit or for punishment! (That'll get their goats!).

More Responses Following Marilyn's Response:

You're in error, but Albert Einstein earned a dearer place in the hearts of people after he admitted his errors.

Xxx, Ph.D.
University of Michigan

I have been a faithful reader of your column, and I have not, until now, had any reason to doubt you. However, in this matter (for which I do have expertise), your answer is clearly at odds with the truth.

xxxx, Ph.D.
Millikin University

May I suggest that you obtain and refer to a standard textbook on probability before you try to answer a question of this type again?

Xxx, Ph.D.
University of Florida

I am sure you will receive many letters on this topic from high school and college students. Perhaps you should keep a few addresses for help with future columns.

xxxx, Ph.D.
Georgia State University

You are utterly incorrect about the game show question, and I hope this controversy will call some public attention to the serious national crisis in mathematical education. If you can admit your error, you will have contributed constructively towards the solution of a deplorable situation. How many irate mathematicians are needed to get you to change your mind?

xxxx, Ph.D.
Georgetown University

I am in shock that after being corrected by at least three mathematicians, you still do not see your mistake.

Xxxx, Ph.D.
Dickinson State University

APPENDIX C Further Discussion Related to the Example: At Least One Boy

7/27/97
Parade
Magazine

I will never read your column again.
—Eric Gatterdam, Tucson, Ariz.

I have lost nearly all my faith in you.
—Douglas Kraft, Notre Dame, Ind.

I can only conclude that you are not woman enough to face the truth and admit your mistake. You are highly intelligent, and that is an admirable quality, but high intelligence coupled with an unwillingness to admit a mistake is unforgivable.
—Leonard Haefele, Overland Park, Kan.

It really puzzles and frustrates me that, despite your great perspicacity, you are unable to see that your answer to the "man and woman each with two children" problem is wrong.
—J.H. Wuller, St. Louis, Mo.

You are wrong. This is borne out by the application of Bayes' rule to the probability structure you imposed, and in the inner refinement functionality as given in the Dempster-Shafer theory of evidential reasoning.
—Dave Ferkinhoff, Middletown, R.I.

I was horrified to read that one of your few supporters was an engineer responsible for assessing risks in the operation of nuclear power plants. I sometimes wonder why critics of IQ testing don't point to some of your work as vivid examples of the vast difference between IQ and logic.
—Robert Williamson, Knoxville, Tenn.

As an anti-nuclear activist, I find it both scary and humorous that a person with a Ph.D. in nuclear engineering who once managed the performance of probabilistic safety analyses of nuclear power plants thinks you are correct.
—Ben Davis Jr., Sacramento, Calif.

I guess the real hope is that the nuclear engineer wasn't paying very close attention when she offered her assent, or else the next problem will involve three-eyed children.
—Jason Zeamon, White Bear Lake, Minn.

That question about a woman and a man, each with two children, is causing controversy again. But this time our women readers are asked to participate, and $1000 is on the line.

I am writing to the Nuclear Regulatory Commission to suggest that any power plants approved for operation by Jennifer Adams be closed immediately.
—Russell Redgate, Marstons Mills, Mass.

This is not going to go away until you admit that you are wrong, wrong, wrong!!!
—Pearl Meibos, Salt Lake City, Utah

You are not the only genius to base logic on a faulty major premise. Einstein did it more than once.
—Margaret-Mary del Tufo, North Myrtle Beach, S.C.

Even the Bulls lose one every once in a while.
—Chris Rowley, Frisco, Tex.

I will send $1000 to your favorite charity if you can prove me wrong. The chances of both the woman and the man having two boys are equal.
—Eldon Moritz, Arlington, Tex.

You're on, Eldon! If you are wrong, you'll donate $1000 to the American Heart Association. If I'm wrong, I'll donate $1000 to that association. Rather than explain my reasoning again, let's just put it to the test. Here's the original problem:

"A woman and a man (unrelated) each have two children. At least one of the woman's children is a boy, and the man's older child is a boy. Do the chances that the woman has two boys equal the chances that the man has two boys?"

I said the chances that the woman has two boys are 1 in 3 and the chances that the man has two boys are 1 in 2. The letter-writers agree with me about the man. But they disagree with me about the woman. Instead, they say the chances that the woman has two boys are 1 in 2 (just like the man's chances).

Readers, here's how you can help prove which answer about the woman is correct. To my women readers: If you have exactly two children (no more), and at least one of them is a boy (either child or both of them), write —or send e-mail—and tell me the sex of both of your children. Don't consider their ages.

In other words, it's fine to write if your older child is a boy and your younger child is a girl. It's also fine to write if your older child is a girl and your younger child is a boy. And it's fine to write if both of your children are boys. I need to hear from all of you (but only if you have two children and no more).

We'll publish the results in an upcoming column.

PROBLEMS

Problems marked with an asterisk (*) are more challenging.
Problems marked with a dagger (†) are considered quantitative.

1. George flips two coins. He believes that both have a fifty-fifty chance of landing heads, and that the results of the flips are irrelevant to one another given his background state of information. The clairvoyant tells George that at least one of the coins has landed heads. What probability should George then assign to the other coin landing heads?

a. $1/4$
b. $1/3$
c. $1/2$
d. You do not have enough information to solve this problem.

2. ABC Insurance Co. estimates that 80% of drivers wear seat belts regularly. They also estimate that 50% of drivers are over age 35. A study showed that 40% of those drivers who wear seat belts regularly are over age 35. Given these estimates, what is the probability that a driver who is over age 35 wears a seat belt regularly?

a. 0.40
b. 0.64
c. 0.96
d. None of the above

3. A friend of yours has two children. You know that at least one of his children is a boy. What is the probability that both children are boys? (You believe that the chance of having a boy is the same as the chance of having a girl regardless of the number and gender of any previous children).

a. $1/4$
b. $1/3$
c. $1/2$
d. None of the above.

†4. Jack believes that the chance of an earthquake happening at Stanford and his car being totally damaged is 10%. If his car is totally damaged, the chance that there has been an earthquake is 80% and if his car is not totally damaged, he believes the chance of an earthquake is 15%. Which of the following intervals covers the chance of his car being totally damaged given there is an earthquake (denoted as p)?

a. $0 \leq p < 0.25$
b. $0.25 \leq p < 0.5$
c. $0.5 \leq p < 0.75$
d. $0.75 \leq p < 1$

5. A recent article in *Fortune* stated that 70% of the billionaires in the United States were school dropouts. Caroline, who lives in the United States, wants to be a billionaire. Which of the following statements can you infer from this information?

a. She has higher chance to become a billionaire if she drops out from school, compared with her staying in school.
b. She has higher chance to become a billionaire if she stays in school, compared with her dropping out from school.
c. Dropping out of school or not does not affect her chance of becoming a billionaire.
d. You do not have enough information to solve this problem.

†**6.** Sara has three coins which you believe to be fair, and are labeled #1, #2, and #3. Suppose Sara flips all three coins and tells you that **at least** 2 of them landed heads.

Given that you believe each coin's probability of landing heads is irrelevant to how the other coins land, what is the probability that coin #1 landed on Head?

a. $1/2$
b. $2/3$
c. $3/4$
d. $4/5$

7. In front of a group of students, Aykut tossed a coin, saw the outcome, and quickly covered it before any of them could see how it landed. Is the outcome of this coin toss still an uncertainty to the other students?

a. Yes, because no student except Aykut knows the result.
b. No, because the coin has already landed. The students just do not know the result.
c. No. Because Aykut knows the result, it can no longer be uncertain.
d. Yes, because it could be a two-headed coin.

†**8.** On Mount Olympus, Zeus is about to announce the name of the winner of the "Greatest Hero in Greek Mythology" award. The finalists are Theseus, Jason and Herakles. Theseus believes that he has a 25% chance of winning and that Herakles is twice as likely to win as is Jason. To build suspense, Zeus is going to call out the names of the two finalists who didn't win before naming the winner, like this: "The second runner-up is *X*, the runner-up is *Y*, and the winner of the award is *Z*." Theseus believes that the announcer is equally likely to name either Jason or Herakles first if Theseus himself is to be named the winner. Then Theseus happens to overhear a conversation between Zeus and Aphrodite, and learns that Zeus will name Jason before Herakles ("Jason, Theseus, Herakles"). We will use p to denote the probability that Theseus assigns to his being the winner of the award. In what range does p lie after hearing this new information?

Hint: When drawing your tree, use the distinctions "Winner" and "Jason's Name Announced Before Herakles"

a. $0 \le p < 0.1$
b. $0.1 \le p < 0.2$
c. $0.2 \le p < 0.3$
d. $0.3 \le p$

†**9.** Jennifer decided to have a barbecue. She invited John over and offered him a piece of wild boar. Unfortunately, the boar was not cooked all the way through, and John became worried that he may have contracted a disease which is known to be transmitted by exposure to raw boars. Having talked to an expert on infectious diseases, John believes that he has a 1 in 50 chance of having the disease. John is a bit worried and goes to his doctor to take a test known to be 90% accurate in predicting this disease. In the doctor's office, John finds out that the test came back positive. Given his other beliefs, with what probability does John believe the (positive) test report is correct?

†**10.** Modified version of Let's Make a Deal: Suppose you believe that there is a 0.1 probability that the car is behind door 1, a 0.3 probability that it is behind door 2, and a 0.6 probability that it is behind door 3. Which door should you choose initially? Should you switch when the host shows you a goat?

†**11.** Doug has defined the event "It will rain" like this: Today, it will either rain or it will not, but not both. He calls the weather service and they say that it will rain today with probability 0.6, which Doug, for the moment, accepts. Doug knows that Cathy is extremely inaccurate about predicting the weather. He believes that if it is not going to rain on a given day, Cathy will say so with a probability of only 0.1. Similarly, if it is going to rain, he feels Cathy will say so with probability 0.2.

a. Assuming that today it will either rain or it will not (but not both), construct the probability tree for Doug's state of belief.
b. Cathy confidently proclaims, "It's not going to rain today." Given Cathy's statement, what probability does Doug assign that it will rain today?

*†**12.** Three prisoners—*A*, *B*, and *C*—are in a prison camp. They all know that one of them is to be executed and the other two are to be set free. However, the warden has not announced who will be executed. Being curious, prisoner *A* asks the warden, who is an honest man, to tell her which of the other two prisoners is to be set free; she already knows that at least one of the other two will be set free.

a. Suppose that prisoner *A* believes that it is equally likely that any one of the three prisoners will be executed. She also thinks that if both prisoners *B* and *C* are to be set free, the warden is equally likely to name either one. The warden quietly tells her that prisoner *B* is to be set free and she agrees not to tell this to the other two prisoners. In light of the warden's remarks, what probability should prisoner *A* assign to the event of being executed herself? Does this probability make sense to you? (Adapted from Frederick Mosteller, Fifty Challenging Problems in Probability, Dover, New York, 1965.)

b. Prisoner *C* believes that he is twice as likely as *B* to be the one executed, and *B* and *A* are equally likely to be the unfortunate one. He asks the warden the same question that *A* asked. He believes that if both prisoners *B* and *A* are to be set free, the warden is twice as likely to name *A* as *B*. If the warden tells him B is to be set free, what probability should prisoner *C* assign to the event of *A* being executed?

c. A new prisoner, named *D*, enters the camp and prisoner *A* tells him that either prisoner *A* (herself) or prisoner *C* will be executed. Prisoner *D* believes that the warden favors neither *A* nor *C*. What probability should prisoner *D* assign to the event of prisoner *A* being executed? Compare this to your answer for part (a) and explain.

*†**13.** Years later, there are three other prisoners, also named *A*, *B*, and *C*. They all know that two of them are about to be set free and that the third is to be kept in prison. Prisoner *A* is in his cell, pondering which one will be kept in prison. Because their crimes are different and prisoner *A* believes that this has been taken into consideration, he assigns different probabilities that each will be the one kept in prison. To prisoner *C*, an inside trader, prisoner *A* assigns probability $1/2$; to prisoner *B*, a check forger, he assigns probability $3/8$; and to himself probability $1/8$.

Being curious, prisoner *A* asks the warden to tell him which of the other two prisoners is to be set free. The warden agrees to tell the truth.

a. Let p be prisoner *A*'s probability that the warden will say "*B* goes free," given that both *B* and *C* are to be released. For what value of p will the warden's response be irrelevant to *A*'s beliefs about whether he himself will stay in prison?

b. Now suppose that, given both *B* and *C* are to be set free, prisoner *A* believes there is a $1/3$ chance the warden will say "*B* goes free;" otherwise, he'll say "*C* goes free."

   1. According to *A*, what is the probability that the warden will say that *B* goes free? That *C* goes free?
   2. Suppose the warden tells prisoner *A* that *B* will go free. Now what should prisoner *A* assign as the probability that he'll stay in jail, given the warden's statement?
   3. Suppose instead that the warden says that *C* will go free. What should prisoner *A* assign as the probability that he'll stay in jail?

*†**14.** The winner of this year's No-Bull prize is about to be announced. The finalists are Danny, George and Sam. Danny believes that exactly one of them will win. Furthermore, Danny believes that he has a 10 percent chance of winning and that Sam is twice as likely as George to win. In order to build suspense, the announcer is going to call out the names of the two finalists who didn't win before naming the winner, as in the example: "The second runner-up is *X*, the runner-up is *Y*, and the winner of the No-Bull prize is *Z*."

Danny believes that the announcer is equally likely to name either George or Sam first if Danny himself is to be named the winner.

Then Danny happens to overhear a conversation between the announcer and a previous No-Bull prize winner, Maggie, and learns that the announcer will name George before Sam. Now what probability does Danny assign to himself being the winner of the No-Bull prize?

†**15.** Suppose you are given a die with 6 sides, marked 1 through 6.

a. For one roll of the die, list all outcomes you consider possibilities. Assign a probability to each of the possible outcomes. Express your beliefs in a probability tree.

b. Graph your probability distribution, cumulative probability distribution, and excess probability distribution over the outcomes.

*†**16.** Your friend takes a coin out of his pocket and tosses it three times. You know that he had a two-headed coin and a regular coin in is pocket, and you believe that it is equally likely for him to have chosen either coin. The outcome of the tosses comes up heads, heads, heads. What is the chance that your friend originally chose the two-headed coin?

17. Reverse the order of the trees in Figure 6.40.

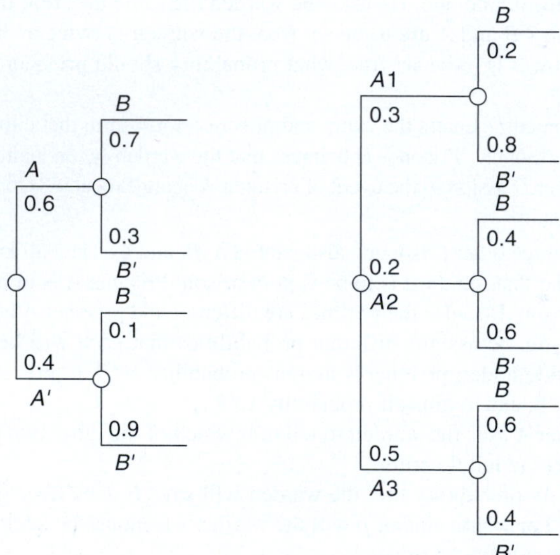

FIGURE 6.40

CHAPTER

7

Relevance

CHAPTER CONCEPTS

After reading this chapter, you will be able to explain the following concepts:

- Relevance
- Relevance is mutual
- Relevance vs. causality
- Relevance diagram
- Reversing arrows in a relevance diagram
- Associative logic errors
- The third factor

7.1 INTRODUCTION

We have seen that our probability assignment for a certain event can change when we know that another event has occurred. For example, knowing that a person has "at least one boy" updates out belief about the second child being a boy. When this change in belief about a distinction occurs after observing the outcome of another distinction, we say that the two distinctions are "relevant." This chapter discusses the properties of relevance relationships and their use in updating belief. New diagrams to represent relevance relations and to facilitate the probability assignment are introduced.

7.2 RELEVANCE WITH SIMPLE DISTINCTIONS

In the last chapter, we discussed assigning probabilities to the two distinctions Beer Drinker and College Graduate. We saw that if we receive new information that a person entering a room is a College Graduate, then that information might change the probability we believe that the person is also a Beer Drinker. In that example, knowing that a person is a Beer Drinker decreased the probability of his (or her) being a College Graduate from $\{G|\&\} = 0.6$ (relying only on the state of information) to $\{G|B,\&\} = 0.2$, knowing the state of information and the person is a Beer Drinker. On the other hand, knowing the person is not a Beer Drinker increased the probability of their being a College Graduate to $\{G|B',\&\} = 0.7$.

> We say that a simple (binary) distinction with outcomes E,E' is **relevant** to another simple distinction with two outcomes, F,F' ***if*** the probability of F given that we know E is *not* equal to the probability of F given that we know E', such as
>
> $$\{F|E,\&\} \neq \{F|E',\&\}$$
>
> We say two distinctions are relevant if: Given your information &, knowing the resolution of one distinction will change the probabilities you assign to the other distinction.

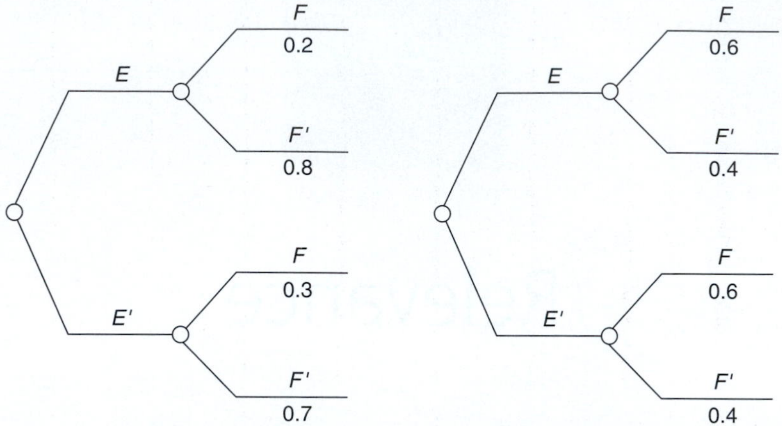

FIGURE 7.1 Left *E* is Relevant to *F*; Right *E* is Irrelevant to *F*

Figure 7.1 shows two trees. The tree on the left hand side shows two relevant distinctions *E* & *F* and the tree on the right hand side shows two distinction where *E* is irrelevant to *F*.

On reflection, we see that this definition corresponds to our basic understanding of relevance. For example, we say that education is relevant to income because if we know the person is well educated, we assign higher probabilities to higher incomes. We say that flipping a coin is not relevant to the weather because we would assign the same probabilities to different weather states regardless of whether we flipped heads or tails.

7.3 IS RELEVANCE MUTUAL?

Now we have defined what we mean by relevance, think about the following question.

> *"If distinction E is relevant to distinction F, is distinction F relevant to distinction E?"*

After reflecting on the question, take a look at Figure 7.2. On the left of the figure, we have a probability tree for the distinctions *E* and *F*. The elemental probabilities, are represented by positive numbers *a*, *b*, *c*, and *d* as shown. Their sum is one.

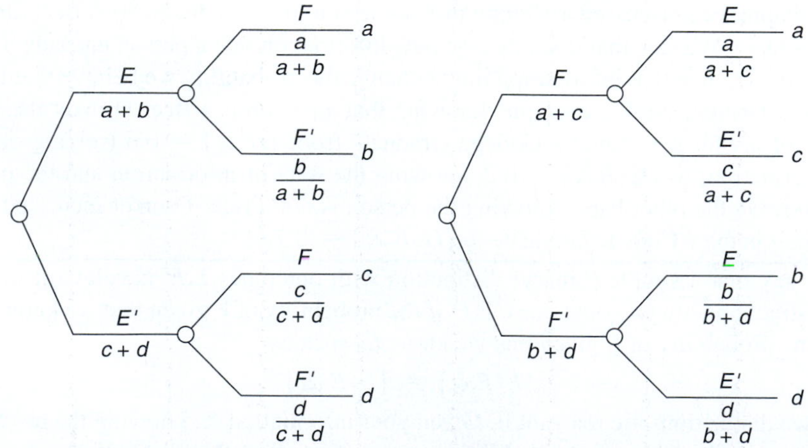

FIGURE 7.2 Two Simple Distinctions with their Elemental Probabilities

If the simple distinction E is relevant to the distinction F, then the probability of F given E must be different from the probability of F given E', $\{F|E,\&\} \neq \{F|E',\&\}$. Since $\{F|E,\&\} = \frac{a}{a+b}$ and $\{F|E',\&\} = \frac{c}{c+d}$, this means that

$$\frac{a}{(a+b)} \neq \frac{c}{(c+d)}$$

Multiplying both sides by $(a+b)(c+d)$ implies

$$ac + ad \neq ac + bc$$

and therefore the two simple distinctions are relevant if

$$ad \neq bc$$

Next, examine the reversed probability tree on the right side of the figure. The only difference is that the elemental probabilities assessed on the two middle elemental possibilities are reversed and are back in their proper positions. If F is relevant to E, then,

$$\frac{a}{(a+c)} \neq \frac{b}{(b+d)}$$

which implies (by multiplying both sides by $(a+c)(b+d)$) that

$$ad \neq bc$$

This is the same condition of relevance required for the tree on the lefthand side of the figure.

Therefore, we have proved that if one simple distinction is relevant to another, then the second is relevant to the first. The same applies to distinctions with multiple degrees. Relevance is thus a **mutual property**.

Observe that the only way you can determine whether distinction E is relevant to distinction F in a probability tree is by examining the numbers on the tree in detail to see whether the conditional probabilities on F are the same given either E or E'. The probability tree is not the most convenient graphical representation of relevance especially when the number of degrees of each distinction is large.

7.3.1 Benefits of Irrelevance Relations for Probability Elicitation

When two distinctions are irrelevant given our state of information, both the construction and the reversal of the probability tree are significantly simplified. When two distinctions, E and F, are irrelevant, then

$$\{E|F,\&\} = \{E|\&\}$$

Therefore, we can assess the probabilities for each distinction without conditioning (such as, $\{E|\&\} = 0.6$ and $\{F|\&\} = 0.3$). We can also draw the full probability tree from these assessments, without conditioning on either distinction as shown in Figure 7.3.

Since irrelevance is a mutual property, we can also draw the tree in the reverse order by using the same probabilities (as shown in Figure 7.4). We leave it as an exercise for the reader to show that this tree is in fact the same tree we would obtain using the steps of tree reversals that we discussed in the last chapter.

In the tree of Figure 7.3 or Figure 7.4, we need only two probability assessments to construct the whole tree (for example, $\{E|\&\}$ and $\{F|\&\}$). If the distinctions were relevant, however, then we would need three assessments (for example $\{F|\&\}$,$\{E|F,\&\}$ and $\{E|F',\&\}$). To generalize, if we

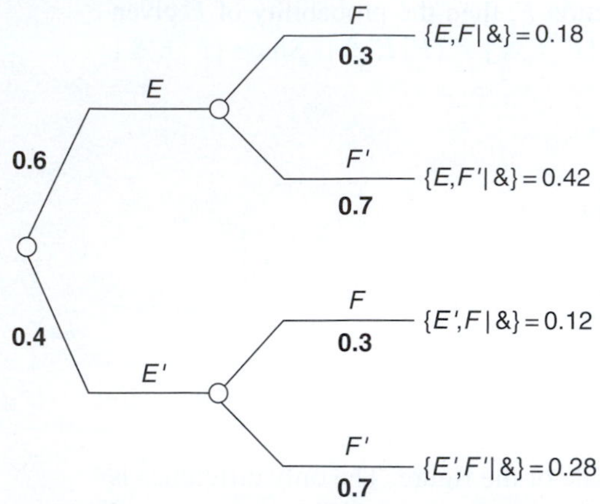

FIGURE 7.3 Probability Tree for Two Irrelevant Distinctions

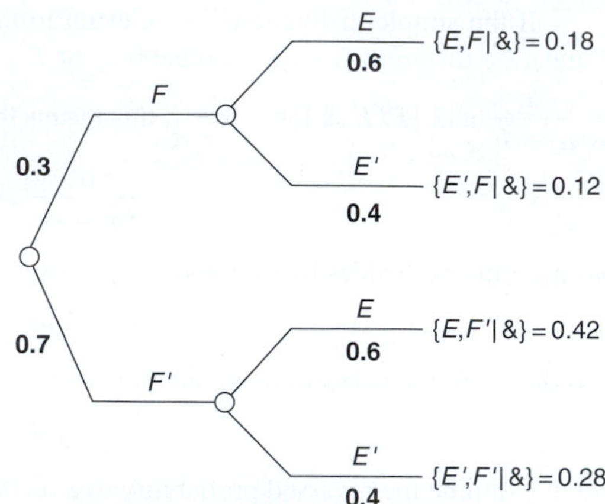

FIGURE 7.4 Probability Tree in Reverse Order $\{E|F,\&\} = \{E|\&\}$ and $\{F|E,\&\} = \{F|\&\}$

can create simple distinctions that are all mutually irrelevant, then we would need an assessment of n probabilities for the n simple distinctions instead of $2^n - 1$ if they were relevant. Furthermore, we would not need to make conditional probability assessments except on the state of information, &.

Once we recognize that distinctions are irrelevant, our tree reversal analysis is significantly simplified. Furthermore, we no longer have to condition our probability assessments on the different degrees of previous distinctions to construct the probability tree.

7.3.2 Relevance, Not Causality

We use the word ***relevance*** because it signifies an informational connection, not a causal connection. There is no notion of causality in our probabilistic descriptions. In many probability classes, the terms *dependence* and *relevance* are used interchangeably. This is misleading. Unlike the word *relevance*, the word *dependence* contains a notion of causality. For example, if we ask whether knowing that people are carrying umbrellas depends on whether it is raining, most people will immediately answer, "Yes." However, if we ask whether rain depends on whether people carry umbrellas, they will answer, "No."

In Figure 6.6, when we assess the conditional probability that the next person entering the room will be a College Graduate, given we know he or she is a Beer Drinker, we are not implying a causal relationship. We are only considering if there is an informational relationship, without considering any notion of causality. We want to be able to consider, with equal ease, the reversed situation. The distinction people carrying umbrellas is relevant to rain and rain is relevant to people carrying umbrellas.

7.4 RELEVANCE DIAGRAMS

As the number of uncertainties in the decision situation increases, probability trees become less efficient for representing the relevance and irrelevance relations between the uncertainties present, as shown in Figure 7.5.

Relevance diagrams provide a simpler and more compact representation of the irrelevance relations between the uncertainties. As we shall see, relevance diagrams contain less information than probability trees as they do not represent the number of degrees of each distinction and they do not contain any numerical probabilities. However, they do efficiently highlight the irrelevance relations.

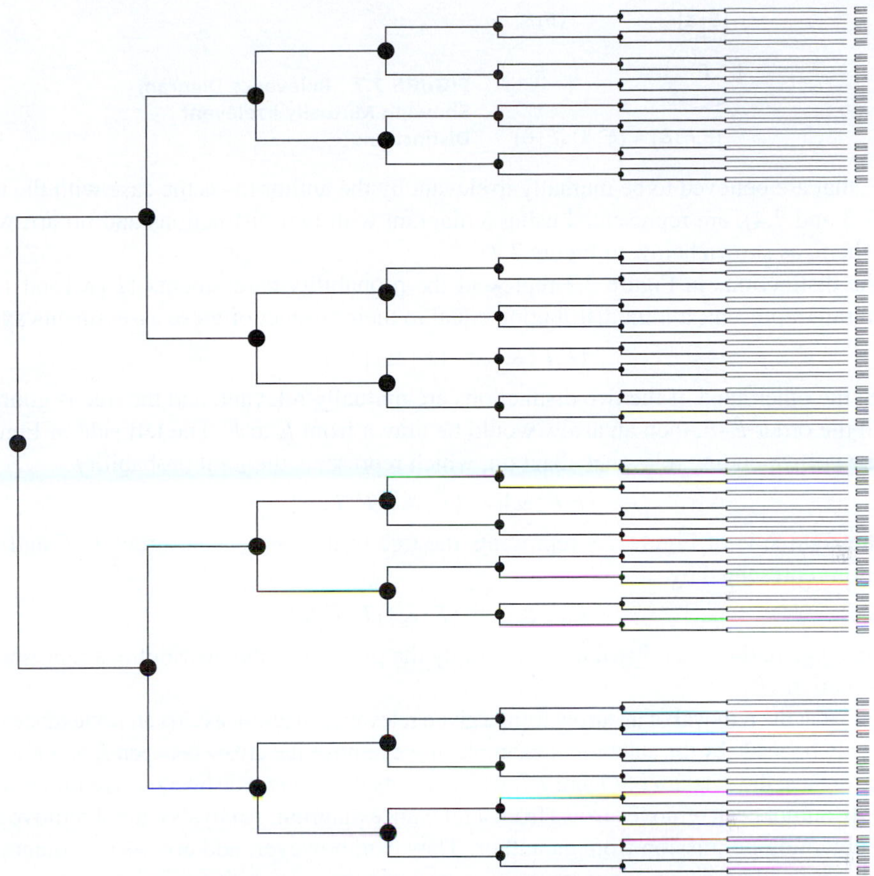

FIGURE 7.5 Trees Growing Exponentially with Uncertainties

Relevance diagrams represent the knowledge of an author based on that author's current state of information, &. The author is responsible for making any assertions relating to relevance of distinctions based on &, and for assigning any probabilities required by the diagram. Relevance diagrams consist of two components, as shown in Figure 7.6:

- **Uncertainties** are depicted by circles or ovals. They represent the distinctions, but they do not carry information about the number of degrees of the distinction or the probability assignments.
- **Arrows** are drawn between the distinctions to represent the possibility of relevance between them. Missing arrows are assertions of irrelevance by the author.

The rule for relevance diagrams is that the arrows entering a distinction show the distinctions upon which the probability assignment for that distinction is conditioned. If a distinction has no input arrows, then it is conditioned only on the state of information, &. Two distinctions,

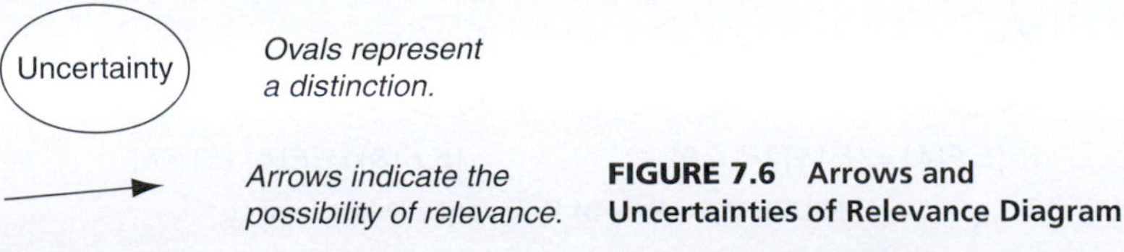
Uncertainty
Ovals represent a distinction.
Arrows indicate the possibility of relevance.

FIGURE 7.6 Arrows and Uncertainties of Relevance Diagram

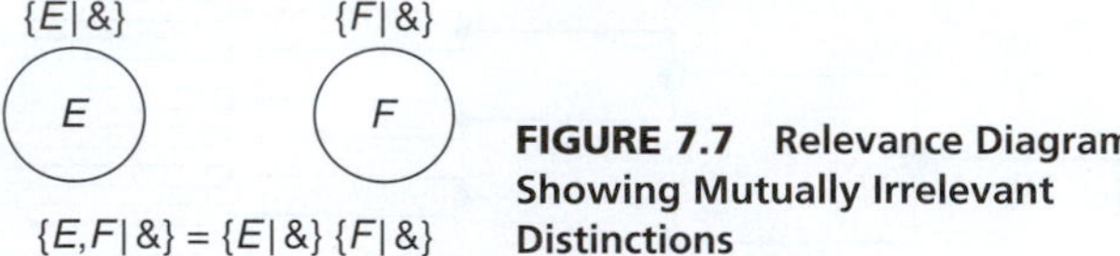

FIGURE 7.7 Relevance Diagram Showing Mutually Irrelevant Distinctions

E and *F*, that are believed to be mutually irrelevant by the author (as is the case with the trees in Figure 7.3 and 7.4), are represented using a diagram with two distinctions and no arrows connecting them, as shown below in Figure 7.7.

The distinctions in Figure 7.7 represent the probability assessments $\{E|\&\}$and $\{F|\&\}$. This diagram represents a joint distribution equal to their product of these assessments as

$$\{E,F|\&\} = \{E|\&\}\{F|\&\}$$

On the other hand, if the two distinctions are mutually relevant, and the tree is going to be drawn in the order *E–F*, then an arrow would be drawn from *E* to *F*. The left side of Figure 7.8 shows the corresponding relevance diagram, which represents the joint probability

$$\{E,F|\&\} = \{E|\&\}\{F|E,\&\}$$

The right side of Figure 7.8 represents the tree in the assessment order *F–E* and a joint distribution represented by

$$\{E,F|\&\} = \{F|\&\}\{E|F,\&\}$$

Once again, this joint distribution is simply the product of the distributions represented by each distinction.

Note that the removal of an arrow from a given relevance diagram asserts an irrelevance relation that can only be made by the author. For example, if we remove the arrow between *E* and *F* in Figure 7.8, then we have the diagram of Figure 7.7, which asserts that the two distinctions are irrelevant.

Only authors can remove arrows from a relevance diagram. Analysts cannot remove arrows without obtaining permission from an author. They can, however, add arrows to a diagram, but making such an addition may unnecessarily complicate the probability assessment process.

For example, if we add an arrow between *E* and *F* in Figure 7.7, it would imply the possibility of relevance between these two distinctions. This would also imply that we would need to assess $\{E|\&\}$ and the conditional probability $\{F|E,\&\}$, instead of just $\{E|\&\}$ and $\{F|\&\}$, as we discussed earlier. During the probability assessment, however, we would find that the conditional probability assessment for $\{F|E,\&\}$ does not change with any degree of *E* and we could then remove the arrow and assert irrelevance based on the author's probability assignment.

7.4.1 Reversing Arrows on a Two-Distinction Diagram

In correspondence with our discussion of tree reversals, if we have a relevance diagram in the form shown on the left of Figure 7.8, we can always convert it into the form shown on the right. This means that the first distinction in the probability tree will be *F*, and the second will be *E*. The probability calculations required to produce consistency in the two representations are identical to those performed in constructing the probability tree of Figure 6.8.

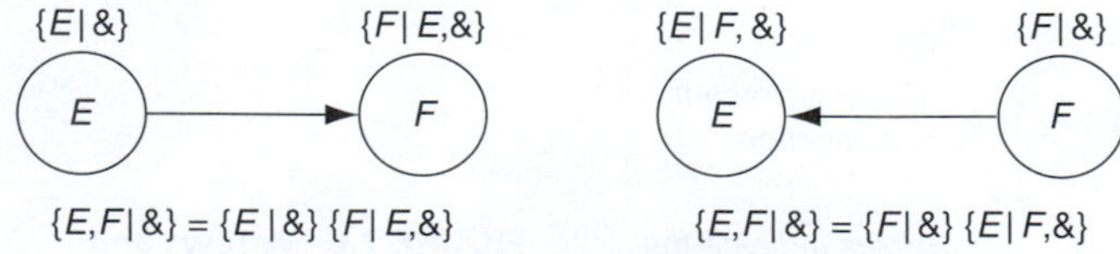

FIGURE 7.8 Pair of Relevance Diagrams

FIGURE 7.9 Relevance Diagrams for Beer Drinker/College Graduate

The relevance diagram is an extremely useful tool for representing the irrelevance assertions made in probability assignment. As we continue our discussions, we shall use larger, more detailed diagrams as powerful tools for representing problems of uncertainty and decision. However, for the Beer Drinker—College Graduate distinctions, the relevance diagram has either of the simpler forms shown in Figure 7.9.

As in the previous example, complications may arise should the analyst decide to add an arrow to a relevance diagram. Each of the relevance diagrams in Figure 7.9 represents a possible assessment order for the probability tree. However, adding an arrow from G to B in the diagram on the left, or from B to G in the diagram on the right, would create relevance diagrams that do not correspond to any probability tree. There is no notion of causality in our probabilistic descriptions. Just like trees in nature, probability trees begin with trunks and end up in smaller and smaller branches. They do not have loops or cycles. When drawing a relevance diagram, we cannot add a conditioning arrow that would create a loop, where the same distinction could be reached by following an arrow leading from it to other distinctions and then back to itself.

7.4.2 Relevance Diagrams with more than Two Distinctions

If there are more than two distinctions to consider, then we need additional distinctions in the diagram. For example, we can now represent the probability tree in Figure 6.18 by the relevance diagram in Figure 7.10. This diagram has three distinctions, indicating the relation

$$\{B,G,S|\&\} = \{B|\&\}\{G|B,\&\}\{S|B,G,\&\}$$

In Figure 7.10, we see that the Beer Drinker distinction has no input, and, therefore, its probability assignment has no conditions except &. The College Graduate distinction is conditioned on the Beer Drinker distinction and &. The sex distinction is conditioned on both the Beer Drinker and College Graduate distinction, in addition to &. This conditioning corresponds exactly to that used in assigning probabilities in the probability tree of Figure 6.19. Once we assess these probabilities, the joint distribution represented by the diagram is simply their product, such as,

$$\{B,G,S|\&\} = \{B|\&\}\{G|B,\&\}\{S|B,G,\&\}$$

The rule for arrow reversals changes slightly if two or more distinctions appear in a relevance diagram. For example, in the case of two distinctions, E and F:

> *"We can reverse the arrow between two distinctions, E and F, only if they are conditioned on the same state of information."*

(if E has a conditioning arrow from any distinction in the diagram, then F must also have a conditioning arrow from the same distinction).

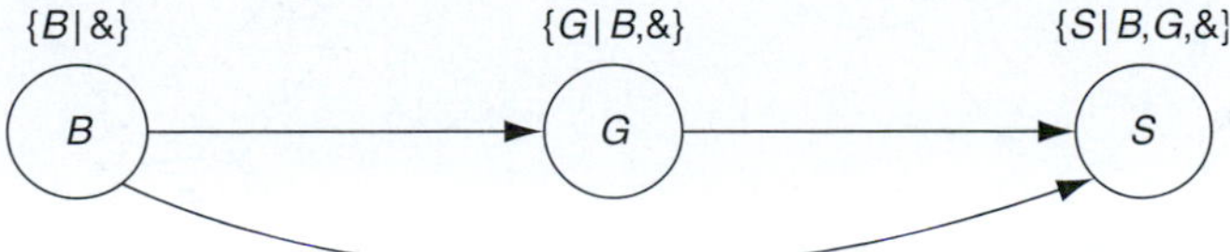

FIGURE 7.10 A Relevance Diagram Indicating $\{B,G,S|\&\} = \{B|\&\}\{G|B,\&\}\{S|B,G,\&\}$

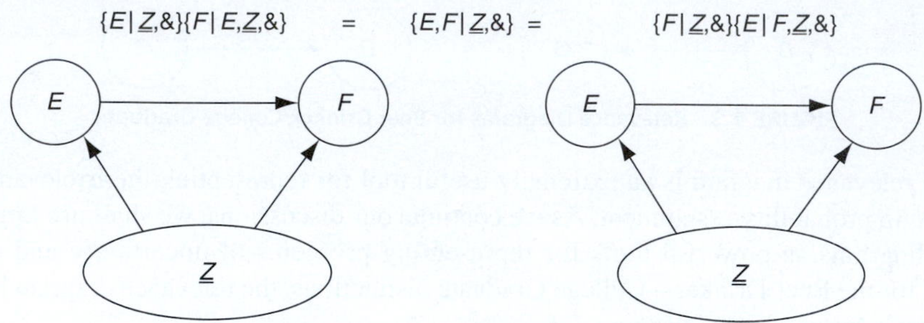

FIGURE 7.11 Arrow Reversals Between Two Distinctions

Figure 7.11 shows an example in which both distinctions, E and F, are conditioned on the same set of distinctions, Z. Therefore, we can reverse the arrow between E and F. The equation below shows the correctness of this reversal,

$$\{E \mid \underline{Z},\&\}\{F \mid E,\underline{Z},\&\} = \{E,F \mid \underline{Z},\&\} = \{F \mid \underline{Z},\&\}\{E \mid F,\underline{Z},\&\}$$

because it is equivalent to that of Figure 7.8, where and & is the common state of information rather than &.

7.5 ALTERNATE ASSESSMENT ORDERS

There are N! orders for drawing a possibility tree with N distinctions. There are also N! possible ways to draw relevance diagrams with N distinctions.

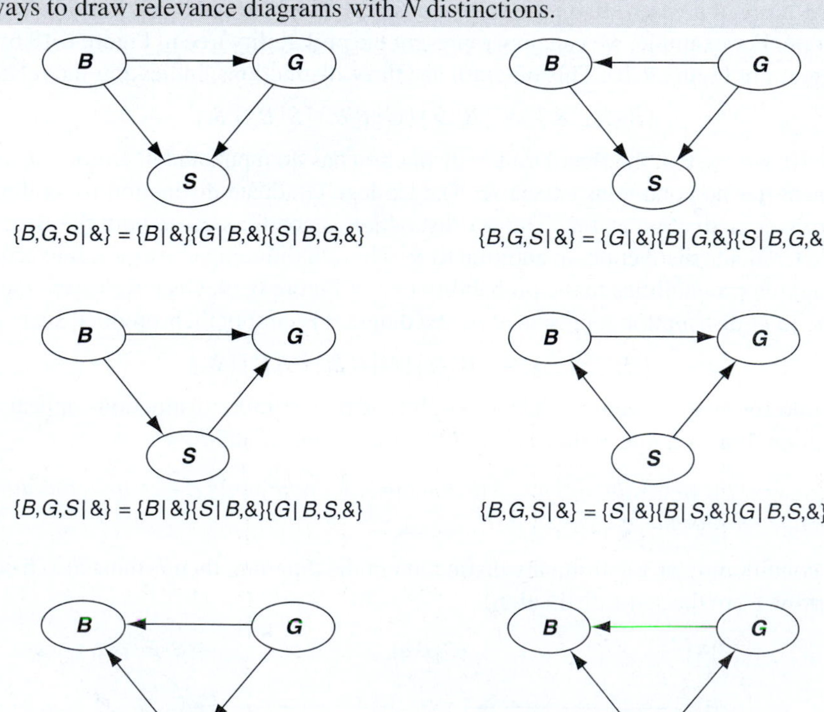

FIGURE 7.12 Relevance Diagrams for Three Distinctions

Therefore, the distinctions B, G, S can be ordered in six ways. Figure 7.12 shows the six possible associated relevance diagrams.

The relevance diagram in the upper left corresponds to the probability tree of Figure 6.18. Once we have assigned the probabilities for any one of these diagrams, we can deduce the probabilities for any other. We will convert the diagram we have assigned to the relevance diagram that appears on the right of the second row in Figure 7.12. The assessment order for this diagram is first Sex, then Beer Drinker—Sex, and finally College Graduate—Beer Drinker—Sex.

To convert the first figure into the second, we first reverse the arrow from S to G. We can do this, since both S and G are conditioned on the same state of information (both have an arrow from distinction B). We now have the diagram on the left hand side of the second row. Next, we reverse the other arrow from B to S. Once again, we can perform this operation, since distinctions B and S are conditioned on the same state of information (neither of them has an arrow from any other distinction in the diagram). We now have the diagram in the assessment order of S–B–G. This diagram implies the possibility of relevance among all distinctions. For example, knowing a person is a Beer Drinker updates our probability of his being both a College Graduate and a Male. The probability tree corresponding to this diagram is in Figure 6.19.

EXAMPLE 7.1

When relevance diagrams have missing arrows, we can determine the irrelevance relations quite easily. Consider for example, Figure 7.13.

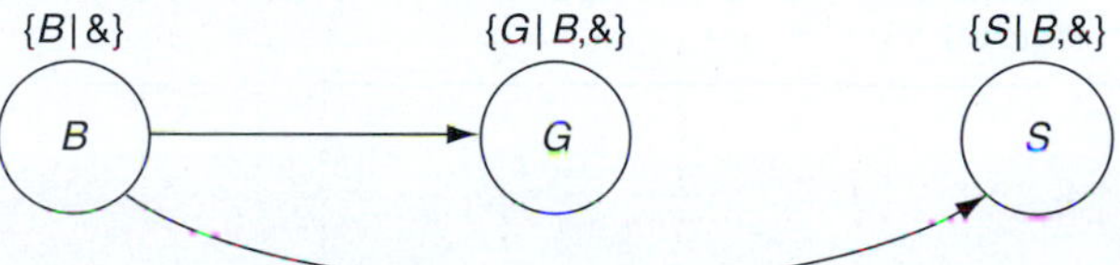

FIGURE 7.13 Relevance Diagram Showing Sex is Irrelevant to College Graduate Given Beer Drinker

This diagram asserts that College Graduate and Sex are mutually irrelevant given we know whether the person is a Beer Drinker. That is, our conditional probability assignment for $\{G|B,\&\}$ would not change if we knew the person is a Male, hence

$$\{G|B,\&\} = \{G|B,M,\&\}$$

Similarly, our conditional probability assignment for $\{M|B,\&\}$ would not change if we knew the person was a College Graduate,

$$\{M|B,\&\} = \{M|G,B,\&\}$$

The diagram represents an assessment order and expresses the relation between the distinctions B,G, and S as

$$\{B,G,S|\&\} = \{B|\&\}\{G|B,\&\}\{S|B,\&\}$$

or equivalently

$$\{B,G,S|\&\} = \{B|\&\}\{S|B,\&\}\{G|B,\&\}$$

When diagrams have missing arrows, they can be represented by more than one probability tree order. For example, Figure 7.13 represents the assessment order Beer Drinker—College

Graduate—Sex or the order Beer Drinker—Sex—College Graduate. Figure 7.14 shows an example of a probability tree corresponding to the diagram in Figure 7.13. As shown, the conditional probability assessments $\{M|G,B,\&\} = \{M|G',B,\&\}$ and $\{M|G,B',\&\} = \{M|G',B',\&\}$ assert irrelevance between Sex and College Graduation given Beer Drinker.

We leave it as an exercise to the reader to reverse the tree of Figure 7.14 in the order Beer Drinker—Sex—College Graduate and verify the relations $\{G|M,B,\&\} = \{G|M',\&\}$ and $\{G|M,B',\&\} = \{G|M',B',\&\}$.

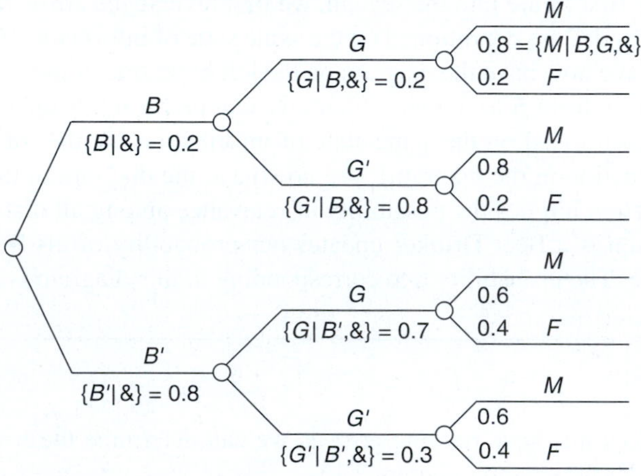

FIGURE 7.14 Tree Corresponding to Relevance Diagram in Figure 7.13

Reflection

Label the distinctions and write the joint probability distributions for the relevance diagrams of Figure 7.15. Mention any irrelevance relations that exist. After doing that, answer the following questions:

1. In the first diagram, can we assert that *G* and *S* are irrelevant given &?
2. In the second diagram, can we assert *B* and *S* are irrelevant given &?

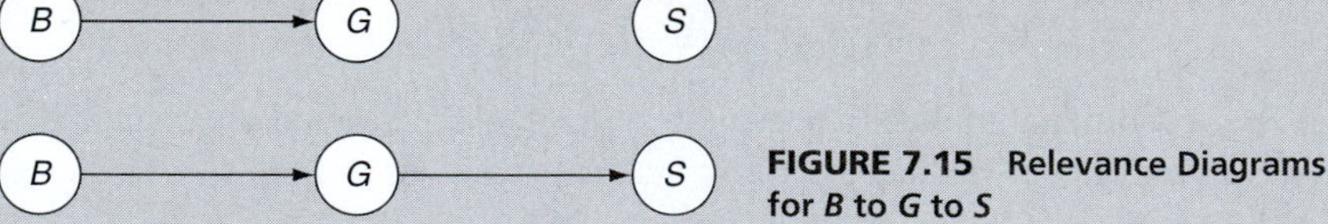

FIGURE 7.15 Relevance Diagrams for *B* to *G* to *S*

7.6 RELEVANCE DEPENDS ON KNOWLEDGE

Relevance is not a matter of logic, but of information. Two distinctions may be mutually relevant to some people and not to others. Some people may think that graduating from college and drinking beer are relevant, while others may believe these are mutually irrelevant. You may disagree with another person's probability assignment, but in no sense is the other person's probability "wrong." The purpose of eliciting probability is to accurately represent a person's beliefs—whatever they might be.

We cannot overemphasize the following: Relevance rests on your **state of information** at a particular time, and not on the **definition of the events**. As people learn, their beliefs and hence probability assignments may, and usually do, change. Our beliefs can change, even though the definitions of distinctions do not. From experience, most errors in dealing with uncertainty are the result of misunderstanding or misapplying the concept of relevance.

7.6.1 Example: Relevance is a Matter of Information

Assume you have a coin with two sides, heads and tails, and you believe the probability of a toss producing either side is 1/2. You toss the coin twice, and you believe the outcome of each toss is irrelevant to the other: Knowing the outcome of one toss does not tell us anything about the outcome of the other. Examine the relevance diagram for these distinctions shown in Figure 7.16. The two distinctions are irrelevant given this current state of information: there is no arrow between them.

FIGURE 7.16 Irrelevance Relations for Head on Toss 1 and Toss 2

Now, suppose we learn that for each of the two tosses, at least one came up as heads. We add this new distinction to the relevance diagram. We draw an arrow from each of the original distinctions to this new one, because knowing the original distinctions will determine whether at least one came up as heads. Figure 7.17 shows the relevance diagram for the distinctions.

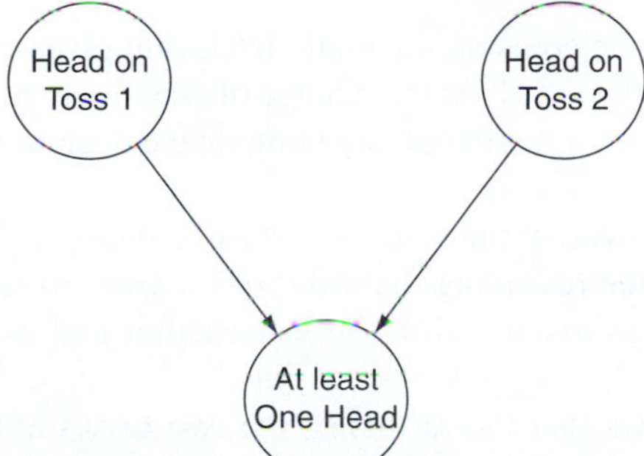

FIGURE 7.17 Relevance Diagram for Three Distinctions

To see how this information affects our beliefs about the results of each of the two tosses, we need to draw the relevance diagram in the order At Least One Head—Toss 1—Toss 2. To do this, we first reverse the arrow between At Least One Head and Head on Toss 2. However, these two distinctions are not conditioned on the same state of information. We can condition them on the same state of information by adding an arrow from Head on Toss 1 to Head on Toss 2, as shown in Figure 7.18. Remember, adding an arrow does not place any restrictions on the diagram, as long as the result is a possible assessment order. The added arrow represents only a *possibility* of relevance.

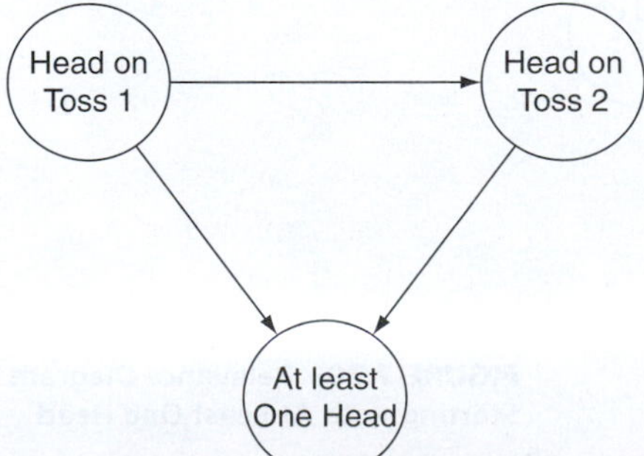

FIGURE 7.18 Adding an Arrow to the Relevance Diagram to Allow for an Arrow Reversal

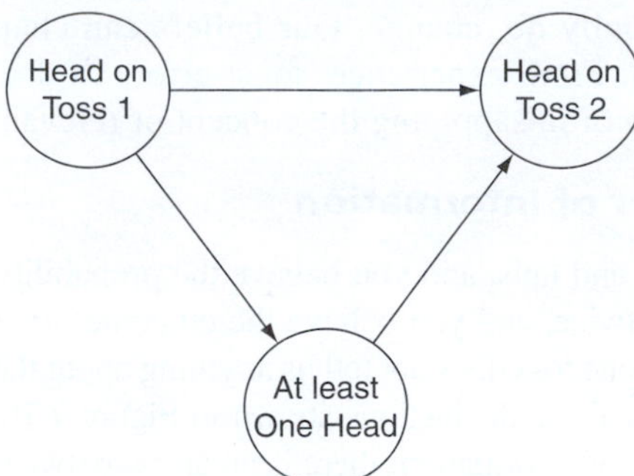

FIGURE 7.19 Reversing the Arrow and Changing the Diagram Order

Next, we can reverse the arrow between At Least One Head and Head on Toss 2, as shown in Figure 7.19.

Finally, we need to reverse the arrow between At Least One Head and Head on Toss 1. We can perform this operation directly, since both distinctions in Figure 7.19 are conditioned on the same state of information. See Figure 7.20.

We now have the relevance diagram in the order we want, and observe that the distinctions Head on Toss 1 and Head on Toss 2 may now be relevant since we know that at least one toss came up as heads.

Consequently, we say the outcomes of the coins are mutually irrelevant given our current state of information but [possibly] mutually relevant given knowledge of whether or not At Least One Head has occurred. We have reached this result without any numerical calculations, relying only on the rules of arrow reversals.

Using probability trees, the reader can verify the same assertion, creating a distinction of the event $I =$ At Least One Head, and then reversing the order of the trees to start with I, then Head on Toss 1 and Head on Toss 2, as discussed in the last chapter and as shown in Figure 7.21.

The left hand side of Figure 7.21 shows that the results of the two tosses are mutually irrelevant given &. The right hand side shows that given I and &, the results of the two tosses are indeed relevant. If you know that at least one toss came up heads, then you would assign probability $2/3$ to Toss 1 producing a head, and probability $1/2$ to Toss 2 producing a head if you knew that Toss 1 produced a head. This is a very important point as it demonstrates that probability rests not in the coins themselves, but in our knowledge. Relevance rests on information, not on logic.

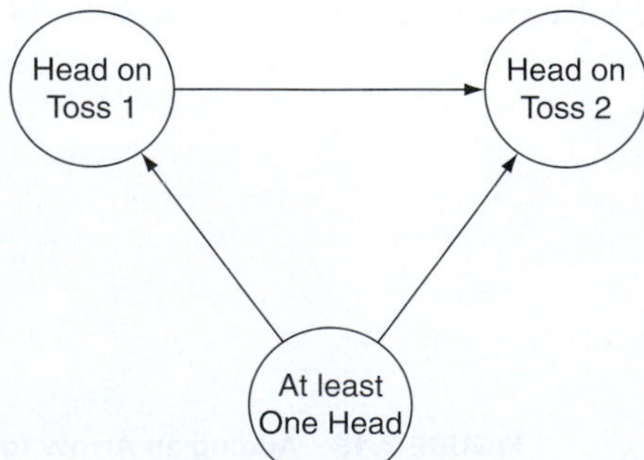

FIGURE 7.20 Relevance Diagram Starting with At Least One Head

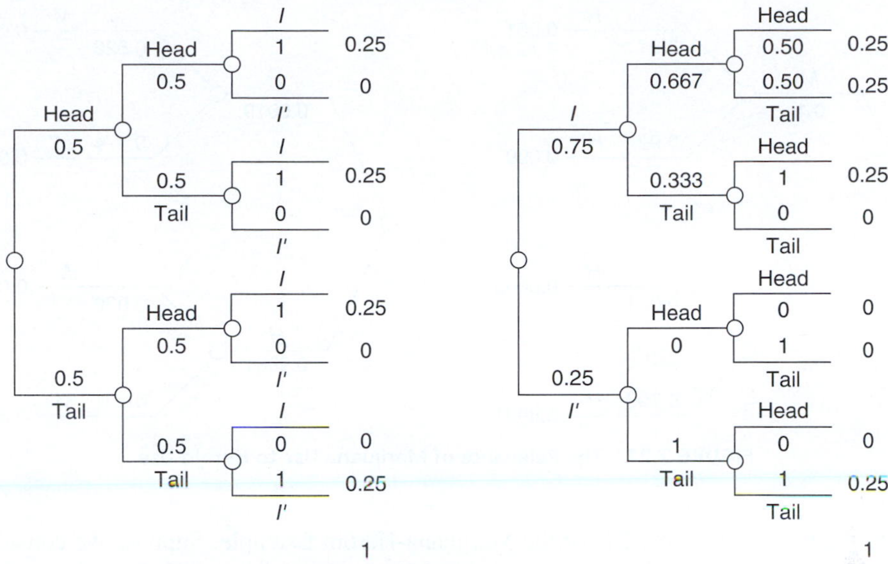

FIGURE 7.21 Probability Trees for Coin Toss Example

7.6.2 Example: The Relevance of Marijuana Use to Heroin Use

The question of relevance can have important implications in everyday life. Several years ago, a defendant convicted of selling marijuana appeared before a judge for sentencing. The judge said,

> *"I have observed in my many years on the bench that most heroin addicts began by using marijuana. Since the drug you sell leads on to heroin addiction, I am going to sentence you as if you were a heroin dealer."*

The judge then gave the convict a long sentence.

EXAMINING THE REASONING USING RELEVANCE RELATIONS Now, we will examine the judge's reasoning in greater detail. We will include as part of our state of information & that we are talking about American adults, those over 18 years of age. Let us define *M* as the event that a person is a Serious Marijuana User using the criterion that he or she must have consumed at least 100 marijuana joints during his or her lifetime. Let us define *H* as the event that the person is a Serious Heroin User, specified as having used at least 20 doses of heroin over a lifetime.

ASSIGNING PROBABILITY We then construct the probability tree in the order *M–H* as shown on the left side of Figure 7.22.

The probability assignments shown are those typical of a group in which this situation is discussed. By our criterion, we see that there is a probability 0.1 that an American adult will be or have been a Serious Marijuana User, and if the person is a Serious Marijuana User then there is a probability of 1 in 100 of being a Serious Heroin User. However, if, by our criterion, the person is not a Serious Marijuana User, then the probability of being a Serious Heroin User is 1 in 1,000. The four elemental probabilities implied by the tree are, as usual, shown at the ends of the branches. We see, for example, that the elemental possibility *MH* has a probability of 1 in 1,000.

VISUALIZING CONDITIONAL PROBABILITIES The difficulties people have in dealing with conditional probability may be alleviated by presenting the concept in another form. Let us

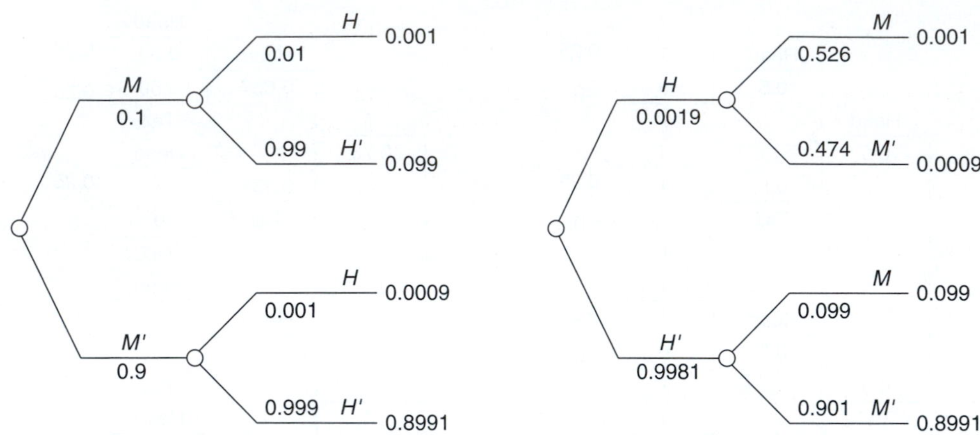

FIGURE 7.22 The Relevance of Marijuana Use to Heroin Use

return to the trees of Figure 7.21 for the Marijuana-Heroin Example. Suppose we consider the trees as representing the results of a survey of 10,000 people as shown in Table 7.1.

Of the 10,000 people, 10 were *MH*, 990 were *MH'*, 9 were *M'H*, and 8991 were *M'H'*. The entries in the table are the same for either the left tree or the right tree of Figure 7.21 because they are constructed from the endpoint probabilities, the same in both forms. The row sums 1000 and 9000, divided by10000, are the probabilities $\{M|\&\}$ and $\{M'|\&\}$. The column sums, 19 and 9981, divided by 10000 are the probabilities $\{H|\&\}$ and $\{H'|\&\}$. All conditional probabilities such as $\{M|H'\&\}$, are obtained by dividing the entry for *MH'*, 990, by the column sum of *H'*, 9981, or 0.099.

Visualizing trees using a table may make the results more intuitive. You may wish to create a similar table for the Smoking-Lung Cancer Example of Figure 7.22.

REVERSING THE TREE Next, we reverse the probability tree, as shown on the right side of Figure 7.22. With elemental probabilities recorded in their appropriate position, we can compute the probability assignments in this tree that will make it consistent with the original one. The probability of the event *H* is the sum of the probabilities of the elemental possibilities *HM* and *HM'* or .0019; the event *H'* has probability .9981. We obtain the conditional probabilities from the ratio of the elemental probabilities and the requirement that they add to 1. Therefore, the probability of *M* given *H*, $\{M|H,\&\} = 0.0010/(0.0010 + 0.0009)$ or 0.526, and the conditional probability $\{M'|H,\&\} = 0.474$. Similarly we find $\{M|H',\&\} = 0.099$ and $\{M'|H',\&\} = 0.901$.

INTERPRETATION We can now interpret what we have computed. First, we see that by our definition, the probability that an adult American will be a Serious Heroin User is 0.0019, or

TABLE 7.1 Marijuana-Heroin Example as Results of a Survey

	H	H'	Sum
M	10	990	1000
M'	9	8991	9000
Sum	19	9981	10000

about two thousandths. Given that the person is a Serious Heroin User, the probability of their being a Serious Marijuana User is 0.526. This is consistent with the judge's observation that when he interviewed heroin addicts, he found that many of them were or had been heavy users of marijuana. Given heroin use, the probability of marijuana use is sizeable.

However, from this observation, the judge had concluded that marijuana use leads to heroin use. In our example this clearly is not the case, since the probability $\{H|M,\&\}$ shows only a 1 in 100 chance of a Serious Marijuana User becoming a Serious Heroin User. The fact that the probability of M given H is large does not imply that the probability of H given M is large. The judge observed correctly that a majority of heroin users were, or had also been, marijuana users, but incorrectly concluded that marijuana use is likely to lead to heroin use.

7.7 DISTINCTIVE VS. ASSOCIATIVE LOGIC

It is both common and easy to reason improperly by confusing the order of conditional probabilities. Without the benefit of this foundation in logic, it is difficult for a person to see and understand why he has made a logical error. A simple associative, non-directional logic might lead someone to conclude that heroin use "goes with" marijuana use, and to believe that both of the conditional probabilities—M given H and H given M—are large. Using the more precise, distinctive logic described here helps us to understand and appreciate the fallacy of these assertions.

7.7.1 Males and Hemophilia

To drive home the fact that a conditional probability and its inverse can be very different, we will now consider two events, M and H, with different definitions. Let M be the event that the person is a male, and H be the event that the person is a hemophiliac. The probability of H given M, that a person will be a hemophiliac given that he is a male, is very small, 1 in 1,000 or less. However, the probability of M given H is virtually 1, since only males appear to express the disease of hemophilia.

7.7.2 Smoking and Lung Cancer

We once presented both the marijuana/heroin and male/hemophilia examples to a group of medical doctors who had returned to the university for an advanced fellowship. One doctor remarked that while this was certainly an important logical observation, such errors in thought would not be common in people like themselves, thoroughly trained in diagnosis and treatment.

To see if this was so, we examined with the group the relationship between smoking cigarettes and contracting lung cancer. We defined the event being a Heavy Smoker, S, as having smoked at least two packs of cigarettes per day for a period of at least 10 years during a lifetime. We defined the event of contracting Lung Cancer, L, according to the usual medical definition, well understood by this group. Then one of the doctors, not a specialist in the treatment of lung cancer, assigned the probabilities shown in the probability tree at the left side of Figure 7.23, with the understanding that we were talking about adult Americans, as before. He assigned 1 chance in 4 that a person would be a Heavy Smoker, and if he or she were a Heavy Smoker, S, that the chance of getting Lung Cancer was about 0.1. If the person was not a Heavy Smoker, S', then the chance of getting Lung Cancer was 1 in 100. We obtained the elemental probabilities by multiplying the probabilities along each branch and recording as shown on the left side of Figure 7.23. When transferred to the tree in reverse order, as shown on the right side of Figure 7.23, they produced about a 3 percent chance of Lung Cancer and a 0.769 chance of being a Heavy Smoker, given that a person did get Lung Cancer.

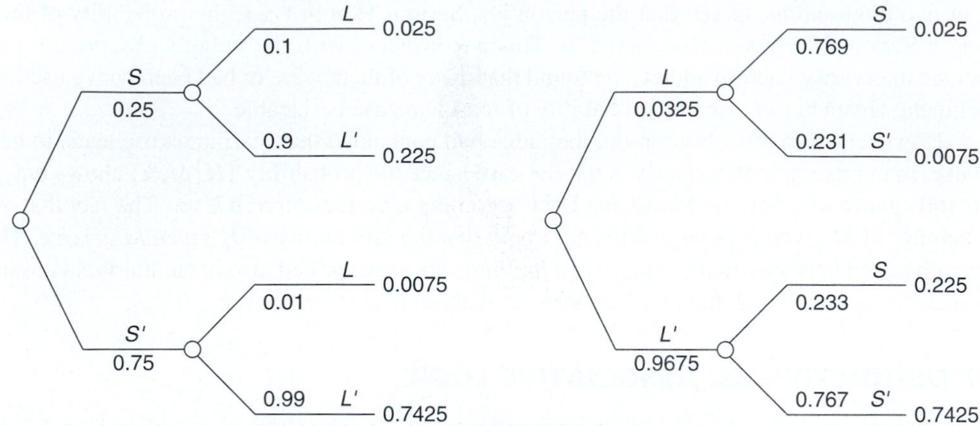

FIGURE 7.23 Relevance of Smoking to Lung Cancer

INTERPRETATION The group then discussed the doctor's assignments. A lung cancer specialist remarked that he thought the probability of *L* given *S*, assigned at 0.1, was much too low. When asked why, he responded, "because when I visit my lung cancer ward, it is full of smokers." This observation, of course, merely reflects what the first doctor had said. The probability of *S* given *L* is, in fact, 0.769, so he, too, would expect that a majority of patients in a lung cancer ward would, by our definition, be heavy smokers. In fact, the probability of getting Lung Cancer if you are a Heavy Smoker is nowhere near as high as the probability of being or having been a Heavy Smoker given that you have Lung Cancer.

When even highly trained people fall victim to inferior reasoning, we know that we all must be careful. No one knows how many people are spending extra years in jail, or perhaps, being improperly treated in hospitals because of a failure to grasp the difference between distinctive and associative logic.

The same inferior reasoning may explain why mechanics believe that most people do not take proper care of their cars, or why lawyers believe that most people do not pay enough attention to legal matters. Like the doctor, in their practices they are much more likely to see people who have these tendencies.

7.8 THE THIRD FACTOR

The notions of relevance and relevance diagrams are useful not only in specifying probability assignments, but in clarifying our thinking. A particular example of this use is the understanding of so-called "third factor" problems. An example of these is the historical story of the connection between stork nests and human births in Copenhagen.

7.8.1 The Relevance between Storks' Nests and Human Births

The story is that when stork nests are more prevalent, human births are higher. In relevance diagram terms, this means that we draw a relevance arrow from stork nests to human births as shown in Figure 7.24.

FIGURE 7.24 Relevance Between Stork Nests and Human Births

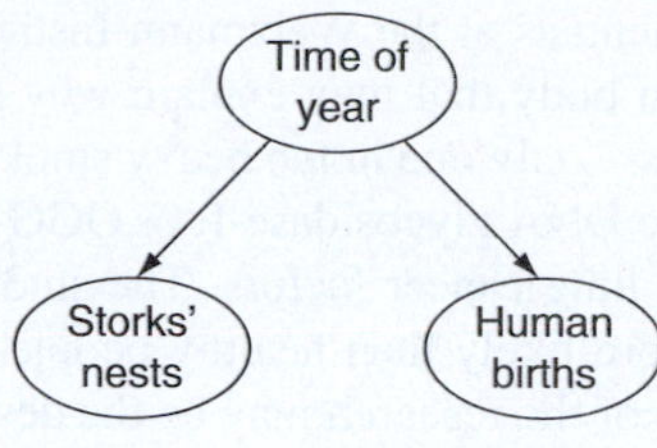

FIGURE 7.25 The Third Factor Explaining Relevance Between Storks' Nests and Human Births

In fact, if you had to bet on the number of human births at any time you would find it useful to know the number of stork nests. Is there any truth to the story that storks deliver babies? It turns out that if we condition our knowledge on time, such as the month of the year, we can solve the mystery.

As shown in the relevance diagram in Figure 7.25, the time of year is relevant to both the number of stork nests and the number of human births. At certain times of the year, storks are more likely to nest in the chimneys of Copenhagen, and babies are more likely to be born. However, if we know the time of the year, there is no relevance of stork nests to human births.

In Figure 7.25, we notice arrows connecting time of year to both stork nests and human births, but we do not see any arrow connecting storks' nests and human births. If you do not know the time of year, stork nests are relevant to human births. If you do, however, they are completely irrelevant. This example is one of many for which the relevance between two distinctions may become irrelevant once a new distinction becomes known. This is also another example of why the concept of relevance depends on your state of information.

7.8.2 The Relevance between Smoking and Lung Cancer

We have already discussed the relevance of heavy cigarette smoking to lung cancer. We represent this relevance using a relevance diagram, such as the one seen in Figure 7.26.

However, cigarette manufacturers have traditionally said that this is *not* a causal relationship. Rather, they say, the relevance can be explained by the existence of a third factor, *X*, that makes people both more likely to smoke and more likely to get lung cancer. In other words, they postulate a relevance diagram like the one shown in Figure 7.27. Of course, since no one at present knows whether a factor *X* exists, or whether they have it, it is prudent for most of us to avoid cigarette smoking.

On the other hand, perhaps factor *X* does exist. The following article from *The Economist* discusses a possible third factor for the relevance between smoking and lung cancer.

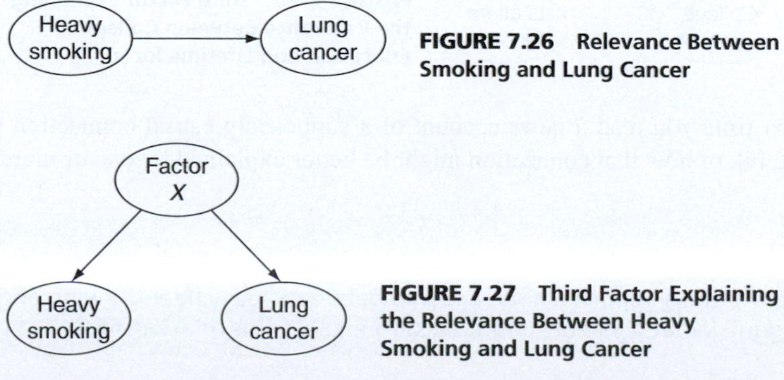

FIGURE 7.26 Relevance Between Smoking and Lung Cancer

FIGURE 7.27 Third Factor Explaining the Relevance Between Heavy Smoking and Lung Cancer

> On Tuesday, the NCI published a study by scientists at the Weizmann Institute in Israel, identifying a substance produced in the human body that may explain why some people are much more prone to lung cancer than others—only one in ten heavy smokers develops the disease. The substance, called 8-oxoguanine DNA glycosidase 1, or OGG1, repairs the genetic damage caused by smoking and other lung-cancer factors. The study found that those suffering from lung cancer are much more likely than healthy people to have low levels of the compound. One possible outcome of the research may be the development of a blood test to identify those most at risk from lung cancer.
>
> "THE BATTLES TO COME," *The Economist:* 9/5/2003
>

7.8.3 The Relevance between College Education and Lifetime Income

As a third example, consider the distinctions of college education and lifetime income. There appears to be relevance between college education and lifetime income. If you know that someone has a college education as opposed to no education, you would assign a higher probability of higher lifetime income. The associated relevance diagram is the one shown in Figure 7.28.

FIGURE 7.28 Relevance Between College Education and Lifetime Income

Professors like to think that this relevance is the result of the wonderful educational experience they provide students.

However, there may be another factor, *X*, that can explain this relevance even if the college education is useless. Such a relevance diagram appears in Figure 7.29. In fact, many factors exist that could, either individually or collectively, serve as factor *X*. Some possibilities are intelligence, industriousness, and motivation of the student, as well as the social status and wealth of the student's family. Of course, professors like to think that even though some relevance may be due to such factors as *X*, there is additional relevance from the effect of the college education itself.

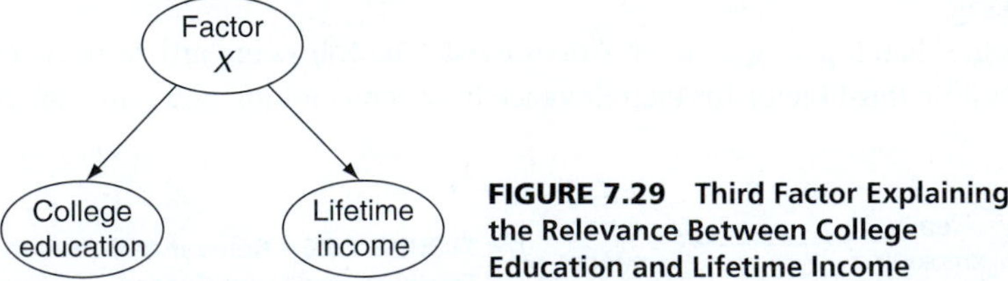

FIGURE 7.29 Third Factor Explaining the Relevance Between College Education and Lifetime Income

The next time you read a news account of a supposedly causal connection between two distinctions, think of how that connection might be better explained by one or more *X* factors.

Reflection

A recent medical story reported a relationship between heavy beer drinking and the development of gout. What *X* factors might exist to explain this relationship?

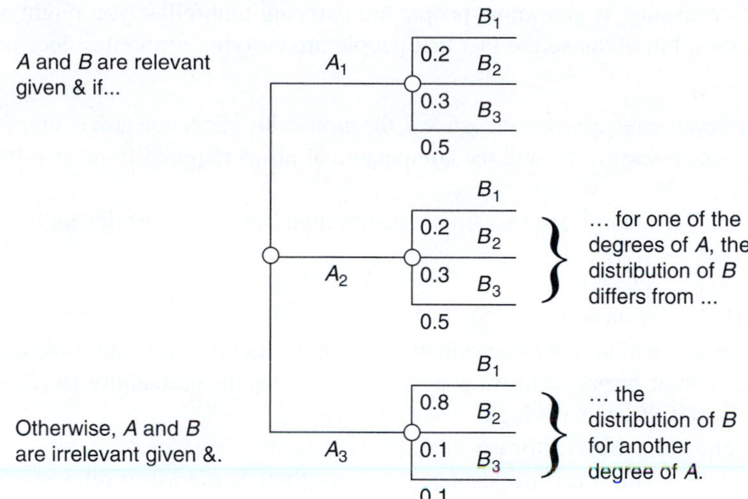

FIGURE 7.30 Relevance with Three Degrees

7.9 MULTI-DEGREE RELEVANCE

The existence of multiple degrees of distinctions will allow us to refine the definition of relevance we used for simple distinctions. Consider a distinction named A with 3 degrees A_1, A_2, A_3, and another distinction named B with 3 degrees B_1, B_2, B_3 as shown in Figure 7.30.

A is relevant to B if there is at least one degree of A, say A_3, such that the conditional probability $\{B|A_3,\&\}$ is not equal to the conditional probability of B when conditioned on other degrees of A.

In Figure 7.30, $\{B|A_3,\&\} \neq \{B|A_2,\&\}$ and so A and B are mutually relevant, even if $\{B|A_2,\&\} = \{B|A_1,\&\}$. It suffices to find one degree of A such that the conditional probability of B is different.

An alternate definition of relevance is to say A is relevant to B if for at least one degree of distinction A_i and at least one degree of distinction B_j.

$$\{B_j|A_i,\&\} \neq \{B_j|\&\}$$

To show that relevance is still mutual for multiple degrees (if A is relevant to B, then B is relevant to A), consider one such pair A_i and B_j. Let A_i be our earlier simple distinction E, and all other degrees of distinction of A be E'. Similarly, let B_j be our earlier simple distinction F, and all other degrees of distinction of B be F′. Then, by our earlier proof, since $\{F|E,\&\} \neq \{F|\&\}$, then $\{E|F,\&\} \neq \{E|\&\}$. Therefore, $\{A_i|B_j,\&\} \neq \{A_i|\&\}$, and we have shown that for multiple degrees of distinction, if one distinction is relevant to another, then the second is relevant to the first.

7.10 SUMMARY

Relevance is a matter of information not logic. We have seen that the outcome of a second coin toss was relevant to the first if we knew "there was at least one Head". But the outcomes of the first and second tosses were irrelevant if we did not know there was at least one Head. Relevance does not imply causality. If you know people are carrying umbrellas, you might assign a high probability of rain, but ofcourse the fact that people are carrying umbrellas does not cause rain to occur.

- When irrelevance relations are recognized, the probability elicitation task is simplified (because we need less assessments) and the computational effort required to reverse the tree is also simplified.
- In particular, If E and F are two simple distinctions, and if E is irrelevant to F, Then
  - $\{E|F, \&\} = \{E|F; \&\}$
  - $\{E|F, \&\} = \{E|\&\}$
  - F is also irrelevant to E.
  - Relevance is a mutual relationship even if the distinctions have multiple degrees.
- Associative logic errors occur when people believe that the probability $\{A|B, \&\}$ or $\{A|\&\}$ must be high if $\{B|A, \&\}$ is high.
- The presence of a third factor can explain away many relevance relations.
- Relevance diagrams assert irrelevance relations; they do not assert relevance relations.

KEY TERMS

- Relevance
- Mutual property
- Relevace diagrams
- Uncertainties
- State of information
- Definition of events
- Associative logic errors
- Third factor

APPENDIX A More on Relevance Diagrams and Arrow Reversals

Relevance diagrams represent a joint probability distribution given by the product of the distributions assigned to each node or distinction. Each distinction in a relevance diagram represents a probability distribution assigned by the author conditional on the author's current state of information &, and knowledge of any other distinctions with arrows that lead to this distinction.

For example, distinction A in the following diagram represents the probability distribution $\{A \mid C,D,\&\}$ since it has two arrows from C and D entering into distinction A, while the distribution for distinction E is simply $\{E \mid \&\}$ since there are no arrows entering into E.

The joint distribution represented by the relevance diagram in Figure 7.31 is

$$\{A,B,C,D,E \mid \&\} = \{C \mid \&\}\{D \mid \&\}\{E \mid \&\}\{A \mid C,D,\&\}\{B \mid D,E,\&\}$$

As we shall see, relevance diagrams are powerful tools for inferring irrelevance relations between distinctions using certain arrow manipulations. These manipulations are equivalent to flipping the trees. To deduce those irrelevance relations graphically, in a way that is consistent with tree-flipping, we discuss the rules for arrow flipping below.

Rule # 1: You can add an arrow to the relevance diagram as long as it does not create a *cycle*.

A cycle exists in a relevance diagram when you can begin with any distinction, traverse an arrow leaving it to another distinction, continue this sequence, and arrive back at the distinction where you started Figure 7.32. A cycle is not allowed because it does not correspond to any possible assessment order.

First, we recall that an arrow implies the possibility of relevance between two uncertainties and does not assert relevance. For example, if we add an arrow from distinction C to distinction D in the relevance diagram shown in Figure 7.31, we have a new representation for distinction D given by $\{D \mid C, \&\}$. Observe that if C and D are in fact irrelevant given &, then this representation is in fact correct, since in this case $\{D \mid C,\&\} = \{D \mid \&\}$. However, we lose this information in the revised relevance diagram. Therefore, the new relevance diagram will require more assessments but will yield the same result as the earlier diagram of Figure 7.31.

Rule # 2: You can reverse arrows between two uncertainties if they are conditioned on the same state of information.

If you decide to do this, it is good to draw a "virtual" rectangle around the arrow you reverse and to include the uncertainties at both ends within the rectangle. Then we make sure that any arrow coming from a distinction into one of the two uncertainties in the rectangle will have another arrow coming from the same distinction to the second uncertainty in the rectangle. If this is the case, then we can reverse the arrow. If this is not the case, then we go back to rule #1 and see if we can add an arrow to make both uncertainties within the rectangle conditioned on the same state of information. Remember, we can add any arrow provided it does not create a cycle.

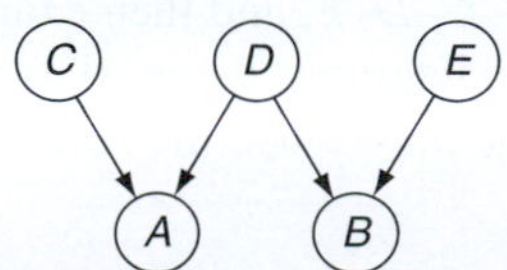

FIGURE 7.31 Relevance Diagram with Five Distinctions

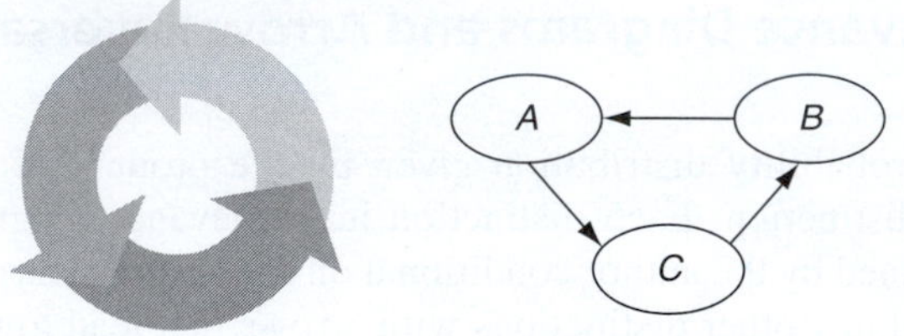

FIGURE 7.32 Adding Arrows without Cycles

Relevance diagrams provide many insights on the nature of the irrelevance relations that exist between the uncertainties present. This representation provides a convenient graphical expression that may not be as clear using the probability tree.

EXAMPLE 7.2 Chain Diagram

Consider the relevance diagram shown below, which has three uncertainties, *A*, *B*, and *C*. In Figure 7.33 (a), we can reverse the arrow between *A* and *B* to yield the diagram of Figure 7.33 (b) and see immediately that *C* is irrelevant to *A* given *B* and our state of information, &.

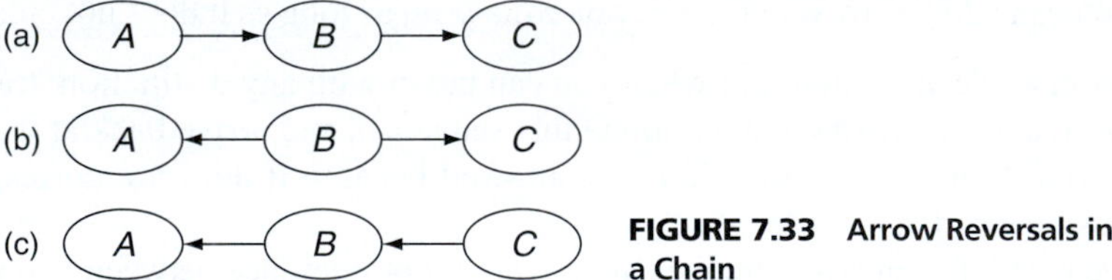

FIGURE 7.33 Arrow Reversals in a Chain

Starting from Figure 7.33 (b), we can also reverse the arrow from *B* to *C* to get Figure 7.33 (c). Comparing Figures 7.33 (a) and 7.33 (c) shows we have obtained a chain in the reverse direction that is consistent with Figure 7.33 (a). For example, we have shown that if the author of the diagram asserts the irrelevances (missing arrows) in Figure 7.33 (a), then those in Figure 7.33 (c) are implied. We derived this result graphically, without using any numbers.

Following is another example of arrow reversals in relevance diagrams.

EXAMPLE 7.3 Relevance Diagram

Let us reconsider the probability trees implied by the relevance diagram of Figure 7.31. One possible tree order is *C–D–E–A–B*.

Can we deduce that *A* is irrelevant to *B* given &? To answer this question using decision trees, we would flip the tree into the order *A–B–C–D–E*, and then examine the first two layers. To answer the same question using relevance diagrams, we need to reverse all arrows in the relevance diagram that are pointing into *A* or *B*. We use the rules of arrow flipping described above and the following order of reversals, adding arrows where necessary. The arrows reversed in Figure 7.34 appear dotted for ease of representation.

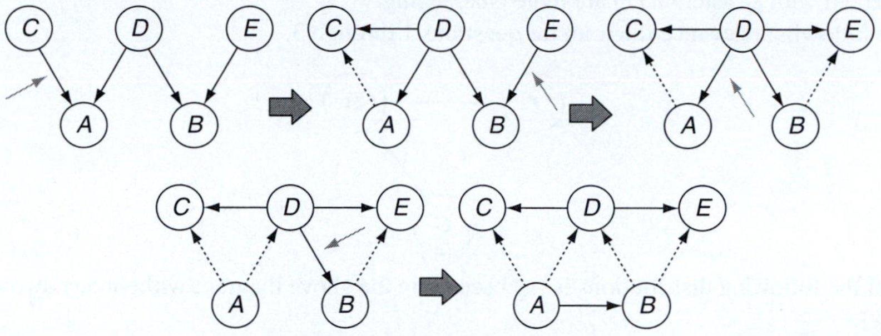

FIGURE 7.34 Arrow Reversals

Since an arrow exists between *A* and *B* in the final diagram, when both *A* and *B* are conditioned on the state of information, &, we cannot say they are irrelevant given &. The probability assignment tree orders of the final relevance diagram are either *A–B–D–C–E* or *A–B–D–E–C*.

PROBLEMS

Problems marked with an asterisk (*) are more challenging.
Examine the following relevance diagram for questions 1 through 3.

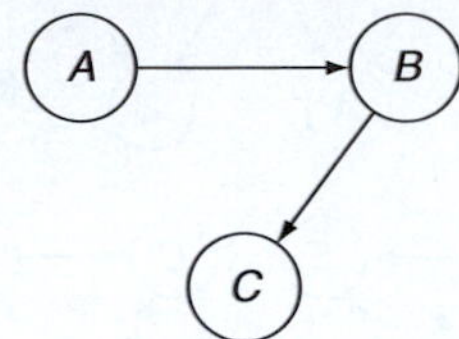

1. Which of the following distributions is represented in the above diagram without any arrow reversals?
   a. $\{B|\&\}$
   b. $\{B|A,\&\}$
   c. $\{B|C,\&\}$
   d. $\{B|A,C,\&\}$
2. Which of the following statements follow from the diagram above?
   I. *A* and *B* are relevant given &.
   II. *A* and *B* are irrelevant given &.
      a. I only
      b. II only
      c. Both I and II
      d. Neither I nor II

*3. Which of the following statements follow from the diagram above?
   I. *A* and *C* are irrelevant given &.
   II. *A* and *C* are irrelevant given *B* and &.
      a. I only
      b. II only
      c. Both I and II
      d. Neither I and II

*4. There are three distinctions: *A*, *B*, and *C*. Consider the following statements:
   I. $\{A|B, C, \&\} = \{A|\&\}$
   II. *B* is irrelevant to *C* given &.
   III. *B* is irrelevant to *A* given &.
   IV. *C* is irrelevant to *A* given &.

   Which one of the following statements is true?
   a. Given statements I, II, and III, statement IV is never true.
   b. Given statements I, II, and III, statement IV is sometimes true.
   c. Given statements I, II, and III, statement IV is always true.
   d. Statements I, II, and III can never all be true.

***5.** Consider three distinctions, A (with degrees a_1 and a_2), B (with degrees b_1 and b_2), and C (with degrees c_1 and c_2), and the probabilities (given &) of the compound distinctions given in the following table.

$a_1b_1c_1$	$a_1b_1c_2$	$a_1b_2c_1$	$a_1b_2c_2$	$a_2b_1c_1$	$a_2b_1c_2$	$a_2b_2c_1$	$a_2b_2c_2$
0.030	.180	.120	.120	.090	.060	.360	.04

Which of the following statements are true?

I. A is relevant to B given &.

II. A is relevant to B given C and &.

a. I only
b. II only
c. Both I and II
d. Neither I nor II

***6.** Which of the following statements best corresponds to the following relevance diagram?

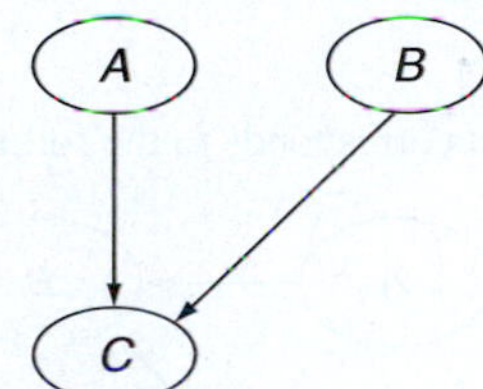

a. $\{A,B,C|\&\} = \{A|\&\}\{B|\&\}\{C|\&\}$
b. $\{A,B,C|\&\} = \{B|C,\&\}\{A|B,\&\}\{C|A,\&\}$
c. $\{A,B,C|\&\} = \{C|A,\&\}\{C|B,\&\}\{A|\&\}\{B|\&\}$
d. $\{A,B,C|\&\} = \{A|\&\}\{B|\&\}\{C|A,B,\&\}$

7. How many of the following statements are correct?

I. Relevance diagrams contain more information than probability trees.

II. With a relevance diagram, one can quickly see if two distinctions are relevant given a certain state of information, whereas relevance between distinctions might not be immediately obvious from a probability tree.

III. Relevance diagrams show the cause and effect relationships between different distinctions, whereas probability trees do not show these relationships immediately.

a. 0
b. 1
c. 2
d. 3

8. Consider the following relevance diagram:

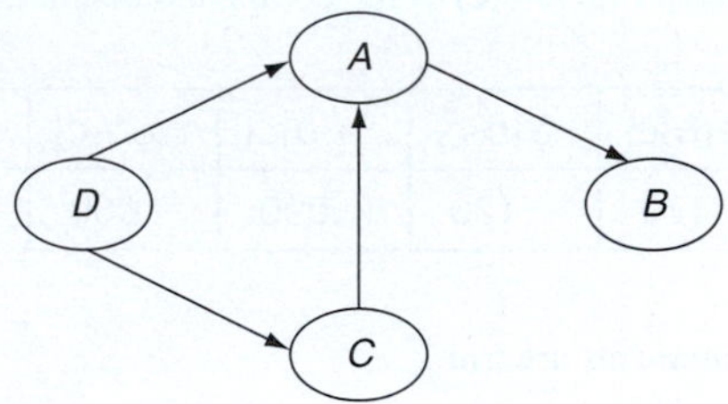

How many of the following statements must be true (after any necessary arrow manipulations are performed)?

I. *B* is irrelevant to *C* given *A, D,* and &.
II. You may add an arrow from *B* to *D*.
III. *B* is irrelevant to *C* given &.

a. 0
b. 1
c. 2
d. 3

9. Which of the following equations best corresponds to the following relevance diagram?

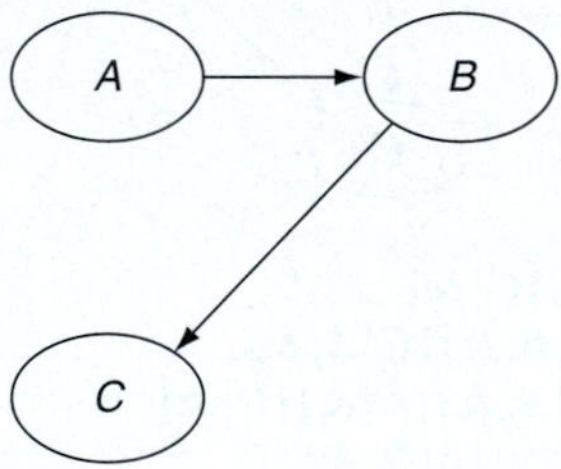

I. $\{A,B,C|\&\} = \{C|A,\&\}\{C|B,\&\}\{A|\&\}\{B|\&\}$
II. $\{A,B,C|\&\} = \{A|\&\}\{B|\&\}\{C|B,\&\}$
III. $\{A,B,C|\&\} = \{A|\&\}\{B|A,\&\}\{C|B,\&\}$
IV. $\{A,B,C|\&\} = \{A|\&\}\{B|\&\}\{C|A,B,\&\}$

10. Consider the following three distinctions:

- *Class Difficulty* (*X*), with degrees "Easy" and "Difficult;"
- *Student Effort* (*Y*), with degrees "Lazy" and "Hard Working;"
- *Student Grade* (*Z*), with degrees "*A*" and "*B* or less."

The following table provides a joint probability distribution over these distinctions:

Class Difficulty – Student Effort – Student Grade	Probability
Difficult – Lazy – *B* or Less	0.19
Difficult – Lazy – *A*	0.01
Difficult – Hard-Working – *B* or Less	0.15
Difficult – Hard-Working – *A*	0.15
Easy – Lazy – *B* or Less	0.10
Easy – Lazy – *A*	0.10
Easy – Hard-Working – *B* or Less	0.05
Easy – Hard-Working – *A*	0.25

How many of the following statements are true?

I. Given just &, a student is more likely to be lazy than hard working;

II. X is irrelevant to Y given &;

III. A hard-working student in a difficult class is more likely to get an A than a lazy student in an easy class;

IV. X is relevant to Y given Z and &.

a. 0 or 4

b. 1

c. 2

d. 3

11. Which of the following relevance diagrams best describes the relationships between distinctions X, Y, and Z as characterized in the Question 10?

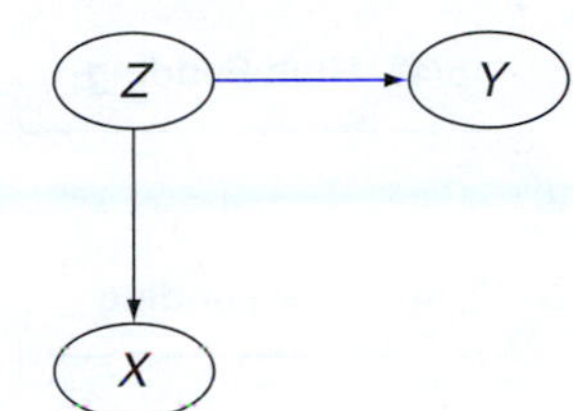

a.

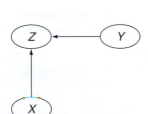

c.

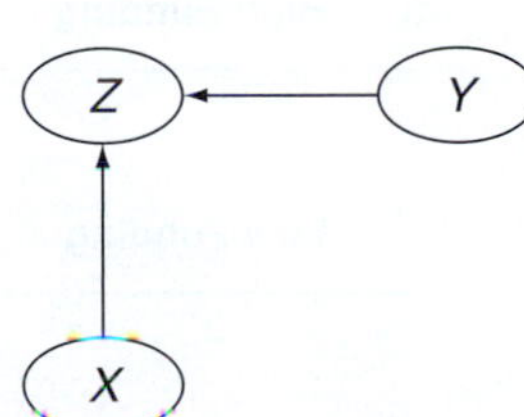

b.

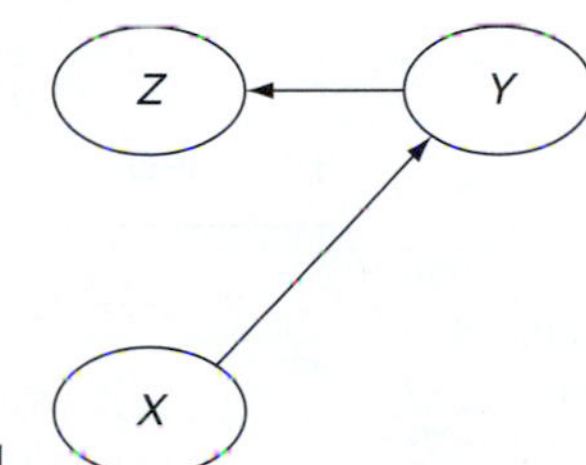

d.

12. Which the following statements are true?

I. Adding an arrow (without creating a cycle) preserves the irrelevance statements graphically shown in the relevance diagram.

II. A relevance diagram communicates the degrees of the uncertainties that it depicts.

a. Only I

b. Only II

c. Both I and II

d. Neither I nor II

13. Rohit is doing research on entrepreneurs. The distinctions he is currently considering are whether a founder of a company has a Stanford degree, whether the company succeeds in having an initial public offering (IPO), and whether the funding of the company was high or low. He describes his beliefs about this subject in the following tree:

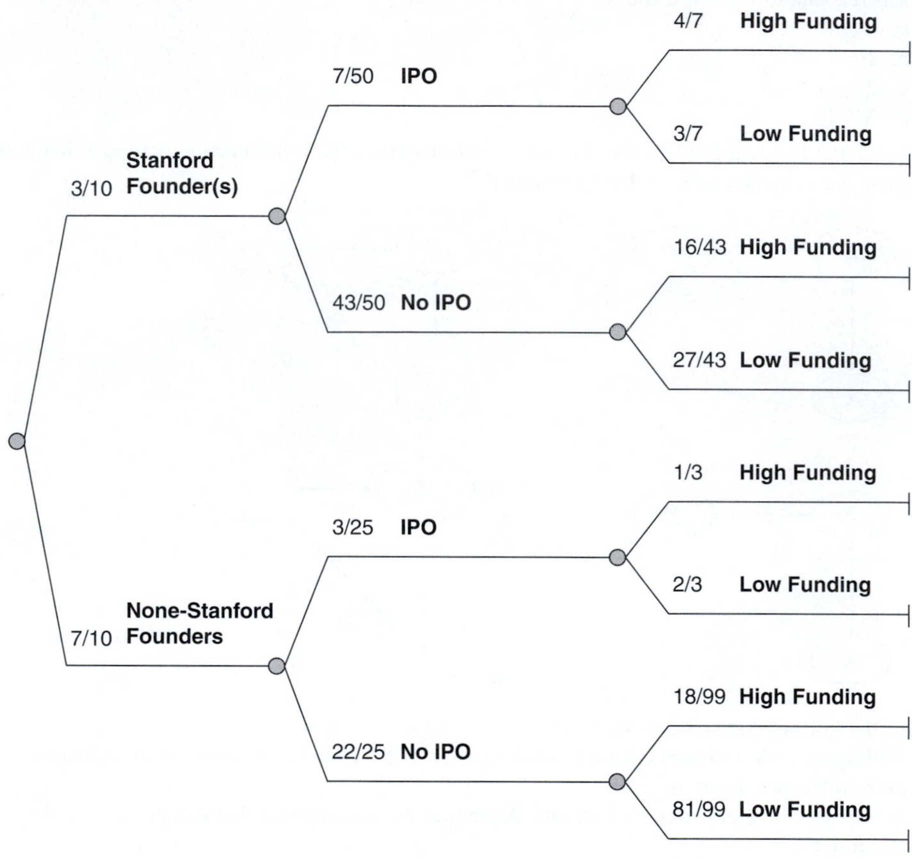

Given this information, how many of the following statements are true?

I. *Founders* is irrelevant to *Funding* given &.
II. *Founders* is irrelevant to *IPO* given *Funding* and &.
III. *Founders* is irrelevant to *Funding* given *IPO* and &.

a. 0 c. 2
b. 1 d. 3

14. Which one of the following relevance diagrams is both consistent with and **captures all of the irrelevance** relationships of the tree shown in the previous problem?

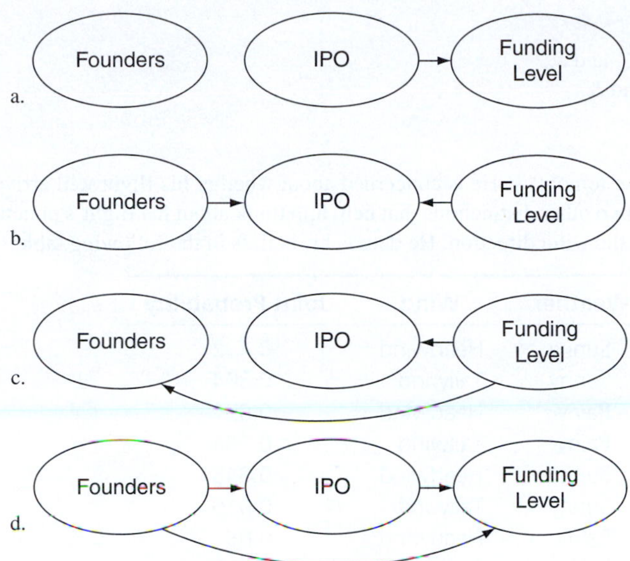

15. Consider the following relevance diagram:

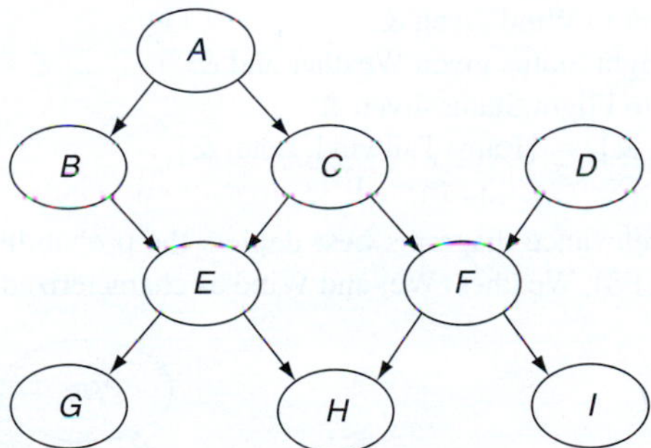

How many of the following operations can you (validly) perform on this relevance diagram *without any further manipulation*?

I. Reverse arrow between *A* and *B*.
II. Reverse arrow between *D* and *F*.
III. Add arrow from *G* to *A*.
IV. Add arrow from *B* to *C*.

a. 0 c. 2
b. 1 d. 3

16. Based on the relevance diagram shown in the previous question, how many of the following statements can you infer *after performing any valid and necessary manipulations*?

I. *E* is irrelevant to *F* given &.
II. *B* is relevant to *C* given &.
III. *G* is irrelevant to *H* given *B*, *C*, and &.
IV. *B* is irrelevant to *D* given *E* and &.

a. 0 or 4 c. 2
b. 1 d. 3

17. Casey will be traveling by airplane tomorrow. He is concerned about whether his flight will arrive to its destination on time. There are two other distinctions that help him think about his flight's punctuality: the weather at the airport and the wind direction. He details his beliefs in the following table:

Flight Status	Weather	Wind	Joint Probability
On Time	Sunny	Headwind	0.432
On Time	Sunny	Tailwind	0.304
On Time	Rainy	Headwind	0.09
On Time	Rainy	Tailwind	0.064
Late	Sunny	Headwind	0.048
Late	Sunny	Tailwind	0.016
Late	Rainy	Headwind	0.03
Late	Rainy	Tailwind	0.016

Given this information, how many of the following statements are true?

I. Flight Status is relevant to Wind given &.
II. Wind is relevant to Flight Status given Weather and &.
III. Weather is irrelevant to Flight Status given &.
IV. $\{\text{Sun} \mid \text{Tailwind, Late, \&}\} = \{\text{Rain} \mid \text{Tailwind, Late, \&}\}$.

a. 0 or 4 b. 1 c. 2 d. 3

18. Which of the following relevance diagrams **best** depicts the probabilistic relationships between the distinctions Flight Status (FS), Weather (We) and Wind as characterized in Question 17?

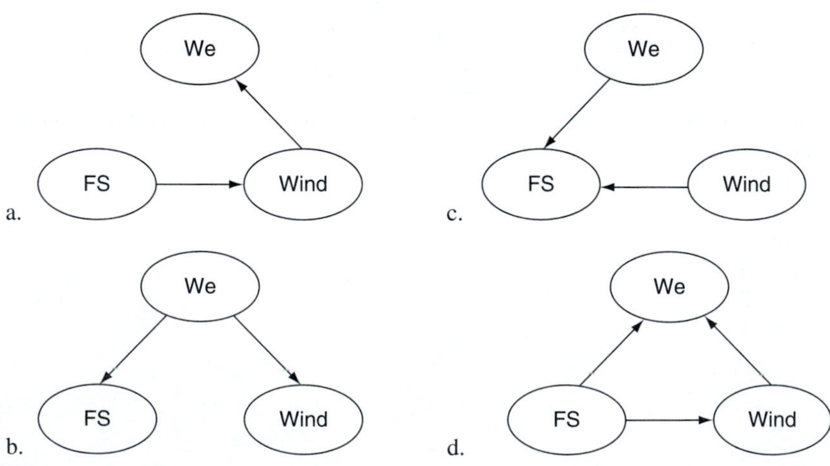

19. Consider the following relevance diagram:

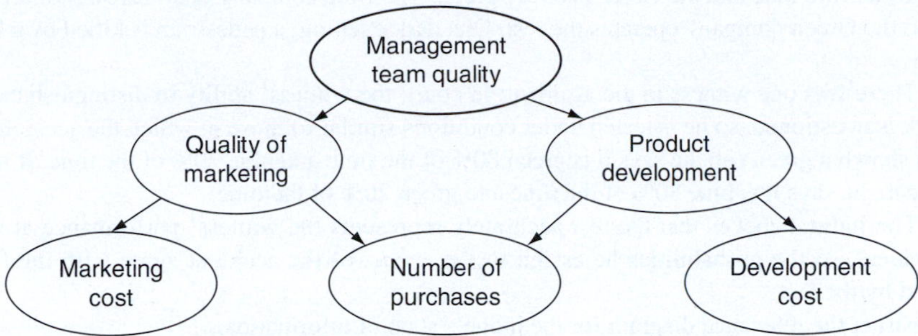

Based on the above relevance diagram, how many of the following statements can you infer?

I. *"Marketing Cost"* and *"Quality of Marketing"* are relevant given &.

II. *"Marketing Cost"* and *"Number of Purchases"* are irrelevant given *"Management Team Quality"* and &.

III. *"Marketing Cost"* and *"Development Cost"* are irrelevant given &.

IV. *"Development Cost"* and *"Number of Purchases"* are irrelevant given *"Management Team Quality"*, *"Product Development"* and &.

a. 0 or 4 c. 2
b. 1 d. 3

20. Consider the following statements about distinctions A, B, and C:

I. $\{A \mid B, C, \&\} = \{A \mid \&\}$.

II. C is irrelevant to A given &.

Which one of the following statements is true?

a. Given statement I, statement II is never true.
b. Given statement I, statement II is sometimes true.
c. Given statement I, statement II is always true.
d. We need to know whether B is irrelevant to C given & to answer this question.

21. Draw the relevance diagram corresponding to the tree below. Then reorder the distinctions so that they are in the order C, A, B and draw its corresponding relevance diagram. Are events A and B relevant to one another given &? How about C and A?

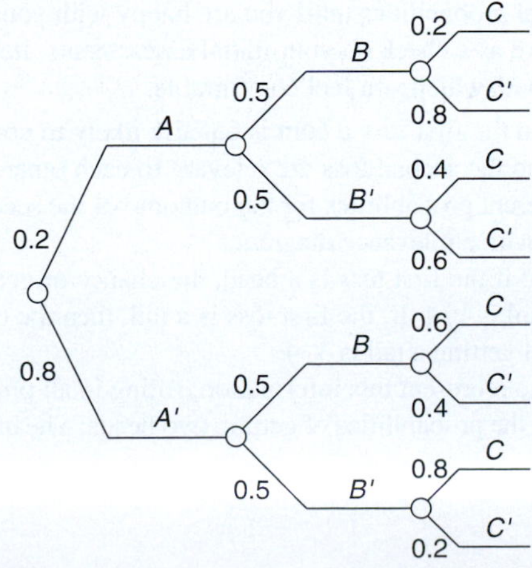

22. In the city of Bedford, there are two taxicab companies, the Blue and the Green. As you may suppose, the Blue cabs are blue and the Green cabs are green. The Blue company operates 90% of all cabs in the city and the Green company operates the rest. One dark evening, a pedestrian is killed by a hit-and-run taxicab.

There was one witness to the accident. In court, the witness' ability to distinguish cab colors in the dark is questioned, so he is tested under conditions similar to those in which the accident occurred. If he is shown a green cab, he says it is green 80% of the time and blue 20% of the time. If he is shown a blue cab, he says it is blue 80% of the time and green 20% of the time.

The judge believes that the test accurately represents the witness' performance at the time of the accident, so the probabilities he assign to the events of the accident agree with the frequencies reported by the test.

a. Construct the relevance diagram for the judge's state of information.
b. Construct the probability tree representing the judge's state of information. Label all endpoints, supply all branch probabilities, and calculate and label all endpoint probabilities.
c. Flip the tree. Label all endpoints, supply all branch probabilities, and calculate and label all endpoint probabilities.
d. Draw a relevance diagram corresponding to your result in Part (c).
e. If the witness says "The cab involved in the accident was green," what probability should the judge assign to the cab involved in the accident being green?
f. How does the answer to Part (e) compare to the witness' accuracy on the test? Does this result seem surprising? Why or why not?

23. Consider what the weather might be on October 20 of next year. If we are considering planning an outdoor birthday party for this day, we may think that two distinctions about the weather are important: whether or not it rains, and what the temperature is. We may even believe that they are relevant to each other.

a. Let R designate the event that it rains. Formulate the distinctions R and R' (where R' means "Not R") so that they are mutually exclusive and collectively exhaustive. Subject your distinctions to the clarity test. Let W represent that the same day is warm and W represent not warm. Formulate this distinction clearly. Without attaching any numbers, draw a tree and label all the possibilities created by these two distinctions.
b. Let & stand for your background state of information. In words define the following: $\{R|\&\}$, $\{W|R, \&\}$, $\{R|W, \&\}$, $\{R,W|\&\}$
c. Discuss the relevance between R and W. Are they both caused by the same weather systems? Does a warm day cause rain? Does rain cool down the temperature?
d. Fill in the probabilities on a tree with R and R' on the first set of branches and with W and W' on the second set. Flip the tree and note the resultant probabilities that this assessment produces for W and W. Adjust the conditional probabilities until you are happy with your overall assessment. Note that flipping the tree can serve as a check on your initial assessments. Iterating a few times will usually produce an assessment with which you feel comfortable.

24. Suppose you believe that on the first toss a coin is equally likely to come up heads or tails. You also believe that the first toss and the second toss are relevant to each other: depending on the outcome of the first toss, you have different probabilities for the outcome of the second toss.

a. Represent this situation with a relevance diagram.
b. Suppose you believe that if the first toss is a head, the chance of getting a head on the second toss is $3/4$ and of getting a tail is $1/4$. If, the first toss is a tail, then the chance of getting a head on the second toss is $1/4$ and of getting a tail is $3/4$.
c. Draw a probability tree to represent this information, filling in all probabilities.
d. From the tree, determine the probabilities of getting two heads; a head followed by a tail; and exactly one head in two tosses.

25. Xi, Ahren and Witi work together at DAcorp. All three travel frequently for business. The following three probabilities represent Xi's beliefs on whether his colleagues will be in the office on any given workday.

{Ahren is in the office | &} = 0.5

{Witi is in the office | &} = 0.5

{Both Ahren and Witi are in the office | &} = 0.3

Given this information, which of the following statements are true for Xi on a given workday?

I. Whether Ahren is in the office is **irrelevant** to whether Witi is in the office.

II. On a given workday, it is less likely for both Ahren and Witi to be in the office than it is for neither Ahren nor Witi to be in the office.

a. I only
b. II only
c. Both I and II
d. Neither I nor II

26. (Continued from previous problem) Ahren and Witi both love coffee. When Ahren is in the office, he drinks one cup of coffee with probability 0.6 and no cups of coffee with probability 0.4. When Witi is in the office, he drinks two cups of coffee with probability 0.5 and one cup with probability 0.5. Ahren and Witi always discard their empty coffee cups in the office wastebasket, and they are the only people to do so.

Xi asserts the following:

- The number of cups of coffee that Witi drinks at the office is irrelevant to Ahren's presence at the office.
- The number of cups of coffee that Ahren drinks at the office is irrelevant to Witi's presence at the office.
- The number of cups of coffee that Ahren drinks at the office is irrelevant to the number of cups of coffee that Witi drinks at the office.

Xi returns late from a client visit and finds that the office is empty. He doesn't know if Ahren, Witi, both, or neither had come to the office that day, but he recalls that the wastebasket was empty early in the morning, before any employees had arrived.

Given this information, how many of the following statements are true?

I. If Xi sees 1 cup in the wastebasket, then his probability that **both** Ahren and Witi were in the office that workday is greater than 0.1.

II. If Xi sees 2 cups in the wastebasket, then his probability that **only** Witi came to the office that workday is 0.4.

a. I only
b. II only
c. Both I and II
d. Neither I nor II

CHAPTER

8

Rules of Actional Thought

CHAPTER CONCEPTS

After reading this chapter, you will be able to explain the following concepts:

- The five rules of actional thought:
  - Probability rule
  - Order rule
  - Equivalence rule
  - Substitution rule
  - Choice rule
- The difference between a prospect of a decision and an outcome of an uncertainty
- A stated preference probability
- The money pump argument

8.1 INTRODUCTION

In previous chapters, we discussed methods of characterizing a decision situation by creating useful distinctions. In this chapter, we present the basis by which we should actually make our decisions and finally answer the question posed in Chapter 1:

What is a good decision?

To answer this question, we shall present the rules of thought that help us think clearly about any decision we face, known as the five Rules of Actional Thought. We can think of these rules as the "logic seat" of the stool metaphor of Figure 1.19 or the "sound reasoning" element of the decision chain in Figure 1.10.

8.2 USING RULES FOR DECISION MAKING

Most people have lived their entire lives without being able to define what makes something a good decision. As we demonstrated in many of the previous examples, such as *Let's Make a Deal* or the associative logic errors, relying on intuition alone is insufficient and often misleading. A clear set of rules can help us make solid decisions in an uncertain world. For almost two centuries, we have understood the inadequacy of trying to use standard logic to deal with uncertainty:

> *"They say that understanding ought to work by the rules of right reason. These rules are, or ought to be, contained in logic; but the actual science of logic is conversant at present only with things either certain, impossible, or entirely doubtful, none of which*

(fortunately) we have to reason on. Therefore, the true logic for this world is the calculus of probabilities, which takes into account the magnitude of the probability which is, or ought to be, in a reasonable man's mind."

JAMES CLERK MAXWELL (1831–1879)

Maxwell, the originator of Maxwell's equations in electrical engineering, did not present the rules as they appear here, but he did recognize the need for a new way of thinking: You do not need to waste time thinking about things that are either certain or impossible. Instead, the challenge in understanding the world and in decision making lies in facing *uncertainty*. As you read the rules of actional thought presented later in this chapter, consider whether you wish to shape your thinking by conforming to these rules. The rules may seem straightforward, but, as we shall see, their implications are profound.

8.2.1 Considerations

In choosing any set of rules to guide logical thought and action when facing uncertainty, the most general rules are

1. The rules should be valid when applied to simple situations.
2. The rules should not break down when extended to the larger situations in which they will be used.

For example, we test the rules of arithmetic by checking that 3 plus 4 is indeed 7 by the using the test of counting. As we extend our use of arithmetic to larger and larger problems, we note that the extension is seamless—there is no apparent point at which dealing with larger or more numbers causes the process to break down.

These same considerations apply to choosing rules for actional thought. Our goal is to have them be *transparent*—that is, they should describe what you would actually do in the simple situation to which they apply. In fact, they should be transparent enough to border on the obvious.

As we combine these simple rules, however, you will obtain power over complex decisions—just as the simple rule of arithmetic gives you power over numbers. When you face an opaque decision situation, with many alternatives, several sources of uncertainty, and complex preferences, you will be able to combine these transparent rules to transform the opaque decision into a transparent one. You will thereby attain your goal of achieving clarity of action.

8.2.2 Rules, Not Axioms

In a decision theory class, we might choose to use the term "axiom," instead of using the term "rule." In decision analysis, however, we are concerned with the *practical application* of decision theory. For this purpose, "rule" is a more powerful description than "axiom" because it is more accessible and just as precise to many people.

For most people, *axiom* is a word that induces anxiety. Graduate engineering students can cope with axioms, but most people have an aversion to math. Their aversion centers around words like "axiom" with which they did not have a happy relationship in high school. For many people, the term "axiom" creates a needless problem in understanding.

In contrast, everyone is familiar with *rules*. There are rules of football. There are rules of bridge. There are rules for every game. In fact, without rules, there is no game.

For example, suppose you are playing checkers with someone who starts moving not just on the black squares, but on the red squares as well. You say "you can't do that!" They reply "what do you mean I can't? There is plenty of room on the red squares. Watch me." As soon as you give up on the rules, the game is destroyed. If you can have many strikes as you like, you are not playing baseball. The rules make the game.

The rules of actional thought are both *personal* and *proactive*. Following the rules requires you to do things such as to order your preferences on possible futures. For individual decisions, there is no outside party to enforce the rules. The appropriate application of the rules is determined by your judgment.

8.3 THE DECISION SITUATION

The rules of thought apply to a particular decision situation at an instant in time. In a decision situation, you face choices among different alternatives with different allocations of resources. Each alternative produces a **deal**. A deal is composed of various future prospects whose realization may be uncertain. Therefore, we think of a decision as a choice between uncertain futures—as a choice of deals with uncertain prospects.

8.3.1 From Possibilities to Prospects

Possibilities are also sometimes referred to as events or outcomes. If we were teaching a probability class, we would need to consider only possibilities. Problems about cards, dice, and balls in urns deal with the possibilities that may occur. But in addition to the possibilities that might ensue, the likelihood of those possibilities, and what those possibilities mean to you in the context of your future life, decision making is also about the alternatives you choose.

Prospects characterize the future created by the choice of a particular decision alternative and any one of its possibilities. The fundamental difference between a prospect and a possibility is that a prospect depends on the decision alternative chosen as well as on the resolution of an uncertainty. Prospects describe your actions and their consequences.

Consider, for example, the uncertainty about the increase in profit of a company in the next year of operation. One possibility for this uncertainty could be $25 million, and this possibility could follow different decision alternatives. One alternative could be the decision to have layoffs, leading to a reduction in operating costs and an increase in profit. Another alternative could be to lower prices, leading to expansion in sales, increased revenue, and increased profit. The $25 million increased profit may be the same for both alternatives, but the prospects are different because the possibility is associated with different decision alternatives.

A prospect is not simply a future possibility following an alternative—it represents the future life associated with having that possibility. Suppose, for example, you were given a chance to flip a coin for $1,000,000. What does winning $1,000,000 mean to you? At one level, it is an additional $1,000,000 in your bank account. At another level, it may change your lifestyle, your job, your attitude, or even your relationships. If so, the prospect of winning $1,000,000 is not adequately described by simply measuring your bank account.

The description of a prospect only needs to be as complete as you need it to be to make your decision. The prospect of winning $100 from tossing a coin may be adequately described by merely measuring your bank account. But, imagine that you are suddenly faced with making a medical decision for which one choice is to have an operation. If you have the operation, one possible side effect is restricted arm motion. What is the prospect "restricted arm motion?" The possibility "restricted arm motion" may be easy to describe. It means you cannot move your arm above shoulder height without great pain. The prospect "have operation followed by restricted arm motion" may be much more difficult to describe, especially if you have limited related experiences. To describe this prospect, you must be able to imagine what it means to you to have restricted arm motion in your future life. Perhaps it will mean that you can no longer play tennis or rock climb, but you will be able to garden and play the piano.

8.3.2 An Instant in Time

All of these thoughts about a decision occur at this particular instant. All notions of consistency apply to your thoughts at this instant—not necessarily to your preferences over time. Just because you prefer chocolate ice cream to vanilla ice cream today does not mean you will prefer it tomorrow. As Emerson[1] said, "a foolish consistency is the hobgoblin of simple minds." The consistency we seek is the consistency of our rules of thought, not the foolish consistency of constancy of preference.

However, there is nothing in the rules to prevent you from being as consistent in your preferences over time as you like. In fact, most of us will prefer to be relatively consistent in our preferences over time, so that we are not buying tomorrow what we gave away today.

8.4 THE FIVE RULES OF ACTIONAL THOUGHT

If you adopt the following five rules, you can achieve high quality actional thought in even the most difficult decision situations.

8.4.1 The Probability Rule

The **probability rule** has two parts and the following requirements.

1. You create prospects: You characterize each alternative and its possibilities as prospects.
2. You assign probabilities: You assign probabilities to each of the prospects following each alternative. In essence, the probability rule requires that you characterize deals using the concepts and procedures from the previous chapters. We can think of each decision alternative as a deal described by a probability tree, perhaps with several measures specified at the endpoints.

Contrast the requirements of this rule with what you learn by taking a class in probability. First, the professor creates the distinctions, such as the toss of a coin, selecting a card drawn from a deck, or drawing a ball from an urn. Then the professor specifies the degrees of each distinction to create possibilities: the coin shows heads, the card is a heart, the ball is red. The task of assigning probabilities to these possibilities is either done by the professor or by some simple rule that implicitly provides them, such as "all degrees are equally likely".

For example, if the professor says the coin is "fair" you are expected to assign equal probability to heads and tails on a toss; if the professor says the deck of 52 cards is well shuffled, you are expected to assign a probability of 1/52 to drawing each card; and if drawing balls from an urn the professor says has some fraction of balls of each color, you are expected to assign that fraction as the probability that a ball of each color will be drawn. For example, in drawing a red ball from an urn that contains 50 red balls and 50 blue balls, you would be expected to assign a probability of 0.5.

To contrast this with the requirements of the probability rule, if you looked at the urn and saw that the great majority of red balls were at the top of the urn and great majority of blue balls

---

[1]Ralph Waldo Emerson was born on May 25, 1803 in Boston, Massachusetts and is widely regarded as one of America's most influential authors, philosophers, and thinkers.

were at the bottom, you would assign a probability much larger than 0.5 to reaching into the urn, even blindfolded, and drawing a red ball. We doubt you would assign a probability of only 0.5.

Decision analysis requires a more fundamental and realistic perspective. Most of the challenge is to create the right distinctions in the first place. As we demonstrated in defining Beer Drinker, creating clear distinctions can be hard work. Creating useful distinctions is even more challenging. Clear and useful distinctions provide much of the value in decision analysis. They create the possibilities that allow careful assigning probabilities based on your knowledge.

The probability rule requires that the characterization be adequate to describe all the important thoughts you would like to have about your decision situation. Who determines if it is adequate? You do. This rule says you will create distinctions and assign probabilities until you have characterized the decision situation to your satisfaction. The question you need to answer in accepting this rule for yourself is "Am I ready to perform such characterization when I face a serious decision problem?"

In applying the probability rule, one important consideration will be deciding what you choose to characterize as an uncertainty. At some level, everything is uncertain. For example, do you really know who your biological parents are? Babies get switched in the hospital, and mothers make admissions to their husbands late in life. What you choose to consider as certain and uncertain will depend on the decision. Making the choice is part of the art of decision analysis.

8.4.2 The Order Rule

The **order rule** requires that you can arrange prospects that are the combination of an alternative and any one of its possibilities in a list of your preferences from best to worst. If you prefer prospect A to prospect B (written $A > B$), then A would appear higher in the list than B. Two prospects to which you are indifferent can appear at the same level in a list, such as a condition described by $A = B$.

Suppose you had three prospects you were facing: receiving the Hope diamond,[2] having a ton of parmesan cheese delivered to your door, and getting a case of pneumonia that would confine you to a hospital bed (See Figure 8.1). Most people would prefer these prospects in that order. However, if you believe, as some do, that the Hope diamond is cursed, you might make that your least preferred prospect.

Notice that the ordering of the prospects requires only the creation of distinctions, and, therefore, the possibilities for each alternative. This ordering describes your preferences for

FIGURE 8.1 Three Prospects (Diamond: Smithsonian Institution/Corbis; Cheese: Marco Mayer/Shutterstock; Patient: Tyler Olson/Shutterstock)

[2]The Hope diamond is the world's largest deep blue diamond, and is more than a billion years old. It was carried from within the earth by a volcanic eruption to a surface in what is now India. According to a legend, a curse foretold bad luck and death to anyone who touched the diamond. Whether or not you believe in the curse, the Hope diamond has intrigued people for centuries because of its quality, size, and rare color.

prospects in a world without uncertainty. You may perform this ordering without having to have done the probability assignments required for a complete characterization of deals.

If you have trouble ordering prospects in a list, it may be because you have not created enough distinctions. For example, if you have difficulty saying whether you would prefer your life with the gift of a new stereo or the gift of a dog, it may be because you do not know how much care a dog requires. You might prefer the stereo to the dog if the dog requires ten hours of care per week, but the dog to the stereo if the dog required only one hour of care per week. By introducing the additional kind of distinction—hours of care for the dog—with sufficient degrees to describe the possibilities, you have created additional prospects.

You could then arrange your expanded list of prospects in order of preference. The order rule allows you to introduce as many kinds and degrees of distinction as necessary to enable you to rank prospects. By accepting the order rule, you affirm your belief that you can always create the distinctions necessary to express a preference between prospects.

CONSISTENCY OF ORDER Consistency in satisfying the order rule prohibits a prospect from appearing at two different levels in the list. This is clear for two prospects: If A is preferred to B, and hence is higher in the list, then B cannot be higher in the list than A. The rule also requires that if A is above B, $A > B$; if B is above C, $B > C$; and if A is above C, $A > C$. If A is preferred to B and B is preferred to C, then A is preferred to C. If you are tempted to violate this requirement, it is usually because you are thinking of different attributes in making the comparison of prospects.

For example, suppose you are given a choice of three cars: The Ajax, the Rex, and the Luxo. Upon reflection, you find that you prefer the Ajax to the Rex because of its acceleration; you prefer the Rex to the Luxo because of its reliability; and you prefer the Luxo to the Ajax because of its luxury. It appears as if you are having trouble in creating an ordered list of preferences. But the question is, all attributes considered, if one of these cars were waiting for you as a free gift in the parking lot, which would you choose? Whichever it is should be highest on the list. Then if that car were not available, which of the remaining two would you choose? That car would rank below the first on the list; finally, the remaining car would be below the other two. Remember that we are ranking prospects where each is your whole life as described by an elemental possibility and its associated measures.

THE MONEY PUMP ARGUMENT Violating the order rule has one major disadvantage. Suppose you have your original confused preferences about the three cars, and you have just been given a Luxo. Then someone could come up to you and say, "I know that you prefer the Rex to the Luxo because of its reliability. If you give me some money, I will exchange your Luxo for a Rex." You agree, and make the exchange. A little later, the person reappears and says "I also know that you prefer an Ajax to a Rex because of its acceleration, and so if you give me a little more money, I will exchange your Rex for an Ajax." Once again, you agree, and hand over some more money. But we are not through. He comes back and says, "I know you prefer the Luxo to an Ajax because of its luxury. Give me some more money and I will exchange your Ajax for a Luxo." You agree yet again, and hand over even more money. Now you are back where you started, having disbursed money three times. There is nothing to prevent this person from coming back and taking you around the cycle again. This is sometimes called being made a **money pump**. Anyone who violates the order rule is vulnerable to a money pumper.

In most cases, you will not have any trouble satisfying the order rule. If someone asks if you would rather have $100, $50, or nothing, you will not hesitate to assign your preference. Even with more complex descriptions of prospects, the choice will usually be simple. For example, suppose that your local daily newspaper is sponsoring a drawing to increase circulation. Three of the prizes in the newspaper drawing are

1. a two-week trip to Tahiti for two people;
2. a $500 gift certificate at the largest department store in town; and
3. a free one-year subscription to the newspaper.

Most of us would have little trouble in establishing the preference order $A > B > C$, specifying that *A* is above *B* and *B* is above *C* in our preferentially ordered list of prospects. However, any ordering is allowed: someone who dislikes air travel might not prefer the trip to the gift certificate.

8.4.3 The Equivalence Rule

Suppose that you prefer prospect *A* to prospect to *B*, and prospect *B* to prospect *C*: $A > B > C$. This means that *A* is above *B* and *B* is above *C* in the preference list. The **equivalence rule** requires that you can state a number, *p*, so that you will be indifferent between receiving prospect *B* for sure and receiving a deal with probability *p* of prospect *A* and probability $1 - p$ of prospect *C*. Figure 8.2 shows the requirement of the equivalence rule.

You have adjusted the probability *p* in the probability tree on the right of the figure until you are indifferent between receiving the deal described by that tree and the deal described by receiving prospect *B* with probability 1. You are the sole arbiter on what the setting of *p* should be to achieve your indifference. The number *p* can be any number between 0 and 1, but cannot be exactly equal to either. If you set *p* to 1, you would be saying that you are indifferent between *A* and *B*, which is not so. If you set *p* to 0, you would be saying that you are indifferent between *B* and *C*, which is also not the case.

Suppose that when you are asked to do this for a newspaper drawing for a Tahiti trip, a gift certificate, or an annual subscription you encounter a problem. You say, "I know I prefer the Tahiti trip to the other two prospects, but before I can state *p*, I have to know more specifics. Is it the deluxe trip with first class airfare and hotel, all expenses paid or the economy trip with a charter flight, youth hostel, and no payment of expenses?" If you have this type of difficulty, the probability rule requires that you introduce distinctions about the type of trip, redo the order rule with the expanded list of prospects, and then come back to the equivalence rule with clearly defined prospects.

PREFERENCE PROBABILITY When you have stated that the *p* that makes you indifferent, we call it the **preference probability** for this situation. We call it a preference probability rather than simply a probability because it is not the probability assignment on any distinction whose occurrence could be resolved by a clairvoyant. We reserve the term "state a preference probability" for the equivalence rule and "assign a probability" for the probability rule.

STATING PREFERENCE PROBABILITIES USING THE PROBABILITY WHEEL The equivalence rule requires you to create a hypothetical deal involving prospects *A* and *C* that you find equally

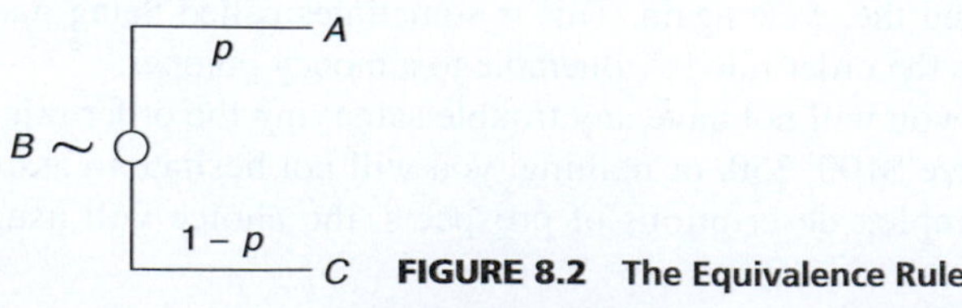

FIGURE 8.2 The Equivalence Rule

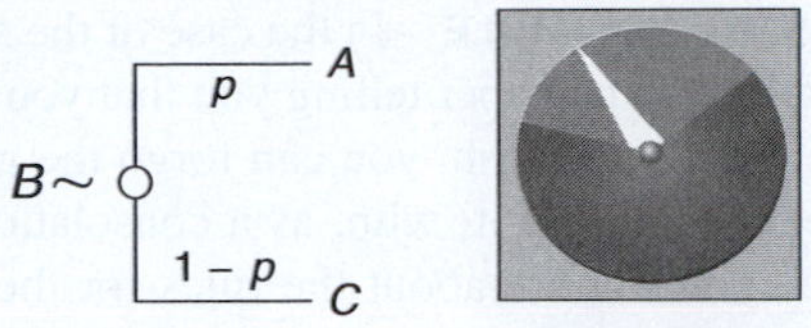

$B \sim$ [p: A; $1-p$: C]

The Wheel

FIGURE 8.3 Preference Probability and the Probability Wheel

attractive to the prospect *B*. One way to think of this is to imagine using the probability wheel of Figure 8.3 to describe the deal, with one color corresponding to receiving prospect *A* and the other to receiving prospect *C*. If, for the particular wheel setting in front of you, you prefer the wheel deal to the prospect *B*, then you must decrease the color corresponding to *A*. If you prefer prospect *B*, you must increase the fraction of the wheel corresponding to *A*. When you are indifferent, the fraction of the wheel corresponding to *A* is the preference probability *p* that you are stating in the equivalence rule.

By accepting the equivalence rule, you assert that you can state a preference probability for any three prospects at different levels of preference as specified by the order rule. Some of these prospects may involve life and death, consequences of surgery such as blindness and paralysis, and other prospects such as your life while undergoing chemotherapy or radiotherapy.

Just as the order rule describes preferences in a world without uncertainty, the equivalence rule describes preferences in a world with uncertainty.

CERTAIN EQUIVALENT For the *p* stated in the equivalence rule, *B* is called the **certain equivalent** of the *A*–*C* deal, and *p* is the stated preference probability of the three ordered prospects. To illustrate these concepts, let us return to the newspaper drawing. Suppose that, after reflection and with the use of the wheel, you are just indifferent between receiving, on the one hand, the $500 gift certificate at the major department store, and on the other, a deal consisting of a 15% chance of the deluxe trip to Tahiti and an 85 percent chance of a free one-year subscription to the newspaper (Figure 8.4). Then you would be stating that the $500 gift certificate is the certain equivalent of a deal consisting of a 0.15 chance of the deluxe trip to Tahiti and a 0.85 chance of a free one-year subscription to the newspaper. You would also be saying that 0.15 is your stated preference probability for the three ordered prospects in the newspaper drawing.

8.4.4 The Substitution Rule

The Substitution Rule requires that whenever you face a prospect (or prospects) for which you have stated a preference probability using the Equivalence Rule, you would be indifferent between your decision facing these prospects or an equivalent decision for which you have replaced any number of prospects with their equivalent binary deals using the Equivalence Rule. The Substitution Rule also works in the reverse direction: you would be indifferent between a decision that contains a binary deal with a given probability of receiving the best and worst prospects and a decision that replaces that binary deal with an equivalent prospect which has the same preference probability as the probability of receiving the best prospect in the binary deal.

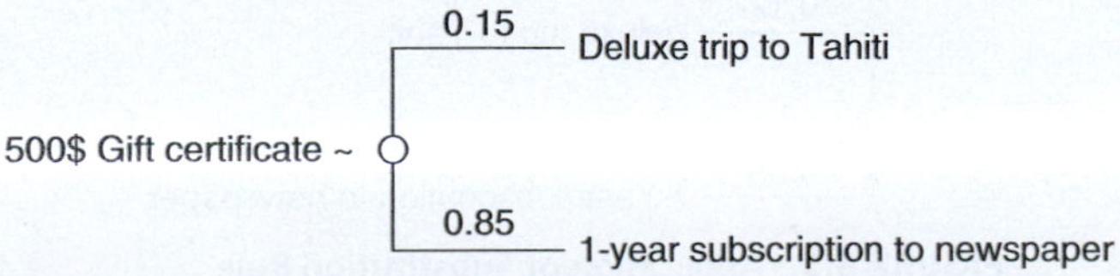

500$ Gift certificate ~
0.15 Deluxe trip to Tahiti
0.85 1-year subscription to newspaper

FIGURE 8.4 Certain Equivalent Deal

LET'S USE AN EXAMPLE TO EXPLAIN THE SUBSTITUTION RULE In the case of the newspaper drawing example, suppose that you received a call from the paper telling you that you have won the $500 gift certificate. The caller informs you that if you wish, you can forgo the gift certificate and enter a drawing for the deluxe two-week trip to Tahiti with, as a consolation prize, a free one-year subscription to the paper. If, when you inquire about the rules for the drawing, you assign a 0.15 probability of winning the deluxe Tahiti trip, you must be indifferent between exchanging and not exchanging the gift certificate for the drawing.

This means that when faced with an actual uncertain deal on *A* and *C* where you have assigned a probability of winning *A* equal to your stated preference probability, you will be indifferent to either receiving that deal or receiving *B*. In other words, your preference probabilities will describe your actions when they are equal to probabilities you assign. In short, it means that you will treat your stated preference probabilities as probabilities in analyzing decision situations.

APPLICATION Figure 8.5 shows an application of the substitution rule. Suppose in the equivalence rule, as in Figure 8.4, you had stated that the $500 gift certificate was the certain equivalent of a deal with 0.15 chance of the deluxe trip to Tahiti and a 0.85 chance of an annual subscription to the local newspaper. Suppose, further, you now actually face a deal where you assign a 0.80 probability of the $500 gift certificate and a 0.20 probability of an annual subscription to the local newspaper. By accepting the substitution rule, you must be ready to evaluate this deal by substituting for its $500 prospect the deal for which this is the certain equivalent determined in the equivalence rule. The substitution reveals that the deal you actually face is equivalent to a 0.12–0.88 chance of getting the trip to Tahiti or the annual subscription.

8.4.5 The Choice Rule

The Choice Rule requires the following. Suppose that you have a choice of two deals that have the same two prospects, you prefer one of the prospects to the other, and you have assigned the probability of receiving that prospect in each deal. The **choice rule** requires that you choose the deal with the higher probability of the prospect you prefer.

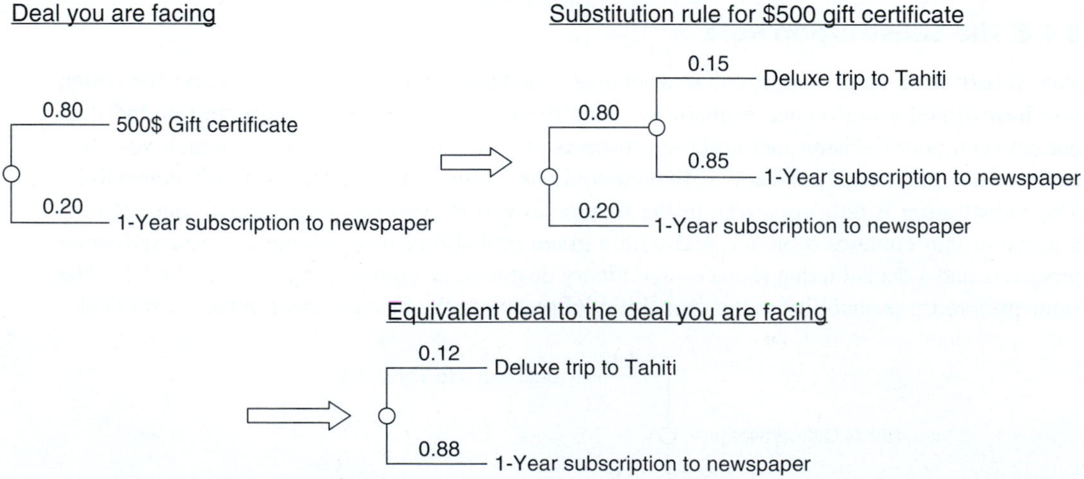

Deal you are facing

0.80 500$ Gift certificate

0.20 1-Year subscription to newspaper

Substitution rule for $500 gift certificate

0.80

0.15 Deluxe trip to Tahiti

0.85 1-Year subscription to newspaper

0.20 1-Year subscription to newspaper

Equivalent deal to the deal you are facing

0.12 Deluxe trip to Tahiti

0.88 1-Year subscription to newspaper

FIGURE 8.5 Application of Substitution Rule

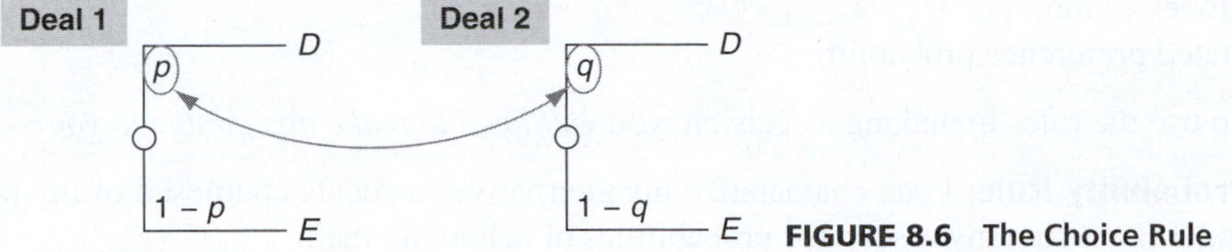

FIGURE 8.6 The Choice Rule

Figure 8.6 illustrates the choice you face between Deal 1 and Deal 2. The deals have the same two prospects D and E, but they may have different probabilities of achieving them. Suppose you prefer prospect D to prospect E, so $D > E$, meaning that D is above E in your preference list. Suppose you have also assigned p and q as the probability of winning D in each deal. The choice rule requires, if $p > q$, you must choose Deal 1.

This is the last rule, and it is the only rule that specifies the action that you must take. The rule says that you must choose the deal with the higher probability of the prospect you like better.

This rule is so transparent that, if you announced you were violating it, people would think you were joking. Suppose you prefer \$100 to nothing and you were given a choice between flipping a coin to win \$100, to which you assign a probability of 1/2 of winning, and drawing a heart from a deck of cards for the \$100, to which you assign a probability 1/4 of winning. If you told a friend that you had chosen the card game, he would think that he had misunderstood you. Notice that someone who is a gambler and who enjoys risk will still prefer the coin deal because he believes he has a better chance of winning.

In the newspaper drawing example, the choice rule requires that if you are given a choice of participating in different drawings with only two possible prizes, the Tahiti trip or the free one-year subscription to the newspaper, then you must prefer the drawing with the best chance of winning the Tahiti trip, presumably the one with the fewest participants.

The choice rule says nothing about your preferences, but only that you must follow them. Even masochists can follow the choice rule. If you prefer being beaten up in a bar to not being beaten up in a bar, then the choice rule requires you to prefer the bar with the higher chance of getting beaten up. *De gustibus non disputandum est*—there is no arguing about taste.

8.5 SUMMARY

Our task in decision analysis is to create representations and reduce them using the rules until you are left with a choice between two alternatives with the same two prospects where you have assigned a higher probability to one prospect than to the other. Sometimes we say our task is to take an opaque decision situation, one where you do not know what to do, and reduce it by a series of transparent steps into something as simple as the choice rule.

We created new distinctions in this chapter:

- Prospect vs. outcome
- Five rules of actional thought
  - Probability rule;
  - Order rule;
  - Equivalence rule;
  - Substitution rule;
  - Choice rule

- Money pump
- Stated preference probability

To use the rules in making a decision, you *will need to make this pledge to yourself*:

Probability Rule: I can characterize my alternatives as deals composed of my prospects and I can assign my associated probabilities of achieving them.

Order Rule: I can order my entire list of prospects from best to worst; two or more prospects may be at the same level.

Equivalence Rule: Given three prospects at different levels, I can state a preference probability of receiving the best vs. the worst prospect that would make me just indifferent to receiving the middle prospect. The middle prospect is the certain equivalent for the two-prospect deal with probability, p, of the better prospect and $1 - p$ of the worse prospect.

Substitution Rule: If my decision contains a prospect (or prospects) that I consider the certain equivalent of a better and worse prospect by stating a *preference probability* p in the equivalence rule, I remain indifferent to receiving that prospect or a deal where I *assign a probability* p of the better prospect and $1 - p$ of the worse prospect.

Choice Rule: Given a choice between two deals involving the same two prospects at different levels in the order rule, I will choose the deal with the higher probability of the better prospect.

If there were no uncertainty present, we would only need the probability rule (to characterize the problem), the order rule (to order the prospects), and the choice rule (to make a selection). We would not need the equivalence or the substitution rules.

In future chapters, we will demonstrate that by repeatedly applying these five rules, we can make virtually any decision. For convenience, we use the word ***rulesperson*** to identify a person who agrees to satisfy the rules.

We will now point out some general properties of this normative form of decision making. For example, we note that the choice between two alternatives depends only on the prospects of these alternatives and not on the prospects of other alternatives. Your preferences are on the future prospects you actually receive, not on the ones you might have received. At this instant in time, the removal (or the addition) of an alternative that you did not choose cannot change the order of your preferences among those that remain provided it does not give away any information that is relevant to the remaining alternatives. These observations may seem strange in the abstract, but they are no more than formalized common sense.

For example, consider the story of the man who entered an ice cream parlor and asked what flavors were available. The server replied, "Vanilla, chocolate, and strawberry." The customer said, "I'll have vanilla." A moment later the server returns saying, "I'm sorry, we have no strawberry." The customer says, "In that case, I'll have chocolate." That this story seems to be a joke shows the extent to which we have internalized the rules of thought.

KEY TERMS

- Certain equivalent
- Choice rule
- Deal
- Equivalence rule
- Money pump
- Order rule
- Possibilities
- Preference probability
- Probability rule
- Prospects

PROBLEMS

Problems marked with a dagger (†) are considered quantitative.

1. Which of the following describe violations of the order rule of the five Rules of Actional Thought?
   I. Joy is indifferent between having an extra \$10 or having the latest CD from U2. Someone offers Joy the latest CD from U2 to Joy for \$5, but she decides not to buy it.
   II. Aaron prefers to fly with Qantas rather than Alaska Airlines because of their better safety record. He prefers Jet Blue to Qantas because they have inflight television. However, he would rather fly Alaska than Jet Blue because they have a better customer satisfaction rating.
   a. I only
   b. II only
   c. Both I and II
   d. Neither I nor II
2. You graduate and get hired at a management-consulting firm. On your first project, the decision maker says "I cannot assign a probability to the outcomes of this uncertainty, but my trusted friend can. Also, I can arrange the prospects in order of preference, but I cannot state a preference probability on either of them."
   You think of the following possible solutions:
   a. Explain the five rules to him and try to get him to state these probabilities.
   b. Tell him it is ok to use his friend's probability assessment for outcomes of the uncertainty and the preference probability of the prospects if he shares his friend's belief.
   c. Tell him it is ok to use his friend's probability assessment for outcomes of the uncertainty if he shares his friend's belief, but he must also share his friend's preferences if he wishes to use the preference probabilities as well.
   Which would you chose to do?
   i. (a) and (b)
   ii. (a) and (c)
   iii. (b) and (c)
   iv. None of the above.
3. Chantal is going to her friend Gregory's house for a dinner party. Gregory had asked Chantal to bring a bottle of wine to drink with the meal. When Chantal goes to the grocery store to pick up a bottle, though, she realizes that she can't remember if Gregory said he was serving beef, chicken, or fish for dinner. She now needs to decide whether to buy a bottle of red or white wine. Her preferences for the various outcomes can be arranged as follows:

Wine	Food	Preference Probability
Red	Beef	1.0
White	Fish	0.8
White	Chicken	0.7
Red	Chicken	0.5
Red	Fish	0.3
White	Beef	0.0

   While thinking about this decision, she bumps into her friend Shelli, who is also picking up some things for the same dinner party. Shelli doesn't know what is being served either, but she tells Chantal that it couldn't be beef because Gregory's girlfriend doesn't eat red meat. Therefore, Chantal decides to recalibrate the table above with the beef options eliminated (so that "white wine and fish" is the best outcome and "red wine and fish" is the worst outcome). What should Chantal's new preference probability be for the outcome "white wine and chicken?"
   a. 0.7
   b. 0.8
   c. 0.9
   d. There is not enough information to decide.

4. (Continued from the previous problem) Shelli also mentions that she thought Gregory might also make vegetarian lasagna, since that is his favorite recipe. Chantal's preferences for the two new outcomes generated by this information are as follows:

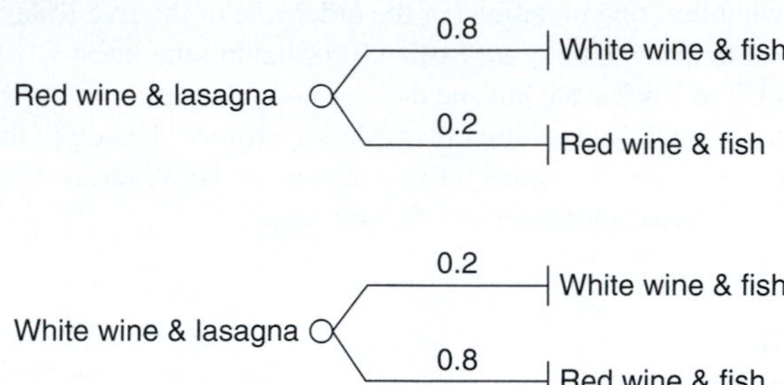

Based on what Shelli told Chantal, how can we characterize Chantal's preference for "white wine and chicken" and "red wine and lasagna"?

a. She prefers "white wine and chicken."
b. She prefers "red wine and lasagna."
c. She is indifferent between the two.
d. There is not enough information to decide.

5. How many of the following statements necessarily violate the five Rules of Actional Thought?

I. Onder is risk averse over some monetary deals and risk neutral over others.
II. Jeremy prefers a latte to a cappuccino, but he prefers a cappuccino with a biscotto to a latte with a biscotto.
III. Yesterday, David assigned a probability of 0.5 to a thumbtack landing heads. Today, David assigns a probability of 0.6 to a thumbtack landing heads.

a. 0　　c. 2
b. 1　　d. 3

6. Hannah follows the five Rules of Actional Thought, prefers more money to less, and has stated three certain equivalents for the deals shown below:

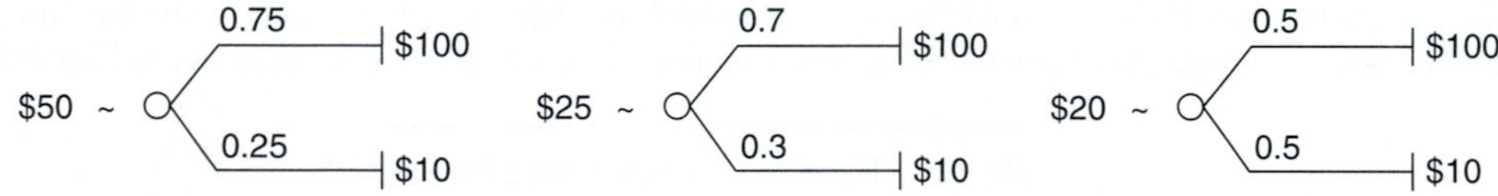

What can we say about Hannah's certain equivalent (CE) for the following deal?

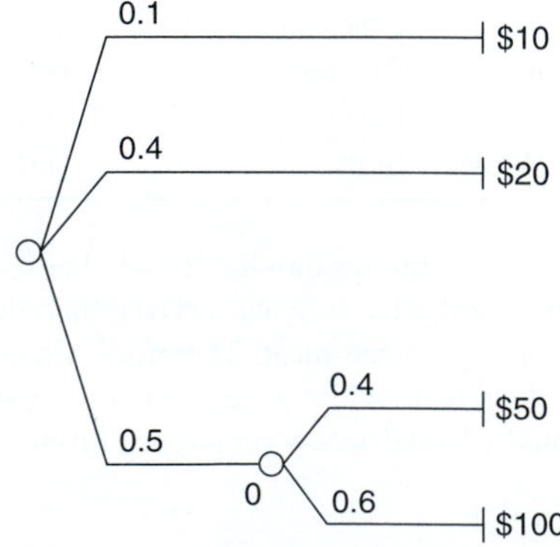

a. $CE \leq \$20$ c. $\$25 < CE \leq \$50$
b. $\$20 < CE \leq \$25$ d. $CE > \$50$

7. Fiona is indifferent between a deal of $60 for certain, and an uncertain deal with a 0.5 probability of $100 and a 0.5 probability of $50. Fiona follows the five Rules of Actional Thought and prefers more money to less. What is Fiona's preference order for deals *X*, *Y*, and *Z*?

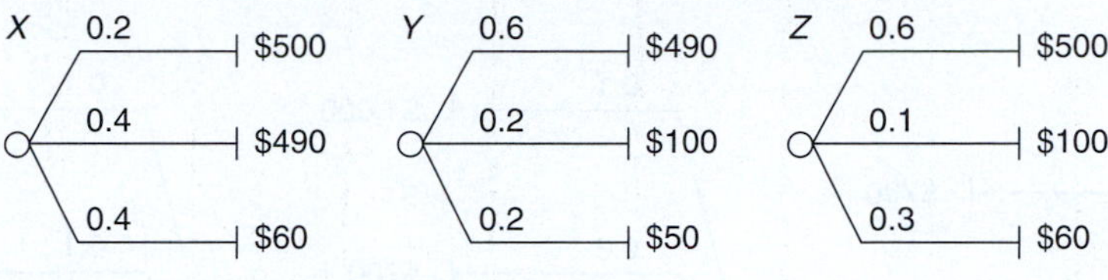

a. $X < Y < Z$ c. $Y < Z < X$
b. $X < Z < Y$ d. $Y < X < Z$

8. Leland follows the five Rules of Actional Thought and plans to travel from Sacramento to Reno. When planning his trip, he considers various modes of transportation. He prefers traveling via steam locomotive to traveling via a high-wheel bicycle. He also asserts the following preferences:

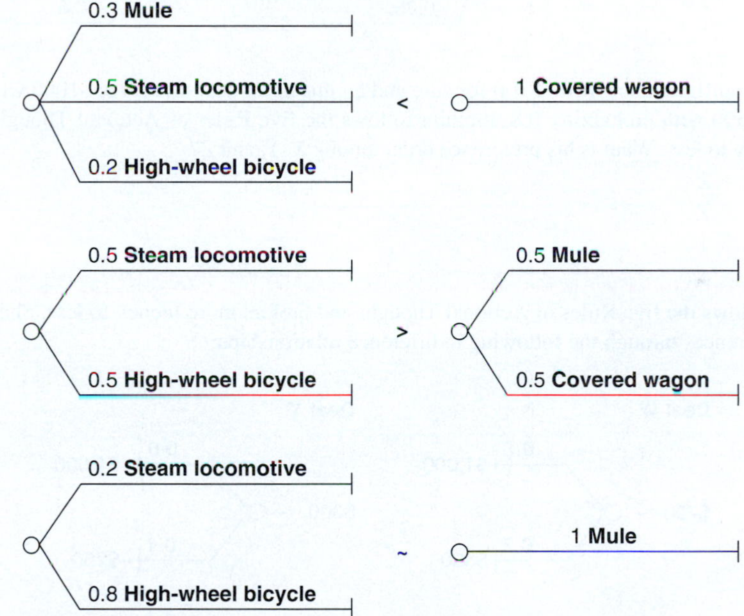

Given this information, which of the following statements are true for Leland?

a.

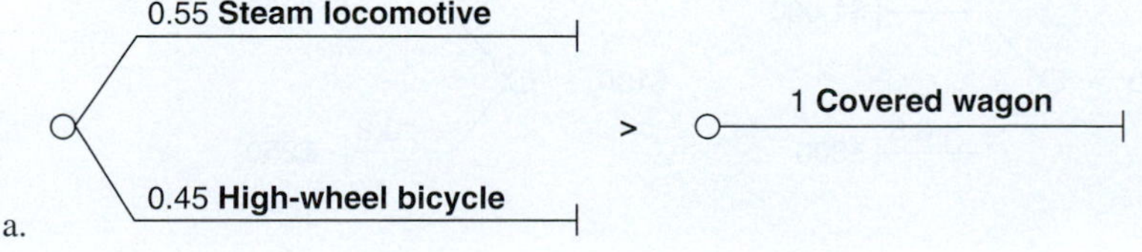

b. High-wheel bicycle < mule < covered wagon < steam locomotive
   a. I only
   b. II only
   c. Both I and II
   d. Neither I nor II

9. Ibrahim faces the following three deals:

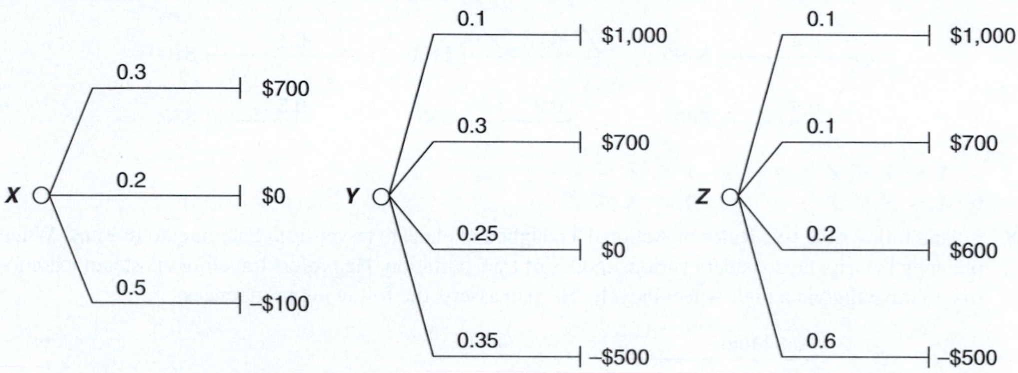

Ibrahim is indifferent between –$100 for sure and an uncertain deal that gives $1000 with probability 0.2 and –$500 with probability 0.8. Ibrahim follows the five Rules of Actional Thought and prefers more money to less. What is his preference order among *X*, *Y*, and *Z*?

a. $X > Y > Z$
b. $Y > X > Z$
c. $X > Z > Y$
d. $Z > X > Y$

10. Rashmi follows the five Rules of Actional Thought and prefers more money to less. She asserts some of his preferences through the following indifference relationships:

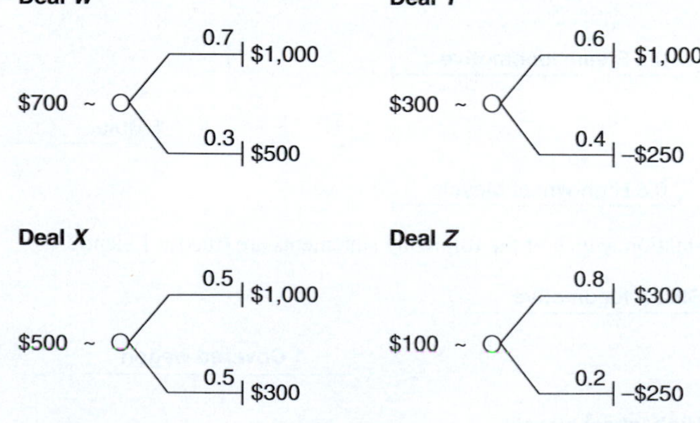

What can we say about Rashmi's equivalent (*CE*) for the following deal lie?

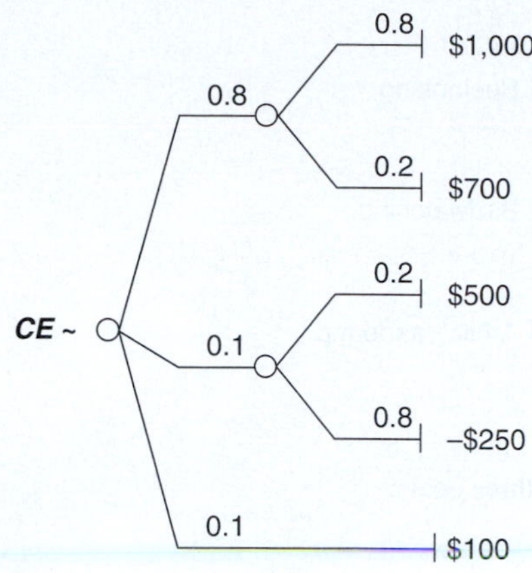

a. $\$100 < CE \leq \$500$
b. $\$300 < CE \leq \$500$
c. $\$500 < CE \leq \$700$
d. either $CE \leq -\$250$ or $CE > \$700$

11. Lionel likes extreme sports and follows the five Rules of Actional Thought. He is indifferent between spelunking and a deal with a 5/6 chance of an Arctic canoe trip and a 1/6 chance of a game of Jai Alai. He is also indifferent between a game of Jai Alai and a deal with a 4/9 chance at spelunking and a 5/9 chance at a birdwatching outing. Finally, he is indifferent between cliff diving and a deal with a 2/9 chance of spelunking and a 7/9 chance of a birdwatching outing.

Given this information, how many of the following three relationships are consistent with Lionel's preferences?

I.
0.8 Spelunking
0.2 Cliff diving
~
0.6 Arctic canoe trip
0.4 Jai Alai

II.
0.3 Jai Alai
0.4 Spelunking
0.3 Cliff diving
>
0.5 Arctic canoe trip
0.5 Birdwatching

Jai Alai > 0.1 Cliff diving; 0.2 Spelunking; 0.3 Birdwatching; 0.4 Arctic canoe trip

III.

a. 0 c. 2
b. 1 d. 3

12. Amit is facing the following three deals:

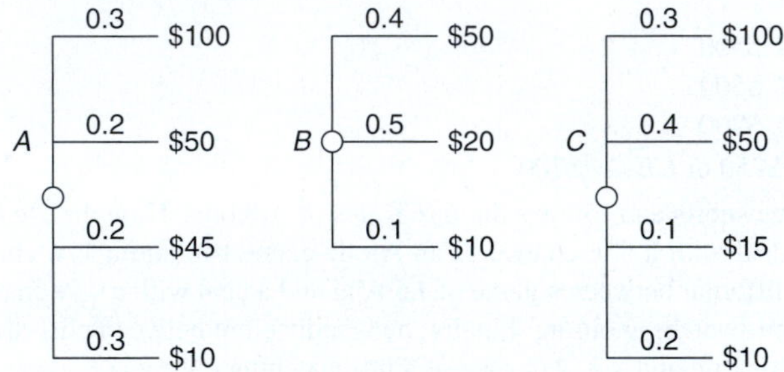

Amit is indifferent between a deal that gives \$20 for sure, and a deal that gives \$100 with probability 0.6 and \$15 with probability 0.4. Amit follows the five Rules of Actional Thought. What is Amit's preference order among *A*, *B*, and *C*?

a. $A > B > C$
b. $B > A > C$
c. $C > B > A$
d. $B > C > A$

13. Salvador has stated the following certain equivalents for the deals shown:

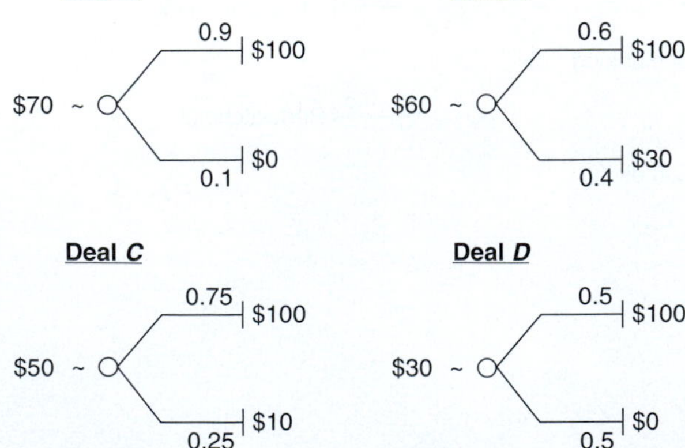

Assuming that he follows the five Rules of Actional Thought, what can we say about his certain equivalent (*CE*) for the following deal?

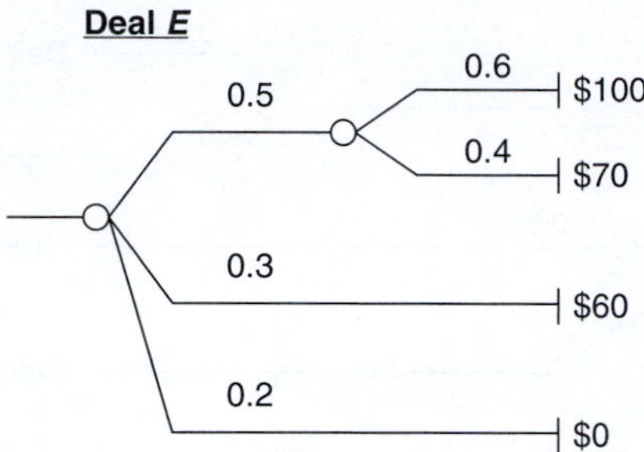

a. $30 < *CE* < $50
b. $50 ≤ *CE* ≤ $60
c. $60 < *CE*
d. There is not enough information to decide.

†**14.** What is the difference between a probability assigned in the probability rule and a preference probability stated in the equivalence rule? What is the clairvoyant's role with respect to each of these?

†**15.** Chris prefers a sled to a harmonica and specifies the following equivalence relations:

Book ~ 0.7 Sled, 0.3 Harmonica

Sweater ~ 0.2 Sled, 0.8 Harmonica

Ball ~ 0.85 Sled, 0.15 Harmonica

a. What is Chris' preference ordering for the sled, the sarmonica, the book, the sweater, and the ball?
b. What is Chris' preference probability (the probability of a sled versus a harmonica in a hypothetical deal) that is equivalent to the following deal?

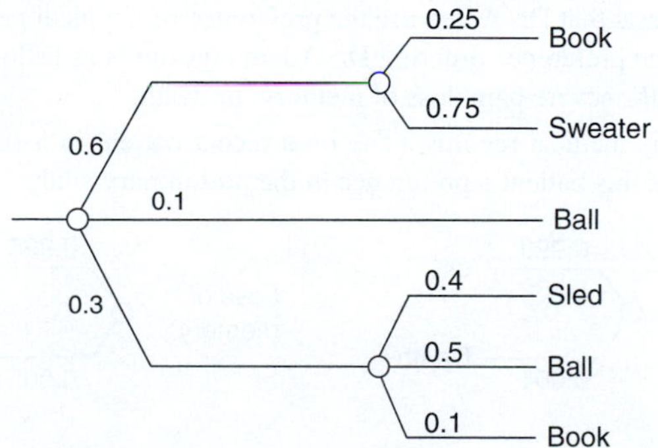

c. Does Chris prefer a book or the following deal? (***Hint***: First, express this deal as a probability of a sled versus a harmonica.)

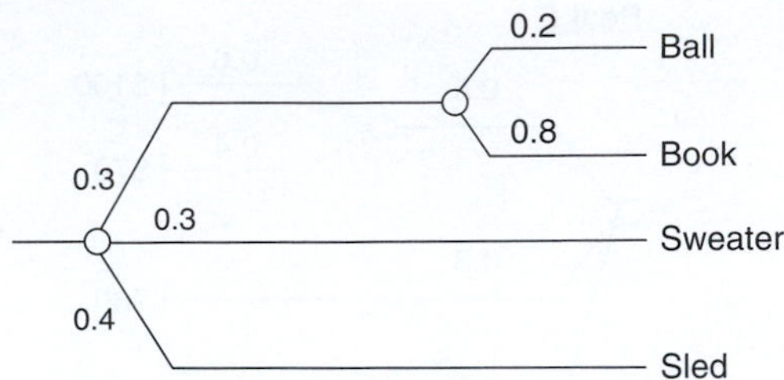

d. What can we infer about Chris' preference for four sweaters versus one book from his indifference statements presented in this problem?

†**16.** Dr. Ben Adam, an eminent brain surgeon, follows the five Rules of Actional Thought. One day, he has a patient, Mr. Smith, who is unconscious. In considering medical procedures to use for Mr. Smith, he creates the following tree about state-of-the-art medical technology and its impact on patients.

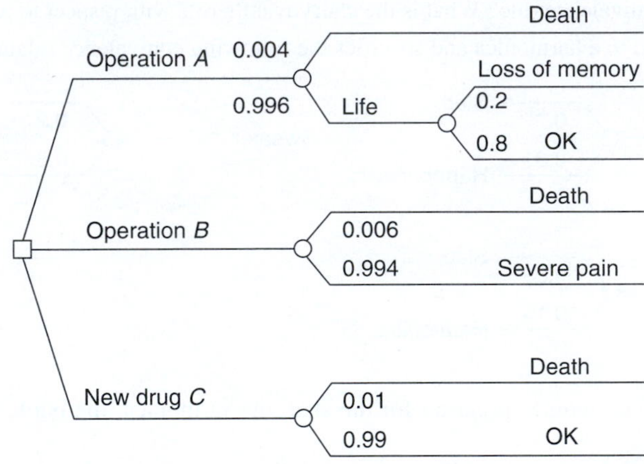

Since Mr. Smith is unconscious, Dr. Adam cannot assess his preference over the possible outcomes. One of the nurses suggests that Dr. Adam use the preference of a typical patient as he sees appropriate from his experience. The preference ordering Dr. Adam lays out is as follows from the most preferred to the least preferred: OK, severe pain, loss of memory, or death.

He pulls from his medical records a file on a recent patient in a very similar situation as Mr. Smith; he intends to use this patient's preference in the preliminary analysis for Mr. Smith's case.

Severe pain ~ [0.999 OK; 0.001 Death]

Loss of memory ~ [0.995 OK; 0.005 Death]

a. According to this information, what is Dr. Adam's best decision?

b. Just as Dr. Adam is thinking about how to improve this decision, Mr. Smith regains consciousness. Mr. Smith, who also considers the five Rules of Actional Thought as his norm for decision making, starts reviewing Dr. Adam's decision process. He agrees with Dr. Adam's ordering and assessments, except that he feels memory is less valuable to him than Dr. Adam supposed. The following is Mr. Smith's modified preference assessment.

Severe pain ~ [0.999: OK; 0.001: Death]

Loss of memory ~ [0.98: OK; 0.02: Death]

What is Mr. Smith's best decision?

c. Suppose Mr. Smith gives the following preference probabilities instead of those above.

Severe pain ~ [$1-p$: OK; p: Loss of memory]

Loss of memory ~ [q: Severe pain; $1-q$: Death]

where $p = 0.05$ and $q = 0.981$. Is the above information sufficient for you to recommend a decision? If so, give your recommendation; if not, explain.

CHAPTER

9

The Party Problem

CHAPTER CONCEPTS

After reading this chapter, you will be able to explain the following concepts:

- Decision tree
- *E*-operation and *e*-value
- Preference probability of an alternative
- The best decision alternative
- The wizard
- Hamlet's dilemma framed as a decision

9.1 INTRODUCTION

In the last chapter, we presented the five Rules of Actional Thought and discussed their importance in decision making. The best way to appreciate these five rules is to see them in action in a decision. In this chapter, we present an application of the five rules in a problem known as the party problem. We shall use the party problem as an extensive example to demonstrate how the rules should be applied. Further, we will examine how to integrate many other principles of decision analysis into this problem.

9.2 THE PARTY PROBLEM

Suppose that Kim has invited her teenage friends to a house party. However, she seems unsettled. She explains, "I'm confused about where to have this party." We say, "Since you're perplexed about a decision, perhaps we can help. Before we can, though, you have to read and agree to follow the rules of actional thought." She looks them over and readily agrees, and we tell her that she is now a *rulesperson*.

THE PROBABILITY RULE (FIRST PART) Following the first part of the probability rule, we start by identifying alternatives and possibilities and then characterizing the prospects Kim faces. We ask her what alternatives she is considering for the venue, and she says, "There are three possible locations: I can have it outdoors on the lawn, indoors in the family room, or on the porch, which is covered, but open to the weather on the sides The problem is, I don't know what the weather is going to be like. If it rains and I have the party outdoors, we could be miserable, but if it is sunny and I have it indoors, I'm going to wish we were outdoors."

9.2.1 The Decision Tree

At this point we say, "Let us help you with your decision." We sketch Figure 9.1.

The **decision tree** for Kim begins with a square decision node where the three alternatives emanating from it are labeled as *O* (outdoors), *P* (porch), and *I* (indoors). Then, for each alternative, we show two possibilities for the weather labeled: (*S*) sunshine and (*R*) rain.

9.2.2 The Clarity Test

She begins, "I see why you are offering two possibilities for the weather, but what exactly do you mean by sunshine and rain?" We explain, "As the decision maker, you need define those terms in a way that meets the standards of the clarity test and is also useful in ranking your prospects." We talk for 20 minutes to determine the best definitions of sunshine and rain, which we will not recount here (in the clarity test definition of the Beer Drinker, we have already shown how to conduct this discussion). Even to arrive at this simple binary distinction, we have to address multiple issues, such issues as cloudiness, precipitation, and intermittent clearing with precipitation.

Finally, though, Kim is satisfied that she has made this distinction clearly and in the most useful way for her purposes. We have now applied the first part of the probability rule and have characterized Kim's prospects. Next, we consider the remaining parts of the five Rules of Actional Thought and apply them specifically to help Kim make a good decision.

9.2.3 Using The Rules in The Party Problem

THE ORDER RULE We now mention the order rule, and point out to Kim that her decision tree has six *prospects* that involve the combinations of her actions and the possibilities that may arise from them.

1. Having an outdoor party in the sunshine, *O–S*
2. Having an outdoor party in the rain, *O–R*
3. Having a party on the porch in the sunshine, *P–S*

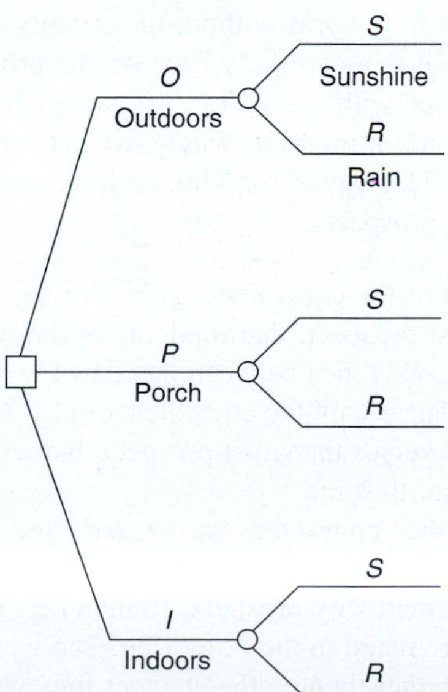

FIGURE 9.1 The Party Decision Tree

O–S	B: Best
P–S	
I–R	
I–S	
P–R	
O–R	W: Worst

FIGURE 9.2 Kim's Preferentially Ordered Prospect List

4. Having a party on the porch in the rain, *P–R*
5. Having a party indoors in the sunshine, *I–S*
6. Having a party indoors in the rain, *I–R*.

We then ask her to identify the best prospect, and she picks *O–S*. She identifies *O–R* as the worst prospect.

Next, we ask her to make a list of the six prospects with the best prospect (*O–S*) at the top, and the worst prospect (*O–R*) at the bottom. She constructs the ranking shown in Figure 9.2.

At the top is *O–S*, which is labeled *B* as the best prospect; at the bottom is *O–R*, which is labeled *W* for the worst prospect. The other four prospects—in order of preference—are *P–S*, *I–R*, *I–S*, and *P–R*. In reviewing the list, we note that she has ranked an indoor party in the rain, *I–R*, higher than an indoor party in the sunshine, *I–S*, and we ask her if this really reflects her preference. She says, "Yes. An indoor party in the rain could still be quite nice, but an indoor party in the sunshine might well make people wish they were outdoors." Just to be sure, we ask her if there are any two prospects on this list to which she is, in fact, indifferent. She replies, "No, I like each one better than any below it." We tell her she has now satisfied the order rule.

The order rule describes Kim's preferences in a world without uncertainty. If Kim has trouble assessing which prospect she likes better, it is most likely because the prospects have some important uncharacterized properties associated with them. For example, if she is not sure how much she would like the porch prospects because it might be windy, we need to go back to the probability rule and add the distinction "windy." However, for Kim, each prospect is characterized sufficiently to enable her to easily order all prospects.

THE EQUIVALENCE RULE Now we remind Kim of the equivalence rule. We say, "You have said that outdoors-in-the-sunshine, *O–S*, is the best prospect: that outdoors in the rain, *O–R*, is the worst prospect: and that porch in the sunshine, *P–S*, lies between these two in your preference." She says, "Right." Then we say, "In accordance with the equivalence rule, we should be able to offer you some chance of the best prospect versus the worst prospect that will make you just indifferent to having a party on the porch in the sunshine."

To help with this task, we introduce another character, the ***wizard***. See Figure 9.3. Unlike the clairvoyant, who can tell you what will happen, the wizard can make things happen in the present by magically waving his wand to create any prospect. Remember, that there is supposed to be *no uncertainty* in prospects that are listed in the order rule. The wizard assures that you focus on your preference for prospects, rather than on the chances they will happen.

FIGURE 9.3 The Wizard Can Make Anything Happen

While we are saying this, we set up the probability wheel, and tell her that if one color appears, the wizard will magically give her an outdoor party in the sunshine, and if the other color appears, he will magically give her an outdoor party in the rain. We ask her if she would be just indifferent between spinning the wheel on the one hand or receiving a party on the porch in the sunshine, also provided by the wizard, for sure. She says that in fact, a porch party in the sunshine could be very nice, so she would have to have a probability much closer to one than to zero of winning the best party before she would be indifferent.

After adjusting the colors, she establishes that if she had a 95% chance of an outdoor party in the sunshine and a 5% chance of an outdoor party in the rain, she would be just indifferent between this deal and a sure party on the porch with sunshine. We remind her that this 0.95 is her **preference probability** between best and worst corresponding to the prospect *P–S*; and that *P–S* is her certain equivalent for a 95% chance of the best as opposed to the worst prospect.

We then ask her to repeat this process for all prospects and she produces Figure 9.4.

For completeness, we have shown that the outdoor-in-the-sunshine party is equivalent to probability 1 of the best prospect, and the outdoor-in-the-rain party is equivalent to probability 1 of the worst prospect. As we go down her prospect preference list, the preference probabilities for best versus worst steadily decrease: 0.95 for *P–S*, 0.67 for *I–R*, 0.57 for *I–S*, and 0.32 for *P–R*. If the preference probabilities had not decreased, it would mean the prospects were not correctly ordered and that she should revisit the ordering process. You should note that these preference probabilities tell us nothing about Kim's beliefs about the chance of sunshine or rain; instead, they reflect the strength of her preference for the various deterministic prospects.

THE PROBABILITY RULE (PART 2) At this point, we return to the issue of information using the probability rule. Specifically, we observe that although we have completed the structure of the decision tree in Figure 9.1, we have not yet supplied Kim's probability assignment on the event in sunshine. We then begin a discussion on probability assignment, and we ask Kim to gather the information about tomorrow's weather. She watches television weather reports, reads the newspaper, looks at the sky, and finally sits down to have her probability encoded

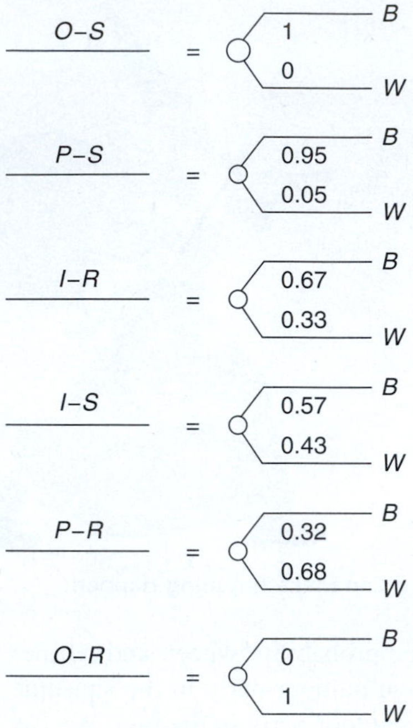

Prospect		B	W
O–S	=	1	0
P–S	=	0.95	0.05
I–R	=	0.67	0.33
I–S	=	0.57	0.43
P–R	=	0.32	0.68
O–R	=	0	1

FIGURE 9.4 Creating Best and Worst Deals

using the processes we have previously discussed. She assigns a probability of 0.4 to Sun and, consequently, a probability of 0.6 to Rain. The decision tree with these probabilities appears in Figure 9.5.

To demonstrate the right decision for Kim, we will combine the alternatives and information for Figure 9.5 with the preference information in Figure 9.4. We assure her that from this point on, as long as she follows the rules of thought, finding the best alternative is just a matter of calculation.

THE SUBSTITUTION RULE Next, we remind Kim of the substitution rule stating that, for any prospect, we can substitute a deal involving the best and worst prospect for which it is the certain equivalent. We perform this operation in Figure 9.6.

In Figure 9.6, for each prospect of the decision tree, we have substituted the equivalent deal from Figure 9.4. For example, if Kim picks the Porch alternative and the weather is Sunny,

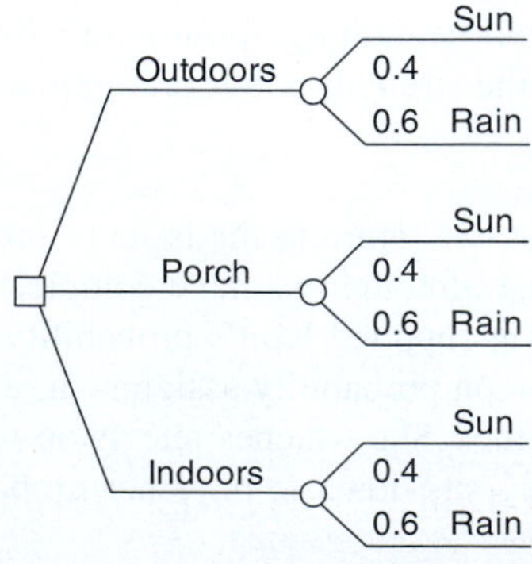

FIGURE 9.5 Decision Tree with Probabilities

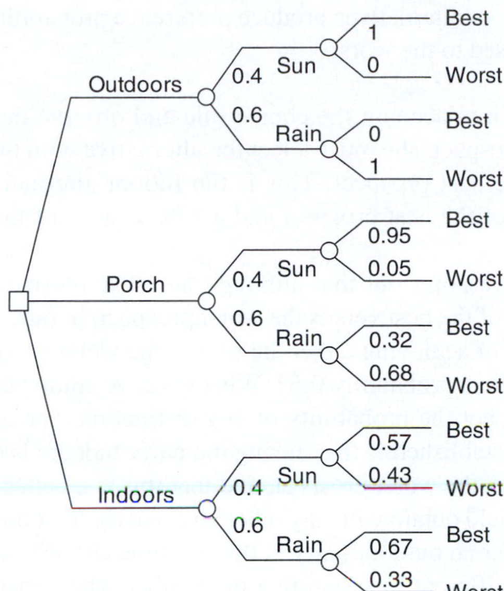

FIGURE 9.6 Substitution of Equivalent Deals for Each Prospect

she will be in a situation that she regards as equivalent to one with a 95% chance of the best prospect, *O–S*, and a 5% chance of the worst prospect, *O–R*.

When we examine Figure 9.6, we note that for each alternative, *O*, *P*, and *I*, there are only two possible prospects at the end of the tree: The best prospect, *B*, and the worst prospect, *W*. Since the substitution rule states that there is no difference between probabilities and preference probabilities for decision purposes, we can compute the preference probability of achieving the best prospect for each alternative by multiplying together the conditional probabilities on the branches of each path that leads to the best prospect and then adding over all paths.

In Figure 9.6, the preference probability of the Outdoor alternative is therefore

$$0.4 \times 1 + 0.6 \times 0 = 0.4$$

The porch alternative has a preference probability of

$$0.4 \times 0.95 + 0.6 \times 0.32 = 0.57$$

The Indoor alternative has a preference probability of

$$0.4 \times 0.57 + 0.6 \times 0.67 = 0.63$$

This is shown in the simplified tree of Figure 9.7.

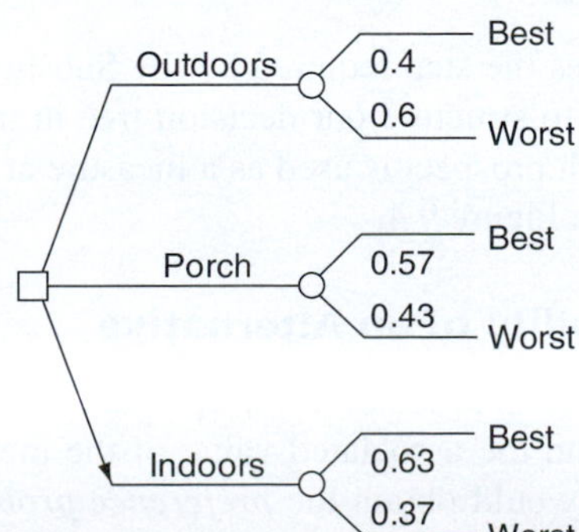

FIGURE 9.7 Preference Probabilities of Each Alternative in Terms of the Best and Worst Prospects

Note that the Outdoor, Porch, and Indoor alternatives produce preference probabilities 0.4, 0.57, and 0.63 of the best prospect, as opposed to the worst.

THE CHOICE RULE Next, Kim must take into account the choice rule and observe that since she prefers the best prospect to the worst prospect, she must select the alternative with the highest preference probability of achieving the best prospect. This is the Indoor alternative. We note that it is equivalent to a 63% chance of the best prospect and a 37% chance of the worst prospect.

In interpreting this result for Kim, we point out that although her best alternative, the Indoor party, is equivalent to a 63% chance of the best versus the worst prospect, in fact, following this alternative leaves her no possibility of achieving either the best or the worst prospect. In this problem, there is no real world event with a probability 0.63. What we have computed is her preference probability for each alternative, not the probability of any distinction that could be resolved by the clairvoyant. What she has established is that having the party indoors is equivalent to a 63 percent chance of the best versus the worst prospect, and that this is a better chance of the best versus worst prospect than she could obtain with any other alternative. If, at this point, the wizard were standing by to produce either an outdoor party in the sunshine (B) or an outdoor party in the rain (W), she would be just indifferent to spinning a probability wheel that would result in B with probability 0.63, or W with probability 0.37.

APPLYING THE RULES You can make any decision by applying the five rules. Think about all the decisions you could potentially face. Suppose, for example, you are blind and are considering an operation that could either cure you or kill you. Your three prospects are alive and sighted, alive and blind, and dead, usually preferred in this order. To help you make your decision, you would apply the rules just as we did in the previous example, helping Kim plan her party.

When it comes to more complex problems, strictly following the rules may seem like an impossible challenge. Think of all ways three prospects at different levels could be chosen from the prospect list in the order rule, and realize that, in principle, you would have to provide a preference probability in each case. For example, the party problem ended up having six prospects, and there are $6!/(3!3!) = 20$ different potential equivalence rule questions. We asked only four.

Fortunately, we will see that simpler methods exist for carrying out the analysis. The logic we use—the seat of the stool—rests on the rules, but they are in the background, just as the rules of plane geometry are in the background of a carpenter building a house.

9.3 SIMPLIFYING THE RULES: *E*-VALUE

By defining a new operation, we can save ourselves the step required by the Substitution Rule in deriving the best alternative. One alternative is to structure our decision tree in the form of Figure 9.8, where the preference probability of each prospect is used as a measure at the end of the tree. The preference probability was obtained in Figure 9.4.

9.3.1 Calculating the Preference Probability of an Alternative

Consider a probability tree with a **measure**. If, for each alternative and for each elemental possibility, we multiplied its elemental probability and the associated value of the measure, and then summed over all elemental possibilities, we would obtain the *preference probability* for

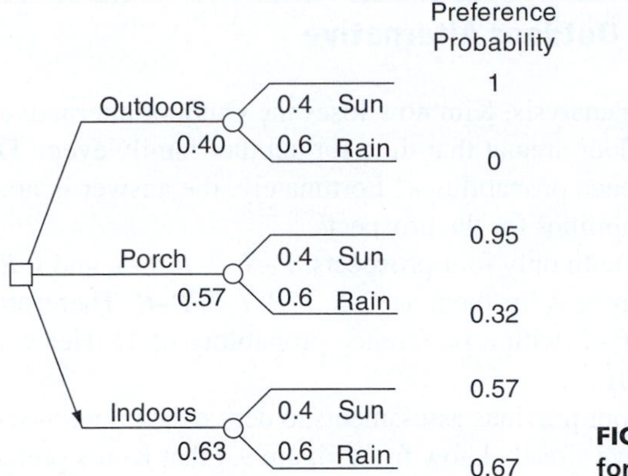

FIGURE 9.8 Preference Probabilities for Each Alternative

the alternative, as shown in Figure 9.7. For example, the preference probability of the Porch alternative is

$$0.4 * 0.95 + 0.6 * 0.32 = 0.57.$$

In Figure 9.8, these preference probabilities are underlined. We call the number we have computed the ***e*-value** of the measure. With this *e*-operation defined, we can now say that

The preference probability for any alternative is the e-value of the preference probabilities of its prospects.

As we proceed, we will discover other uses for the *e*-operation.

9.3.2 Implications of the Rules

As we have just learned, the preference probability for any alternative is the *e*-value of the preference probabilities for each of its prospects. This new observation has several implications. First, using this operation means that we do not need to perform the full set of substitutions for each tree, as we did in Figure 9.6. Instead, we simply draw the tree with its preference probabilities as measures, and calculate the *e*-value of the preference probabilities for each alternative. The alternative with the highest *e*-value of preference probabilities is the one that is consistent with the five rules.

A second, and more important implication, is that if you wish to follow the five Rules of Actional Thought, then the method we have just described is the only way you can make decisions:

Choose the alternative with the highest e-value of preference probabilities.

Any other method—such as maximizing the probability of the best outcome, minimizing the probability of the worst, or taking benefit–cost ratios—can violate the five Rules of Actional Thought. If you are tempted to make a choice that is inconsistent with the result of using the five rules, then you must ask yourself which of the five rules you are willing to violate, and why? Remember that you control the frame and all three legs of the stool. You declare the decision, specify the alternatives, assign your probabilities, and state your preferences. The rules describe the logic of the seat that implies a course of action.

As we discussed, if we have no uncertainty, we would only need the probability rule (to characterize the problem), the order rule (to order the prospects), and the choice rule to make a selection. We would need neither the equivalence nor the substitution rule.

EXAMPLE 9.1 Kim Loses the Outdoor Alternative

Suppose that, after carrying out all this analysis, Kim now loses the Outdoor alternative. Her father tells her that they need the Outdoor area at that time for another family event. Do we have to go back and assess her preference probabilities? Fortunately, the answer is no. Kim has already stated her preference probabilities for the prospects.

In this new situation, Kim is left with only four prospects: *P–S*, *P–R*, *I–S*, and *I–R*. She has already stated her order rule preferences for them as *P–S*, *I–R*, *I–S*, *P–R*. Therefore, her best prospect in this new situation is *P–S* (with a preference probability of 1). Her worst is *P–R* (with a preference probability of 0).

In this new situation, we can use our previous assessments to derive the preference probabilities for *I–R* and *I–S*. For example, we already know from Figure 9.4 that Kim's preference probability for *I–R*, in terms of *O–S* and *O–R*, is 0.67. Since she no longer has *O–S* or *O–R* prospects, what we are interested in now is her preference probability expressed in terms of the probability of getting a porch–sunny (*P–S*) or a porch–rainy (*P–R*) party. This is the preference probability, p, shown on the right side of Figure 9.9.

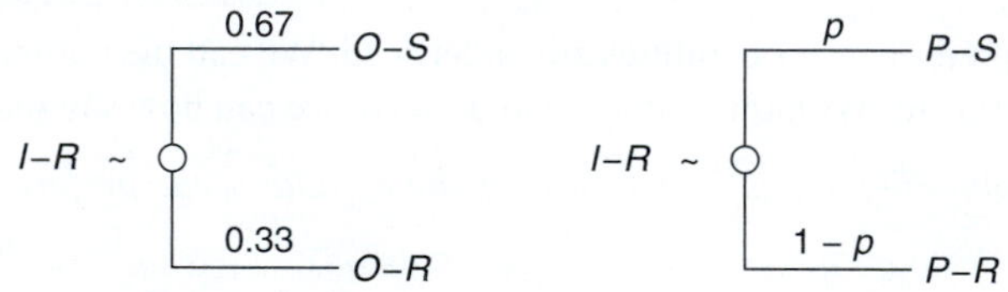

FIGURE 9.9 Preference Probabilities for Indoor-Rainy with (Left) Outdoor Alternative and (Right) No Outdoor Alternative

To derive the preference probability, p, of Figure 9.9, we need to convert this figure into prospects involving *O–S* and *O–R*, and compare the corresponding values, as shown in Figure 9.10. To do this, we refer back to Figure 9.4 and use the substitution rule and original preference probabilities of *P–S* and *P–R*: 0.95 and 0.32.

By equating the preference probability of *I–R* in the right-hand side of Figure 9.10 to that of Figure 9.4, we get the equation

$$0.95p + 0.32(1 - p) = 0.67$$

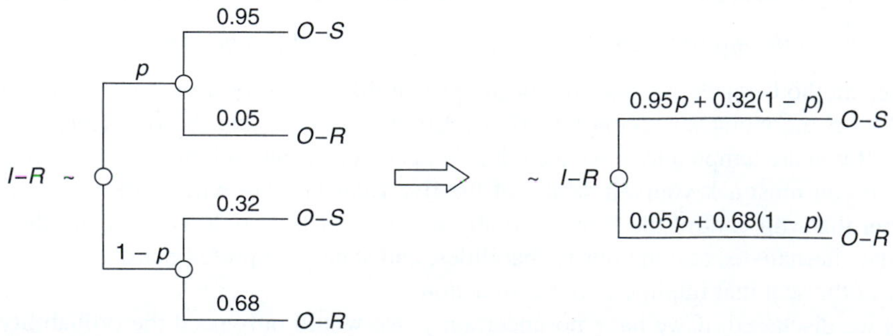

FIGURE 9.10 Deriving the Preference Probability for *I–R* After Removing *O* Alternative

from which we get $p = 0.56$, the preference probability for *I–R* after removing the Outdoor alternative. In a similar way, we can deduce the preference probability for *I–S*, q, after removal of the Outdoor alternative by solving for

$$0.95q + 0.32(1 - q) = 0.57$$

to find $q = 0.40$.

To compute Kim's calculated preference probability for the Indoor alternative without the Outdoor alternative, we perform the *e*-operation on the new preference probabilities, $0.4 \times 0.4 + 0.6 \times 0.56 = 0.5$. Her calculated preference probability for the Porch alternative is now $0.4 \times 1 + 0.6 \times 0 = 0.4$. Therefore, Kim still prefers the Indoor alternative to the Porch alternative.

This is an important observation: *the removal of a decision alternative does not change the preference order for any of the remaining alternatives.*

We did not have to redo any of the calculations to know that Kim would prefer the Porch alternative if she lost the Indoor alternative.

Note that we could conduct the previous analysis even if Kim did not lose the Outdoors alternative. Given the preference probabilities assessed for the best and the worst prospects, we can determine the preference probabilities for any other combinations of the desired prospects. We can also use any prospects to represent the best and the worst prospects for the preference probability assessment, even if they are not included in the actual decision problem.

EXAMPLE 9.2 Kim Gets a Hayride Alternative

Kim has a friend, Allison, whose dad has a hayride business. At the time of the party, regardless of the weather, he offers her a free hayride around town with her friends. Kim feels that if the weather is sunny, the Hayride Sunny prospect (*H–S*) would be better than her original Outdoors-Sunny (*O–S*) party. But if the weather is rainy, the Hayride Rainy (*H–R*) prospect would be the worst prospect, since she and her friends will be exposed to rain for a few miles before the ride is over. With this new alternative, do we need to re-assess Kim's preference probabilities for all the prospects? Once again, the answer is no.

Kim now faces eight preferred prospects in the following order, from ascending to descending: *H–S*, *O–S*, *P–S*, *I–R*, *I–S*, *P–R*, *O–R*, *H–R*. In this new situation, *H–S* has a preference probability of 1, and *H–R* has a preference probability of 0. All we need to do is assess her preference probabilities for two prospects (say *O–S* and *O–R*) in terms of *H–S* and *H–R*, and we can deduce all the remaining preference probabilities.

For example, suppose Kim states that her preference probabilities for *O–S* and *O–R* were 0.9 and 0.05, respectively, as shown in Figure 9.11.

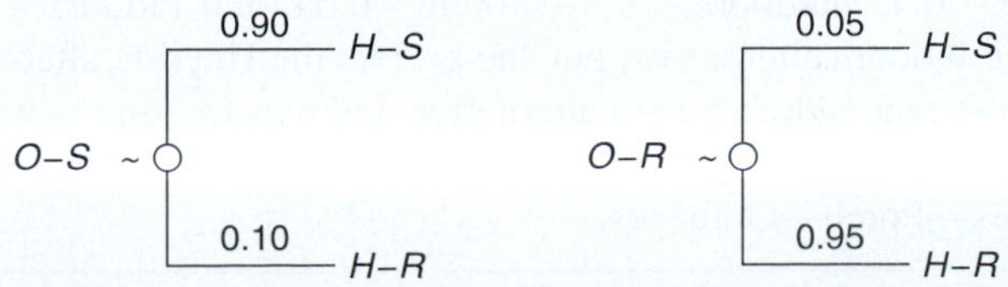

FIGURE 9.11 Preference Probability Assessments for Two Prospects

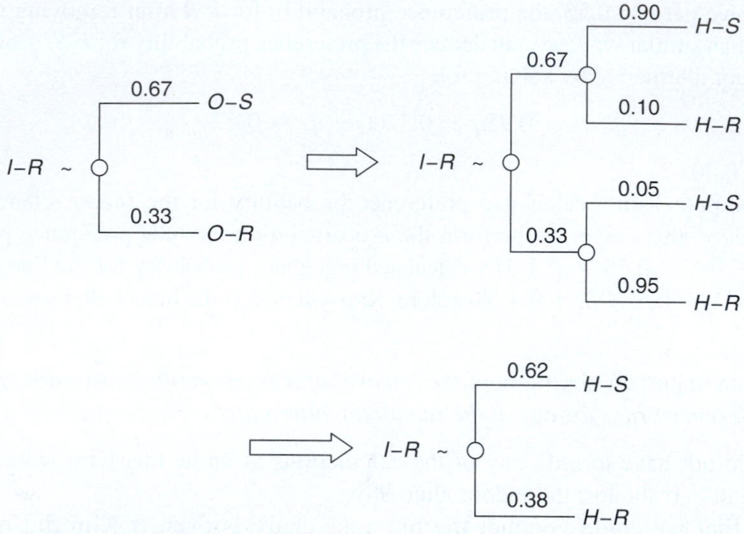

FIGURE 9.12 Deriving Preference Probability for *I–R* Using *H–S* and *H–R*

Kim's preference probabilities for any of the prospects can be expressed in terms of the *H–S* and *H–R* prospects using the substitution rule. For example, her preference probability for the *I–R* prospect can be calculated using her assessments in Figure 9.4 and 9.12 to yield 0.62, as shown in Figure 9.12.

The same procedure can be repeated for all other prospects to the get the preference probabilities in Figure 9.13.

H–S	1
O–S	0.9
P–S	0.86
I–R	0.62
I–S	0.53
P–R	0.32
O–R	0.05
H–R	0

FIGURE 9.13 Derived Preference Probabilities

We find the preference probabilities for each alternative by calculating the *e*-values of the preference probabilities: Hayride = 0.4, Outdoors = 0.39, Porch = 0.54, and Indoors = 0.58. Kim's best alternative is still the Indoors alternative, but she prefers the Hayride alternative to the Outdoors alternative. Note that adding a new alternative and constructing new preference probabilities did not change the original preference order of the alternatives that Kim has previously calculated: Indoors—Porch—Outdoors.

9.4 UNDERSTANDING THE VALUE OF THE PARTY PROBLEM

For several reasons, the party problem is a useful pedagogical paradigm. First, it is clearly contrived, so students spend little energy deciding whether it mirrors a real-life decision. If we used a more practical problem in a popular field, such as investing, satellite design, or research and development, someone would question our accuracy, and we would have to spend time explaining particulars irrelevant to the principles we are trying to illustrate.

Second, the party problem structure can represent many common decisions. As an example, we once used the party problem to illustrate the principles of decision analysis to a group of nuclear engineers who built power plants. Their response was that this was all very interesting but "We don't face this kind of decision." So, we spent the next part of the presentation changing some of the terminology of the party problem to see how it might fit one of their most important decisions: How much to retrofit their nuclear power plants to face seismic events? Rain became "a major seismic event in the operating life of the plant," while the sun did not. While there were many other uncertainties, a major seismic event was by far the most important. Outdoors became "no structural modifications for this risk;" Indoors became the alternative of extensive, expensive, and effective modifications; Porch became a limited and less effective modification alternative. If they spent a lot of money on retrofitting (the Indoor alternative), they do not really care much if there is an earthquake (rain), because they are defended against it. If they do not spend the money (the Outdoor alternative), they are hoping no major seismic event will happen (rain), because that would precipitate a disaster. The nuclear engineers were quick to say that these were exactly the kind of decisions they were facing on a regular basis. By incorporating suitable changes in distinctions, probabilities, and values, the necessary process of analysis is much the same.

For example, the party problem structure also applies to the purchase of medical insurance. Sunny would correspond to no serious medical expenses during the year, and Rainy the opposite. Outdoor would be purchasing no medical insurance, Porch to buying a high deductible policy, and Indoors to paying the higher premium of a low deductible policy. The same reasoning would apply to the purchase of any type of casualty insurance.

This structure is also quite useful in the medical domain. For example, someone with cancer may face three treatment choices: surgery, chemotherapy, or do nothing. The key uncertainty could be characterized as cured (i.e., cancer has gone into remission) or not. Of course, there are many other issues to potentially consider, including such factors as quality of life, cost, and discomfort during treatment. As we proceed with future chapters, we shall see that incorporating such additional features is simple once the basic foundations are understood. We shall also see that the party problem structure, though simple, has many properties that challenge our intuition.

DECISION ANALYSIS IN THE LIBERAL ARTS: HAMLET'S DILEMMA This section shows that the five Rules of Actional Though can be used to explain many decisions, even ones that have puzzled characters in famous novels. As an illustration, we consider Hamlet's dilemma from the famous play by William Shakespeare.

Plot Summary of Hamlet:

The play starts when King Hamlet of Denmark killed the King of Norway, Fortinbras, and took his lands. King Hamlet is now dead and his son, Prince Hamlet, is living in his father's castle. Prince Fortinbras, the son of the dead King of Norway, has sworn to attack Denmark and retake the lost lands. Hamlet asks guards to watch for this attack, and while watching, the guards see a ghostly figure. Horatio, Prince Hamlet's friend, sees the ghost, and identifies it as the ghost of King Hamlet. He rushes off to tell Prince Hamlet that he had seen his father's ghost.

Meanwhile, Prince Hamlet's father has been succeeded by his brother, Claudius, as the new King of Denmark. Claudius has married the former king's wife (Prince Hamlet's mother), Gertrude. Hamlet is very suspicious about his mother's quick remarriage to his uncle, Claudius, after his father's death.

Later, outside the castle, the ghost of King Hamlet speaks to young Hamlet, confirming Hamlet's suspicions that Claudius murdered his father. Hamlet is troubled. He pretends to be mad and contemplates suicide in his famous soliloquy.

> "**To be, or not to be: That is the question:**
> *Whether 'tis nobler in the mind to suffer*
> *The slings and arrows of outrageous fortune,*
> *Or to take arms against a sea of troubles,*
> *And by opposing end them?*"

This is the decision that Hamlet needs to resolve and is the cause of his worries: "To be or not to be;" to commit suicide; or to do nothing. Should he bear the "slings and arrows" of his current life or "take arms against a sea of troubles" and kill himself?

> "*To die: to sleep;*
> *No more; and, by a sleep to say we end*
> *The heartache and the thousand natural shocks*
> *That flesh is heir to, 'tis a consummation*
> *Devoutly to be wish'd. To die, to sleep;*
> *To sleep: perchance to dream: ay, there's the rub;*"

Hamlet would like to die and just sleep or rest in peace "'tis a consummation devoutly to be wished". But he realizes there is an uncertainty that may affect him if he decides to commit suicide; the possibility of "to dream" in the afterlife.

> "*For in that sleep of death what dreams may come,*
> *When we have shuffled off this mortal coil,*
> *Must give us pause. There's the respect*
> *That makes calamity of so long life;*"

Hamlet is not sure what the afterlife might be like. This is the cause of his dilemma.

> "*For who would bear the whips and scorns of time,*
> *The oppressor's wrong, the proud man's contumely,*
> *The pangs of dispriz'd love, the law's delay,*
> *The insolence of office, and the spurns*
> *That patient merit of the unworthy takes,*
> *When he himself might his quietus make*
> *With a bare bodkin? Who would fardels bear,*
> *To grunt and sweat under a weary life,*
> *But that the dread of something after death,*
> *The undiscover'd country from whose bourn*

No traveler returns, puzzles the will,
And makes us rather bear those ills we have
Than fly to others that we know not of?"

Even though the current life has several unpleasant experiences "whips and scorns of time, the oppressor's wrong" to name only a few, the dread of what will happen after death makes us bear with life's pains rather than committing suicide with a bodkin (dagger).

"Thus conscience does make cowards of us all;
And thus the native hue of resolution
Is sicklied o'er with the pale cast of thought,
And enterprises of great pith and moment
With this regard their currents turn awry
And lose the name of action."

—Shakespeare, *Hamlet,* Act III, Scene 1, 56

The decision tree for Hamlet's situation is shown in Figure 9.14.

Suppose Hamlet had asked you to help him with his dilemma using the five Rules of Actional Thought. The first step would be to characterize the decision situation. First, we note that in his soliloquy Hamlet is deciding whether or not he should commit suicide. He has two alternatives: To Be or Not to Be.

If we apply the five Rules of Actional Thought to help Hamlet, we proceed as follows.

1. **Probability Rule**
   Hamlet must create distinctions that characterize his decision situation. Furthermore, should assign a probability to each prospect. If we look at Figure 9.14, we see that Hamlet has two alternatives: To Be or Not to Be. The alternative Not to Be has a distinction: *Dream–Not Dream,* or *Afterlife–No Afterlife.*

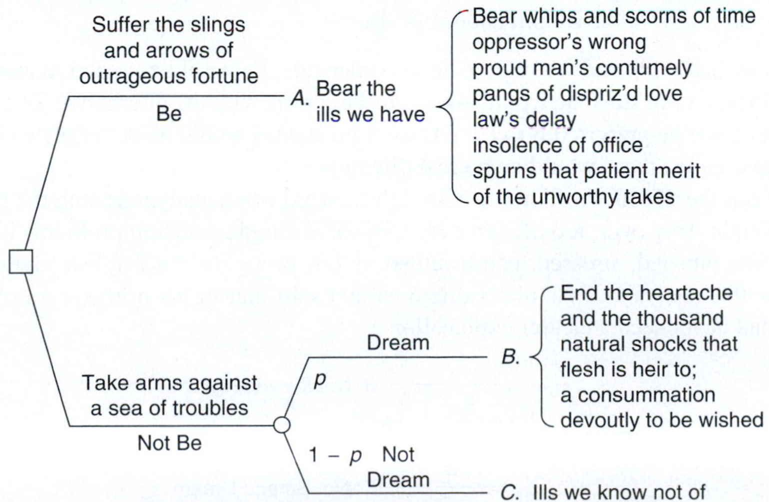

FIGURE 9.14 Decision Tree for Hamlet's Dilemma

Hamlet faces three prospects:

A: Be: Do not commit suicide, and bear the ills that he has.

B: Not Be and Not Dream: This prospect that "ends the heartache and the thousand natural shocks that flesh is heir to." He also refers to it as a "consummation devoutly to be wished."

C: Not Be and Dream: This is a prospect that Hamlet fears. He has seen his father's restless ghost. He knows the church's teaching that suicide is a sin.

To complete the probability rule, Hamlet then assigns a probability, p, for Not Dream after suicide; that is, the probability of no life after death. If necessary, we can assist him in this task using the probability wheel. The reason Hamlet assigns a non-zero probability is that he has seen his dead father's troubled ghost appear to him and talk to him.

2. **Order Rule**
Now, Hamlet orders the prospects from best to worst. We note that if he preferred *A* to both *B* and *C*, then he would not have a dilemma; he would simply "Be." Conversely, if he preferred both *B* and *C* to *A*, then his resolution of the dilemma would be "Not to Be." From the passage, we understand that Hamlet, in fact, prefers *B* to *A* and *A* to *C*: $B > A > C$. This is the source of his dilemma.

3. **Equivalence Rule**
Next, Hamlet states a preference probability q for the prospect *B* versus *C* that would make him just indifferent to receiving *A* for sure. We could have him use a probability wheel to state a probability where he would be just indifferent to receiving prospect *A* for sure, or he would a deal with a probability q of getting *B* and $1 - q$ of getting *C*. Note that if Hamlet does indeed follow the five Rules of Actional Thought, he should be able to state this preference probability. Figure 9.15 shows the assessment needed for the equivalence rule.

4. **Substitution and Choice Rules**
The substitution rule allows us to move directly to the choice rule: since replacing prospect *A* with the equivalent stated preference probability of q, we have the same two prospects for alternatives Be and Not Be, as shown in Figure 9.16. Hamlet must choose the alternative with the higher probability of the prospect he prefers. If $p > q$, he should choose Not to Be, and if $p < q$, he should choose To Be.

If Hamlet had preferred *A* to *B* to *C* in his order rule, he would not need to assess either of the probabilities, p or q, since he would always be better off with the alternative To Be. This also would be the case if he preferred *B* to *C* to *A*, since he always would be better off with Not to Be. In both of these cases, there would be no real dilemma.

As we can see, Hamlet's dilemma is straightforward when analyzed using the five Rules of Actional Thought. However, too often we try to resolve simple decision problems in our minds, and we become puzzled, stressed, and troubled by our thoughts. An English literature professor once saw this analysis of Hamlet's dilemma and said that in his thirty years of explaining Hamlet, he had never seen a clearer explanation.

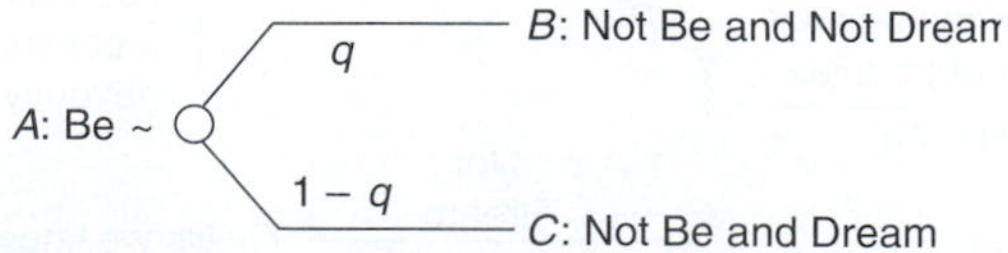

FIGURE 9.15 Equivalence Rule for Hamlet's Dilemma

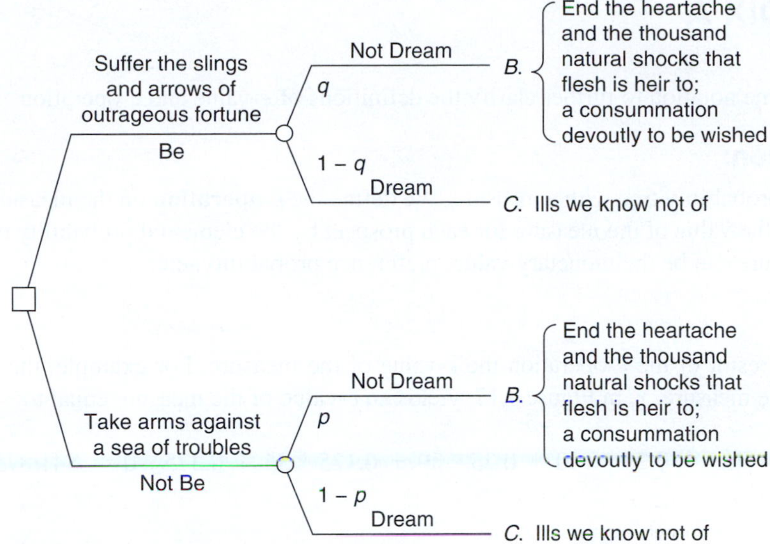

FIGURE 9.16 Hamlet's Dilemma Reduced to Identical Prospects for Each Alternative

9.5 SUMMARY

You can make any decision by applying the five Rules of Actional Thought.

- There is a difference between a *prospect of a decision* and an *outcome of an uncertainty.*
- The ***preference probability of a prospect is*** obtained using the equivalence rule stated preference probability in terms of the best and worst prospects.
- We can calculate the preference probability for a prospect in terms of the preference probabilities of any other two prospects (one more preferred and one less preferred) introduced in the tree.
- The ***preference probability of an alternative is*** the *e*-value of the preference probabilities of its prospects.
- The best decision alternative is the one with the highest preference probability.
- The addition or removal of an uninformative alternative or a prospect does not change the preference order of the remaining alternatives.
- Multiplying preference probabilities by any positive number or adding a constant to them does not change the ranking of the decision alternatives.

KEY TERMS

- Decision tree
- *e*-operation
- *e*-value
- Measure
- Preference probability of an alternative

APPENDIX A

Let's use some notation to further clarify the definitions of *e*-value and e-operation.

e-operation:

Consider a probability tree with a measure. We define the ***e*-operation** on the measure as simply multiplying the value of the measure for each prospect by the elemental probability of achieving it. The measure can be the monetary value, preference probability, etc.

e-value:

We call the result of the *e*-operation the *e*-value of the measure. For example, the *e*-operation applied to the measure, *x*, in Figure 9.17. yields an *e*-value of the measure equal to:

$$e\text{-value} = 0.25 * 100 + 0.25 * 50 + 0.125 * 35 + 0.375 * 0 = \$41.875$$

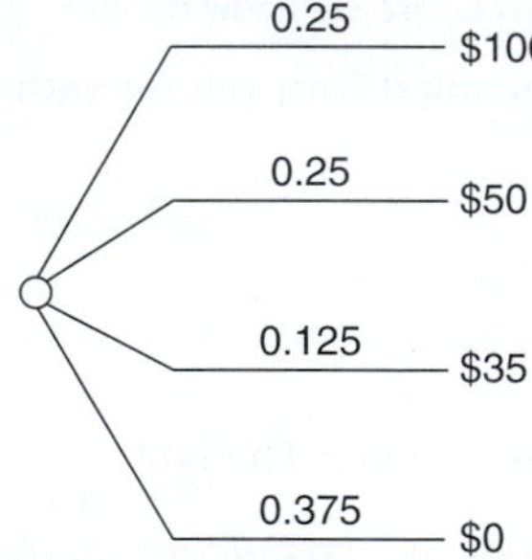

FIGURE 9.17 Example of a Tree with Monetary Measure

From here on, we will denote the *e*-value of a measure *x* by the symbol $<x|\&>$. For short, we will often express the *e*-operation using the symbol $\sum_{i-1}^{n} p_i x_i$. For example, given a deal with prospects x_1, x_2, x_3 with corresponding probabilities p_1, p_2, p_3, we have

$$e\text{-value of } x = <x|\&> = \sum_{i=1}^{3} p_i x_i = p_1 x_1 + p_2 x_2 + p_3 x_3.$$

Preference Probability of an Alternative:

The preference probability of an alternative is the *e*-value of the preference probability of its prospects.

The Best Decision Alternative:

The best decision alternative is the one with the highest preference probability.

PROBLEMS

Problems marked with an asterisk (*) are considered more challenging.

1. How many of the following statements are true?

I. The five Rules of Actional Thought only apply to situations where there is money or to business decisions, but they do not apply to medical decisions.

II. The five Rules of Actional Thought are normative, and they describe how you should make a decision not how people actually make decisions. Therefore, they should not be used in real world decisions.

III. The inclusion of additional prospects requires again eliciting all of the preference probabilities from the decision maker.

a. 0 c. 2
b. 1 d. 3

2. How many of the following statements are true?

I. The *e*-operation applied to the preference probabilities of an alternative yields the same *e*-value as using the substitution rule for the prospects of the alternative and reducing its prospects in terms of the best and the worst.

II. The best decision alternative is the one with the highest preference probability.

III. The best decision alternative is the one with the highest benefit–cost ratio.

a. 0 c. 2
b. 1 d. 3

***3.** Which of the following represents Hamlet's Order rule preferences?

a. To Be > Not to Be
b. Not to Be > To Be
c. Be > Not Be and Not Dream > Not Be and Dream
d. Not Be and Not Dream > Be > Not Be and Dream

***4.** Which of the following best explains why the best decision alternative is the one with the highest preference probability?

a. This corresponds to the alternative with the highest probability of getting a good outcome
b. It is a result of applying the five rules of actional thought
c. It corresponds to the alternative with the lowest probability of a good outcome
d. It leads to minimum regret

***5.** If we remove two prospects from the set of prospects, how many preference probabilities do we need to re-assess?

a. 0
b. 1
c. 2
d. The number of remaining prospects

***6.** If we add two prospects to the set of prospects, how many preference probabilities do we need to re-assess?

a. 0
b. 1
c. 2
d. The new number of remaining prospects

CHAPTER

10

Using a Value Measure

CHAPTER CONCEPTS

After reading this chapter, you will be able to explain the following concepts:

- The advantages of using money as a value measure
- Assessing a *u*-curve over money
- Material distinction
- The value of clairvoyance
- The certain equivalent

10.1 INTRODUCTION

The last chapter provided many examples of the five Rules of Actional Thought as applied to actual decisions, both—simple and more complex. The procedure we described in the last chapter can be used to analyze any decision even if the prospects are not described by a value measure—we simply use the preference probability as the value measure and calculate the *e*-value of the preference probabilities to determine the best alternative. It is often useful to introduce money as a measure of the desirability of prospects. We shall explain the important reasons for using money in decision analysis, and we shall illustrate its usefulness in assigning values to the alternatives of the party problem introduced in the last chapter. We shall also discuss the use of money in determining the value of information.

10.2 MONEY AS A VALUE MEASURE

There are several important reasons for introducing money as a value measure in decision analysis. First, money is **familiar**. We have a lifetime of experience in using a monetary measure in valuation. We have developed an internal scale that allows us to assess the value of things on a common basis. When you see a pen, for example, you can often tell whether it is a one dollar pen or a hundred dollar pen. When you say it is a hundred dollar pen, most people will understand that there must be something special about it to justify an unusually high value. We are, therefore, much more confident about our assessments when we use the concept of money in the process.

A second, equally important reason for using money is that it is both **fungible** (meaning there is no preference for any unit of this measure over another unit) and **divisible** (meaning it can be divided into smaller units as practically necessary). Are there other value measures besides money? Some might suggest time or time saved by some alternative. The problem with time as a value measure is that units of time are not fungible. If someone says an alternative can save you one hour, you would want to know which hour—an hour of your busiest time or an hour you might have spent relaxing.

A third reason for introducing money as a value measure is that it is **alienable**, meaning we can transfer it to other people in exchange for goods or service. In fact, our whole discussion in Chapter 3 of Personal Indifferent Buying Price (PIBP) and Personal Indifferent Selling Price (PISP) rests on the notion of a *fungible, divisible*, and *alienable* value measure. Since the value of clairvoyance is the PIBP for the information, we can compute the value of information in a decision situation only when we have introduced a value measure. In the party problem as we have framed it so far, there is no way to discern Kim's value of information on the weather.

A final reason for using money is **simplicity**: You can often dramatically simplify the application of the rules. As you will soon see, using money may allow you to assess one curve that can be used to compute all preference probabilities, saving the work of all of the individual assessments.

Some people become uncomfortable when money is introduced into a decision situation that seems distinctly non-monetary, as if discussing money were demeaning. You have heard people say, "money is the root of all evil," but the actual teaching is "the love of money is the root of all evil." Money is just money; a measure of worldly resources. In fact, even Mother Teresa probably would have preferred more money to less, because she would have been able to use it to help more needy people.

10.2.1 Bring On the Wizard

One way we introduce money is to ask the decision maker for his willingness to pay for changes from one prospect to another. To free the decision maker's mind from questions about the likelihood of such changes, we employ the wizard to effect the changes and ask only about his willingness to pay for the wizard's services. We illustrate this procedure in the party problem by determining Kim's value in dollars for each of the party prospects.

KIM'S PARTY PROSPECT VALUES Kim owns her alternatives and her prospects (as discussed in Chapter 1, every decision maker must own his or her alternatives). We therefore can ask Kim what her PISP would be for any potential prospect. Suppose, for example, when we ask her about the outdoor party in the rain, she says that she would sell this prospect for nothing and she would not pay to get rid of it: She is just indifferent between having the outdoor party and not having it. Her PISP for this prospect, then, is zero. She explains that if she ended up with an outdoor party in the rain they would probably call the whole thing off and go to the movies; the whole notion of the party would simply have served as a way to get together.

Next, we ask Kim about the prospect of a sunny outdoor party. How much would we have to pay her to give it up? After some reflection she answers, "$100." In other words, her PISP for the *O–S* party is $100. What she means by this statement is that she is indifferent between having an outdoor party in the sunshine with her present bank account and having no party with $100 more in her bank account. Continuing with the same questions, we find that her PISPs for the remaining prospects are $90 for Porch–Sunny (*P–S*), $50 for Indoor–Rainy (*I–R*), $40 for Indoor–Sunny (*I–S*), and $20 for Porch–Rainy (*P–R*).

PROSPECT DOLLAR VALUES Kim's PISPs for the party prospects are recorded in Figure 10.1. The amounts Kim will pay to change any prospect into any other prospect are the differences in the values of the prospects. Figure 10.2 shows the preferentially ordered list of prospects with columns of preference probabilities and dollar values.

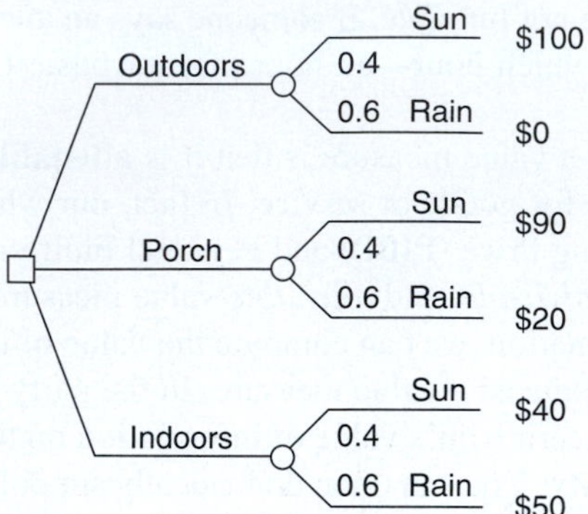

FIGURE 10.1 Kim's Dollar Values of Each Prospect

Note: The higher the value of a prospect, the higher its position in the list in Figure 10.2, and the higher its preference probability.

Prospect	Preference Probability	Dollar Value
Outdoors, Sunshine	1	$100
Porch, Sunshine	0.95	$ 90
Indoors, Rain	0.67	$ 50
Indoors, Sunshine	0.57	$ 40
Porch, Rain	0.32	$ 20
Outdoors, Rain	0	$ 0

FIGURE 10.2 Prospect Preference Probability and Dollar Value Measure

For these six prospects, we can now plot preference probability versus dollar value, as shown in Figure 10.3. We can think of this plot as a transformation from dollars to preference probabilities. If we had had this plot for Kim for the dollar values she assigned for the prospects,

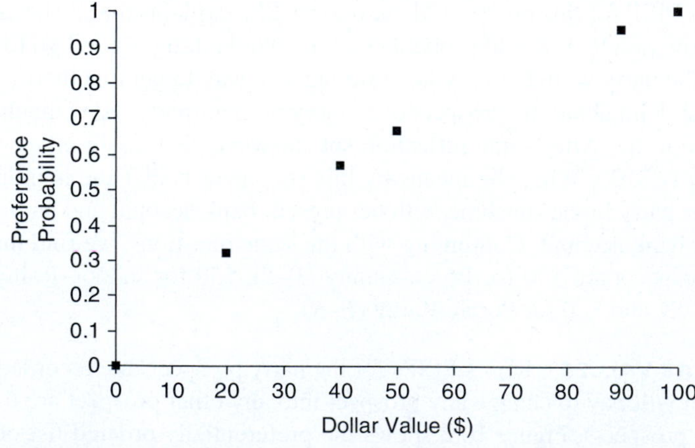

FIGURE 10.3 Plot of Preference Probabilities vs. Dollar Values

we could have used it to obtain the preference probabilities for Figures 9.4, and we could have found the preference probability for each alternative.

Suppose we would like to have more points in this plot to represent more dollar values and more preference probabilities. All we have to do is propose additional prospects to Kim that are intermediate in desirability between the best party and the worst party, and then ask her for both a PISP and the preference probability for each prospect.

For example, suppose Kim is offered a card that would allow her to view movies at her favorite online website for one month. She might have a PISP of $30 for this card and a preference probability of 0.45 between best and worst party. This answer would allow us to plot the corresponding point in Figure 10.3. Since $30 lies between the dollar values of $50 and $40 corresponding to *I–R* and *I–S* in Figure 10.2, consistency would require Kim to state a preference probability for this card between the corresponding preference probability values of 0.67 and 0.57. If Kim states a PISP of $30 and a preference probability for the card between best and worst party of 0.75, for example, then we can alert her to this inconsistency and reassess this preference probability or dollar value.

To develop the whole curve, we would have to repeat this process several times until the curve was complete. We can extend the dollar axis beyond the $0 to $100 range by considering better and worse prospects for the party, like those for the Hayride alternative discussed in the last chapter.

If you are concerned that this process would prove to be exhausting, you should know that in practice we do not have to follow it, as we shall see later in this and following chapters.

10.3 *u*-CURVES

The curve formed by plotting preference probability versus monetary amounts is called a ***u*-curve**. If you can convert the prospects of your decision into equivalent, purely monetary values—that is, establish a PISP for each prospect—the ***u*-curve** summarizes all preference information necessary for you to make your decision. In other words, a *u*-curve gives you preference probabilities, thus doing the work of the equivalence rule.

> *"If any two prospects have the same monetary value, they must have the same preference probability and to be consistent: you must be indifferent between them. Therefore, the u-curve must apply to any monetary prospects for which it is drawn."*

For purposes of making the decision, the vertical scale of Figure 10.3 is irrelevant. If we had multiplied all of the preference probabilities by 10 and then carried out the *e*-operation of Figure 9.8, the preference probabilities for each alternative would merely be multiplied by 10. The one that was largest before the multiplication by 10 still would be largest after the multiplication. The same thing would be true if we added or subtracted any number from all of the preferences probabilities in the plot. The same number would merely be added to or subtracted from the calculated preference probabilities for each alternative; the alternative with the largest preference probability would be unchanged. On the other hand, if (for any reason) you multiplied all of preference probabilities by a negative amount and then carried out the *e*-operation, the preference probabilities for each alternative would be multiplied by the same negative amount, and the best alternative would be the one with the lowest preference probability.

While such transformations would no longer permit the interpretation of the results as preference probabilities, they would still lead to the selection of the correct alternative. It is sometimes numerically convenient to perform such a transformation and to forgo the preference probability interpretation. If at any time it is desired to interpret the numeric results as a preference probability, it is merely a matter of defining a best and a worst prospect and of rescaling. Therefore, we refer to a point on the *u*-curve as the ***u*-value**, rather than a preference probability, keeping in mind that the preference probability interpretation can be obtained immediately by normalizing the *u*-curve to range from zero to one. Think of a plot like Figure 10.3 as a thermometer for taking the temperature of deals. It does not matter whether the thermometer measures temperature in Celsius, Fahrenheit, or any other scale with equal degrees where hotter is higher. So long as you use the same thermometer, you can have confidence that you will always find the hottest (best) deals.

10.3.1 Kim's *u*-curve

After consideration of many prospects, Kim's preference probability plot of Figure 10.3 has been completed in Figure 10.4.

Notice that the plot extends beyond the value 1 and below the value 0; we have forgone the interpretation as preference probability outside the range from \$0 to \$100.

The *u*-curve as plotted is adequate for the following discussion. However, if you wish to confirm our calculations, you should know that Kim's particular *u*-curve miraculously has the form

$$u(x) = \frac{4}{3}\left[1 - \left(\frac{1}{2}\right)^{\frac{x}{50}}\right].$$

This means that to find the *u*-value corresponding to any dollar measure *x*, we raise 1/2 to the power *x* divided by 50, subtract the result from 1, and then multiply by 4/3. Once we have Kim's *u*-curve and her dollar valuation on prospects, we use the *u*-curve to obtain the *u*-value of each prospect

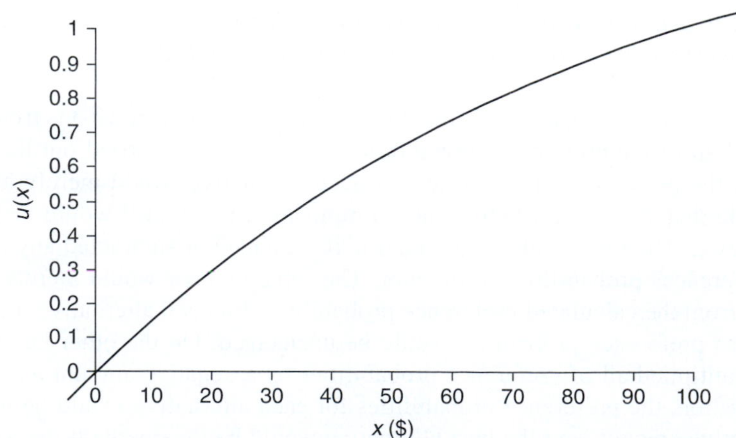

FIGURE 10.4 Kim's *u*-curve: Preference Probabilities vs. Dollar Values

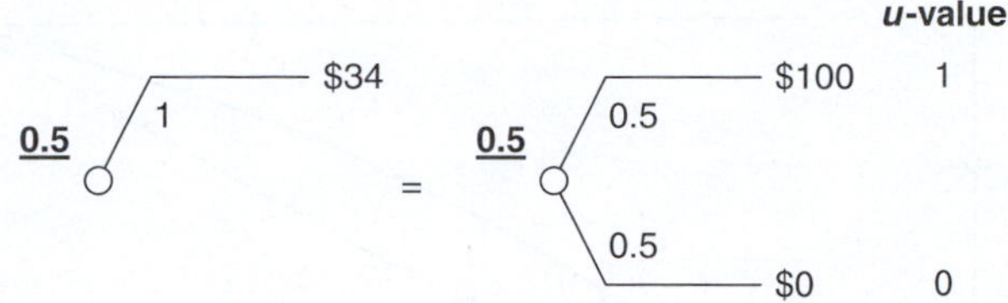

FIGURE 10.5 Kim's Certain Equivalent for a 50–50 Chance of $100

and then find the best alternative for her by finding the alternative with the highest *e*-value of the *u*-value. It is clear that this procedure will result in the Indoor alternative in the party problem.

Your u-curve works for all decisions whose prospects are described by a value measure.

As we said, your *u*-curve can be used to obtain your certain equivalent for any monetary deal that is within its scope. In addition, a *u*-curve only changes infrequently, and then usually due to a major change in your wealth. For example, suppose Kim is offered a deal which she assesses as a 50–50 chance of winning $100 as opposed to nothing. What should her certain equivalent in dollars for this deal be? Once we have her *u*-curve from Figure 10.4, we can answer this question without consulting her.

We know that her certain equivalent will be the amount of money she would have to receive for sure to be indifferent between that prospect and the deal with a 50 percent chance of winning $100 and a 50 percent chance of winning nothing. We know that if she is to be indifferent, the certain equivalent and the deal must have the same preference probability or ***u*-value**. The ***u*-value of the deal** is the *e*-value of the *u*-values of its prospects. The calculation is shown in Figure 10.5.

As recorded in Figure 10.5, the *u*-curve of Figure 10.4 shows that the prospect $100 has a *u*-value of 1; the prospect of $0 has a *u*-value of 0. The *e*-value of these *u*-values is $0.5 \times 1 + 0.5 \times 0 = 0.5$, which we refer to as the *u-value of the deal*. If she is to be indifferent to this deal and its certain equivalent, the certain equivalent must also have a *u*-value of 0.5. By referring to the *u*-curve of Figure 10.4, we find that a *u*-value of 0.5 corresponds to a dollar value of about $34. Consequently, the monetary amount $34 and the uncertain deal both have the same *u*-value of 0.5. This means that $34 is Kim's certain equivalent for the 50–50 chance at $100.

Even if you multiplied her *u*-curve by a negative sign, if you are consistent in using her *u*-curve and its inverse, you still get the correct certain equivalent calculation. In the extreme, suppose we measured Kim's *u*-curve and the vertical scale got washed off in the rain. No problem: We can mark off any equal-division vertical scale and calculate the correct certain equivalent for each alternative. The alternative with the highest certain equivalent is the one consistent with the five rules.

10.3.2 Graphical Interpretation

When a deal has only two possible monetary payoffs, we can find its certain equivalent from the *u*-curve by using a simple construction that implements this set of calculations. First, we draw a straight line between the two points on the *u*-curve corresponding to the *u*-values of the payoffs. Next, we find the point on this line, such that the distance from the lower *u*-value to this point divided by the total length of the line is equal to the probability of the larger payoff. Finally, we

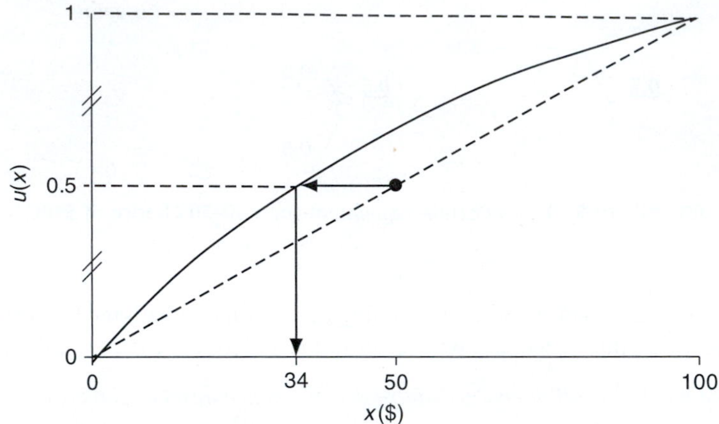

FIGURE 10.6 Graphical Construction to Find Certain Equivalent

move horizontally to the left from this point until we meet the u-curve and drop down to read the certain equivalent. Figure 10.6 illustrates the procedure for the 50–50 chance at $100.

The dashed straight line connects the points on the u-curve corresponding to 0 and 100 dollars. Since the probability of winning the higher amount of $100 is 1/2 starting from the lower point, we proceed 1/2 the way along the line and locate the point shown as a large dot. The height of this point is the u-value of the deal. Then we move left from this point until we encounter the u-curve and read off the corresponding certain equivalent of $34 on the dollar scale.

The u-curve also answers other questions about Kim's preferences, such as her stated preference probability for a deal paying two different dollar amounts with a specified certain equivalent between them. Suppose Kim has a choice between receiving a sure $50 or a deal that will give her $100 with probability p and nothing with probability 1-p. What probability p would make her indifferent? We calculate it from her u-curve, since the u-value of $50 must be equal to the u-value of the deal, which is in turn p times the u-value of $100 (1) plus 1-$p$ times the u-value of $0 (0), or simply p. Her preference probability is the u-value of $50 from the curve or 2/3. Kim would require two to one odds of winning to be indifferent.

The u-curve also allows us to compute the certain equivalent of deals with any number of prospects. For example, suppose Kim is considering a deal that pays her $100 with probability 0.2; $50 with probability 0.3, and $0 with probability 0.5.

The u-values of 100, 50, and 0 are 1, 2/3, and 0 respectively, and so the u-value for this deal is

$$u\text{-value} = 1 \times 0.2 + \frac{2}{3} \times 0.3 + 0 \times 0.5 = 0.4$$

The certain equivalent for this deal is the dollar value that has a u-value of 0.4 in Figure 10.4, $25.73. We can also state the Kim is indifferent between receiving this three-prospect deal and receiving a two-prospect deal of $100 and $0 with probability 0.4 of winning, since both deals have the same u-value of 0.4.

10.3.3 Using Kim's *u*-curve in the Party Problem

Suppose Kim already had had her u-curve when she first faced the party problem. How would her analysis have changed? Most likely, she first would have created the decision tree of Figure 10.1 to represent her alternatives, weather probabilities, and prospect values. Next, she would use her

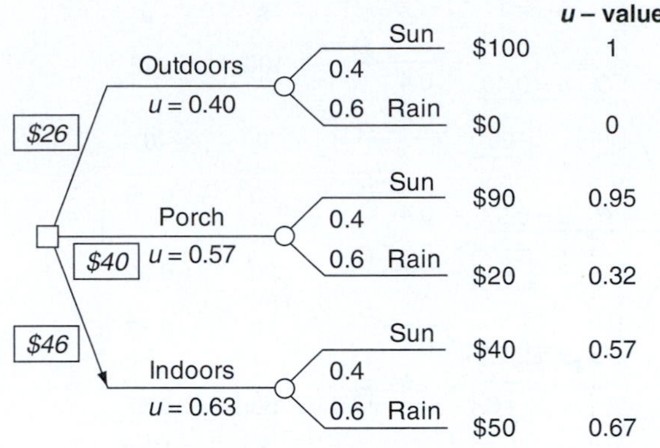

FIGURE 10.7 Kim's Certain Equivalents for Party Problem Alternatives

u-curve to find the *u*-value of each prospect, as shown in Figure 10.7. The *e*-values of the *u*-values for the Outdoor, Porch, and Indoor alternatives correspond to the *u*-values of 0.40, 0.57, and 0.63. By referring to the *u*-curve of Figure 10.4, we find that the dollar certain equivalents corresponding to these *u*-values are \$26, \$40, and \$46 (rounded to the nearest dollar), respectively. These certain equivalents are recorded in boxes adjacent to each alternative.

Kim would be just indifferent between following the Outdoor alternative and receiving \$26; to following the Porch alternative and receiving \$40; and to following the Indoor alternative and receiving \$46. These certain equivalents give us a feeling for her strength of preference among the three alternatives.

Since Kim will always prefer more money to less (she can always give it away), Kim's *u*-curve must increase as dollars increase. This means that the higher her *u*-value for an alternative, the higher her certain equivalent for that alternative. Consequently, she can determine the best alternative to follow either by choosing the one with the highest preference probability—*u*-value—or by choosing the alternative with the highest certain equivalent.

10.4 VALUING CLAIRVOYANCE

Suppose that at this juncture in Kim's saga, the clairvoyant arrived with an offer of help. Recall from our introduction in Chapter 2 that the clairvoyant is competent and trustworthy—he knows the future weather, and he will tell the truth if we engage him.

However, unlike the wizard, he cannot change anything. The clairvoyant offers to tell whether the weather will be Sunny (*S*) or Rainy (*R*). Notice that he does not describe the weather at the time of the party in some general sense but only in terms of a distinction that meets the clarity test.

The good news is that the clairvoyant has arrived; the bad news is that he wants to be paid. His fee for his service is \$15. Should Kim hire him? The question is whether she would be better off by doing the best she can without him or by hiring him for \$15 and then deciding on the party location with the knowledge of whether *S* or *R* will occur. We represent Kim's decision on buying clairvoyance at a price of \$15 by the decision tree of Figure 10.8.

The upper part of the tree represents the situation where Kim does not buy clairvoyance. She then faces the same choices for party location with the same consequences shown in Figure 10.7. Her best alternative is to have the party indoors with a *u*-value of 0.63 and a certain equivalent of \$46.

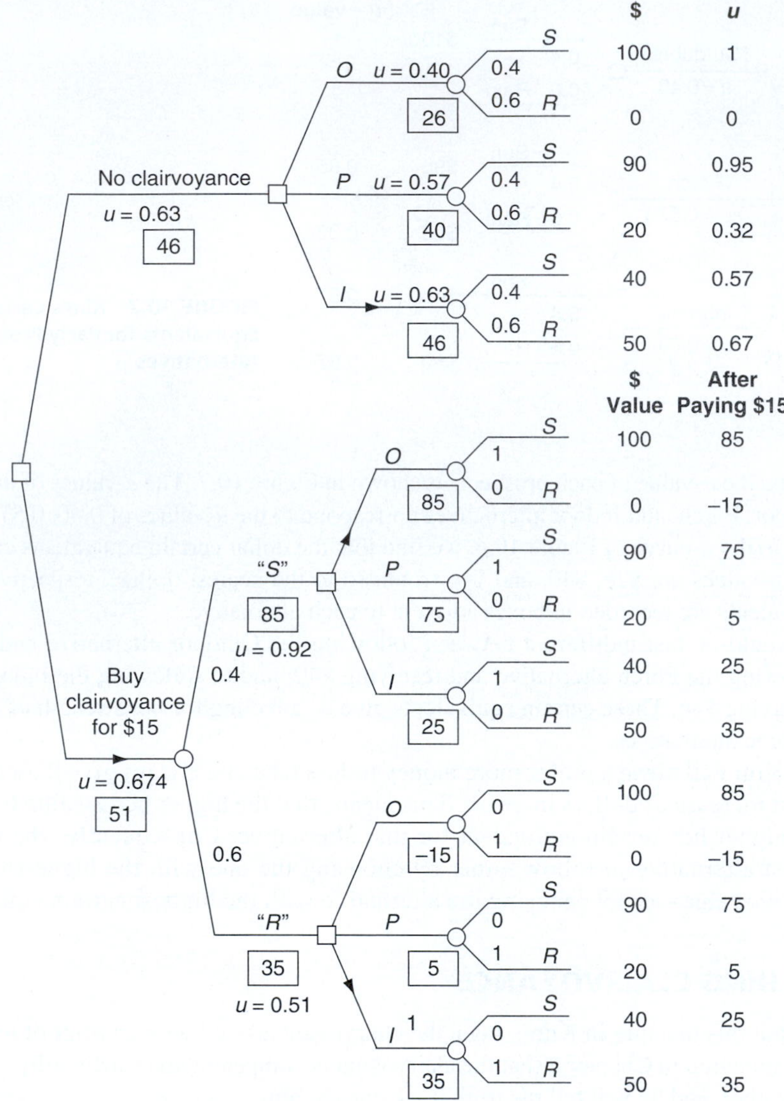

FIGURE 10.8 Determining the Value of Clairvoyance

The lower portion of the tree in Figure 10.8 represents the clairvoyance deal at a price of \$15. If she takes this deal, the first thing that will happen—after committing herself to the payment of \$15—is receiving the report of the clairvoyant on what will happen. We let "*S*" represent the clairvoyant's report that *S* will occur and "*R*" be his report that it will not. We are uncertain about what he will say because we are uncertain about the weather. Since he will report "*S*" when and only when *S* will occur, "*S*" and *S* represent the same distinction and must be assigned the same probability; in this case, 0.4, with 0.6 probability assigned to "*R*." Once the clairvoyant's report is known, Kim will be able to choose a party location: Outdoors, Porch, or Indoors. Then the weather will actually turn out to be sunny (*S*) or rainy (*R*), and at the end of the tree, we show the measure she has attached to each of these prospects. The dollar measure of each prospect is computed by subtracting the \$15 that must be paid to the clairvoyant from the measures that we have been using for each prospect. Thus, for example, if she has an outdoor party in the sunshine, her value is not $100, as it was before; due to the clairvoyant's payment, her value is now $85. All the dollar values at the end of the tree have been reduced by $15 to reflect this commitment.

We have now completed the structure of the tree and the assignments it requires, except for noting that when the clairvoyant says "*S*" of course the weather will be *S* with probability 1 and that when he says "*R*" it will be *R* with probability 1. Therefore, when the clairvoyant reports *S*, Kim is looking forward to Sunshine and must select between party locations that will pay her the equivalent of $85, $75, and $25 for the locations Outdoors, Porch, and Indoors. Clearly, in this case, she will choose to have her party Outdoors and obtain the $85 payoff. This is shown by the arrow on the Outdoor branch of the location decision following the report "*S*."

If the clairvoyant reports "*R*," she faces rainy weather and must choose among the three locations that will then pay –$15, $5, and $35. Again, she will choose the one that gives her the highest payoff; she will have the party indoors.

We thus see that the alternative of purchasing clairvoyance at a price of $15 reduces to selecting a deal with a 0.4 chance of an $85 payoff and a 0.6 chance of a $35 payoff. From the *u*-curve of Figure 10.4, the *u*-values of $85 and $35 are 0.92 and 0.51, as shown adjacent to the "*S*" and "*R*" branches. The *u*-value of buying the clairvoyance is then $0.4 \times 0.92 + 0.6 \times 0.51$ or 0.674. From the *u*-curve, we find that the certain equivalent of this alternative is $51.

Thus, we have determined that buying the clairvoyance at this price is a better deal for Kim than her original deal without it, because the *u*-value of 0.674 is higher than the *u*-value of 0.63, or equivalently, the certain equivalent of the clairvoyant's deal is $51 compared to the certain equivalent of $46 from proceeding without clairvoyance. Kim would be well advised to buy the clairvoyance at a price of $15.

10.4.1 Material Information

We define **material information** as information that has the possibility of changing your decision. For example, Kim might end up changing her course of action based on the clairvoyant's report. If he reports that the weather will be rainy, then she will have an indoor party just as she would have had if she had never consulted him. If, on the other hand, the clairvoyant's report is that the weather will be sunny, then she will change from an indoor party to an outdoor party.

The clairvoyant's information is *material* since the *possibility* exists that the information may change her decision. In this case, there is a 40% chance that she will change her course of action as the result of engaging the clairvoyant. If no possible report of information will change your decision, then that has no value in the context of this decision and is **immaterial information**. The value of clairvoyance on any immaterial distinction is zero.

10.4.2 Kim's Value of Clairvoyance

We have found that Kim should buy clairvoyance at a price of $15, but what is the value of clairvoyance—the amount she would have to pay for it so that she would be indifferent to buying or not buying? The value of clairvoyance is the threshold-buying price for clairvoyance: Her PIBP for the clairvoyant's services.

To compute the value of clairvoyance, we must increase the cost of clairvoyance until Kim is just indifferent between buying it and not. Thus, in the tree of Figure 10.8, we would consider successively larger values of the cost of clairvoyance, such as $16, $17, and $18, and determine whether it was still a good idea to buy clairvoyance. When we have increased the price to the point where the certain equivalent of the alternative of purchasing clairvoyance is equal to the certain equivalent of not purchasing it, then we will have determined her PIBP of clairvoyance. You may wish to verify that at a price of $20 Kim will find that she does not care whether or not she buys clairvoyance, since she will face the same $46 certain equivalent regardless of whether she buys it.

We observe that the increase in price from $15 to $20 required to find the value of clairvoyance is just equal to the difference in the certain equivalent of the clairvoyance and no clairvoyance alternatives of Figure 10.8, which is $51 less $46. While this result is true in our example, it is not generally true for any *u*-curve that a person might actually have. The only general way to compute the value of clairvoyance is to increase its price iteratively until the point of indifference between buying it and not buying it is reached. We shall soon discuss the special conditions that the *u*-curve must satisfy if we are to be able to compute the value of clairvoyance simply by computing the certain equivalent of the deal with free clairvoyance and then subtracting the certain equivalent of the deal without clairvoyance.

10.4.3 Graphical Procedures

We can graphically pursue the issue of valuing clairvoyance, as shown in Figure 10.9. If Kim pays $15 for clairvoyance as we have shown, she will face a deal with a 40% chance of an $85 payoff and a 60% chance of a $35 payoff. To compute the certain equivalent of this deal, we connect the points on the *u*-curve corresponding to $35 and $85 by the dashed straight line. Next, we locate a point that is 40% of the way up this line and move left to the point *C*. Point *C* represents the *u*-value of clairvoyance at this price on the *u* scale, namely 0.674, and the certain equivalent of clairvoyance at this price on the dollar scale, namely $51. Since the point marked *I* for indoor alternative is otherwise her best choice and this point has a *u*-value of 0.63 and a certain equivalent of $46, the clairvoyant's deal is clearly superior. This merely confirms what we found in Figure 10.8.

To visualize computing the value of clairvoyance from Figure 10.9, we imagine moving the two ends of the dashed line by equal dollar amounts to the left and observing the motion of the point *C*. We see that, as the dashed line moves down the *u*-curve in this way, the point *C* will slide down until for some total payment to the clairvoyant the *u*-value with clairvoyance will coincide with the *u*-value of the Indoor alternative. A careful construction will show that this point will be reached when the total payment for clairvoyance is about $20.

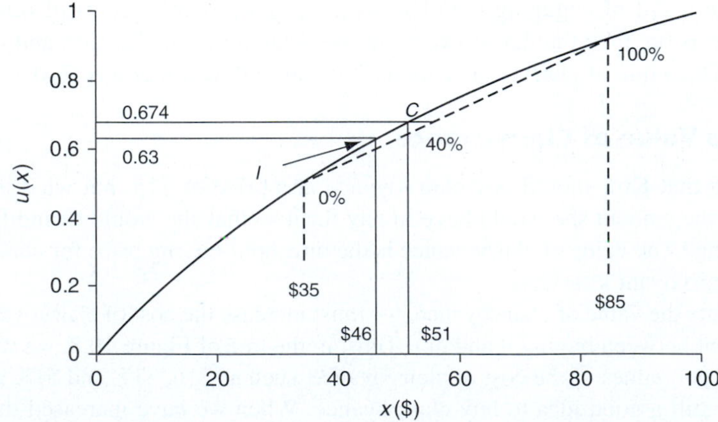

FIGURE 10.9 Graphing the Value of Clairvoyance

10.4.4 Interpretation

The value of computing the value of clairvoyance does not rest in any way on the existence of an entity like the clairvoyant. Its value comes from quickly giving you intuition about where to focus your attention in a decision problem. It can help you identify information gathering activities that you may want to explore. More importantly, it helps you identify information-gathering activities that would be a waste of your time and money.

The value of clairvoyance represents the most that a decision maker should pay for *any* kind of information on an uncertain distinction. No device, person, survey, or other information-gathering process possibly can be worth more than the value of clairvoyance. Knowing the value of clairvoyance, the decision maker has a benchmark against which to compare any information-gathering scheme. If the cost of the scheme exceeds the value of clairvoyance, there is no need to examine the scheme in any further detail. In Kim's case, therefore, no source of information about the distinction *S*/*R* for weather in the party problem could be worth more than $20.

10.4.5 Can the Value of Clairvoyance be Negative?

The value of clairvoyance may be zero, but it can never be negative in any decision. Regardless of what the clairvoyant reports, you would choose the same alternative that you would have chosen without the aid of the clairvoyant and then the value of clairvoyance will be zero. However, if for the least one possible report you will be able to improve your deal by choosing a different alternative, the report will be material and clairvoyance will have a positive value.

Notice that we prefaced these comments by saying "in any decision." In other circumstances, you may not want to hear what the clairvoyant has to say. For example, if you were reading a wonderful mystery novel and the clairvoyant (or a friend) offered to reveal the murderer, you would likely decline the offer because the premature solution of the mystery would diminish your enjoyment of the book. One of the longest running plays in London is an excellent mystery, *The Mousetrap*, written by Agatha Christie. It is said that if you do not tip the taxi driver who takes you to the theater to see the play, he might say, "Look out for (name of character) who plays a very special role."

You will have a negative value for reports on any activity where finding the answer is what gives you pleasure. As an example, if you like crossword puzzles and you turned to one in an airline magazine only to find it completely filled out, you would not say "Lucky me!" Rather, you would feel disappointed that you would not have the opportunity to solve it yourself.

A couple expecting a child might wish to be surprised by whether it will be a boy or a girl. However, if they were more concerned about being able to decorate the baby's room appropriately before the birth, they would have a positive value for knowing the sex of the child.

10.5 JANE'S PARTY PROBLEM

To illustrate how another person might deal with a decision problem about the party, let us consider it from the point of view of Kim's friend Jane. We will assume that Jane is similar to Kim in many ways. She faces the same party problem, and has the same information about the weather, leading her to assign a probability of 0.4 to the event *S*. However, Jane's preferences with respect to party prospects may be different from Kim's.

Since Jane has also agreed to follow our five Rules of Actional Thought, we begin to check her preferences by asking her to order the six prospects for the party. She produces the same ordering as Kim did with an outdoor party in the sunshine *O*–*S* being the best prospect and an outdoor party in the rain *O*–*R* being the worst. Then, using her agreement to follow the equivalence rule, we ask her to assign to each of the four intermediate prospects the preference

Prospect	Jane's Preference Probability
Outdoors, Sunshine	1
Porch, Sunshine	0.90
Indoors, Rain	0.50
Indoors, Sunshine	0.40
Porch, Rain	0.20
Outdoors, Rain	0

FIGURE 10.10 Jane's Preference Probabilities

probability for the best prospect versus the worst to which she is indifferent, just as we did for Kim in Figure 9.4. The result is shown in Figure 10.10.

We observe that Jane's preference probabilities are different from Kim's. For example, while Kim said that a porch party in the sunshine was equivalent to a 95% preference probability of the best party, Jane says that it is only equivalent to a 90% preference probability of the best party. By comparing the two figures, we find that Jane's preference probabilities are always less than Kim's for the same prospect, except, of course, for the best and the worst.

We can determine the effect of these different preference probabilities on the choice that Jane would make for party location. We do this by constructing the decision tree of Figure 10.11 and by recording at the endpoint representing each prospect Jane's preference probability for that prospect as a measure.

Then, following our usual procedure, we find the preference probability for each of the location alternatives by computing the *e*-value of this measure for each alternative. For the Outdoor alternative, we record $0.4 \times 1 + 0.6 \times 0 = 0.40$; while for the Porch and Indoor alternatives, we record 0.48 and 0.46. Note that the highest preference probability is the 0.48 for the Porch alternative. Therefore, by the choice rule, Jane will choose to have her party on the porch. Recall that Kim's best decision was to have the party indoors. Thus, the effect of the difference between Jane's and Kim's preference probabilities for prospects is to cause Jane to follow a different alternative for the party location.

Just like Kim, Jane would like to have the advantages of using a monetary measure. Therefore, we determine Jane's PISP for each party prospect. We find that Jane's PISP in every case is exactly equal to Kim's, so that the dollar measures she assigns to each prospect are

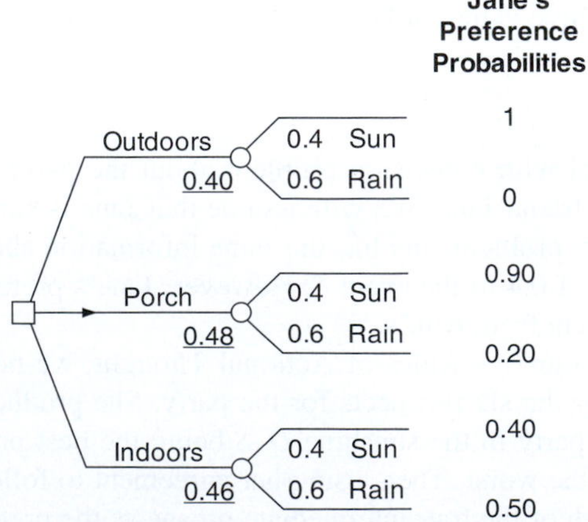

FIGURE 10.11 Jane's Decision Tree

Prospect	Jane's Preference Probability	Jane's Dollar Value
Outdoors, Sunshine	1	$100
Porch, Sunshine	0.90	$90
Indoors, Rain	0.50	$50
Indoors, Sunshine	0.40	$40
Porch, Rain	0.20	$20
Outdoors, Rain	0	$0

FIGURE 10.12 Jane's Preference Probabilities and Dollar Values

exactly the same as Kim's. We record Jane's preference probabilities and corresponding dollar values of each prospect in the table shown in Figure 10.12, just as we did for Kim in Figure 10.2.

We note that Jane's dollar values are directly proportional to her preference probabilities; in fact, they are just the preference probabilities multiplied by 100. This means that if we make a plot of preference probability versus dollar value for Jane, all the points will lie on a straight line. If we assume that this relationship will apply for any point, then we can plot the *u*-curve for Jane as the straight line shown in Figure 10.13.

Kim's *u*-curve is also shown for convenience and comparison. We can use this *u*-curve for Jane to answer the same type of questions we explored using Kim's *u*-curve. For example, to find Jane's certain equivalent for a 50–50 chance at $100 versus nothing, we would proceed as shown in Figure 10.14.

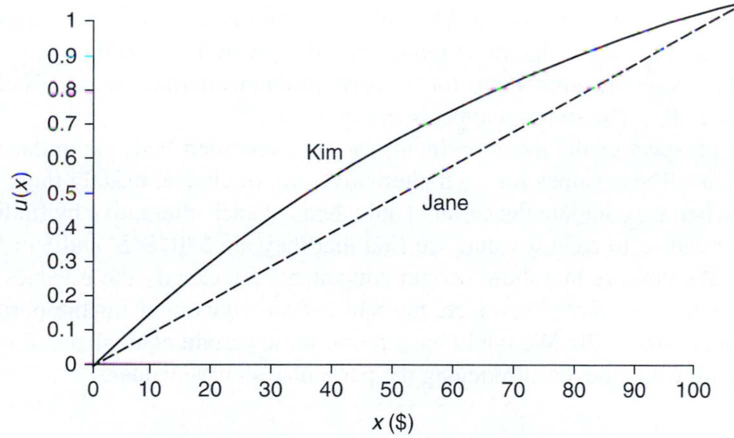

FIGURE 10.13 Kim's and Jane's *u*-Curves

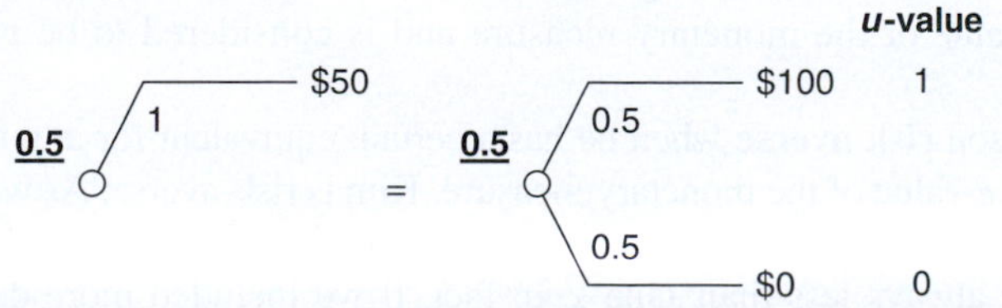

FIGURE 10.14 Jane's Certain Equivalent for 50–50 Chance of $100

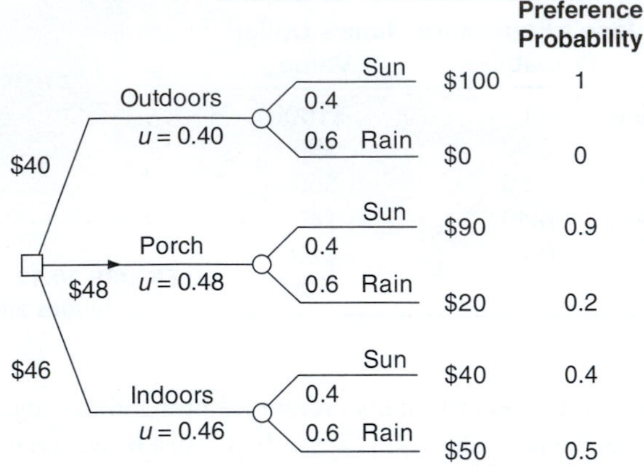

FIGURE 10.15 Jane's Certain Equivalents for Party Problem Alternatives

We know that for Jane to be indifferent between the deal and the certain equivalent, they both must have the same *u*-value. On the right side of the figure, we compute the *u*-value of the deal by first finding the *u*-value of its prospects \$100 and \$0 as 1 and 0 by consulting Jane's *u*-curve. Then, we find the *u*-value of the deal by finding the *e*-value of these *u*-values using $0.5 \times 1 + 0.5 \times 0 = 0.5$. The certain equivalent, therefore, must also have a *u*-value of 0.5. By examining Jane's *u*-curve, we find that \$50 has a *u*-value of 0.5. Therefore, \$50 is her certain equivalent of a 50–50 chance at \$100. We observe that Jane's certain equivalent for the deal is, in fact, equal to the *e*-value of its monetary values with $0.5 \times 100 + 0.5 \times 0$.

To find Jane's certain equivalents for the party problem alternatives, we proceed as we did for Kim in Figure 10.7. The solution appears as Figure 10.15.

For each prospect of the party problem, we have recorded both the dollar value and its associated *u*-value. The *u*-values for each alternative are, of course, exactly those computed in Figure 10.11. When we compute the certain equivalents of each alternative by finding the dollar amounts corresponding to each *u*-value, we find that they are \$40, \$48, and \$46 for *O*, *P*, and *I*, respectively. We observe that these certain equivalents are exactly the *e*-values of the dollar measures for each alternative. Therefore, the \$48 certain equivalent for the porch alternative equals $0.4 \times 90 + 0.6 \times 20$. We could have found these certain equivalents directly from the monetary measure without ever introducing the particular *u*-curve values.

10.6 ATTITUDES TOWARD RISK

A person with a *u*-curve that is a straight line will have a certain equivalent for any monetary deal that is the *e*-value of the monetary measure and is considered to be **risk-neutral**. Jane is risk-neutral.

We call a person **risk averse** when he has a certain equivalent for any monetary deal that is always less than the *e*-value of the monetary measure. Kim is risk-averse. As we saw in Figure 10.7, Kim's certain equivalents for the three party locations Outdoors, Porch, and Indoors are \$26, \$40, and \$46, which are always less than Jane's. In fact, if we included more decimal places in our calculation of the certain equivalent for the Indoor alternative, we would see than Kim's certain

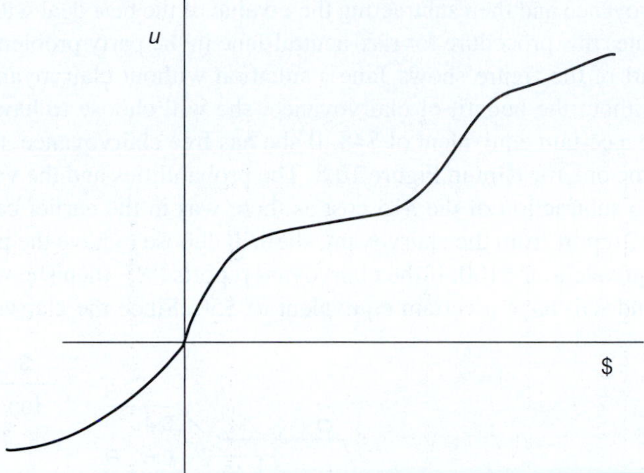

FIGURE 10.16 A Possible *u*-Curve Permissible by the Rules

equivalent for the Indoor alternative is $45.83, which is slightly less than Jane's at exactly $46. Since Kim's certain equivalents are always less than Jane's for the same deals we know in this situation, we can call Kim risk-averse. In fact, since her curve shown in Figure 10.4 is always concave downwards, Kim's certain equivalents always will be less as long as uncertainty is present.

If a person had a *u*-curve that was always concave upward, that person's certain equivalents always would be greater than the *e*-values of the monetary measure. We would call such a person **risk preferring**.

Every rulesperson must be able to create a personal *u*-curve to describe his or her risk preference for uncertain monetary deals. The only requirement is that the *u*-value increase as the value measure increases. Figure 10.16 shows a possible *u*-curve permissible under the rules. The curve continually increases as the value measure increases. Notice that as it bends it exhibits risk aversion in some regions and risk preference in other regions. By requiring other properties of our *u*-curve such that it never is risk preferring, we can try to eliminate such perverse behavior.

10.6.1 Computational Advantages of Risk Neutrality

A person who is risk-neutral for a monetary deal derives a considerable practical advantage in being able to compute certain equivalents as *e*-values of money. Note that if we increase the payoffs of the uncertain deal in Figure 10.14 by $10 to $110 and $10, their *u*-values would increase to 1.1 and 0.1, and yield a deal with *e*-value of *u*-values of $0.5 \times 1.1 + 0.5 \times 0.1$ or 0.6. The corresponding certain equivalent would be $60, which is the *e*-value of the payoffs and an increase of $10 over the original. Changing all the payoffs up or down by the same amount causes the same change in the certain equivalent.

This property is especially useful when we wish to compute the value of clairvoyance for a risk neutral person. Recall that to compute the value of clairvoyance in general, we had to subtract larger and larger costs of clairvoyance from all of the prospect measures until the person was indifferent between buying the clairvoyance at that price and not buying it. In the case of a risk-neutral person, we know that the certain equivalent for any deal will be its *e*-value of the monetary measure and, therefore, that any common addition or subtraction adjustment to the measure will cause the adjustment in the certain equivalent. The practical importance of this is that we can find the value of clairvoyance for risk-neutral persons by computing the *e*-value of the clairvoyance deal with free clairvoyance and then subtracting the *e*-value of the best deal without clairvoyance. Figure 10.17 illustrates this procedure for risk-neutral Jane in the party problem.

The upper part of the figure shows Jane's situation without clairvoyance, as determined in Figure 10.15. Without the benefit of clairvoyance, she will choose to have the party on the porch and will have a certain equivalent of $48. If she has free clairvoyance, the structure of the tree is exactly like the one for Kim in Figure 10.8. The probabilities and the values are identical; however, there is no subtraction of the $15 cost as there was in the earlier case. We see that if Jane receives the "*S*" report from the clairvoyant, she will choose to have the party Outdoors and achieve a certain equivalent of $100. If the clairvoyant reports "*R*," then she will choose to have the party Indoors and will have a certain equivalent of $50. Since the clairvoyant will say "*S*"

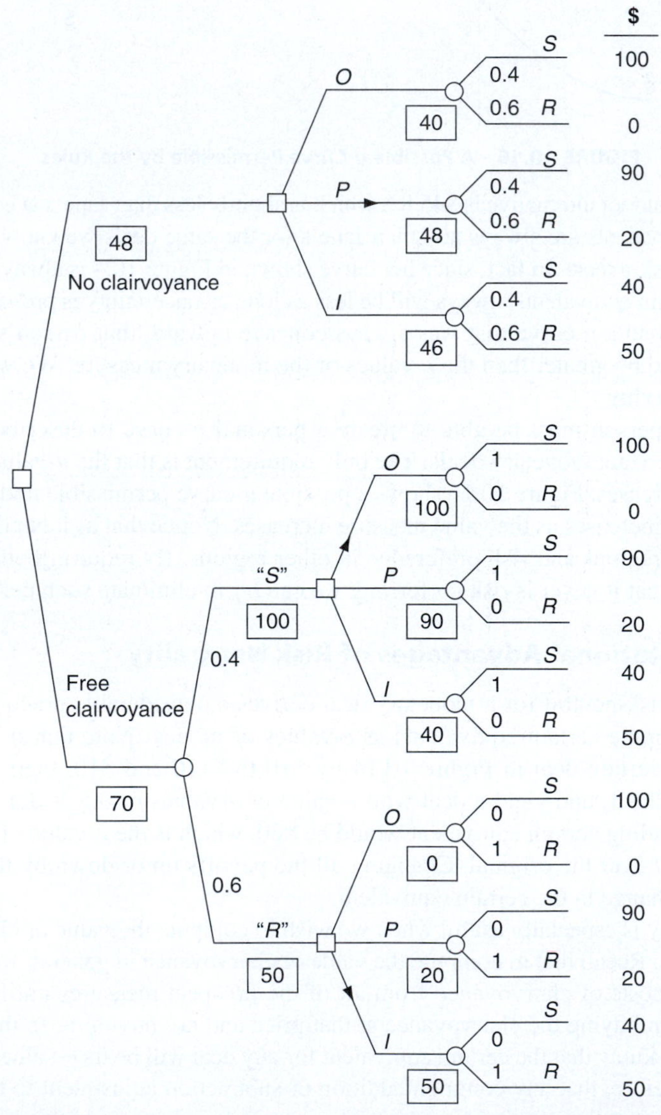

FIGURE 10.17 Determining Jane's Value of Free Clairvoyance on S/R

with probability 0.4 and "R" with probability 0.6, her certain equivalent with free clairvoyance is $0.4 \times 100 + 0.6 \times 50$ or \$70. We know that if she had to pay \$70 − \$48 or \$22 for clairvoyance, then her certain equivalent would be \$48 regardless of whether she bought clairvoyance. Therefore, Jane's value of clairvoyance on the distinction S/R in the party problem is \$22.

Note that either clairvoyant statement will change Jane's initial decision to have the party on the Porch; both reports are material. As we have seen, it is much simpler to calculate the value of clairvoyance without iteration. We might ask ourselves whether this simplification is available only for a risk-neutral person. The answer is no, as we shall soon see.

10.7 MARY'S PARTY PROBLEM

We shall now consider Mary, who is another friend of both Kim and Jane. With respect to the party, she shares certain preferences with each of them and assigns the same weather probabilities that they do. Mary has the same preference probabilities for party prospects as does Kim; namely, those shown in Figure 9.4. However, she has the same risk-neutral u-curve for money as does Jane in Figure 10.13. This means that her dollar values will be proportional to her preference probabilities as were Jane's in Figure 10.12, but they will be different dollar values from the ones shared by Kim and Jane.

Mary's preference probabilities and PISP dollar values appear in Figure 10.18. Although Mary's risk attitude toward money is different from Kim's, she makes exactly the same decision in this situation as does Kim because she has the same preference probabilities. In fact, the operations performed for Kim in Figures 9.6 and 9.7 are indistinguishable from those for Mary: Both women will choose to have the party indoors. As we can see from these figures, Kim and Mary will make the same decision on party location for any probability of S as long as they agree on that probability. Therefore, we see that we cannot tell whether a person is risk-neutral or risk-averse by observing the decision made in any situation where prospects are not described by a value measure.

When will Mary's actions differ from Kim's? They will differ when the clairvoyant arrives. We compute Mary's value of clairvoyance on S using the decision tree in Figure 10.19 just as we did for Jane in Figure 10.16. Since Mary is risk-neutral, we can use the same simplification of first finding the e-value of the deal without clairvoyance and then subtracting that from the e-value of the deal with free clairvoyance.

In the upper part of the decision tree, we review her decision without clairvoyance. Since she is risk neutral like Jane, she enters her dollar values for each prospect and computes the e-value of each alternative as its certain equivalent. The Indoor alternative has the highest certain equivalent of \$63; she will take the action we predicted.

Prospect	Mary's Preference Probability	Mary's Dollar Value
Outdoors, Sunshine	1	\$100
Porch, Sunshine	0.95	\$ 90
Indoors, Rain	0.67	\$ 50
Indoors, Sunshine	0.57	\$ 40
Porch, Rain	0.32	\$ 20
Outdoors, Rain	0	\$ 0

FIGURE 10.18 Mary's Preference Probabilities and Dollar Values

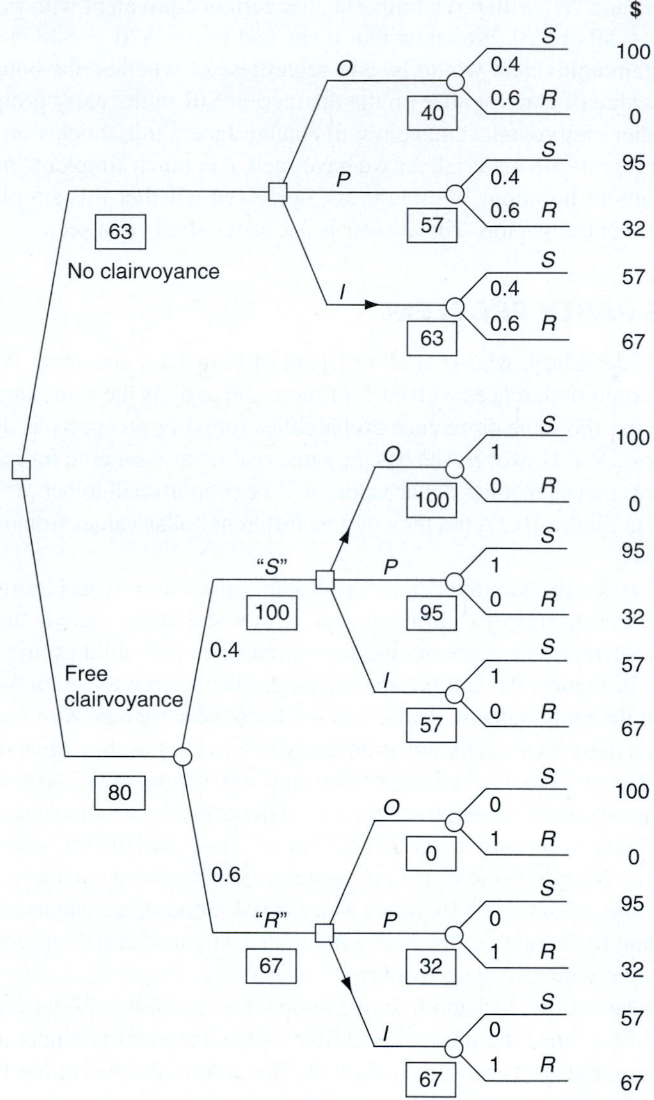

FIGURE 10.19 Determining Mary's Value of Free Clairvoyance on Weather

In the lower part of the tree, we compute Mary's value of free clairvoyance. Like Jane and Kim, she will have the party outdoors when the clairvoyant says "*S*" and indoors when he says "*R*." As both Kim and Jane had done, she assigns a \$100 value to the report "*S*;" however, unlike both Kim and Jane, the value she assigns to the report "*R*" is \$67. Her certain equivalent for the party with free clairvoyance is then $0.4 + 100 + 0.6 \times 67$ or \$80. Since Mary is risk-neutral, we can compute her value of clairvoyance on *S* by subtracting her certain equivalent without clairvoyance from her certain equivalent with free clairvoyance as $\$80 - \$63 = \$17$.

We have thus found that Kim, Jane, and Mary will each pay different amounts for clairvoyance: \$20, \$22, and \$17, respectively. These differences arise solely from differences in preference—not from differences in alternatives or information.

10.8 SUMMARY

You do not need money to determine the best decision alternative. However, there are many advantages to using money as a value measure in decision analysis:

1. You can calculate a certain equivalent for each alternative.
2. You can calculate the indifference amount required to give up one alternative for another.
3. You can calculate the value of information on one or more uncertainties.

The concepts of *u*-values and *u*-curves are used to compile the preference information used to make a decision under uncertainty, we discussed.

- Calculating certain equivalents of deals using *u*-curves.
- Implications of additive/multiplicative transformations of the *u*-curve.
- Different people may have different values for the same information when facing the same decision, based on their particular *u*-curves.
- Using certain equivalent (in dollars) to select the best decision alternative.
- The value of clairvoyance can be zero, but it is not negative in making decisions.

KEY TERMS

- Alienable
- Divisible
- Fungible
- Immaterial information
- Material information
- Risk-averse
- Risk-neutral
- *u*-curve
- *u*-value
- *u*-value of the deal

PROBLEMS

Problems marked with a dagger (†) are considered quantitative.

1. Which of the following statements about Kim's Indoor alternative are true?
   a. 0.63 is the probability Kim will get her best prospect if she sets up her party indoors
   b. 0.63 is the preference probability of the *e*-value of the monetary prospects
   c. 0.63 is her preference probability for the Indoor deal
   d. 0.63 is the *e*-value of the preference probabilities for the Indoor prospects
      i. a and c iii. b and c
      ii. a and d iv. c and d

2. In general, the value of clairvoyance is defined as
   a. The difference between the certain equivalent of a deal with free clairvoyance and the certain equivalent of that deal without the clairvoyance.
   b. The minimum market selling price for that information.
   c. The decision maker's Personal Indifferent Buying Price (PIBP) for that information.
   d. The decision maker's Personal Indifferent Selling Price (PISP) for that information.

   Which of the above statements is/are true?
      i. d only iii. a and b only
      ii. c only iv. a and c only

3. Samantha owns the given decision opportunity and has the following preference probabilities for deals with $200 as the best outcome and $0 as the worst outcome.

Value ($)	Preference Probability
0	0.00
10	0.17
20	0.32
40	0.57
50	0.67
80	0.89
90	0.95
200	1.00

What is Samantha's certain equivalent for the following decision opportunity?

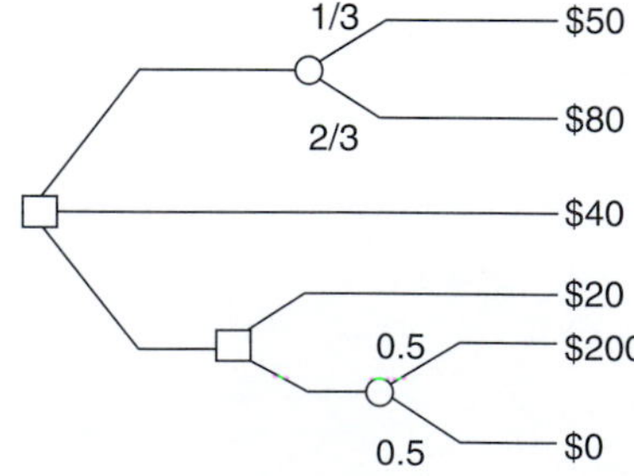

   a. Between $20 and $50
   b. Between $50 and $90
   c. Between $90 and $200
   d. There is not enough information to solve the problem

4. Consider the four statements below.
   I. You should never pay more than the value of clairvoyance for any information gathering process for an uncertain distinction.
   II. The five Rules of Actional Thought prove that computing the *e*-value of the monetary worth of our prospects will always give us the preferred alternative.
   III. A person's preference probability for a given outcome will always be greater than his/her assessed probability for the outcome given clairvoyance.
   IV. The three legs of the decision quality stool are preferences, information, and alternatives.

   Which of the preceding statements is true?
   a. I and III only
   b. I and IV only
   c. II and IV only
   d. I, II, and IV only
5. Sam has just paid $215 for a certificate that allows him to call the flip of an "unfair" coin. If he calls it correctly, he will win $500. If he calls it incorrectly, he will win $0. His PISP for the certificate is $315. The clairvoyant offers to tell him the outcome of the impending coin toss. What is Sam's PIBP for the clairvoyant's information? You do not know anything about Sam's risk attitude.
   a. $185
   b. Need Sam's *u*-curve to determine his PIBP for the information
   c. $285
   d. Need Sam's *u*-curve and the probability he assigns to the "unfair" coin landing Heads (or Tails) to determine his PIBP for the information

For Problems 6, 7, and 8, consider the decision situation given in the following decision tree where the dollar values are winnings and the probabilities are the assessments of the deal's owner. Suppose Stuart, who is risk-neutral, owned the deal below.

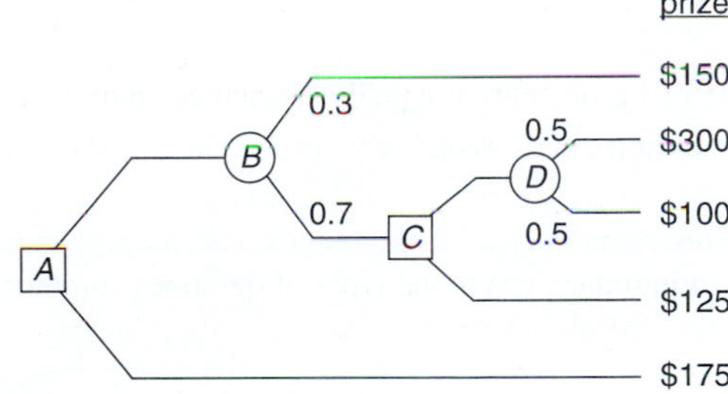

6. What is Stuart's certain equivalent for the deal?
   a. 175 c. 215
   b. 200 d. 185
7. What is the most Stuart should pay for clairvoyance on uncertainty *D* before making decision *A*?
   a. 215 c. 30
   b. 50 d. 185
8. What is the most Stuart should pay for clairvoyance on uncertainty *B* before making decision *A*?
   a. 25.5 c. 192.5
   b. 7.5 d. 185
9. Consider the following five statements.
   a. The difference between the certain equivalent of a deal with free clairvoyance and the certain equivalent of that deal without the clairvoyance represents the value of clairvoyance for any deal.
   b. The value of clairvoyance for any deal is equal to the PISP of clairvoyance for that deal.
   c. The five Rules of Actional Thought prove that it is not logical to be risk seeking.

d. A risk-averse individual never places a lower value on a deal than a risk-neutral individual, assuming they agree on probabilities and the value of the prospects.

How many of the above statements are false?

i. 1 iii. 3
ii. 2 iv. 4

†**10.** Ellen is a risk-neutral test taker who has been studying for her Decision Analysis midterm. The midterm has fifteen questions, and each question has four possible answers. The midterm is different from any she has taken before. Instead of just marking the answer she thinks is correct—*a*, *b*, *c*, or *d*—she assigns a probability to the chance that each answer is correct. The sum of the probabilities she assigns for each question must equal one. The following formula is used to score each question (normalized for a possible 100 points on the exam):

$$\text{Score} = \frac{100}{15} + \frac{100 \times \ln p}{15 \ln 4}$$

where p is the probability she assigns to the correct answer.

Suppose she believes that $\{a \text{ correct} | \&\} = 0.7$ and $\{b \text{ correct} | \&\} = \{c \text{ correct} | \&\} = \{d \text{ correct} | \& = 0.1$. This can be represented as the following deal.

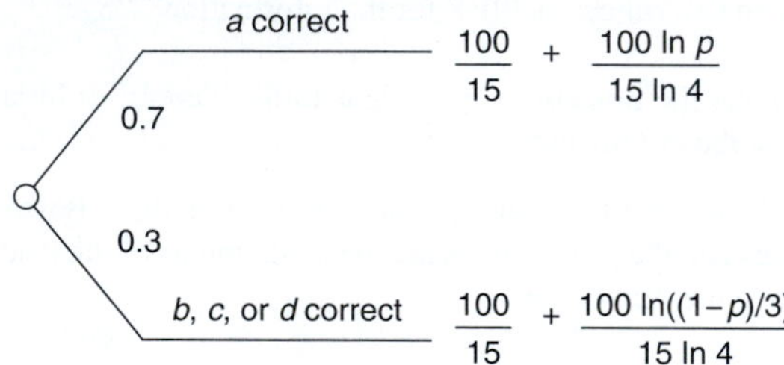

For simplicity, we assume that probability is equally distributed among choices *b*, *c*, and *d*.

If Ellen wants to maximize her *e*-score, what probability, *p*, should she write down for answer (*a*)? (Show your reason.)

Note: This question requires some calculus background whose purpose is to induce you to assign your true belief. It is intended to introduce you to the types of questions commonly seen in probabilistic test taking.

CHAPTER

11

Risk Attitude

CHAPTER CONCEPTS

After reading this chapter, you will be able to explain the following concepts:

- Equality of personal indifferent buying and selling prices of uncertain deals around a cycle of ownership
- The delta property:
  - *u*-curves satisfying the delta property
  - Implications for value of clairvoyance
- Assessing risk odds for a deltaperson
- Risk tolerance and risk aversion coefficient for a deltaperson

11.1 INTRODUCTION

To begin, we refer back to Chapter 10—specifically, to Figure 10.4—where, a *u*-curve displays preference probabilities versus dollar values. Graphing *u*-curves like that shown in Figure 10.4, and showing the dollar values with a range from \$0 to \$100 is possibly misleading. These dollar amounts are supposed to represent **prospects**—that is, our current state of wealth is the starting point, and any additional amounts are to be added. This means that zero dollars, as shown in Figure 11.1, is actually not a value of zero. Instead, it is w_0 whatever our present wealth happens to be. Any additional amount, x, received by the uncertain deal should be added to w_0. We will denote w as the total wealth and denote w_0 as the person's present wealth. Figure 11.1 represents the *u*-curve on total wealth.

11.2 WEALTH RISK ATTITUDE

Everything we have done up to this point is consistent with the *u*-curve on total wealth interpretation. For example, recall for a moment the party problem in Chapter 10. If at the outset Kim's total wealth was \$13,000. We have been using the portion of her *u*-curve on total wealth that is between \$13,000 and \$13,100 to do our calculations.

If a person is given a monetary deal at no cost, we can use the *u*-curve on total wealth to calculate the certain equivalent of total wealth, $\tilde{w}$, that the person will now have—as well as his certain equivalent for the deal, $\tilde{x}$. The present wealth w_0, the certain equivalent of total wealth $\tilde{w}$, and the certain equivalent of the deal $\tilde{x}$ are related by

$$\tilde{w} = \tilde{x} + w_0$$

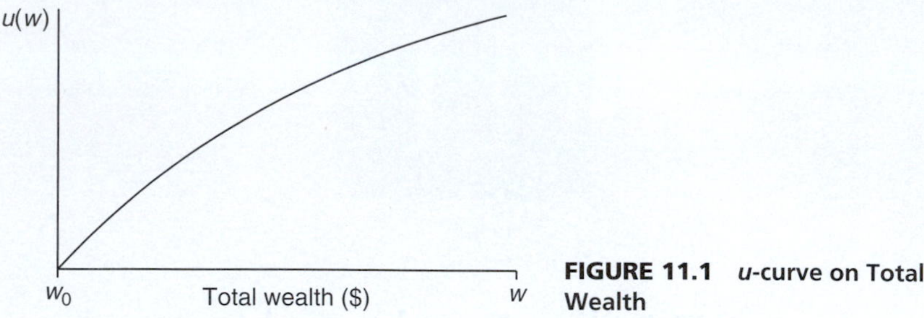

FIGURE 11.1 *u*-curve on Total Wealth

To illustrate the calculation of the certain equivalent $\tilde{x}$, let the uncertain deal x be represented by the possibility of receiving different amounts x_i with corresponding probability p_i. Since the person should be indifferent between the certain equivalent of the deal and the uncertain deal, his u-value of the certain equivalent on total wealth, $u(\tilde{w})$, must equal his e-value of u-values of total wealth that result from the deal. Therefore,

$$u(\tilde{w}) = \sum_i p_i u(w_0 + x_i)$$

The certain equivalent of the deal $\tilde{x}$ is

$$\tilde{x} = \tilde{w} - w_0$$

The amount $\tilde{x}$ is the lowest amount a person would be willing to accept to give up an uncertain deal he currently owns. This is also his Personal Indifferent Selling Price (PISP) for the uncertain deal.

11.3 BUYING AND SELLING A DEAL AROUND A CYCLE OF OWNERSHIP

In Chapter 3, we discussed Personal Indifferent Buying Prices (PIBPs) and Personal Indifferent Selling Prices (PISPs) for deterministic deals (deals with no uncertainty). As we learned, your PIBP of a deal you do not own is equal to your Personal Indifferent Selling Price (PISP) of the deal if you acquired the deal at your PIBP. If you buy a deal you do not own at your PIBP and sell it instantaneously at your PISP, then this transaction comprises a **cycle of ownership**.

Similarly, your PISP of a deal you own is equal to your PIBP of the deal if you sold it for your PISP and are now thinking of buying it back. This transaction also comprises a cycle of ownership. Now, we show that even in uncertain deals, the notion of a cycle of ownership still applies.

Case 1: Selling, then Buying, a Deal You Own

In addition to a wealth w_0, suppose a person also owns an uncertain deal and then sells it. In this case, let the PISP for the uncertain deal be s^*. This situation is shown in Figure 11.2. The left-hand side of Figure 11.2 shows the possible outcomes the person will get from the uncertain deal if he does not sell it with his wealth, w_0, added to each outcome. The right-hand side shows his wealth and a deterministic amount, s^*. If he is indifferent between the two states, then s^* is his PISP for the uncertain deal. Note that the PISP for an owned deal is the same as the certain equivalent $\tilde{x}$ for the deal defined in the preceding section. Similar to the case of a deterministic deal, the PISP might change with the wealth level.

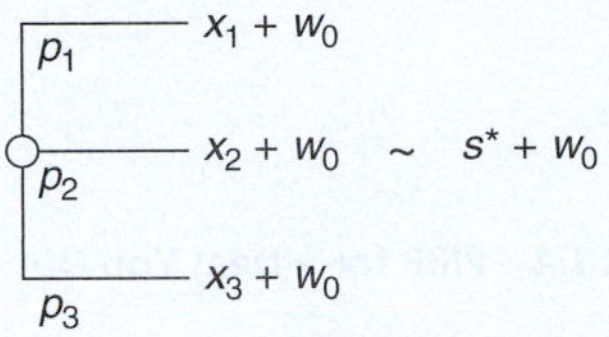

$$\begin{array}{l} p_1 \quad x_1 + w_0 \\ p_2 \quad x_2 + w_0 \quad \sim \quad s^* + w_0 \\ p_3 \quad x_3 + w_0 \end{array}$$

FIGURE 11.2 PISP for an Uncertain Deal You Own

Since the person is indifferent between the two cases, his *u*-values for both cases must be the same. Equating the *u*-values of both sides of Figure 11.2 gives

$$\sum_i p_i u(x_i + w_o) = u(s^* + w_o)$$

After selling the deal for s^*, the PIBP at which he would buy it back, b^*, would be defined at the same instant by the relation shown in Figure 11.3. The left-hand side of the figure shows his state with the deterministic amount $w_0 + s^*$, which is his state after selling the deal for the PISP. The right hand side of the figure shows the possible outcomes when he buys back the deal for an amount b^*. If he is indifferent between the two states, then b^* is his PIBP for the deal at the wealth level $w_0 + s^*$.

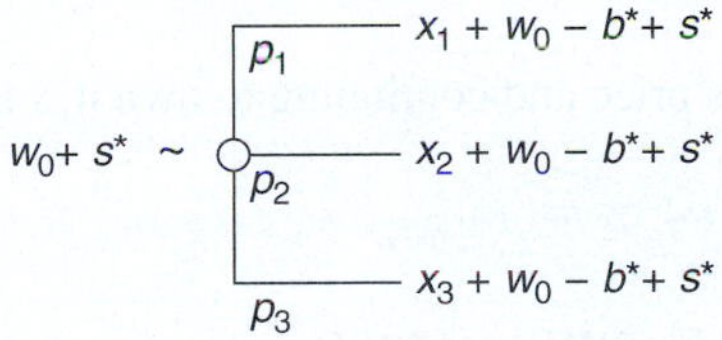

$$\begin{array}{l} \qquad\qquad\quad p_1 \quad x_1 + w_0 - b^* + s^* \\ w_0 + s^* \sim \quad p_2 \quad x_2 + w_0 - b^* + s^* \\ \qquad\qquad\quad p_3 \quad x_3 + w_0 - b^* + s^* \end{array}$$

FIGURE 11.3 Buying the Deal Again

Equating the *u*-values for both sides of Figure 11.3 gives

$$u(w_0 + s^*) = \sum_i p_i u(w_0 + s^* - b^* + x_i).$$

The only solution to this equation is $s^* = b^*$: equality of the PISP to the subsequent PIBP.

> *"If a person owns an uncertain deal, the personal indifferent selling (PISP) must be equal to the personal indifferent buying price (PIBP) at which he is just willing to buy it back after he sells it."*

Note *This result does not depend on the u-curve of the person facing this deal.*

This result is equivalent to the concept of a cycle of ownership for deterministic deals that we presented in Chapter 3. We have now illustrated that it also applies to uncertain deals, and it applies to any *u*-curve. Next, we see that it also applies in the reverse direction if the person first buys a deal he does not own and then sells it back.

Case 2: Buying, Then Selling, a Deal You Do Not Own

Suppose that another man with wealth w_0 does not own the deal x but is thinking of buying it. What is his PIBP, b, for it? Since when he pays b he is indifferent between buying the deal at that price and not having it, this situation is represented by Figure 11.4.

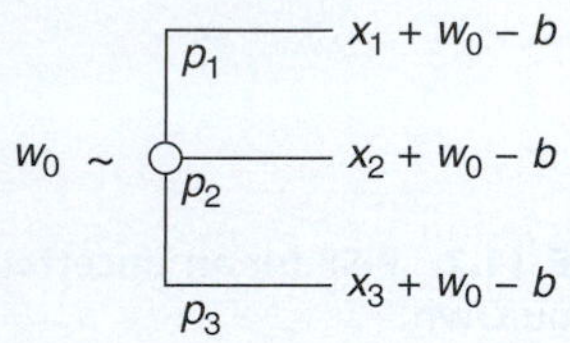

p_1 — $x_1 + w_0 - b$

w_0 ~ p_2 — $x_2 + w_0 - b$

p_3 — $x_3 + w_0 - b$

FIGURE 11.4 PIBP for a Deal You Do Not Own

If the person is indifferent between the two sides of Figure 11.4, his u-values for both cases must be the same. Equating the u-values of both sides of the figure results in

$$u(w_0) = \sum_i p_i u(w_0 + x_i - b)$$

Therefore, solving this equation will determine b, which is his PIBP for the deal. Note that b does not need to equal b^* of the previous case, because the wealth state is different when he is buying it. (In the previous case, the wealth was $w_0 + s^*$, and in this case, the wealth state is w_0.)

Suppose now that having bought the deal for this PIBP, the man wants to figure out his PISP for the deal. Since selling at his PISP should bring him back to the initial state of not owning the deal at wealth w_0, we have a cycle of ownership. However, now there is no uncertainty, since he will always have $w_0 + s - b$. This is why in Figure 11.5 all prospects are the same on the left-hand side of the tree.

Since he is indifferent between selling it at this price and continuing to own it, s is defined by

$$\sum_i p_i u(w_0 + s - b) = u(w_0)$$

The solution to this equation is $s = b$; equality of the PIBP and PISP.

The amount for which a person is willing to sell the uncertain deal he just bought (his PISP) must be equal to the amount he was just indifferent to paying for it (his PIBP) at that instant in time. This is a logical tautology, since he is returning to the state of not having the deal after paying and then receiving the same amount of money.

Once again, $b = s$ around this cycle of ownership, and once again, this applies to any u-curve.

IDENTIFYING TWO DIFFERENT STATES The two cycles of ownership that we have just discussed take place in different worlds. In the first world, the man has wealth w_0 and also owns a deal. The PIBP and PISP were b^* and s^*, respectively, and $b^* = s^*$. In the second world, the man has wealth w_0, but he does not own the deal. The resulting PIBP and PISP in this state are b and s respectively with $b = s$.

We now pose the following question.

"For any wealth level w_0 and any uncertain deal, is it necessary that $s = s^*$ and $b = b^*$?"

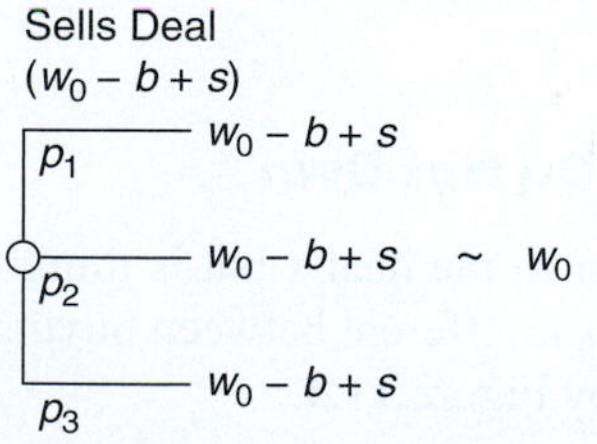

Sells Deal
($w_0 - b + s$)

p_1 — $w_0 - b + s$

p_2 — $w_0 - b + s$ ~ w_0

p_3 — $w_0 - b + s$

FIGURE 11.5 PISP for the Deal after Purchase

The answer, as we might expect, is no. Although the PIBP and PISP must be the same around each cycle of ownership, it does not follow that the buying prices or selling prices need to be the same in both worlds with different wealth levels. It is very possible that $s \neq s^*$ and $b \neq b^*$. The price at which he is just willing to buy something if he owns a similar deal does not need to equal the price he is willing to buy it if he does not own a similar deal. Therefore, your PIBP for a deal you do not own (when you have initial wealth w_0) is not necessarily equal to your PISP for the same deal if you owned it in addition to your initial wealth of w_0. This result also corresponds to the discussion of the Rolls-Royce example in Chapter 3, where your PIBP for a Rolls-Royce is not necessarily the same as your PISP for the Rolls-Royce if you got it for free, since wealth can change your cycle of ownership.

If a person was to have $s = s^*$ and $b = b^*$ for all uncertain deals and all wealth levels, what would have to be true about the person's preferences? In the next section, we will explore this question.

11.4 THE DELTA PROPERTY

Suppose you own a deal x and someone offers to increase each and every payoff x_i by the same dollar amount Δ. What should happen to your certain equivalent (PISP) for the deal?

For example, in Chapter 10, we discovered that Kim's certain equivalent for a 50–50 chance at \$0 versus \$100 was \$34. Imagine now that someone sweetens the deal so that at worst she will receive \$20 and at best she will receive \$120. In other words, she is guaranteed to receive \$20 more than the previous case. For the revised deal, what should Kim's certain equivalent be?

Many would say that her certain equivalent should increase by the same amount that the payoffs have increased—namely, by \$20—to a new certain equivalent of \$54. The idea is that the \$20 is "money in the bank" for which there is no uncertainty. If this is the case, we say Kim follows the "delta property" for this \$20 range. The delta property is illustrated graphically in Figure 11.6.

On the left side of the figure, the person has expressed a certain equivalent $\tilde{x}$ for a three branch deal. On the right side of the figure, the payoffs have changed by Δ (positive or negative), and the certain equivalent has changed by the same amount. If this describes the person's wishes for any monetary deal within a range of payoffs, then we say that the person satisfies the **delta property** within that range. Within the range where it is valid, we shall call such a person a *deltaperson*.

The delta property makes the person's certain equivalent for any deal independent of the person's initial wealth: Adding the same amount to the payoffs of any deal he owns will increase his wealth certain equivalent by that amount. This is an important property that you may want to think about in your valuation of uncertain deals.

The delta property allows us to remove the initial wealth, w_0, from our calculations of buying and selling prices for a deltaperson, since the actual value of w_0 does not matter, and we can consider it to be zero.

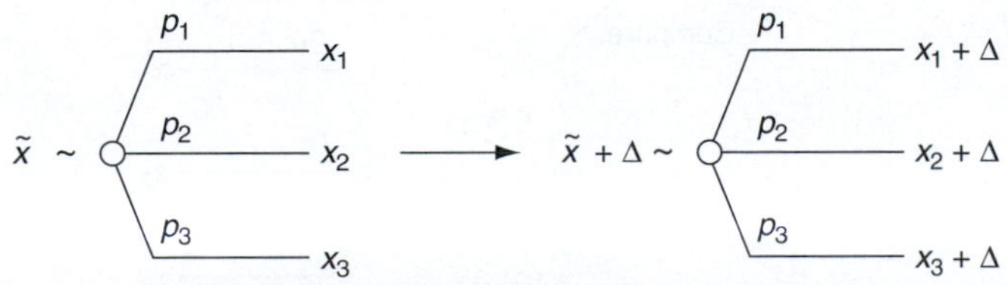

p_1 — x_1; p_2 — x_2; p_3 — x_3: $\tilde{x} \sim$ deal $\longrightarrow$ $\tilde{x} + \Delta \sim$ deal with p_1 — $x_1 + \Delta$; p_2 — $x_2 + \Delta$; p_3 — $x_3 + \Delta$

FIGURE 11.6 The Delta Property

Therefore, if a person follows the delta property over a range of payoffs, then for any wealth level w_0 and any uncertain deal he faces along this range, we should expect that $s = s^*$ and $b = b^*$. We will illustrate this further in the next section.

For some range of payoffs, the delta property is very attractive to almost everyone. However, when Δ is very large and positive, some people say that possession of the deal changes their attitude toward risk so that they are now willing to assess certain equivalents that are much closer to the *e*-value of payoffs. When Δ is large and negative, others are equally concerned that they will not be sufficiently risk-averse.

You may find it useful to think of the delta property as a convenient approximation to any *u*-curve for a sufficiently small range of payoffs. For some people, this range may in fact be very large. On the other hand, a risk-neutral person will automatically satisfy the delta property.

11.4.1 Equality of Buying and Selling Prices

Accepting the delta property has strong implications for decision making about monetary deals. To illustrate, consider the three-branch deal shown in Figure 11.7, representing buying prices for an unowned deal and selling prices for a deal you already own. For a deltaperson, we know the valuation of the deals is independent of the initial wealth. Therefore, we have removed the initial wealth w_0 from all sides in the figure.

On the left side of Figure 11.7, a woman owns the deal and establishes a PISP $\tilde{x} = s^*$ using the idea that she is indifferent between having s^* for sure and having the payoff from the deal. On the right side, she does not own the deal and establishes a PIBP, b, by finding the amount she would pay for the deal before she would be indifferent to owning it. If she satisfies the delta property, then we know that if we add b to all of the payoffs on the top right, the resulting deal shown on the bottom right must have certain equivalent, b, since the certain equivalent must change by the same amount as the payoffs.

However, we already know from the top left that this deal has certain equivalent s^*; therefore, $s^* = b$. Thus, we have discovered that if a person satisfies the delta property, her selling price for an owned deal will be the same as her buying price for an unowned deal. We now understand part of what needs to be true about a person's preferences so that $s^* = b$. We

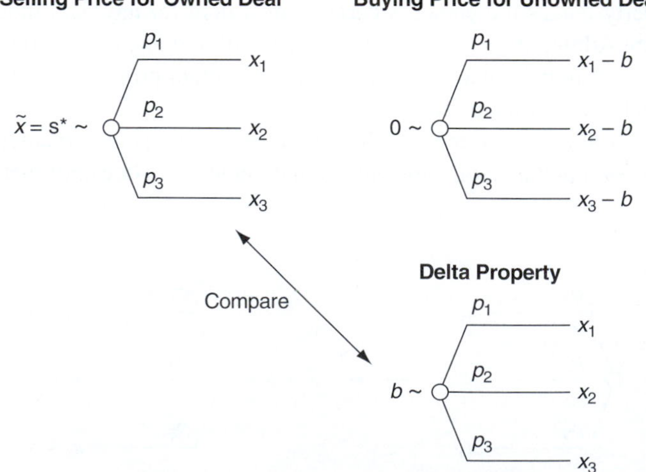

FIGURE 11.7 Delta Property Creates Equal Buying and Selling Prices

now know the person needs to satisfy the delta property, but we still do not know the person's u-curve.

Let's now summarize what we have learned so far. We know that, around a cycle of ownership at a particular instant, whether we are talking about shirts or uncertain deals, PIBP and PISP must be the same ($s = b$ and $s^* = b^*$). For any u-curve, the equality of buying and selling prices for a deal around a cycle of ownership will be true.

In general, however, it will not be true that your buying price for an uncertain deal you do not own will be equal to your selling price for the same deal if you owned it. This will be true only if you satisfy the delta property.

11.4.2 Implications for Value of Clairvoyance

Someone who satisfies the delta property will have a major simplification when computing his value of clairvoyance. The certain equivalent of a deal for a deltaperson will undergo the same change made to all the payoffs of the deal. This means that a deltaperson can find the certain equivalent of a decision with free clairvoyance and subtract from it the certain equivalent of the decision without clairvoyance to find the price of clairvoyance that would lead to indifference—*the value of clairvoyance*.

Value of clairvoyance for a deltaperson =
Value of decision with free clairvoyance − Value of decision with no clairvoyance

This also means that a deltaperson will have no need to use the iterative or graphical techniques for evaluating the value of clairvoyance that we discussed earlier.

This simplification is of such great practical importance that, in many cases, it is wise to assume that the delta property holds exactly when it is even close to being acceptable.

At this point, we know that a person who satisfies the delta property must:

- Have the same buying and selling prices for an uncertain monetary deal, regardless of ownership, and
- His value of clairvoyance on any uncertainty is equal to his value of the decision with free clairvoyance less his value of the decision with no clairvoyance.

However, we still do not know the form of his u-curve.

11.4.3 Forms of *u*-Curve Required by Delta Property

The delta property puts stringent requirements on the shape of the u-curve. If the person has wealth w_0 then the implication of Figure 11.5 is that for any deal and any Δ. Thus,

$$u(w_0 + \tilde{x} + \Delta) = \sum_i p_i u(w_0 + x_i + \Delta)$$

We shall now discuss the only two utility functions that satisfy this equation.

1. STRAIGHT LINE Suppose that for any monetary amount y, the u-curve has the form of a straight line,

$$u(y) = a + by$$

where a and b are constants with b positive, so that increasing y causes increasing u. Then,

$$a + b(w_0 + \widetilde{x} + \Delta) = \sum_i p_i[a + b(w_0 + x_i + \Delta)]$$

and since $\sum_i p_i = 1$, we find,

$$\widetilde{x} = \sum_i p_i x_i = < x|\& > = e\text{-value of dollars}$$

Therefore, a person who is risk neutral will satisfy the delta property, and his certain equivalent for any uncertain deal is equal to the e-value of the dollar value measure, regardless of the person's wealth w_0. Recall from Chapter 10 that Jane and Mary, both of whom were risk-neutral, satisfied the delta property.

Are there other forms for the u-curve that satisfy the delta property? Yes. By methods that lie outside our scope, it can be shown that there is only one other form of u-curve that satisfies the delta property, as we discuss next.

2. EXPONENTIAL The exponential form of u-curve is the only other u-curve besides the linear u-curve that will satisfy the delta property. The exponential u-curve has the form

$$u(y) = a - br^y$$

where a, b, and r are constants and y is the wealth . If the u-curve increases with wealth, y, then certain conditions are imposed on the parameter values: if $r < 1, b$ must be negative, and if $r > 1, b$ must be positive. In the case of $r = 1$, we change the form to the linear case discussed in the previous section. We shall see that risk averse deltapersons have $r > 1$.

Note that the whole exponential u-curve is conveniently described by a single number, r. The u-curve we used for Kim was in the exponential form, so

$$u(x) = \frac{4}{3}\left[1 - \left(\frac{1}{2}\right)^{\frac{x}{50}}\right] = a - br^{-x}$$

Here, $a = b = \frac{4}{3}$ and $r = \left(\frac{1}{2}\right)^{-\frac{1}{50}} = 2^{\left(\frac{1}{50}\right)} = 1.014$. Therefore, Kim satisfies the delta property.

In the previous chapter, our calculations did not include Kim's, Jane's, or Mary's initial wealth, but now we know that since they were delta people, all of our calculations were valid. This is also why their value of clairvoyance was equal to their value of the deal with free clairvoyance minus their value of the deal with no clairvoyance. Given we now know that Kim is a deltaperson, future calculations involving her case will be much simpler.

11.5 RISK ODDS

It is useful to provide a direct interpretation for the constant r in the exponential form discussed in the previous section. This will help with assessing its value directly to determine the u-curve and will provide a better understanding of its magnitude.

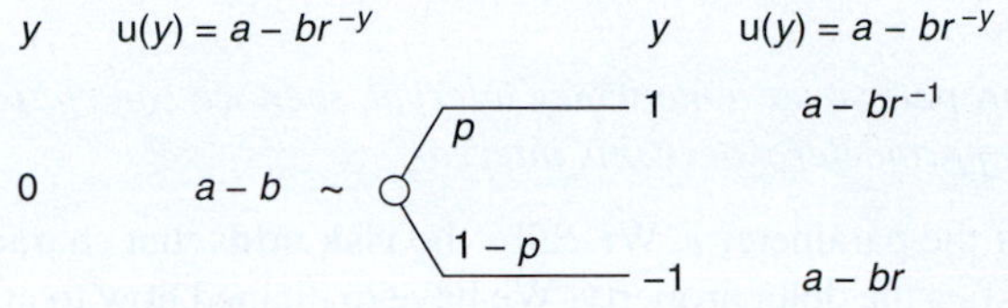

y	$u(y) = a - br^{-y}$		y	$u(y) = a - br^{-y}$
		p	1	$a - br^{-1}$
0	$a - b$ ~			
		$1 - p$	−1	$a - br$

FIGURE 11.8 Indifference to Deal Involving Winning or Losing One Monetary Unit

Suppose that someone satisfying the delta property and, therefore, having a u-curve in this form, faces a deal with a probability p of winning one monetary unit and a probability $1 - p$ of losing one monetary unit, as shown in Figure 11.8.

The monetary unit could be any size, like \$1000. The question is, for a given r, what p will create indifference between accepting the deal and not accepting it. The **Equivalence Rule** requires that any rulesperson must be able to state a preference probability p. The right side of Figure 11.8 shows the monetary values and u-values associated with accepting the deal. The left side shows the monetary values and u-values associated with not accepting the deal (a deal payoff of \$0). To be indifferent, a person must have the same e-value of u-values for accepting and not accepting the deal. The u-value of the left side is $u(0) = a - b$, and the u-value of the right side is the e-value of the u-values. We have

$$a - b = p(a - br^{-1}) + (1 - p)(a - br)$$
$$= a - b[pr^{-1} + (1 - p)r]$$

Now we subtract a and divide by b from both sides to get

$$1 = pr^{-1} + (1 - p)r$$
$$(1 - p)r^2 - r + p = 0$$

Solving for r, we have,

$$r = \frac{1 \pm \sqrt{1 - 4p(1 - p)}}{2(1 - p)} = \frac{1 \pm \sqrt{1 - 4p + 4p^2}}{2(1 - p)}$$
$$= \frac{1 \pm (1 - 2p)}{2(1 - p)}$$

Or r can take the two values of

$$r = 1, \frac{p}{1 - p}$$

The solution $r = 1$ is a limiting case that corresponds to the risk-neutral u-curve we have discussed previously: $u(y) = a - by$.

The solution $r = \dfrac{p}{(1 - p)}$ means that, when r is equal to the odds of winning one monetary unit versus losing one monetary unit, the person is indifferent between accepting and rejecting the deal.

This is an important result:

> *"If a person satisfies the delta property over a monetary interval, then we can characterize his whole u-curve using just one parameter r over this interval."*

We now have an interpretation for the parameter r. We call r the **risk odds** that characterize the risk attitude of a person who satisfies the delta property. We have explained how to assess r using the indifference probability of Figure 11.8 for 1 monetary unit. We now show how to assess r for any monetary unit.

11.5.1 Using Risk Odds

The constant r can be put to immediate use. Suppose, as shown in Figure 11.9, a person faces a deal of winning or losing m monetary units.

What is the probability of winning q that will just create indifference? We proceed, as before, by equating the u-values of both sides. Thus,

$$\begin{aligned} a - b &= q(a - br^{-m}) + (1 - q)(a - br^{m}) \\ &= a - b[qr^{-m} + (1 - q)r^{-m}] \end{aligned}$$

then

$$\begin{gathered} 1 = qr^{-m} + (1 - q)r^{m} \\ (1 - q)r^{2m} - r^{m} + q = 0 \end{gathered}$$

By comparing this equation with our earlier development, we see that the solution for r to the power m must be either $r = 1$ or $r^m = \dfrac{q}{1 - q}$, or since $r = \dfrac{p}{1 - p}$,

$$\frac{q}{1 - q} = \left[\frac{p}{1 - p}\right]^{m}$$

The odds required to be indifferent for stakes m are the same as the odds required to be indifferent for unit stakes raised to the mth power. For an exponential deltaperson, this is true for any value of m. For example, if the stakes were one-half a monetary unit, then the odds on winning for indifference would be just the square root of the risk odds calculated for that monetary unit.

We can now relate the risk odds for different monetary amounts x and y. Let us denote your risk odds for gaining or losing x dollars as $r(x)$, and your risk odds for gaining or losing y dollars as $r(y)$. If you are a deltaperson along this range of dollar values, then

$$r(x) = (r(y))^{\frac{x}{y}}$$

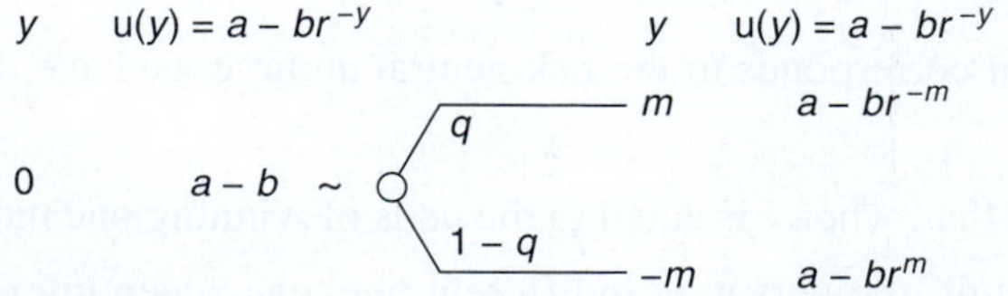

y	$u(y) = a - br^{-y}$			y	$u(y) = a - br^{-y}$
			q	m	$a - br^{-m}$
0	$a - b$	~	○		
			$1 - q$	$-m$	$a - br^{m}$

FIGURE 11.9 Indifference to a Deal of Winning or Losing m Monetary Units

We illustrate this relation further through the following example:

EXAMPLE 11.1 Using Risk Odds

A person satisfies the delta property and has a risk odds for \$100 is $r(100) = 1.5$. As we discussed, this means that he is indifferent between getting \$0, and a binary deal that gives him 0.6 probability of payoff \$100 and 0.4 probability of payoff \$−100.

Note *As with our earlier discussion of odds in Chapter 6, risk odds for \$100 is the ratio* 0.6/0.4 = 1.5.

What can we say about the risk odds of $r(1000)$ for a binary deal with pay offs \$1000 and \$−1000?

SOLUTION

We have

$$r(1000) = \left(r(100)\right)^{\frac{1000}{100}} = (1.5)^{10} = 57.665$$

This means the person would be indifferent to receiving either 0 dollars or a binary deal with a probability $\dfrac{57.665}{1 + 57.665} = .983$ for a payoff of \$1000 and a probability of $1 - 0.983 = 0.017$ for a payout of \$−1000.

Note *Recall from Chapter 6, we obtain the corresponding probabilities from the odds using the relation* $p = \dfrac{r}{1 + r}$.

11.5.2 Kim's Risk Odds

Let us test these ideas using Kim's case, which first was presented in Chapter 10. We have already found that Kim's risk odds for a \$1 monetary unit is

$$r(1) = \left(\frac{1}{2}\right)^{-\frac{1}{50}} = 2^{\frac{1}{50}} = 1.014$$

This means that Kim would have to have 1.014 odds of winning versus losing \$1 before she would be indifferent.

More interesting, however, are the risk odds of $r(100)$ for winning versus losing \$100 that would just make her indifferent.

$$r(100) = \left(2^{\frac{1}{50}}\right)^{100} = 2^2 = 4$$

To become indifferent to winning versus losing \$100, Kim would require 4:1 odds. This means that she would have to have a 0.8 chance of winning and 0.2 chance of losing.

11.5.3 Assessing Risk Odds

We have now found one way to assess Kim's *u*-curve. First, we could ask questions confirming that she wanted to satisfy the delta property for gains and losses in the \$100 range. Next, we could ask her what chance of winning versus losing \$100 would have made her indifferent to playing.

When she answered 80% or 4:1 odds, we could use this answer to construct her u-curve and answer any questions regarding risky choice that she might face in the specified range.

From our discussion, it follows that Kim is just indifferent to 2:1 odds of winning versus losing \$50, so that $r(50) = 2$. Note the relative ease of using this method to establish Kim's u-curve instead of having to determine the equivalence of preference probabilities and dollar amounts, as was described in Chapter 9.

We can use the same method to assess the u-curve for anyone who accepts the delta property. The best way is to select a large enough monetary unit to capture the person's interest. Depending on the person, the unit could be relatively small—like Kim's \$100—or much larger. Next, you determine the odds of winning versus losing that would create their indifference. You can check for consistency by doubling or halving the size of the monetary unit and then repeating the question.

If the person always assigns a probability of $q = 0.5$ in Figure 11.9, his risk odds are always equal to one, since $r(1) = \frac{0.5}{0.5} = 1$. In this case, we are also observing risk neutrality, and we use the u-curve form $u(y) = a - by$. This person values the deal at its e-value for the monetary prospects, since $0.5(1) + 0.5(-1) = 0$.

On the other hand, if the person assigns a probability of $p < 0.5$ in Figure 11.9, then his risk odds must be less than one. For example, if $p = 0.4$, then $r = \frac{0.4}{0.6} = \frac{2}{3}$. This person must also value the deal higher than its e-value for monetary prospects, since he values the deal at 0, but the e-value is $0.4(1) + 0.6(-1) = -0.2$. Therefore, with $p < 0.5$, we are observing risk-preferring behavior.

To summarize, we can determine whether a decision maker is risk preferring, risk averse or risk neutral from the risk odds as

Risk Attitude Relations

$r = 1$	Risk Neutral
$r > 1$	Risk Averse
$r < 1$	Risk Preferring

***u*-CURVE EQUATION FROM THE RISK ODDS** Once we know the risk odds, the expression for the u-curve is

$$u(x) = a - b\big(r(1)\big)^{-x}$$

where x is in dollars.

As we discussed earlier, the parameters a and b are arbitrary, since a linear transformation on the u-values does not change the optimal decision alternative. However, to get an increasing u-curve with x when $r < 1$, b should be positive, and to get an increasing u-curve with x when $r > 1$, b should be negative.

The r and the x values have to be in the same monetary unit. For example, if x is in one thousand dollar units, then we use $r(1000)$. We can, therefore, write a more general formula for the u-curve as

$$u(x) = a - b\big(r(m)\big)^{-x}$$

where x is expressed in units of m dollars.

11.6 DELTA PROPERTY SIMPLIFICATIONS

Now we will look at some specific examples of how the delta property can be used to simplify risk assessment. For these examples, we ask you to recall the party problem presented in Chapter 10.

11.6.1 Kim Loses Some Party Options

Remembering the party problem, imagine that Kim's parents tell her she can no longer have her party outdoors. This shouldn't cause her any problems, since her best alternative was shown to be the Indoor alternative. On the other hand, if her parents ask her not to have the party indoors because of the potential for too much noise, Kim would resort to having the party on the porch. Since her certain equivalent for the Porch party ($46) is $6 less than her certain equivalent for the Indoors party, Kim would be just indifferent to having the party indoors or having the party on the porch and receiving a monetary payment of $6 from her father.

We know this is true because Kim is a deltaperson. If she was not a deltaperson, we would have to draw the tree, add a certain amount of value measure s to each prospect in the Porch alternative, and iterate the value of s until her certain equivalent for the Porch deal with a payment of s is equal to her certain equivalent for the Indoor deal.

11.6.2 Valuing Insurance on the Weather

Suppose Kim is offered insurance on the weather. The insurance company would pay her $50 if it rains, and 0 if it is sunny. Should Kim pay for this insurance? If so, what is the maximum amount, b, that she should pay? To calculate Kim's value of insurance b, we draw the tree of Figure 11.10.

For each value of b, we calculate the certain equivalent of the deal with insurance. The value of b that makes her certain equivalent equal to $46, which is her value of the deal without insurance, is the maximum value she should pay. By repeated iterations, we find this value is equal to $31.35, as shown in Figure 11.11, and given the insurance policy, the best party is the Porch party.

We can also calculate the value of insurance for the particular u-curves of Kim, Mary, and Jane by calculating the value of the party deal with free insurance and subtracting it from the

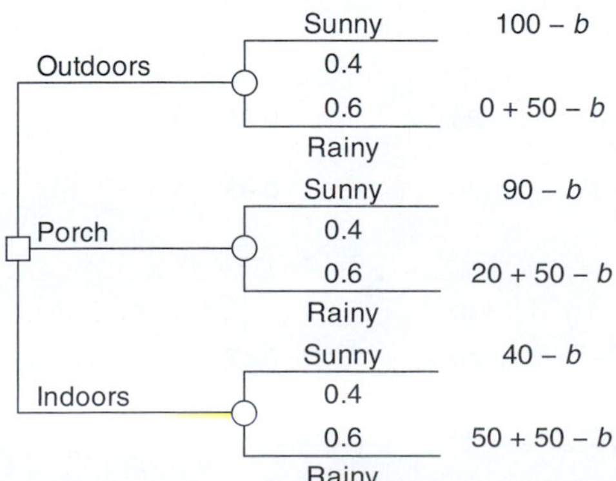

FIGURE 11.10 Kim's Value of Party Insurance

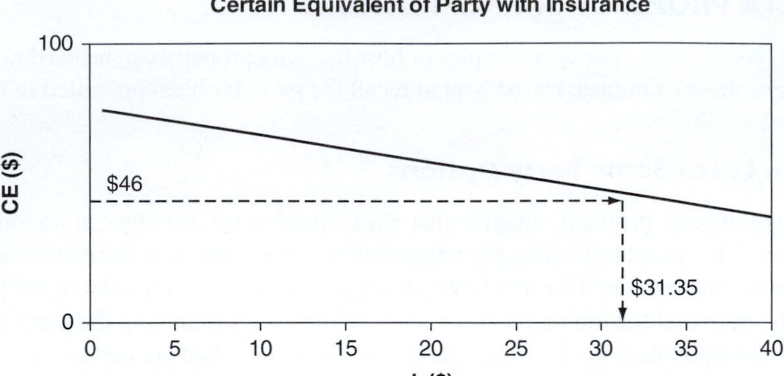

FIGURE 11.11 Kim's Certain Equivalent of Party Deal for *b*

value of the party with no insurance. The tree for the value of free insurance for Kim is shown in Figure 11.12.

In Figure 11.12, we see that the Porch alternative has the highest certain equivalent of \$77.35. Therefore, the value of this insurance policy for Kim is \$77.35 − \$46 = \$31.35.

From the insurance firm's perspective, if they were risk-neutral and also believed that the probability of rain is 0.6, then they are offering a deal with an *e*-value from their own perspective: $-50 \times 0.6 + 0 \times 0.4 = -\$30$.

Furthermore, the insurance company will probably have administrative costs that would make them charge even more than \$31.35 for the insurance certificate. As such, they probably could not offer Kim this insurance certificate for less than her PIBP.

The approach of using iterations to find either the value of clairvoyance or the value of insurance works for all types of *u*-curves.

Alternative	Weather	Probability	Value with Free Insurance ($)	*u*-value
Outdoors (*u*-value = 0.8, CE = \$66.10)	Sunny	0.4	100	1.00
	Rainy	0.6	50	0.67
Porch (*u*-value = 0.87703, CE = \$77.35)	Sunny	0.4	90	0.95
	Rainy	0.6	70	0.83
Indoors (*u*-value = 0.827014, CE = \$69.85)	Sunny	0.4	40	0.57
	Rainy	0.6	100	1.00

FIGURE 11.12 Kim's Party Deal with Free Insurance

11.7 OTHER FORMS OF EXPONENTIAL *u*-CURVE

In decision analysis, there is tremendous usefulness in the exponential risk attitude associated with satisfying the delta property and being a deltaperson. In fact, this usefulness is so significant that in this section we will explore the features of this family of *u*-curves in even greater detail.

We sometimes prefer to write the exponential in forms other than the one we have used here. For example, if we let $r(1) = e^{\gamma}$, we can place the *u*-curve in the form

$$u(x) = a - b(r(1))^{-x} = a - b(e^{\gamma})^{-x} = a - be^{-\gamma x}$$

The quantity

$$\gamma = \ln(r(1))$$

is called the **risk-aversion coefficient**; its unit is the reciprocal of the unit of the value measure. Its reciprocal, ρ, is known as the **risk tolerance** and is given by

$$\rho = \frac{1}{\gamma} = \frac{1}{\ln r(1)}$$

Once again, the risk tolerance is expressed in the same monetary units as the risk odds. If x is in units of thousands of dollars and we use $r(1000)$, then ρ is also expressed in thousands of dollars.

From the previous discussion, we can now summarize the different attitudes towards risk in terms of the values of the risk odds, risk-aversion coefficient, and the risk tolerance as shown in Figure 11.13.

11.7.1 Kim's Risk Aversion Coefficient

We can now calculate Kim's risk tolerance and risk-aversion coefficient using her risk odds $r(1)$ as

$$\rho = \frac{1}{\ln(r(1))} = \frac{1}{\ln(2^{\frac{1}{50}})} = \$72.13$$

and $\gamma = \ln(2^{\frac{1}{50}}) = 0.01386$. As we will see, γ and ρ can both be useful in both practical and theoretical discussions of risk attitude.

If we wish to define the *u*-curve $u(x) = a - be^{-\gamma x}$, such that it is normalized with

$$u(0) = 0 \text{ and } u(1) = 1$$

then we have

$$u(0) = 0 = a - b$$

	Risk Preferring	Risk Neutral	Risk Averse
Risk Odds, r	$r < 1$	$r = 1$	$r > 1$
Risk Aversion, γ	$\gamma < 0$	$\gamma = 0$	$\gamma > 0$
Risk Tolerance, ρ	$\rho < 0$	$\rho = \infty$	$\rho > 0$

FIGURE 11.13 Risk Relations

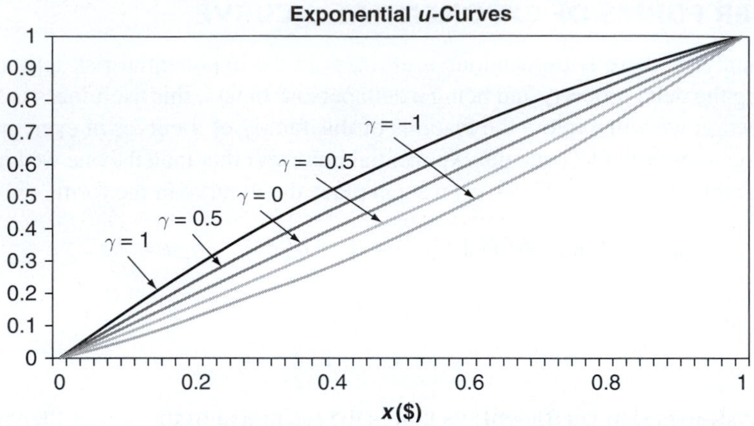

FIGURE 11.14 **Exponential *u*-Curves for Different Risk-Aversion Levels**

and

$$u(1) = 1 = a - be^{-\gamma}$$

Solving these two equations gives

$$a = b = \frac{1}{1 - e^{-\gamma}}$$

so that

$$u(x) = \frac{1 - e^{-\gamma x}}{1 - e^{-\gamma}}$$

Using this equation, we can now plot the exponential *u*-curves on a domain [0,1] for deltapeople with different risk-aversion coefficients, as shown in Figure 11.14. Kim's *u*-curve would be described by a gamma of 1.386 in $100 units. It would be more risk averse than the one for gamma equals one in Figure 11.14.

The risk-aversion coefficient, risk tolerance, and risk odds each characterize the risk attitude of deltapeople. We do not recommend, however, that a person be risk preferring value, because his certain equivalent for a deal will be larger than its mean value.

11.8 DIRECT ASSESSMENT OF RISK TOLERANCE

We can also assess the risk tolerance, ρ, directly from the decision maker without having to assess the risk odds first. Of course any deal where the deltaperson can provide a certain equivalent provides one equation in one unknown, which in this case is the value of the risk tolerance.

METHOD 1: RISK TOLERANCE ASSESSMENT When we wish to assess a deltaperson's risk tolerance, we begin by asking a question like,

"Would you be willing to accept a deal where you were equally likely to win $100 and lose $50?"

The chances are equal and the loss amount is always chosen to be half the gain amount. If the answer is "yes," then the stakes are raised. The next question might become, "How about winning $300, losing $150?" If the answer is "yes," then we might try "Win $1,000, lose $500?"

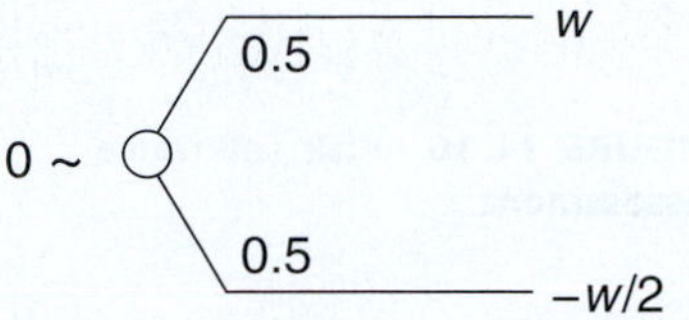

$$0 \sim \begin{cases} 0.5 & w \\ 0.5 & -w/2 \end{cases}$$

FIGURE 11.15 Risk Tolerance Assessment

At some point, everyone will become uneasy, although the question might have to be, "Win a million dollars, lose half a million dollars?" or in the case of a company, "Win a billion dollars, lose half a billion dollars?"

The aim is to find the point where the person is indifferent—not happy to take the risk. Thus, if the person says "That deal would be just OK", it becomes a little more extreme until the person does not care whether he receives the deal or not.

At the point of balance, the person is indifferent to investing w for equal chances of doubling his money or losing half of it.

Suppose that the deltaperson has an exponential u-function where we have chosen the constants for convenience to be $a = 0$ and $b = 1$. Then we have

$$u(x) = -e^{-\gamma x}$$

To determine the risk tolerance, we first assess w for which the decision maker expresses the indifference in Figure 11.15.

Indifference implies we equate the u-value of each side of Figure 11.15. We have

$$U(0) = \frac{1}{2}u(w) + \frac{1}{2}u\left(-\frac{w}{2}\right)$$

Substituting gives,

$$\begin{aligned} -1 &= 0.5(-e^{-\gamma w}) + 0.5(-e^{\gamma \frac{w}{2}}) \\ 2 &= e^{-\gamma w} + e^{\frac{\gamma w}{2}} \end{aligned}$$

Solving gives $\gamma w = 0.96$. Thus,

$$\gamma = \frac{0.96}{w}$$

and

$$\rho = \frac{1}{\gamma} = \frac{w}{0.962} = 1.04w$$

Approximately, $\rho = w$. Better, $\rho = 1.04w$. (You can add 4% to the value of w that you use in evaluating the deal, although that is seldom necessary.)

METHOD 2: RISK TOLERANCE ASSESSMENT Occasionally, people object to this assessment procedure because they say that in their business, they never face even odds and they don't like to think about even–odds deals. If they are still deltapeople, we can accommodate them reasonably

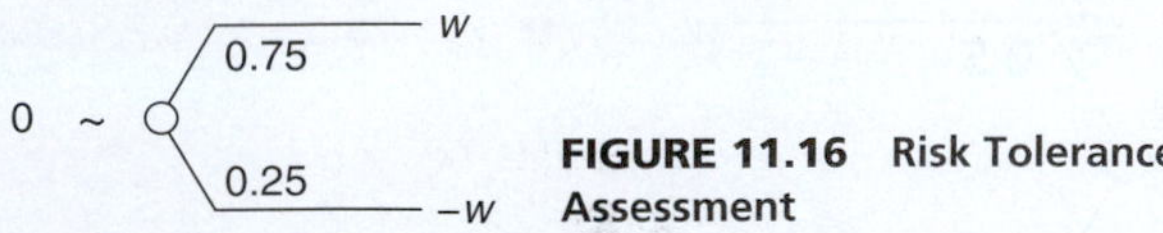

FIGURE 11.16 Risk Tolerance Assessment

well with another form of assessment question, which asks whether they are indifferent between receiving nothing and receiving a deal that has a 3 to 1 chance of gaining as opposed to losing w.

Suppose we have the deal shown in Figure 11.16. Again, we find the value of w that makes the decision maker indifferent. Note that this time, the decision maker has a higher probability of getting a positive reward, but at the same time, he may lose w instead of $w/2$ as compared to Figure 11.15.

We now equate the u-values for both sides of Figure 11.16 to get

$$u(0) = 0.75u(w) + 0.25u(-w)$$
$$-1 = -0.75e^{-w/\rho} - 0.25e^{w/\rho}$$
$$4 = 3e^{-w/\rho} + e^{w/\rho}$$

If we make the substitution $\alpha = \dfrac{w}{\rho}$, we have

$$4 = 3e^{-\alpha} + e^{\alpha}$$

Solving gives

$$\alpha = \frac{w}{\rho} = 1.09861$$

$$\rho = \frac{w}{1.09861} = 0.91024w$$

$$\rho \approx 0.91w \approx w$$

11.8.1 Risk Odds or Risk Tolerance

We have now discussed two basically different ways of obtaining the parameter required to describe the exponential u-curve. The risk-odds method requires a preference probability assignment for specified monetary payoffs. The risk-tolerance methods specify the probabilities and requirements in achieving difference to the scale of a deal that is increasing in its certain equivalent to a risk-neutral person—but with greater and greater possibility of loss. Having the person think about greater and greater losses seems to cause a focus on the idea of "what can I afford to lose," especially since the result is a number called "risk tolerance" that is measured in monetary terms. Sometimes people incorrectly think of the risk tolerance as the amount that they can "afford to lose."

The purpose of the assessment is to obtain an initial u-curve that the person can use to test whether his attitude toward monetary risk is adequately represented. You can think of this exercise like trying on a suit of clothes for fit and then making adjustments to suit your tastes. In later discussions, we will have the opportunity to consider other forms of u-curves. However, the exponential is an excellent and convenient first choice to learn about the power of certain equivalent.

EXAMPLE 11.2 Comparing Risk Odds and Risk Tolerance

A person who satisfies the delta property has risk odds of 2 for a binary deal with payoffs \$250 and \$ −250.

1. Calculate his risk tolerance.
2. Calculate his risk aversion coefficient.
3. Calculate his preference probability for \$0, expressed in terms of \$250 and \$−250.
4. Calculate his risk odds for a binary deal with \$500, \$−500.
5. By inspection, can you tell approximately the value of w that makes him indifferent to the following deal in Figure 11.17?

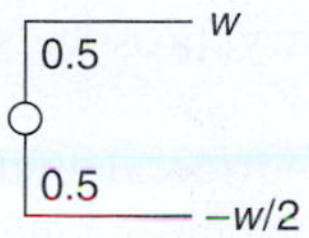

0.5 — w
0.5 — $-w/2$

FIGURE 11.17 Value of w for Indifference

The decision maker now faces the following deal of Figure 11.18.

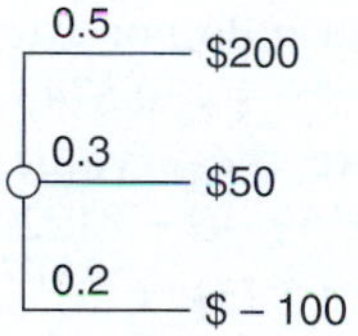

0.5 — \$200
0.3 — \$50
0.2 — \$ − 100

FIGURE 11.18 Uncertain Deal

6. Calculate the e-value of the deal in Figure 11.18.
7. If he owns the deal in Figure 11.18 and his initial wealth is \$500, find his certain equivalent for this deal.
8. If he does not own this deal and his initial wealth is \$300, find his PIBP for the deal.
9. Repeat Parts (1) through (3) for a decision maker who follows the delta property and has a risk odds of 0.8 for a binary deal with payoffs \$250 and \$ −250.

SOLUTIONS

1. We know the risk odds for \$250 are $r(250) = 2$. This means that the risk odds for one monetary unit are $r(1) = (r(250))^{\frac{1}{250}} = (2)^{\frac{1}{250}} = 1.00276$. The risk tolerance expressed in dollars is related to r_1 by

$$\rho = \frac{1}{\ln(r(1))} = \frac{1}{\ln(1.00276)} = \$360.67$$

2. The risk-aversion coefficient is the reciprocal of the risk tolerance. So

$$\gamma = \frac{1}{\rho} = \frac{1}{360.67} = 0.002773$$

3. His preference probability for 0 dollars is expressed in terms of \$250 and \$−250 where the probability makes him indifferent as to getting this binary deal or getting 0. By definition of the risk-odds calculation, this is also the probability that corresponds to his risk odds for \$250. His preference probability is

$$p = \frac{r(250)}{1 + r(250)} = \frac{2}{3}$$

4. Let his risk odds for \$500 be $r(500)$. We have $r(500) = \left(r(250)\right)^{\frac{500}{250}} = 2^2 = 4$.
5. Since this corresponds to a risk-tolerance assessment, we know the value of w that makes him indifferent is approximately equal to his risk tolerance. Therefore, $w \approx \$360.67$. The exact value of w can be determined by equating $u(0) = 0.5u(w) + 0.5u(-w/2)$ for a risk odds of $r(1) = 1.00276$ gives $w = \$348$.
6. The e-value is the e-operation on the monetary prospects $= 0.5*200 + 0.3*50 + 0.2*(-100) = \$95$.
7. To calculate the certain equivalent, we first need to calculate the u-values. Since the person satisfies the delta property, (i) he has an exponential or linear u-cuve and (ii) his certain equivalent is independent of his wealth level. We can, therefore, remove the initial wealth from the calculations. Let us choose an expression for the u-curve as $u(x) = a - be^{-\gamma x} = -e^{-0.002773x}$. Here, we set $a = 0$ and $b = 1$. As discussed, the actual values of a and b do not affect certain equivalent calculation. We have $u(200) = -e^{-0.002773(200)} = -0.5743$, $u(50) = -.87053$, and $u(-100) = -1.31956$. We now calculate the e-value of the u-values as $0.5(-0.5743) + 0.3(-.87053) + 0.2(-1.31956) = -.8122$. To calculate the certain equivalent, we get the inverse of this $u(\tilde{x}) = -e^{-0.002773\tilde{x}} = -.8122$, which implies the certain equivalent of $\tilde{x} = \$75$. Notice that for this decision maker, the certain equivalent of the uncertain deal is less than the e-value of the uncertain deal. This is true for all risk-averse decision makers.
8. If he does not own this deal, then his PIBP is equal to his certain equivalent for it if he bought it and decided to sell it. Furthermore, since he is a deltaperson, the initial wealth does not matter in the certain equivalent calculation. Therefore, regardless of his initial wealth of \$300, his PIBP is still equal to \$75.
9. If the risk odds for \$250 are 0.8, then $r(250) = 0.8$, and $r(1) = (0.8)^{\frac{1}{250}} = 0.9991$.

   Note that the risk odds are less than one, so he is risk seeking. (i) His risk tolerance is $\rho = \frac{1}{\ln r(1)} = \frac{1}{\ln(0.9991)} = -1120.36$. The risk tolerance is negative, as we discussed, since the decision maker is risk seeking. (ii) The risk-aversion coefficient is $\gamma = \frac{1}{\rho} = \frac{1}{-1120.36} = -0.00089$. (iii) His preference probability for zero, which is expressed in terms of \$250 and \$−250, is $p = \frac{r(250)}{1 + r(250)} = 0.444$. (iv) His risk odds for a binary deal of \$500 and \$−500 are $r(500) = (r(250))^2 = 0.64$. (v) The approximate value of w is −1120.36, which is the risk tolerance. (vi) The e-value of the deal does not change and is still \$95. (vi) Let us now use the other expression $u(x) = -r(1)^{-x} = -(0.8)^{-\frac{x}{250}}$. (We could have equally well just used the exponential form with e and note the risk-aversion coefficient is negative. We use the risk odds form for this part to illustrate its use numerically). The u-values are $u(200) = -r(1)^{-200} = -(0.9991)^{-200} = -1.19544$, $u(50) = -1.0456$, and $u(-100) = -.9146$. Taking the e-value of the u-values gives -1.09433. To calculate the certain equivalent, we equate $u(\tilde{x}) - (0.8)^{\frac{\tilde{x}}{250}} = -1.09433$, which gives $\tilde{x} = \$100.99$. Notice that the certain equivalent of the uncertain deal is now higher than its e-value. This is true for all risk-preferring people.

To recap some learning from this example, risk-averse people value deals less than their e-value, while risk-preferring people value deals higher than their e-value. You can satisfy the delta property and be risk preferring, risk-averse, or even risk neutral.

11.8.2 Commentary: The Meaning of Risk Aversion

UNDERSTANDING PROSPECTS Now that we have had this discussion of attitude toward risk, we return to the distinction of a *prospect* because it is so important in achieving clarity of thought.

Suppose that Lenny, who is risk-averse, faces a situation where he owes a loan shark \$1000, but now has only \$800. The loan shark says that if Lenny does not come up with the \$1000 within a few minutes, the loan shark will kill him—a threat that Lenny doesn't doubt. A witness to this conversation says that he will give Lenny a chance to call the toss of a coin for \$1000 if Lenny pays him the \$800 he currently has. In the desperate hope that he will successfully make the call, Lenny agrees. In the face of an extreme situation, is he suddenly becoming risk-preferring?

Recall that the u-curve only applies to prospects described completely in terms of a value measure. In this case, Lenny faces three prospects. First, he could be a dead man with \$800 in his pocket, until the loan shark takes it. Second, he could be dead, having spent his \$800 for the coin deal and lost. Third, he could have spent \$800 for the coin deal and then called it correctly; he is now without any money, but still alive.

In all of these prospects, the possibility of being alive or dead will likely be much more significant to Lenny than any amount of money. In this case, the prospects are not completely described by a value measure. For a purely financial deal, Lenny still may be as risk-averse as he was before; he would not normally spend \$800 for the chance to call a coin toss for \$1000.

PROSPECTS AND GAMBLING This notion of prospect may help us to understand the popularity of gambling. We know it is financially lucrative to offer to the public a series of chance deals with negative monetary means. Why would a risk averse person succumb?

If you think about the Las Vegas gambling experience, it involves not just financial costs and payoffs, but also such attributes as free drinks, exciting locales, and the possibility of new social encounters. These prospects are only partially described by a value measure. Perhaps this explains why a prospective business venture that we call Compu-gamble would be doomed to fail.

Suppose that a person who wants to gamble tells us his complete gambling strategy, including the games he would play and how much he would bet conditional on winning and losing. In other words, he would specify a detailed decision tree showing how he would make his gambling choices. Then, in just a few moments, the Compu-gamble company would simulate the strategy on a computer using the actual probabilities of the games of chance he is playing and simply tell the customer what he had won or lost. At that point, he would then either pay the bill or receive

FIGURE 11.19 **Evel Knievel**
(AP Images)

Evel Knievel died of natural causes in 2007.

a payment. There would be no need for him to travel to Vegas or even take the time to actually gamble. What could be more efficient? If gambling prospects were just the monetary results, Compu-gamble might be a very successful business.

Why do we find that people with very limited means will buy government lottery tickets even though they are poor investments? For one, the purchaser may view the ticket expenditure as a tiny decrease in his current unattractive circumstances—compared to the hope of even a miniscule chance of what he thinks will be a transformed life. The government television advertisement showing the wonderful experience of being a winner does not allude, even slightly, to the possibility of loss. If instead that advertisement spent the same 30 seconds showing each of the millions of people who bought a ticket and lost, the campaign would undoubtedly be much less successful in attracting new ticket buyers.

JUDGING THE RISKY BEHAVIOR OF OTHERS Sometimes, in everyday conversation, you hear someone describe another person as being risk-averse or risk seeking. Do these descriptions correspond to our present discussion of risk attitude? The answer is no. The *u*-curve specifies a person's preferences only on uncertain prospects that are describable entirely by a value measure. When we observe someone else's decision, we know neither the non-monetary benefits nor the disadvantages that the person may have considered. Furthermore, we do not know the probabilities the person assigned to the prospects.

Consider the famous motorcycle stunt performer Evel Knievel, pictured in Figure 11.19. Some called him a daredevil, but for all we know, he assigned much higher probabilities to successfully performing a stunt than did others, and he could well have been extremely risk-averse in his financial dealings. The risk attitude you state following the principles in this chapter is the only one on which you can be an authority.

11.9 SUMMARY

In general, when we calculate the *u*-value of a monetary prospect, we include the payoff of the deal plus initial wealth. If a person is a deltaperson, we do not need to include the initial wealth in the calculations.

For *any u*-curve, the buying and selling prices for a deal around a cycle of ownership are equal. In general, however, it is not necessarily that the buying price for an uncertain deal you do not own will be equal to your selling price for the same deal if you owned it. This will be true only if you satisfy the delta property.

Here we summarize the properties of a deltaperson:

- The PIBP of a deal you do not own is equal to the PISP for the deal if you owned it. This property can be useful when we wish to calculate the PIBP of an uncertain deal since we can instead calculate the PISP, which is easier.

- Value of clairvoyance = value of deal with free clairvoyance minus value of deal with no clairvoyance. This property is also very useful, as it considerably simplifies the calculation of the value of information for deltapeople.
- You have an exponential or a linear *u*-curve.
- We can characterize your *u*-curve by a single parameter—either the risk odds, the risk tolerance, or the risk-aversion coefficient.
- Certain equivalent of a deal does not depend on the initial wealth or on any other irrelevant deals owned by the decision maker. This property is also useful because thinking about the exact amount of your initial wealth can be difficult.
- A person can be risk preferring, risk-averse, or risk-neutral and still satisfy the delta property.
- By observing the risk odds, we can determine whether the decision maker is risk-averse, risk-neutral, or risk preferring:

Risk Attitude Relations

$r = 1$	*Risk Neutral*
$r > 1$	*Risk Averse*
$r < 1$	*Risk Preferring*

- We can assess the risk odds directly in one monetary unit and relate it to the risk odds for another monetary unit $r_m = (r_n)^{\frac{m}{n}}$.
- We can relate the risk odds to the probabilities of indifference $p = \dfrac{r}{1+r}, r = \dfrac{p}{1-p}$.
- We can also relate the risk odds to the risk-aversion coefficient and risk tolerance $\gamma = \ln r(1)$, $\rho = \dfrac{1}{\gamma} = \dfrac{1}{\ln r(1)}$.
- Risk-averse people value deals less than their *e*-value, while risk-preferring people value deals higher than their *e*-value. This is true for delta people and for non-delta people.

KEY TERMS

- Cycle of ownership
- Delta property
- Prospects
- Risk-aversion coefficient
- Risk odds
- Risk tolerance

PROBLEMS

Problems marked with a dagger (†) are considered quantitative.

1. Ron of RonCo Rain Detectors is considering rolling out his line of rain-detecting equipment nationwide. Two rollout alternatives are under consideration: national and regional. Ron is uncertain how large the rain-detecting market might be. After speaking with many experts, Ron believes no matter which alternative he chooses, the probability of a Large Market is 0.7 while the probability of a Small Market is 0.3. Ron evaluates the four prospects as follows.

Prospect	Value (thousands of $)
National and Large Market	450
National and Small Market	−300
Regional and Large Market	300
Regional and Small Market	−100

Mark of Mark's Market Testing offers to sell Ron a market detector. If the market will be large, the detector will indicate "Large" 60% of the time. If the market will be small the detector will indicate "Small" 95% of the time. What is Ron's value of clairvoyance (Vα) on the market detector indication? (You may assume that RonCo is risk-neutral for the range of prospects in this problem. Values are rounded to the nearest thousands of dollars.)

a. 15
b. Ron would not buy clairvoyance because he is risk-neutral.
c. 25
d. 0

2. Consider the following decision tree from Kim's party problem after she has applied the first four Rules of Actional Thought:

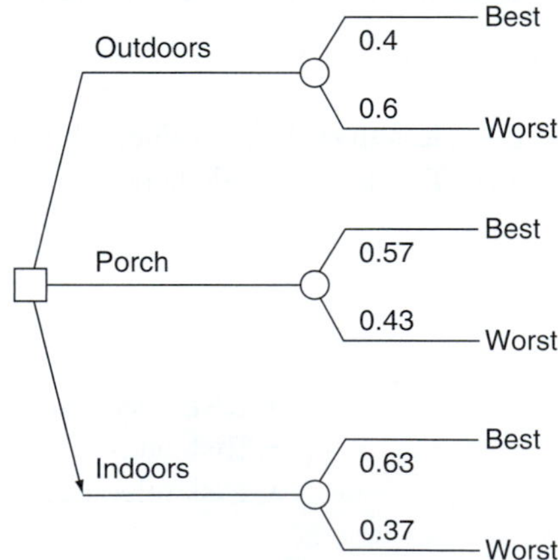

Which of the following statements about Kim's Indoors alternative are true?

a. 0.63 is the probability Kim will get her best prospect if she sets up her party Indoors
b. 0.63 is the preference probability of the *e*-value of the monetary prospects

c. 0.63 is her preference probability for the Indoor deal
d. 0.63 is the *e*-value of the preference probabilities for the Indoor prospects
  i. (a) and (c)
  ii. (a) and (d)
  iii. (b) and (c)
  iv. (c) and (d)

3. Samantha owns the decision opportunity given here and has the following preference probabilities for deals with $200 as the best outcome and 0 as the worst outcome.

Value ($)	Preference Probability
0	0.00
10	0.17
20	0.32
40	0.57
50	0.67
80	0.89
90	0.95
200	1.00

What is Samantha's certain equivalent for the following decision opportunity?

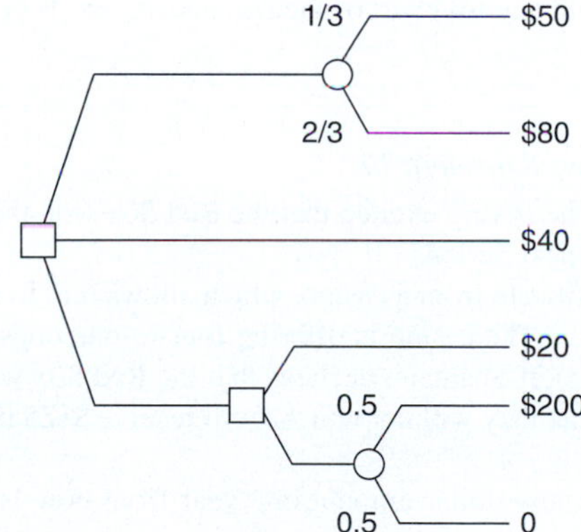

a. Between $20 and $50
b. Between $50 and $90
c. Between $90 and $200
d. Not enough information given to solve the problem.

4. Sam has just paid $215 for a certificate that allows him to call the flip of an "unfair" coin. If he calls it correctly, he will win $500. If he calls it incorrectly, he will win $0. His PISP for the certificate is $315. The clairvoyant offers to tell him the outcome of the impending coin toss. What is Sam's PIBP for the clairvoyant's information? You do not know anything about Sam's risk attitude.
a. $185
b. Need Sam's *u*-curve to determine his PIBP for the information.
c. $285
d. Need Sam's *u*-curve and the probability he assigns to the "unfair" coin landing Heads (or Tails) to determine his PIBP for the information.

For Problems 5, 6, and 7 consider the decision situation below where the dollar values are winnings and the probabilities are the assessments of the deal's owner. Suppose Stuart, who is risk-neutral, owned the deal given here.

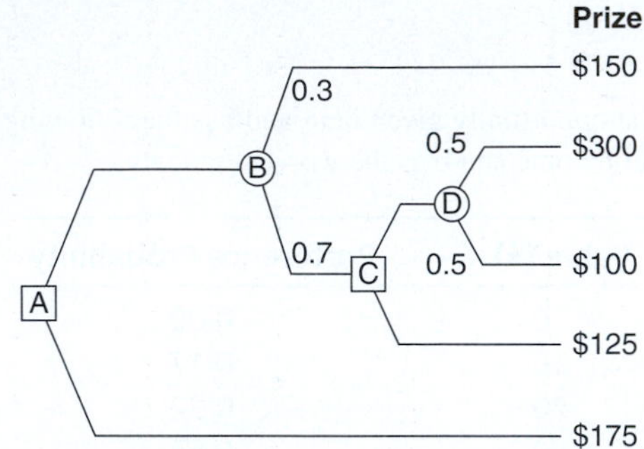

5. What is Stuart's certain equivalent for the deal?
   a. 175 c. 215
   b. 200 d. 185
6. What is the most Stuart should pay for clairvoyance on uncertainty D before making decision A?
   a. 215 c. 30
   b. 50 d. 185
7. What is the most Stuart should pay for clairvoyance on uncertainty B before making decision A?
   a. 25.5 c. 192.5
   b. 7.5 d. 185

Use this information for Problems 8 through 10.

Shantanu is a huge baseball fan; he is very excited that the Red Sox won the 2007 series, so he decided to travel to Las Vegas to bet on the next Series.

For $100, he bought a certificate from a casino, which allows him to stake that $100 on whether the Red Sox will win the 2008 Series. The casino is offering four-to-one odds for the Red Sox not to win a second time consecutively. That is, if Shantanu declares that the Red Sox will win, he will receive $500 if he is correct, and if he declares that they will not win, he will receive $125 if he is correct. In either case, if he is incorrect, he gets nothing.

Assume that to Shantanu, any dollar amount one year from now is worth as much as any dollar amount today.

8. Suppose that Shantanu thinks that the Red Sox have a 40% chance of winning the 2008 Series and that the clairvoyant comes to Shantanu and offers to tell him who is going to win, for a fee. Assuming that Shantanu is risk neutral, in what range does his value of clairvoyant lie with respect to the deal he faces?
   a. Greater than or equal to $0, but strictly less than $50
   b. Greater than or equal to $50, but strictly less than $100
   c. Greater than or equal to $100, but strictly less than $150
   d. Greater than or equal to $150
9. Now suppose Shantanu thinks that the Red Sox do have a four-to-one chance of losing their title next year; that is, he agrees with the casino's assessment of the odds and believes that there is a 0.8 chance that the Red Sox will lose next year. Unfortunately, you do not know whether Shantanu follows the delta property, and you don't even know what his initial wealth is—you just know that he follows the five Rules of Actional Thought and prefers more money to less. How should Shantanu bet?
   a. There is not enough information to answer this question.
   b. He should bet on the Red Sox winning if he is risk-averse and on their losing if he is risk preferring.

c. He should bet on the Red Sox losing if he is risk-averse and on their winning if he is risk preferring.
d. No matter what his risk attitude is, he should be indifferent between the two alternatives.

10. Going back to the case in which Shantanu follows the delta property, how many of the following could possibly change his choice of a best alternative? (Shantanu has to bet.)
a. His *u*-values for all prospects get divided by 2.
b. He reads in the newspaper that the World Series might be open to more teams in 2008 than in 2007.
c. Before making up his mind on what to bet on, he loses $100 at the roulette table.
d. He learns that no matter what he bets on, he will be charged a $2 processing fee for placing his bet.
i. (a) or (c)
ii. (a)
iii. (b)
iv. (c)

11. Minying has just bought a single share of stock from DA Corporation for $50, and she is considering buying a put option for the stock, which would cost her an additional $10. A put option would allow her to sell the stock one year from now at the current price of $50. If she does not buy the option, then she would sell the share one year from now at its market price. Minying believes that the price of the share will go up to $100 with a probability of 0.6 and go down to $30 with a probability of 0.4. Minying follows the five Rules of Actional Thought and prefers more money to less. She follows the delta property for prospects up to $200, and she is indifferent between receiving $30 for sure and a deal with a .6 chance of getting $50 and a .4 chance of getting $10.
What should Minying do? (Do not worry about discounting for time.)
a. Not buy the option, which means that she will sell it in one year at market price.
b. Buy the option and use the option only if the market price goes up to $100.
c. Buy the option and use the option only if the market price goes down to $30.
d. Buy the option and use it in either case.

12. This question uses the information given in the Problem 11. Which of the following statements MUST be true about Minying's value of clairvoyance (VOC) on the future price of the share before she decides to buy the put option?
a. VOC $= 0$
b. $\$0 < \text{VOC} \leq \$5$
c. $\$5 < \text{VOC} \leq \$15$
d. $\$15 < \text{VOC}$

Use this information for Problems 13 through 16.

Ibrahim is debating whether to purchase a mortgage note that is being marketed by a troubled bank. The mortgage looks good. It is backed by a piece of valuable commercial real estate and MallCo, the borrower, has a good payment history. The two distinctions relevant to Ibrahim's value for this mortgage note are MallCo's sales in the coming year (Strong or Weak) and whether they will default (Yes or No) at any point over the coming year. Ibrahim values the prospect of not purchasing the loan at 0. If Ibrahim decides to purchase the loan, he faces the prospects tabulated below. The dollar values of these prospects include the purchase price of the mortgage note and appropriate time discounting.

Sales	Default	Value [$millions]
Strong	No	10
Strong	Yes	−30
Weak	No	7
Weak	Yes	−35

Ibrahim's probabilities are as $\{\text{Strong Sales}|\&\} = 0.9$ $\{\text{Default}|\text{Strong Sales}, \&\} = 0.05$, and $\{\text{Default}|\text{Weak Sales}, \&\} = 0.35$. Ibrahim is risk-neutral over the dollar value prospects of this decision, and he follows the five Rules of Actional Thought.

13. Ibrahim's certain equivalent (CE) for the mortgage note lies in what range?
   a. $-\$2.5\text{ million} < \text{CE} \leq 0$
   b. $0 < \text{CE} \leq \$2.5\text{ million}$
   c. $\$2.5\text{ million} < \text{CE} \leq \$5\text{ million}$
   d. $\$5\text{ million} < \text{CE}$

14. Sara is also considering purchasing the mortgage note. Her probabilities and prospect dollar values are the same as Ibrahim's. Like Ibrahim, Sara follows the delta property over the prospect dollar values, as well as the five Rules of Actional Thought. Unlike Ibrahim, Sara has a risk tolerance (ρ) of \$50 million. Given this information, Sara's value of clairvoyance (VOC) on the sales distinction lies in what range?
   a. $0 < \text{VOC} \leq \$2\text{ million}$
   b. $\$2\text{ million} < \text{VOC} \leq \$4\text{ million}$
   c. $\$4\text{ million} < \text{VOC} \leq \$6\text{ million}$
   d. $\$6\text{ million} < \text{VOC}$

15. Sara's value of clairvoyance (VOC) on the default distinction lies in what range?
   a. $0 < \text{VOC} \leq \$2\text{ million}$
   b. $\$2\text{ million} < \text{VOC} \leq \$4\text{ million}$
   c. $\$4\text{ million} < \text{VOC} \leq \$6\text{ million}$
   d. $\$6\text{ million} < \text{VOC}$

16. How many of the following statements about the mortgage note purchase decision are true?
   I. Ibrahim's VOC on Sales is less than Sara's VOC on Sales.
   II. Ibrahim's VOC on Default is less than Sara's VOC on Default.
   III. Sara's VOC on Default equals her VOC on Sales and Default together.
      a. 0 c. 2
      b. 1 d. 3

17. Consider the following two deals faced by Matt, a person who follows the five Rules of Actional Thought, prefers more money to less, and has a consistent risk attitude (either risk-averse, risk-neutral, or risk seeking) over all dollar-valued prospects:

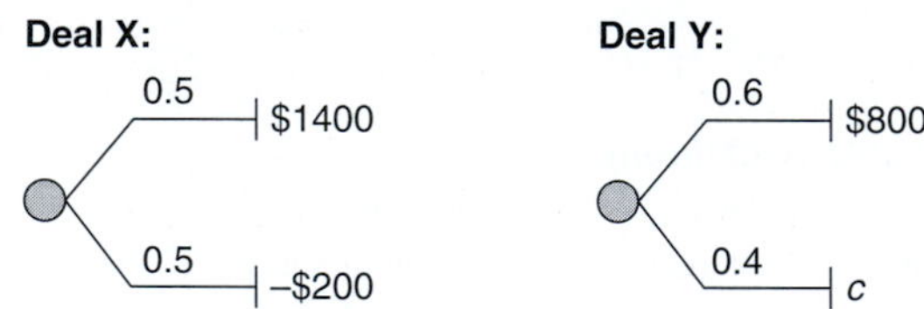

   I. If $c = \$200$ and Matt's certain equivalent for Deal Y equals \$500, he must be risk seeking.
   II. If $c = \$300$ and Matt is risk-neutral, he must prefer Deal X to Deal Y.
   III. If Matt prefers \$700 for certain to Deal X, he must be risk-averse.

   How many of the previous statements are true?
      a. 0 c. 2
      b. 1 d. 3

18. Louis follows the delta property for deals with prospects between $-\$3000$ and \$3000, and he prefers more money to less. His certain equivalent is \$500 for a deal with a 0.8 chance at \$2000 and a 0.2 chance at $-\$1000$. Now consider the following u-curves:
   I. $u(x) = 4 - 4^{-x/1500}$
   II. $u(x) = 4^{-x/1500}$
   III. $u(x) = -(1/4)^{x/1500}$
   IV. $u(x) = 8 - 8^{-x/1500}$

If x is measured in dollars, how many of the following u-curves are consistent with Louis's preferences?

a. 0 or 4 c. 2

b. 1 d. 3

19. Yongkyun follows the five Rules of Actional Thought, and he faces the following deals:

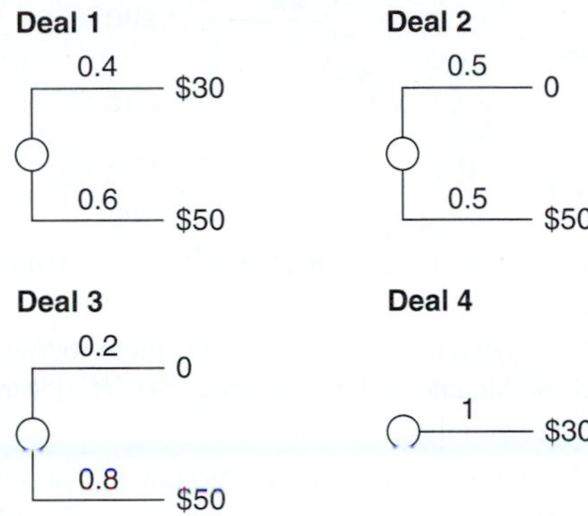

Now consider this statement:

"If Yongkyun prefers Deal 1 over Deal 3, then he prefers Deal 2 over Deal 4."

Which of the following best describes the statement?

a. Never true
b. True only if Yongkyun is risk seeking
c. True only if Yongkyun is not a deltaperson
d. Always true

20. James is thinking about betting on the outcome of the next presidential election. An online betting service offers him a deal where he can bet $1000 on a Democratic, Republican, or third-party victory. If he bets on a Democratic victory, he will win $1500 if a Democratic candidate wins the election. If he bets on a Republican victory, he will win $2000 if a Republican candidate wins. If he bets on a third-party victory, he will win $20,000 if a third-party candidate wins. In any case, if another type of candidate wins, the one James bet on, he receives $0.

James believes that there are 70%, 25%, and 5% chances of Democratic, Republican, and third-party victories respectively.

James values a dollar today the same way he will after the election, so the time value of money is not an issue.

You don't know James's risk attitude or u-curve, however you do know that he follows the five Rules of Actional Thought and prefers more money to less.

Which of the following statements MUST be true?

a. James should bet on a Democratic victory if he is risk-averse and on a third-party victory otherwise.
b. Regardless of his u-curve, James should never bet on a third-party victory.
c. Regardless of his u-curve, James should never bet on a Republican victory.
d. None of the above statements MUST be true.

†**21.** Risk odds: Wendy adheres to the delta property and has risk odds of 2:1 for a deal where she could win or lose $100.

a. What are her risk odds for a deal where she could win or lose $50?
b. What is her preference probability for 0 when the best and worst prospects are winning and losing $1000?

†22. Joe follows the Rules of Actional Thought, the delta property, and has 3:1 risk odds for $\pm$\$1000. Create an equation that could describe Joe's u-curve.

†23. Consider the following decision situation:

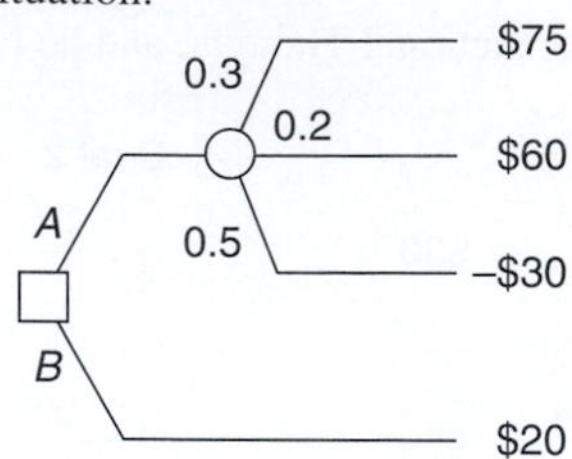

a. Using the u-curve $u(x) = 0.2x + 5$, what is the u-value of the preferred alternative for this decision?

b. Transform the u-values for the dollar amounts in this problem so that the highest dollar amount has a u-value of 1 and the lowest dollar amount has a u-value of 0. (*Hint*: First perform an additive operation to get a u-value of 0 for the lowest dollar amount.)

24. Suppose Alejandro follows the five Rules of Actional Thought and prefers more money to less. Suppose $0 < p < 1, 0 < q < 1$, and $0 < r < 1$. Consider Deals W, X, Y, and Z below.

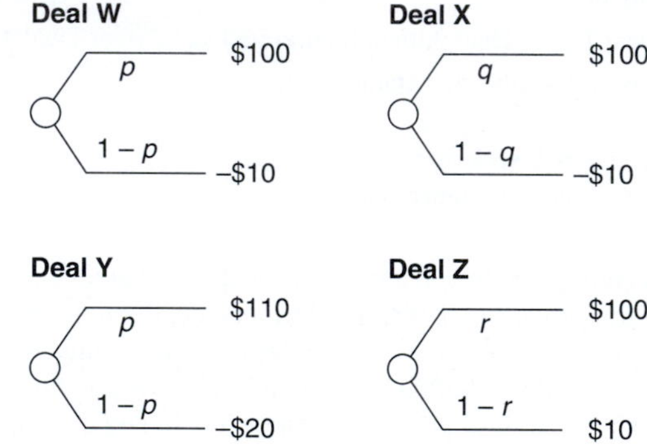

Consider the following three statements:

i. If $p > q$, then Alejandro prefers Deal W to Deal X.
ii. If Alejandro is a "deltaperson", his certain equivalent for Deal Y is \$10 more than his certain equivalent for Deal W.
iii. If $p > r$, then Alejandro prefers Deal W to Deal Z.

Which of the above statements is/are true?

a. 0
b. 1
c. 2
d. 3

CHAPTER

12

Sensitivity Analysis

CHAPTER CONCEPTS

After reading this chapter, you will be able to explain the following concepts:

- Sensitivity analysis
- Risk sensitivity profile
- Value of clairvoyance sensitivity

12.1 INTRODUCTION

Sensitivity analysis enables us to investigate how the decision will change if we change certain numbers in the decision basis. By doing that, we can determine whether additional effort should be expended in increasing the precision of the numbers we have used. **Sensitivity analysis** is an important feature of professional decision analysis, where we must continually refine and focus our attention on the important aspects of the problem. We shall discuss the subject of sensitivity analysis in further sections. However, let us now illustrate its use in the party problem.

12.2 KIM'S SENSITIVITY TO PROBABILITY OF SUNSHINE

Suppose that Kim is concerned about how sensitive her decision is to the probability of Sunshine she assigns; she has currently assigned 0.4. She may want to know whether small changes in this probability will change the best alternative and/or substantially change the value of the party to her. She also may be interested in a sensitivity analysis to this probability because she expects to receive additional information and wants to know how she should adapt her party strategy in the face of a new probability of Sunshine. Knowing how sensitive the decision is to this probability also helps Kim to decide how hard she needs to work on the probability assignment to accurately represent her belief.

To explore this issue, we let p be the probability of sunshine that she assigns and consider how the analysis of the problem will depend on p. In Figure 12.1, we draw Kim's decision tree for a probability of sunshine p replacing the value of 0.4 we used in Figure 9.5.

The u-values of the three alternatives are easily computed. The u-value of the Outdoor alternative is simply p.

The u-value of the Porch alternative is

$$0.950p + 0.323(1 - p) = 0.323 + 0.627p$$

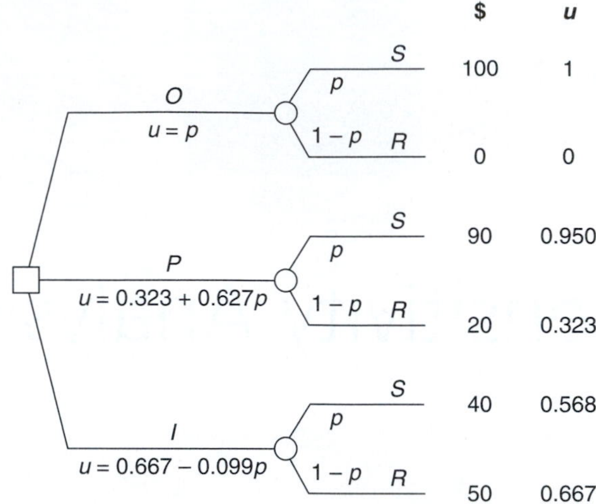

FIGURE 12.1 Kim's Decision Tree for *p*

The u-value of the Indoor alternative is

$$0.568p + 0.667(I - p) = 0.667 - 0.099p$$

The u-value for each alternative is shown by the branch for that alternative in the figure. Figure 12.2 shows a plot of these u-values versus p, which is the probability of sunshine as p ranges from 0 to 1.

Each alternative plots as a straight line that joins the u-value of the alternative with $p = 0$ to the u-value of the alternative with $p = 1$. The best alternative to follow is the one with the highest u-value for the particular p. Thus, we see that the best alternative when $p = 0$ is the Indoor alternative. The Indoor line remains highest until it crosses the Porch line at a value of $p = 0.47$. Then the Porch line is highest until it crosses the Outdoor line at $p = 0.87$. So, for probabilities of sunshine less than 0.47, the Indoor alternative is best; for probabilities of sunshine greater than 0.87, the Outdoor alternative is best; and otherwise the Porch alternative is best. Note that the value of probability of sunshine that Kim originally assigned, 0.4, leads her to follow the Indoor alternative, as we found in Figure 10.7. You can see that Kim will only move to the Outdoor alternative when she is very sure that the weather will be sunny.

One easy way to draw Figure 12.2 is to create on the right-hand side of the figure a dollar scale that is distorted according to Kim's u-curve of Figure 10.4. For example, we plot the point corresponding to \$50 on the right of Figure 12.2 by observing from Figure 10.4 that \$50 corresponds to a u-value of 0.667. Once we have constructed this distorted dollar scale, we can draw the lines corresponding to each alternative by simply connecting the end points that correspond to the dollar values. For example, the Porch line connects the point corresponding to \$20 when $p = 0$ to the point corresponding to \$90 when $p = 1$.

We can also use this figure to show the u-value of free clairvoyance. Recall that if Kim receives free clairvoyance on the weather and finds that the weather will be sunny, then she will have a party outdoors with a value of \$100. If she finds that the weather will be rainy, then she will have her party indoors with a value of \$50. Consequently, if the probability of sunshine is p,

The u-value of free clairvoyance is

$$pu(100) + (1 - p)u(50)$$

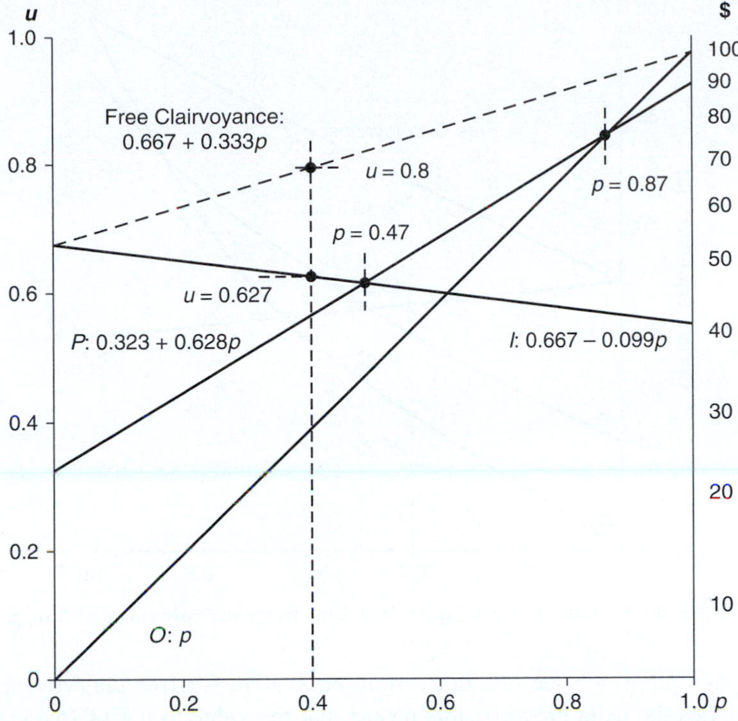

FIGURE 12.2 **Kim's *u*-Value Sensitivity to Probability of Sun, *p***

This is the line that connects the point corresponding to \$50 when $p = 0$ to the point corresponding to \$100 when $p = 1$. In Figure 12.2, it is shown as a dotted line. Note that it connects the highest point on the $p = 0$ axis to the highest point on the $p = 1$ axis. It is clear that the alternative consisting of making a decision *after* receiving free clairvoyance is better than any of the other alternatives in the problem without clairvoyance, unless $p = 0$ or $p = 1$.

12.3 CERTAIN EQUIVALENT SENSITIVITY

From Figure 12.2, we can determine the *u*-value for any probability of sunshine for any given alternative. By reading from the distorted dollar scale, we can also obtain the corresponding **certain equivalent**.

However, it will be more convenient to perform this operation on the whole figure, so we obtain a plot of Kim's certain equivalent sensitivity for probability of sunshine *p*, as shown in Figure 12.3.

Note that these plots showing how the certain equivalent of each alternative depends on the probability of sunshine are not straight lines because of the distortion in the dollar scale of Figure 12.2. We see, for example, that in accordance with Figure 10.7, for Kim's original probability of sunshine $p = 0.4$, the Indoor, Porch, and Outdoor alternatives have certain equivalents of \$45.83, \$40.60, and \$25.73, respectively.

Figure 12.3 also shows that, as the probability of sunshine varies from 0.2 to 0.6, Kim's certain equivalent will be much less sensitive to this change if she follows the Indoor alternative than it would be if she followed the Porch alternative. For the Indoor alternative, there is only about a \$4 difference in certain equivalents over this range, while for the Porch alternative there is a \$24 difference.

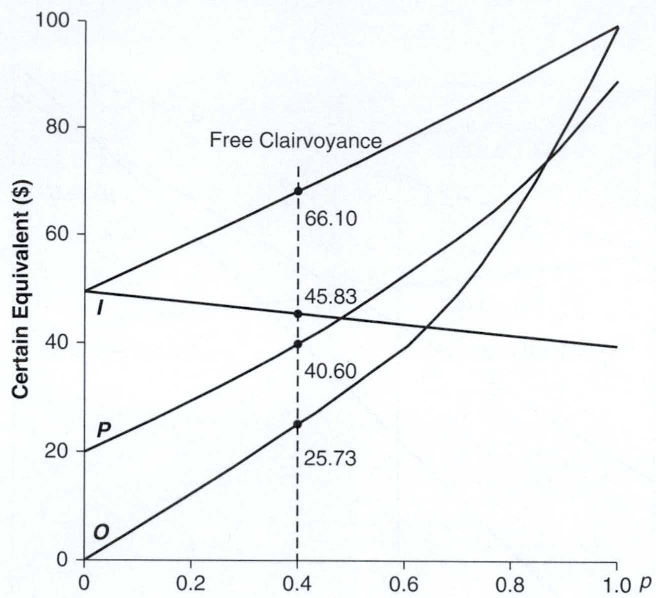

FIGURE 12.3 Kim's Certain Equivalent Sensitivity to Probability of Sun, *p*

Also, at this point, we see that her certain equivalent for free clairvoyance is \$66.10. Since Kim satisfies the delta property, this means that the value to her of free clairvoyance is \$66.10 − \$45.83, or \$20.27, which is in agreement with our earlier results.

12.4 VALUE OF CLAIRVOYANCE SENSITIVITY TO PROBABILITY OF SUNSHINE

Because Kim wishes to accept the delta property, we can refer to Figure 12.3 to define her value of clairvoyance (VOC) for any probability of sunshine p. Specifically, we need to compute the difference between the certain equivalent for free clairvoyance and the certain equivalent for the best alternative. The result of the computation appears in Figure 12.4.

We observe that there will be no value of clairvoyance when p is either 0 or 1, because at these points, Kim would have clairvoyance. Otherwise, in this problem, there will be a positive value to free knowledge of the weather.

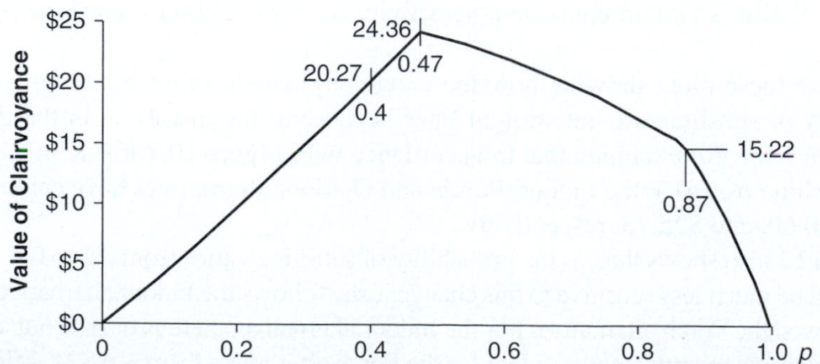

FIGURE 12.4 Kim's Value of Clairvoyance Sensitivity to Probability of Sun, *p*

As p increases from 0, the value of clairvoyance also increases, reaching a peak of about \$24.36 at $p = 0.47$. Note that the value is \$20.27 at $p = 0.4$. When p exceeds 0.47, the value of clairvoyance falls, and when p exceeds 0.87, it falls even more rapidly. For probabilities of sunshine that range from about 0.3 to 0.85, we can see that clairvoyance will be worth at least \$15.

12.5 JANE'S SENSITIVITY TO PROBABILITY OF SUNSHINE

We can perform this analysis even more easily for the risk-neutral Jane. Figure 12.5 shows Jane's decision tree for the general probability of sunshine p, in correspondence with Figure 10.11.

Since Jane's certain equivalents are simply her e-values of monetary values, we find her certain equivalent for the Outdoor alternative is simply $100p$.

For the Porch alternative, her certain equivalent is

$$90p + 20(1 - p) = 20 + 70p$$

For the Indoor alternative, it is

$$40p + 50(1 - p) = 50 - 10p$$

We plot these certain equivalents in Figure 12.6.

The certain equivalent for each alternative is a straight line connecting the dollar values for $p = 0$ and $p = 1$. Again, as p increases from 0 through 1, the best alternative will change from Indoor to Porch to Outdoor; however, the points of change are different from those for Kim. Jane will change from the Indoor to the Porch alternative when p crosses 0.375, and from the Porch to the Outdoor alternative when p crosses 0.667.

Thus, she is more willing than Kim to follow the more exposed alternatives for lower probabilities of sunshine. We see that for Jane, the value of $p = 0.4$ corresponds to the Porch region. Her certain equivalents are shown for the Indoor, Porch, and Outdoor alternatives for $p = 0.4$ as \$48, \$46, and \$40.

If Jane is given free clairvoyance, her certain equivalent is represented in Figure 12.6 by the dotted line that connects the \$100 value she will receive when $p = 1$ by having the party outdoors, and the \$50 value she will receive when $p = 0$ by having the party indoors. When $p = 0.4$, this line shows that Jane's certain equivalent of free clairvoyance will be \$70. Since

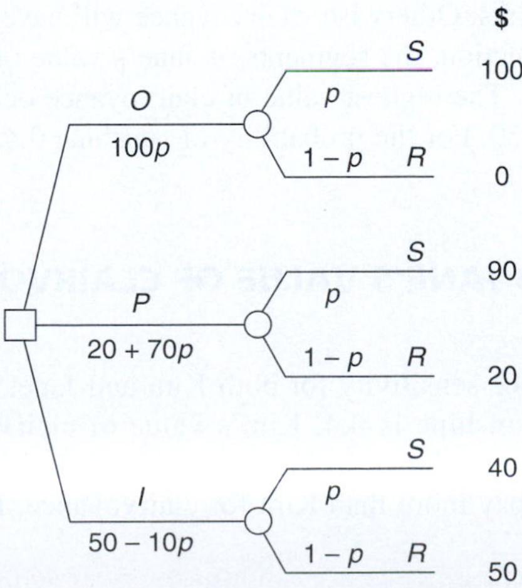

FIGURE 12.5 Jane's Decision Tree for General Probability of Sun, *p*

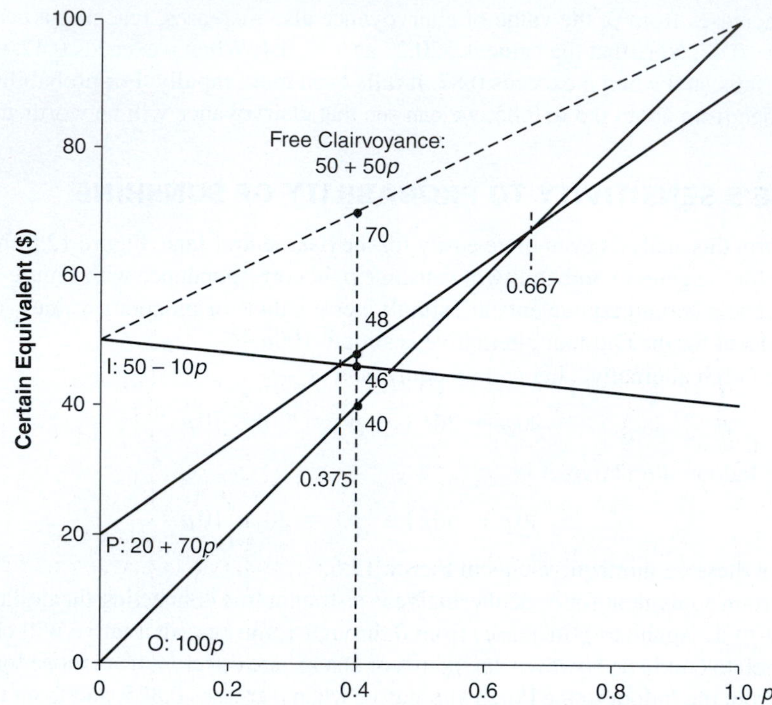

FIGURE 12.6 Jane's Certain Equivalent Sensitivity to Probability of Sun, *p*

Jane is risk-neutral, she satisfies the delta property. Consequently, we can obtain her value of clairvoyance by subtracting the $48 certain equivalent of her otherwise best alternative, having the party on the porch, from the $70 certain equivalent of having free clairvoyance. The result is the $22 value of clairvoyance for Jane that we computed originally in Figure 10.16, at $p = 0.4$.

To obtain Jane's value of clairvoyance sensitivity to p, as shown in Figure 12.7, we can perform this same computation. Specifically, we subtract the certain equivalent of the best alternative for a particular p from the certain equivalent of having free clairvoyance at the same p.

Just as for Kim, Jane will have no value of clairvoyance when $p = 0$ or $p = 1$, because she will already have clairvoyance at these points. Otherwise, clairvoyance will have a positive value to Jane. As you can see from the construction, the segments of Jane's value of clairvoyance sensitivity curve will all be straight lines. The highest value of clairvoyance occurs at the crossover point $p = 0.375$, and here it is $22.50. For the probability of sunshine 0.4, the value of clairvoyance is $22, as we have observed.

12.6 COMPARISON OF KIM'S AND JANE'S VALUE OF CLAIRVOYANCE SENSITIVITIES

We can now compare the value of clairvoyance sensitivity for both Kim and Jane. Note that, from Figure 12.4, when the probability of sunshine is 0.4, Kim's value of clairvoyance is $20.27, whereas Jane's value of clairvoyance, as shown in Figure 12.7, is $22. In this case, Jane, as the risk-neutral person, is willing to pay more than Kim for clairvoyance. But would this always hold true?

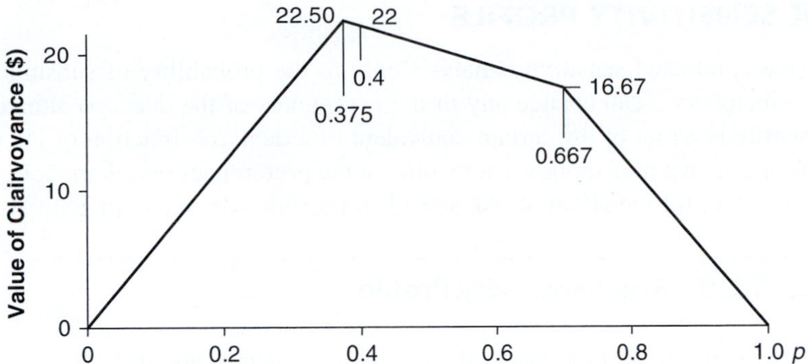

FIGURE 12.7 **Jane's Value of Clairvoyance Sensitivity to Probability of Sun, *p***

While many people think so, the answer is no. Compare the two figures when $p = 0.5$. At this point Kim's value of clairvoyance is about $24, while Jane's is $20. The order of their values is reversed for this different probability of sun.

When $p = 0.5$, both Kim and Jane will follow the Porch alternative and be more exposed to the weather. Kim, as a risk-averse person, is more concerned about this and, consequently, is willing to pay more for clairvoyance regarding the weather. However, when $p = 0.4$, Kim has already switched to the Indoor alternative, while Jane is still following the Porch alternative. Kim, as we found in the sensitivity analysis, is now much less exposed to the weather, so her value of clairvoyance becomes less than Jane's. Yet we note that when $p = 0.3$, both Kim and Jane are following the Indoor alternative—the least exposed—yet Jane is willing to pay more ($18) for clairvoyance than Kim ($14.87). Even in such a simple problem as this, intuition can often be misleading.

By superimposing the value of clairvoyance sensitivities for both Kim and Jane, as shown in Figure 12.8, we observe that these two people who share everything but risk attitude will, in fact, have quite different dependence of value of clairvoyance on probability of the Sun p.

Note: *In Figure 12.8, we see that Jane will be willing to pay more for clairvoyance for values of p below 0.423, and that Kim will be willing to pay more for clairvoyance for values of p above 0.423. It is clear that even in this simplest nontrivial decision problem with three alternatives, you cannot make any generalizations about whether increasing risk-aversion will increase or decrease the value of clairvoyance.*

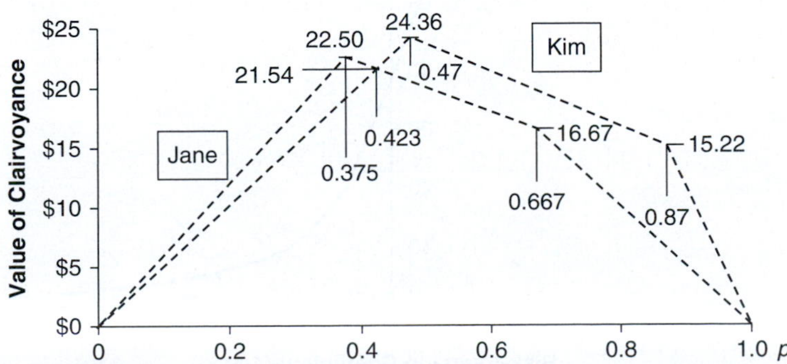

FIGURE 12.8 **Value of Clairvoyance Sensitivity to *p* for Kim and Jane**

12.7 RISK SENSITIVITY PROFILE

So far, we have conducted sensitivity analysis only to the probability of sunshine in the party problem. In principle, we can change any of the parameters of the decision situation. The **risk sensitivity profile** is a plot of the certain equivalent of a deal as a function of the risk-aversion coefficient. While such a plot applies strictly only to the preferences of deltapeople, it is still useful in obtaining a feel for the effect of risk attitude on certain equivalent in general.

EXAMPLE 12.1 Risk Sensitivity Profile

To examine the concept, let us consider the two deals shown in Figure 12.9.

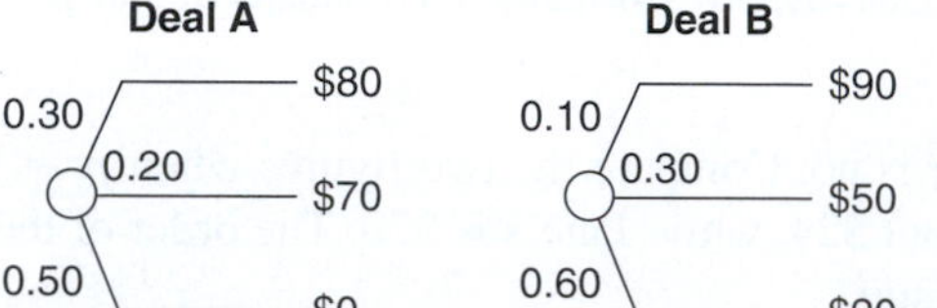

FIGURE 12.9 Examining Risk Sensitivity Profiles

Deal A is a 30% chance of $80, a 20% chance of $70, and otherwise nothing. Deal B is a 10% chance of $90, a 30% chance of $50, and a 60% chance of $20. If someone were given the choice between the deals, we can imagine a thought that might arise. "With Deal A, I have a 50% chance of nothing, but with deal B, I am assured of at least $20 and may win as much as $90." When we examine the deals from the point of view of a risk-neutral person, we find that Deal A has a certain equivalent (*e*-value of dollars) of $38, whereas Deal B has a certain equivalent of $36. Therefore, as the risk-aversion coefficient increases, it will be interesting to see if and when Deal B becomes preferable to Deal A.

The risk sensitivity profile of Figure 12.10 reveals the answer.

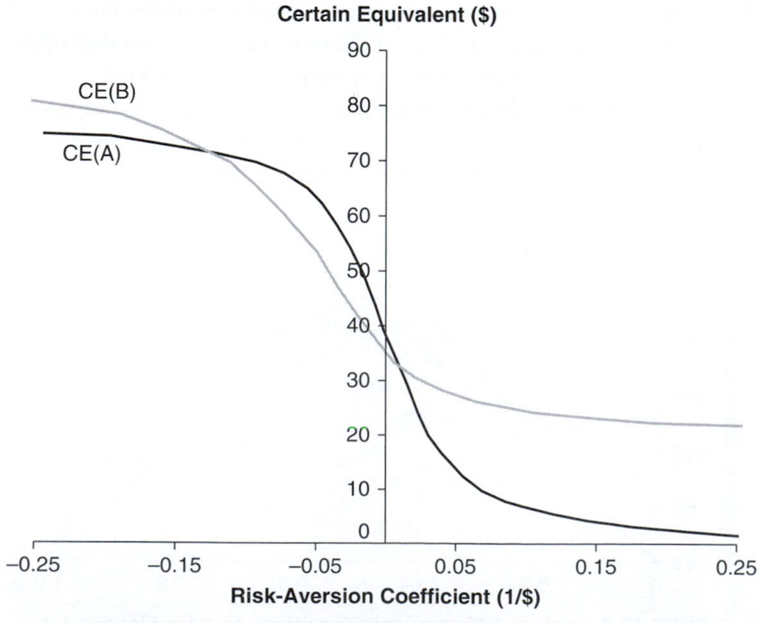

FIGURE 12.10 Risk Sensitivity Profiles with Deals A and B

In Figure 12.10, we have plotted the certain equivalents of the deals against both negative and positive values of the risk-aversion coefficient. We recall that a risk-aversion coefficient of 0 corresponds to risk-neutrality, and we observe that the curves representing each deal are equal to the *e*-values of the deals when $\gamma = 0$. As γ increases above 0, the certain equivalents of the deals fall. We find that the curves cross and the deals have the same certain equivalent of \$35.00 when $\gamma = 0.00415$ or ρ equals \$241; for positive values of ρ larger than \$241, Deal A will be preferred.

Note *When the risk-aversion coefficient* γ *becomes negative, the certain equivalent of the deal increases, the more negative it becomes.*

Of special interest are the limits that the risk sensitivity profiles for each deal approach as γ becomes large and positive or large and negative.

As γ becomes larger and larger in the positive direction, the certain equivalent approaches the smallest payoff in the deal, regardless of its probability.

Extreme risk averters act as if they were sure to receive the worst result. Therefore, in this case, the risk sensitivity profile for deal A approaches 0 as γ becomes large and positive, whereas for Deal B, the profile approaches \$20.

Similarly, as γ becomes large and negative—regardless of probability—the certain equivalent of a deal approaches the largest payoff.

Extreme risk preferrers act as if they are sure to win the highest payoff. The risk sensitivity profiles for large and negative γ show that the certain equivalent of Deal A approaches \$80 and that of Deal B approaches \$90.

12.7.1 Risk Sensitivity in the Party Problem

Speaking of Kim, as we show in Figure 12.11, we can construct the risk sensitivity profile for the deals provided by the location alternatives in the party problem. This figure only shows the

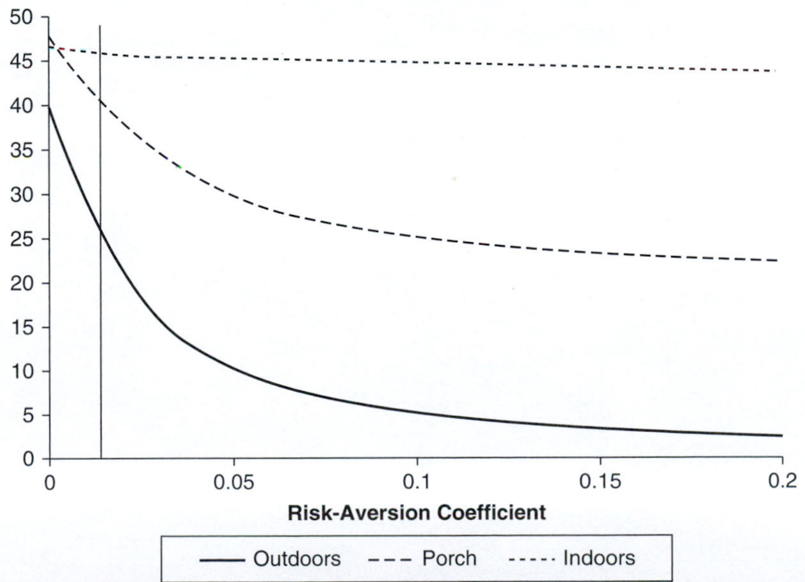

FIGURE 12.11 Kim's Risk Sensitivity Profile for Party Problem

portion of the profile representing risk-aversion, since this is the region of practical interest. When $\gamma = 0$, the curves for the certain equivalents of the Porch, Indoor, and Outdoor alternatives have the values \$48, \$46, and \$40 corresponding to the *e*-values of dollars of risk-neutral Jane. The vertical line at $\gamma = 0.01386$ or $\gamma = \$72.13$ represents the portion of the profile describing Kim's preferences. Here, the alternatives are ranked in the order Indoor, Porch, and Outdoor with certain equivalents \$45.83, \$40, and \$25, respectively. From the profile, we observe that no risk-averse deltaperson will ever prefer the Outdoor alternative to one of the other two, as long as all agree on dollar values and probabilities.

Risk sensitivity profiles are one way to explore the possibilities for creating deals that two or more people will find satisfying. Thus, we see from the profile that although Jane and Kim disagree on where to have the party, Jane would agree with Kim to have the party Indoors instead of on the Porch if Kim paid her about \$2, and that such a payment would still leave Kim with a certain equivalent of about \$44—three dollars more than if she had to settle for a Porch party. Of course, such calculations are only a guide to the possibilities. They rely on complete candor in information and preference, both of which are often absent during actual negotiations.

12.8 SUMMARY

Sensitivity analysis is a useful tool that determines how the decision may change if we change certain numbers of the problem.

KEY TERMS

- Sensitivity analysis
- Certain equivalent
- Risk sensitivity profile

PROBLEMS

Problems marked with a dagger (†) are considered quantitative.

1. Josephine needs a new car for her job that delivers meals to the elderly. She needs to buy the car now, but she knows that in a month she will be assigned by her employer to a new delivery route, which will either be in the country or in the city. She does not yet know which route will be assigned. She is considering three types of cars:

   I. a large four-wheel drive SUV.
   II. a mid-size sedan.
   III. a small economy car.

   Her preference probabilities for the various prospects associated with this decision are as follows:

Buy SUV, Country route	0.8
Buy SUV, City route	0.2
Buy Sedan, Country route	0.4
Buy Sedan, City route	0.8
Buy Economy, Country route	0
Buy Economy, City route	1

   Josephine assigns a probability p that she will be assigned to the city route. The following graph illustrates her sensitivity to p for the preference probabilities of each of the three alternatives:

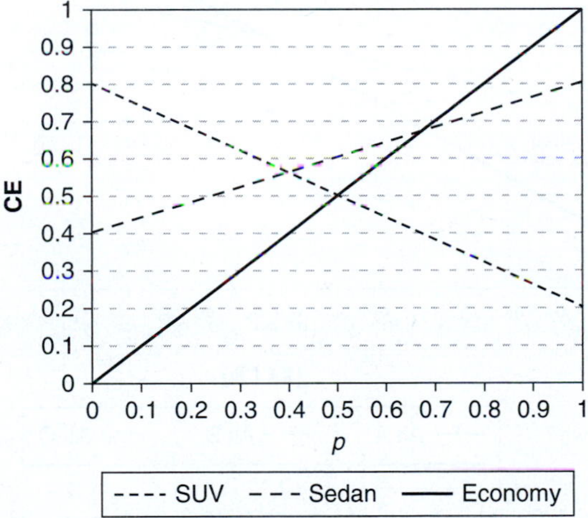

   Which of the following values of p is closest to the one with the *greatest* value of free clairvoyance on the outcome of the delivery route assignment for Josephine?

   a. 0 c. 0.55
   b. 0.4 d. 0.7

2. For Kim's friend, Mary, whom we introduced in the party problem and has preference probabilities and dollar values presented in Figure 10.18, plot the sensitivity of the value of clairvoyance (VOC) to the probability p of Sunshine. How many of the following statements are true?
   I. For any value of p, Mary should be willing to pay at least as much as Kim for clairvoyance.
   II. The difference between the value of the deal with free clairvoyance and the value of the deal with no clairvoyance is greatest when p is approximately 0.87.
   III. Mary's VOC sensitivity curve has the same switch points as Jane's. (By "switch points," we mean the values of p at which the decision changes).
   a. 0 c. 2
   b. 1 d. 3

3. Sarah, a decision maker who follows the five Rules of Actional Thought, drew the following sensitivity plot during her decision analysis. By visual inspection, you may assume that the u-values of the end points of the curves shown below are (0,0), (1,20); (0,6), (1,14); (0,8), (1,0) for alternatives A, B, and C, respectively. Please do not assume anything else about the plot.

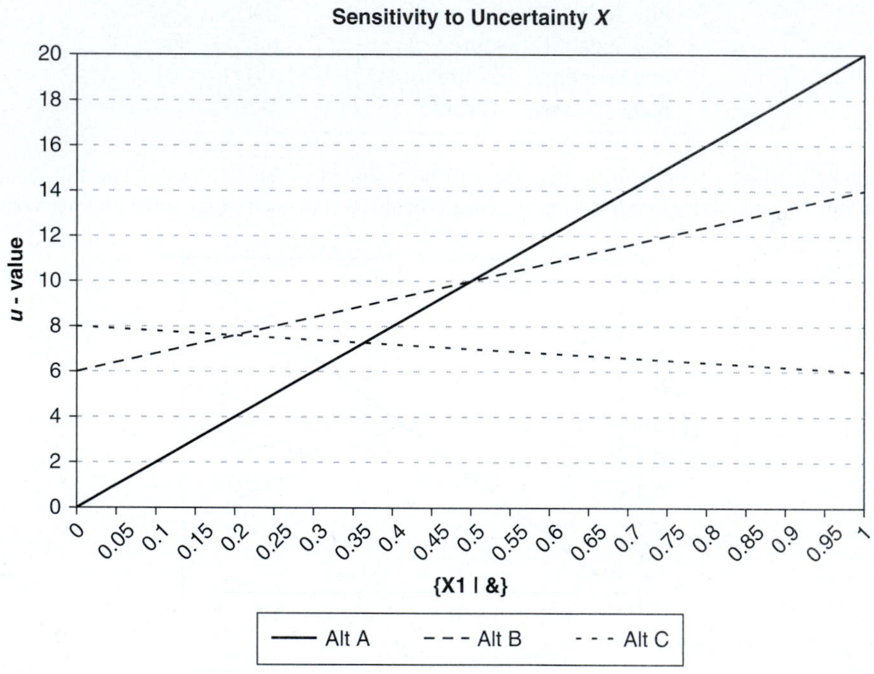

Let Sarah's u-curve be defined by the function u (.), and it's inverse as $u^{-1}(.)$. Sarah has two degrees for distinction X ($X1$ and $X2$), and her current belief is that $\{X1|\&\} = 0.75$. X is the only distinction in her decision.

I. The "curves" in this plot are straight lines.
II. The alternatives in Sarah's plot of CE versus $\{X1|\&\}$ are three straight lines.
III. If Sarah owned alternative B, she would pay $u^{-1}(3)$ to receive alternative A.

How many of the following statements must be true?

a. 0 c. 2
b. 1 d. 3

Refer to the information here to answer Problems 4, 5 and 6. Davood is the CEO of a major Silicon Valley company, Joojle. His company has decided to expand into the mobile advertising market. To implement the company's expansion policy, he is faced with two strategic alternatives. The first is to acquire MobAd, a company that has developed the required technology and is gaining a large customer base. The second alternative is to build up the technology internally. The acquisition is very appealing to Davood because it will ensure Joojle's faster entrance to the market. However, the acquisition deal has to be approved by the FTC. Davood is very uncertain about the FTC decision. The given decision tree represents Davood's decision situation.

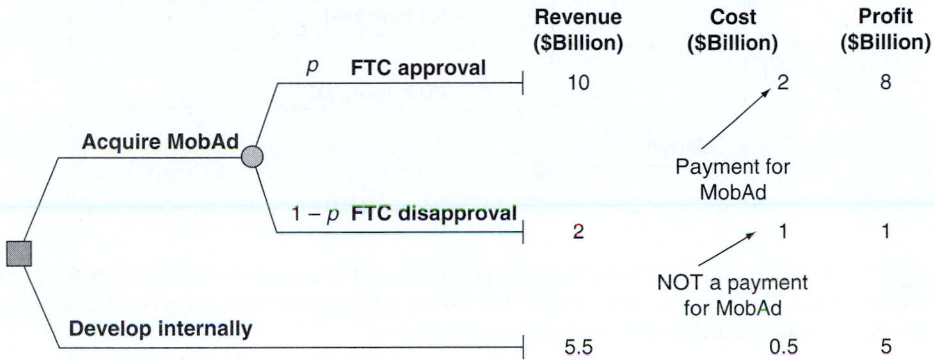

4. The Joojle DA team has not yet elicited Davood's probability of FTC approval which is denoted by p in the decision tree. However, they know that he is risk-neutral in the range of this deal. Denote p_i as the probability of FTC approval in which Davood is indifferent between acquiring MobAd and developing the technology internally. In what range does p_i lie?
   a. $p_i \leq 0.5$
   b. $0.55 < p_i \leq 0.65$
   c. $0.65 < p_i \leq 0.75$
   d. $p_i > 0.75$
5. The DA team has now assessed the probability of FTC approval from Davood to be $p = 0.5$. Currently, MobAd is asking for \$2 Billion to sell itself. Joojle rejects the offer at this price, but Davood would like to know the maximum amount that Joojle should be willing to pay for MobAd.
   a. \$1500 M
   b. \$1000 M
   c. \$750 M
   d. \$500 M
6. Davood wants to see how the value of information moves with his belief about the FTC approval. Given that MobAd is still asking for \$2 billion, what does the probability of approval p have to be for his VOC to be its highest?
   a. $0 \leq p \leq 0.1$
   b. $0.4 \leq p < 0.6$
   c. $0.6 \leq p < 0.8$
   d. $0.8 \leq p < 1$

Use this information to solve Problems 7, 8 and 9.

This situation of concerns Kelton, the CEO of Pistell-Feiers Squib Pharmaceuticals (PFS). He faces a decision about his newest and most promising drug candidate, Smartiva. Clinical trials have been completed, and the team awaits an approval decision from the U.S. Food and Drug Administration (FDA) before it can be marketed. Kelton was planning to market Smartiva, but he receives an offer from Eli Willy Pharmaceuticals to license Smartiva from PFS. Kelton's decision situation is described in the following tree:

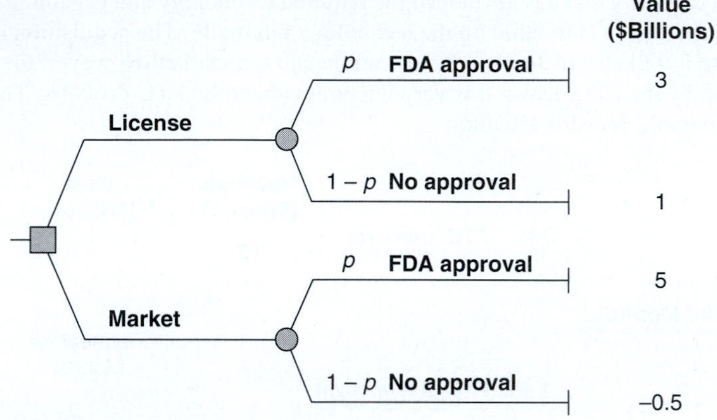

The PFS DA team has not yet elicited Kelton's probability of FDA approval, which is denoted by "p" in the decision tree. However, they know that he follows the delta property over prospects of $\pm$ \$10 Billion. Also, they recently elicited the following deal from Kelton.

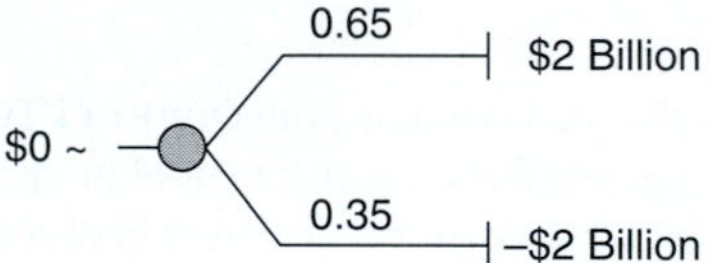

7. Denote p_i as the probability in which Kelton is indifferent between licensing Smartiva and marketing it. In what range does p_i lie?

a. $p_i \leq 0.5$
b. $0.55 < p_i \leq 0.65$
c. $0.65 < p_i \leq 0.75$
d. $p_i > 0.75$

8. Before Kelton makes his decision, AstroZene Co. makes an offer to purchase the rights to Smartiva for \$1.5 B. His decision situation now changes to that shown here.

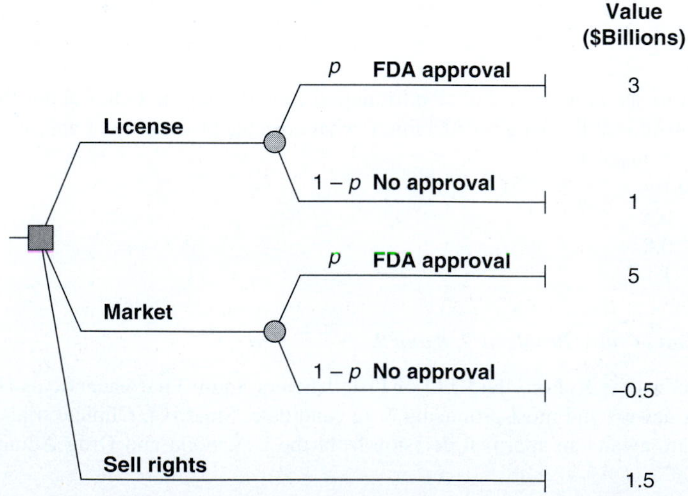

Consider the following statements about Kelton's new decision situation.

I. If $p = 0.26$, he should choose to sell the rights to Smartiva.

II. His decision to sell the rights to Smartiva is insensitive to p.

III. The probability at which he is indifferent between marketing and selling lies between 0.55 and 0.75.

IV. When he is indifferent between marketing and selling, his best alternative is to license Smartiva.

V. When he is indifferent between licensing and marketing, his best alternative is to sell the rights to Smartiva.

VI. He should always choose to market when $p > 0.8$.

VII. If $p = 0.65$, the VOC on FDA approval is greater than it is when $p = 0.8$.

How many of the above statements are true?

a. 0 or 4 c. 2 or 6
b. 1 or 5 d. 3 or 7

9. The PFS decision analysis team elicits the probability of FDA approval from Kelton. However, due to the recent fallout from a class-action lawsuit, the value of PFS stock has sunk by 30%. As such, the DA team will consider the same decision as shown before but thinks it wise to reassess Kelton's risk-aversion coefficient, γ. Before doing so, they construct the following graphic.

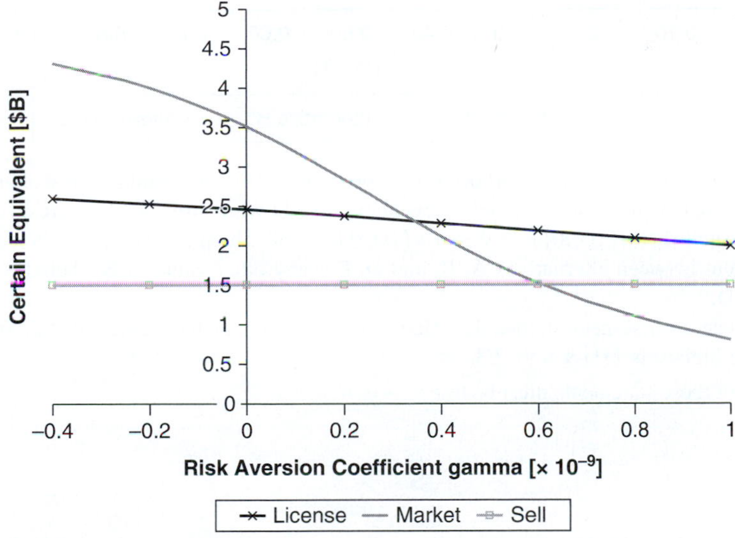

Consider the following statements about Kelton's decision problem:

I. If he is risk seeking, he should choose to market.

II. There is a degree of risk-aversion where his best alternative would be to sell the rights to Smartiva.

III. His assessed probability of the FDA approval for Smartiva is greater than 0.70.

How many of the above statements are true?

a. none c. 2
b. 1 d. 3

10. Aishwarya follows the five Rules of Actional Thought and prefers more money to less. She faces one decision with three alternatives of A, B, and C and one uncertainty with degrees $i0$ and $i1$. In other words, the alternative she chooses and the degree of the uncertainty she later observes fully describes her prospects.

This plot details her u-value for each of the three alternatives as a function of her probability that degree $i1$ will occur.

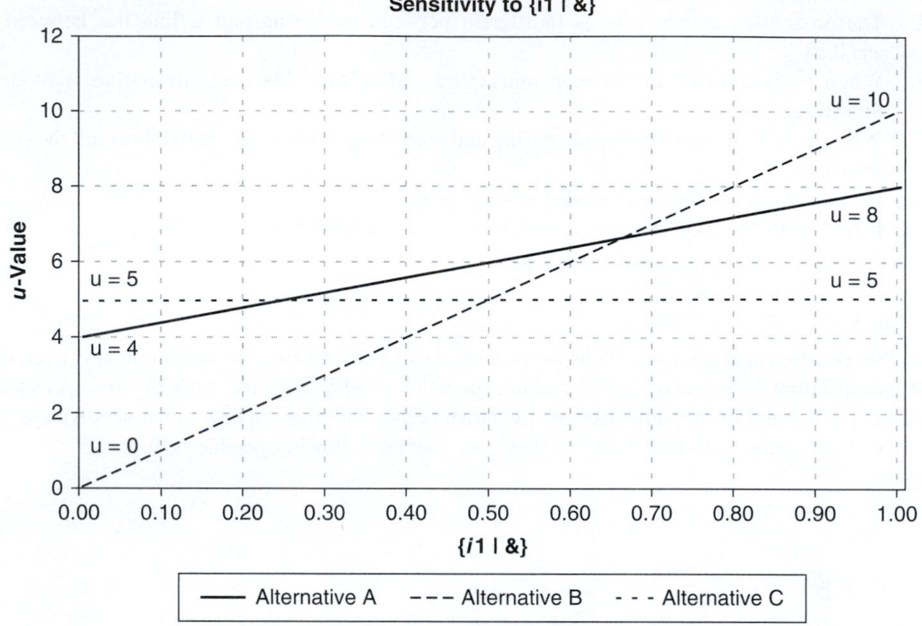

I. If Aishwarya believes that $i1$ will occur with probability 0.7, she should choose alternative B.

II. Suppose that Aishwarya is offered a new alternative D with corresponding prospects [D, i0] and [D, $i1$] such that $u([\text{D}, i0]) = 2$ and $u([\text{D}, i1]) = 9$. Then, if $\{i1|\&\} = 2/3$, she should be indifferent between alternatives A, B, and D. For any other value of her belief, she should not choose D.

III. If Aishwarya is risk-neutral, then the value of $\{i1|\&\}$ for which her value of clairvoyance on Part. I will be highest is $\{i1|\&\} = 2/3$.

How many of these statements must be true?

a. 0
b. 1
c. 2
d. 3

11. Xena does not follow the delta property. She has just paid \$100 for a certificate that allows her to call the flip of a coin. If she calls it correctly, she wins \$400: otherwise she gets nothing. Her PISP for the certificate is \$250.

A clairvoyant offers to tell Xena the outcome of the impending coin toss. What is her PIBP for the clairvoyant's information?

a. \$150
b. \$300
c. We need Xena's u-curve to determine her PIBP for the clairvoyant's information.
d. We need both Xena's u-curve and the probability she assigns to the coin landing "Heads" to answer this question.

†12. The figure shows a sensitivity analysis of value of information to the probability of Sunshine, p, for both Kim and Jane in the party problem.

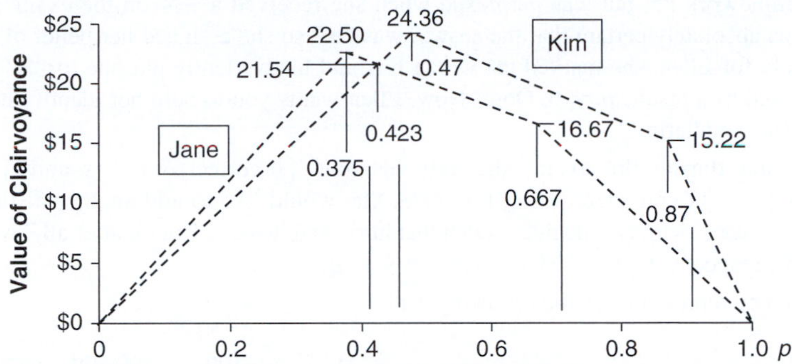

On this figure, plot the sensitivity of Mary's value of clairvoyance to p.

At a consulting project, you hear the phrase "Risk-averse people always pay more for clairvoyance than risk-neutral people." From the curves of Kim, Jane, and Mary, how would you respond to this statement?

†13. If the monetary prospects and the probabilities are the same for a risk-averse decision maker and a risk-neutral decision maker, what can you say about the statement in Problem 12.

Use the information presented here to answer Problems 14 through 17 regarding sensitivity to probability. Then use the information presented to answer Problems 18 through 22 regarding value of clairvoyance sensitivity.

Eduardo, a confused graduate student, has just won the weekly Bingo prize, and now he has to choose among three alternatives. Although he follows the five Rules of Actional Thought, he is having trouble picking and he wants your help. In the first alternative, he is given a jar of balls. Some are red, and the others are white. If Eduardo picks a red ball, he wins \$2500; if he picks a white ball, he wins \$1000. The second alternative is a similar deal. However, if he picks a red ball from the second jar, he wins \$3000; if not, he wins nothing. The third alternative is a sure \$1500. Assume Eduardo is risk-neutral.

†14. Assuming there will be the ratio of red balls in both jars, what percentage of red balls must there be for Eduardo to be indifferent between alternative 1 and alternative 2?

†15. What percentage of red balls must be in jar one for him to be indifferent between alternative one and alternative three?

†16. What percentage of red balls must be in jar two for him to be indifferent between alternative two and alternative three?

†17. Draw a sensitivity to probability graph and explain what the values, axes, and lines represent. Show where Eduardo's best alternative is represented on the graph and how it changes as the number of red balls in the jars changes. Also show the points of indifference that you found in both alternatives.

†18. On a similar graph, show how you would represent free clairvoyance.

†19. Calculate the certain equivalents for each alternative when the fraction of red balls is equal to $1/6$ and then when the fraction is equal to $5/6$. Show these values on the graph.

†20. In each case, determine Eduardo's best alternative and explain why they might be different.

†21. What is Eduardo's value of clairvoyance at these two points?

†22. Construct your own graph of the sensitivity of the value of clairvoyance to probability for Eduardo's decision problem as done for the party problem in Chapter 9. Be sure to label the axes and include the values that would be important to Eduardo in making his decision.

Use the information to solve Problems 23 through 27.

Ellen, a risk-neutral test taker, just finished taking her decision analysis midterm. She understood the scoring rule from Homework #5, but was perplexed when she received a $-\infty$ on the exam. On one of the questions she was absolutely certain that the answer was "b," so she assigned her belief of $p = s = 1$ to "b." Unfortunately for Ellen, she marked the wrong box and inadvertently put one for "c" and a zero for everything else, and as a result, $p = 0$. Oops. Now, Ellen wants you to help her identify how she should approach next year's midterm.

Suppose that during the exam, she believes that $\{\text{b correct}|\&\} = s$ and $\{\text{a correct}|\&\} = \{\text{c correct}|\&\} = \{\text{d correct}|\&\} = (1 - s)/3$. She would like to add another distinction, named "oops, I made a mistake" with two degrees "ouch that hurt" and "whew, I'm ok after all," which we'll just denote M and M′, respectively. Let $\{\text{M}|\text{b correct}, \&\} = q$.

This all can be represented as the following deal.

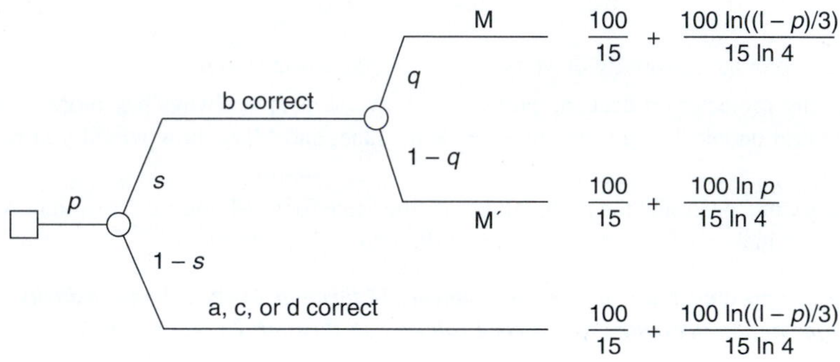

For simplicity, assume that she equally distributes the probability weights among the other choices (a, c, and d.)

†**23.** Given that $s = 1$ and $p = 1$, compute Ellen's sensitivity to the probability q. What is her e-value for $q = 1$? What about for $q = 0.00001$?

†**24.** Suppose $s = 0.7$ and $p = 0.7$. Compute Ellen's sensitivity to the probability q. What is her e-value for $q = 1$? What about for $q = 0.00001$?

†**25.** Now, include a decision about p, where p can be one of three possible values $\{0.25, 0.7, 1\}$. Suppose $s = 1$. Compute Ellen's sensitivity to the probability q and represent your results in a graph. Repeat for the case when $s = 0.7$.

†**26.** Suppose Ellen is really a deltaperson with risk odds $r = 2$. Repeat Problem 26. How do your results change? Given her new-found attitude toward risk, do the results make sense?

†**27.** Suppose Ellen is a deltaperson with risk odds $r = 1/2$. Repeat Problem 25 again. How do your results change? Do they make sense given this attitude toward risk?

CHAPTER

13

Basic Information Gathering

CHAPTER CONCEPTS

After reading this chapter, you will be able to explain the following concepts:

- Assessed and inferential forms of relevance diagrams
- Prior, likelihood, pre-posterior, posterior
- Important principles of experimentation:
  - Relevant
  - Material
  - Economic
- Sensitivity and specificity of detector
- The Value of Information
- The equivalent probability of clairvoyance

13.1 INTRODUCTION

One of the most useful features of decision analysis is its ability to distinguish between constructive and wasteful information gathering and, furthermore, its ability to place a value on information. We have already discussed the value of clairvoyance in a decision and discussed the ease of calculating this value for a deltaperson. But what if clairvoyance is not available in a given decision? Suppose, instead, that imperfect information is available that provides an indication with a certain level of accuracy. Is this information useful? Can we calculate the value of information in this case? This chapter illustrates how to calculate the value of (imperfect) information on a distinction of interest. In Chapter 18, we will extend this analysis to the value of multiple sources of information, and in Chapter 20, we will discuss the value of information from a source that provides multiple indications.

13.2 THE VALUE OF INFORMATION

To understand the notion of information gathering, we start with a pair of **relevance diagrams**, shown in Figure 13.1, that can be used for a typical information gathering or experimental process.

Each diagram has two distinctions: A **distinction of interest** and another related **observed distinction** that will be determined by the result of the experimental process. The value of the observed distinction, in fact, is derived from the relevance between it and the distinction of interest that is indicated by the relevance arrow.

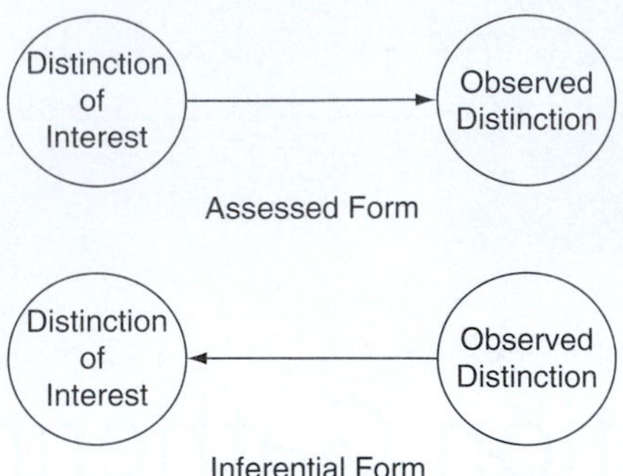

FIGURE 13.1 Relevance Diagrams for Information Gathering

The relevance diagrams in Figure 13.1 apply to many types of authentic information gathering activities. For example, the distinction of interest could be the amount of oil in a particular underground structure, and the observed distinction could be the result of seismic tests. Or, the distinction of interest could be future product sales for a company, and the observed distinction could be the results of a market survey. We could even be dealing with a medical situation, such as the presence of a disease (distinction of interest) and the results of a medical test (the observed distinction). Yet another example is in-service manufacturing costs and pilot plant manufacturing costs. As you can see, information gathering activities are a part of most decisions worthy of analysis.

Notice that which distinction we regard as the *distinction of interest* and which we regard as the *observed distinction* is somewhat arbitrary, since relevance is a symmetric property and knowing the degree of one distinction will update our belief about the other.

For example, if we refer back to our earlier example of Beer Drinker and College Graduate, we could consider the extent of beer drinking as the result of a test that is relevant to the amount of education, or the amount of education as a test that is relevant to the extent of beer drinking. The context of the decision problem will usually indicate which of the uncertainties is to be regarded as the distinction of interest and which as the observed distinction. A good way to find the distinction of interest is to ask which question you would want to ask the clairvoyant if given that chance.

Reflection

Think of other information gathering activities and characterize them using two relevant distinctions: a distinction of interest and an observed distinction.

13.2.1 Assessed Form

The relevance diagram in the upper half of Figure 13.1 is known as an **assessed form**. Probabilities representing our information are usually assessed in this way. The arrow from the distinction of interest to the observed distinction represents the conditional probability of the observed distinction given the distinction of interest. This is often called the **likelihood**. As we shall see, the likelihood probability assessment describes the accuracy of information gathering.

The other probability assessment needed to complete this form is the probability distribution on the distinction of interest. It is called the **prior** because it is assessed before the results from the experimental process are known.

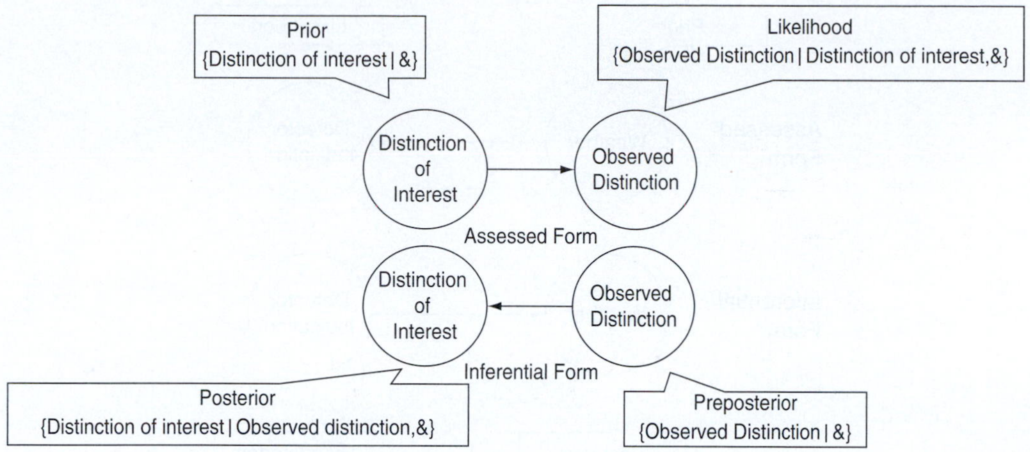

FIGURE 13.2 Relevance Diagrams for Various Distributions

13.2.2 Inferential Form

If we are to obtain information from the experimental process in its most useful form, the information contained in this relevance diagram must be placed in the **inferential form**, which is shown in the lower half of Figure 13.1. Here, the arrow from observed distinction to distinction of interest shows us the probability distribution that we will have to assign to the distinction of interest when the observed distinction is known. This updated probability is usually called a **posterior distribution** because it is our belief about the distinction of interest after the results of the experiment are known.

The other probability distribution implied by the inferential form is the probability distribution on what will be observed in the experiment—the probability distribution on the observed distinction. This is the **preposterior distribution**. Note that if there is no uncertainty in what will be observed, there is no point in performing the experiment.

We summarize the notation and terminology for experimentation in Figure 13.2.

The question we ask now is "How much is information provided by an experiment worth in making a decision?" The nature and value of this entire experimental process becomes more apparent when we apply it to the party problem that was introduced in Chapter 9.

13.3 THE ACME RAIN DETECTOR

Suppose that, as Kim is preparing to decide on her party location, she is approached by someone who offers her the services of an Acme Rain Detector. The detector will provide either an indication of Sunshine, S, or of Rain, R. Extensive testing of the detector's operation has shown that, when the weather will actually be sunny, there is an 80 percent chance that the detector will indicate sunshine, and that when the weather will actually be rainy, there is an 80 percent chance that the detector will indicate rain. Therefore, the detector's indication is a useful, but not infallible, prognostication of the weather. These conditional probabilities for detector indication given the weather are the likelihood probabilities in this experiment.

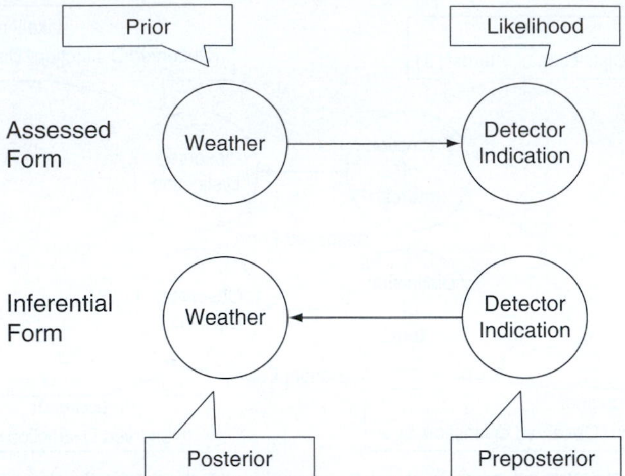

FIGURE 13.3 Relevance Diagrams for Acme Rain Detector

The experimental relevance diagrams for the Acme Rain Detector appear in Figure 13.3.

Kim has already assessed her probabilities on the weather as Sunshine S or Rain R: They constitute her prior. As shown in the assessed form, the weather is relevant to the detector's indication. When this likelihood is assessed—in this case, as an 80% chance of a correct indication—we can change the diagram into the inferential form shown in the lower part of Figure 13.3. Here, we can find the posterior probability on the weather given the detector's indication as well as the preposterior probability that the detector will produce either possible indication.

The person offering the rain detector service, however, is not an altruist. He seeks a fee of \$10 for the detector's services. So, Kim's decision has now expanded to whether she should pay \$10 to use the detector. Before we proceed any further, Kim should review the implications of her value of clairvoyance (VOC) on the weather. We found that she would be willing to pay \$20.27 to predict the weather. If the Acme service had requested more than this amount, then she could reject it outright. However, its price is about half her VOC, so there is at least a possibility that using it may be a good idea.

13.3.1 Assessed Form

To be sure, Kim will have to do some further analysis. She begins by representing the likelihood for the detector in a tree form, as shown in Figure 13.4.

Notice that there is, indeed, relevance between the two distinctions of weather and rain indication in Figure 13.4. We see that when the weather will, in fact, be rainy, there is an 80% chance of a detector indication of Rain. In some settings, such as medical testing, we also refer to this conditional probability of a detector indication as the detector's **sensitivity**. This is the chance of the detector's indicating the bad result (rain or disease) given that the bad result occurs.

Conversely, if the weather will be sunny, there is an 80% chance that the detector will indicate Sun. **Specificity** is the other conditional probability of a detector indication. This is the chance of the detector's indicating the good result (sunshine or good health) given that the good result occurs. In this simple case, the sensitivity and specificity of the Acme Rain Detector

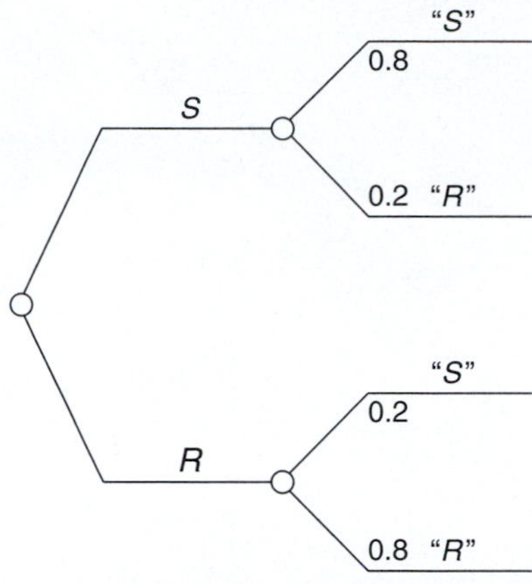

FIGURE 13.4 Assessment of Detector Properties

completely specify its operation. For simplicity, we have made the device symmetric with equal sensitivity and specificity. In medical and in most other settings, this would rarely be the case.

Kim must supplement these likelihood probabilities with her prior probabilities of the weather—0.4 for sunshine and 0.6 for Rain—to describe the experiment completely in the tree form of Figure 13.5.

We know that to use the information this tree contains we will have to reverse its order. As a first step, we compute the probabilities of each of the four elemental possibilities implied by the tree by multiplying together the probabilities along each path leading to each elemental possibility.

Reflection

If the detector sensitivity and specificity in Figure 13.5 were both equal to 100%, what would be the posterior probability of Sun for each possible detector indication? How much would the detector be worth?

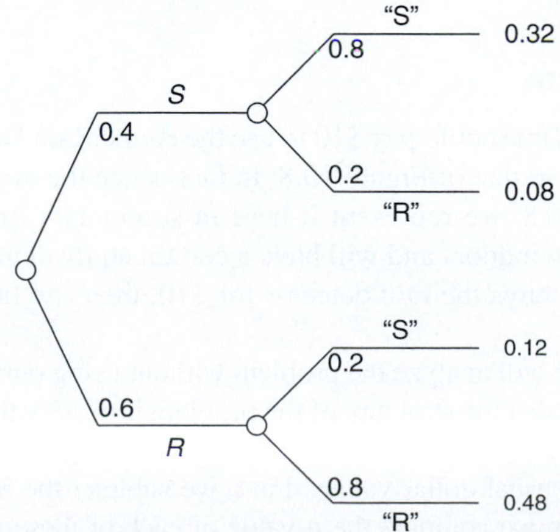

FIGURE 13.5 Complete Assessment of Experimental Situation

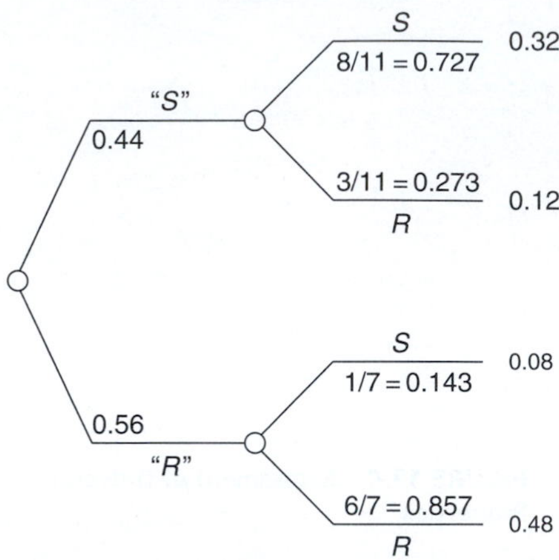

FIGURE 13.6 Experimental Inference

13.3.2 Inferential Form

In Figure 13.6, we have reversed the tree order, using our customary procedure.

We have copied the probabilities of the elemental possibilities to the correct end point of the reversed tree and then used our standard rules to distribute probabilities among the branches. We see, for example, that there is a 0.44 chance that the detector will indicate Sunny, *S*, before it is turned on; a result obtained from adding together 0.32 and 0.12. This is the preposterior probability of *S*, because it indicates the chance of the experimental indication before the experiment is performed.

The posterior probabilities on the weather, given the experimental indication, are those shown at the second layer of the tree. Since the odds are 32 to 12 that it will be sunny when the detector says Sunny "*S*", there is a probability of $8/11 = 0.727$ of sunshine following such an indication. However, if the detector indicates Rainy "*R*", there are 8 to 48 odds or probability $1/7 = 0.143$ of sunshine.

Note that after Kim turns on the detector, the one thing of which she will be certain is that her probability of sunshine will no longer be 0.4. It will become either 0.727 if the detector indicates Sunny or 0.143 if the detector indicates Rainy. Therefore, 0.727 and 0.143 are her posterior probabilities of Sunshine given the detector indications *S* and *R*, respectively. The probabilities that the detector will indicate S and R are 0.44 and 0.56, respectively (the pre-posterior probabilities in this information gathering activity).

13.3.3 Detector Use Analysis for Kim

Now we are ready to discover whether or not Kim should pay $10 to use the Acme Rain Detector. To begin, we construct a decision tree similar to that of Figure 10.8; in fact, since the upper part of the figure is identical to that of Figure 10.8, we represent it here in summary form only. Recall that, in this case, she will have the party indoors and will have a certain equivalent of $46 ($45.83 to be more precise). However, if she buys the rain detector for $10, then she faces the tree shown in the lower part of Figure 13.7.

In order to observe a general analysis, we will analyze the problem without using our knowledge that Kim satisfies the delta property. Note that the structure of the problem is exactly the same as that for clairvoyance; the only difference is in the probabilities that we will be assigning.

The values at the tips of the tree are the usual dollar values. First, we subtract the $10 cost of the experiment from each of the values, then we compute the *u*-value of each of these reduced

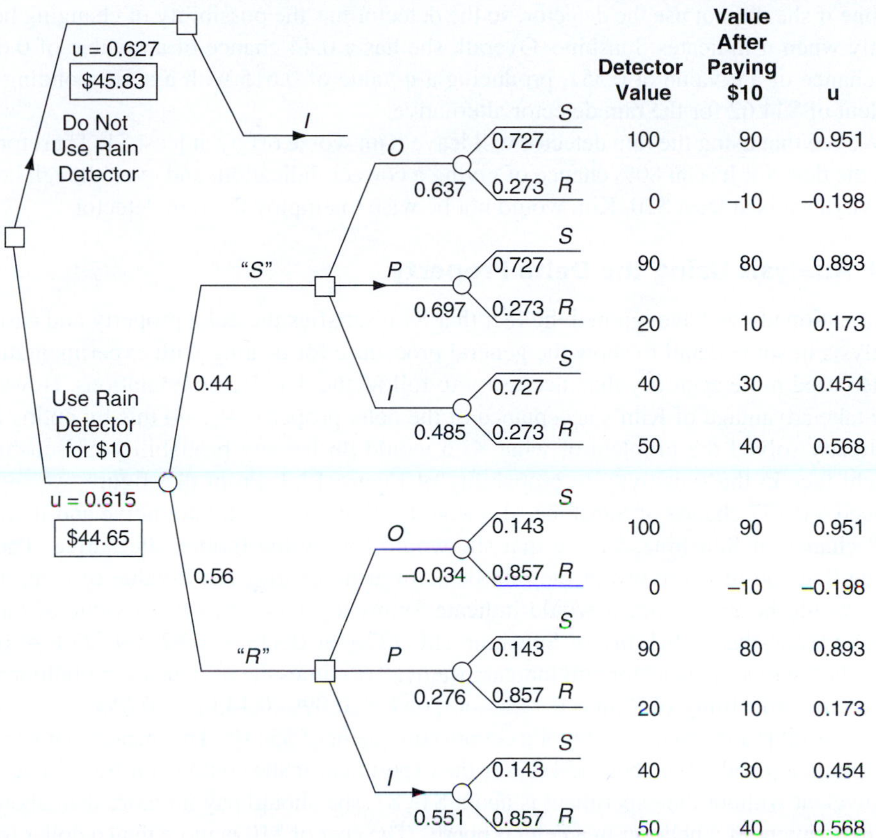

FIGURE 13.7 Kim's Detector Use Analysis

amounts, as shown in the column at the far right of the figure. Next, we copy the appropriate probability assignments into the figure—for example, the preposterior probabilities of 0.44 and 0.56 that the detector will indicate Sunny and Rainy, respectively. Following each indication, we then record the proper conditional probability of sunshine and rain, as computed in Figure 13.6. Therefore, following the Sunny indication, there will be a 0.727 chance of sunshine and, following the Rainy indication, there will be a 0.143 chance of sunshine. Now, we can roll back this tree to see whether the *u*-value of buying the rain detector service is greater than the *u*-value of 0.63 of doing no experiment or, conversely, whether the certain equivalent of buying the rain detector at a cost of $10 is greater or less than $45.83, which is the certain equivalent of performing no experiment. 0.2

As we proceed, we find that following the branch of a Sunny indication *S*, the decision to have the party outdoors will produce a 0.727 chance of a 0.951 *u*-value and 73 chance of a −0.198 *u*-value, so a *u*-value for this alternative is 0.637. The Porch alternative in this situation produces a *u*-value of 0.697, and the Indoor alternative has a *u*-value of 0.485.

Consequently, the Porch alternative is best and, if the detector indicates Sunny, Kim will decide to have the party on the porch. Notice that this is a change from the decision she would make otherwise to have it indoors; the detector's indication is material to her decision.

The alternatives following the Rainy indication have *u*-values of −0.034, 0.276, and 0.551. Clearly, the Indoor alternative is best when the detector indicates Rainy. This is what Kim would have done if she did not use the detector, so the detector has the possibility of changing her decision only when it indicates Sunshine. Overall, she has a 0.44 chance of a u-value of 0.697 and a 0.56 chance of a u-value of 0.551, producing a u-value of 0.615 with a corresponding certain equivalent of $44.62 for the rain detector alternative.

We see that using the rain detector will leave Kim worse off by at least $1. Therefore, even though the detector has an 80% chance of giving a correct indication, and even though her value of clairvoyance is at least $20, Kim would not be wise to employ the rain detector.

13.3.4 Analysis Using the Delta Property

As we mentioned, we have ignored the fact that Kim satisfies the delta property and carried out the analysis in some detail to show the general procedure for dealing with experimentation. We have also used more accuracy than necessary to follow the detailed computations. However, let us now take advantage of Kim's acceptance of the delta property. We do this by noting that we have already solved the problem of what Kim would do for any probability of Sunshine that she might face in the sensitivity to probability of Figure 12.2. From that figure, we see that if Kim faced a 0.727 chance of Sunshine, she would follow the Porch alternative and if she faced a 0.143 chance of Sunshine, we see that she would follow the Indoor alternative. Therefore, assuming that she received use of the rain detector at no charge, her u-value of using the free detector would be the chance it would indicate Sunny at, 0.44 times the u-value of the Porch alternative when the probability of Sunshine is 0.727—or $0.323 + 0.628(0.727) = 0.780$—plus the chance that the detector will indicate Rainy, 0.56 times the u-value of the Indoor alternative when the probability of Rain is 0.143—or $0.667 - 0.099(0.143) = 0.653$.

The result is a u-value of 0.709 or a certain equivalent of $54.65. This means that Kim would have a certain equivalent of about $54.65 for the experiment if she could get it free. Since her certain equivalent without the experiment is about $45.83, she should pay no more than about $8.82 for the experiment as a believer in a delta property. The cost of $10 is more than a dollar too high. Relying on intuition, we might at first be tempted to think that the value of an 80% accurate detector is 80% of the VOC. As we have seen, however, this is not the case.

> **Note**
>
> The worst possible detector we can face is one with a 50% accuracy, as it is irrelevant to the distinction of interest.

13.3.5 Detector Use Analysis for Jane

We can use this same approach to determine rather quickly whether the detector is a good idea for Jane. Recall that Jane's value of clairvoyance was $22.00. We use Figure 12.6 to compute Jane's certain equivalents for each possible detector reading. If the detector reads "S," the probability of Sunshine will be 0.727, and she will choose to have the party outdoors with a certain equivalent of $100p = \$72.70$. If the detector reads "R," then the probability of Sunshine will be 0.143, and she will choose to have the party indoors with a certain equivalent of $50 - 10p$ or $50 - 10(0.143) = \$48.57$. Using the probabilities of these indications, we obtain the certain equivalent of a free detector to the risk-neutral Jane as $0.44(72.70) + 0.56(48.57) = \$59.19$. Since Jane's certain equivalent of the party without using the detector is $48.00 obtained by having the party on the porch, the value of the detector to her is $11.19. If the detector is offered at a cost of $10.00, Jane will make a profit of over $1.00 by using it.

Notice the difference in the materiality of the detector for Kim and for Jane. Jane's decision will change from her current alternative of having the party on the porch to either having the party outdoors for a Sunshine indication or indoors for a Rain indication. Therefore, Jane will benefit from both indications of the detector, while Kim will benefit from only one. This is one reason that using the Acme Rain Detector at a cost of $10 is a good deal for Jane, but a bad deal for Kim.

13.4 GENERAL OBSERVATIONS ON EXPERIMENTS

Experimental decisions—no matter how complicated—can be handled by exactly the same principles and using exactly the same tools illustrated in the party problem.

13.4.1 Important Principles of Experimentation

There are three important principles that guide experimentation or information gathering: First, the relevance of the result to the distinctions of interest; second, the possibility that the result will change the decision; and third, whether it is economic—worth the cost.

RELEVANCE First, an information gathering process is worth considering only if its results are relevant to the distinctions of interest in the decision problem. If we were considering an investment decision in an oil company, for example, it would not be prudent to spend money or time on an experiment to determine the outcome of a coin toss, since we believe the coin toss to be irrelevant to the distinction of interest.

MATERIAL Second, an information gathering process is worth considering only if the observations it produces have the possibility of changing the decision that is made. We say that the process must be material to the decision.

For example, suppose we conducted a sensitivity analysis to the sensitivity and specificity values of the detector in Figure 13.4 but kept their values equal:

$$\text{Sensitivity} = \{\text{"S"}|S,\&\} = \{\text{"R"}|R,\&\} = \text{Specificity} = p$$

The previous analysis focused on the case where $p = 0.8$. Figure 13.8 conducts a sensitivity analysis to this value and shows the posterior probabilities of Sun, $\{S|\text{"S"},\&\}$ and $\{S|\text{"R"},\&\}$, for different values of p, when the prior probability of Sunshine $\{S|\&\}$ remains equal to 0.4.

We observe several important points from Figure 13.8. First, when the detector accuracy is 0.5 $(\text{Sensitivity} = \text{Specificity} = 0.5)$, Kim's (or Jane's) posterior probability of Sun is equal to her prior probability, 0.4, no matter what the detector says. The rationale, of course, is that the detector is irrelevant to the weather in this case, so she would not update her belief about the weather when she gets this indication.

We also note that the higher the detector accuracy, the higher is the posterior probability of Sun when the detector says "S," and the lower is the posterior probability when the detector says "R." This makes intuitive sense, since she would be more likely to believe a more accurate detector.

Note that, as the accuracy becomes $p = 1$, the detector indication is now clairvoyance and her posterior probability is either 0 or 1, depending on the detector indication. Note further that Kim's posterior probability is also 0 or 1 when the detector has a 0% accuracy $(p = 0)$. Of course, if you knew the detector always gives a wrong indication, it is just as good as a detector that always gives a true indication, since you will just believe the opposite of its indication.

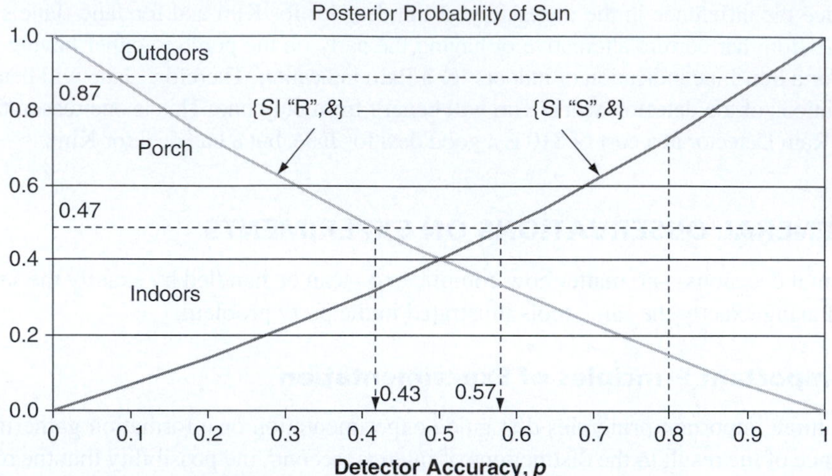

FIGURE 13.8 Sensitivity to Detector Accuracy

Recall from Figure 12.2 that when Kim's probability of Sun is less than 0.47, she would go indoors; when it is between 0.47 and 0.87, she would go on the porch; and when it is higher than 0.87, she would go outdoors. The posterior probabilities of 0.47 and 0.87 are presented as dashed lines in Figure 13.8. Since Kim will have the party indoors without the detector indication, what values of p will make her change her decision for some detector indication, and what values will not change her decision?

From Figure 13.8, we see that if the detector accuracy is anywhere between $p = 0.43$ and $p = 0.57$, Kim will always go indoors no matter what the detector says, since her posterior probabilities of Sun, in this case, will always be less than 0.47. We call such a detector an **immaterial detector** for this decision. Of course, it may be material for other decisions depending on the probabilities and preferences involved.

Figure 13.8 also shows the posterior probability values when $p = 0.8$ for the Acme detector, where $\{S|\text{"S"},\&\} = 0.727$ and $\{S|\text{"R"},7\} = 0.143$ matching our previous analysis, and asserting that this detector is a material detector for the given prior probability of Sun.

ECONOMICAL Third, an information gathering process is worth considering only if using it is economical in terms of the overall decision. The Acme Rain Detector was relevant and material for both Kim and Jane, but economical only for Jane.

13.4.2 Experiments and the Value of Clairvoyance (VOC)

The idea of the value of an experiment already resides in the value of clairvoyance (VOC). To illustrate this relation, note that the value of an experiment is simply the value of clairvoyance on the observations of the experiment. Therefore, to determine the value of an Acme Rain Detector, we consider the detector indication itself as a distinction in the diagram. We then calculate the value of clairvoyance for this observed distinction. The value of clairvoyance on the detector indication distinction is equal to the value of the detector.

Every information gathering process, regardless of its form, logically requires the steps that we described in Section 13.2.1. That is, we first must describe the experiment in terms of the likelihood, and next, we must describe of our original knowledge in terms of the prior. Then, we must process this information to provide both the posterior distribution, which shows what we will learn from different results of the experiment, and the preposterior distribution, which will show us the chance of obtaining each of these results.

It is important to note that the detector accuracy (sensitivity and specificity) alone is not sufficient to determine the quality of a test. We need also to consider the prior probabilities. We illustrate this notion through the following example.

EXAMPLE 13.1 Testing for a Virus

Suppose you are testing for a virus, and you believe that 1/400th of the population is infected. A medical test is available. If used on an infected person, the test is 99% likely to indicate "positive" The person is indeed infected. If the person is not infected, the test is 99% likely to indicate "negative" Then the person is not infected.

Suppose you took the test and the test indicates "positive" What is the probability that you have the virus? If the test indicates "negative," what is the probability that you have the virus?

To answer these questions, we first think about the problem in assessed form, as shown in Figure 13.9. In the medical domain, the conditional probability {"Pos" | Virus,&} is often referred to as the sensitivity of the test, as we discussed earlier, and {"Neg" | No Virus,&} is its specificity. These conditional probabilities are the likelihood of a test indication given the person has the virus or does not have the virus.

Now we reverse the tree and draw the relevance diagram in inferential form, as shown in Figure 13.10.

Figure 13.10 shows that if the test indicates "positive," the conditional probability that the person has the virus is only {Virus | "Pos",&} = 0.2, while if the test indicates "negative," the probability the person does not have the virus is {No Virus | "Neg",&} = 0.999975.

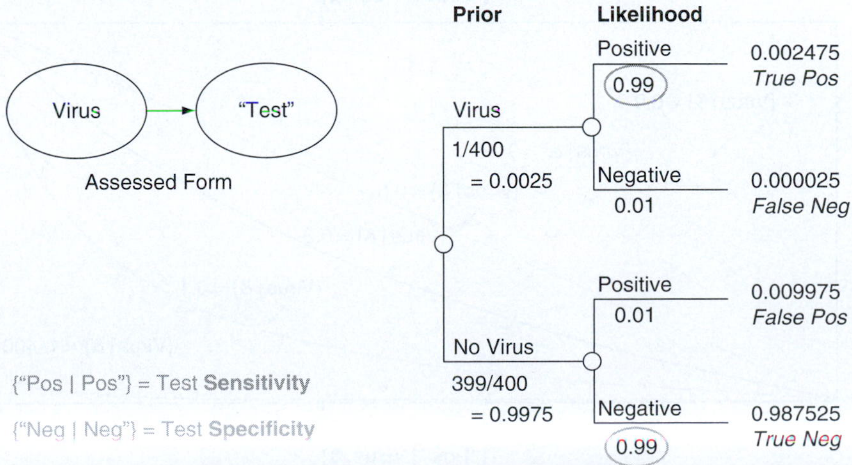

FIGURE 13.9 Assessed Form for Medical Testing

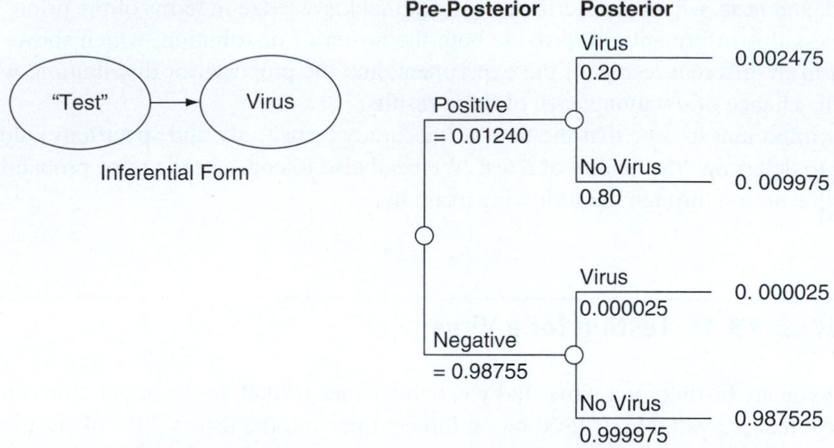

FIGURE 13.10 Inferential Form for Medical Testing

The test is a good test when it gives a negative indication, but it is not a good test when it gives a positive indication. This is because it is not sufficient to mention that a test has a 99% accuracy without comparing it to the actual prior that is being tested, $\{\text{Virus}|\&\} = 1/400$.

To illustrate further, Figures 13.11 and 13.12 show the updated conditional probabilities of $\{\text{Virus}|\text{"Pos"},\&\}$ and $\{\text{No Virus}|\text{"Neg"}|,\&\}$ versus the test accuracy. Each curve represents a different prior probability level for $\{\text{Virus}|\&\}$. If $\{\text{Virus}|\&\} = 0.1$, then a 99% accurate test would give $\{\text{Virus}|\text{"Pos"},\&\} = 0.916$ and $\{\text{No Virus}|\text{"Neg"},\&\} = 0.998$. This is a big improvement for the use of the test upon a positive indication. Notice at $\{\text{Virus}|\&\} = 0.5$, both curves of Figures 13.11 and 13.12 are identical due to the symmetry of the problem, and the curves are mirror images of each other around this value.

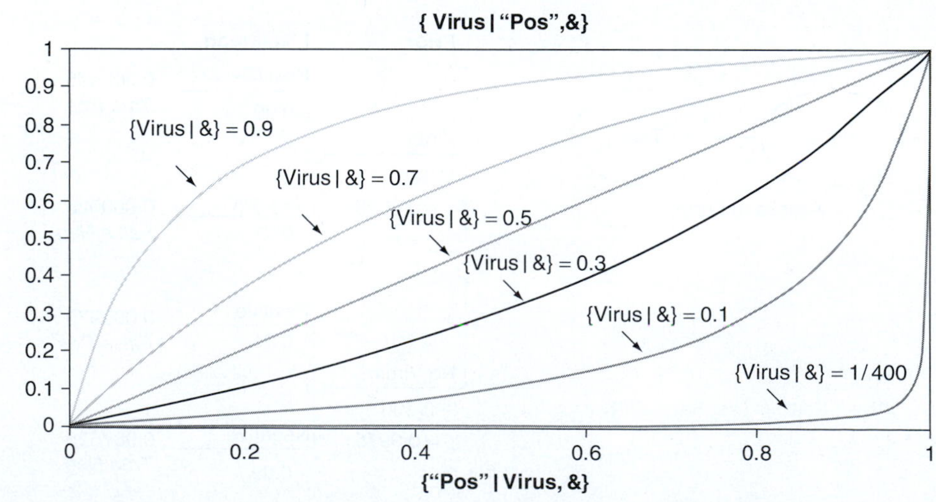

FIGURE 13.11 Plot of {Virus|"Pos",&} with Sensitivity and Prior

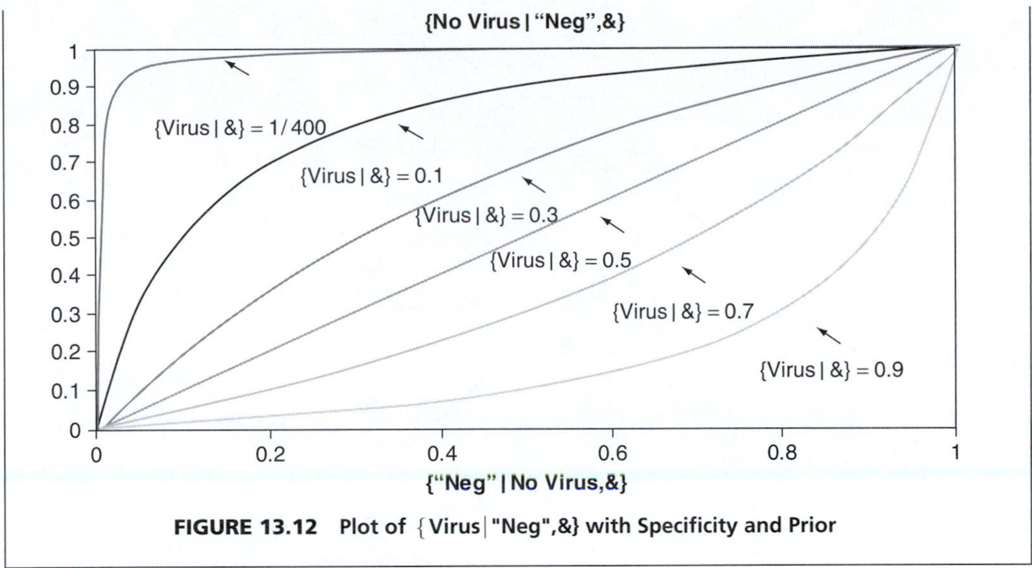

FIGURE 13.12 Plot of {Virus|"Neg",&} with Specificity and Prior

13.5 ASYMMETRIC EXPERIMENTS

The Acme Rain Detector had the same chance of saying Sunny when the weather is going to be sunny as it did of saying Rainy when the weather was going to be rainy: 0.8 in both cases. There is no reason why an experiment need be equally accurate for both degrees of the distinction of interest. In fact, experiments that are asymmetric in this sense are the rule rather than the exception. We can analyze **asymmetric experiments** just as easily as we can symmetric ones. (We can also handle any number of experimental indications—not just two.)

13.5.1 Assessed Form

We can analyze a two-indication asymmetric experiment on the occurrence of a simple dichotomy as shown in Figure 13.13. In the upper part of the figure, we show the experiment in assessed form. The event of interest E will occur with probability p. If it does occur, the experiment will report its occurrence "E" with probability t. We can think of p as representing the probability of the occurrence of the event as truly reported. If the event does not occur, E', then the probability that the experiment will say so, namely, "E'" is f. We can think of f as representing a correct report of the falsity that the event occurred.

Thus, the characteristics of the experiment are described by the two numbers t and f; for the Acme Rain Detector, they will both be 0.8. These numbers are sometimes given special names in particular situations. For example, in the field of medicine, the event E is often thought of as a disease that may be detected by the experiment. In this case, the number t is called the sensitivity of the test and the number f is specificity of the test. All such two-indication experiments on a dichotomous simple event may be analyzed within the structure we have represented.

13.5.2 Inferential Form

The bottom portion of Figure 13.12 shows the flipped probability tree that represents the inferential form of the experiment. The tree is reversed according to our usual procedures.

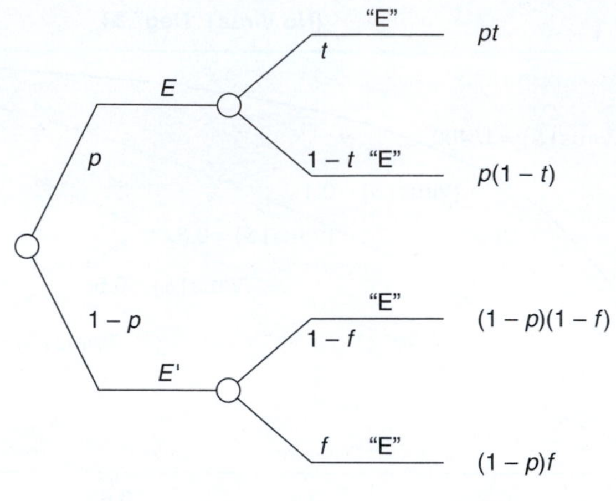

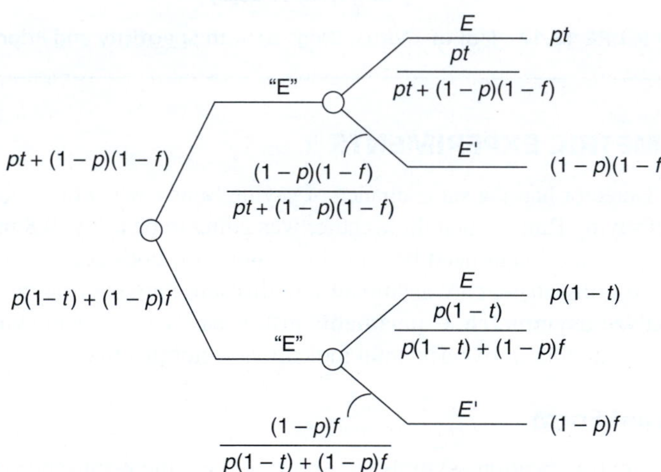

FIGURE 13.13 Tree Analysis of an Asymmetric Experiment

We see that the preposterior probability that the experiment will report "E" is

$$\{\text{"E"}|\&\} = pt + (1 - p)(1 - f)$$

The posterior probability that E has in fact occurred when such a report is given is

$$\{E|\text{"E"},\&\} = \frac{pt}{pt + (1 - p)(1 - f)}$$

If, on the other hand, the nonoccurrence of *E* is reported, "E'", then the probability that *E* has occurred is

$$\{E|\text{"E'"},\&\} = \frac{p(1 - t)}{p(1 - t) + (1 - p)(1 - f)}$$

For any numbers p, t, and f that specify the assessed form of the experiment and the prior information, we can use the inferential form in the lower half of Figure 13.12 to compute both the preposterior and posterior probabilities associated with the experiment or information gathering process.

13.5.3 The Asymmetric Acme Rain Detector

Figure 13.14 uses the results of Figure 13.13 to reproduce and generalize the analysis of the Acme Rain Detector for Kim. To perform this analysis, we note that the event E is S and E' is R. Further, for the Acme Rain Detector, the probability of Sunshine is $p = 0.4$.

Columns 1 and 2 show the probabilities of a true positive and true negative report from the detector, t, f. These values are also the sensitivity and specificity of the test, respectively. Column 3 shows the posterior probability of sunshine given that the detector gives a Sunny indication, S. Column 4 shows Kim's best decision given an indication of S. Column 5 is a u-value associated with this alternative given the detector indication, again from Figure 12.2. Column 6 shows the preposterior probability of a Sunny indication from the detector.

The next four columns are analogous to the preceding four, except they show a quantity associated with a report of Rainy from the detector, R. Column 11 is the u-value of the experiment computed by multiplying the u-value of a sunny report from Column 5 by the probability of such a report from Column 6 and adding the u-value of a Rainy report from Column 9 multiplied by the probability of a Rainy report from Column 10. Finally, Column 12 shows the certain equivalent associated with the u-value of Column 11.

1	2	3	4	5	6
Sensitivity t	Specificity f	Posterior Prob Given "S" {S|"S",&}	Best Party Location Given "S"	u-Value of Party Location Given "S"	Preposterior {"S"|&}
0.8	0.8	0.727	Porch	0.780	0.44
0.9	0.7	0.667	Porch	0.742	0.54
0.7	0.9	0.824	Porch	0.840	0.34
0.6	1.0	1.0	Outdoor	1.0	0.24

7	8	9	10	11	12
Posterior Prob Given "R" {S|"R",&}	Best Party Location Given "R"	u-Value of Party Location Given "R"	Pre-Posterior {"R"|&}	u-Value of Experiment	CE with free Experiment
0.143	Indoor	0.653	0.56	0.709	$54.70
0.087	Indoor	0.658	0.46	0.703	$54.00
0.182	Indoor	0.649	0.66	0.714	$55.30
0.211	Indoor	0.646	0.76	0.731	$57.30

FIGURE 13.14 Sensitivity Analysis of Asymmetric Rain Detector

The first row in Figure reproduces the analysis of the Acme Rain Detector we have previously developed by setting both t and f to 0.8. We see that Kim will have the party on the porch if the detector indicates Sunny, and indoors if the detector indicates Rainy. Her u-value is 0.709 and her certain equivalent for a free detector is $54.7. Recall that Kim's certain equivalent without the detector was almost $46. Consequently, Kim, as a person accepting the delta property, would not pay $10 for the use of the detector.

The second row of the figure analyzes a detector that is more accurate in Sunshine and less accurate in Rain. It has a 0.9 probability of correctly indicating sunshine, but only a 0.7 probability of correctly indicating Rain. We see that this detector yields the same actions in the face of its reports as it does the original Acme Rain Detector. However, its u-value is 0.703; therefore, it is somewhat less valuable to Kim than the original detector.

The third row of the figure illustrates a detector that is asymmetric in the other direction. The probability that it will correctly indicate Sunshine is 0.7; whereas, the probability that it will correctly indicate Rain is 0.9. This detector again causes the same decisions in the face of its reports, but now the u-value is 0.714 with a certain equivalent of $55.3.

We can think of these two detectors that we have just analyzed as elements of a sensitivity analysis to asymmetry in the behavior of the detector. Observe that in Kim's situation, she would prefer a detector that was better at detecting Rain even at the expense of poorer performance in detecting Sunshine.

The last row of the table shows the result of an even greater asymmetry in the detector. Here it detects Rain perfectly: $f = 1$, but the probability that it will correctly detect Sunshine is only 0.6. Note that the effect of this behavior is that the posterior probability of sunshine given a report of Sunny is 1. If the detector says Sunny, you are sure to see sunshine. Note that, in the face of such a report, Kim will change her decision and have the party outdoors. The u-value of the detector is 0.731 with an associated certain equivalent of $57.3. This detector is worth more than $2 more than the Acme Rain Detector of the first row. In fact, Kim would be pleased to pay $10 to use it.

As we have said, any type of information gathering process can be analyzed by the methods we have described. Sensitivity analyses to their characteristics, such as the ones performed in Figure 13.14, provide important indications of the characteristics of a process that would be most effective in helping the decision maker.

We can also conduct a sensitivity of Kim's value of the detector for different values of the detector sensitivity and specificity. This is shown in Figure 13.15.

Note that the value of the detector is zero if both the sensitivity and specificity are equal to 0.5. This is an intuitive result since, as previously discussed, the detector indication of Figure 13.4 would no longer be relevant to the weather. Note also that the value of the detector is equal to the value of clairvoyance when both sensitivity and specificity are equal to one.

13.6 INFORMATION GATHERING EQUIVALENTS

We sometimes find it useful to gain insight into the efficacy of information gathering schemes by comparing them with standard information gathering schemes. One way to do this is to recognize that since any information gathering scheme cannot have a negative value and cannot have a value greater than that of clairvoyance, we can characterize it as being equivalent to some chance that clairvoyance will be available. That is, we can represent any information gathering scheme by some probability p that we receive clairvoyance and a probability $1 - p$ that we will be told no clairvoyance is available and that we will receive no indication or help at all.

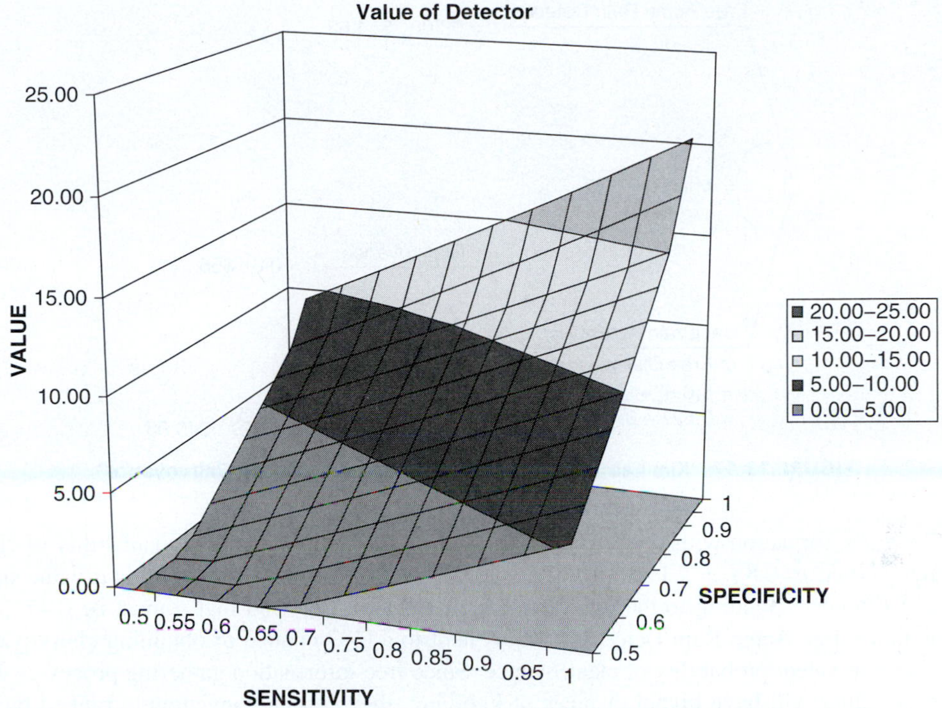

FIGURE 13.15 Kim's Detector Value for Different Sensitivity and Specificity Values

Figure 13.16 shows this representation graphically. We ignore all costs and simply ask, "What chance of receiving clairvoyance would make us just indifferent to the information gathering scheme under consideration?"

13.6.1 Kim's Equivalent Probability of Clairvoyance for Acme Rain Detector

We illustrate the construction of this equivalent probability of clairvoyance by applying it to Kim's party problem when she faces the possible use of the Acme Rain Detector. What chance of free clairvoyance would just make her indifferent to a free Acme Rain Detector?

Figure 13.17 shows the calculations. As we found in our discussion of Figure 10.6, a free Acme Rain Detector has a u-value of 0.708 and a certain equivalent of \$54.62. Figure 12.2 shows that, for her probability of 0.4 of Sunshine, Kim will have a 0.8 u-value for clairvoyance and a

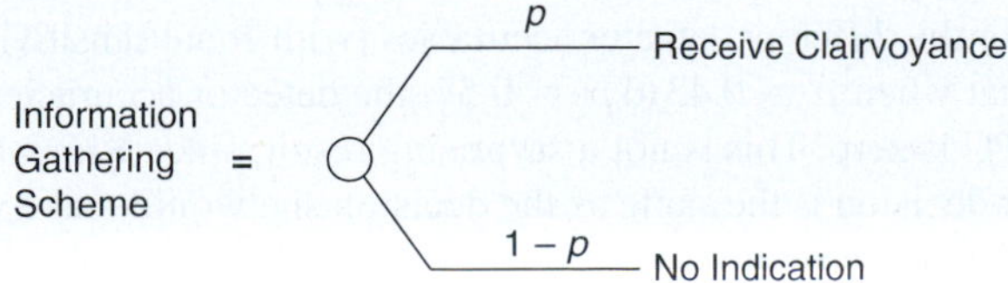

FIGURE 13.16 Information Gathering as Equivalent Probability of Clairvoyance

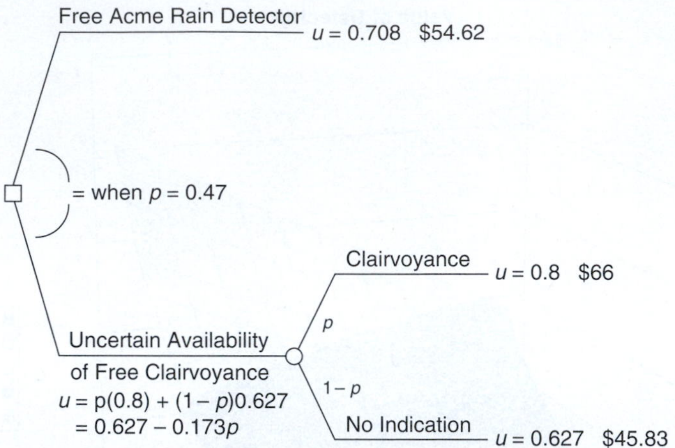

FIGURE 13.17 Kim Calculates the "Equivalent Probability of Clairvoyance"

0.627 u-value for action with no clairvoyance. The u-value of the uncertain availability of clairvoyance is then $p(0.8) + (1 - p)0.627 = 0.627 - 0.173p$. When we set this u-value equal to the 0.708 corresponding to the free Acme Rain Detector, we find that p must be 0.47. This means that a free Acme Rain Detector is equivalent to a 0.47 chance of obtaining clairvoyance or a 0.47 equivalent probability of clairvoyance. Since free information gathering processes with higher u-values will have higher chances of knowing, they can be conveniently ranked by the chance of knowing index.

To reconcile this calculation with our earlier results, recall from Chapter 10 that we found Kim's value of free clairvoyance to be $20.26. The chance she will receive clairvoyance from the detector is the chance of knowing, 0.47. Therefore, she has a 0.47 chance of receiving something worth $20.26, and a 0.53 chance of receiving nothing. For someone with Kim's risk preference (exponential, risk tolerance $72.13), the certain equivalent of this deal is $8.82, just the value of a free detector we found earlier.

In general, the **equivalent probability of clairvoyance (EPC)** of p can be calculated using the u-values of free detector, no detector, and free clairvoyance as

$$\text{EPC} = \frac{u(\text{Free Detector}) - u(\text{No Detector})}{u(\text{Free Clairvoyance}) - u(\text{No Detector})}$$

Notice that if we multiplied all u-values by a constant, a, or added a constant, b, the equivalent probability of clairvoyance remains the same. Furthermore, a detector with a higher accuracy will have a higher equivalent probability of clairvoyance.

For example, Figure 13.18 plots a sensitivity analysis for the equivalent probability of clairvoyance (EPC) for the different detector accuracies (with equal sensitivity and specificity) of Figure 13.17. Note that when $p = 0.43$ to $p = 0.57$, the detector accuracies that make it immaterial: the value of EPC is zero. This is not a surprising result, since Kim would go indoors with any indication, so her decision is the same as the decision she would make without a detector.

We can also calculate the equivalent probability of clairvoyance for the asymmetric detectors. For example, Figure 13.19 shows Kim's equivalent probability of clairvoyance for the asymmetric detectors of Figure 13.14.

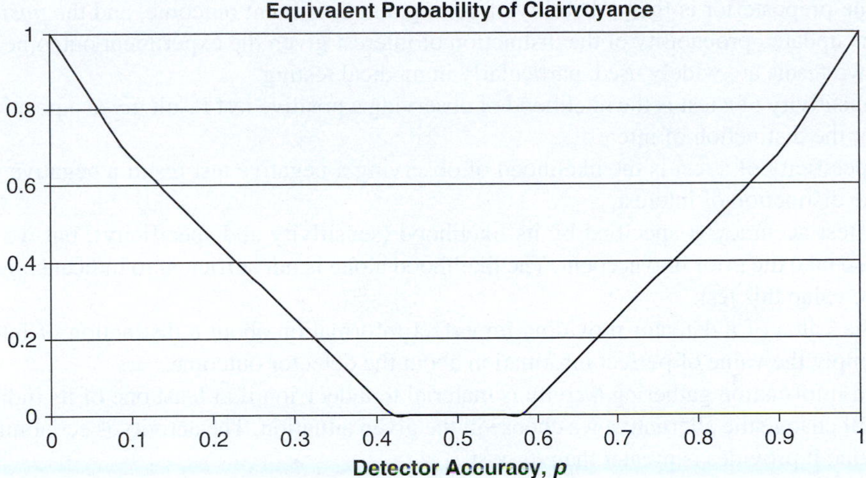

FIGURE 13.18 Sensitivity Analysis

1	2	3	4
Sensitivity *t*	Specificity *f*	*u*-Value of Experiment	EPC
0.8	0.8	0.709	0.47
0.9	0.7	0.703	0.44
0.7	0.9	0.714	0.50
0.6	1.0	0.731	0.60

FIGURE 13.19 Equivalent Probability of Clairvoyance

Note that the equivalent probability of clairvoyance does not require money. We could have determined it directly for Kim in the party problem in Chapter 10 using her preference probabilities. However, the equivalent probability of clairvoyance does not tell you what to pay for the detector indications. If the costs of gathering information are included, the chance of knowing index is still useful: The values of any experiments must be ordered the same as their equivalent probability of clairvoyance, and we would never choose a more expensive experiment over one with a higher epc.

13.7 SUMMARY

- To determine how an information gathering activity (or experiment) can be used to update belief, we first think of the assessed form and consider the conditional probabilities of the experimental outcomes for each value of the distinction of interest. Next, we reverse the order of the tree and think of inferential form—how our belief changes given the results of the experiment.
- The ***prior*** is the probability we assign to the distinction of interest. The likelihood is the conditional probability of getting an experiment outcome given the distinction of interest.

The preposterior is the probability of getting an experiment outcome, and the ***posterior*** is the updated probability of the distinction of interest given the experiment outcome.

- Two terms are widely used, particularly in medical testing.
- Sensitivity of a test is the likelihood of observing a positive test result given a positive state for the distinction of interest.
- Specificity of a test is the likelihood of observing a negative test result a negative state of the distinction of interest.
- A test accuracy is specified by its likelihood (sensitivity and specificity), but we should also take the prior into account. The likelihood alone is not sufficient to indicate how much we value this test.
- The value of a detector providing imperfect information about a distinction of interest is simply the value of perfect information about the detector outcome.
- An information gathering activity is material to a decision if at least one of its indications will change the alternative we choose in the given situation. The activity is economic if the value it provides is greater than its cost.
- The value of an information gathering activity can also be expressed in terms of an equivalent probability of clairvoyance (EPC) index.
- For a detector (or any information gathering activity) to be used in a decision, it should be:
    - Relevant to the distinction of interest.
    - Material to the decision situation.
    - Economic: its value is higher than its cost.

PROBLEMS

Problems marked with an asterisk (*) are considered more challenging.

Use the following information to answer Problems 1 and 2:

Eudaemonic Pie, Inc. sells computers which predict the outcome of a spin of a roulette wheel. If someone enters the position of the ball and the approximate speed of the wheel at a given instant in time, the computer will tell you whether to bet on "red," "black," or "green." Normally, red, black, and green arise as the outcomes of any given spin with probabilities 18/38, 18/38, and 2/38, respectively. The company rigorously tests their computer system and finds that given the outcome, the probabilities of each computer reading are as follows.

Computer Indication **Actual outcome**	**"Red"**	**"Black"**	**"Green"**
Red	0.6	0.2	0.2
Black	0.2	0.6	0.2
Green	0.2	0.2	0.6

At the roulette table where you are sitting, $10 bets are taken for each spin, and you can stake your bet on the outcome being red, black, or green. The payoff matrix is as follows.

Bet **Outcome**	**"Red"**	**"Black"**	**"Green"**
Red	$20	$0	$0
Black	$0	$20	$0
Green	$0	$0	$180

You always have the option not to place a bet.

1. Suppose you are risk-neutral and that the company is offering you the use of their computer for $5 per bet. What should you say to the company?
   a. "Thanks, I'll accept your offer, and place a bet."
   b. "No thanks, your offer is too expensive; I'll go ahead and bet without using your computer."
   c. "No thanks; although using your computer at the price you quote would be to my advantage if I were to place a bet, I'm still better off not betting at all."
   d. "No thanks; using your computer at the price you quote would not be to my advantage if I were to place a bet. I choose not to bet at all."

2. Now suppose that you are a deltaperson with risk odds of 3 on a deal where you can win or lose $10. The company offers the use of their computer to you for $5 per bet. What should you say to the company?
   a. "Thanks, I'll accept your offer, and place a bet."
   b. "No thanks, your offer is too expensive; I'll go ahead and bet without using your computer."
   c. "No thanks; although using your computer at the price you quote would be to my advantage if I were to place a bet, I'm still better off not betting at all."
   d. "No thanks; using your computer at the price you quote would not be to my advantage if I were to place a bet. I choose not to bet at all."

Use the following information to answer Problems 3 through 5.

Eudaemonic Pie, Inc. has just come out with Model II of their roulette-predicting computer. They are now able to tune their machine so that its predictions are as accurate as one wants them to be. The accuracy of the Model II is controlled by setting a parameter p which ranges in the interval $[0, 1]$. The accuracies as a function of p are represented in the following chart.

Reading Outcome	"Red"	"Black"	"Green"
Red	$(3 + 2p)/5$	$(1 - p)/5$	$(1 - p)/5$
Black	$(1 - p)/5$	$(3 + 2p)/5$	$(1 - p)/5$
Green	$(1 - p)/5$	$(1 - p)/5$	$(3 + 2p)/5$

3. Suppose that you are risk neutral and the company offers the use of their Model II machine to you for one bet for a price of $\$5 + 10p$, for any p you choose in $[0, 1]$. What should be your response to the company?
   a. "Thanks, I'll buy it with $p = 1$, and place a bet."
   b. "Thanks, I'll buy it with $p = 0$, and place a bet."
   c. "No thanks; although using your computer at the price you quote for some p would be to my advantage if I were to place a bet, I'm still better off not betting at all."
   d. "No thanks; using your computer at the price you quote for any p would not be to my advantage if I were to place a bet. I choose not to bet at all."
4. Suppose Justin is a deltaperson with risk odds of 3 on a deal where he can win or lose \$10. Suppose that Justin has bought the rights to use the Model II machine for the next roulette spin, at some value of p. Now the company offers Justin an "upgrade" to change p to $p + 0.1$. Which of the following statements are true?
   a. Justin's PIBP for the upgrade would be lower if $p = 0.1$ than if $p = 0.9$.
   b. Justin's PIBP for the upgrade would be higher if $p = 0.1$ than if $p = 0.9$.
   c. Justin's PIBP for the upgrade would be the same regardless of the value of p.
   d. You need to know how much the rights to use the computer cost in order to compute Justin's PIBP for an upgrade.
5. Suppose that you are a deltaperson with risk odds of 3 on a deal where you can win or lose \$10. The company offers the use of their Model II computer to you for one bet at a price of $\$5 + 10p$ for any p that you choose in $[0, 1]$. What should you say to the company?
   a. "Thanks, I'll buy it with $p = 1$, and place a bet."
   b. "Thanks, I'll buy it with $p = 0$, and place a bet."
   c. "No thanks; although using your computer at the price you quote for some p would be to my advantage if I were to place a bet, I'm still better off not betting at all."
   d. "No thanks; using your computer at the price you quote for any p would not be to my advantage if I were to place a bet. I choose not to bet at all."
6. Recall Kim's party decision. What is the minimum "accuracy" that a rain detector would need for Kim to be willing to pay \$10 for its use in helping her make her party decision? By "accuracy," we mean the probability that the detector indicates Sunshine when it will actually be sunny, which equals the probability that the detector indicates Rain when it will actually be rainy.
   a. 90%
   b. 83%
   c. 57%
   d. None of the above
7. Suppose that Kim is offered not a "weather detector" but a "sunshine detector" to help her make a decision on the party problem—when the weather is actually going to be sunny, the detector is 99% likely to indicate Sun, but when the weather is going to be rainy, the detector is equally likely to indicate Sun or Rain.

Which of the following is closest to the most Kim should be willing to pay to acquire this detector?

a. \$0 c. \$5
b. \$3 d. \$10

8. Which of the following statements must be true?
   I. Since Mary has the same preference probabilities as Kim for the different prospects in the party problem, if we plot the value of clairvoyance as a function of the probability of sunshine for Mary and for Kim, the two curves will have the same breakpoints.
   II. If your PIBP for the Clairvoyant's services on the answer to a probabilistic question is strictly greater than 0, it means that you are not 100% sure as to which answer is the correct answer.
   a. Neither I nor II
   b. I only
   c. II only
   d. Both I and II

9. Consider the following statements about tests and test results:
   I. A test must be observable to be useful.
   II. A test may be material but not relevant.
   III. A test may be relevant but not economic.

   Which of the above statements are true?
   a. II only
   b. I and III only
   c. II and III only
   d. I, II, and III

10. RegionBank, a commercial real estate lender, has fallen on financial difficulty and a government takeover is a possibility. Pierre, the CEO of CashDollarCo, one of RegionBank's minor creditors, is working on a strategy for restructuring its portion of RegionBank's debt. Pierre presents his board with four alternatives that are detailed in the graph given here. CashDollarCo is risk-neutral in the range between −\$200 Million and \$400 Million.

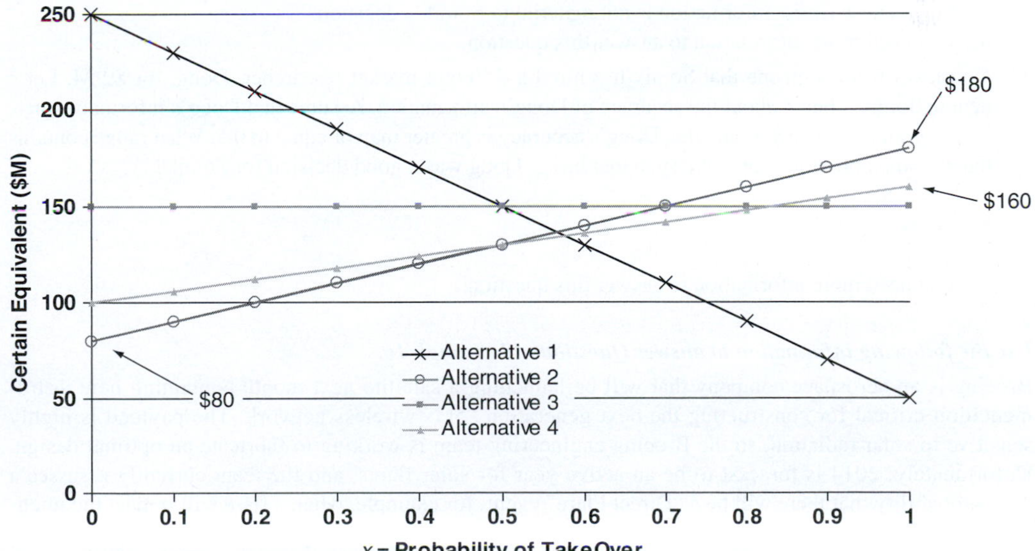

How many of the following statements are true?

I. Whether Pierre should choose Alternative 4 is insensitive to the value of x that he assigns.
II. If $x = 0.2$ and one of RegionBank's major creditors were to offer Pierre \$50 M to choose Alternative 3, he should refuse.
III. If Pierre were to gain (perfect) insider knowledge that the government was planning a takeover, he would choose Alternative 2.
IV. The value of clairvoyance on a government takeover is greatest when $x = 0.5$.
V. If $x = 0.4$, the value of clairvoyance on a government takeover is less than \$50 M.
VI. If $x = 0.4$, Pierre should purchase a \$10 M symmetric detector that predicted a government takeover with accuracy 0.75 if it is his best offer for new information.
VII. If $x = 0.4$ and Pierre knew of a detector with specificity 0.75, he should pay nothing for that detector if its sensitivity were below 0.45.

a. 0 or 4
b. 1 or 5
c. 2 or 6
d. 3 or 7

Use the following information to answer Problems 11 and 12.

Vasanth is an enthusiastic entrepreneur that just launched DATown.com as an online social network. For funding, Vasanth approached Somik, which is an aspiring venture capitalist. Somik believes that Vasanth's site needs an investment of \$12 M to be fully functional.

If DATown succeeds, Somik's stake in the company will be worth \$100 M, but if it fails, he will get nothing. Somik believes that DATown has a 1 in 4 chance of success. Somik asks you to advise him with this decision. You assess his risk attitude and find that he follows the delta property and that he has a risk tolerance of \$200 M.

11. Chris, a market researcher, offers Somik his services for \$1 M. Somik believes that Chris' information has an accuracy of 60% and is symmetric. Should Somik hire Chris for his services?
a. Yes, because Chris' information is economic.
b. No, but he would hire Chris if the charge were \$500 K.
c. No, because Chris' information is not material to Somik's decision.
d. You need more information to answer this question.

12. You learn from someone that Somik has hired a different market researcher, Doug, for \$2 M. Let c denote Somik's belief about the accuracy of Doug's information. Assume that Doug's information acts like a symmetric detector, and that Doug's accuracy is greater than or equal to 0.5. What range contains the *minimum* possible value of c such that hiring Doug was a good decision for Somik?
a. $c < 0.8$
b. $0.8 \leq c < 0.9$
c. $c \geq 0.9$
d. You need more information to answer this question.

Use the following information to answer Questions 13 through 16.

Broeing is an aerospace company that will be launching a satellite next month containing new instrumentation critical for constructing the next generation "5G" wireless network. The payload is highly sensitive to solar radiation, so the Broeing engineering team is working to fabricate an optimal design. Unfortunately, 2014 is forecast to be an active year for solar flares, and the team currently assesses a 20% probability that there will be a "Direct Flare" event; for example, when a flare will impact the integrity of the space systems.

Broeing's u-curve is assessed at $u(x) = 1 - \exp(-x/\rho)$.

The Broeing team draws the following decision tree, whose prospects are the value measure x that is used in Broeing's u-curve.

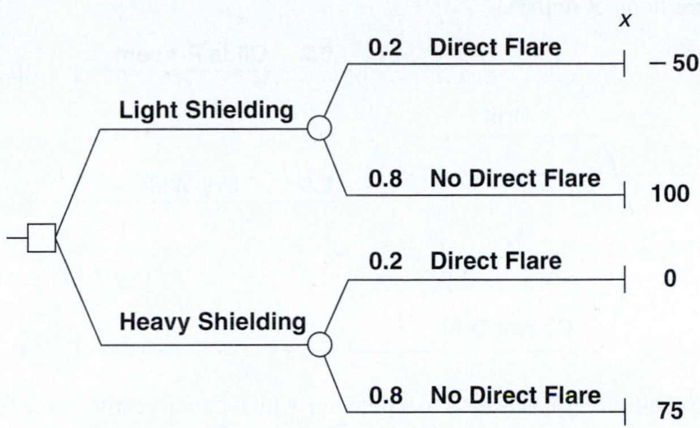

13. Which range contains the risk tolerance ρ in which Broeing is indifferent between Light Shielding and Heavy Shielding?

a. $0 \leq \rho < 50$　　c. $100 \leq \rho < 150$
b. $50 \leq \rho < 100$　　d. $\rho \geq 150$

***14.** Broeing has the option to hire graduate students to conduct sunspot research to better understand the timing and directionality of next year's solar events. After consulting with the students, Broeing assesses the following probabilities for the research program:

$$\{\text{Alert!} \mid \text{Direct Flare}, \&\} = 0.8 \qquad (\text{sensitivity})$$
$$\{\text{No Alert} \mid \text{No Direct Flare}, \&\} = 0.7 \qquad (\text{specificity})$$

Let x be Broeing's PIBP for the graduate research, and assume $p = 100$.
Which range correctly contains b, which is Broeing's PIBP for the graduate student research?

a. $b = 0$　　c. $5 < b \leq 10$
b. $0 < b \leq 5$　　d. $b > 10$

Please assume Broeing is risk-neutral in completing problems 15 and 16.
(*Hint*: You can "approximate" this in your model by setting ρ to an appropriate level.)

***15.** After consulting with the students, Broeing negotiates for the right to pay an additional sum of money to hire an extra student for a portion of the research work. Depending on what the student investigates, Broeing translates this to incrementally increasing either the sensitivity or specificity. Broeing must choose which to increase before beginning the study.

Which of the following statements should be Broeing's response for the research directive?

a. "We would like to increase the specificity of the test."
b. "We would like to increase the sensitivity of the test."
c. "We would like to hire the student, but are indifferent to her work topic."
d. "Hiring the extra student would not provide value to Broeing."

***16.** Let V_1 be Broeing's value of clairvoyance for the flare event before any research is conducted. After conducting their research, the graduate students report an "Alert!" to Broeing. Let V_2 be Broeing's value of clairvoyance for the flare event now that it has the new, relevant information.

Which of the following is true?

a. $V_1 < V_2$
b. $V_1 = V_2$
c. $V_1 > V_2$
d. Cannot be determined from the information given.

***17.** Mr. Schellburger, the CEO of Schellburger Oil, is facing a dilemma—he is not sure whether or not he should drill an oil well. His decision situation is captured by the following decision tree (all prospects are expressed in millions of dollars):

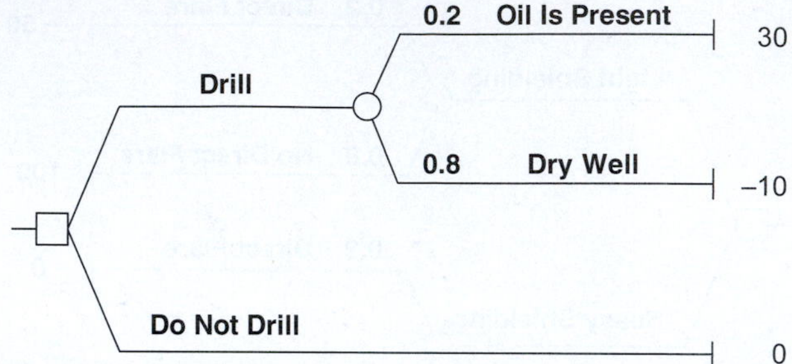

Fortunately, Mr. Schellburger has access to a detector which can give him some further information on whether or not there the site contains oil. The detector is symmetric and has accuracy 0.6.

Mr. Schellburger's engineering team tells him that he can use the detector as it is, or they can improve it for him in one of three ways before he uses it:

I. They can improve $\{$"Oil Is Present"$|$Oil Is Present, &$\}$ to be equal to 1, at a cost of $3 M.

II. They can improve $\{$"Dry Well"$|$Dry Well, &$\}$ to be equal to 1, at a cost of $3 M.

III. Or, they can improve both of those probabilities to be equal to 1, at a cost of $6 M.

Mr. Schellburger follows the delta property and his risk tolerance is $200 M. What should he do?

a. Use the detector in its present state, without any improvements.

b. Improve $\{$"Oil Is Present"$|$Oil Is Present, &$\}$ before using the detector.

c. Improve $\{$"Dry Well"$|$Dry Well, &$\}$ before using the detector.

d. Improve both probabilities before using the detector.

***18.** Katie follows the five Rules of Actional Thought and prefers more money to less. She faces one decision with three alternatives, *A*, *B*, and *C* and one uncertainty with degrees i_0 and i_1. In other words, the alternative she chooses and the degree of the uncertainty she later observes fully describes her prospects.

This plot details her *u*-value for each of the three alternatives as a function of her probability that degree i_1 will occur:

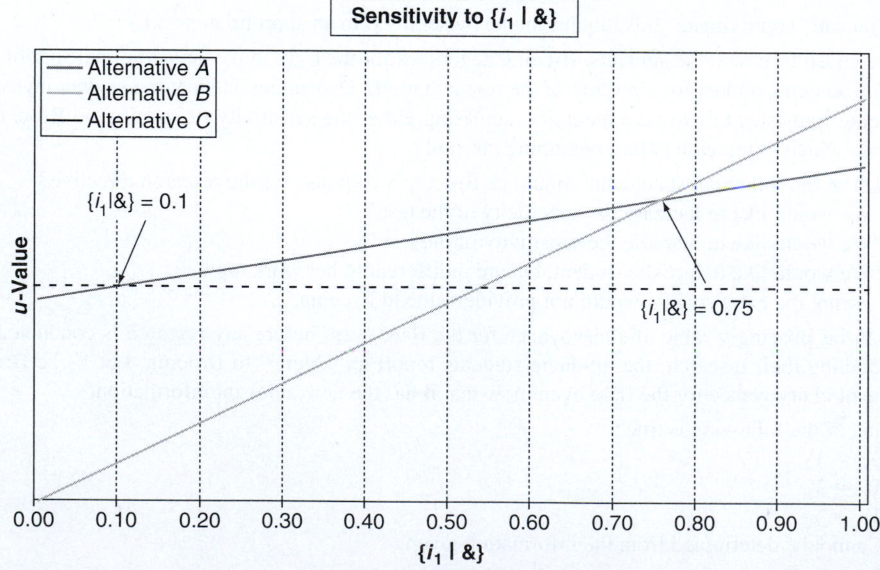

Katie also states that $\{ i_1 | \& \} = 0.4$.

Katie can perform a test that could give her more information on I. Her beliefs are $\{\text{“}i_1\text{”} | i_1, \&\} = p$, and $\{\text{“}i_0\text{”} | i_0, \&\} = q$.

How many of the following statements must be true?

I. If $p = 0.5$ and $q = 0.9$, the test is both relevant and material.

II. If $p = 0.5$ and $q = 0.9$, Katie should choose Alternative B if the test says “i_1” and Alternative C if the test says “i_0”.

III. If the test is symmetric and if $p \geq 0.5$, the minimum accuracy p that will make the test material for Katie is approximately 79% (rounding to the nearest percent).

a. 0 c. 2
b. 1 d. 3

19. Kate wonders if she has a certain disease named SDIA, and she initially believes that the chance of having it is $1/1000$. She went to a hospital to take a medical test for SDIA, and she believes the false positive chance is 5% ($\{\text{“}+\text{”} | \text{“}-\text{”}, \&\} = 5\%$). After the test, she received a positive test result. What can she infer about her chance of having SDIA (denoted as p)?

a. $p \leq 1/50$
b. $1/20 \leq p < 1/10$
c. $1/10 \leq p < 1/2$
d. There is not enough information to answer this question.

Problems 20 through 23 deal with the same situation.

20. Tom believes that there is a 99.99% chance that he is *not infected* with the HIV virus. Tom takes an HIV test which he believes has an accuracy of 99% and is symmetric (when the subject has HIV, the test flags presence of HIV correctly 99% of the time, and when the subject does not have HIV, the test flags absence of HIV correctly 99% of the time). The test comes back positive, indicating he is infected. Given his beliefs, in what range is p, the probability Tom assigns to HIV infection?

a. $98\% \leq p < 100\%$
b. $0\% \leq p < 2\%$
c. $8\% \leq p < 10\%$
d. Cannot be determined with the given information.

21. Tom takes another HIV test. Tom believes that this test is 99% accurate, symmetric (per definition in previous question), and is irrelevant to the first test given his state of infection. Given his beliefs and the information from the previous problem, in what range is p, the probability Tom assigns that the test will come back positive?

a. $1\% \leq p < 3\%$
b. $4\% \leq p < 6\%$
c. $98\% \leq p < 100\%$
d. Cannot be determined with the given information.

22. Unfortunately for Tom, the second test also came back positive. Given his beliefs and the information given in the previous two problems, in what range is p, the probability Tom assigns that he **is infected** with the HIV virus?

a. $98\% \leq p < 100\%$
b. $49\% \leq p < 51\%$
c. $0\% \leq p < 2\%$
d. Cannot be determined with the given information.

23. Tom takes two more tests that he believes are 99% accurate, symmetric (per previous definition), and are irrelevant to the other tests (including each other), given the state of infection. These two tests come back negative. Given Tom's beliefs, what is p, the probability Tom assigns that he *is infected*?

a. $49\% \leq p < 51\%$
b. $0\% \leq p < 2\%$
c. $10\% \leq p < 12\%$
d. Cannot be determined with the given information.

CHAPTER

14

Decision Diagrams

CHAPTER CONCEPTS

After reading this chapter, you will be able to explain the following concepts:

- Decision diagrams
- Nodes in a decision diagram
- Arrows in a decision diagram
- Decision diagram for a detector use decision

14.1 INTRODUCTION

As we learned in Chapter 7, *relevance diagrams* provide a compact representation of the uncertainties in a decision situation and the possibility of relevance between them. This chapter extends the concept of relevance diagrams to decision diagrams that provide a graphical representation of the whole decision situation—not just the uncertainties present.

Decision diagrams are communication tools useful to improve the clarity of thought in decision analysis and also to provide a more compact representation than a decision tree. As we might expect, decision diagrams involve more types of nodes than those used in relevance diagrams, since they capture the whole decision situation. When we incorporate these additional nodes in the decision diagram, the arrows in the diagram will have different interpretations based on the types of nodes they connect. We discuss the interpretations of these new arrows and nodes and how to capture the decision situation using decision diagrams.

14.2 NODES IN THE DECISION DIAGRAM

We now discuss the additional types of nodes in decision diagrams: Decision, uncertainty, deterministic, and value nodes.

14.2.1 Decision Node

Decision nodes are depicted by a rectangle, as shown in Figure 14.1, and represent the decisions faced by the decision maker. A decision node can be thought of as a high-level view that contains additional embedded information. The complete specification of each decision node requires a name, a rough definition, an explanation about the importance of creating this decision node, a list of the alternatives considered, a clear definition of each alternative, the author of the node, and the time of its creation. The minimum description would be the name.

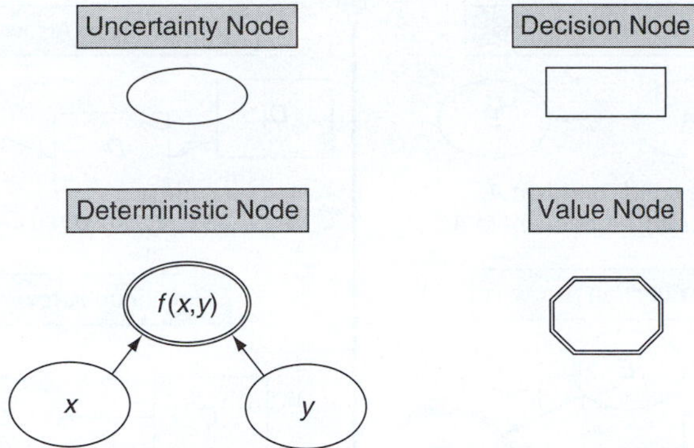

FIGURE 14.1 Nodes in a Decision Diagram

14.2.2 Uncertainty Node

The **uncertainty node** is similar to the same component of a relevance diagram. It has an oval shape that represents the uncertainty. Associated with each uncertainty node is the name of the distinction we have created, a rough definition of the distinction, the clarity test definition, the importance of the distinction, the degrees in the distinction, the name for each degree, the probability assessments made, and the author's time stamp for this information. The distribution for each node represents the distribution for that distinction conditioned on our state of information and all other nodes that have arrows entering into this node. Just as for the decision node, the minimum description would simply include the name and a rough definition.

14.2.3 Deterministic Node

A **deterministic node** is graphically represented by two concentric ovals as a deterministic function of its inputs. For example, if the deterministic node is profit and its inputs are revenue and cost, the deterministic node is a subtraction operation of cost from revenue. If we know the inputs to a deterministic node, there is no further uncertainty about it. For example, while profit may be uncertain, if we know the revenue and cost, it is no longer uncertain. Therefore, unlike an uncertainty node, there is no uncertainty about a deterministic node once its inputs are specified.

14.2.4 Value Node

The **value node** is also a deterministic node that is completely determined by its inputs. However, it has a special status in that it represents the value that the decision maker is seeking to maximize. For example, a value node can be profit or safety. To differentiate the value node from a general deterministic node, we represent it with either a hexagon or an octagon.

14.3 ARROWS IN DECISION DIAGRAMS

Now we discuss the arrows used in decision diagrams: Relevance, information, influence, function, and direct value. Examples of decision diagram arrows are shown in Figure 14.2.

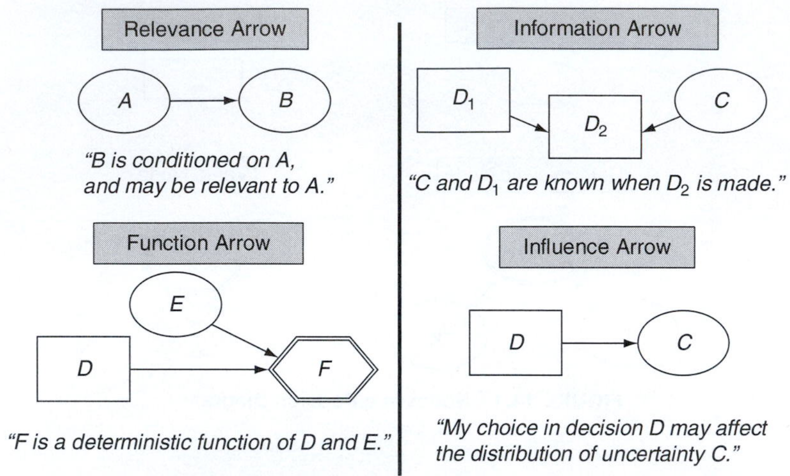

FIGURE 14.2 Arrows in Decision Diagram

14.3.1 Relevance Arrows

Relevance arrows exist between two uncertainties and have the same meaning as the arrows in a relevance diagram. Again, the absence of an arrow asserts irrelevance between the uncertainties, and the presence of an arrow implies possible relevance between them conditioned on their state of information.

14.3.2 Information Arrows

Information arrows enter into a decision node from either another decision node, a deterministic node, or an uncertainty node. Information arrows represent knowledge of what is at the other end of the arrow before making the decision they enter. An arrow from an uncertainty into a decision node means we know the outcome of the uncertainty before making the decision. This implies perfect information about the uncertainty. An arrow from a decision node into another decision node is a "no-forgetting" arrow, which means that we remember the decision made previously before making the current decision.

14.3.3 Influence Arrows

Influence arrows exist from a decision node into an uncertainty node. They signify that the decision we are making influences the probability distribution of the uncertainty.

For example, a budget director needs to decide whether to spend money on product marketing or technical feature enhancement. The budget allocation decision may change the probability distribution for marketing success or technical success of the product. Therefore, we have an arrow from the budget allocation node to the marketing success and technical success nodes.

14.3.4 Function Arrows

Function arrows enter into either a deterministic node or a value node, since the deterministic node is a function of its inputs.

14.3.5 Direct Value Arrows

Direct value arrows are function arrows that enter into the value node.

EXAMPLE 14.1 Decision Diagrams

The simplest decision diagram consists of a single decision node and a single value node, as shown in Figure 14.3. This diagram implies a decision situation consisting of selecting the alternative with the highest value. Note that there is no uncertainty in this diagram. This diagram could represent a decision situation such as "Should I choose to receive $5 or $10?"

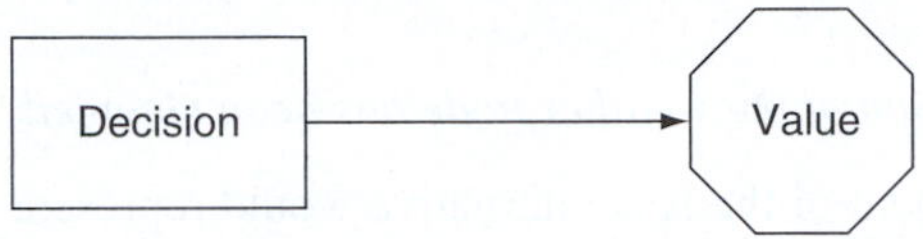

FIGURE 14.3 The Simplest Decision Diagram

Most interesting decision problems have at least one uncertain node and possibly more than one decision node. For example, in the party problem introduced in Chapter 9, Kim must make a decision about the party location when she is uncertain about the weather. Her value for the party depends on both the location (Outdoors, Porch, or Indoors), and on the weather (Sunshine or Rain). The corresponding decision diagram for Kim's problem appears as Figure 14.4.

Notice that we have two function (direct value) arrows entering into the Value node indicating that the value of the party is directly determined by both the weather and the location. There is no arrow entering into the Weather node from the Decision node, because that would imply that the location decision influences the probability distribution on weather. Furthermore, there is no arrow from the Weather node to the Location node, since this would imply perfect information about the weather when making the party location decision. As we know, Kim does not know the outcome of the weather before making her decision. This is a common subtlety that arises in practice. The fact that we have thought about the weather during our analysis does not mean that we know its outcome.

Note, also, that we could have drawn this decision diagram for Kim's problem before we had made any judgments about the number of alternatives for location, the number of degrees of the weather distinction, or its clarity test definition. We could draw it without specifying probability assignments on weather or particular value numbers associated with combinations of location and weather. In fact, drawing the decision diagram of Figure 14.4 would have been the most convenient place to begin our discussion of the party problem with Kim. Once we had agreed on its basic structure, we could have proceeded to our more detailed analysis. Note, also, that this diagram would work just as well to help both Jane and Mary.

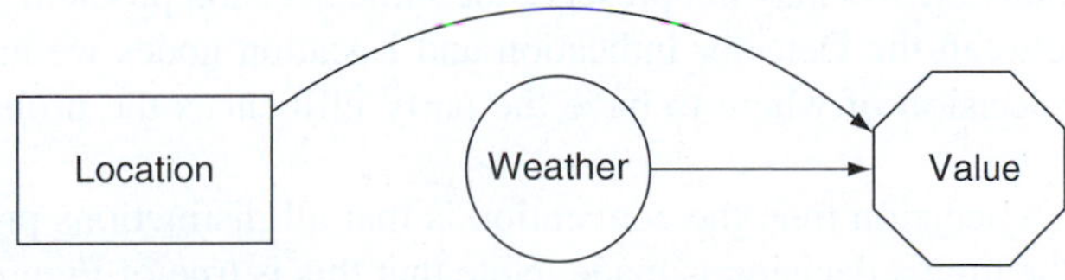

FIGURE 14.4 Decision Diagram for Party Problem

14.4 VALUE OF CLAIRVOYANCE

As we have said, an arrow into a decision node means that the information at the other end of the arrow is known at the time of the decision. For example, if we wish to consider the value of clairvoyance (VOC) on weather, all we need to do is construct the decision diagram with an arrow connecting Weather to Location, as shown in Figure 14.5.

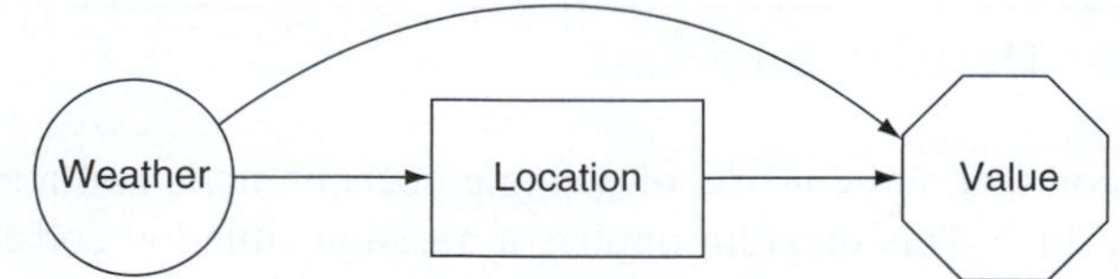

FIGURE 14.5 Decision diagram with Free Clairvoyance on Weather

Note *For the sake of convenience, the location of the weather node has been changed.*

In this decision diagram, the certain equivalent of the best alternative would represent the certain equivalent of the deal with free clairvoyance on the weather. If Kim satisfies the delta property from this quantity—as she does—we would only have to subtract the certain equivalent of the best deal without clairvoyance on the weather, as computed from Figure 14.4, to find the value of free clairvoyance on weather to Kim.

14.5 IMPERFECT INFORMATION

If Kim did not receive free clairvoyance on the weather but instead received free Acme Rain Detector service, her decision diagram would appear as in Figure 14.6. Here, we do not have an arrow from the Weather node to the Location node, since Kim will not know the weather for certain before making the location decision. However, there is a new Detector Indication node that is relevant to the weather, and an information arrow from the Detector Indication node to the Location node. The reason is that Kim will observe a distinction (the Detector Indication) before making the location decision.

14.6 DECISION TREE ORDER

To use the detector information in tree form, we need to reverse the arrow between the Weather and Detector Indication nodes shown at the bottom of Figure 14.6. This change in the decision diagram is reflected in Figure 14.7. Note that this reversal can be performed directly, since both weather and detector information are conditioned on the same state of information, &. Relevance arrow reversals between uncertainties, therefore, extend to decision diagrams. However, reversing the direction of other arrows may not preserve the same decision problem. For example, if we reverse the arrow between the Detector Indication and Location nodes we have a new diagram that means that the decision of where to have the party influences the probability of Detector Indication.

Recall that, in a decision tree, the convention is that all distinctions preceding a decision node are known at the time a decision is made. Note that this is true of Figure 14.7, but it is *not*

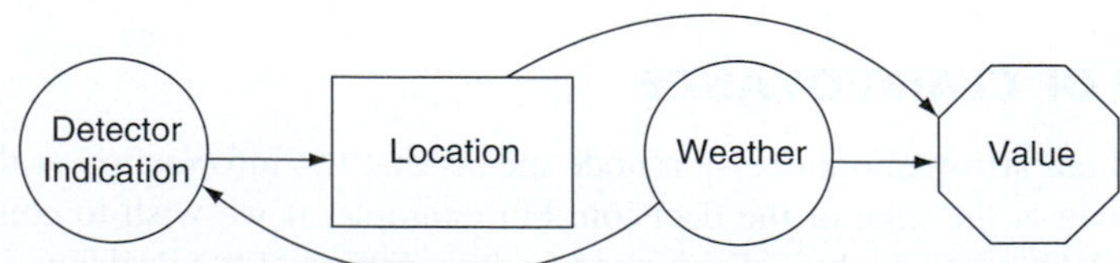

FIGURE 14.6 Decision Diagram with Free Detector in Assessed Form

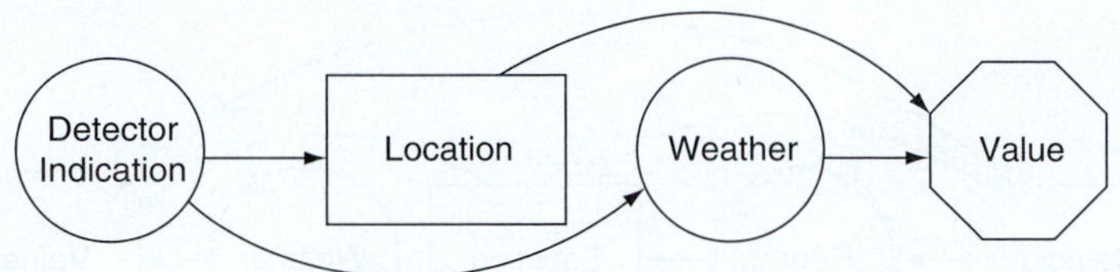

FIGURE 14.7 Decision Diagram with Free Detector in Decision Tree Form

true of Figure 14.6. If we attempt to draw a decision tree from Figure 14.6, we would have to start out with Weather, then Detector Indication, then Location, and, lastly, Value. However, this would not be correct because the tree structure implies that weather is known when location is chosen, and that is not so with Figure 14.6. In Figure 14.7, however, the order would be Detector Indication, Location, Weather, and Value. The tree can be drawn in this order, and its certain equivalent would be Kim's certain equivalent for free use of the detector.

We can describe quite succinctly what is necessary to place a decision diagram in decision tree order by thinking of each arrow in a diagram as representing a parent to child relationship, using conventional genealogical terms. For example, in Figure 14.6, we can say that the Location node has two ancestors: Detector Indication and Weather. However, it has only one parent: Detector Indication. To place a decision diagram in decision tree order, we must reverse the relevance arrows until the only ancestors of any decision node are its parents. Notice that this is the case for Figure 14.7.

14.7 DETECTOR USE DECISION

In most cases, information is not free. If we wish to get the results of the Detector Indication, we may need to pay for it. In these cases, we need to think about whether or not we want to purchase the Detector Indication. This is a new decision. To help Kim determine whether she should use the detector at a specified price, we might begin by drawing a decision diagram like that shown in Figure 14.8.

We have now added a decision node to help her decide whether or not to purchase the detector, with three arrows emanating from the Detector decision. One arrow goes to the Value node to show the effect on value of the cost of using the detector. A second arrow goes to Location node to show that, when Kim makes the Location decision, she will know whether she decided to purchase the detector. As in all decision diagrams, we presume that decisions previously made are never forgotten at the time of future decisions. Finally, an arrow from the

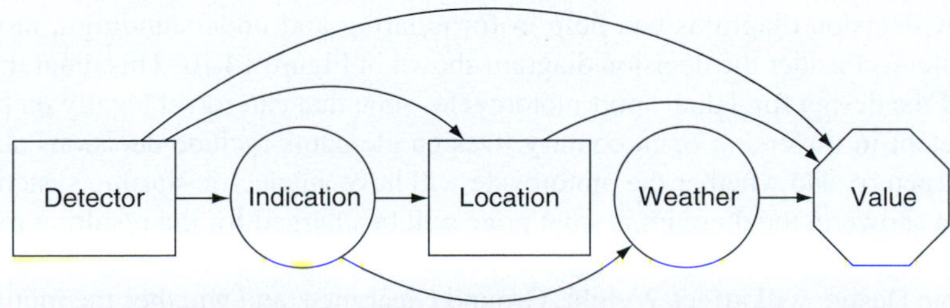

FIGURE 14.8 Decision Diagram for Party Problem with Detector

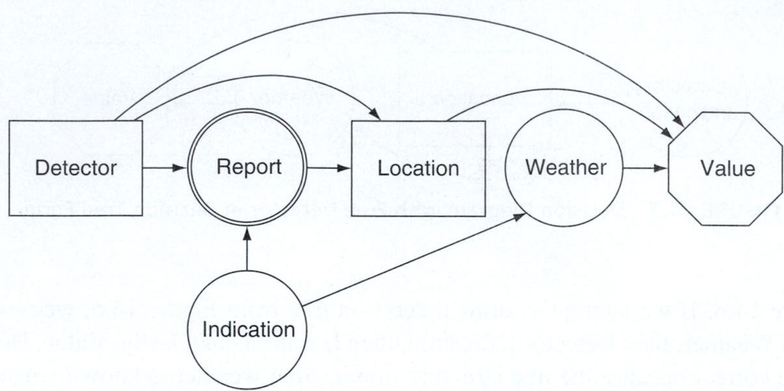

FIGURE 14.9 Accurate Decision Diagram for Party Problem with Detector

Detector node enters the Indication node to show that the indication observed would depend on whether or not she uses the detector.

While this diagram could be used with care, it can create some confusion because the arrow between the Detector node and the Indication node makes it look as if the actual indication of the detector may be influenced by whether or not it is used, and that is not our intent. Recall that arrows from decisions into uncertainties are influence arrows that can change the probability distribution of the uncertainty based on the alternative that is chosen.

A more accurate decision diagram to represent the possibility of purchasing the use of the detector is shown in Figure 14.9.

In Figure 14.9, we have added a deterministic node called Report, and the Detector node is now an input to the Report node. The Report node has a double circular border to indicate that it is a deterministic node—that is, its outputs are known when its inputs are known.

This means that the detector report that will be available at the time the location decision is made will depend on both the detector purchase decision and on the detector indication obtained if it is going to be used. One possible detector report is "no indication" as the result of a decision not to buy the detector service. This decision diagram would look very much like the decision tree in Figure 10.6 that was used to determine whether Kim should use the rain detector for a cost of, for example, $10.

As we have seen, one or more relevance diagrams may form a major portion of decision diagrams. Everything that we have learned about the manipulation of relevance diagrams is directly applicable when they appear in decision diagrams.

14.7.1 Dual Sport Motorcycle Design Decision Diagram

To show how decision diagrams can help in formulating and understanding a more realistic decision problem, consider the decision diagram shown in Figure 14.10. This diagram is the representation of the design for a dual sport motorcycle—one that can travel legally on paved roads and is competent in traversing open country. Design elements include decisions about engine size, frame structure, and whether the motorcycle will have an electric starting system. The only other decision shown in the diagram is what price will be charged for the resulting motorcycle.

We can begin to examine the diagram at any node. Engine Size will affect Power and Weight: Frame Design will affect Weight, Ground Clearance, and whether the motorcycle will have Capacity for Two—the ability to carry two people. Power and Weight affect both Road

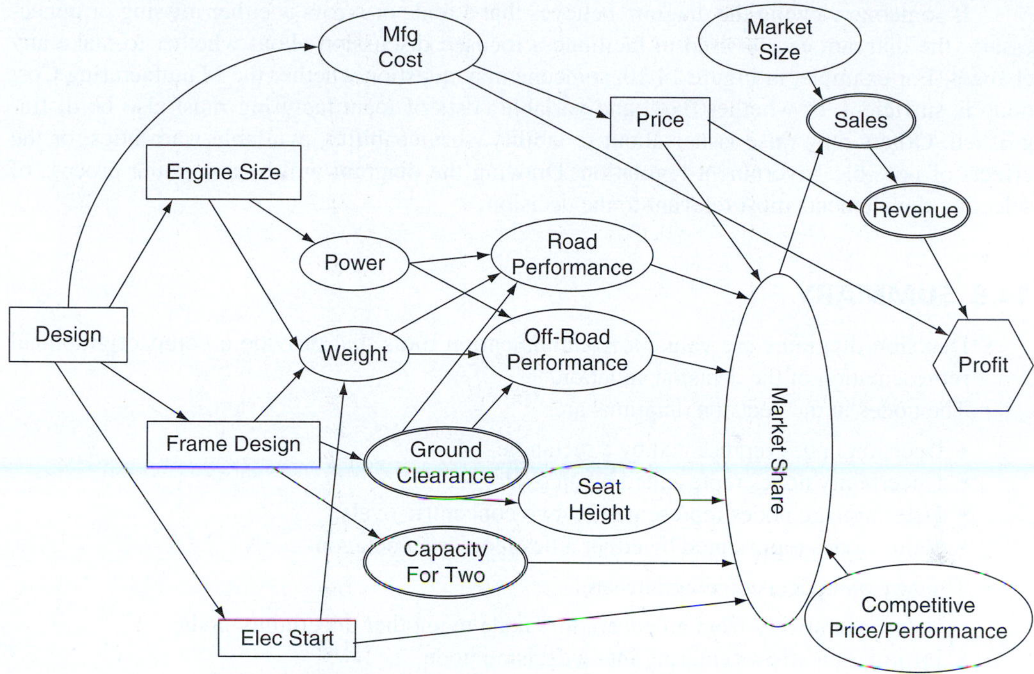

FIGURE 14.10 Dual Sport Motorcycle Design

and Off-Road Performance, because heavy motorcycle weight is a major disadvantage in off-road travel. Ground Clearance affects both Road and Off-Road performance; High Ground Clearance favors Off-Road Performance and negatively affects Road Performance; High Ground Clearance raises Seat Height, which may make the motorcycle undesirable for shorter people. As built, the motorcycle will be characterized by Road and Off-Road Performance, Seat Height, Capacity for Two, and whether it has Electric Start. Electric starting can be a very desirable on-road feature, but, it can be even more desirable off-road where you may have difficulty trying to kickstart the engine. However, Electric Start does increase the overall weight of the motorcycle.

Design will affect the Manufacturing Cost which will be known when the decision Price is made. The prospective customer will know the price and feature set of the motorcycle, as well as those quantities for other competitors, represented by the Competitive Price/Performance node. The result will be the Market Share of the company. When combined with Market Size, Market Share produces the company's Sales. Sales and Price yield Revenue, and Revenue, together with Manufacturing Cost, creates Profit. Ultimately, this diagram may help this company realize their greatest possible profits.

This diagram is a very simplified version of what would be needed to describe the entire decision process. However, it serves to illustrate the two complementary requirements of a correctly drawn decision diagram.

1. Any thought relevant to the decision must be represented by an element in the diagram.
2. Every element in the diagram must correspond to a thought essential to the decision.

If someone viewing the diagram believes that a node or arrow is either missing or unnecessary, the diagram can be used to facilitate a focused discussion about whether to make any changes. For example, in Figure 14.10, someone may question whether the Manufacturing Cost node is sufficient, or whether fixed and variable costs of manufacturing must also be distinguished. Others may raise issues about reliability, serviceability, available warranties, or the effects of possible government regulation. Drawing the diagram will be a ongoing process of selecting the elements most relevant to the decision.

14.8 SUMMARY

- Decision diagrams are valuable communication tools that provide a compact graphical representation of the decision situation.
- The nodes in the decision diagrams are:
  - Decision nodes represented by a rectangle.
  - Uncertainty nodes represented by an oval.
  - Deterministic nodes represented by two concentric ovals.
  - Value nodes represented by either a hexagon or an octagon.
- The arrows in decision diagrams are:
  - Relevance arrows from an uncertainty node to another uncertainty node.
  - Information arrows entering into a decision node.
  - Function arrows entering into a deterministic node.
  - Influence arrows from a decision node to an uncertainty node.

KEY TERMS

- Decision diagram
- Decision node
- Uncertainty node
- Deterministic node
- Value node
- Relevance arrows
- Information arrows
- Influence arrows
- Function arrows
- Direct value arrows

PROBLEMS

Problems marked with an asterisk (*) are considered more challenging.
Problems marked with a dagger (†) are considered quantitative.

1. Which of the following statements can you conclude about the decision situation depicted in the decision diagram below?

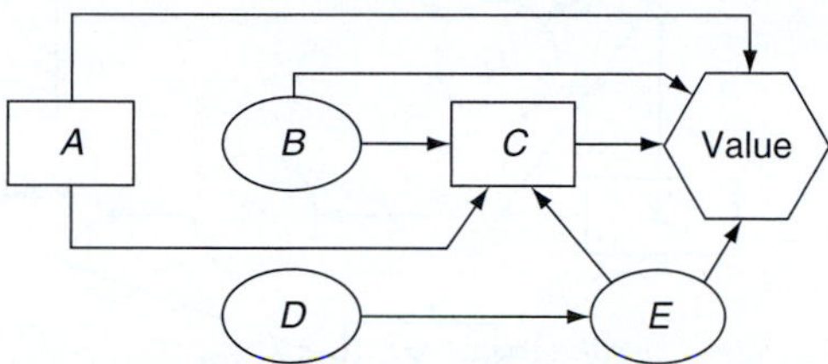

I. The decision maker observes Uncertainty *E* before making Decision *A*
II. The decision maker observes Uncertainty *E* before making Decision *C*, but after making Decision *A*.
III. Uncertainty *B* is irrelevant to Uncertainty *D*, given &.
  a. I only
  b. III only
  c. I and II only
  d. II and III only

Refer to the following decision diagram when answering Problems 2 and 3.

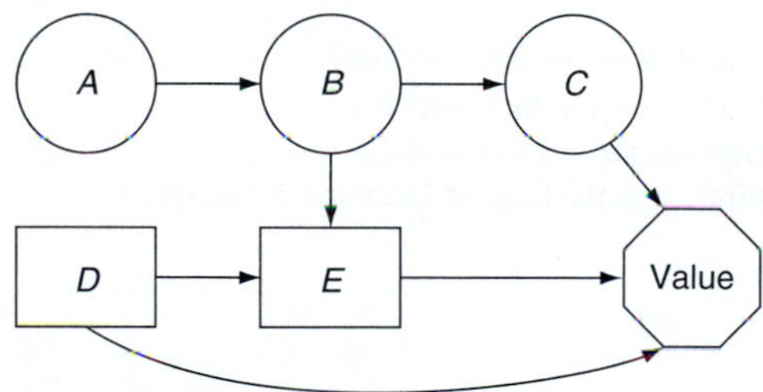

2. How many of the following statements must be true?
I. Uncertainty *B* has been observed at the time of Decision *D*.
II. Uncertainty *B* has been observed at the time of Decision *E*.
III. Uncertainty *A* has been observed at the time of Decision *E*.
IV. Uncertainty *C* has been observed at the time of Decision *E*.
  a. 0 or 4
  b. 1
  c. 2
  d. 3

3. How many of the following statements must be true? (VOC = value of clairvoyance)
I. The VOC on Uncertainty *A* at the time of Decision *E* equals zero.
II. The VOC on Uncertainty *B* at the time of Decision *D* equals zero.
  a. I only
  b. II only
  c. Both I and II
  d. Neither I nor II

Refer to the following decision diagram to answer Problems 4 and 5.

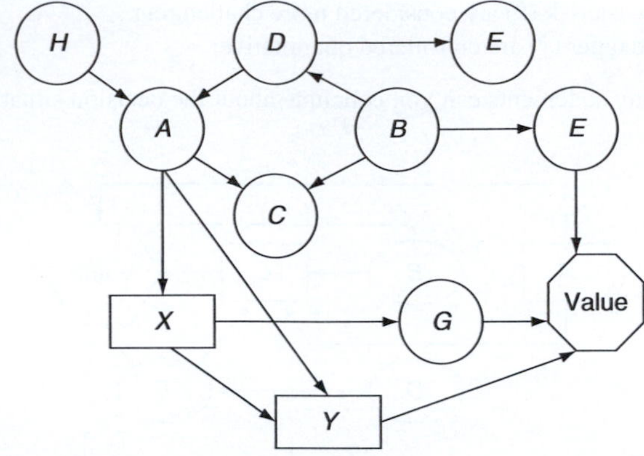

4. How many of the following statements must be true?
I. Uncertainty *A* is irrelevant to Uncertainty *E* given Uncertainty *D* and &.
II. Uncertainty *A* is irrelevant to Uncertainty *F* given &.
III. Uncertainty *A* is irrelevant to Uncertainty *F* given Uncertainty *D* and &.
a. 0
b. 1
c. 2
d. 3

5. How many of the following statements must be true?
I. There are three function arrows in this diagram.
II. There are three information arrows in this diagram.
III. The VOC of Uncertainty *H* at the time of Decision *Y* equals zero.
a. 0
b. 1
c. 2
d. 3

Use the following decision diagram for Problems 6, 7, and 8.

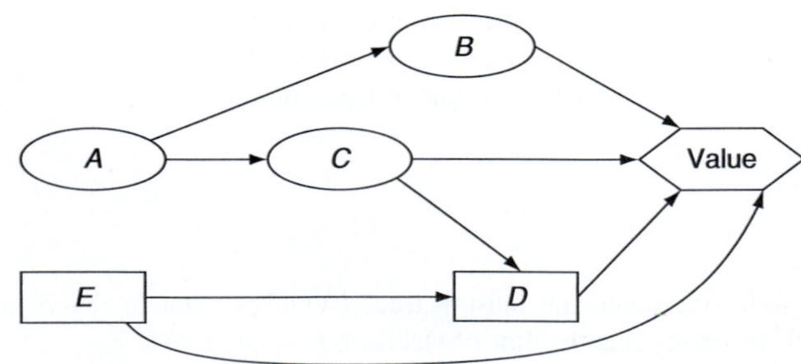

6. Which of the following statements are correct?
I. *A* and *B* are irrelevant given *C* and &.
II. *B* and *C* are irrelevant given &.

III. Once B and C are both observed, and Decision D is made, the decision maker's value is fully determined.

a. I only
b. I and II
c. II and III
d. III only

7. How many of the following statements are correct?

I. Uncertainty A and Uncertainty C have both been observed when Decision D is made.
II. Decision E is known before Decision D is made.
III. At the time Decision D is made, the value of clairvoyance on Uncertainty A is less than or equal to the value of clairvoyance on Uncertainty B.

a. 0
b. 1
c. 2
d. 3

8. Which of the following statements are correct?

I. The arrow between nodes C and D is an influence arrow.
II. The number of relevance arrows and information arrows are equal.
III. The arrow between nodes B and V is of the same type as the arrow between D and V.

a. I and II
b. I and III
c. II and III
d. I, II and III

9. Consider the following decision diagram:

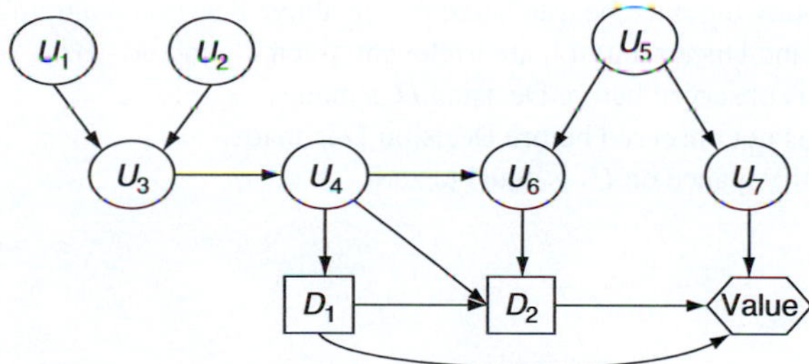

How many of the following statements are necessarily true?

I. There are two influence arrows in the diagram.
II. U_3 is observed before D_1 is made.
III. U_4 is observed before D_2 is made.
IV. U_1 is relevant to U_2 given U_3.
V. U_3 is irrelevant to U_6 given U_4.
VI. U_6 is irrelevant to U_7 given U_5.
VII. The value of clairvoyance on U_7 is zero.

a. 0 or 4
b. 1 or 5
c. 2 or 6
d. 3 or 7

10. Which of the following statements can you conclude about the decision situation depicted in the decision diagram below?

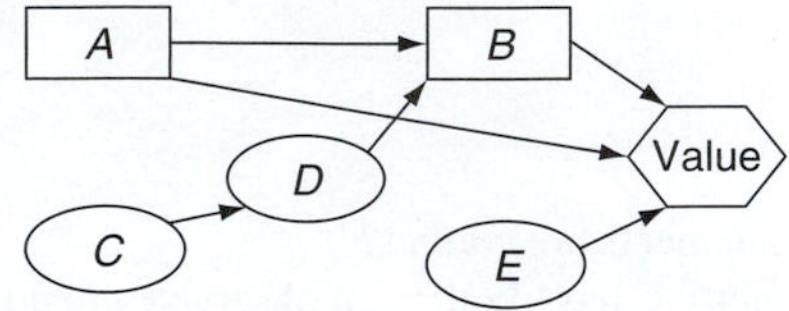

I. Uncertainty *E* is observed before making Decision *A*.
II. Uncertainty *D* is observed before making Decision *B*, but after making Decision A.
III. Uncertainty *C* is irrelevant to Decision *E*, given &.
   a. I only
   b. III only
   c. I and II only
   d. II and III only

11. Consider the following decision diagram:

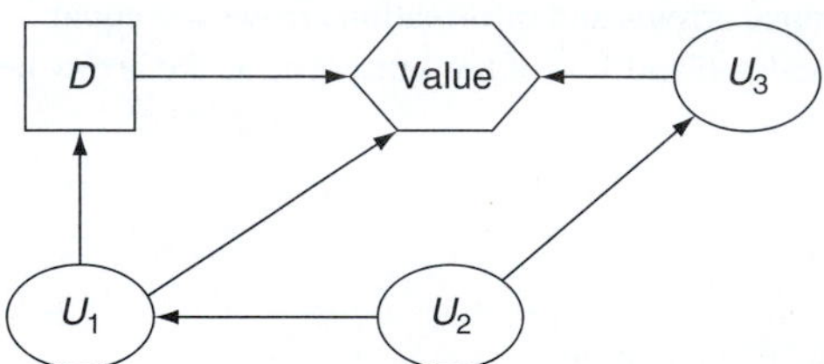

How many of the following must be true based on the above decision diagram?
I. Uncertainty U_1 and Uncertainty U_3 are irrelevant given U_2 and &.
II. Uncertainty U_1 is observed before Decision *D* is made.
III. Uncertainty U_2 is not observed before Decision *D* is made.
IV. The value of clairvoyance on U_3 is equal to zero.
   a. 0 or 4
   b. 1
   c. 2
   d. 3

12. Consider the following decision diagram.

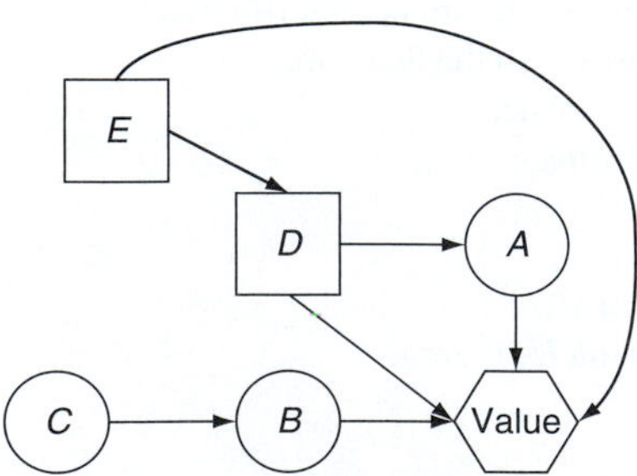

How many of the following statements must be true?
I. There are as many information arrows in this decision diagram as there are influence arrows.
II. Uncertainty *C* is observed at the time Decision *D* is made.

III. At the time Decision E is made, the value of clairvoyance on C is less than or equal to the value of clairvoyance on B.
  a. 0
  b. 1
  c. 2
  d. 3

*†**13.** Name the types of arrows in the following diagrams, and explain the meaning of each diagram.

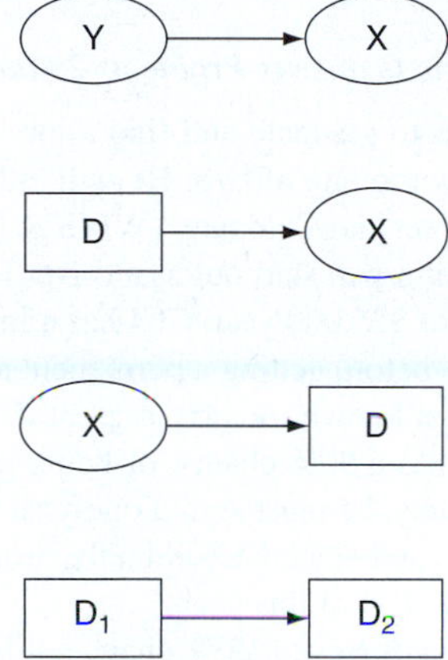

†**14.** Think of a decision situation you are experiencing now or have experienced recently. Identify the decisions, uncertainties, and values. Draw the decision diagram for your decision situation.

An Aircraft Company: Use this scenario to answer the Problems 15 through 18.

An aircraft company is considering using a new light-weight fastener, called the Superlight, to rivet certain sheet metal parts together. These particular items are not critical to safety, and so the decision is purely financial. Using Superlights instead of the basic fastener would save a value of $100 K to the aircraft manufacturer.

The Superlight fastener has never been used before on a production aircraft, and engineers are wary of its durability. From what little they know of the design and initial lab tests, they have high uncertainty about the number of Superlight fasteners that would need replacement during routine maintenance. They translate this to costs of $300 K with probability of 0.3, $120 K with probability 0.5, and $75 K with probability 0.2. (Label these possible cases High, Moderate, and Low repair cost, for future reference.) These costs include both parts and labor.

The alternative to the Superlight is the Basic fastener, which, although heavier than the Superlight, has been used extensively on past aircraft. Stress engineers feel that the Basic has a "tried and true" performance record: They conclude from past statistics that the repair cost associated with Basic fastener failure is $125 K.

In these dollar ranges, assume that the aircraft company is a risk-neutral decision maker.

†**15.** What is the certain equivalent for the best alternative?

†**16.** What is the value of clairvoyance on the repair costs for the Superlights?

†**17.** What probability of high repair costs would make the aircraft company indifferent between the two alternatives? When you vary the probability, keep the ratio of the other two probabilities constant.

†**18.** What probability of high repair costs leads to the highest value of clairvoyance on repair costs? Again, when you vary the probability, keep the ratio of the other two probabilities constant.

The Test Engineering group could run a batch of Superlights through a test program to determine their strength and durability characteristics under simulated flight conditions. Given what the stress group has told him regarding the Superlight, the Test Engineering group manager claims that his test program would be of little value. Further, he says that for Superlights with high repair cost, the test batch will pass his test with a likelihood of only 30%. If the Superlights have moderate repair cost, they will pass the test with a probability of 80%, and if they have a low repair cost, he estimates a 90% passing probability. The manager also cautions that a lot of time and hardware would be needed to conduct the test.

†**19.** What is the value of the Test Engineering group's test program?

†**20.** Draw the decision diagram for the aircraft company's decision.

†**21.** Draw the decision diagram for the aircraft company's decision to buy clairvoyance.

†**22.** Draw the decision diagram for the aircraft company's decision to use the Test Engineering group's test program.

Konstantin's Decision: Use this scenario to answer Problems 23 through 28.

At the end of the year, Konstantin plans to graduate and find a job. He is currently evaluating several job prospects, and his choice has come down to one of two. He will either work for Clairvoyant Consultants, Inc. (CCI), or for the Wizard of Wall Street and Company (WWS & Co.).

If he goes to work for CCI, Konstantin will start out as a driver for their famous Field Action Minivan. He has been promised a starting salary of $25,000/year. CCI has a firm rule that everyone must start out as a driver in the Bay Area for three weeks before getting a permanent assignment.

The position of driver has long been known to carry a great deal of prestige at CCI, and Konstantin figures that starting out as a driver, he has a 90% chance of being promoted to some better position after only three weeks on the job. Unfortunately, he must sign a one-year contract with CCI before he finds out whether or not he'll get the promotion, and since CCI is badly in need of drivers, there is still that 10% chance that they'll keep him there for the rest of this year.

Despite that, Konstantin suspects that there is a 30% chance of being promoted to Associate Clairvoyant. Given that he is promoted, he suspects there is a 50–50 chance that he will be sent to either CCI's San Francisco office or their San Jose office. The San Francisco office pays its Associate Clairvoyants $42,000/year, while the San Jose office pays its Associate Clairvoyants on a very strange commission schedule. If they bring in a new client, San Jose's Associate Clairvoyants are paid $53,000/year, but if they do not bring in a new client, they are only paid $32,000/year. Konstantin figures there is a 40% chance that, as an Associate Clairvoyant, he could bring in a new client.

Having listened closely to the rumor mills, Konstantin discovers that CCI is desperately in need of people to start up a new office in Osaka, Japan. Being fluent in Japanese, he suspects there is a 60% chance that CCI will choose to send him to Japan rather than keeping him in the bay area. But once he gets there, he doesn't know whether he will be an Associate Clairvoyant or a Senior Clairvoyant. After consulting with Blair's probability wheel, Konstantin determines that if he is sent to Osaka, there is a 85% chance that he'll be made a Senior Clairvoyant, and only a 15% chance that he will be an Associate Clairvoyant. Japanese Senior Clairvoyants are paid $95,000/year, but Associate Clairvoyants are only paid $35,000/year.

The Wizard of Wall Street & Company has offered Konstantin $60,000/year to work in New York City. However, there is no chance of promotion once he gets there.

Konstantin decides that he would be equally happy working in any of these cities. His only concern is for his salary level at the end of the first year. He doesn't care about anything after the first year—he plans to join a Tibetan monastery and will relinquish all worldly possessions at that time.

After pondering his predicament for several hours, Konstantin comes up with the following information. He wishes to follow the delta property over all the prospects in his problem. He considers a deal for equal chances of either $100,000 or nothing to be worth about $38,000. He considers that same deal for $100,000 or nothing, with a 75% probability of getting $100,000, to be worth about $64,000.

†**23.** Draw a decision diagram for Konstantin's problem.

†**24.** Find Konstantin's best decision.

†**25.** How much would WWS & Co. need to offer Konstantin for him to be indifferent between the two offers?

†**26.** Konstantin decides that he wasn't quite sure of his risk preference. What risk tolerance would make him indifferent between the two offers?

†27. Konstantin meets secretly with his friend the clairvoyant, who will reveal the outcome of any uncertainty Konstantin chooses. Draw Konstantin's new decision diagram. How much should Konstantin pay for his friend's information?

†28. Konstantin's friend has changed his mind; she will not allow Konstantin to choose which uncertainty she reveals. Konstantin is uncertain which of his uncertainties his friend will reveal, and assigns equal probabilities to each of his four uncertainties (Driver/Bay Area/Japan, San Jose/San Francisco, Finding a new client in San Francisco, Being promoted to Senior Clairvoyant in Japan). Draw his new decision diagram, and determine his PIBP for his friend's services.

†29. Describe the decision situation represented by the following decision diagram.

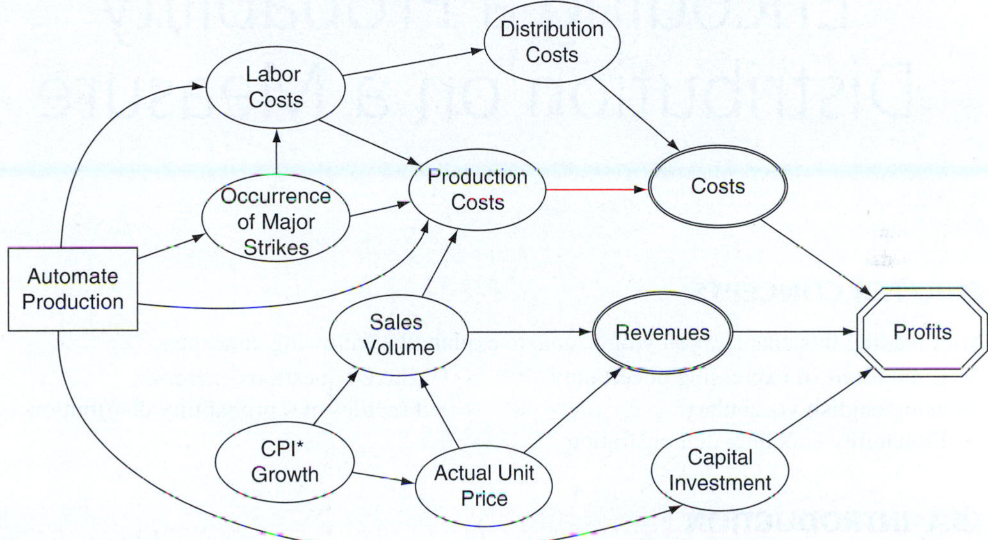

CHAPTER

15

Encoding a Probability Distribution on a Measure

CHAPTER CONCEPTS

After reading this chapter, you will be able to explain the following concepts:

- Limitations of expressing uncertainty using english vocabulary
- Probability encoding demonstration
- The 20 questions exercise
- Fractiles of a probability distribution

15.1 INTRODUCTION

In Chapter 6, we demonstrated how to assign a probability to the degrees of the Beer Drinker–College Graduate distinctions. This probability assignment reflected our belief that the next person entering the room was a Beer Drinker, and the conditional probability that he (or she) would be a College Graduate given the person was a Beer Drinker. The idea of probability being a degree of belief and its use in decision making dates back to Laplace's *Philosophical Essay on Probabilities*, Chapter 18:

> *By this theory, we learn to appreciate precisely what a sound mind feels through a kind of intuition often without realizing it. The theory leaves nothing arbitrary in choosing opinions or in making decisions, and we can always select, with the help of this theory, the most advantageous choice on our own. It is a refreshing supplement to the ignorance and feebleness of the human mind.*
>
> *If we consider the analytic methods brought out by this theory, the truth of its basic principles, the fine and delicate logic called for in solving problems, the establishments of public utility that rest on this theory, and its extension in the past and future by its application to the most important problems of natural philosophy and moral science, and if we observe that even when dealing with things that cannot be subjected to this calculus, the theory gives the surest insight that can guide us in our judgment and teaches us to keep ourselves from the illusions that often mislead us, we will then realize that there is no other science that is more worthy of our meditation.*
>
> — Laplace, 1812

Using probability to describe our beliefs is essential to clarity of thought and action, as we have shown. Since we cannot rely on language alone to represent belief, we need probability. To demonstrate the inadequacy of words, we often ask students to fill in Table 15.1 with the lowest and highest probability values they mean when they use the corresponding words.

Reflection

Fill in the lowest and highest numbers that best represent what you mean when you say Probably, Likely, Unlikely, and Almost Certainly.

TABLE 15.1 Representing Belief using English Vocabulary

Fill in the lowest and highest values you mean when you say the following:

	Lowest	Highest
Probably	______	______
Likely	______	______
Unlikely	______	______
Almost certainly	______	______

The following example, based on Table 15.1, illustrates a discussion a class might have if comparing their results. "I" is the instructor; "C" is the class; various class members are referred to by name.

I: Tom, what did you assign for Lowest when you use the word Probably?

Tom: I put a probability of 0.6. When I say this will probably happen, the lowest I mean is 0.6.

I: Did anyone put a higher number than 0.6 for Lowest when they say the word Probably?

Mary: I put a probability of 0.7.

I: Anyone higher than 0.7? No. OK, I will note Mary's number, 0.7. Mary, what did you put for Highest when you use the word Probably?

Mary: I put 0.9.

I: Anyone put lower than 0.9?

John: Yes, I put 0.6 to be the highest I mean when I use the word Probably.

I: Anyone lower than 0.6? No. Ok, John, I will note your number.

The instructor then records the highest value anyone in the class put down for Lowest and the lowest value anyone in the class put down for Highest. Table 15.2 shows some typical values obtained during this exercise for words including Probably, Likely, Unlikely and Almost Certainly:

As we might expect, terms such as Likely, Unlikely, and Almost certainly mean different things to different people. For example, suppose that Mary and John have a conversation about an event using the word Probably. Mary would think the event has probability of at least 0.7,

TABLE 15.2 Class Results obtained by taking the highest "Lowest" and lowest "Highest" values

	Highest "Lowest"	Lowest "Highest"
Probably	0.7	0.6
Likely	0.85	0.6
Unlikely	0.4	0.2
Almost Certainly	0.75	0.99

while John would assign a probability no higher than 0.6. This difference in meaning may lead to misunderstanding, frustration, or even lack of trust.

For example, a manager like John, who says to an employee like Mary "you will probably get promoted" may have the employee feeling the lowest probability this could happen is 0.7, while the manager thinks the highest probability this will happen is 0.6.

The results also show that two people in the class can have a confusing conversation. For example, John asks his friend, "Can you give me a ride home tonight?" The friend replies "Likely," thinking he was committing to only a 0.6 chance that this will happen, while John might think his friend has committed to at least a 0.85 chance of providing the ride. These examples illustrate why words alone are insufficient for representing uncertainty.

15.2 PROBABILITY ENCODING

We now present a **probability encoding** session and demonstrate the good practices that should be followed during the elicitation.

15.2.1 The Certificate

The need to express probability by assigning a number becomes particularly important when a decision must be based on the uncertainty. To illustrate, suppose a philanthropist has provided the sum of $10,000 for use in a class exercise. Your instructor decides to use it as follows: A certificate is prepared stating that the bearer will receive $10,000 if the weight of the instructor's desk, which is in plain view of all students, meets a certain **Requirement**. The certificate appears in Figure 15.1. Note that the requirement has not yet been filled in.

Next, suppose the instructor asks you to select someone from the class as your agent in making decisions about the certificate. Your agent will then enter a room, and at that time the

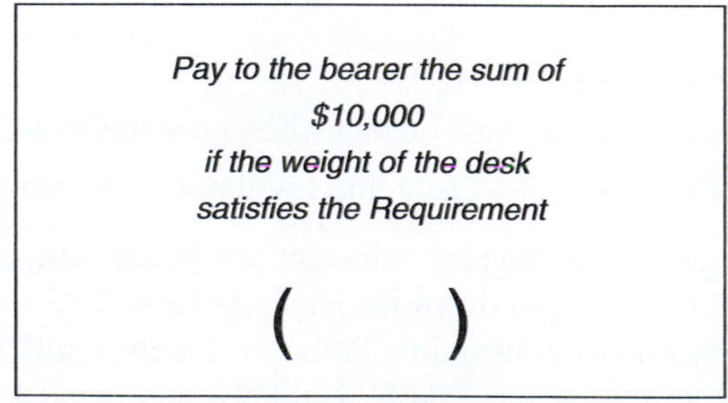

Pay to the bearer the sum of
$10,000
if the weight of the desk
satisfies the Requirement
( )

FIGURE 15.1 The Certificate

Requirement on the weight of the desk will be written on the certificate. The agent may be asked to pay some of your money to obtain the certificate, or may instead be given the choice of taking the certificate, or receiving a particular sum of money for you. Your agent will have your signed check, suitable for making any payment that is in your interest to make, and can receive on your behalf any payments.

After your agent has made a decision for you, the desk will be weighed on a printing scale, thereby determining whether you receive the $10,000, or absolutely nothing in addition to other receipts and payments by your agent.

Next, you must decide what instructions to give your agent to ensure that he or she will make the same decision you would make regarding the certificate if you were the one in that room. You should assume that your agent is competent in all the concepts and methods we have discussed in class and will have any needed computational aid, but will have no other source of information than your instructions.

I: Suggestions for the instructions?

C: (Various shout-outs) What you could afford to lose ... what you think the desk weighs ...

Think back to the three-legged Decision Stool first introduced in Chapter 1. Until the certificate's requirement is filled in, you can't specify the alternatives leg of your stool. The other two legs of the stool are information and preference. What do you need to describe preference? Since the prospects are described by money, your preference will be specified by your *u*-curve, as we have discussed, so you would give that to your agent.

Since you want the agent to use your information, you must give the agent your probability distribution on the weight of the desk. No matter how the certificate requirement is filled in, your probability distribution and *u*-curve will allow your agent to make the same decision you would make if you were present.

The certificate example requires that you delegate your decision making and simply provide your information and preferences. This is not uncommon. For example, if you are hiring someone to handle your investments, you would normally expect only to have to provide your *u*-curve to enable that person to make financial determinations. Now, we proceed to encoding the probability distribution on the weight of the desk.

PROBABILITY ENCODING DEMONSTRATION The following demonstration is an example of a probability encoding session for a continuous measure. We present the demonstration in the form of a conversation between an instructor and a student named Mary. The instructor is encoding the student's probability distribution for the weight of a desk.

I: Mary, I would like you to help me encode your knowledge about the weight of this desk in the front of the room. Let's agree that everything will be taken out of the desk and off the top, and that it will be weighed right here on an accurate printing scale. You can look at the desk from where you are over there, but do not touch it. John, my assistant, will be recording what you tell me, but he will remain out of sight so that you will not be concerned about your consistency with previous replies.
Let's begin. Mary, would you say this is a light desk or a heavy desk?

Mary: It looks heavy to me.

I: How much is "heavy?"

Mary: 100 to 200 pounds.

I: Mary, I have here in my hand a wheel, which I call the probability wheel. (For a view of the probability wheel, refer to Figure 6.4.) Would you rather receive $10,000 if the desk weighs more than 150 pounds, or if the wheel comes up orange?

Mary: I'll take the wheel.

I: In that case, I am slightly reducing the amount of orange a little. How about now?

Mary: I will still take the wheel.

I: OK, you get less orange. (*He repeats this several times.*)

Mary: Now I am indifferent.

I: Fine. I will just show the back of the wheel to John, indicating the fraction of orange. (*Instructor then changes the fraction of orange on the probability wheel.*)

I: Now, would you rather be paid if the wheel comes up orange, or if the desk weighs less than 70 pounds?

Mary: I really like the wheel. (*Successive adjustments lead Mary to be indifferent at a small fraction of orange. Then the instructor once more shows the back of the wheel to John. Instructor changes fraction of orange again.*)

I: Now, would you rather be paid if the wheel comes up orange, or if the desk weighs more than 100 pounds? (*Once again the demonstration continues till Mary is indifferent.*)

The instructor repeats the exercise for all of the following possible desk weights: Less than 220 pounds; more than 130 pounds; less than 180 pounds; more than 250 pounds; less than 100 pounds; more than 50 pounds.

As each indifference point is reached, the instructor stops and shows it to the assistant, John.

I: Mary, I think we are done. Thank you for your help. Now, let's look at John's plot in Figure 15.2.

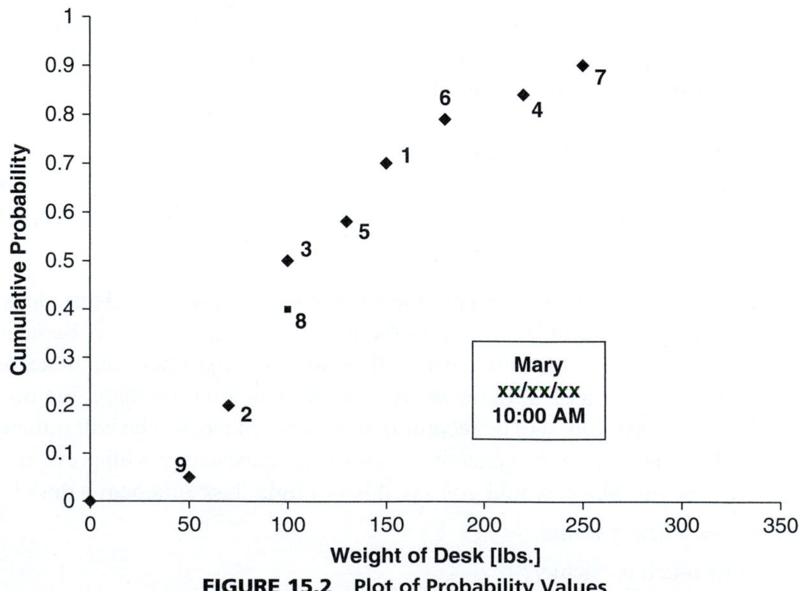

FIGURE 15.2 Plot of Probability Values

1) $\{weight \geq 150 | \&\} = 0.7$
2) $\{weight \leq 70 | \&\} = 0.2$
3) $\{weight \geq 100 | \&\} = 0.5$
4) $\{weight \leq 220 | \&\} = 0.84$
5) $\{weight \geq 130 | \&\} = 0.58$
6) $\{weight \leq 180 | \&\} = 0.79$
7) $\{weight \geq 250 | \&\} = 0.9$
8) $\{weight \leq 100 | \&\} = 0.4$
9) $\{weight \geq 50 | \&\} = 0.05$

FIGURE 15.3 John's Probability Encoding List

I: We see that at 100 pounds, there was a little inconsistency. Note that John has numbered your answers in the correct chronological order. Do you think your earlier or later answers are more or less reliable?

I: John, show us what you were writing. (*John holds up a list of numbers with notations as shown in Figure 15.3.*)

I: Mary, how did you feel about the process?

Mary: I found I really got into it after a while. I feel better about my later answers.

I: John has, in fact, sketched a smooth cumulative probability distribution through your answers. (*He holds up Figure 15.4.*)

He has drawn the curve much closer to your later answer of 100 pounds. Now Mary, is there anything you would like to change about your curve?

Mary: No, that is the best I can do with my present information. Would it be OK if I lifted the desk a bit? I want to see if my impression of its weight is accurate.

I: Not just yet, if you don't mind. John has labeled the plot with the measure being assessed, weight of the desk, and with your name, the date, and the time. Why is the time stamp important, do you think?

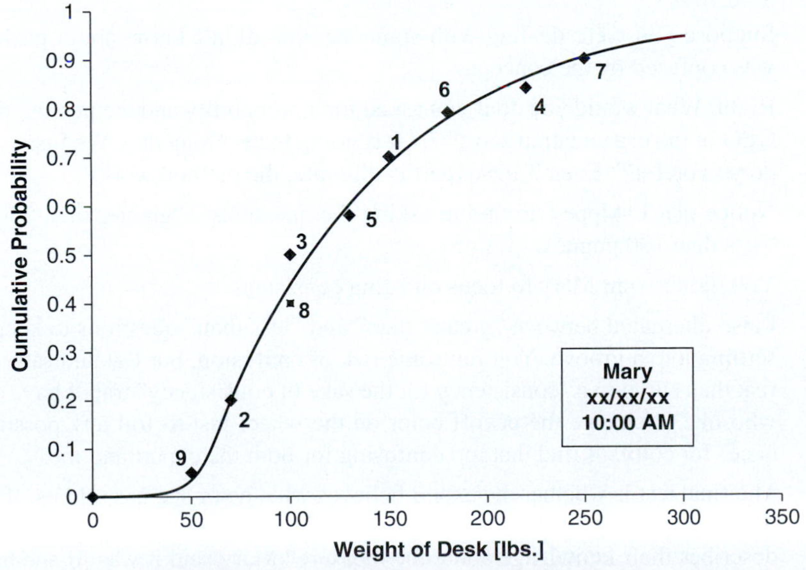

FIGURE 15.4 Cumulative Probability Distribution

Mary: Well, if I lifted the desk my curve is likely to change.

I: Right, and so we have taken a snapshot of your knowledge, or ignorance, before it is modified by your lifting it, or by someone in the class yelling out a comment like "that desk is a lot heavier (or lighter) than you think." Now let us look into the assessment procedure. What was the first question I asked Mary about the weight of the desk?

C: Is the desk heavy or light?

I: Why did I ask that question?

C: That let you see the range of weights she was thinking about without planting an idea in her head.

I: What else?

C: She said 100-200 pounds, which means she thinks in pounds.

I: Right. Suppose she were from Europe?

C: She would probably have answered in kilograms, unless she thought you wanted the weight in pounds.

I: What unit would you want to use in an assessment?

C: The unit she thinks of naturally.

I: Right. That is why we don't ask oil experts to encode oil reserves in liters. We want to use the unit that is familiar and customary for the expert. We want to perform any necessary conversions rather than have the expert trying to do them as we speak.

The first question asked allowed us to set up both the scale of the measure and the unit. John could now put appropriate unit marks on the horizontal axis.

I: Did I ever use the word "probability?"

C: No.

I: Why not?

C: Suppose you were dealing with someone who didn't know about probability, or was confused by the concept.

I: Right. What would you do if you asked for a probability and the person, perhaps the CEO of the organization, says "7." It is going to be a long day. We just ask, "Which do you prefer?" Even if the expert is illiterate, the method works.

I: Notice that I skipped around in asking the questions…"greater than 250 pounds," "less than 100 pounds." Why?

C: You didn't want Mary to focus on being consistent.

I: I also alternated between "greater than" and "less than" questions to keep her from settling into a groove. You run some risk of confusion, but I would rather face that risk than slip into a "consistency for the sake of consistency" trap. I have colleagues who like to change the payoff color on the wheel just to foil any possible preferences for colors. I find that too confusing for both the expert and me.

I: The final test is whether the expert believes what he or she has told us. If the expert had to act now on the basis of current information, would this curve be the one that describes their knowledge about the measure? Mary said it was. If she had wanted an adjustment, we would have made it.

Mary: Can I try to pick up the desk now?

I: Just one second. Class, raise your hands if you think the desk weighs more than Mary thinks it does; now raise your hands if you think it is lighter. (*Lots of hands are raised for both answers.*) It appears there is opinion on both sides of yours, Mary. OK, pick it up.

Mary: This thing is heavier than I thought.

I: John, you can tear up that distribution. It is now as useless as a cancelled check, and no longer represents Mary's state of information. This is why we have a time stamp on each probability assessment.

RETURNING TO THE CERTIFICATE PROBLEM Now that we know how to obtain your probability distribution on the weight of the desk, how would your agent use it in combination with your *u*-curve to make the right decision for you?

The first step would be to use the specified requirement written into the certificate to compute your probability of winning the \$10,000 {Win | Requirement,&}. This probability would produce the certificate deal representation shown in Figure 15.5. This probability tree shows receiving \$10,000 with probability {Win | Requirement,&}, and receiving \$0 with probability 1 − {Win | Requirement,&}.

Your *u*-curve would then allow your agent to compute your certain equivalent, or PISP, for this deal in dollars. If your agent was asked to choose for you between receiving a sum of money or the deal, your agent would keep the deal for any offered sum less than your PISP.

If, instead, your agent was offered the opportunity to buy the deal, the agent would compute your PIBP for the deal by subtracting a number b from both payoffs and then determining the value of b that would make the certain equivalent of the resulting deal equal to zero. Your agent would then pay any amount less than your PIBP to receive the deal, the lower the payment, the better.

Now, let us return to the question of how to compute {Win | Requirement,&} for various Requirements. Suppose the requirement was weight greater than 150 pounds? The probability of winning the \$10,000 would be

$$\{\text{Win} \mid \text{Requirement}, \&_{\text{Mary}}\} = \{\text{weight} \geq 150 \mid \&_{\text{Mary}}\} = 1 - \{\text{weight} < 150 \mid \&_{\text{Mary}}\}$$
$$= 1 - 0.69 = 0.31$$

a point on the cumulative probability distribution. If the requirement was weight less than 200 pounds, then the probability of winning would be

$$\{\text{Win} \mid \text{Requirement}, \&_{\text{Mary}}\} = \{\text{weight} < 200 \mid \&_{\text{Mary}}\} = 0.82$$

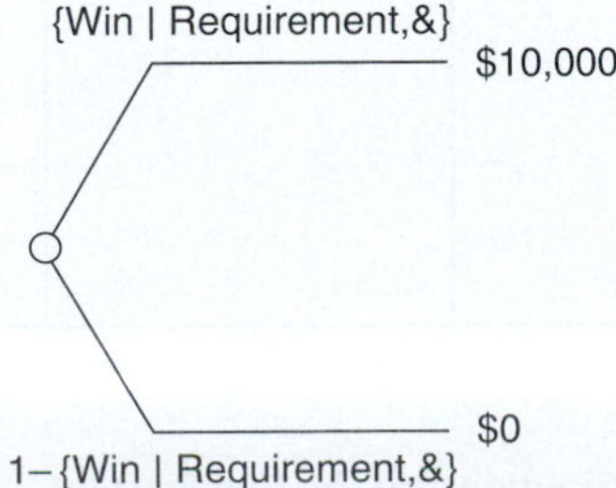

FIGURE 15.5 Probability of Winning Given Requirement

a point on the excess probability distribution. If the requirement was that the weight be at least 150 pounds, but less than 200 pounds, then the probability of winning would be

$$\{\text{Win} \mid \text{Requirement}, \&_{\text{Mary}}\} = \{\text{weight} < 200 \mid \&_{\text{Mary}}\} - \{\text{weight} < 150 \mid \&_{\text{Mary}}\}$$
$$= 0.82 - 0.69 = 0.13$$

Finally, just to illustrate the generality of the representation, suppose that the requirement was that the difference between the weight of the desk and 100 pounds be less than 30 pounds. This means Mary will win if the desk weighs more than 70 pounds, but not more than 130 pounds. Her probability of winning is

$$\{\text{Win} \mid \text{Requirement}, \&_{\text{Mary}}\} = \{\text{weight} > 70 \mid \&_{\text{Mary}}\} - \{\text{weight} > 130 \mid \&_{\text{Mary}}\}$$

or equivalently,

$$\{\text{Win} \mid \text{Requirement}, \&_{\text{Mary}}\} = \{\text{weight} \le 130 \mid \&_{\text{Mary}}\} - \{\text{weight} \le 70 \mid \&_{\text{Mary}}\}$$

From Figure 15.4 showing Mary's cumulative distribution, this difference is approximately $0.43 - 0.23 = 0.2$. Therefore, Mary's agent would know that her probability of winning the \$10,000 would be 0.2 in Figure 15.5. The agent can evaluate any deal offered regarding the certificate using her *u*-curve and her probability distribution on the weight of the desk.

15.3 FRACTILES OF A PROBABILITY DISTRIBUTION

Rather than encoding a whole distribution, we can use a small number of probability assessments to create a compact representation that often provides a sufficient approximation of the full distribution. We discuss this method in more detail here, as we refer back to the instructor demonstration.

Now, as you have seen, probability encoding can be a time-consuming process. Sometimes it is helpful to elicit only certain fractiles of a distribution. **Fractiles** correspond to the values of the measure at certain percentiles of the probability distribution. For example, some of the common fractiles are the 1%, 25%, 50%, 75%, and the 99% fractiles (Figure 15.6).

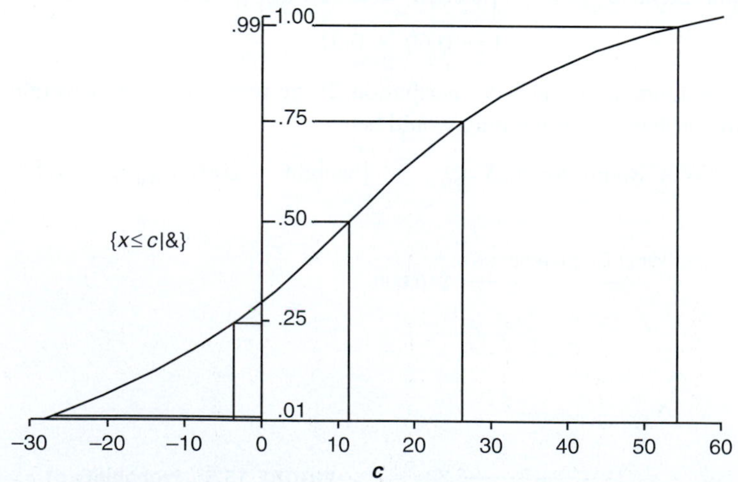

FIGURE 15.6 Cumulative Distribution and the Five Fractiles

TABLE 15.3 Five Fractiles of Distribution for Figure 15.6

f	$\leq x(f)$
0.01	−26
0.25	−4
0.50	11
0.75	27
0.99	54

Fractiles provide a compact description of the probability distribution. If we are given a cumulative distribution, we can read the fractiles directly from the graph. For example, the five fractiles 1%, 25%, 50%, 75%, and 99% for the distribution of Figure 15.6 are shown in Table 15.3.

The **mid-range value** is what we call the interval on the measure between the 25% fractile and the 75% fractile. In Table 15.3, the mid-range is the interval of the measure from −4 to 27. The mid-range has two interesting properties:

1. it is equally likely that the measure will be above or below it (since the probabilities of the measure lying above the mid-range and below it are both equal to 0.25)
2. it is equally likely that the measure will be inside or outside this mid-range (since values within the mid-range have a probability of occurrence equal to 0.5).

The **median value** is what we call the value of the measure that corresponds to the 50% fractile. Note that it is equally likely that the measure will be above or below the median. Finally, the measure values that fall below the 1% fractile and above the 99% fractile are referred to as **surprises**. The total probability of the surprise regions is equal to 0.02.

The five fractiles thus present a compact representation of the probability distribution. They can be elicited directly from the decision maker. When eliciting fractiles of a distribution, it is good practice to start with the surprises, then the mid-range, and finally, the median.

To assess surprises, we ask ourselves, "For what value of the measure do we believe there is only a 1% chance that the measure will fall below that value?" and then, "For what value of the measure do we believe there is only a 1% chance that the measure will exceed it?" These two values, respectively, are what will determine the lower and upper surprise fractiles.

Next, we assess the mid-range of the distribution. We ask, "What value of the measure do we believe it is 25% likely to be below?" and, "What value of the measure do we believe it is 25% likely to be above?" These questions determine the 25% and 75% fractiles, respectively. These two fractiles constitute the mid-range.

As a consistency check, we now use the properties of the mid-range to ask, "Do we believe that the measure is equally likely to be inside or outside the mid-range?" Also, "Do we believe that the measure is equally likely to be above or below the mid-range?" If we answer "no" to either of these questions, then we need to reassess our mid-range values. If we answer "yes" to both of them, then we move on to assess the median values of the measure.

Finally, to assess the median, we ask ourselves, "What value of the measure is the median equally likely to be above and below?"

A very common bias for assessors is to assign overly narrow mid-ranges, thereby demonstrating overconfidence in their estimates. Sometimes, people don't know as much as they think they do. Often, this happens because of excessive adherence to an initially assessed central value, such as the median. To avoid this tendency, try to elicit first the surprises, then the mid-range, and finally, the median.

It is also very common to hear comments during the assessment like, "we have no idea about this distribution" or "we cannot give fractiles." A bit of reflection reveals that people actually know more than they think they do.

For example, suppose we are interested in the fractiles of a probability distribution for the number of pounds of commercial dog food production in the United States in 2003, as reported in an almanac. While we may believe we know very little about this subject, there are some distinctions can help us think about the fractiles.

- Number of U.S. households that have dogs
- Average dog food consumption per dog, per day
- Shelf space for dog food in grocery stores
- Number of TV ads about dog food
- Percentage of commercial dog food not eaten by dogs
- Percentage of dogs that eat other food besides dog food (for example, food scraps)

Reviewing the list, we realize that we know more about the commercial production of dog food than we originally thought. We have prior knowledge about many distinctions that can inform us about the fractiles to be assessed. If, instead, we truly have little knowledge about this measure, then we should assign a wide mid-range that reflects our lack of knowledge.

Figure 15.7 shows an example of a relevance diagram that may help us think about the production of dog food.

15.3.1 The 20 Questions Exercise

The following exercise teaches several lessons about how to make effective probability assignments. First, students fill out the questionnaire in Figure 15.8, answering either "yes" or "no" for each question without further explanation of its meaning.

When the questionnaires are collected, we record the fraction of responses that answered "yes" and "no" for each question. This recorded information is not yet revealed to the students.

Next, we ask the students to assign their fractiles for the 20 questions shown in Table 15.4. Five of the questions refer to the survey in Figure 15.8. The remaining 15 questions are what we might call "almanac" questions, since the answers appear in an almanac or a similiar book of facts.

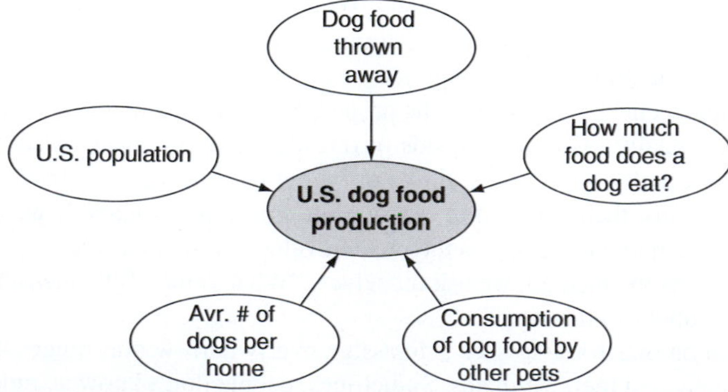

FIGURE 15.7 Thoughts for U.S. Dog Food Production

	YES	NO
1. Do you favor abolishing laws regarding "victimless" crimes?		
2. Do you consider yourself a member of a religion?		
3. Have you ever served in the armed forces of any country?		
4. Have you ever ridden a motorcycle (solo)?		
5. Would you accept a gamble where you were equally likely to lose $50 or win $100?		

FIGURE 15.8 Five Question Survey

The purpose of this exercise is, of course, not for the students to look up the correct answers on the internet or in a library, but to assign fractiles based on their beliefs about these quantities. We remind them of our earlier discussion about eliciting fractiles of a distribution: Start with the surprises, then move to the mid-range, then, finally, the median.

Reflection

Using our earlier discussion about assigning fractiles, assign your own fractiles for the 20 questions in Table 15.4. Note that questions 4, 8, 12, 16, and 20 are the "internal" questions, and refer to the fractiles for the five questions in the distributed questionnaire. This information is only known to the instructor at the time and cannot be found anywhere else. When you answer these questions, think of fractiles that represent a group of first-year graduate students taking a decision analysis class. Note that Table 15.4 has six columns in addition to the columns for the fractiles. For the moment, leave blank the columns numbered 1 through 6.

Answers to the 20 questions, including answers to the internal questions taken from one decision analysis class, appear in Table 15.5.

Seeing the answers to the 20 questions allows us to analyze the class responses. To analyze your results, put a checkmark in one of the Columns 1 through 6 in Table 15.4 according to how the actual answer compared with your assigned fractiles. For example, if your fractiles for the first question were as shown in Table 15.6, you would place a check mark in Column 3 for the first question, since the actual answer (7,631,000) lies between your 25% fractile and your 50% fractile. Furthermore, you would place a check mark in Column 6 for Question 2, since the correct answer (21,366) lies in the surprise region above your 99% fractile. You also would place a checkmark in Column 2 for Question 3, since the correct answer (5,855) lies between the 1% fractiles and the 25% fractiles. Although this question will rarely arise, recall that the fractiles represent a point on the cumulative probability distribution—the probability that the measure will be less than or equal to the fractile. For example, you would place a check mark in Column 4 if your answer were greater than the 50% fractile and less than or equal to the 75% fractile.

Now complete the number checked for the 20 questions based on the fractiles you have assigned. If you have checkmarks in Columns 1 or 6, you have surprises, and if you have checkmarks in Columns 3 or 4, then the correct answer lies within your mid-range. Record

TABLE 15.4 The 20 Questions

	1	1%	2	25%	3	50%	4	75%	5	99%	6
Population of Cuba, 1965											
1965 imports of U.S. ($ million)											
Airline distance in statute miles from San Francisco to Moscow											
Fraction of group favoring abolition of "victimless" crime laws											
The closing Dow Jones Industrials average for May 26, 1969											
The number of pages in the Wall Street Journal of May 27, 1969											
Birthday of Arnold Palmer (month/day/year)											
Fraction of group considering themselves a member of a religion											
American battle deaths in Revolutionary War											
Number of labor strikes in U.S. during WW II (Pearl Harbor - VJ Day)											
Height of Hoover Dam (feet)											
Fraction of group that has served in the Armed Forces											
Length of Broadway run of "Oklahoma" (days)											
Gross tonnage of liner "United States"											
U.S. whiskey production (legal) in 1965 (thousands of gallons)											
Fraction of group that has ridden a motorcycle (solo)											
Length of Danube River (miles)											
Popular votes cast for Eisenhower in 1956 (millions)											
Worldwide airplane accident deaths on scheduled flights during 1960											
Fraction of group willing to accept (−$50, $100) gamble											

Note: Do not turn the page until you have assigned your fractiles.

your number of surprises and mid-range answers. What do you notice? If your belief was an accurate representation of your knowledge, then you would have only 1 to 2 of the 20 questions in the surprise region, and you should have about 8 to 12 answers within the mid-range. If you have too many surprises and too many answers outside the mid-range, you have shown that you are overconfident in your ability to make probability assignments: Your distributions are too narrow.

TABLE 15.5 Answers to the 20 Questions

Question	Answer
1. Population of Cuba, 1965	7,631,000
2. 1965 imports of U.S. ($ million)	21,366
3. Airline distance in statute miles from San Francisco to Moscow	5,855
4. Fraction of group favoring abolition of "victimless" crime laws	***26.83%***
5. The closing Dow Jones Industrials average for May 26, 1969	946.9
6. The number of pages in the Wall Street Journal of May 27, 1969	36
7. Birthday of Arnold Palmer (month/day/year)	SEPT. 10. 1929
8. Fraction of group considering themselves a member of a religion	***54.88%***
9. American battle deaths in Revolutionary War	4,435
10. Number of labor strikes in U.S. during WW II (Pearl Harbor - VJ Day)	14,371
11. Height of Hoover Dam (feet)	726
12. Fraction of group that has served in the Armed Forces	***12.20%***
13. Length of Broadway run of "Oklahoma" (days)	2,246
14. Gross tonnage of liner "United States"	52,072
15. U.S. whiskey production (legal) in 1965 (thousands of gallons)	117,930
16. Fraction of group that has ridden a motorcycle (solo)	***32.93%***
17. Length of Danube River (miles)	1,770
18. Popular votes cast for Eisenhower in 1956 (millions)	35.6
19. Worldwide airplane accident deaths on scheduled flights during 1960	307
20. Fraction of group willing to accept (−$50, $100) gamble	***75.61%***

Table 15.7 presents sample responses obtained from 82 respondents in a decision analysis graduate class. The figure shows the number of responses obtained in each column for each of the 20 questions. Note that the percentage of surprises is 44% (11% + 33%), indicating very narrow distributions. Note, also, that 69% of the responses were outside the mid-range (100% − 16% − 15%).

DISCUSSION OF RESULTS Before making any comments on the results of the 20 questionnaire exercise, we need to note that our observations will be on the overall performance of the group. Some individuals may do an excellent job of assessing the fractiles even when the group performance is not ideal.

TABLE 15.6 Sample Responses and Fractiles and Number Assignments

	1	1%	2	25%	3	50%	4	75%	5	99%	6
1.		2,200,000		6,000,000	x	9,000,000		14,000,000		27,000,000	
2.		100		300		700		900		3000	x
3.		4000	x	7000		9000		9500		9700	

TABLE 15.7 Sample Graduate Student Responses from 82 Participants

	Answer	1	2	3	4	5	6
1. Population of Cuba, 1965	7,631,000	8	11	24	17	10	12
2. 1965 imports of U.S. ($ million)	21,366	6	7	10	4	7	48
3. Airline distance in statute miles from San Francisco to Moscow	5,855	26	23	12	13	4	4
4. Fraction of group favoring abolition of "victimless" crime laws	***26.83%***	7	19	32	14	4	6
5. The closing Dow Jones Industrials average for May 26, 1969	946.9	23	12	15	12	15	5
6. The number of pages in the Wall Street Journal of May 27, 1969	36	2	8	18	14	13	27
7. Birthday of Arnold Palmer (month/day/year)	SEPT. 10. 1929	14	8	3	7	30	20
8. Fraction of group considering themselves a member of a religion	***54.88%***	6	8	22	37	3	6
9. American battle deaths in Revolutionary War	4,435	48	20	7	4	2	1
10. Number of labor strikes in U.S. during WW II (Pearl Harbor - VJ Day)	14,371	0	1	0	0	1	80
11. Height of Hoover Dam (feet)	726	6	14	15	11	13	23
12. Fraction of group that has served in the Armed Forces	***12.20%***	8	20	20	18	8	8
13. Length of Broadway run of "Oklahoma" (days)	2,246	0	2	6	9	7	58
14. Gross tonnage of liner "United States"	52,072	10	2	8	2	6	54
15. U.S. whiskey production (legal) in 1965 (thousands of gallons)	117,930	0	1	0	5	4	72
16. Fraction of group that has ridden a motorcycle (solo)	***32.93%***	4	4	18	17	19	20
17. Length of Danube River (miles)	1,770	0	7	6	9	20	40
18. Popular votes cast for Eisenhower in 1956 (millions)	35.6	8	16	18	16	12	12
19. Worldwide airplane accident deaths on scheduled flights during 1960	307	9	12	25	15	6	15
20. Fraction of group willing to accept (−$50, $100) gamble	***75.61%***	3	2	10	29	26	12
	Average	9.4	9.9	13.5	12.7	10.5	26.2
	Percent	11%	12%	16%	15%	13%	32%

When we assign the 20 questions exercise to the students, we tell them that when we review their overall performance we will be able to demonstrate two things. First, that they think they know more then they do, and second, that they know more then they think they do.

To establish that they think they know more then they do, we simply refer to their responses, which are typically like those shown in Table 15.7. Here, we see that 31% of the answers, and not the expected 50%, were in the mid-range. Similarly, when we look at surprises, we find that 43% of the answers, not the desired 2%, were. These results are typical not only of student groups, but of all groups—including experienced professionals. What these results mean is that the distributions assigned are much too narrow: People think they know more than they do.

To show that they know more then they think they do, we direct attention to the almanac questions. We begin by asking what an almanac is. People know that it is a book of facts on many topics that would be of interest to a general reader. They quickly affirm that almanac compilers do not have extensive research staffs, and that they gather their information from the reports of others, including many government reports.

Starting with Question 1, the population of Cuba, we can see that this number was derived from some type of census reported in an official document. The same would be true for Question 2 on U.S. imports. When we come to Question 3, we see that the airline distance from San Francisco to Moscow is to be measured in "statute miles." Why does this question mention statute miles when Question 17 refers only to "miles?" Some people may know that airline distances are usually measured in nautical miles, a unit useful for measuring distances on the globe. This, in turn, triggers thoughts about the maximum distance between any two points on the globe, and may provide new information in assessing the fractiles.

Continuing our discussion of the almanac, we note that it is usually composed of many tables of facts that are extreme in nature. This observation is relevant to the questions on the height of Hoover Dam, the length of the Broadway run of "Oklahoma," the gross tonnage of the liner "United States," and the length of the Danube River. In each case, we can imagine the table in which this information might first have appeared. Students quickly come up with table titles, like "Tallest Dams in the World," "Longest-Running Broadway Plays," "Largest Passenger Ships Ever Built," and "Longest Rivers of the World." Of course, many variations are possible: The Danube River could appear in a table entitled "Longest Rivers of Europe."

The point is that information appearing in an almanac is typically extreme in nature, and that knowledge is available to you in making your assessments. Since most people have heard of the musical "Oklahoma," it is very unlikely to appear in a table entitled "Shortest Running Broadway Plays." Consider this available information when you look at the number of surprises on these questions. In summary, most students agree that they failed to utilize all of the available information.

Questions 9 and 10 deserve special discussion. To most people, the number of Revolutionary War deaths recorded in the almanac is surprisingly low. To explain this, we should remember that the number in the almanac comes from a government report, perhaps related to the number of official graves. However, at that time in history, it would not be surprising if many wounded fighters withdrew to their homes, and later died without any official report. Furthermore, the importance of this war in American history may lead us to believe that it was more costly in human lives than was actually the case.

Many are surprised by the high number of labor strikes during World War II reported in Question 10. This number is the result of a government report. At that time, a labor strike anywhere in the economy might well have an impact on the war effort. Employers would have been required to report strikes so that corrective action could be taken. In view of the natural reluctance to prepare a report unless it is absolutely necessary, many borderline work interruptions that could possibly be classified as strikes might not have been reported. Therefore, the official number in the almanac is likely to be too low rather than too high. The surprise may come from the documentaries we have seen about the U.S. home front during World War II. We have all seen pictures of "Rosie the Riveter" zealously advancing the war effort. On the other hand, it's hard to find documentaries that discuss black markets and profiteering that might have happened during the war.

Students usually do better in assessing fractiles on the internal questions from the class survey. Perhaps this is because they have learned something about each other, but more likely it is because all fractiles for these questions are limited to the range from 0 to 100%.

THE ALMANAC GAME To help you improve your ability to assign fractiles for uncertain measures, we recommend a very useful exercise called the Almanac game. To play this game, we need three players. Player *A* opens an almanac (a book of reported "facts") or its internet equivalent and specifies the measure to be assessed. Player *B* assigns a mid range to that chosen measure, and Player *C* then bets on whether the measure lies inside or outside Player *B*'s mid-range.

Note: Player *B* should be indifferent to this bet.

If Player *C* chooses inside the mid-range and the report does lie within the mid-range, then Player *C* wins. Otherwise, Player *C* loses. Sometimes, the loser must pay for a round of beverages. Then the Players rotate roles and the game is repeated. In many cases where this carried out in the classroom or in professional executive education settings, people tend to place very narrow mid-ranges, exhibiting, once again, overconfidence in their assignments. If Player *C* always bets outside *B*'s mid-range and wins, Player *B* should learn to widen the assigned mid-ranges.

15.4 SUMMARY

- Probability distributions can be encoded from decision-makers.
- Fractiles of a probability distribution provide a compact representation of the distribution.
- The 20 Questions Exercise showed that (as a group) people tend to use narrow intervals for the mid-range. There were also too many "surprises" in the responses.
- The Almanac game can help you train to provide better fractiles.

KEY TERMS

- Probability encoding
- Five fractiles; 1%, 25%, 50%, 75%, 99%
- Median value
- Surprises
- Mid-range

PROBLEMS

1. Assign your five fractiles for the following distinctions:
   - Population of Turkey in July 2008
   - Birthday of Arnold Schwarzenegger
   - Estimated population of the United States in July 2008
   - Height of Eiffel tower
   - Height of Empire State building
   - Closing of the Dow Jones Industrial Average (DJIA) on December 30th 2008
   - Zero-fuel weight of an empty A330-200 Airbus
   - Length of an A330-200 Airbus

Answers are on the next page. Please do not turn the page until you have assigned your fractiles.

2. The following probability tree characterizes prospects that may occur after choosing a given alternative. The profit (expressed in thousands of dollars) is presented at the end of the tree.

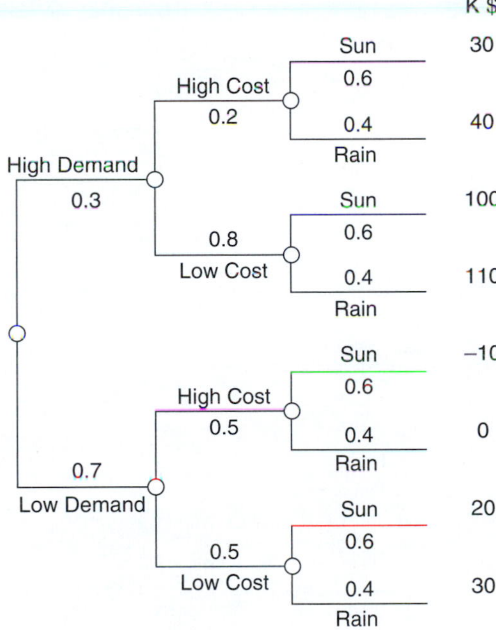

   a. Plot the probability mass function for profit
   b. Plot the cumulative probability for profit.
   c. Plot the excess cumulative probability for profit.
   d. Determine:
      {Profit <= 80 K$ | &}
      {20 <= Profit <= 70 | &}
      {Profit >= 30 K$ | &}

Answers to Problem 1

Population of Turkey in July 2008 (**Answer:** 71,892,808)
Birthday of Arnold Schwarzenegger (**Answer:** July 30th 1947)
Estimated population of the United States in July 2008 (**Answer:** 303, 824, 640)
Height of Eiffel tower (**Answer:** 324 meters, 1063 ft)
Height of Empire State building (**Answer:** 381 meters, 1250 ft)
Closing of the Dow Jones Industrial Average (DJIA) on December 30th 2008 (**Answer:** 8668.39)
Zero-fuel weight of an empty A330-200 Airbus (**Answer:** 168 Tons)
Length of an A330-200 Airbus (**Answer:** 58.8 m, 192 ft 11 in)

CHAPTER

16

From Phenomenon to Assessment

CHAPTER CONCEPTS

After reading this chapter, you will be able to explain the following concepts:

- Information transmission
- Perception
- Cognition and cognitive biases:
  - Availability bias
  - Representativeness
  - Anchoring and adjustment
  - Implicit conditioning
  - Mental accounting
- Motivational biases

16.1 INTRODUCTION

The path is complex between a phenomenon existing in the world and our personal assessment of that phenomenon. The complexity exists both in our own thoughts and in any assessment process designed to produce an assessment of the phenomenon. Figure 16.1 shows the assessment process as a diagram and is useful in understanding both our thoughts and the nature of assessment.

We shall see that there are several stages that intervene between the phenomenon and the assessment. We will discuss each of these in turn, learn the effects they have in distorting assessments, and investigate ways to compensate for these effects.

16.2 INFORMATION TRANSMISSION

Information transmission refers to the process of amplification, or attenuation, that transforms relevant information about the phenomenon into what we are able to actually observe. Some characteristics of the information transmission process are easy to understand. For example, consider the nature of news whether in print, on the radio, on television, or on the Internet. Only the unusual is newsworthy. Remember the adage, "Dog bites man, that is not news; man bites dog, that is news." If a family of four is felled by botulism after eating canned salmon, the story is very likely to appear on the news. However, four older men dying of emphysema in the local Veterans Administration hospital is not likely to make the front page.

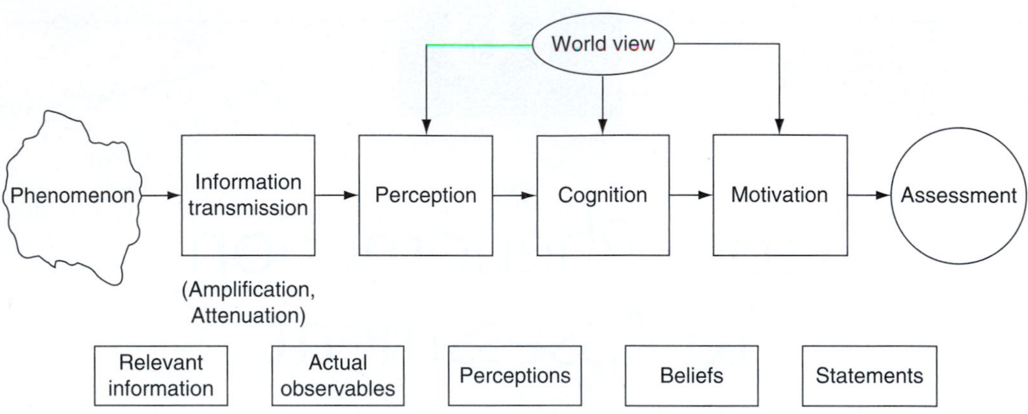

FIGURE 16.1 The Assessment Process

The fact that news reports the unusual makes people believe that rare events are much more likely to occur than is, in fact, the case. If newspaper stories reflected reality, there would be millions of stories like "Joe got up this morning, brushed his teeth, had a quick cup of coffee, drove in heavy traffic to work, spent the day at his desk, drove home, had a microwaved dinner, watched TV, and went to bed." You are unlikely to see a story of this kind anywhere except maybe in the Sunday lifestyles section.

Also, you are also unlikely to see news stories that go against commonly accepted public views on matters of behavior. For example, can you imagine seeing this headline in a mainstream newspaper: "Try Recreational Drugs—40 Million Americans Can't be Wrong!" Instead, you are much more likely to see a story on how the "War on Drugs" is being waged.

Sometimes, the agencies of information transmission have ends other than just simple newsworthiness. For example, we have all seen scenes of the great homefront American support of World War II: Rosie the riveter, victory gardens, scrap metal drives. Seldom will you see a discussion of the extensive black market for rationed goods—gasoline, meat, canned goods—that existed during the war. Consider that the typical book selection committees for history books in public high schools are likely to look unfavorably on a book not suitably patriotic.

In company or school newsletters, purposeful omission is also to be expected. How often, for example, will your company newsletter feature a story about a competitor's new product, emphasizing the many improvements over your company's current product? How often will you read in a university newspaper a story extolling the major achievements of a rival?

If we are aware of these distortions and omissions, we can compensate for them by expanding our sources of information.

16.3 PERCEPTION

The perception process transforms what we are actually able to observe into our personal perceptions. Unfortunately, our senses may be faulty. We have all seen optical illusions like the one in Figure 16.2.

Here, even though the horizontal lines are exactly the same length, the lower one appears longer. Other effects are more subtle. For example, since we are wired to use clarity as a cue for

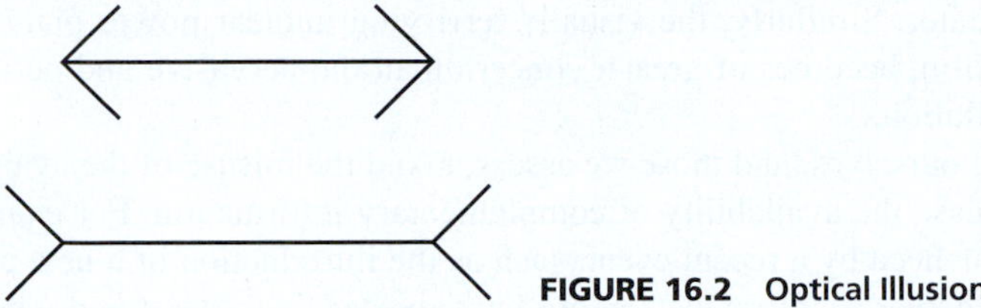

FIGURE 16.2 Optical Illusion

distance, sharply etched objects appear closer to us than they really are. We encounter this effect when driving in the desert and those mountains that seem "right over there" are still "right over there" after another half-hour of driving.

Our world view can also affect our perception. When people are given newspapers containing stories with equal numbers of stories in favor of drug legalization and opposed to it, they tend to perceive only the ones that are consistent with their pre-existing views on the subject.

We ask the class how many have seen the Man in the Moon: About half say yes and the other half do not know what I am talking about. Many people can see the Big Dipper or Great Bear in the sky. Yet these are just interpretations that we are placing on observables to create a perception. As any magician will tell you, "Seeing is believing." This may lead you to error.

16.4 COGNITION

Cognition is the process that turns perceptions into beliefs. **Cognitive biases** are thought process errors that cause a person's beliefs to improperly reflect his or her perceptions. One cause of such errors is **wishful thinking**: Forming beliefs based on what should be, rather than on evidence or actual experience. In this way, we let our world view affect our thinking process.

For example, when asked the chance that a particular presidential candidate might be elected, one young woman in class assessed a probability of 90%—a number far higher than any other class estimates. When asked why she had offered such a high probability, she said, "It needs to be high. I'm working for his campaign." In this case, she had confused her desire for an event to occur with the probability of the event actually occurring.

Another cause of problems in cognition is the improper use of **cognitive heuristics**. In many situations, heuristics can be useful, but often they can also lead us astray. The use of these heuristics is unconscious and uncontrollable, but it is also correctable once we are aware of their existence. We will now discuss several of these heuristics in detail, explaining how they function and what can be done to compensate for them.

16.4.1 Availability

Simply stated, the **availability heuristic** claims "the easier it is to think of an event happening, the more likely it is to happen." Information that is more ordered, dramatic, recent, or imaginable is used as the basis of making events loom larger.

For example, when people are asked whether they can form from a group of 10 more committees of two or of eight, they typically answer that they can create more committees of two. Yet, every committee of two that is selected corresponds to a committee of eight that was not selected; the number of possible committees is the same in each case.

The drama of drowning in a car far outweighs the gradual onset of Type 2 diabetes, making many people become more concerned about the first situation—even though the chances of the second are far greater. Similarly, the visually terrifying nuclear power plant accident, which we have seen on film, becomes of greater concern than the pervasive and permanent effects of environmental pollution.

We can help ourselves, and those we assess, avoid the misuse of the availability heuristic. First, we can increase the availability of complementary information. For example, if a person seems unduly influenced by a recent event, such as the introduction of a new competitive product, we can ask whether such products might have existed in the past and whether, ultimately, they had any impact. Sometimes, having the person try to view the situation retrospectively from the future can also be helpful.

16.4.2 Representativeness

Representativeness is the thinking error based on discarding previous experience—even in the face of worthless evidence—and estimating probabilities based on often incorrect similarity judgments that misinterpret the effect of uncertainty.

The following exercise illustrates insensitivity to previous experience. Subjects are told a group consists of 70 lawyers and 30 engineers. They are then given the following sketch: "Dick is a 30-year old man. He is married with no children. A man of high ability and motivation, he is quite successful in his field. He is well liked by his colleagues." After listening, the subjects decide that the chance that Dick is an engineer is 0.5. In other words, they have disregarded the composition of group, 70 lawyers and 30 engineers, in arriving at their assessment. They did not put enough weight on the prior: the group composition.

In addition, people often have misconceptions about chance. They think that in six tosses of a coin equally likely to fall heads and tails, the pattern HTHTTH is more likely than the pattern HHHTTT because the latter is "not random." They also think that the pattern HTHTTH is more likely than the pattern HHHHTH because the latter is "not fair." Yet in truth, all patterns are equally likely.

At any time, varying process extremes of high or low is likely to be followed by a return toward the average. This is the principle of **regression**, a process we frequently observe, but often forget. One example of regression involves a flight instructor who said he believed that punishment was much better than reward in training new pilots. When asked why, he said he had observed that, whenever he praised a pilot who performed extremely well, the next day the pilot performed worse. However, when he chastised a pilot following a poor performance, the next day that pilot would do better.

Our academic department fell into the same trap. Depending on the exam scores of applicants, the department received a number of fellowships. One particular year, scores were exceptionally high, and we received a large number of fellowships. Naturally, we congratulated ourselves on the rising attractiveness of our department. The next year scores returned to their normal range, and we sought to find where we had gone astray. Regression had struck again.

Many people readily misinterpret the effects of **uncertainty**. A famous example is provided by the *book bag problem*. Two indistinguishable book bags are filled with red and blue balls. The red book bag contains 70% red balls; the blue book bag contains 70% blue balls. One of the book bags is selected at random and balls are sequentially drawn from it and then replaced. The sequence of draws is red, blue, red, red. What is the probability that this is the red book bag? Usually, people assign a probability like 60 or 70 percent, yet the actual answer is 49/58, or 84 percent.

In another example, subjects are asked to consider two hospitals. On average, the larger hospital has 45 births per day; the smaller, 15. The subjects are asked which hospital has a greater number of days per year with more than 60% boy births. Often, people choose the larger hospital, but the correct answer is the smaller hospital. The larger the hospital, the more closely the number of births on any day will tend toward 50% boys. The chance that the smaller hospital will have more than 60% boys on any given day is greater than 15%; the larger hospital has less than a 7% chance of the same result.

To avoid representativeness errors, we must first separate prior information from new evidence, and then process our information using probability theory.

16.4.3 Anchoring and Adjustment

Another cognitive bias is caused when we form our beliefs prematurely. We fall victim to this when our initial ideas play too large a role in determining our final assessments, because we have prematurely fixated on an initial assessment without adequate adjustment. This is referred to as **anchoring.**

The following exercise illustrates the phenomenon. A group of MBA students was asked to write down the last three digits of their home phone numbers, for example, *XYZ*. The instructor then told them that he was going to add 400 to this number, forming $XYZ + 400$. Then, he asked each of them together to write down whether they thought Attila the Hun was defeated in Europe before or after A.D. $XYZ + 400$. Finally, he asked each of them to write down their best guess for the year in which Attila the Hun was actually defeated. In fact, that year was A.D. 451. The results of their estimates are shown in Table 16.1.

Note that the higher their initial anchor, the higher their average guess of the year of Attila's defeat. Although everyone in the group knew that the original anchor had nothing to do with the question, the group still fell subject to the anchor.

Recall that, in the 20 questions exercise, the general tendency is to assign overly narrow distributions. This could be caused by anchoring on an initial number, such as the median, and then failing to adequately adjust. To minimize this tendency, we advise the students to assign first the surprises, then the mid range, and finally the median. Yet, most people still displayed overconfidence in their assessments.

Another consequence of anchoring occurs in assessing the probability of an event that requires many steps to occur. For example, if people are told that going to Mars will require the success of 20 steps, each with 0.95 probability of success, they assign that the probability of success of the Martian mission to be quite high, perhaps because they anchor on the number 0.95. If they are told that going to Mars will require the success of 20 steps each with 0.05 chance of failure, they assess the probability of success to be quite low, perhaps because they are anchoring

TABLE 16.1 Estimates for Attila the Hun's Defeat

Range of the Initial Anchor (Last 3 digits of phone number plus 400)	Average Estimate of the Year of Attila Defeat
400 to 599	629
600 to 799	680
800 to 999	789
1000 to 1199	885
1200 to 1399	988

on the number 0.05. If this seems strange, remember the Atilla the Hun example. In fact the probability of success of the Martian mission for both descriptions is $(0.95)^\wedge 20 = 0.358$.

The remedy for anchoring is to obtain an initial assessment in a form that does not commit the subject. Decision analysts must avoid anchors at all costs.

16.4.4 Implicit Conditioning

Implicit conditioning arises when the probability assessed is implicitly conditional on the occurrence of uncertain events. Without realizing it, subjects may be conditioning on things such as "No war, no strike, no fire, no flood, no price control, no revolution, no new competitor, no failure of design approach." Once the conditioning is discovered, the subject can make an assessment based on each possibility for the condition, and then assign a probability to those possibilities. Making the implicit conditions explicit avoids a very inappropriate assessment.

One way to discover the implicit conditioning is to pose an extreme outcome to the subject and ask for an explanation. For example, if you are encoding a distribution on next year's sales, and you suspect that the distribution is too narrow, you can ask the subject, "Suppose I told you that the sales were going to be x (a very low number). How would you explain that?" Suppose the subject tells you that would mean that would indicate a strike. You then ask, "Did you include the possibility of a strike in your original assessment?" If the subject says, "No," then you would ask the subject to refigure his assessment accordingly. Alternatively, you can ask for a distribution on sales given strike and given no strike, ask for the probability of a strike, and then compute the resulting unconditional distribution.

16.4.5 Mental Accounting

Mental accounting causes a final type of cognitive error. One form of mental accounting is **labeling**, where we attach special characteristics to assets in our possession that are not apparent to others. For example, some people consider gambling winnings as money that is somehow less valuable that what they get in their paychecks, and, therefore, can be treated with more abandon. Imagine that a person is pickpocketed of $100 in gambling winnings in a state where such a loss would constitute a grand theft. Could the thief's lawyer claim that less than $100 was stolen, since these were only measly gambling winnings?

I once had a client who worked for a major New York bank and talked about "pictures of George," by which he meant one dollar bills. What counts is how many "pictures of George" you have, and not the tales you tell about them.

This point was driven home to me when my wife was left some stock by her aunt and she asked our financial manager if he could handle them in our portfolio. He said, "Yes, if they are just shares of stock, but not if they are Auntie Con's shares." In other words, he did not know how to handle shares with "labels."

Another form of mental accounting arises when we deal with percentages rather than "pictures of George." In one demonstration of this, two similar groups were asked two different questions.

People in the first group were told, "You are about to buy a calculator for $15 when the salesclerk tells you it is on sale at their branch store across town for $10." They were then individually asked, "Would you go to the branch store to buy it?"

People in the second group were told, "You are about to buy a jacket for $115 when the salesclerk tells you it is on sale at their branch store across town for $110." They were then individually asked, "Would you go to the branch store to buy it?"

A much higher fraction of people were willing to drive across town to save $5 on the calculator than $5 on the jacket.

I find that when I ask such questions of people who have studied decision analysis, their first thought is how much would I have to be paid to drive across town?

16.5 MOTIVATION

The final link in the chain from phenomenon to assessment is **motivation**. The statements that people make may differ from their beliefs for motivational reasons. Some people may misrepresent their beliefs for reasons of self-interest, including promise of reward or fear of punishment.

Usually, in assessment, we have less concern with outright attempts to mislead than with the hidden effects of motivation. As an example, a colleague had just completed an assessment of next year's sales by one of the firm's leading salesmen. As he was leaving, he saw a plaque on the wall announcing that the salesman was a member of the 400% Club. When the colleague asked about the plaque, the salesman said that it was awarded to salesmen who had exceeded their annual estimates by a factor of four and that he had won it several times. The colleague realized that the assessment he had obtained was virtually worthless.

16.6 SUMMARY

- Cognition is the process that turns perception into beliefs
- Numerous cognitive biases exist. Identifying them can help us become better at providing our beliefs for use in decision analysis
- Motivational biases may cause people to make statements that are different from their beliefs. It is important to recognize such biases when conducting a probability encoding excercise.

KEY TERMS

- Anchoring
- Availability heuristic
- Cognition
- Cognitive biases
- Cognitive heuristics
- Implicit conditioning
- Information transmission
- Labeling
- Mental accounting
- Motivation
- Regression
- Representativeness
- Uncertainty
- Wishful thinking

CHAPTER

17

Framing a Decision

CHAPTER CONCEPTS

After reading this chapter, you will be able to explain the following concepts:

- Selecting a frame
- Zooming in and out of the decision
- The decision hierarchy:
  - What is taken as given
  - What is to be decided now
  - What is to be decided later

17.1 INTRODUCTION

In Chapter 1, we mentioned that decisions are not found, but rather, are declared. Up to this point, our discussions have focused on making a decision once we have declared one. But how should we go about declaring a decision? Recall that in our stool decision metaphor in Chapter 1, the location of the stool is the frame. The **frame** determines what you have declared to be within the bounds of your present decision. The purpose of this chapter is to provide guidance on the choice of the frame and to emphasize its importance in decision making.

17.2 MAKING A DECISION

The word "decision" comes from the same Latin root as "scissors." A **decision**, by its nature, cuts off any other course of action but the one you decide to follow. Simply because you can analyze a declared decision does not mean that it is the decision you should be declaring.

Reconsider Kim's party problem, first introduced in Chapter 9 and revisited throughout this book. Kim has identified three possible party locations and has declared her final decision. But now let us ask some fundamental questions. Why should Kim have a party in the first place? Why have it now? Why not choose some other form of leisure? Why not choose another way to spend her time, like studying her schoolwork or even visiting a sick relative? In the party problem, we were given a frame that Kim has decided to have a party and the she has three alternatives from which to choose. This is why these questions did not come up earlier. However, when making a decision, Kim would need to answer these questions to decide if she even needs to think about the party in the first place or the best way to spend her time. This task is achieved by selecting an appropriate frame for the decision.

FIGURE 17.1 Limits of Using a Frame

17.3 SELECTING A FRAME

The frames used to help us make decisions function in the same way as the frames we use for photographs. In deciding to record any scene, the photographer must decide what to include within the image and, therefore, what to exclude (Figure 17.1). Just as the quality of the resulting photograph depends strongly on its frame, so does the quality of a decision.

Take a moment to think about Figure 17.1. Focusing on only a small portion of the picture will certainly provide focus and will filter out extraneous details. However, you also will not get the whole picture, and at some point, that detail you lose might become important.

Sometimes your decision frame is conscious, but often it is not. Selecting a frame is the most important aspect of making good decisions, since solving the wrong problem correctly leads us to error. Selecting the wrong frame is sometimes described as "making an error of the third kind." It is an error that is easy to make and difficult to correct.

17.3.1 The Six Blind Men and the Elephant

Often we start with a frame that might seem appropriate—given our narrow view (perspective) of the situation. Later, it may become apparent that it is the wrong one and that there are other better frames.

The story of the six blind men and the elephant shows the effect of a limited perspective (Figure 17.2). One of the men felt the elephant's side and said,

"The elephant is like a wall."

The second touched the elephant's tusk and said,

"The elephant is like a spear."

The third handled the trunk and said,

"The elephant is like a snake."

FIGURE 17.2 The Six Blind Men and the Elephant

The fourth encircled the elephant's leg with his arms and said,

> *"The elephant is like a tree."*

The fifth touched the elephant's ear and said,

> *"The elephant is like a fan."*

The sixth held the elephant's tail and said,

> *"The elephant is like a rope."*

All of these men were individually correct, but none had characterized the elephant as it really was. Any point of view provides a frame that is appropriate from that perspective, but it can be perhaps inappropriate considering the whole phenomenon under consideration.

We might say the blind men were led to have different beliefs about the nature of an elephant because they had limited their perspectives to the perception of a single elephant attribute. The same limitation in perspective arises when a person views a situation in terms of the concepts and tools he has learned in developing his professional competence. If you are an expert with a hammer, for example, every problem starts to look like a nail.

Learning how to effectively frame a decision problem is qualitatively different from the kind of analysis we have been discussing in earlier chapters. Once we have a decision basis, we have learned how we can perform a correct analysis using the five Rules of Actional Thought and other tools, such as trees and diagrams.

Learning how to effectively frame decisions is more like taking a class in painting. While knowing the properties of art materials is necessary to produce a high quality painting, it alone is insufficient. Ultimately, it is up to the decision maker to judge whether the chosen frame has produced a masterpiece or a "color by numbers" piece of trivia. As though we are participants in an art class, our discussions in this chapter will be more guidance than instruction. Yet achieving mastery in framing is the most valuable skill in decision making.

17.3.2 The Frame Metaphor–Zooming and Focusing

Just as when we see a scene through a viewfinder, we can distinguish two aspects of framing. One aspect is **zooming**, which refers to how large a view we are taking. The other aspect is **focusing**, referring to which features of the present view attract our attention.

As an illustration of zooming, imagine a slideshow presenting a series of pictures of the same scene, each expanded using a scale 10 times as large as the preceding one. It begins with an overhead view of the backyard of a suburban house. The next scene shows many houses with this backyard in the center. Successive scenes show the neighborhood, the suburbs, the city, and then the region. At some point we see a continent—followed by the earth. Much later, we see the solar system, the galaxy, and even the neighboring galaxies, always with the original backyard in the center.

Each scene might trigger several possible thoughts as we focus on different features. For example, viewing the backyard might cause us to ask whether it has enough room for a swimming pool or to consider installing a sprinkler system to improve the condition of the landscaping. Viewing the city scene might make us think about commuting time. The scene of the region might stimulate thoughts about possibilities for boating or mountain climbing. The earth's scene might trigger thoughts about global issues of environment or conflict. The scene showing neighboring planets could create visions of interplanetary travel, while those of our galaxy could lead us to wonder whether mankind will ever visit another star. The largest scale views may lead us to think of just how small a part of the universe we comprise.

These thoughts are the stimuli that may cause us to declare decisions. The backyard frame might be useful and appropriate for deciding where to plant a fruit tree, but completely inappropriate for finding good boating.

17.3.3 The Thrown Frame

Once we have a frame in mind, we will naturally come to regard the first one that pops up, also known as our **thrown frame**, as the "right" one. From that point on, it will be difficult for us to consider using a different frame that might actually better serve our needs.

COPS AND ROBBERS An example of a thrown frame taken from a newspaper is shown in Figure 17.3a.

Unsung Hero

LONDON: One man in the crowd gripped his walking stick when he saw three men racing down a busy street near Trafalgar Square yesterday [followed by] policemen. He knew where his duty lay if cops were chasing robbers.

He raised his stick, cracked one man over the head, and vanished from the scene. His only desire was to be an unsung hero.

FIGURE 17.3a Thrown Frame, Part 1

Later, we read from the same article the text shown in Figure 17.3b.

Unsung Hero

......

The injured man was taken to the hospital to have his gashed head stitched. Last night, nursing his aching head, 30-year-old actor Michael McStay bemoaned the fact that the movie sequence had proved too realistic. "I suppose this is an occupational hazard," he said, "but I do think he owes me a drink."

FIGURE 17.3b Thrown Frame, Part 2

Think of the frame that the movie director would have while viewing this scene. The director might start by saying "Where did that idiot come from?" Later, he might change his tune to "What can I salvage from today's filming given that I have lost one of my actors?"

A COOKIE STORY A thought-provoking example of the importance of framing concerns the story of a man who was taking a flight from San Francisco to New York with a stop in Chicago. As was his custom, he purchased a bag of his favorite chocolate chip cookies before boarding the plane, knowing that they would make the long flight more pleasant. When he boarded the aircraft, he took his desired window seat and placed his purchase and coat on the seat in the middle of the three seat row. The aisle seat was soon occupied by a woman who also placed her belongings in the seat between them, which they knew at this point was not going to be occupied.

Soon after takeoff, the man fetched one of his cookies from the bag and enjoyed it while reading a book. Some minutes later, he noticed that the woman in the aisle seat had reached over and taken one of his cookies. He was surprised and somewhat annoyed by her thoughtlessness, but took no further notice. A half hour later he enjoyed another of his cookies, and soon afterward was astonished to see the woman help herself again. Now, he believed her to be incredibly rude, and fully aware of the situation. The same sequence occurred a third time, and this time, the bag ended up empty. Though he was now peeved and angry, he decided that bringing attention to her behavior would only make him look petty.

When the plane landed in Chicago, the woman got up, gathered her things, and left without a word of explanation. He sat for a moment, ruminating on the lack of concern for others in modern society when he glanced at the middle seat and saw that when the woman had removed her possessions, the bag of chocolate chip cookies he had bought was exposed, and that it was unopened. Not only had the woman not eaten his cookies, *he* had eaten *hers*.

Here, we see that by unthinkingly creating and accepting decision frames, we often end up making poor judgments regarding our life experiences and the people and situations we encounter.

17.3.4 Zooming: The Decision Hierarchy

The previous examples show how we may often rush to a frame by creating a mental image of the situation in our minds that may be very different from the actual situation at hand. But even if we are fully aware of a given situation, we may still choose different courses of action simply based on the way we happen to focus our attention. Consider the following example (see Figure 17.4).

THE FLAT TIRE A young man leaves work at the end of the day to find that his car has a flat tire. He realizes that he needs to make a decision and thinks to himself, "Should I fix the tire myself, or call a repair service?" He balances in his mind the likely consequences of soiling his clothing if he tries it himself or waiting forever if he calls the repair service. Annoyed with having to deal with this problem, he has the following successive thoughts:

- Rather than driving, I should use public transportation to get to work.
- Regardless of how I get here, I do not like working for this company.
- In fact, I do not like the kind of work I am doing here, or anywhere else.
- I should change the way I am making a living.
- All my life, I have wanted to be an actor. That is what I should do now: Begin an acting career.
- To pursue an acting career, I should move to New York or Los Angeles.
- When I get home, I am going to decide where I should go to pursue my life's dream.
- Now should I fix this tire myself, or call the repair service?

These mental reflections show how this person has indeed recognized that a decision needs to be made about the flat tire, but his thoughts have roamed from how to get the tire fixed, to whether to use public transportation, to whether to change companies, to whether to change fields, to how to pursue the field he desires. Each of these thoughts can lead to a different course of action, and all of them are, in fact, very important, but the original decision about how to fix the tire was never made.

To help provide focus on the actual decision that needs to be made and the choice of the appropriate frame, it is useful to think of a **decision hierarchy** (Figure 17.5).

Figure 17.5 illustrates the role of the frame in focusing on the decision under consideration. The figure represents a decision hierarchy (or pyramid) with the frame in the center. The frame specifies the set of decisions that will be made in a given situation. Above the frame are

FIGURE 17.4 Flat Tire Example
(Kate Bialasiewicz/Shutterstock)

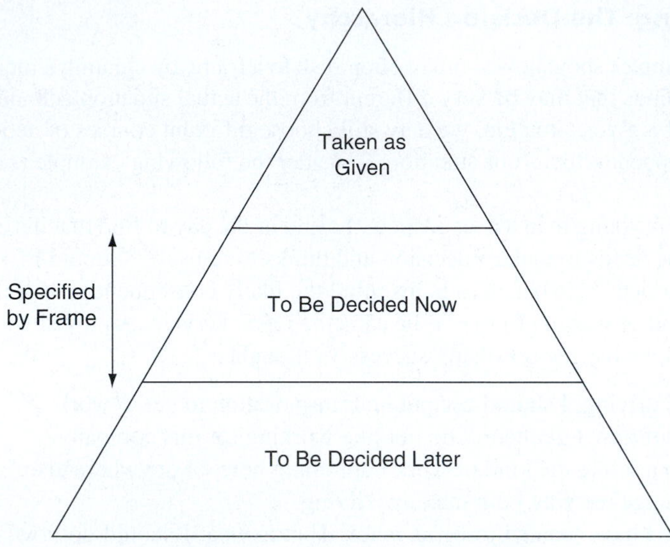

FIGURE 17.5 The Decision Hierarchy

decisions that are taken as a given—not to be questioned at this time. These are often matters of policy. Below the frame are decisions that will be made later—after the decision in the current frame has been made. They include decisions on how the framed decision will be implemented.

THE HOME REMODEL The process of arriving at an appropriate frame is demonstrated by the experience of a professional colleague. He and his wife had lived in the same home for many years raising their children, who were now entering their own careers. They had delayed making

FIGURE 17.6 Home Remodeling Example
(David Sacks/Getty Images)

home improvements, but now it was time. They were going to remodel the kitchen, build a deck with a hot tub, and repaint the interior—all projects they had long anticipated.

As they were about to decide on contractors and color schemes, our colleague began to wonder whether they were thinking broadly enough about their housing choices. Now that the children had left, the attributes they would want in a home might be different. Perhaps it was time to explore selling their home and purchasing a new one. The thought of buying a new home made my friend question how long they might be living in this area, which in turn led him to contemplating the possibility of retirement and perhaps moving to a new and remote location. Each of these successive thoughts caused an increase in the size of the frame for his housing decision.

On further reflection, he realized that retirement and accompanying location changes were far off in the future. Therefore, for the time being, he took it as given that he would continue to live and work in the same general area. The frame, now smaller in size, contained the possibilities of either buying a new home or remodeling their current home. Landscaping and furniture decisions, they realized, could wait until later.

After searching for suitable homes currently on the market, they found one they liked very much and decided to buy. However, before moving, they would still have to sell their current home. To speed up the process, their real estate agent advised them to consider extensive remodeling. Unlike their original and very personal remodeling ideas, these changes would be intended to increase the market value of the home. Their personal preferences were no longer germane to the remodeling decision (see Fig. 17.6).

Figures 17.7a and 17.7b show the initial and final decision hierarchies for the remodeling decision.

This example demonstrates two important concepts of effective framing. First, as shown here, you can expand a frame until it is too large, and then contract it until it is appropriate. Second, you should understand that changing the frame can completely alter the decision basis. Once my colleague and his wife had decided that the exclusive purpose of remodeling their existing home was to maximize profit from the sale, entirely new alternatives, information, and preferences came into play, and the old ones were no longer relevant.

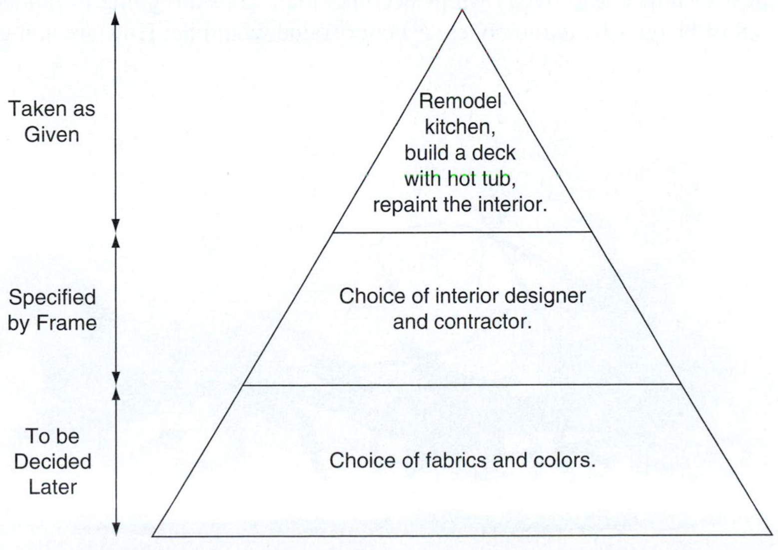

FIGURE 17.7a Home Improvement Initial Decision Hierarchy

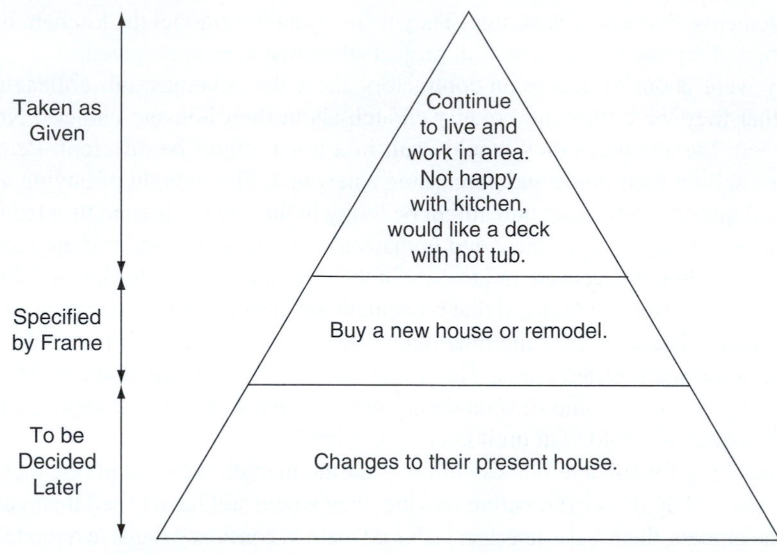

FIGURE 17.7b Home Improvement Final Decision Hierarchy

THE BUGGY WHIP MANUFACTURER Choosing the wrong frame can be disastrous. Consider a buggy whip manufacturer in the early 1900s who finds that sales and profits are continually falling (see Figure 17.8). He thinks the reason is that, because his costs are too high, his prices are too high, and he decides to hire consultants to improve the efficiency of his operations.

The consultants discover ways to lower the cost of raw materials to better organize the factory floor, to eliminate waste in the manufacturing process, and to find cheaper distribution channels. They do an excellent job of "re-engineering." However, they have solved the wrong problem. Figure 17.9 shows the decision hierarchy for the frame they have taken.

Regardless of how cheap buggy whips become, their sales are going to fall as a result of the replacement of buggies by automobiles. A better frame would be: How are you going to use

FIGURE 17.8 Coach Driver using a Buggy Whip (Collection of the authors)

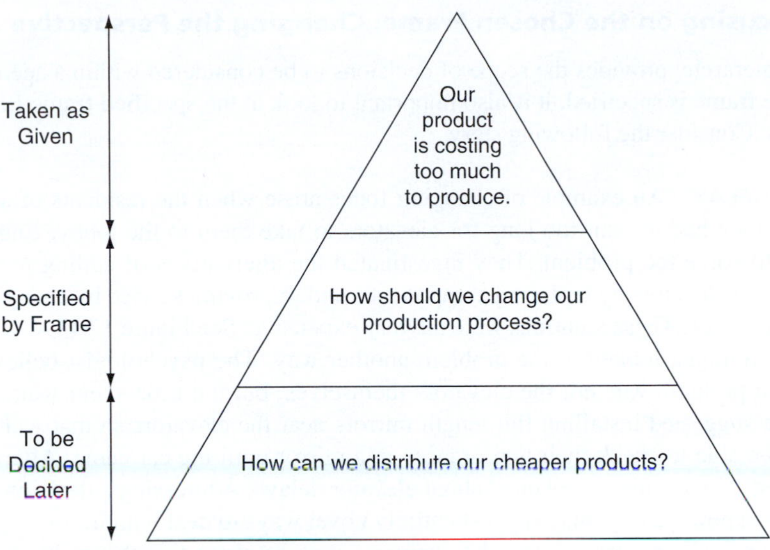
Taken as Given
Specified by Frame
To be Decided Later
Our product is costing too much to produce.
How should we change our production process?
How can we distribute our cheaper products?

FIGURE 17.9 Buggy Whip Inappropriate Decision Hierarchy

the skills of your company in design, production, and distribution to take advantage of the many opportunities provided by the burgeoning market for automobiles? (See Figure 17.10.)

FREUD'S "PRESENTING PROBLEM" Freud used to distinguish between the **presenting problem** and the underlying problem of a given patient. For example, the presenting problem—a symptom of insomnia—might be the product of a more underlying problem—an unresolved and troubling relationship, for example. Finding the proper frame is the corresponding practice of a decision analyst.

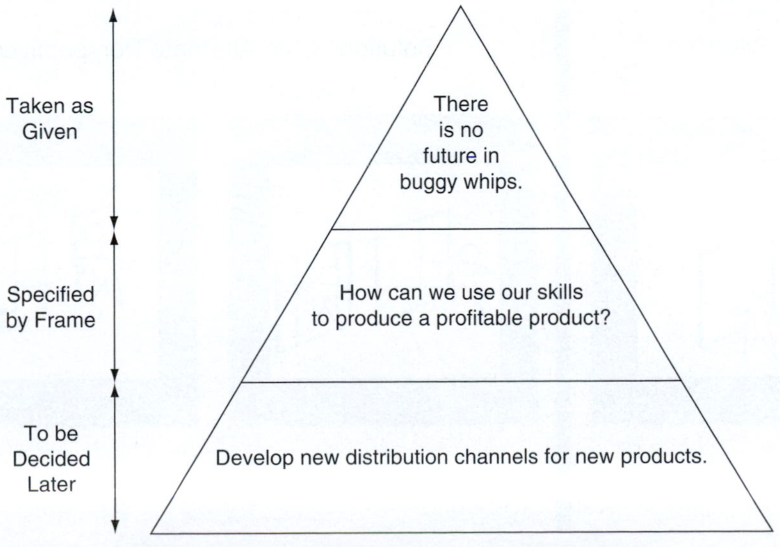
Taken as Given
Specified by Frame
To be Decided Later
There is no future in buggy whips.
How can we use our skills to produce a profitable product?
Develop new distribution channels for new products.

FIGURE 17.10 Buggy Whip Better Decision Hierarchy

17.3.5 Focusing on the Chosen Frame: Changing the Perspective

A decision hierarchy provides the scope of decisions to be considered within a specified frame. But once the frame is specified, it is also important to look at the specified frame from different perspectives. Consider the following story.

ELEVATOR DELAY An example of changing focus arose when the residents of a hotel complained that they had to wait too long for elevators to take them to the lobby. Engineers were summoned to solve the problem. They investigated the alternatives of adding new elevators, speeding up the elevators by replacing components, and improving service by changing the rules for dispatching cars. These solutions were all very expensive. See Figure 17.11.

A psychologist looked at the problem another way. The psychologist believed that the source of the problem was not the elevators themselves, but the time spent waiting for them to arrive. He suggested installing full-length mirrors near the elevators so that waiting passengers would be able to check their appearances in a mirror until the car came. After the mirrors were installed, few people complained about elevator delays. Addressing a decision from a new perspective—a new frame—may suggest entirely novel ways to deal with it.

In dealing with elevator delay, the engineers took as given that the problem required an engineering solution; they framed their decision as determining which engineering solution was best. The psychologist understood that the residents were annoyed by the delay of the elevators, a matter of perception rather than clock time, and he sought an alternative that would decrease this annoyance.

17.3.6 Stakeholders

A useful concept in framing a decision is to identify the **stakeholders** in the decision. We define an **affecting stakeholder** as someone who can affect the decision—for the elevator delay problem, it is the hotel owner. The **affected stakeholders** are the people affected by the decision—in this same example, they are the hotel residents.

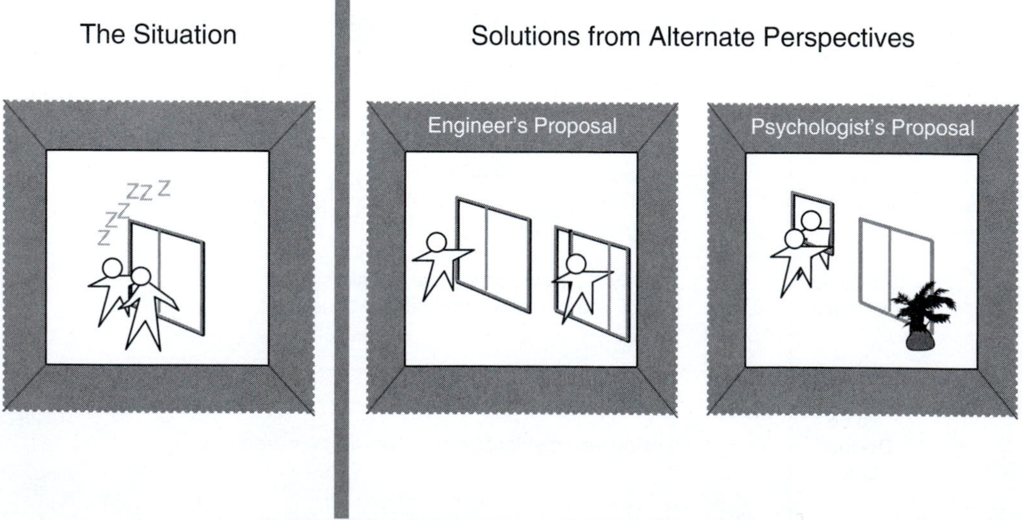

FIGURE 17.11 Same Situation, Different Perspectives

For business decisions, the affecting stakeholders are typically management executives, possibly the board of directors of the corporation, and often government regulators. The affected stakeholders could be current employees, potential and retired employees, vendors, and customers. In some cases, the competitors could fall into either or both categories. In personal decisions, family members may be in both categories.

To illustrate the importance of recognizing stakeholders, consider the Channel tunnel linking England and France. Its construction was a major engineering achievement, yet it has not been a financial success. One reason is that the framing of the original decision did not adequately consider the effect of competitive reaction by other trans-Channel services on the profitability of the enterprise.

17.3.7 Questioning the Frame

You may wish to test a frame by asking yourself these questions:

- What am I taking as given?
- Am I being too narrow or too broad?
- Have I considered both short run and long run consequences?
- Are all stakeholders included?

17.3.8 Re-examining the Frame

A change in the decision situation may require a change in frame. Frame adjustment may be required due to external factors or due to some modification of the decision basis: changes in alternatives, information, or preferences. Both people and organizations may be slow to change frames as they mature and as the world evolves around them. We see major companies become minor through their failure to adapt, while others thrive by taking advantage of the opportunities missed by their competitors.

17.3.9 Framing Workshop

In class, we typically discuss framing by asking each student to think of a personal or professional decision they have made and to prepare to discuss it. We then proceed to inquire about the frame for the decision. Using the notion of the decision hierarchy, we first ask,

> *"What did you take as given?"*

We usually find that some things that had been taken as given need not have been so considered and could have been part of the decision. Further discussion often reveals the existence of factors taken as given without conscious awareness.

We might then ask,

> *"Who were the stakeholders?"*

The results can be funny. One student said that he would be the one deciding the car the family should purchase, but his wife would be the one driving it.

Discussions of framing can sometimes lead to new solutions. On one occasion, for example, a student said that his decision involved how to remove problematic members from a sports club with a minimum of disruption. He had taken as given that he and the other members of a club would have to figure out how to solve this problem. At the next class, he reported a change of frame. He now took as given that the club would form an alumni advisory group of former members. This group would have the responsibility to make decision concerning, separation from membership.

The discussions can sometimes become very personal when people consider a major decision they have made in their lives and its consequences. For example, one woman realized that she had taken as a given that she would attend the graduate school offering the most attractive financial support without giving sufficient consideration to the actual quality of education she would be receiving.

In another example, a man deciding whether to return to his country or to remain in the U.S. to study for a doctorate realized that his major concern was loneliness following a breakup with his girlfriend. He left one class troubled, and arrived at the next class joyful as the result of reflecting on the framing discussion.

Still another example involves a woman trying to decide whether to pursue doctoral education at the British university where she had already received a master's degree, or to return to the U.S. to pursue graduate studies there. She had taken as a given that she would continue working with the same professor in England under whom she was currently studying. Although she very much liked living in England, she did not want to work with that professor anymore and decided to return to the U.S. The discussion seemed very matter of fact until she suddenly got quite emotional. When asked why, she said that until our discussion, she had not realized that she could have continued her English studies under a different professor.

17.4 SUMMARY

- A frame filters what is germane to action from what is not.
- A decision hierarchy sets the frame of the decision by specifying three categories: What is taken as given, what is to be decided by the current analysis, and what is to be decided later.
- Expanding and contracting the frame can help find one that is appropriate.
- The wrong frame will lead to the right action for the wrong problem.
- It is important to look at the specified frame from different perspectives.
- Stakeholders form an important element that needs to be considered within the frame.
- Frames can change with time and need to be periodically re-examined.

KEY TERMS

- Frame
- Decision hierarchy
- Stakeholders

PROBLEMS

1. Think of a decision you are currently facing. Specify the frame for this decision. What are you taking as given? What decisions are you thinking about? What decisions will you decide upon later? Draw the decision hierarchy.
2. Think of a situation you encountered where two people had the following scenarios.
   a. Different thrown frames for the same situation.
   b. Different perspectives for the decision.
3. Select a decision from a news article. Specify the frame for the decision. Draw the decision hierarchy.

CHAPTER

18

Valuing Information from Multiple Sources

CHAPTER CONCEPTS

After reading this chapter, you will be able to explain the following concepts:

- Value of multiple sources of information
- Approaching clairvoyance using multiple detectors

18.1 INTRODUCTION

In many decision situations, we may have multiple sources of information about the distinctions we create. For example, we may have several weather forecasts from several different sources. In oil well exploration, we may have several test measurements that are all relevant to the presence of oil. In the medical domain, we may have several diagnostic tests that provide information relevant to the presence of a disease.

Most of these measurements do not provide clairvoyance on the distinction of interest. Therefore they can be viewed as imperfect sources of information. In this chapter, we shall discuss how to incorporate information from multiple sources into our analysis of a decision. We also discuss the value of information obtained from multiple sources. Once again, we begin our discussion using the party problem from Chapter 9.

18.2 THE BETA RAIN DETECTOR

Refer back to the party problem where Kim is deciding where to have her party. She has three alternatives: Indoors, Outdoors, and the Porch. She also considers an uncertainty. Weather, which could be Sun or Rain. Kim is considering information gathering alternatives about the weather. In Chapter 13, we discussed the Acme Rain Detector and its value to Kim in making this party decision. Suppose that while Kim and the salesman are discussing the possibility of using the Acme Rain Detector for her party, another salesman arrives, representing the Beta Rain Detector. He claims that his Beta detector is more accurate than the Acme and, in fact has a 90% chance of correct indication of both Sunshine and Rain—rather than the 80% obtained by the

Acme detector. Kim can now buy the Acme detector indication (at any time) for $10 or the Beta detector indication (at any time) for $13.

Putting their heads together, the two salesmen even offer to supply the services of both detectors combined for a cost of $16. If Kim chooses this alternative, she can now get one of the following possible joint detector indications:

"Acme says Sunny, Beta says Sunny"
"Acme says Sunny, Beta says Rainy"
"Acme says Rainy, Beta says Sunny"
"Acme says Rainy, Beta says Rainy"

Kim now faces several information gathering alternatives:

- Buying the Acme Detector alone, for $10
- Buying the Beta Detector alone, for $13
- Buying the two detectors combined for $16
- Not buying either one

Should Kim consider any of these deals? Since the highest cost involved in this information gathering activity ($16 for both detectors) is less than her value of clairvoyance on the weather ($20.27), there is at least the possibility that one of these alternatives will be worthwhile. We already know that she should not accept the $10 offer for the Acme detector alone, but what about the Beta detector alone, or the two detectors combined?

VALUE OF JOINT INFORMATION FROM THE TWO DETECTORS Before Kim can consider the question of what the two detectors are worth, she must first think about how they function. Do the detectors operate on identical physical principles or are they different? Is the Beta detector just a more accurate version of the Acme detector, or does it have a distinct principle of operation? Should Kim have a preference for detectors that operate under the same principle? Kim's considerations about the relevance of the different indications and the weather can be summarized in the following relevance diagram.

Note that, as before, Kim still needs to assess the probability of sun or rain and the conditional probability of the different Acme indications given the weather. But now she has to make an additional assessment: the probability of different indications for the Beta detector given its parents. If the Beta assessment requires conditioning on both Weather and Acme's indication, the representation will look like Figure 18.1.

However, if the Beta detector operates on an entirely different physical principle from the Acme detector, then the two detectors' indications will be mutually irrelevant given the weather, and there will be no arrow from the Acme detector indication to the Beta detector indication (see Figure 18.2). This latter representation will simplify the number of probability assessments she needs to make, since she will make the assessment of the Beta detector indication given only the weather.

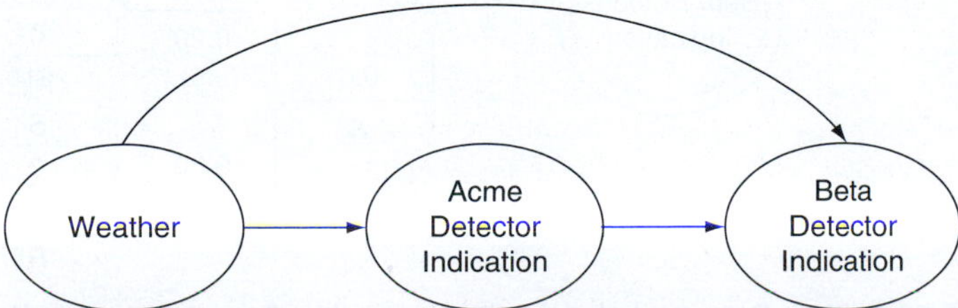

FIGURE 18.1 Assessed Relevance Diagram for Two Rain Detectors

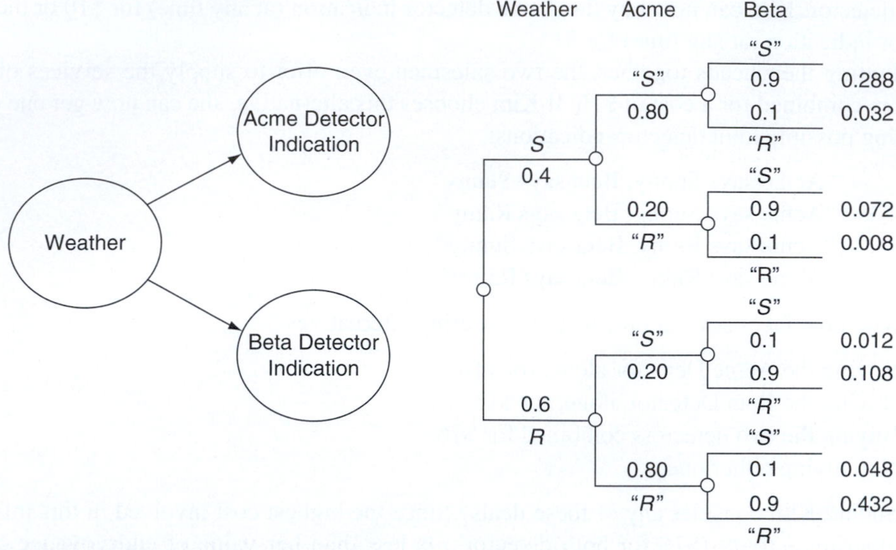

FIGURE 18.2 Two Irrelevant Detectors Given Weather

As another extreme case, if the Beta detector indication is advertised as having the same 80% accuracy as the Acme and operates with the exact same principle of operation, it could, in fact, be identical to the Acme detector indication—just marketed under a different name. In this case, there would be no arrow from weather to the Beta detector indication, but only an arrow from the Acme detection indication to the Beta detection indication, assigning probability 1 to the same indication. In other words, the Beta detector could then be represented by a deterministic node whose value is determined completely by the indication of the Acme detector indication (Figure 18.3). Furthermore, the roles of Acme and Beta could be interchanged.

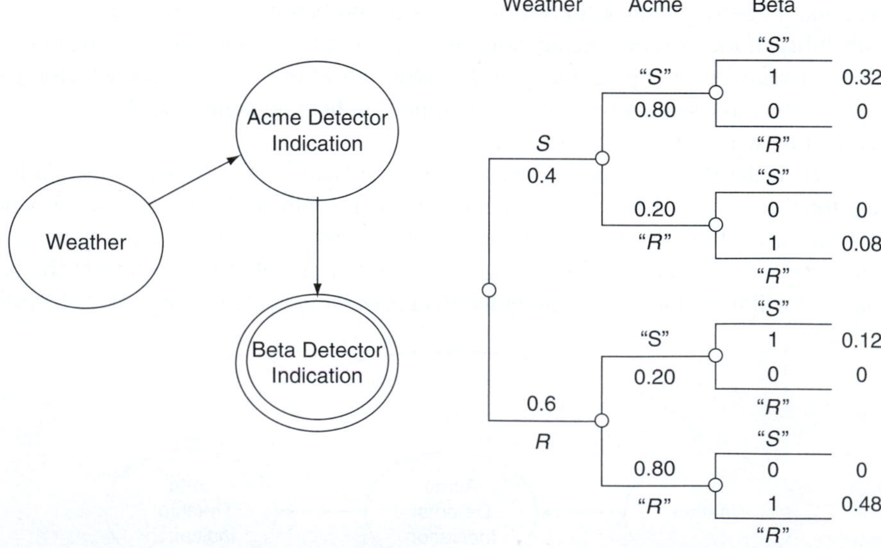

FIGURE 18.3 Beta Detector Indication Determined by Acme Detector Indication

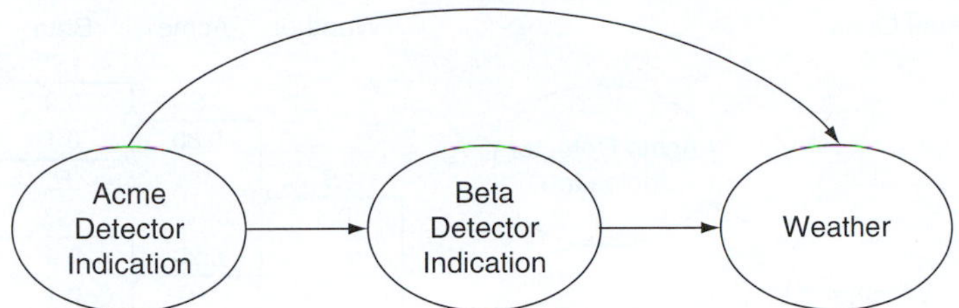

FIGURE 18.4 Relevance Diagram for Two Detectors in Inferential Form

In our case, however, this is not possible, since the accuracies of the detectors are different.

The essence of Kim's problem for the case of two detectors is to provide the probability assignments required by Figure 18.1 after due consideration of her knowledge of the detector mechanisms and representations. Once she has done this, she can change the diagram into its inferential form by reversing the order of the distinctions, as shown in Figure 18.4, using our standard procedures.

The diagram in Figure 18.4 allows her to compute the posterior probability on the weather given any detector indications and the preposterior probabilities of those indications. This is all the probabilistic information she will need to determine her purchasing decision.

Figure 18.5 shows the different orders of assessment and inference when the detectors function using entirely different principles of operation.

Figure 18.6 shows Kim's decision diagram for the use of the two detectors. This diagram incorporates the relevance diagram of Figure 18.4. Depending on Kim's decision about which detector to use, she will obtain different detector reports as a basis for the location decision. The costs associated with her decision on detector use are also reflected in the value node. The decision tree would be constructed in the order: Detector use decision, Acme Rain Detector indication if available, Beta Rain Detector indication if available, Location, Weather, and Value.

You can see that any number of mutually relevant tests could be subject to this type of analysis. Figure 18.7 shows the decision tree that Kim will face if she decides to purchase the two detectors simultaneously. To show the generality of the analysis, we have used the method appropriate for any *u*-curve of subtracting the detector costs from the values at the end of the tree. We used Kim's *u*-curve to obtain the *u*-values.

Since we know that Kim is a deltaperson, we could have solved the problem more simply for her by calculating the value of the party deal with free detector information and subtracting that from the value of the deal with no information to get the value of the detector.

Buying both detectors for \$16 leaves Kim with a certain equivalent of \$44.21, which is less than the \$45.83 certain equivalent if she had bought neither. Therefore, she should not buy both detectors at this price. To determine the most she should pay for both, we decrease the cost to a number below \$16, until she is indifferent to buying them. This cost will be her Personal Indifferent Buying Price (PIBP) for buying both detectors.

As we discussed, since Kim is a deltaperson, we can also calculate this value by finding her certain equivalent of the deal with free use of both detectors and then subtracting from it her certain equivalent with no detectors. By using the tree with zero price for both detectors, we find the certain equivalent of the deal to be \$60.21. Without them, it is \$45.83. Therefore, her PIBP for both detectors is \$14.38.

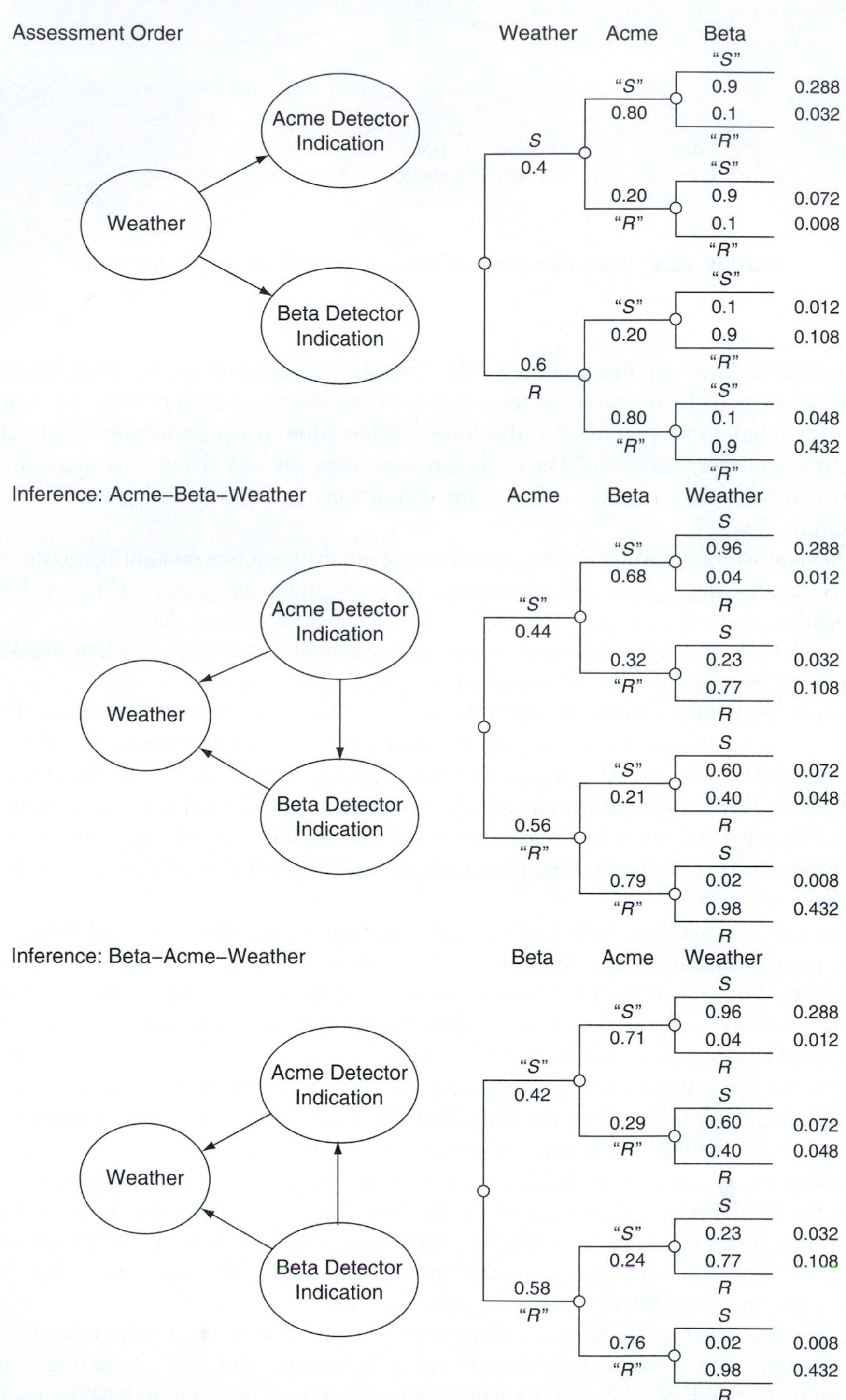

FIGURE 18.5 Orders of Assessment and Inference with Both Detectors

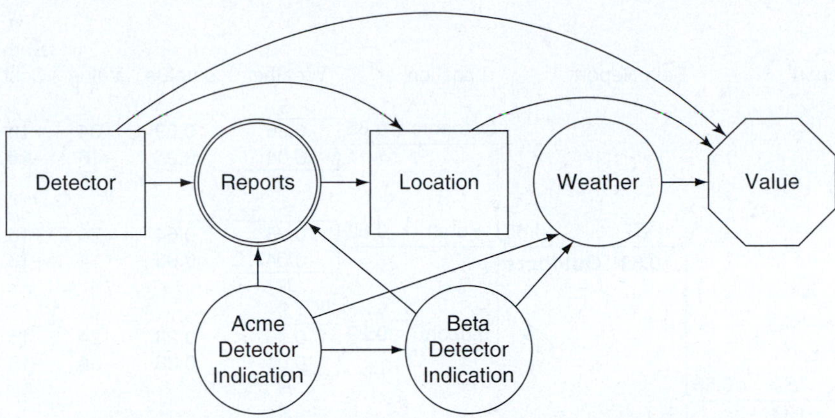

FIGURE 18.6 Decision Diagram for Possible Use of Two Detectors

Using an analysis similar to that conducted in Chapter 13 for the Acme Rain Detector alone, we can also calculate Kim's value of the deal using the Beta Rain Detector alone when she pays $13. Her certain equivalent for the party with this Beta Detector is $45.92, which is $0.09 higher than her value with no information. Given its current price at $13, the Beta detector is just marginally worth buying alone.

Table 18.1 summarizes Kim's certain equivalents for the party deal with the various sources of information she has been offered. Table 18.2 summarizes Kim's value for each of the possible indications.

TABLE 18.1 Certain Equivalent of Party using Different Information Alternatives

Party Deal with Various Information Sources	Certain Equivalent ($)
Party with no detectors	45.83
Party with Acme indication at $10	44.65
Party with Beta indication at $13	45.92
Party with both Acme and Beta indications at $16	44.21

TABLE 18.2 Value of Different Information Alternatives

Value of Various Information Sources	Value of Indication ($)
No detectors	0
Acme indication for free	8.82
Beta indication for free	13.09
Both Acme and Beta indications for free	14.38

Note

The value of the two detectors jointly ($14.38) is not equal to the sum of the individual values of the detectors ($8.82 + $13.09 = $21.91).

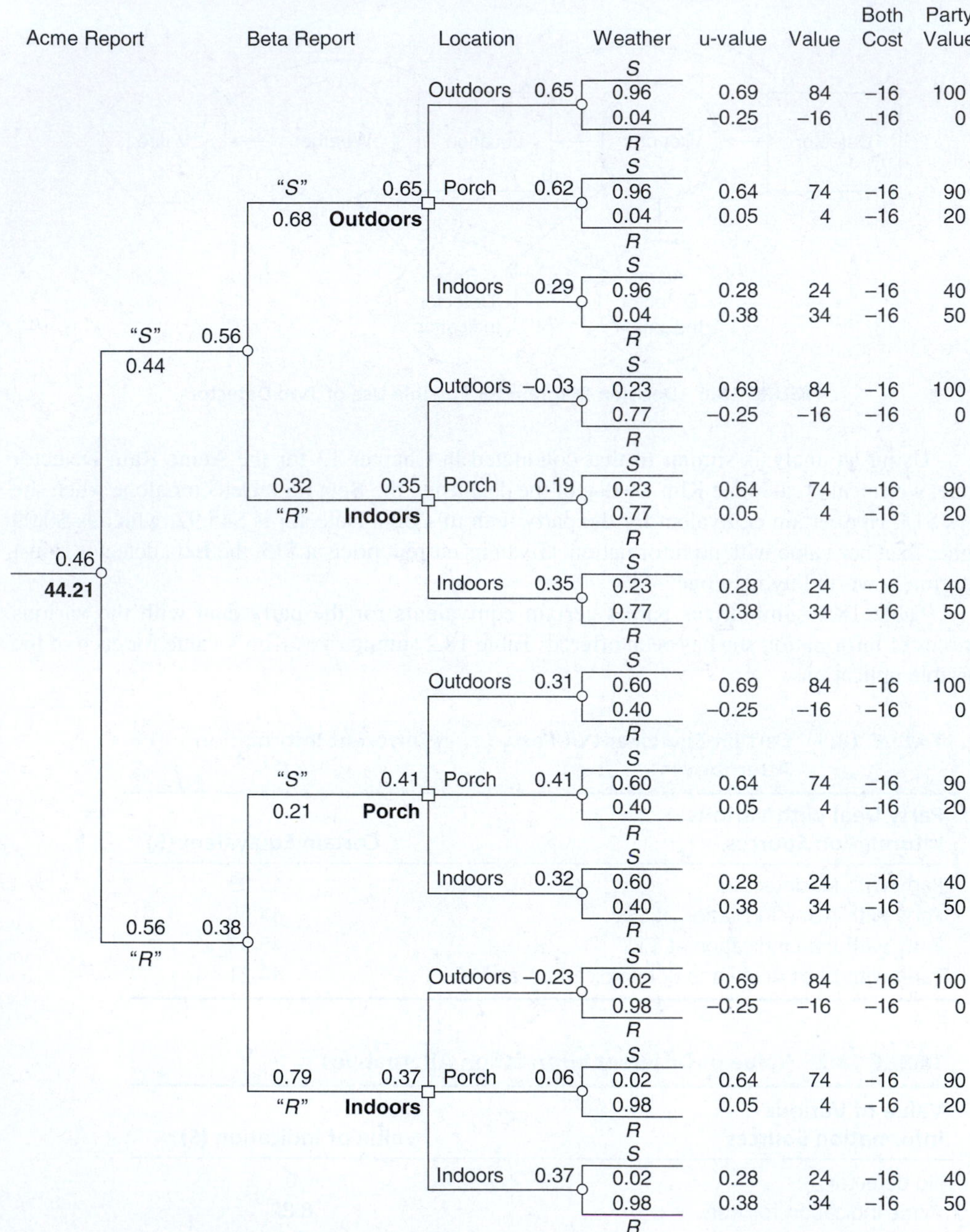

FIGURE 18.7 Decision Tree for Both Detector Indications

18.3 CLARIFYING THE VALUE OF JOINT CLAIRVOYANCE ON TWO DISTINCTIONS

So far, we have seen that the value of the combined Acme and Beta detectors in the party problem is not the sum of the values of the individual detectors. One might argue that we have three distinctions in this case. While the detector indications are mutually irrelevant given knowledge of the weather, they are relevant without knowledge of the weather.

But suppose we have only two distinctions that we would like to know in making a decision, and furthermore, these two distinctions are mutually irrelevant. Does the value of joint clairvoyance on the two irrelevant distinctions have to equal the sum of the values of clairvoyance on the individual distinctions?

To examine this question further, suppose you are considering playing three games based on two tosses of a coin. You believe that the probability of heads on each toss is 0.5 and that the result of one toss tells you nothing about the probability of the results on future tosses. Suppose you are risk-neutral over the range of monetary prospects considered in these games. In each game, we calculate the value of clairvoyance (VOC) for each coin toss and then the value of joint clairvoyance for the two coin tosses.

GAME I In this game, you receive \$200 if you flip the coin twice and the results of both tosses are either two heads or two tails (HH or TT). It costs \$100 to play this game. The probability tree of Figure 18.8 describes the payoffs and probabilities.

Now, we calculate the value of clairvoyance (VOC) on the first toss alone, the second toss alone, and both tosses jointly.

Obtaining clairvoyance on the first toss still leaves you with a 50% chance of winning \$200 or nothing on the second toss. This has an *e*-value of \$100, but since you pay \$100 to play,

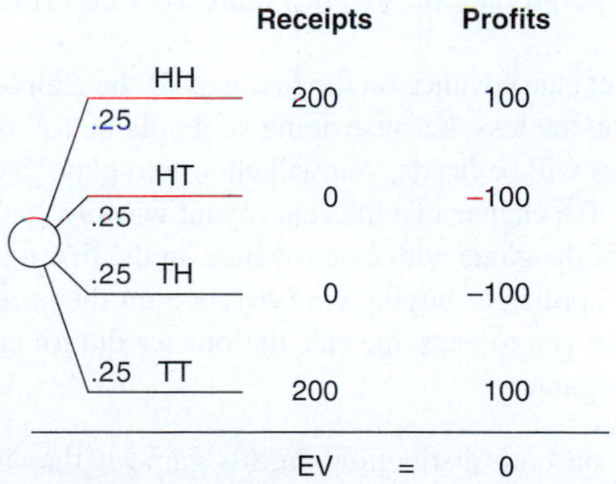

Probability	Outcome	Receipts	Profits
.25	HH	200	100
.25	HT	0	−100
.25	TH	0	−100
.25	TT	200	100
		EV =	0

FIGURE 18.8 Probability Tree for Game I

> **Note**
>
> The certain equivalent of profit for the game if it is played is
>
> $$CE = 0.25 * 100 + 0.25 * (-100) + 0.25 * (-100) + 0.25 * (100) = 0$$
>
> Therefore, you are indifferent to playing or not playing.

	Receipts	Profits
.25 HH	200	100
.25 HT	100	0
.25 TH	100	0
.25 TT	0	−100
EV	=	0

FIGURE 18.9 Probability Tree for Game II

you still have a certain equivalent of 0 for the game. The same reasoning leads us to conclude that the VOC on the second toss alone is 0.

However, if you can receive clairvoyance on the results of both tosses when the clairvoyant says that both tosses will be either heads or tails, you will receive $200 (less the $100 for playing) and still come out $100 ahead by playing the game. If the clairvoyant says that the results of the tosses will be different, you will choose not to play the game. The probability that the clairvoyant will say both tosses will be the same is 0.5, and therefore, the certain equivalent of the game with clairvoyance is $50. This is the VOC on both tosses.

In summary, the VOC on either toss alone is zero, but the VOC on both tosses is $50.

GAME II In this game, you make the same two tosses of the coin, but this time you receive $100 for each head. Furthermore, you now have a choice on whether you will play on each toss; playing a toss costs $50. The probability tree in Figure 18.9 describes the case where you play both tosses.

Suppose you could get clairvoyance on the first toss. If the clairvoyant tells you the first toss is a tail, you will not play on the toss, because doing so would be to lose the $50 cost. If the clairvoyant tells you the first toss will be heads, you will choose to play, pay $50, and receive $100 for a net gain of $50. There is a 0.5 chance that the clairvoyant will say that the first toss will be heads, so your certain equivalent of the game with clairvoyance on the first toss is $25 (See Fig. 18.10).

The same reasoning applies to buying clairvoyance on the second toss. Buying clairvoyance on both tosses will lead you to the same calculations we did for each toss and will make $50 your personal value for the game.

Therefore, the VOC on both distinctions in this game is the sum of the VOCs on each distinction.

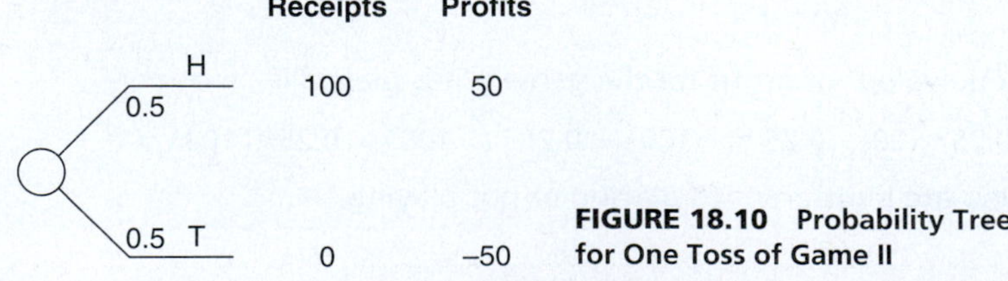

FIGURE 18.10 Probability Tree for One Toss of Game II

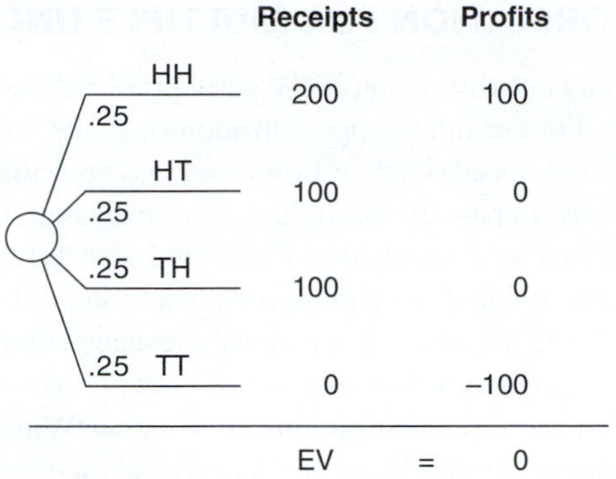

		Receipts	Profits
.25	HH	200	100
.25	HT	100	0
.25	TH	100	0
.25	TT	0	−100
		EV =	0

FIGURE 18.11 Probability Tree for Game III

GAME III In this game, you receive $100 for each head; the game costs $100 to play. Figure 18.11 shows the probability tree for this game.

Your profits will be $100 if both tosses are heads and −$100 if both tosses are tails; each prospect has probability 0.25. With probability 0.5, you will receive nothing. Once again, your certain equivalent for the game is 0. Suppose you get clairvoyance on the first toss. If it is a head and you play, you will receive a profit of either $100 or nothing on the second toss, with certain equivalent $50. If the first toss is a tail, then the profit from the game would be either 0 or −$100 with equal probability if you played the game; therefore, you would not play and would receive zero. Therefore, the VOC on the first toss is $50 with probability 0.5 and 0 with probability 0.5; therefore, it has a certain equivalent of $25. The same reasoning applies to the second toss; it also would have a VOC of $25.

Now, suppose that you considered buying clairvoyance on both tosses. The probability tree of Figure 18.11 shows that you will have a positive profit of $100 from playing the game *only* if clairvoyant indicates that both tosses will be heads. The probability that the clairvoyant will say both tosses will be heads is 0.25; therefore, the certain equivalent of the game with clairvoyance on both tosses is $25.

We see then that the VOC on both tosses is less than the sum of the values of clairvoyance on each toss individually.

Table 18.3 shows the results of our investigation. We have found that the joint value of clairvoyance on two mutually irrelevant distinctions can be greater than, equal to, or less than the sum of their individual values of clairvoyance. This is an important observation that needs to be made about the value of joint clairvoyance: It is not necessarily equal to the sum of values of clairvoyance on the individual distinctions.

TABLE 18.3 Relation of Joint to Individual VOCs

	Game I	Game II	Game III
S_1 = Value of clairvoyanceon first toss:	0	25	25
S_2 = Value of clairvoyanceon second toss:	0	25	25
J = Value of clairvoyanceon both tosses:	50	50	25
Result	$J > S_1 + S_2$	$J = S_1 + S_2$	$J < S_1 + S_2$

18.4 VALUE OF INFORMATION FOR MULTIPLE UNCERTAINTIES

So far, we have discussed only one distinction in the party problem: the Weather. In principle, Kim can create many distinctions. For example, suppose in addition to the Sun/Rain weather distinctions, she also creates two more: Wind Speed (High or Low), and Parents (either at Home or Out). Each of these new distinctions will need to pass the clarity test. Kim feels that if the wind is High, the party outdoors or on the Porch will be less desirable than if the wind were Low. Furthermore, if she has the party Indoors, it will be less desirable if her parents are present since she will be constrained by the amount of noise they will tolerate. However, her parents' presence will not be a problem if the party is on the Porch or Outdoors, since they will not be so bothered by the noise level.

Kim also believes that the three distinctions of Weather/Wind Speed/Parents are potentially mutually relevant, since it is more likely that her parents will be at home if the Weather is Rain than if it is Sun, and the Wind speed is more likely to be High if the Weather is Rain than if it is Sun. Kim draws the decision tree in Figure 18.12 for the modified party problem.

Suppose that Kim can receive clairvoyance about the Wind Speed alone. Is this information useful to her? We know this information would update her belief about the Weather being Sun/Rain, but how does this information update her belief about her Parents being Home or Out? To determine this information, Kim will need to assess the joint distribution for the three distinctions.

Kim may believe that given the Weather, knowing the Wind Speed will not update her information about her Parents being at Home during the party. This assessment is represented by the left-hand diagram in Figure 18.13. Note however, that if she does not know the Weather, knowing the Wind Speed can still update her beliefs about her Parents being at Home during the party. We can verify this using our knowledge of arrow reversals in relevance diagrams. First, we reverse the arrow from Wind Speed to Weather. We can do this immediately because the two nodes are conditioned on the same state of information as represented by the middle diagram in Figure 18.13. Next we reverse the arrow from Weather to Parents at Home. To do this, we need to add an arrow from Wind Speed to Parents at Home to get the diagrams on the right of Figure 18.13.

On further reflection, Kim realizes that knowing the Wind Speed will update her belief about her Parents being at Home even if she knows the Weather, since it will be more likely for them to be at Home than Out if the Wind Speed is High. The relevance diagram with this additional belief appears as Figure 18.14.

Kim decides to describe her party uncertainties by assessing the probability distributions implied by this diagram with the results shown in Figure 18.15. It is of course straightforward to draw the probability tree in any other order, as we discussed earlier. For example, the probability tree corresponding to the order Wind Speed–Weather–Parents at Home is shown in the right-hand side of the figure.

Figure 18.16 shows the decision diagram for the modified party problem with these new distinctions.

Figure 18.17a shows Kim's preference for the prospects of the modified party problem. Kim's best party is Outdoor–Sun–Low Wind Speed. She is indifferent as to whether her parents are at home or out if the party is outdoors because they will not be affected by the noise. Her second best party is Porch–Sun–Low Wind Speed, and she is also indifferent between her parents being at home or not if she is on the Porch. Her third best party is Outdoor–Sun–High Wind Speed and then Porch–Sun–High Wind Speed. Kim then prefers Indoor–Rain–Parents Not

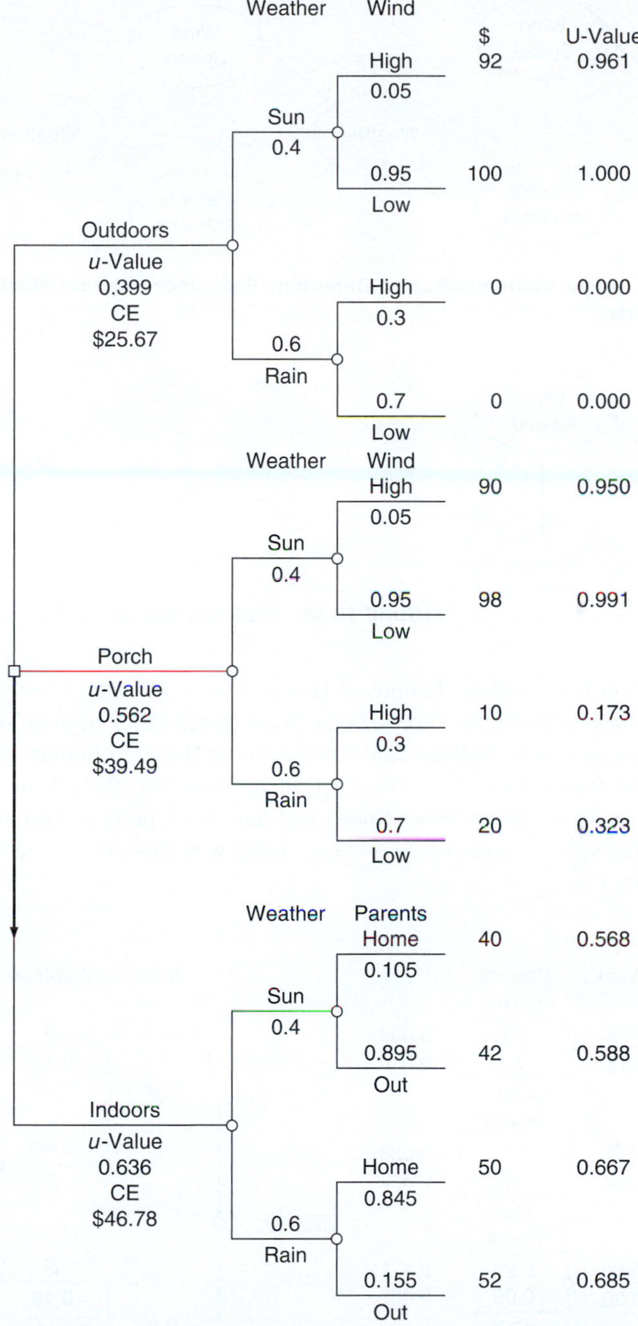

FIGURE 18.12 Decision Tree for Modified Party Problem

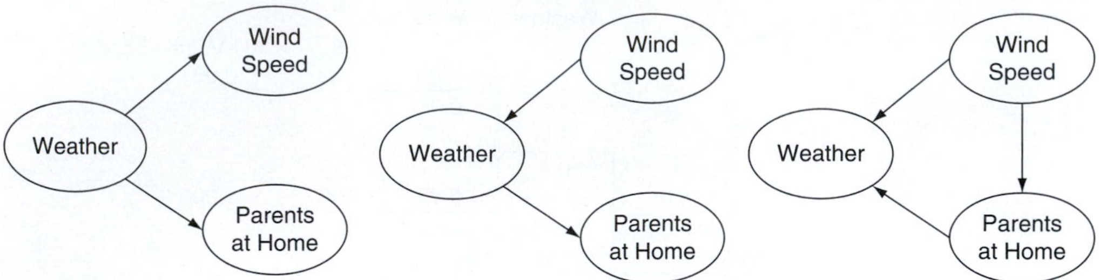

FIGURE 18.13 Arrow Manipulations to Determine Relevance Diagram Wind Speed–Parents at Home–Weather

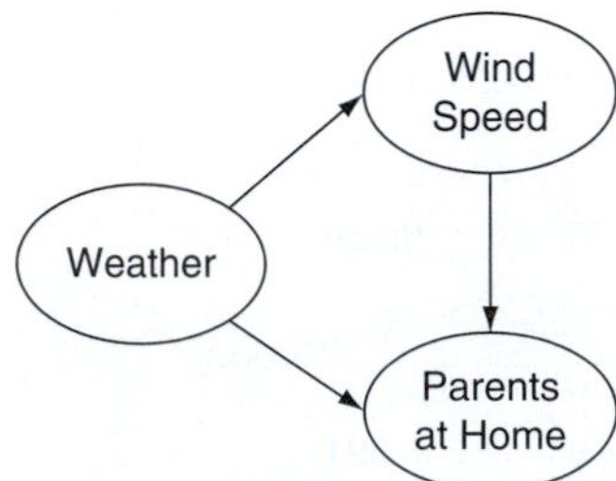

FIGURE 18.14 Modified Relevance Diagram

at Home and then Indoor–Rain–Parents at Home. She would rather her parents be away if the party is Indoors and is indifferent between the Wind Speed being high or low in this Indoor case. Similarly, Kim then prefers Indoor–Sun–Parents Not at Home to Indoor–Rain–Parents at Home. Then she prefers Porch–Rain–Low Wind Speed and Outdoor–Rain–Low Wind Speed. Finally, she prefers Porch–Rain–High Wind Speed and her worst party is Outdoor–Rain–High Wind Speed. The dollar values corresponding to each party with this preference order are shown in the tree in Figure 18.17b.

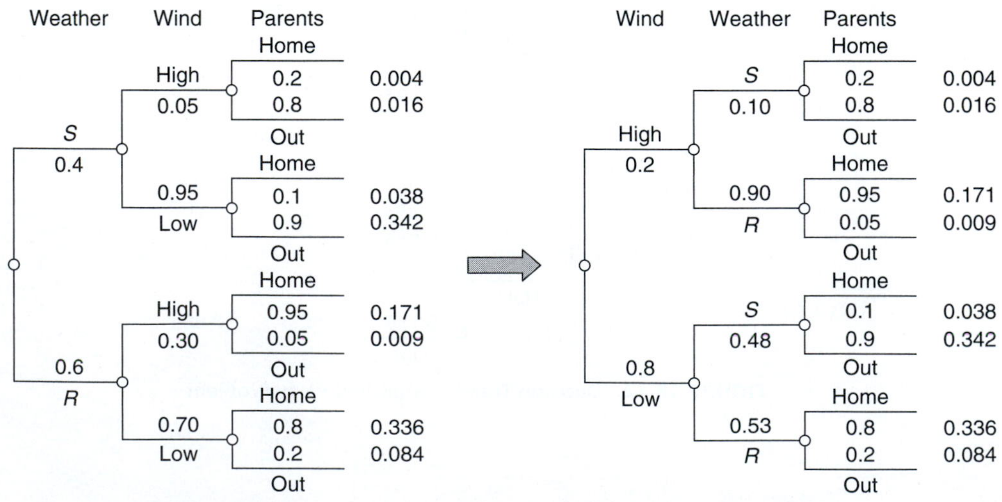

FIGURE 18.15 Probability Trees Corresponding to Figure 18.14

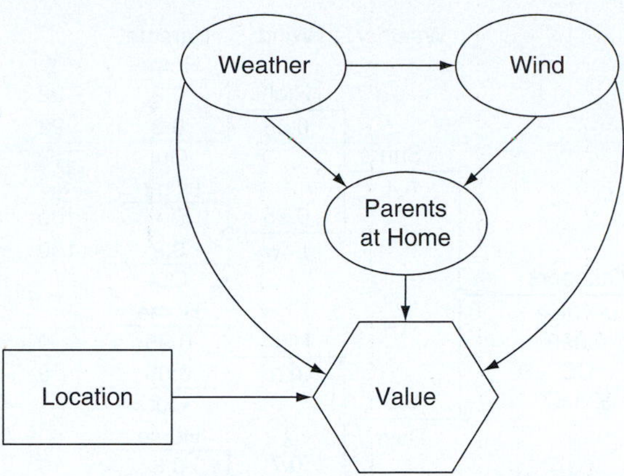

FIGURE 18.16 Decision Diagram for Party Problem with New Distinctions

Location	Weather	Wind	Parents	$
Outdoors	Sun	Low	Out	100
Outdoors	Sun	Low	Home	100
Porch	Sun	Low	Out	98
Porch	Sun	Low	Home	98
Outdoors	Sun	High	Out	92
Outdoors	Sun	High	Home	92
Porch	Sun	High	Out	90
Porch	Sun	High	Home	90
Indoors	Rain	Low	Out	52
Indoors	Rain	High	Out	52
Indoors	Rain	Low	Home	50
Indoors	Rain	High	Home	50
Indoors	Sun	Low	Out	42
Indoors	Sun	High	Out	42
Indoors	Sun	Low	Home	40
Indoors	Sun	High	Home	40
Porch	Rain	Low	Out	20
Porch	Rain	Low	Home	20
Porch	Rain	Low	Out	10
Porch	Rain	Low	Home	10
Outdoors	Rain	Low	Out	0
Outdoors	Rain	Low	Home	0
Outdoors	Rain	Low	Out	0
Outdoors	Rain	Low	Home	0

FIGURE 18.17a Kim's Preference Order and Dollar Values For Modified Party Prospects

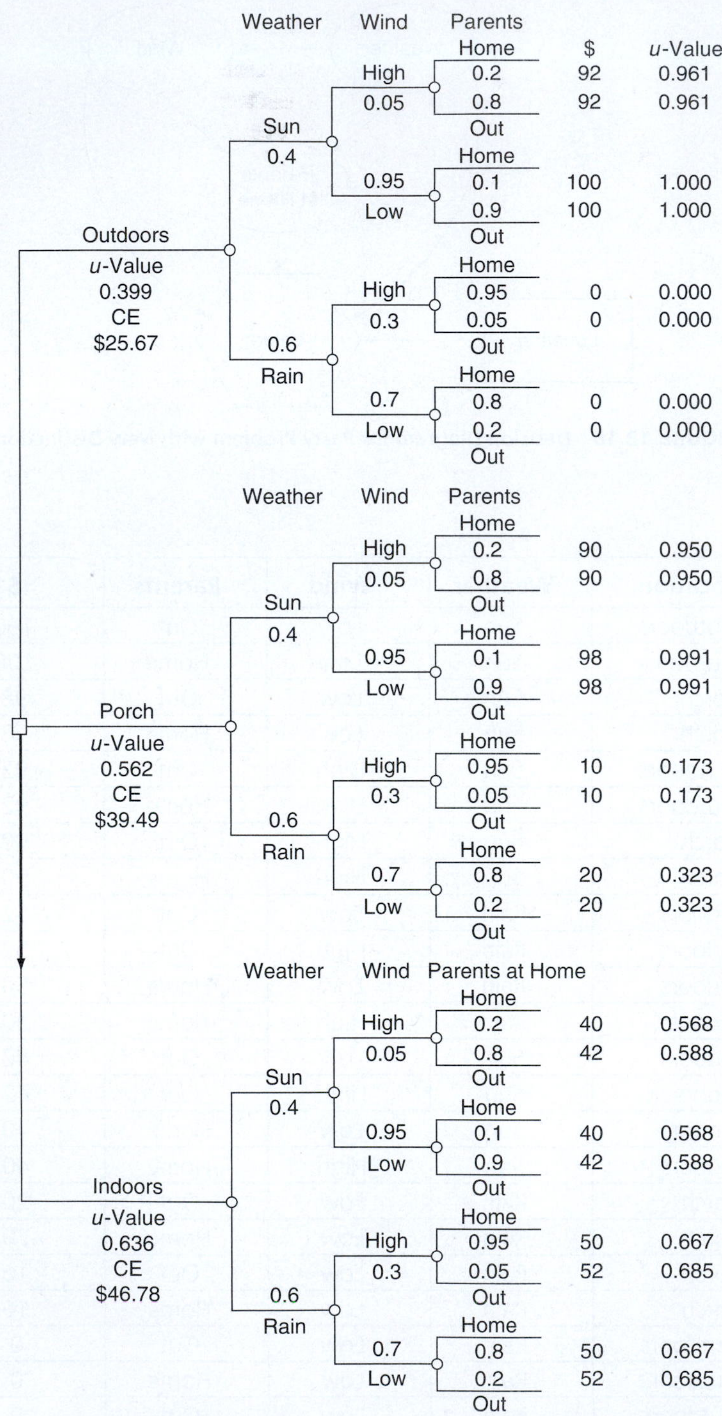

FIGURE 18.17b Complete Decision Tree for Modified Party Problem

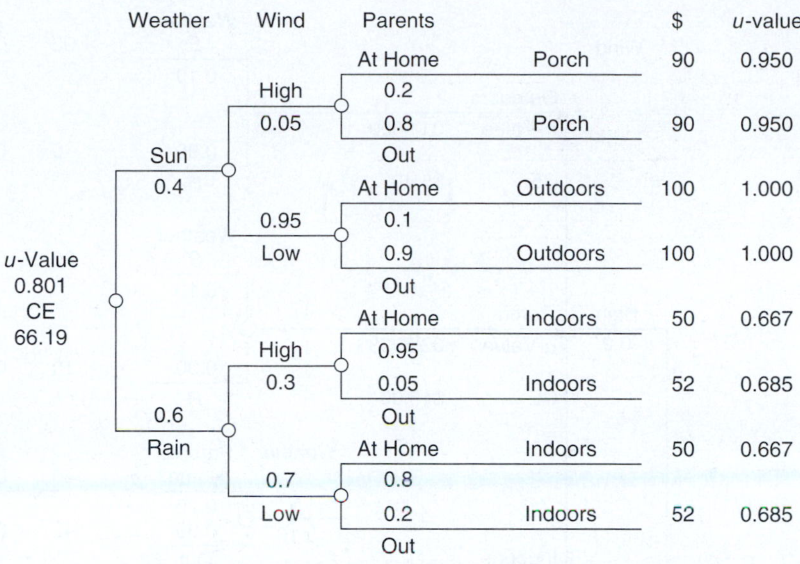

FIGURE 18.18 **Free Clairvoyance on the Three Distinctions**

Kim will choose to have the party Indoors, just as she did in the original party problem; her certain equivalent of the party deal has increased to $46.78 from $45.83.

Kim can use trees to determine the value of clairvoyance on all distinctions, or on a given selection. If she receives free clairvoyance on all distinctions, her value of the deal would be $66.19, as computed in Figure 18.18. Note that Kim will have a Porch party when the Weather is Sun and the Wind is High, an Outdoor party when the Weather is Sun and the Wind is Low, and an Indoor party whenever the Weather is Rain. The presence (or absence) of Parents does not affect the decision.

Since she is a deltaperson, Kim's value of joint clairvoyance for the three distinctions is $66.19 − $46.78 = $19.41, which is less than the $20.27 value of clairvoyance on Weather alone in the original party problem. The net effect of the additional uncertainties is to decrease slightly the value of location alternatives.

To find the value of clairvoyance on Wind Speed alone, Kim draws Figure 18.19. If the Wind Speed is High, the best alternative is for her to go Indoors. If the Wind Speed is Low, Kim would go on the Porch. The value of clairvoyance on Wind Speed alone is not very much, as $47.58 − $46.78 = $0.80.

18.5 APPROACHING CLAIRVOYANCE WITH MULTIPLE ACME DETECTORS

In the detector example, we know that the value of an imperfect detector will be less than the VOC on a distinction of interest. Now, we will show that if we have a sufficient number of detectors that are all mutually irrelevant given the distinction of interest, we can approach clairvoyance on this distinction by observing their results. This result applies even if the detectors are individually immaterial to the distinction of interest. You can think of this as the "wisdom of a large enough crowd."

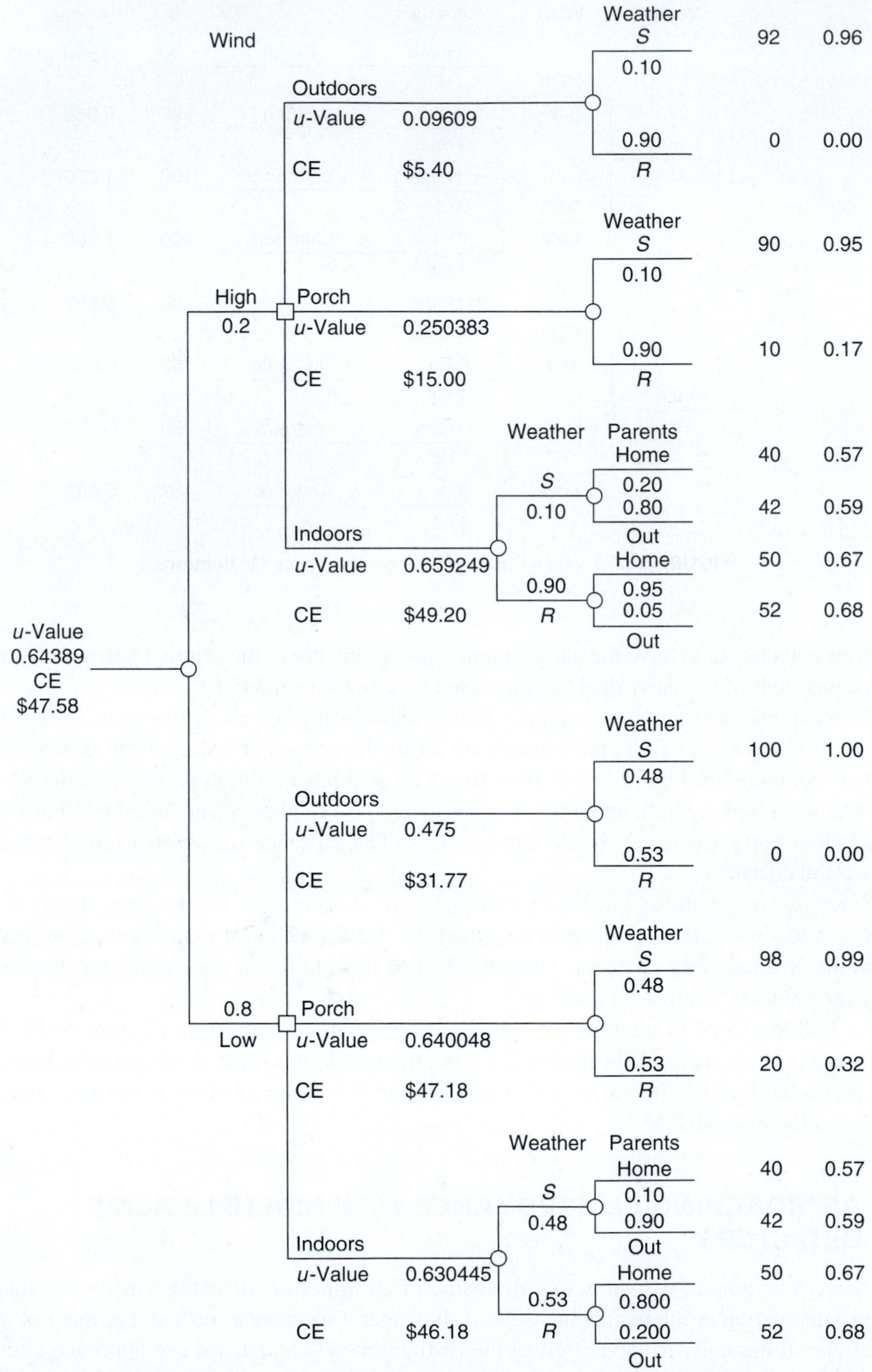

FIGURE 18.19 Calculating Value of Clairvoyance on Wind Speed

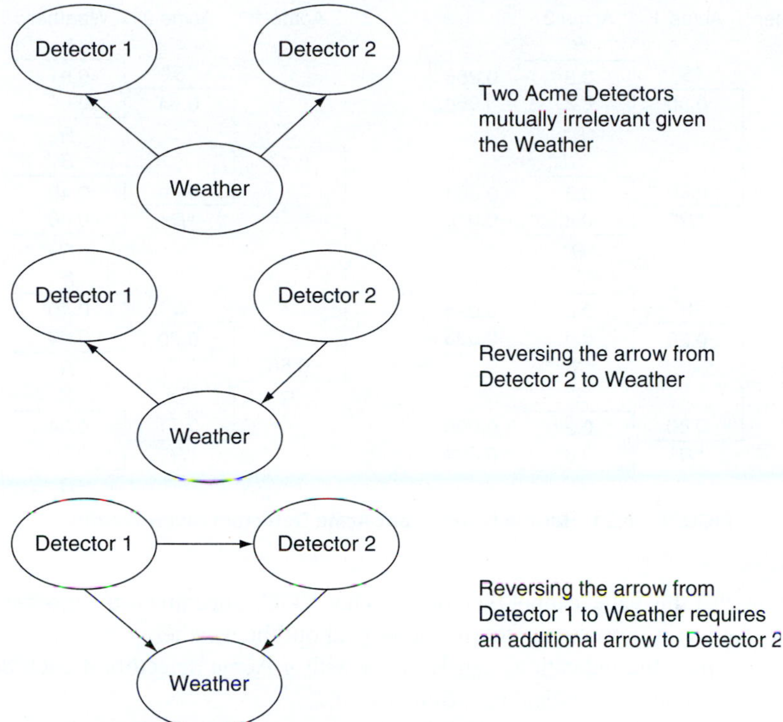

FIGURE 18.20 Relevance Diagrams for Two Acme Detectors

To illustrate, suppose we have two Acme detectors. The detectors might operate using different principles of operation and therefore be irrelevant given the weather (see top diagram of Figure 18.20), but they may still have identical accuracies. If you knew the weather and if, you observed the indication of one of them, you would not learn anything new about what the other detector would say; you would still assign 0.8 to the other's saying Sunny on a sunny day and Rainy on a rainy day. However, if you did not know the weather, knowing the indication of one would tell you something about the indication of the other (see bottom diagram in Figure 18.20).

Figure 18.21 shows the probability tree for two Acme detectors (that are mutually irrelevant given the weather).

Should Kim place a higher value on information received from two detectors than what she receives from only one? Using analysis similar to that conducted for the Acme and Beta Rain Detectors, but using detector accuracy of 0.8 for both detectors, we can find Kim's value of information for two detectors. Direct computation yields a value of $10.38, which is higher than the value of $8.82 for only one Acme indication, but is still less than her VOC, $20.27.

18.5.1 Three Acme Detectors

Now, we can further extend this analysis. Suppose Kim is offered information from three Acme Detectors that are all mutually irrelevant given the weather. What is her value of this information?

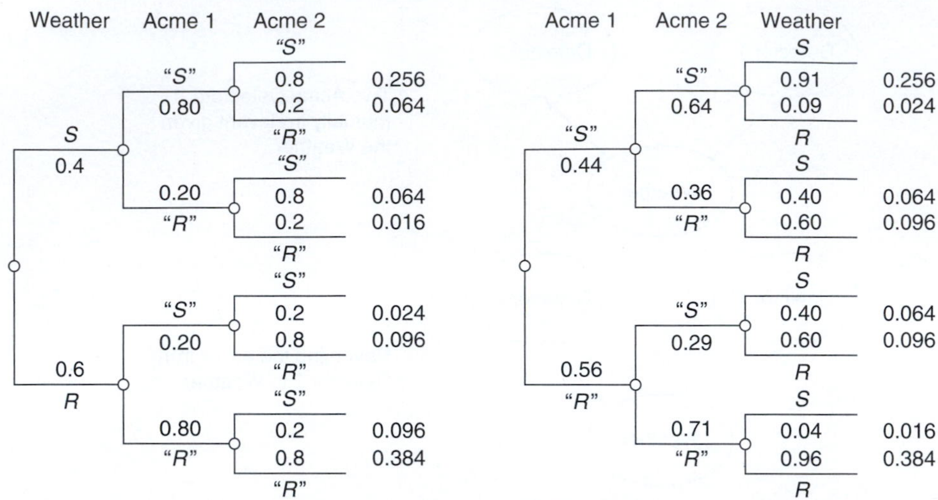

FIGURE 18.21 Mutually Irrelevant Acme Detectors Given Weather

We know that this value cannot be more than her VOC, since this is the most she would be willing to pay for any information gathering activity about the weather.

How many possible indications can Kim get with n Acme detectors if each detector can give only one of two possible indications: Sun or Rain?

By drawing a possibility tree, we find that she can get 2^n possible indications. For example, if Kim has three detectors, she can receive the following $2^3 = 8$ indications, which we write as

"SSS," "SSR," "SRS," "SRR," "RSS," "RSR," "RRS," "RRR"

***Reflection:** Should the order in which Kim receives the indications change her posterior probability of the weather?*

In other words, should Kim have a different belief about the weather if she receives the indications SSR, SRS, or RSS, all of which give the same number of Sun and Rain indications from detectors having the same accuracy? Drawing a probability tree confirms that the probability of any pattern of reports depends only on the number of Sun reports and the number of Rain reports in that pattern. Since all detectors have the same accuracy, it also does not matter which detector yielded a given indication. This is known as **exchangeability**: You don't care which detector generates your report; all you care about is the number of indications of each type. We therefore can simplify the analysis by considering only the number of Sun and Rain indications she receives from the n detectors.

If the weather is Sun, Kim can receive two indications of Sun and one indication of Rain with three detectors in three possible ways: SSR, SRS, or RSS. These possibilities have a probability equal to

$$\{2S, 1R \text{ indications} | S, \&\} =$$
$$\{\text{“SSR”} | S, \&\} + \{\text{“SRS”} | S, \&\} + \{\text{“RSS”} | S, \&\} = 3p^2(1 - p) = 0.384$$

where p is the probability the detector will indicate S when the weather is Sun, which is 0.8 for Acme.

Similarly, if the weather is Sun, she can receive one Sun indication and two Rain indications with the possibilities SRR, RSR, or RRS. These possibilities have a probability equal to

$$\{1S, 2R \text{ indications} | S, \&\} = \{\text{"SRR"} | S, \&\} + \{\text{"RSR"} | S, \&\} + \{\text{"RRS"} | S, \&\}$$
$$= 3p(1-p)^2 = 0.096$$

Furthermore, if the weather is Sun, Kim will receive no Sun indications with the possibility RRR, which has a probability of $\{\text{"RRR"} | S, \&\} = (1-p)^3 = 0.008$, and finally, she can receive three Sun indications with the possibility SSS, which has a probability of $\{\text{"SSS"} | S, \&\} = p^3 = 0.512$. We therefore can draw the simplified probability tree for Kim in assessed form, as shown in Figure 18.22.

The elemental probabilities for the tree in Figure 18.22 are

$$\{0 \text{ Sun Indications,S} | \&\} = 0.4*0.512 = 0.2048$$
$$\{1 \text{ Sun Indication,S} | \&\} = 0.4*0.384 = 0.1536$$
$$\{2 \text{ Sun Indications,S} | \&\} = 0.4*0.096 = 0.0384$$
$$\{3 \text{ Sun Indications,S} | \&\} = 0.4*0.008 = 0.0032$$

and for the lower portion of the tree are

$$\{0 \text{ Sun Indications,R} | \&\} = 0.6*0.008 = 0.0048$$
$$\{1 \text{ Sun Indication,R} | \&\} = 0.6*0.096 = 0.0576$$
$$\{2 \text{ Sun Indications,R} | \&\} = 0.6*0.384 = 0.2304$$
$$\{3 \text{ Sun Indications,R} | \&\} = 0.6*0.512 = 0.3072$$

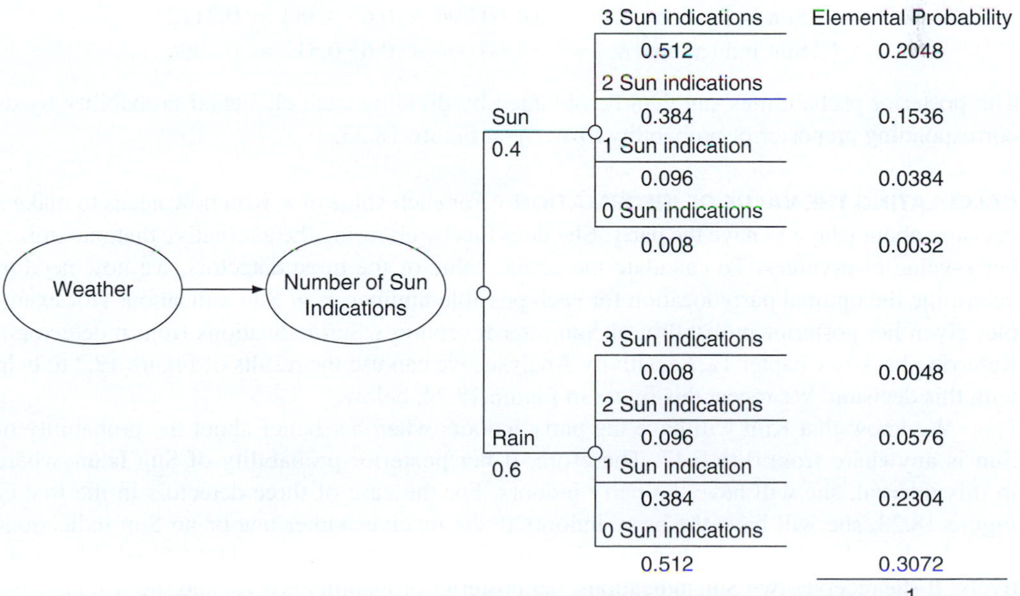

FIGURE 18.22 Simplified Probability Tree/Relevance Diagram (Assessed Form)

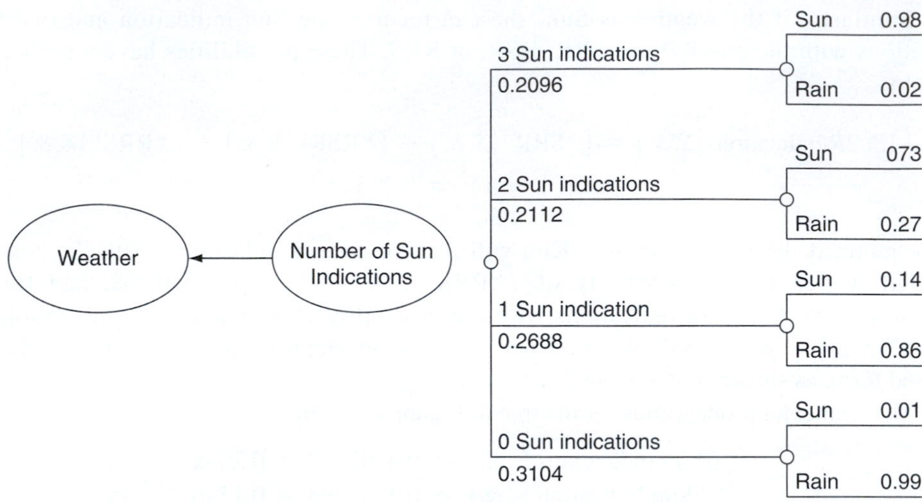

FIGURE 18.23 Probability Tree in Inferential Form

Notice that these eight probabilities must equal one, since they are elemental probabilities of mutually exclusive and collectively exhaustive possibilities.

To reverse the tree, we now calculate the preposterior probability for getting $\{r\,\text{Sun indications} \mid \&\}$. As we discussed earlier, this is obtained by summing all elemental probabilities:

$$\begin{aligned}
\{0\,\text{Sun Indications} \mid \&\} &= 0.4*0.512 + 0.6*0.008 = 0.3104\\
\{1\,\text{Sun Indications} \mid \&\} &= 0.4*0.384 + 0.6*0.096 = 0.2688\\
\{2\,\text{Sun Indications} \mid \&\} &= 0.4*0.096 + 0.6*0.384 = 0.2112\\
\{3\,\text{Sun Indications} \mid \&\} &= 0.4*0.008 + 0.6*0.512 = 0.2096
\end{aligned}$$

The posterior probabilities can then be obtained by dividing each elemental probability by its corresponding preposterior probability to arrive at Figure 18.23.

CALCULATING THE VALUE OF INFORMATION For each value of r, Kim now needs to make a decision about where to have the party. She does this by choosing the alternative that maximizes her e-value of u-values. To calculate the actual value of the three detectors, we now need to determine the optimal party location for each possible number, r, of Sun indications (for example, given her posterior probability of Sun after receiving r Sun indications from n detectors). Referring back to Chapter 12, Sensitivity Analysis, we can use the results of Figure 12.2 to help with this decision. We repeat this figure in Figure 18.24, below.

We know that Kim will have the party indoors when her belief about the probability of Sun is anywhere from 0 to 0.47. Therefore, if her posterior probability of Sun is anywhere in this interval, she will have the party indoors. For the case of three detectors in the tree of Figure 18.23, she will have the party indoors if she receives either one or no Sun indications from the three detectors, since her posterior probabilities of Sun are equal to .01 and .14 respectively. If she receives two Sun indications, her posterior probability is 0.73, and she will have her

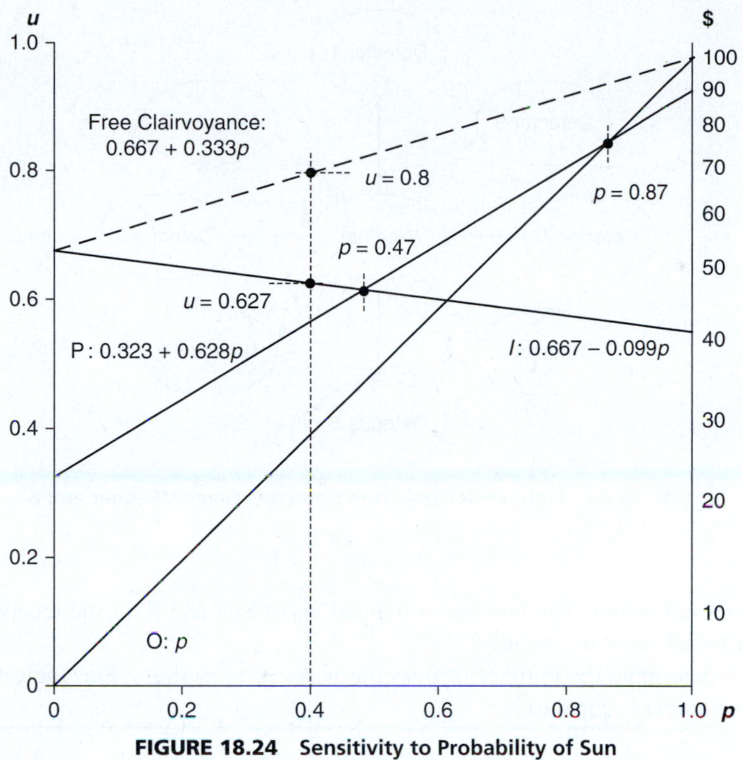

FIGURE 18.24 Sensitivity to Probability of Sun

party on the porch. If she receives three Sun indications, her posterior probability is 0.98, and she will have her party outdoors.

Kim's *e*-value of *u*-values for three free detectors is

$$0.2096 * (0.98 * u(\text{O,S}) + .02 * u(\text{O,R})) + 0.2112 * (0.73 * u(\text{P,S}) + 0.27 * u(\text{P,R})$$
$$+ 0.2688 * (0.14 * u(\text{I,S}) + 0.86 * u(\text{I,R})) + 0.3104 * (0.01 * u(\text{I,S}) + 0.99 * u(\text{I,R}))$$

Kim's certain equivalent for the deal with three free detectors is \$59.80.

Since Kim follows the delta property, we can calculate her value for the three detectors by calculating the difference between her value of three free detectors and her value of no detectors, as \$59.80 − \$45.83 = \$13.97. This value is, of course, higher than the \$10.38 value of two detectors that are irrelevant, given the weather, as computed earlier.

18.5.2 Multiple "*n*" Detectors

We can extend this analysis further. Suppose Kim is offered information from $n > 2$ Acme detectors that are all mutually irrelevant given the weather; Figure 18.25 shows the relevance diagram for eight detectors. What is her value of this information?

Since each detector can give only one of two possible indications: Sun and Rain, Kim can get 2^n possible indications with n Acme detectors. Again, the order in which she receives the

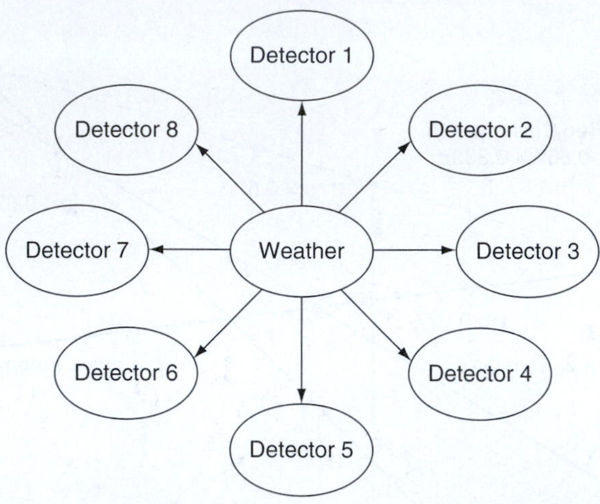

FIGURE 18.25 Eight Irrelevant Acme Detectors Given Weather and &

information does not matter: The number of indications of Sun and Rain she receives is what is relevant to her belief about the weather.

Next, we determine the number of possible ways of receiving r Sun indications from n detectors. This number is equal to

> Number of ways of getting r Sun indications from n detectors $= \binom{n}{r}$
>
> where $\binom{n}{r}$ is equal to
>
> $$\binom{n}{r} = \frac{n(n-1)(n-2)\ldots.(n-r+1)}{r!}$$

For example, if Kim faces ten detectors, she can receive $2^{10} = 1024$ possible indications. She can receive eight Sun indications with as many as

$$\binom{10}{8} = \frac{10.9.8.7.6.5.4.3}{8.7.6.5.4.3.2.1} = 45 \text{ possible ways.}$$

> Since each possible way of getting r correct indications has a probability equal to $p^r(1-p)^{n-r}$, then if the Weather will be Sun, Kim will receive r Sun indications from n detectors with probability equal to the sum of these probabilities,
>
> $$\{r \text{ Sun indications} \mid S, n, \&\} = \binom{n}{r} p^r (1-p)^{n-r}$$

For example, if each detector has a sensitivity and specificity equal to 0.8, as does Acme, then the probability of receiving eight Sun indications is equal to

$$\binom{10}{8}(0.8)^8(0.2)^2 = 0.30199$$

Similarly, the probability of getting r Sun indications with n detectors given that the Weather will be Rain also can be stated as the probability of getting r incorrect indications and $(n - r)$ correct indications given Rain. This has a probability of

$$\{r \text{ Sun indications} | R, n, \&\} = \binom{n}{n-r} p^{n-r}(1-p)^{r}$$

To calculate the elemental probabilities at the end of the tree, we now need to multiply these conditional probabilities by the marginal probabilities to get

$$\{r \text{ Sun indications}, S | n, \&\} = 0.4\binom{n}{r} p^{r}(1-p)^{n-r}$$

$$\{r \text{Sun indications}, R | n, \&\} = 0.6\binom{n}{n-r} p^{n-r}(1-p)^{r}$$

The preposterior probability of receiving r Sun indications given n detectors and & is obtained by summing up the corresponding elemental probabilities:

$$\{r \text{ Sun Indications} | n \text{ detectors}, \&\} = 0.4\binom{n}{r} p^{r}(1-p)^{n-r} + 0.6\binom{n}{n-r}(1-p)^{r}(p)^{n-r}$$

To reverse the tree, we now calculate the preposterior probability for getting $\{r\,S \text{indications} | n, \&\}$. As we discussed earlier, this is obtained by summing all elemental probabilities

For the case of n detectors, the conditional probability of Sun given Kim receives r Sun indications is then

$$\{S | r \text{ Sun indications}, n, \&\} = \frac{0.4\binom{n}{r} p^{r}(1-p)^{n-r}}{0.4\binom{n}{r} p^{r}(1-p)^{n-r} + 0.6\binom{n}{n-r}(1-p)^{r}(p)^{n-r}}$$

For example, Table 18.4 shows the posterior probability of Sun when Kim receives r Sun indications from $n = 10$ detectors.

Figure 18.26 plots this graphically.

Note the large jump in posterior probability from four to six indications of Sun in ten detectors. Also note that when receiving five Sun indications and five Rain indications $(r = 5)$, Kim's posterior probability of Sun is 0.4, which is back to her original prior probability of Sun. This is true whenever she receives an equal number of Sun and Rain indications for any number of detectors.

CALCULATING THE VALUE OF INFORMATION Once again, for each value of r, Kim now needs to make a decision about where to have the party. She does this by choosing the alternative that maximizes her e-value of u-values. To calculate the actual value of the n detectors, we now need to determine the optimal party location for each possible number, r, of Sun indications (i.e., given her posterior probability of Sun after receiving r Sun indications from n detectors).

TABLE 18.4 Posterior Probability of Sun Given *r* Sun Indications $\{S|r, 10, \&\}$

r	$\{S\|r, 10, \&\}$
0	6.35782E-07
1	1.01724E-05
2	0.000162734
3	0.002597403
4	0.04
5	0.4
6	0.914285714
7	0.994174757
8	0.999633923
9	0.999977112
10	0.999998569

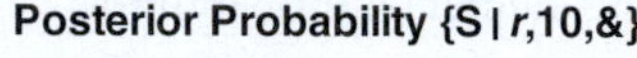

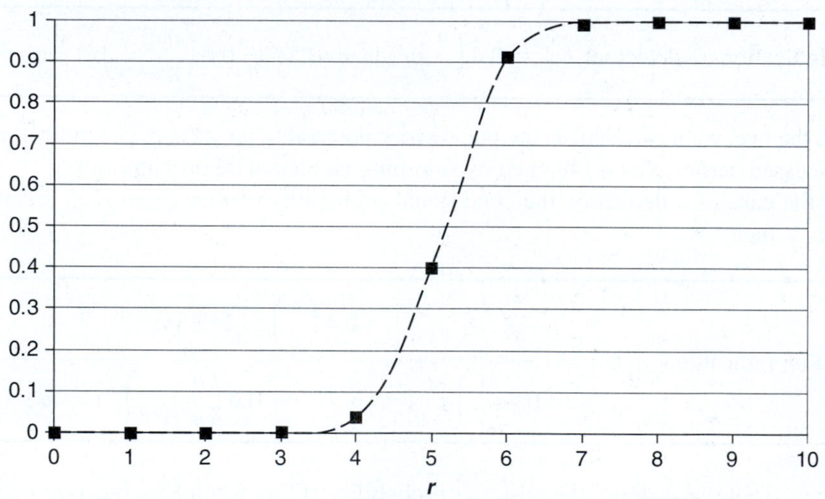

FIGURE 18.26 Posterior Probability of Sun Given *r* Sun Indications

Figure 18.27 plots Kim's value of information for up to ten detectors. Her value for ten detectors is $19.16. Notice that this value approaches but never exceeds her VOC, $22.

18.6 VALUING INDIVIDUALLY IMMATERIAL MULTIPLE DETECTORS

Suppose Kim faces a detector with an accuracy of 0.55. Using the probability tree shown in Figure 18.28, we can see that her posterior probability for the Sun is 0.45 if the detector indicates "S" and 0.35 if the detector indicates "R".

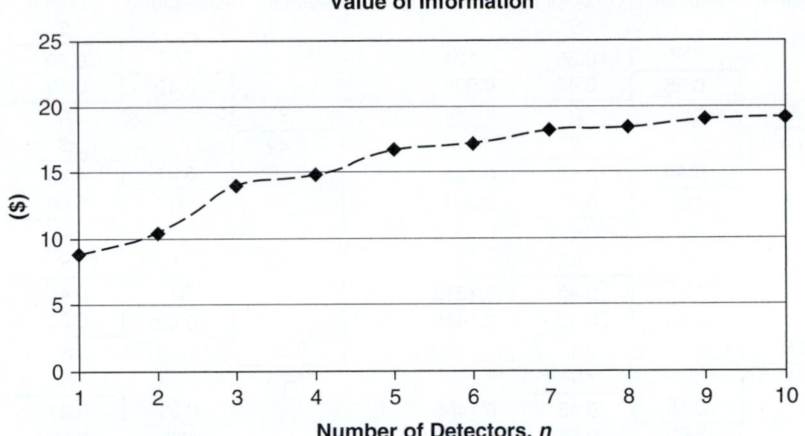

FIGURE 18.27 Value of Information vs. Number of Detectors

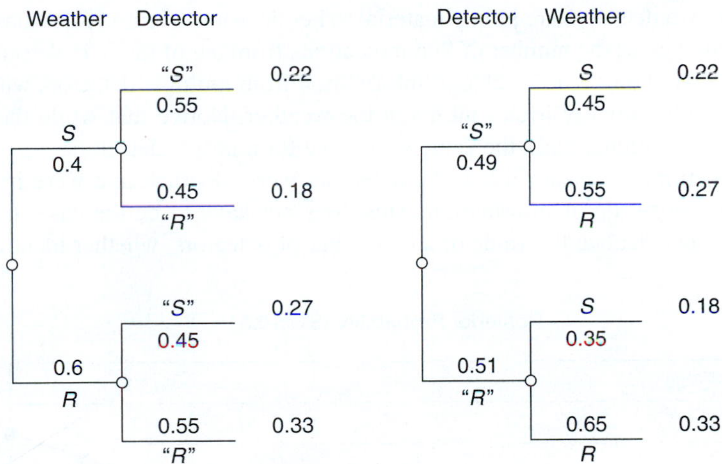

FIGURE 18.28 Detector with 0.55 Accuracy

From the sensitivity analysis curve of Figure 18.24 we observe that Kim will go indoors with both these indications. Therefore, the detector indication is immaterial to her decision, and her value of the detector is 0.

Now suppose Kim is offered two detectors with an accuracy of 0.55 with each relevant to the weather, but the two detectors are mutually irrelevant to each other given the weather (Figure 18.29).

Her posterior probability $\{S|\text{"S","S"},\&\} = 0.5$. From Figure 18.26, this means she will choose a Porch party. For all other indications, her posterior probability is less than 0.47 and so she will have an Indoor party. Thus, the indications from both detectors become material to her decision. Using analysis similar to that conducted with Acme and Beta Rain Detectors, we can show that the value of these two detectors is $0.46. Therefore, while each detector is individually

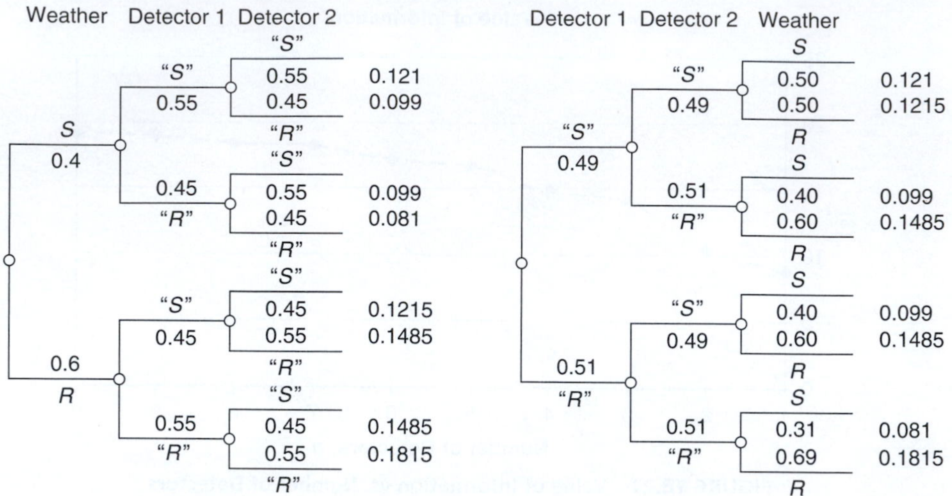

FIGURE 18.29 Two Detectors Each with 0.55 Accuracy

immaterial, the two detectors are jointly material to her decision. Figure 18.30 shows Kim's posterior probability versus the number of Sun indications from ten of the 0.55 detectors.

Figure 18.31 shows Kim's value of information from multiple detectors with an accuracy of 0.55 that are all mutually irrelevant given the weather. Notice that while the value of one detector is 0 (as it is immaterial), the value is positive for multiple detectors.

In this section, we assumed that the detectors were identical and were irrelevant given the weather and our state of information. This does not have to be the case. Using the same procedure, we can calculate the value of any number of detectors, whether identical or not, and

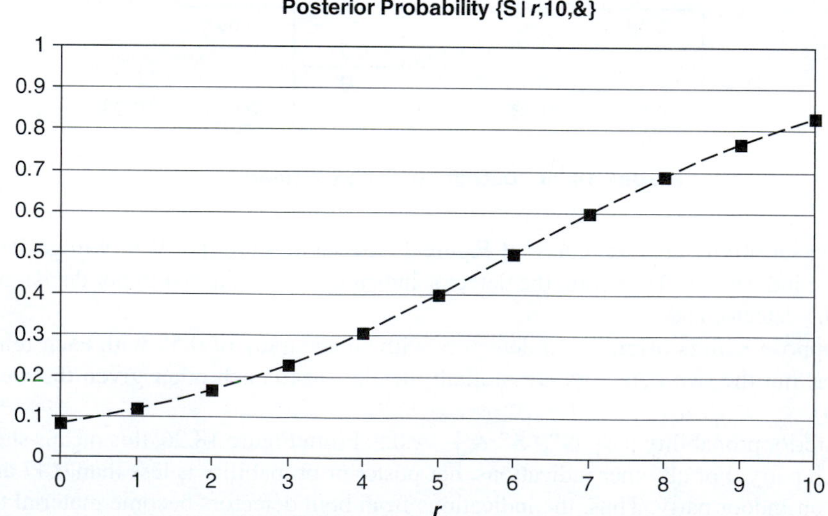

FIGURE 18.30 Posterior Probability vs. Ten Sun Indications

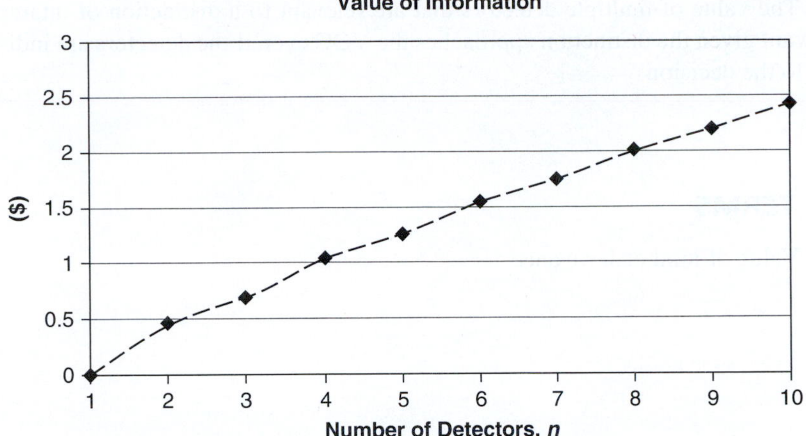

FIGURE 18.31 Value of Information from 0.55 Accuracy Multiple Detectors

whether relevant or not. We first construct the decision tree in assessed form using our belief about the conditional probability of detector indications. We then reverse the tree and calculate the posterior probability of the weather given the detector indications.

18.7 SUMMARY

When multiple sources of information are present, we need to assess the relevance relations between them and draw the tree in assessed form. We then reverse the tree and update our belief based on the detector indication.

The value of joint information can be higher than, less than, or equal to the sum of value of information from the individual sources.

If different distinctions are associated with the different alternatives (such as in the modified party problem):

- We can calculate the alternative with the highest certain equivalent directly.
- We need to assess the relevance relations between the distinctions if we wish to calculate the value of information on any one of the distinctions.

To calculate the value of n detectors:

- Assess the diagram (or tree) in assessed form
- Reverse the order of the arrows (reverse the tree) to get the distinction of interest conditioned on the multiple detectors
- Calculate the optimal alternative for each possible indication of the multiple detectors
- Calculate the value of information from the multiple sources

The value of n detectors cannot be more than the value of clairvoyance.

The order in which we receive information from multiple Acme detectors does not change our posterior probability.

The value of multiple detectors that are relevant to a distinction of interest but mutually irrelevant given the distinction approaches the VOC even if the detectors are individually immaterial to the decision.

KEY TERMS

- Value of joint clairvoyance

PROBLEMS

Problems marked with an asterisk (*) are considered more challenging.

1. Mr. De Cat is the CEO of an oil company. If he decides to drill, he will drill only one well. He believes the outcomes of the drilling are

Dry Well	$0
Small Amount	$25 Million
Large Amount	$60 Million

It costs $ 10 million to drill a well. Mr. De Cat's *u*-curve is

$$u(x) = 56.37^*x - 324.5$$

Mr. De Cat asks the company geologist to assess the probabilities of each possibility. The geologist says this assessment depends on whether or not a dome exists. The geologist says there is a 0.7 chance that a dome does exist. He also provides Mr. De Cat with the following information:

$\{\text{Dry Hole} \mid \text{Dome}, \&\} = 0.5$
$\{\text{Small amount of oil} \mid \text{Dome}, \&\} = 0.3$
$\{\text{Dry Hole} \mid \text{No Dome}, \&\} = 0.7$
$\{\text{Small amount of oil} \mid \text{no dome}, \&\} = 0.25$

a. What is Mr. De Cat's certain equivalent for this drilling site?
b. What is the value of information on the presence of the dome?
c. What is the value of information on the amount of oil present?

2. On your own, repeat the calculations of the value of clairvoyance for
(i) Table 18.2
(ii) Table 18.3
(iii) Kim's modified party problem of Figure 18.12.
(iv) Table 18.4
(v) Figure 18.31

***3.** Kim now believes that the two detector indications are relevant given weather and her background state of information. She still believes that Alpha detector's accuracy equals 0.8 and the Beta detector's accuracy equals 0.9, but she is having great difficulty assessing the relevance between them.

Assume for this analysis that

$$\{\text{Beta says "S"} \mid \text{Alpha says "S"}, \text{Sun}, \&\} = \{\text{Beta says "R"} \mid \text{Alpha says "R"}, \text{Rain}, \&\}$$

Show how Kim's optimal choice might change when the detectors become relevant.

Refer to Figure 18.2 and reverse the tree in the following order:
a. Beta Detector–Weather–Acme Detector
b. Acme Detector–Weather–Beta Detector

***4.** Construct a decision with two sources of information such that:
a. The value of joint clairvoyance is equal to the sum of individual values of clairvoyance
b. The value of joint clairvoyance is less than the sum of individual values of clairvoyance
c. The value of joint clairvoyance is greater than the sum of individual values of clairvoyance

CHAPTER

19

Options

CHAPTER CONCEPTS

After reading this chapter, you will be able to explain the following concepts:

- Contractual and non-contractual options
- The value of an option
- The option price and excercise price
- Value of information as an option
- Sequential detector options

19.1 INTRODUCTION

Too often, individuals focus only on the choice between two or more alternatives without taking enough time to consider additional decision alternatives that they can pursue to enhance the overall value of the decision they face. We have seen that gathering information is an additional alternative that can increase the value of the decision. Different choices can be made following the revelation of the outcome of the information source. Insurance is another example of an alternative that may provide value following the outcome of an uncertainty. In this chapter, our purpose is to increase the awareness about these types of alternatives that can add value by creating new alternatives in the future following the revelation of an uncertainty. We refer to such alternatives as **options** and discuss their value in decision making.

19.2 CONTRACTUAL AND NON-CONTRACTUAL OPTIONS

By an option, we mean not simply any decision alternative, but rather an alternative that permits or may permit a future decision following revelation of some information. This meaning not only encompasses the common options in financial affairs, but it also extends to many other areas of personal and professional decision making. The expanded concept of an option is both one of the most fundamental notions of decision analysis and an important aid to decision synthesis. A decision diagram for the decision to pursue an option is shown in Figure 19.1.

Crucial to the idea of an option is the exercise of human will in the future decision. For example, engaging in research and development in the hope of finding a profitable product is an option, because human will is exercised throughout the development and commercialization process as information is revealed. Purchasing a lottery ticket is an option, since we can also exercise our will and choose not to buy one. Other common options include carrying a pocket

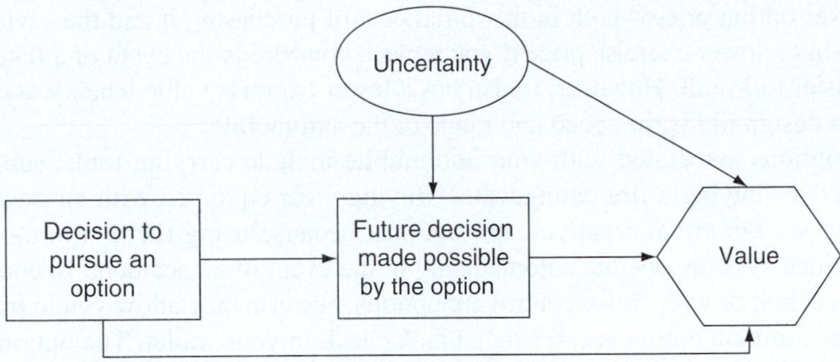

FIGURE 19.1 Decision Diagram for Option Analysis

knife, installing extra wiring for possible future use, and making freeway bridges wider than whatever is currently needed.

Contracting lung cancer after a life of heavy cigarette smoking, however, is not an option, but choosing to smoke is an option that may necessitate a later decision about cancer treatment. Too frequently, we overlook our prerogative to recognize and create options. By doing so, we fail to recognize the sequential nature of many decisions.

Options can be either *contractual* or *non-contractual*. **Contractual** options are agreements with other parties to provide future decision alternatives you otherwise would not have. For example, one type of contractual option is the exclusive right to buy or sell property within a specified time and for a specified price. This option is usually obtained for a fee, called the option price. When buying a house, you can submit an offer to purchase and simultaneously place some amount of money in escrow as a deposit. If the buyer cannot (or does not wish to) fulfill the commitment and close on the house, the seller keeps the deposit. Escrow creation is therefore a contractual option between two parties that allows either one to withdraw from an agreement easily if the other party does not fulfill its commitment.

Non-contractual options exist when we recognize or create the possibility of deciding after the revelation of information with the same or a modified set of alternatives. Non-contractual options may involve the acquisition of contractual options after the new information is revealed. To illustrate, imagine you are buying an option to purchase oil rights to a property, paying for a test drilling program, and deciding whether to buy the oil rights based on the test drilling results. This idea is a non-contractual option that contains a contractual option and will enable you to access information that would otherwise not be available.

19.3 OPTION PRICE, EXERCISE PRICE, AND OPTION VALUE

We have defined a *decision* as an irrevocable allocation of resources. The decision to pursue or to create an option will typically require paying a resource price in some value measure. We call this price the **option price**; it is represented in Figure 19.1 by the arrow from the first decision to the value node. After the revelation of information, choosing an alternative provided by the option may require an additional resource payment. We call this payment the **exercise price** for that alternative in the second decision. Given the exercise price for each alternative, the value of an option is the option price at which the decision maker is indifferent to buying it—the PIBP for the option. In general, the option value increases with uncertainty and an increasing number alternatives.

There are many examples of options in our every day life. For example, you have many options in your automobile. You can carry with you a spare tire or a mini spare tire. The mini tire has a lower option price—both in the initial cost of purchasing it and the saving of storage space. It also has a lower exercise price if you replace your tire in the event of a flat, because it is somewhat easier to install. However, it also has a lower exercise value when you choose to use it, because its design limits the speed and range of the automobile.

Other options associated with your automobile include carrying tools, subscribing to a towing service, or buying a fire extinguisher. Buying a car equipped with air bags or an automatic braking system are alternatives, not options, because using them is not an exercise of human will. Such systems operate automatically in the event of an accident. In contrast, buying a car with a seat belt or with cruise control are options, because they allow you to make a choice.

Another common option is carrying a blank check in your wallet. The option price is low because it occupies only a small space. The exercise price is also low: If you need to use it, you write the check and sign it. However, the value of carrying a blank check may be very high in an emergency or when a credit card cannot be used.

The purchase of insurance is also a very common option. The option price is usually called the premium, and it is the amount you have to pay the insurer up front to acquire the insurance policy. The exercise price may either be fixed—(for example, fixed payments per doctor visit) or they may be a function of the loss—(for example, co-payments for medical services above a predetermined deductible amount). Note that in this particular case, the option will not be exercised—no claim will be filed—unless deductible payments accumulate to satisfy a maximum annual deductible amount.

Not only is buying insurance an option, there are also options to buy insurance. Purchasing an extended warranty on a new car or appliance is buying insurance. Often, the seller will extend the opportunity to buy the insurance for some time period after the original purchase, and usually at no charge. The buyer therefore has the alternative of buying insurance after observing the initial performance of the product: an option on an option. In the same vein, life or health insurance policies may confer options to increase coverage without additional proof of insurability.

Reflection

Think of other options that you can acquire in your daily life. For each, what is the option price? What information is revealed before you decide to exercise the option? What is the exercise price?

19.4 SIMPLE OPTION ANALYSIS

An option affords us at least the possibility of adjusting our decisions in the light of our experience. The value of the option is derived from the likelihood and benefits of such adjustments. Consider the simplified option decision shown in Figure 19.2 that corresponds to the decision diagram of Figure 19.1.

Without the option, suppose you face a deal where some condition, E, will happen with probability q resulting in a monetary payoff b, and some other condition E' may happen with probability $1 - q$ in which case the payoff is a. For example, E may represent an automobile accident in which case b may be the losses due to the accident, and E' may be the condition of no accident in which case the payoff, a, is your current situation with no accident.

Now, suppose you can buy an option for an option price p_o. If you do, then with probability q the exercise condition E will still arise, and with probability $1 - q$ it will not. (There is no reason to believe that buying an option will increase or decrease the probability of E occurring.)

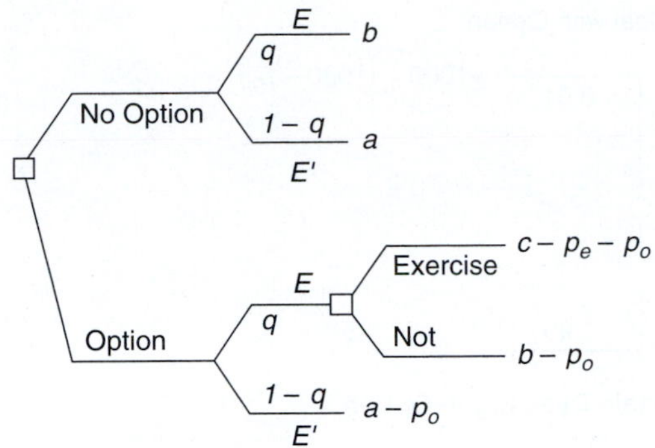

FIGURE 19.2 Simple Option Decision Diagram

If condition E does not occur upon buying the option, the payoff is $a - p_o$, which is what you would get without the option, less the option price. However, if condition E does occur, you now have a second decision to make: whether to exercise the option or not.

If you do not exercise the option, the payoff will be $b - p_o$, which is the payoff you would get with condition E less the option price that you purchased. If you exercise the option, you will incur the additional exercise price p_e and receive the payoff c (due to the usage of the option). For example, c may be some monetary amount added to b or any other payment agreed upon when purchasing the option.

WHEN IS THE RIGHT TIME TO EXERCISE AN OWNED OPTION? The answer is, you exercise the option if the certain equivalent of exercising it is greater than that of not exercising it at the time of that decision (when condition E occurs). In more specific terms, we would exercise the option when

$$c - p_e - p_o > b - p_o$$
$$\Rightarrow c - b > p_e$$

Note that if the exercise price is zero and if $c > b$, you would always exercise the option.

WHAT IS THE VALUE OF THE OPTION IF IT IS FREE? A risk-neutral decision maker, as a deltaperson, can simply subtract the value of the deal without the free option from the value of the deal with the free option.

The value of the deal without the option is simply $[qb + (1 - q)a]$.

If the option is free, $p_o = 0$ and E occurs, the option would be exercised if $c - b > p_e$ with the e-value

$$[q(c - p_e) + (1 - q)a]$$

The value of an option with zero option price for a risk-neutral decision maker is therefore

$$[q(c - p_e) + (1 - q)a] - [qb + (1 - q)a]$$
$$= q(c - b - p_e)$$

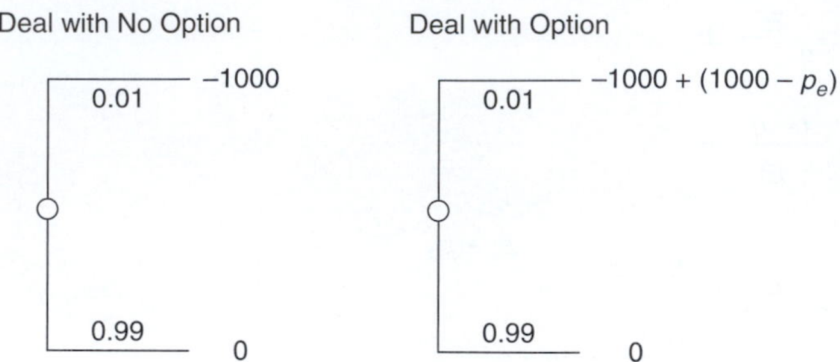

FIGURE 19.3 Left–Uncertain Deal, Right–Option

Since the value of the free option is $q(c - b - p_e)$, it should be bought if this value exceeds its price p_o. Therefore,

$$q(c - b - p_e) > p_o$$

We can also calculate the value of an option for more general *u*-functions. In this case, we would need to change the option price p_o until the certain equivalent of the deal with the option is equal to the certain equivalent of the deal without it. If the calculated *u*-value of the deal with a free option is greater than the calculated *u*-value of the deal without the option, the option price will be positive. Therefore,

$$qu(c - p_e - p_o) + (1 - q)u(a - p_o) > qu(b) + (1 - q)u(a)$$

implies a positive option price.

NUMERIC EXAMPLE Consider a person facing a 1% chance of a \$1000 loss and 99% of no loss. He is offered an option that will reimburse the loss at different exercise prices (Figure 19.3).

For a risk-neutral decision maker, the uncertain deal on the left has a negative certain equivalent of $0.01 * (-1000) + 0.99 * 0 = -\$10$.

For a risk-averse deltaperson decision maker with risk a tolerance of \$1000 and therefore with the exponential *u*-curve of $u(x) = 1 - e^{-\frac{x}{1000}}$, we have $u(0) = 0$, so $u(-1000) = -1.71$, leading to a *u*-value of the deal of $0.5 * 0 + 0.5 * (-1.71) = -0.855$ and a corresponding negative certain equivalent of $-\$17.03$.

At an exercise price of $p_e = 0$, the option value is simply the person's personal indifferent selling price for this undesirable deal, since it will rid him of the possibility of a loss. Therefore, the option value is \$10 for a risk-neutral decision maker and \$17.03 for the deltaperson with risk tolerance \$1000. As we would expect, the option is considerably more valuable to the risk-averse person when the exercise price is low, because without the option, this person has a larger negative certain equivalent of the deal.

As the exercise price increases, the value of the option decreases. We can calculate the value of the option at an exercise price p_e for a deltaperson by calculating the certain equivalent of the deal with the option and subtracting it from the certain equivalent of the deal with no option.

Figure 19.4 shows how the option value will vary with the exercise price if the person is risk-neutral and if the person is a deltaperson with a risk tolerance of \$1000.

The value of the option falls about the same amount for both people with initial increases in exercise price.

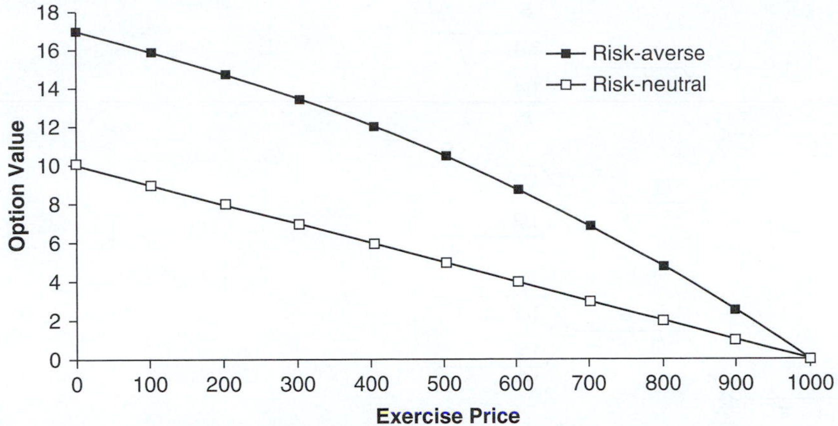

FIGURE 19.4 Effect of Exercise Price on Option Value

Note

When the exercise price p_e is equal to $1000, the decision maker is just indifferent to exercising the option or not, so the option value is zero. Any higher value of an exercise price will result in not exercising the option, so its value remains zero.

19.5 CONSEQUENCES OF FAILURE TO RECOGNIZE OPTIONS

If a sequential decision is wrongly represented as a simple decision, serious decision errors can result. We can demonstrate this with a simple example. Suppose that a risk-neutral person is offered the opportunity to participate in an investment determined by three Success *S*–Failure *F* trials. Observing Successes makes future Successes more likely; observing Failures makes future Successes less likely. The probability of Success on the first trial is $1/2$. As successive trials are observed, the probability of Success on the next trial will be the ratio of (1 plus the number of Successes observed) to (2 plus the number of trials observed). For example, if the first trial is a Success, then the probability of Success on the second trial will be $2/3$. However, if the first trial is a Failure, then the probability of Success on the second trial will be $1/3$. The pattern continues: If the first two trials have been Successes, the probability of Success on the third trial will be $3/4$.

The probability tree for the first three trials appears in Figure 19.5. Note that the probability of Success on the second trial if you do not know the result of the first trial is $1/2 * 2/3 + 1/2 * 1/3 = 1/2$. Likewise, the probability of Success on the third trial if you do not know the result of previous trials is also $1/2$.

The investment pays $100 for a Success and costs $80 for a Failure. Since, there is a 0.5 chance of Success on each of the three trials without observing the results of the other trials, each trial contributes $10 to the *e*-value of the deal for a total value of $30.

Now, suppose that the investor is offered an option to stop any further participation in the trials. What is the value of this option? Figure 19.6 shows the analysis.

We begin by anticipating the third trial. For the four possible prospects that precede the third trial—*SS*, *SF*, *FS*, *FF*—the valuations of the third trial are 55, 10, 10, and −35, respectively.

This means that if the investor arrives at the third trial after observing two failures, he would choose to exercise his option to stop, since engaging in the third trial has a certain

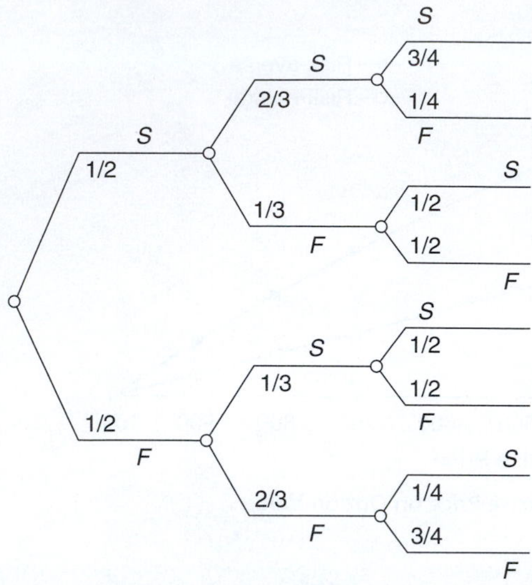

FIGURE 19.5 Probability Tree for Three Trials

equivalent of −\$35. Otherwise, he would continue. The X shown in Figure 19.6 represents this decision to exercise the option and stop.

If he had observed only one trial and it was a Success, he would have a 2/3 chance of receiving 100 + 55 (the EV of the third trial if the second trial is a Success) and a 1/3 chance of receiving −80 + 10 (the EV of the third trial if the second trial is a Failure) for a valuation of 80.

However, if the first trial was a Failure, he would have a 1/3 chance of receiving 100 + 10 and a 2/3 chance of −80 + 0 (since he would then stop and not play the third trial with a negative EV of −35). This valuation is −50/3, and he would again exercise his option to stop.

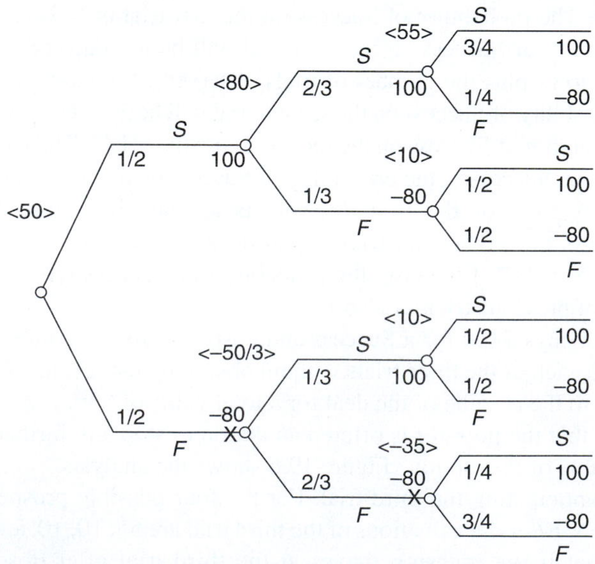

FIGURE 19.6 Decision Tree Evaluation of Investment Option

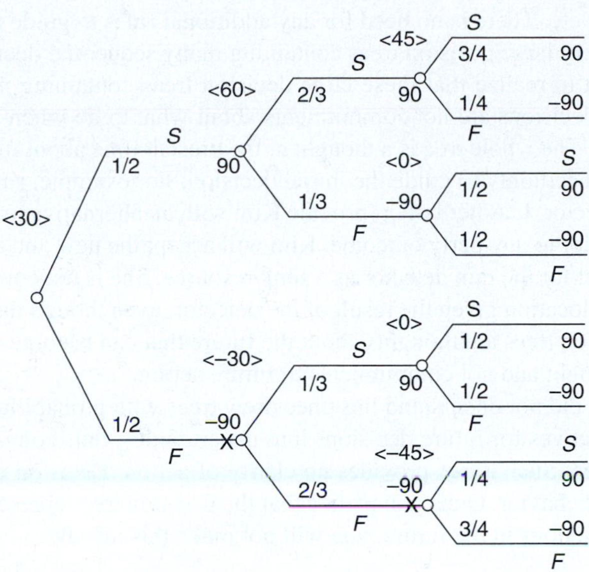

FIGURE 19.7 Decision Tree Evaluation of Modified Investment Option

At the beginning of the process, the overall value is then $(1/2)(100 + 80) + (1/2)(-80) = 50$. The value of the option is, therefore, $50 - 30$, or \$20. In other words, the option to stop participating increases the value of the investment from \$30 to \$50.

If the option allows the investor to pass on any trial, rather than to stop participating, then that option becomes slightly more valuable. The investor would pass on the second trial if the first trial was a Failure, but would participate for the third trial if the second trial was a Success. This would increase the valuation of the process and the value of the option by (1/2)(1/3)(10), or \$1.67.

Now, let's consider the same investment where the payoffs for Success and Failure are each reduced by \$10 to \$90 and −\$90. Without an option, this investment would have no value to a risk neutral person. With an option to stop at any time, the analysis proceeds as before, with the results shown in Figure 19.7.

The option to stop will begin be exercised if there is a Failure on the first trial. The value of the option and therefore the value of the investment is \$30. Note that the option to stop is worth more for this investment than for the original investment (\$20). For this investment, there is no additional value of an option to pass on any trial rather than to stop.

Options that permit someone to cut their losses following revelation of unfavorable information can be extremely valuable. Failure to represent such options in a decision problem can lead to serious error.

As we have seen, using the Five Rules for Sequential Decisions, we often encounter decision situations where the initial decision we make will result in a future decision. For example when Kim is considering buying the Acme rain detector, if she does buy it at an attractive price she will still face the decision of where to have the party. This is an example of a sequential decision: where one decision is followed by another. As we have seen, each prospect of the initial deal is evaluated by a *u*-Value (or certain equivalent) of the subsequent deal. This evaluation is carried out using the same five Rules of Actional Thought. The rules are then applied again to make the initial decision. If we have several sequential decisions, the last one is evaluated first, then the next to last, etc. There is no need for any additional rules to guide the process. Thus we can evaluate even very large decision trees containing many sequential decisions.

It is important to realize that these large decision trees containing many what we might call "downstream" decisions are not commitments about what to do when the situations we are contemplating arise. The whole tree is a thought at the present time about future possible actions and resolutions of uncertainty to guide the initial decision. For example, suppose, after Kim has paid for the rain detector, I, as her father, provide Kim with an alternative to having the party that she regards as superior to any party outcome. Kim will accept the new and better alternative and consider the payment for the rain detector as a sunk resource. She is not committed to having the party in a particular location given the result of the detector, even though that thought was in her decision tree. Decision trees are thoughts about the future that can become useless in a moment. They are aids to thought and not commitments to future action.

Someone who did not understand this once drew trees with probabilities on future actions. He turned the alternatives for future decisions into uncertainties, thus converting a decision tree into a probability tree. Such a tree provides no clarity of action. Decision trees are not descriptive of actual future behavior. Once you realize that the decision tree represents your best current thought about your actions in the future, you will not make this mistake.

19.6 JANE'S PARTY REVISITED

Once again, we refer back to the party problem first introduced in Chapter 9. Think back to Jane's party decision and redraw her decision tree in Figure 19.8 for convenience.

Recall that Jane's risk-neutral certain equivalents of the three alternatives are \$40, \$48, and \$46; therefore she chooses a porch party. If she knew it would be sunny, she would have an outdoor party and make 100; if she knew it would be rainy, she would have an indoor party and make 50. Since the probabilities of Sunny and Rainy are 0.4 and 0.6, free clairvoyance would provide her with a valuation of 70 at an increase of 22. This means that the value of the option represented by clairvoyance on the weather is 22.

Returning to the party problem, note that Jane can create the clairvoyance option by setting up the party both outdoors and indoors. With this setup, if the Weather is Sunny, she will have the party outdoors and if it is Rainy, she will have it indoors. If she can do this setup for an option price of less than 22, she will come out ahead of her current deal.

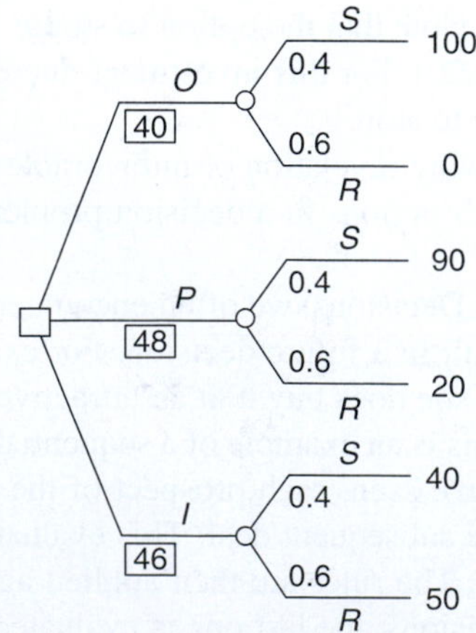

FIGURE 19.8 Jane's Decision Tree for Party Problem

She can also decrease the option price by increasing the exercise price—she could partially set up the party both indoors and outdoors with the rest of the party facilities arranged so they can be moved to the party location once the weather is known. Since rain is more likely, she would set up these moveable features for an indoor party and move them outdoors if necessary. If she could have this partial setup for an option price of \$10 and an exercise price of \$5 given sunshine, the valuation of the option would be $22 - 10 - 0.4(5) = 10$. For Jane, creating this option would raise the value of the party from \$48 to \$58. Thinking of inventing options and then modifying them to construct different option and exercise prices can be extremely helpful in alternative creation.

19.6.1 Jane's Party Problem with Many Days: Sequential Decisions

Jane can further enhance her party deal by creating other options for herself. For example, suppose Jane has an option to have the party on any of the next two days after observing the weather each morning. She assigns the same 0.4 probability to the weather's being sunny on any day regardless of the weather on the preceding day. If the weather is sunny on the first day with a probability of 0.4, she will have her party outdoors with a certain equivalent of \$100. If it is rainy on the first day, she can wait for the second day. If it is sunny on the second day, she will have her party outdoors (with a probability of it being rainy on the first day and sunny on the second day $= 0.6 * 0.4 = 0.24$). Her certain equivalent for this Outdoor sunny party on the second day is still \$100. If it is rainy on the second day, she will have her party indoors with a certain equivalent of \$50 and a probability of $0.6 * 0.6$. Jane's new party deal can now be represented using the following tree in Figure 19.9.

Jane's certain equivalent for this modified deal is now

$$0.4 * 100 + 0.6 * 0.4 * 100 + 0.6 * 0.6 * 50 = \$82.$$

The option to have the party on any of two days has increased her certain equivalent by an option value of $\$82 - \$48 = \$34$.

If Jane can have the party on any of the next $n > 2$ days and she has the same mutually irrelevant probabilities of rain, her option value increases even further. Under these conditions, she will hold her party outdoors on the first sunny day with a value of \$100. If she waits until $n = 1$ on the last day, she will have the party on that day in the location best suited to the day's

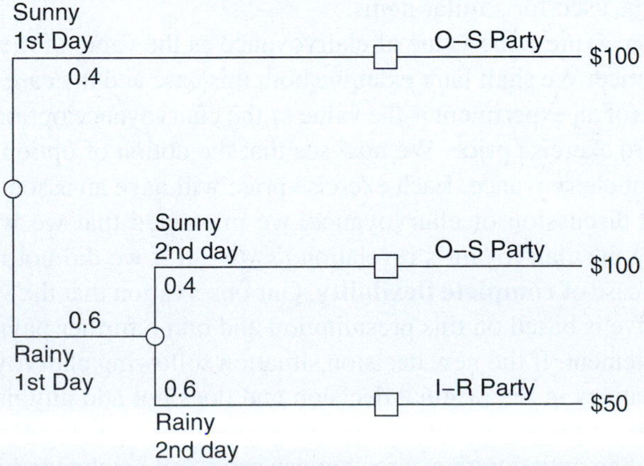

FIGURE 19.9 Jane's Decision Tree with Any Two Days Option

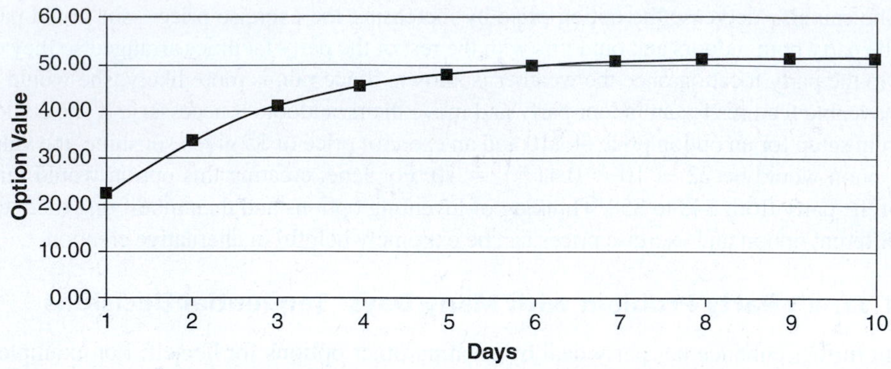

FIGURE 19.10 Change in Option Value with Days Remaining

weather—that is, with clairvoyance. The value of the option when $n = 1$ is therefore 22, as we calculated previously. With this option for n days, Jane will be able to enjoy an outdoor party in the sunshine if at least one of the n days is sunny. This will happen with the probability that not every day is rainy, as $1 - (0.6)^n$.

As Figure 19.10 shows, for n greater than 9, the option is worth about $52, which is the difference between the value of the best party at $100, and the value of the porch party at $48. In other words, if she has 10 or more days in which to have her party, she is virtually sure of having a party in sunshine.

19.7 VALUE OF CLAIRVOYANCE AS AN OPTION

Clairvoyance is an option that permits making decisions after revelation of the outcomes of specified distinctions. Since any experiment or information gathering activity can be viewed as providing clairvoyance on the results of that experiment or activity, any experiment or other information gathering activity can also be viewed as an option. Clairvoyance with limited flexibility is also an option and is the case of receiving information that limits the set of alternatives that can be made after revelation of specified distinctions. Examples of these situations include destructive testing, where you can no longer use the item you have tested, but you have acquired information that can be used for similar items.

We can therefore express the value of clairvoyance as the value of the clairvoyance option with a zero exercise price. We shall later examine both this case and the case of non-zero exercise price. The usual value of an experiment is the value of the clairvoyance option on the results of the experiment with a zero exercise price. We now see that the notion of option allows us to expand the concept of value of clairvoyance. Each exercise price will have an associated option value.

In our original discussion of clairvoyance, we presumed that we would have the same alternatives following the clairvoyant's revelation as we did if we did not receive this information. We call this the case of **complete flexibility**. Our observation that the value of clairvoyance could never be negative is based on this presumption and on no further payment to the clairvoyance based on his statement. If the new decision situation following clairvoyance eliminates one or more of the alternatives in the original decision and does not add any new alternatives, then even with a zero exercise price, the value of clairvoyance given this limited flexibility could be negative. The value of an option with a zero exercise price and no change in alternatives, that is, complete flexibility, cannot be negative.

19.8 SEQUENTIAL INFORMATION OPTIONS

In the last chapter, we discussed Kim's value of information for the Acme and Beta detectors individually and also discussed the value of receiving information from both detectors simultaneously. We also showed that the value of joint clairvoyance is not necessarily equal to the sum of the values of clairvoyance on the individual uncertainties.

We can now discuss other methods by which Kim can receive information. For example, suppose that Kim is offered the Acme detector with the option to receive the Beta indication after she observes the indication of the Acme detector. Should Kim's valuation for the Beta detector now be different? Alternatively, Kim can receive the Beta detector indication with the option of receiving the Acme indication after observing the indication of the Beta detector. In some cases, as in the case of destructive testing, this idea of sequential testing may not be available, since an item will be destroyed after the first test. In other cases, however, this option may be available, and we may be interested in calculating the value of an option to receive clairvoyance after learning the detector indication.

To illustrate the idea of sequential information further, let us now refer back to Kim's Acme Rain detector decision. Figure 19.11 shows the decision tree for using the Rain detector if it is free; arrows indicate the best alternative. Note that Kim will have the party on the porch if the detector says Sunny, but she will stay indoors if the detector says Rainy. Recall that the value of the detector option to her is \$54.65 − 45.83, or \$8.82, and her value of clairvoyance is \$20.

Now suppose Kim is given free use of the rain detector, but she has not yet observed its indication. If the clairvoyant then approaches her and offers his services, what would be her

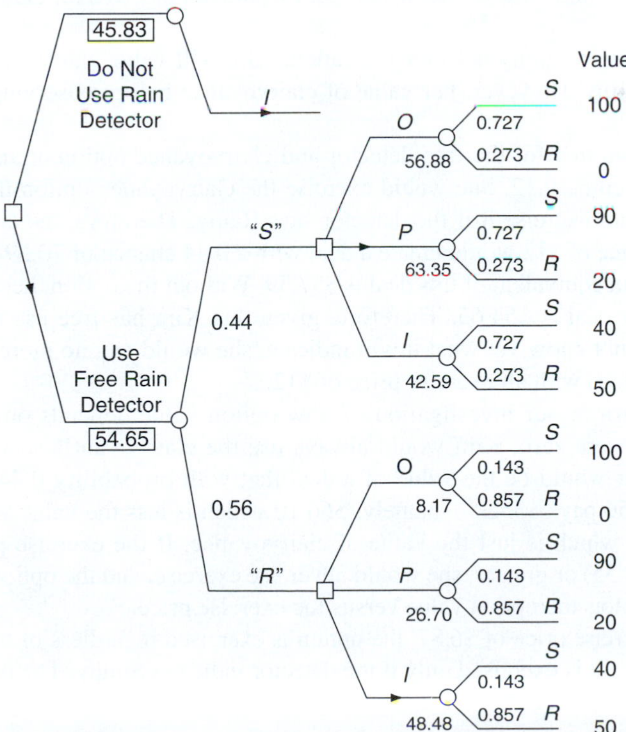

FIGURE 19.11 Kim's Decision Tree for Rain Detector

value of clairvoyance? Since Kim is a deltaperson, her new value of clairvoyance would be the value of the deal with free clairvoyance minus the value of the deal with no clairvoyance but with the free detector. The value of clairvoyance now falls to $66.10 − 54.65, or 11.45. In other words, she would not accept the services of a clairvoyant who charges even $12.

Now let us consider another variation of the clairvoyant option and illustrate how the clairvoyant can charge more for his services. Suppose the clairvoyant asks Kim, "How much would you pay me for an option to use my services at any time for an exercise price of $12?" That is, Kim can first view the free detector indication and then decide whether or not she wants to use the clairvoyant's services. Let us critically examine Kim's analysis of this question.

Kim reasons, "If Acme says Sunny, which it will with probability 0.44, I would have the party on the porch with a certain equivalent of 63.35 and a 0.727 chance of Sunshine (Figure 19.11). If I then obtained clairvoyance, I would have a 0.727 chance that the clairvoyant would say Sunny, giving me a 0.727 chance of an outdoor-Sunny party with a value of 100, and a 0.273 chance the clairvoyant would say Rainy giving me a 0.273 chance of an indoor party with value 50. Such a deal would be worth $82.59 to me, an increase of $19.24 over the $63.35 value I would otherwise face. Therefore $19.24 would be my value of clairvoyance under these circumstances. I would certainly pay $12 for clairvoyance when the detector says Sunny, and obtain a net 7.24 gain.

"On the other hand, if Acme says Rainy, which it will with probability 0.56, I would have the party indoors with a certain equivalent of $48.48 and a 0.143 chance of Sunshine. If I then obtained clairvoyance, I would have a 0.143 chance of an outdoor party with a value of 100 and a 0.857 chance of an indoor party with a value of 50. Such a deal would be worth $55.35 to me, an increase of 6.87 over the $48.48 value I would otherwise face. Therefore 6.87 would be my value of clairvoyance under these circumstances. I would not pay $12 for clairvoyance when the detector says Rainy.

After observing a Sunny detector indication, Kim will value clairvoyance higher than if the indication is Rainy. However, her value of clairvoyance before observing the indication is still $11.45.

Kim's decision tree for the free detector and clairvoyance option at an exercise price of $12 is shown in Figure 19.12. She would exercise the clairvoyance option if the detector says Sunny and would not exercise it if the detector says Rainy. Therefore, the clairvoyance option with an exercise price of $12 would create a deal with a 0.44 chance of 70.59 and a 0.56 chance of 48.48. The certain equivalent of this deal is $57.39. Without the option, her certain equivalent of the free detector deal is $54.65. Therefore, given that Kim has free use of the Acme Rain Detector, but does not know yet what it will indicate, she would pay no more than $2.74 for an option on clairvoyance with an exercise price of $12.

Now we return to our investigation of how option value depends on exercise price. If the exercise price were zero, Kim would always use the clairvoyant's service and the value of the option to her would be the value of a deal that with probability 0.44 pays $82.59 and with probability 0.56 pays $55.35—namely; $66.10 which is less the value without the option, $54.65: or $11.45, which is just the value of clairvoyance. If the exercise price were $19.24 (i.e., $82.59 − $63.35) or greater, she would never use exercise, and the option would be worthless. Figure 19.13 plots the option value versus the exercise price.

Below an exercise price of $6.87, the option is exercised regardless of the detector indication; above this price it is exercised only if the detector indicates Sunny. The option has no value when the exercise price exceeds $19.24, since she would never use the clairvoyant after observing any detector indication.

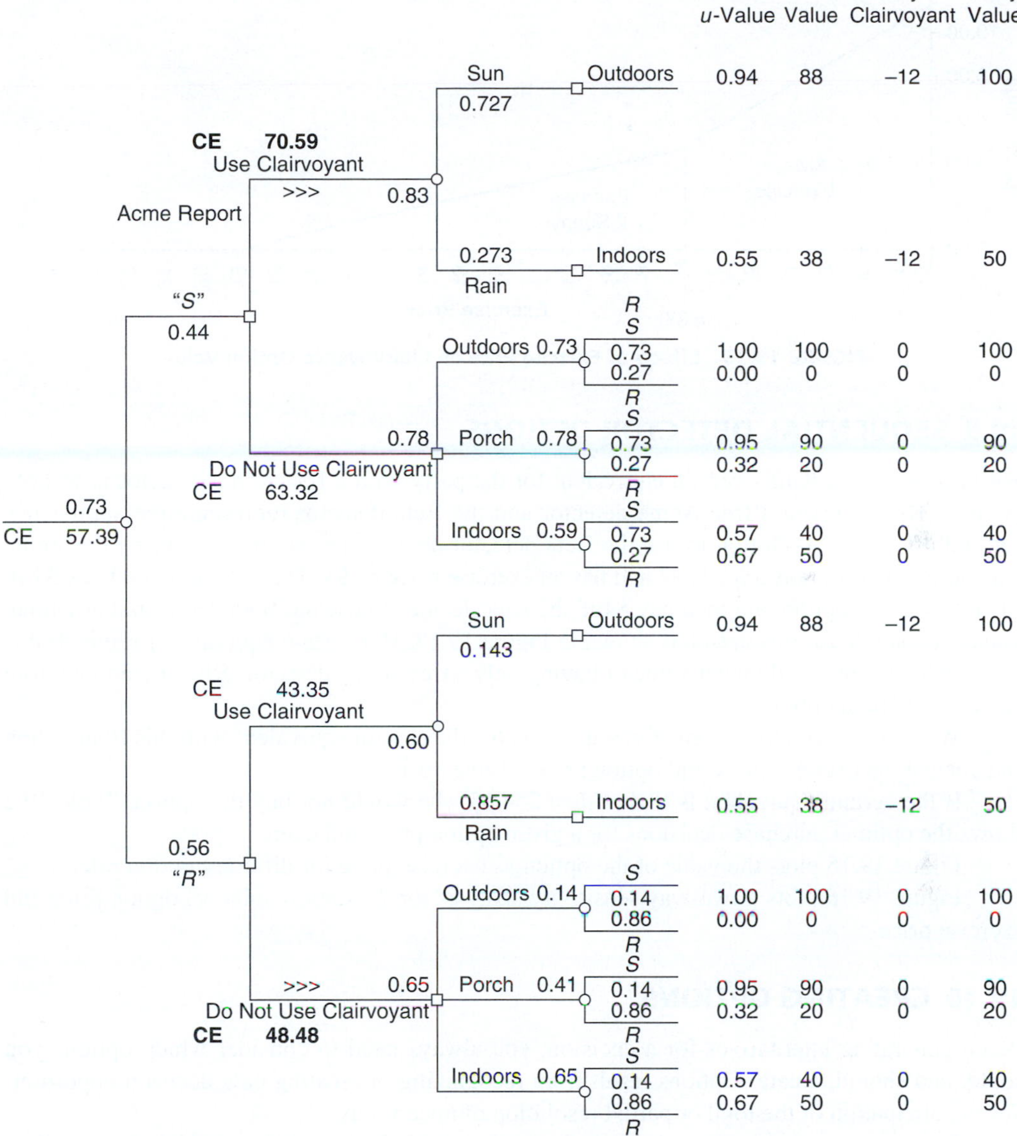

FIGURE 19.12 Decision Tree for Clairvoyance Option with $12 Exercise Price

Calculating the option value is more involved when the decision maker is risk-sensitive, but not a deltaperson. We can no longer find the value of the option by subtracting the certain equivalent of the deal without the option from the certain equivalent of the deal with a free option. However, the method of calculation of Figure 19.12 can be used if you enter different option and exercise prices and then find what combination of them will make the decision maker indifferent to the option. This approach will allow drawing a diagram like Figure 19.13 without assuming the delta property.

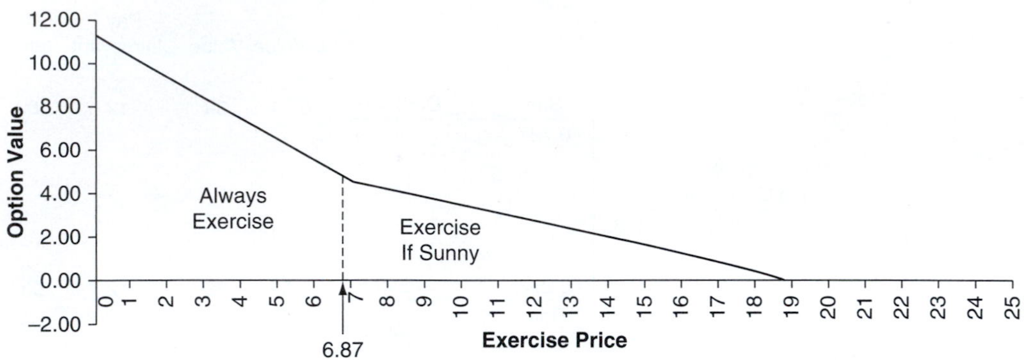

FIGURE 19.13 Effects of Exercise Price on Clairvoyance Option Value

19.9 SEQUENTIAL DETECTOR OPTIONS

We have seen that Kim's certain equivalent for the party with a free Acme detector is $54.65. Suppose Kim is given a free Acme detector and the Beta detector representative approaches her and offers her an Option to use the Beta detector anytime (even after observing the Acme indication). This option price is $2 and has an exercise price of $4. That is Kim would pay $2 in advance for this option and then pay $4 if she later decides to use the Beta detector at any time. The decision tree for this option is shown in Figure 19.14. Her certain equivalent for this deal is $55.4, which is higher than her value of having only a free Acme detector. Kim should therefore agree to this Beta option.

We can also conduct a sensitivity analysis for the certain equivalent with this Beta option and change the exercise price and option price. (Table 19.1)

If the certain equivalent is higher than $54.65, she would not buy the option. Table 19.2 shows the optimal purchase decisions for a given option price and exercise price.

Figure 19.15 plots the value of the option vs exercise price for different option prices.

Figure 19.16 plots a two-way sensitivity analysis for the option value vs option price and exercise price.

19.10 CREATING OPTIONS

When generating alternatives for a decision, you always need to consider which options you could, and should, create. Options result from recognizing or creating new decision opportunities in anticipation of the total or partial resolution of uncertainty.

Suppose you are on a motoring vacation and you are concerned about whether you will be able to find a motel room at your desired destination, though you assign only a 50% chance of reaching it given the highway's bad traffic conditions. You call ahead and find you can reserve a room if you pay for it now by credit card. Your alternatives are to guarantee the reservation or not.

Suppose that you offer to guarantee the innkeeper half the cost of the room (the option price), and then offer to pay an additional $3/4$ of the cost of the room if you arrive (the exercise price). If you are risk-neutral, your evaluation of the cost of the option is $1/2 + (1/2)(3/4)$, or $7/8$ the cost of the room, which is less than the cost of the guarantee. If the innkeeper is also risk-neutral and assigns the same 50% chance of your appearing, the value to the innkeeper is also $7/8$. This option will be a good deal to an innkeeper who assigns less than a $7/8$ probability to renting the room to someone else.

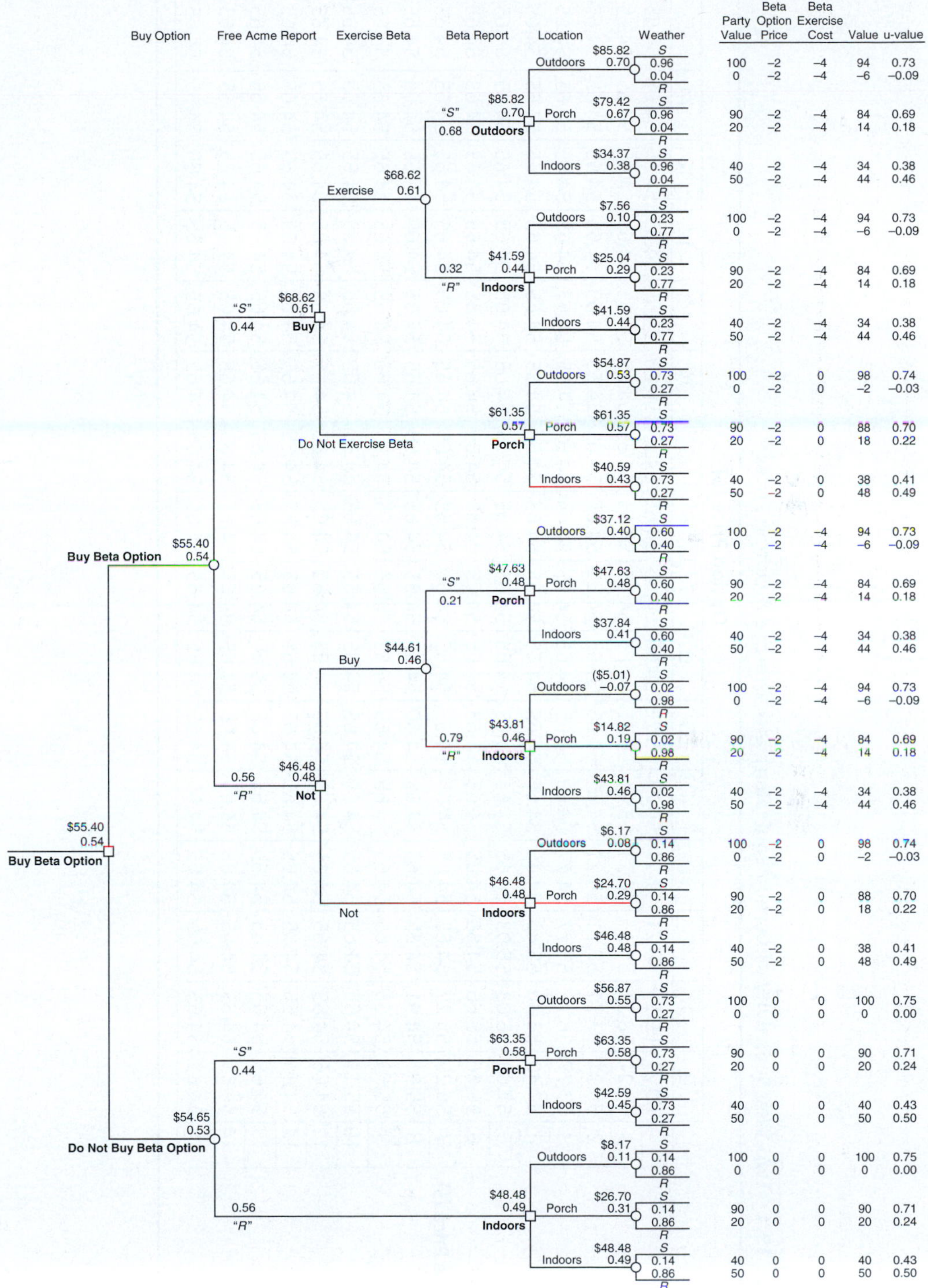

FIGURE 19.14 Decision Tree for Sequential Detector Options

TABLE 19.1 Two-Way Sensitivity of Certain Equivalent with Exercise & Option Prices

Excercise Price ($)	Option Price ($)															
	0	1	2	3	4	5	6	7	8	9	10	11	12	13	14	15
0	60.21	59.21	58.21	57.21	56.21	55.21	54.65	54.65	54.65	54.65	54.65	54.65	54.65	54.65	54.65	54.65
1	59.21	58.21	57.21	56.21	55.21	54.65	54.65	54.65	54.65	54.65	54.65	54.65	54.65	54.65	54.65	54.65
2	58.21	57.21	56.21	55.21	54.65	54.65	54.65	54.65	54.65	54.65	54.65	54.65	54.65	54.65	54.65	54.65
3	57.77	56.77	55.77	54.77	54.65	54.65	54.65	54.65	54.65	54.65	54.65	54.65	54.65	54.65	54.65	54.65
4	57.40	56.40	55.40	54.65	54.65	54.65	54.65	54.65	54.65	54.65	54.65	54.65	54.65	54.65	54.65	54.65
5	57.03	56.03	55.03	54.65	54.65	54.65	54.65	54.65	54.65	54.65	54.65	54.65	54.65	54.65	54.65	54.65
6	56.66	55.66	54.66	54.65	54.65	54.65	54.65	54.65	54.65	54.65	54.65	54.65	54.65	54.65	54.65	54.65
7	56.29	55.29	54.65	54.65	54.65	54.65	54.65	54.65	54.65	54.65	54.65	54.65	54.65	54.65	54.65	54.65
8	55.91	54.91	54.65	54.65	54.65	54.65	54.65	54.65	54.65	54.65	54.65	54.65	54.65	54.65	54.65	54.65
9	55.53	54.65	54.65	54.65	54.65	54.65	54.65	54.65	54.65	54.65	54.65	54.65	54.65	54.65	54.65	54.65
10	55.14	54.65	54.65	54.65	54.65	54.65	54.65	54.65	54.65	54.65	54.65	54.65	54.65	54.65	54.65	54.65
11	54.76	54.65	54.65	54.65	54.65	54.65	54.65	54.65	54.65	54.65	54.65	54.65	54.65	54.65	54.65	54.65
12	54.65	54.65	54.65	54.65	54.65	54.65	54.65	54.65	54.65	54.65	54.65	54.65	54.65	54.65	54.65	54.65
13	54.65	54.65	54.65	54.65	54.65	54.65	54.65	54.65	54.65	54.65	54.65	54.65	54.65	54.65	54.65	54.65
14	54.65	54.65	54.65	54.65	54.65	54.65	54.65	54.65	54.65	54.65	54.65	54.65	54.65	54.65	54.65	54.65
15	54.65	54.65	54.65	54.65	54.65	54.65	54.65	54.65	54.65	54.65	54.65	54.65	54.65	54.65	54.65	54.65

TABLE 19.2 Optimal Purchase Decision Given Option Price and Exercise Price

Excercise Price ($)	Option Price ($)															
	0	1	2	3	4	5	6	7	8	9	10	11	12	13	14	15
0	Buy	Buy	Buy	Buy	Buy	Buy	Not	Not	Not	Not	Not	Not	Not	Not	Not	Not
1	Buy	Buy	Buy	Buy	Buy	Not	Not	Not	Not	Not	Not	Not	Not	Not	Not	Not
2	Buy	Buy	Buy	Buy	Not	Not	Not	Not	Not	Not	Not	Not	Not	Not	Not	Not
3	Buy	Buy	Buy	Buy	Not	Not	Not	Not	Not	Not	Not	Not	Not	Not	Not	Not
4	Buy	Buy	Buy	Not	Not	Not	Not	Not	Not	Not	Not	Not	Not	Not	Not	Not
5	Buy	Buy	Buy	Not	Not	Not	Not	Not	Not	Not	Not	Not	Not	Not	Not	Not
6	Buy	Buy	Buy	Not	Not	Not	Not	Not	Not	Not	Not	Not	Not	Not	Not	Not
7	Buy	Buy	Not	Not	Not	Not	Not	Not	Not	Not	Not	Not	Not	Not	Not	Not
8	Buy	Buy	Not	Not	Not	Not	Not	Not	Not	Not	Not	Not	Not	Not	Not	Not
9	Buy	Not	Not	Not	Not	Not	Not	Not	Not	Not	Not	Not	Not	Not	Not	Not
10	Buy	Not	Not	Not	Not	Not	Not	Not	Not	Not	Not	Not	Not	Not	Not	Not
11	Buy	Not	Not	Not	Not	Not	Not	Not	Not	Not	Not	Not	Not	Not	Not	Not
12	Buy	Not	Not	Not	Not	Not	Not	Not	Not	Not	Not	Not	Not	Not	Not	Not
13	Buy	Not	Not	Not	Not	Not	Not	Not	Not	Not	Not	Not	Not	Not	Not	Not
14	Buy	Not	Not	Not	Not	Not	Not	Not	Not	Not	Not	Not	Not	Not	Not	Not
15	Buy	Not	Not	Not	Not	Not	Not	Not	Not	Not	Not	Not	Not	Not	Not	Not

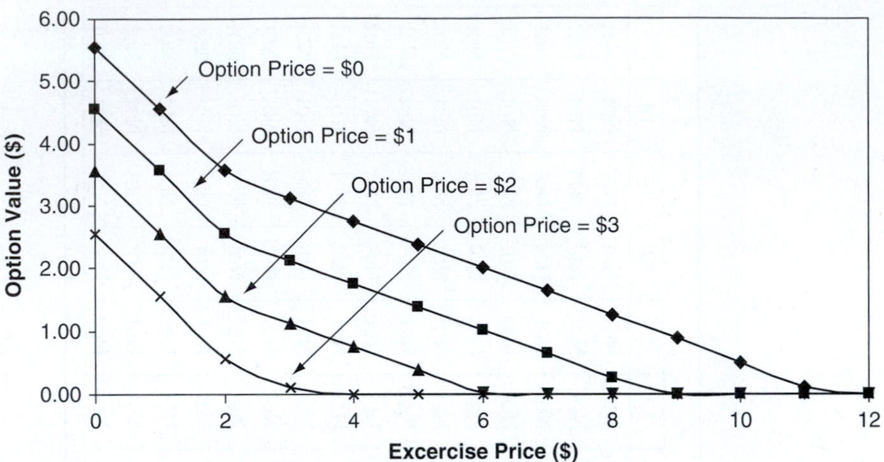

FIGURE 19.15 Option Value vs Exercise Price

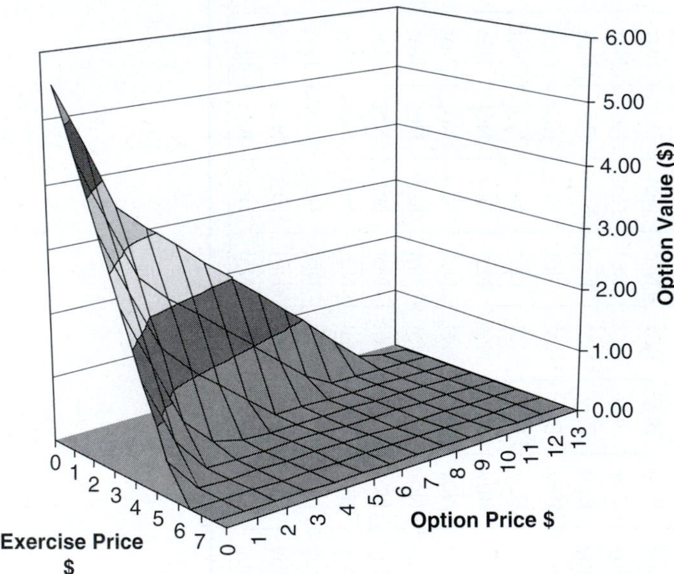

FIGURE 19.16 Sensitivity Analysis Option Value vs. Exercise & Option Price

You can adjust the option to fit other situations. For example, if the exercise price is one-half the room cost, the option cost is 3/4; selling this option will be desirable to the innkeeper if there is otherwise a 75% chance of otherwise renting the room.

19.10.1 Use in Developing Strategies

Options should play an important role in helping people select business strategies. A strategy is a set of alternatives from a group of decisions that a decision maker may decide upon. When defining possible strategies, your first step is normally to identify every decision required to define the strategy and then to specify the alternatives for each decision. Strategies refer to a collection of alternatives for each of those decisions. The strategies to investigate would be the few most promising, most coherent selections of one single alternative for each decision selected.

For example, if you were a cosmetic company thinking of possible strategies for a new perfume, the marketing decision might have alternatives like mass market, medium priced, and very highly priced. The distribution channel decision might have alternatives like discount chains, mall department stores, and exclusive boutiques. A strategy would be a selection of an alternative from each of these decisions. A strategy that matched the very highly priced product with a discount chain distribution would not be a coherent or useful approach.

As helpful as the notion of strategy construction can be, framing of the choices for each decision as alternatives may lead to insufficient consideration of strategies that contain options and hence may prevent the development of sequential decisions.

Before making a decision, one way to uncover new options is to ask yourself whether you could make a better decision if you could first clarify one or more uncertain distinctions. If so, you would ask what options you could create to create the type of decision opportunity you seek. Then, you would evaluate whether those options make analytic sense using the type of analyses we have illustrated. In the perfume example, this might mean paying for an option on manufacturing facilities rather than committing to them before the seeing results of a market survey.

19.11 SUMMARY

An option is an alternative that permits or may permit a future decision following revelation of information. The clairvoyant and any information gathering experiment provide options. An option has a value, an option price, and an exercise price. If you have no decision to make, receiving information is an option. If you do have a decision to make, acquiring information is an option. The option of waiting often adds value. If the option price is 0 and the exercise price is 0, the value of an option, like the value of clairvoyance, must be non-negative. There are many good reasons to generate options in our daily lives.

KEY TERMS

- Value of an option
- Option price
- Exercise price

PROBLEMS

Problems marked with an asterisk (*) are considered more challenging.

1. Think about a decision situation you are currently facing.
- What are your current alternatives?
- What are your uncertainties?
- What options can you create to enable a larger set of alternatives in the future upon revelation of the outcome of an uncertainty?
- Which uncertainty will most likely be revealed?
- What future alternatives will be created by your option?
- Draw the decision tree and calculate the option value.

***2.** This problem focuses on the option to purchase of multiple sources of information with different prices. Two salesmen approach Kim: One is selling an Acme Rain Detector, and the other is selling a Beta Rain Detector. After conferring with one another, they offer Kim the following opportunities.
- Kim can buy the Acme detector (at any time) for $10.
- Kim can buy the Beta detector (at any time) for $13.
- Kim can buy both detectors (at the same time) for $16.

Kim believes that the Acme detector is symmetric, and it has an accuracy of 0.8. In other words,

$$\{``S" \mid S, \&\} = \{``R" \mid R, \&\} = 0.8$$

where "S" is the event that the test indicates Sunny, S is the event that the weather is actually sunny (similarly for rain).

Kim also believes that the Beta detector is symmetric, with accuracy of 0.9, and that the two detector indications are irrelevant given the weather and her background state of information.

a. Determine the different purchase strategies that Kim faces. (***Hint:*** They are not simply Acme, Beta, both, or none).
b. Draw a decision diagram for Kim's detector purchase decision.
c. Redraw the decision diagram in inferential form (weather conditioned on detector indications).
d. Determine the optimal detector purchase strategy for Kim. Please draw the decision tree and show the best decisions on the tree.

***3.** A risk-neutral decision maker carries a stock option for one day. This option pays him $\$(P - 50)$ is the stock price, P, is above 50 and pays him 0 otherwise. The current stock price is $45, and each day it can go up 20% or down 20% with probabilities 0.6 and 0.4, respectively. What is the value of this option for one day? Repeat if the decision maker has a three-day option on this stock, and he can
- Exercise this option only on the third day, or
- Exercise this option at any day during the three days.

***4.** On your own, repeat the analysis of
(i) Figure 19.10
(ii) Figure 19.13
(iii) Table 19.1
(iv) Table 19.2.

CHAPTER

20

Detectors with Multiple Indications

CHAPTER CONCEPTS

After reading this chapter, you will be able to explain the following concepts:

- Valuing detectors with multiple indications
- Valuing a detector indication that is less than or equal to some value
- Valuing a continuous Beta detector
- Discretizing a continuous Beta detector

20.1 INTRODUCTION

In previous chapters, we have discussed situations where multiple sources of information are available. Now, we will consider situations where a detector has multiple degrees. For example, while a temperature sensor might indicate either "Hot" or "Cold," it could also produce a wide range of possible indications of temperature based on its resolution. Is a sensor with a wide range of indications more valuable than a simple indication of "Hot" and "Cold?" Would you prefer a car battery condition indicator that gives you several degrees of the battery condition or just a simple indicator with "good" or "bad?" If you are considering an investment in a stock, would you value knowing whether "Stock will go up" or "Stock will go down" more or less than knowing the actual future stock price?

In this chapter, we will present several concepts. First, we will illustrate how to calculate the value of a detector with multiple indications. The basic principles we discussed for a detector with two indications will also apply here. However, it will be helpful to use tables to organize computations for the case of multiple indications. To maximize the learning from this chapter, we suggest that you conduct the analysis yourself in tabular form. We will also show how to conduct the analysis for detectors providing an indication on a continuous interval. Second, we will introduce beta distributions, which are very useful probability distributions that can be used in a variety of decisions because of the variety of shapes they can have. Third, we will illustrate how distinctions with multiple degrees can be discretized into distinctions with a fewer number of degrees. We will revisit this subject in more detail in Chapter 35.

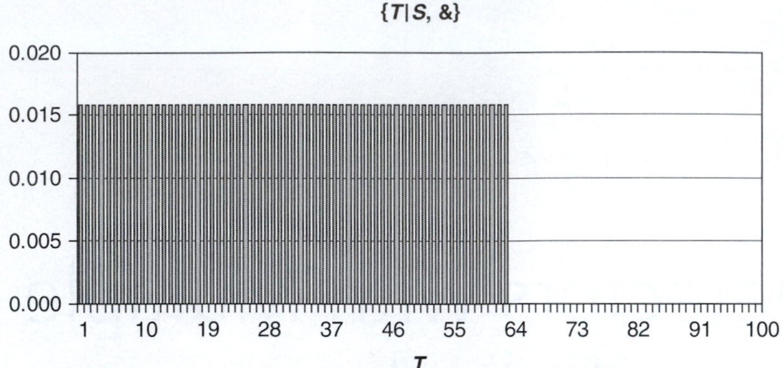

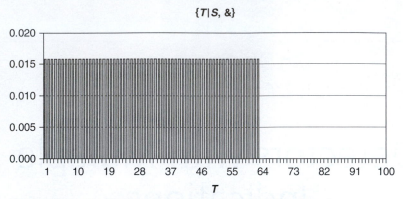

FIGURE 20.1 Possible Detector Indications/Probability Given Sun

20.2 DETECTOR WITH 100 INDICATIONS

Once again, we will revisit the Party Problem introduced in Chapter 9, where Kim is deciding where to have her party and is considering three alternatives. She is uncertain about the weather, which could be Sun or Rain. She considers purchasing detector information. In Chapter 18, we discussed the Acme and Beta Rain Detector purchase decisions. Suppose that while Kim is considering using the Acme Rain Detector for her party, another salesman arrives who offers her a rain detector that provides an indication from 1 to 100. He says that his 100 Degree Detector has a higher resolution than the Acme detector. It provides an integer, *T*, that ranges from 1 (for Sun, no Rain) to 100 (for Rain, no Sun). He tells Kim that the detector will produce a number between 1 and 63 with equal probabilities $(1/63 = 0.0159)$ if the weather is Sun and a number between 38 and 100 with equal probabilities $(1/63 = 0.0159)$ if the weather is Rain. The probability distribution of the **detector indication**, *T*, given the weather is

$$\{T|S,\&\} = \begin{cases} \dfrac{1}{63}, & T = 1, 2, .., 63 \\ 0, & T = 64, 65, \ldots, 100 \end{cases} \qquad \{T|R,\&\} = \begin{cases} 0, & T = 1, 2, .., 37 \\ \dfrac{1}{63}, & T = 38, 39, \ldots, 100 \end{cases}$$

Figures 20.1 and 20.2 show the probability distributions of the detector indication given Sun and Rain respectively.

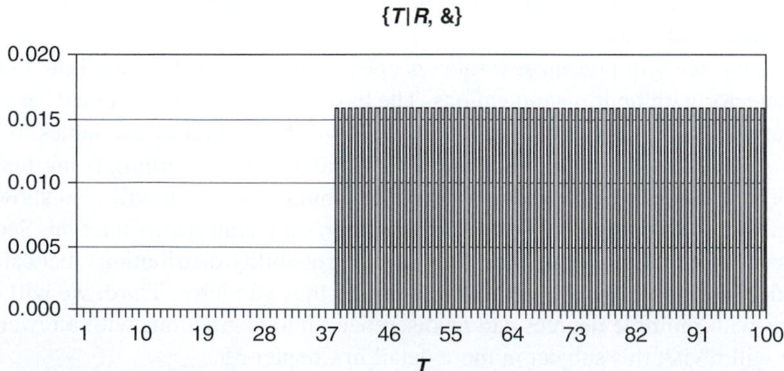

FIGURE 20.2 Possible Detector Indications/Probability Given Rain

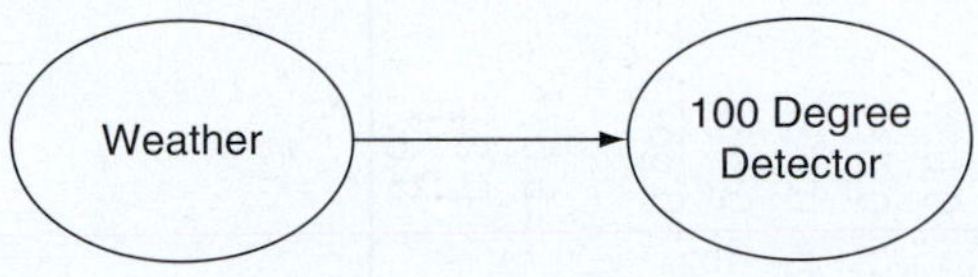

FIGURE 20.3 Relevance Diagram for 100 Degree Detector

Notice that the detector indications are relevant to the weather, because we have a different probability distribution for each degree of the weather distinction. The relevance diagram for this situation is given in Figure 20.3 and is exactly the same as the one used for the simple detector.

The tree representation for this relevance diagram appears in Figure 20.4 and shows both the prior distribution on the weather and the likelihood of the indication given the weather.

To find the implications of using the detector and its value, we need to reverse this tree to create its inferential form. Table 20.1 shows this **tree-reversal** analysis in tabular form.

Column (a) lists the possible indications produced by the detector. Columns (b) and (c) list the conditional probabilities of an indication given the weather is Sun and Rain, respectively, so we have $\{T|S,\&\}$ and $\{T|R,\&\}$. Note that these numbers are either $1/63 = 0.0159$ or 0, as previously seen in Figures 20.1 and 20.2. Further more, there is relevance between the detector indication and the weather, since the conditional probabilities $\{T|S,\&\}$ and $\{T|R,\&\}$ are different. Column (d) shows the joint probability of an indication T and the weather being Sun at $\{S,T|\&\}$ that is obtained by multiplying the number in Column (b) by the probability of Sun at $\{S|\&\} = 0.4$, since

$$\{S,T|\&\} = \{T|S,\&\}\{S|\&\}$$

Column (e) shows the joint probability of an indication, T, and the weather being Rain, $\{R,T|\&\}$, obtained by multiplying the numbers in Column (c) by $\{R|\&\} = 0.6$, since

$$\{R,T|\&\} = \{T|R,\&\}\{R|\&\}$$

Column (f) shows the preposterior probability of an indication $\{T|\&\}$ that is obtained by summing Columns (d) and (e),

$$\{T|\&\} = \{S,T|\&\} + \{R,T|\&\}$$

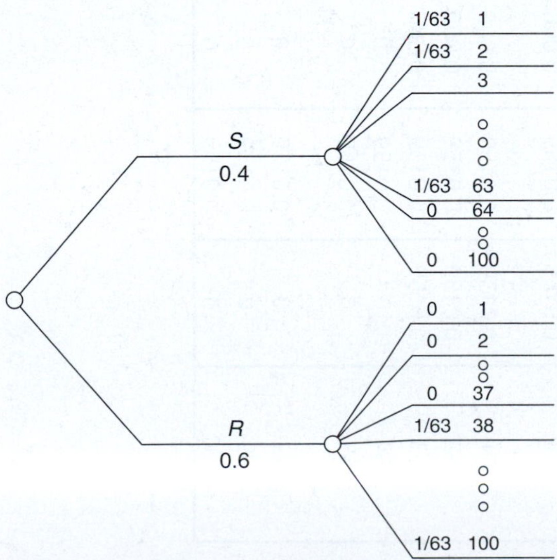

FIGURE 20.4 Assessed Probability Tree for 100 Degree Detector

TABLE 20.1 Inferential Representation of 100 Degree Detector

(a)	(b)	(c)	(d)	(e)	(f)	(g)	(h)	(i)	(j)	(k)
Indication, T	$\{T\|S, \&\}$	$\{T\|R, \&\}$	$\{S, T\|\&\}$	$\{R, T\|\&\}$	$\{T\|\&\}$	$\{S\|T, \&\}$	$\{R\|T, \&\}$	Decision	CE of Party	u-Value of Party
1	0.0159	0	0.006	0	0.006	1	0	O	100	1
2	0.0159	0	0.006	0	0.006	1	0	O	100	1
3	0.0159	0	0.006	0	0.006	1	0	O	100	1
4	0.0159	0	0.006	0	0.006	1	0	O	100	1
…	…	…	…	…	…	…	…	…	…	…
34	0.0159	0	0.006	0	0.006	1	0	O	100	1
35	0.0159	0	0.006	0	0.006	1	0	O	100	1
36	0.0159	0	0.006	0	0.006	1	0	O	100	1
37	0.0159	0	0.006	0	0.006	1	0	O	100	1
38	0.0159	0.0159	0.006	0.010	0.016	0.4	0.6	I	45.83	0.627
…	…	…	…	…	…	…	…	…	…	…
61	0.0159	0.0159	0.006	0.010	0.016	0.4	0.6	I	45.83	0.627
62	0.0159	0.0159	0.006	0.010	0.016	0.4	0.6	I	45.83	0.627
63	0.0159	0.0159	0.006	0.010	0.016	0.4	0.6	I	45.83	0.627
64	0	0.0159	0	0.010	0.010	0	1	I	50	0.667
65	0	0.0159	0	0.010	0.010	0	1	I	50	0.667
…	…	…	…	…	…	…	…	…	…	…
98	0	0.0159	0	0.01	0.01	0	1	I	50	0.67
99	0	0.0159	0	0.01	0.01	0	1	I	50	0.67
100	0	0.0159	0	0.01	0.01	0	1	I	50	0.67

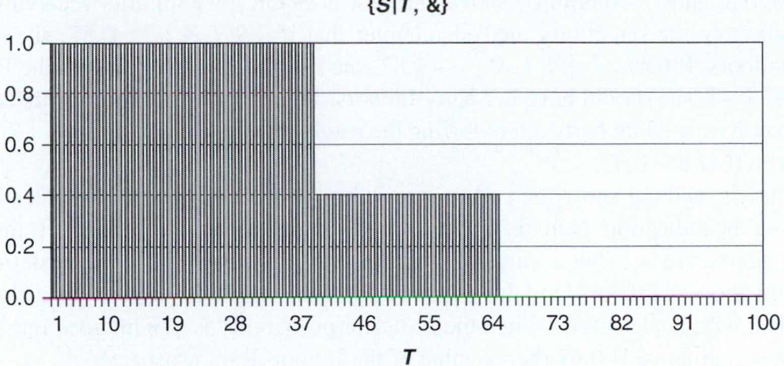

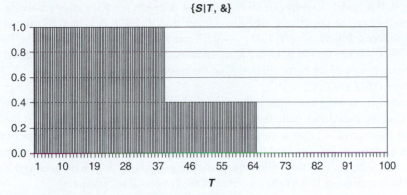

FIGURE 20.5 Posterior Probability, $\{S|T,\&\}$

Column (g) shows the posterior probability of Sun after receiving an indication T, so we have $\{S|T,\&\}$. This is obtained by dividing the corresponding cells in Columns (d) and (f).

$$\{S|T,\&\} = \frac{\{S,T|\&\}}{\{T|\&\}}$$

Column (h) shows the posterior probability of Rain given an indication T, so we have $\{R|T,\&\}$, which is obtained by dividing the corresponding cells in Columns (e) and (f). Thus,

$$\{R|T,\&\} = \frac{\{R,T|\&\}}{\{T|\&\}}$$

The analysis we have done in tabular form corresponds to the tree reversal we conducted earlier. The table format is convenient for the large number of indications. Figures 20.5 and 20.6 plot the posterior probabilities of Sun and Rain for a given indication, T.

Column (g) and Figure 20.5 show that there only three possible values for the probability of Sun after receiving the detector indication: 1, 0.4, and 0. Column (h) and Figure 20.6 show that there are only three possible values for the probability of Rain after receiving the detector

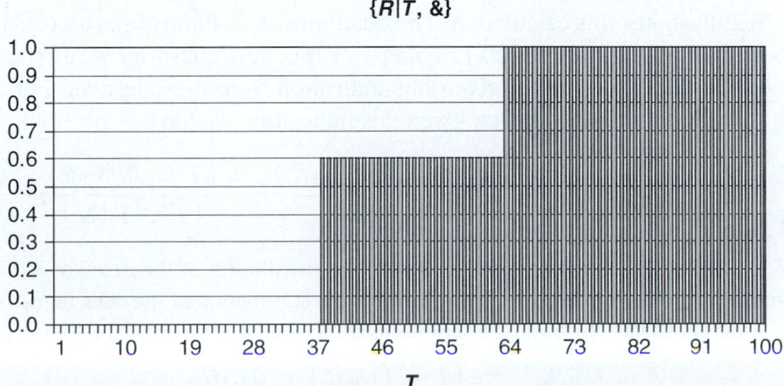

FIGURE 20.6 Posterior Probability, $\{R|T,\&\}$

indication: 0, 0.6, and 1. Column (i) shows the best decision for Kim after receiving an indication T. Recall from the sensitivity analysis chapter that if $\{S|T,\&\} \geq 0.87$, she should have her party Outdoors. If $0.47 \leq \{S|T,\&\} < 0.87$, she should have her party on the Porch, and if $\{S|T,\&\} < 0.47$, she should have her party Indoors. Therefore, we are not surprised to see that Kim will never have a Porch party after viewing the results of the detector because $\{S|T,\&\}$ does not lie between 0.47 and 0.87.

Columns (j) and (k) show the certain equivalent and the *e*-value of *u*-values of the party after receiving the indication. Note that, when the posterior probability of Sun is 1, the *u*-value of the Outdoor alternative is 1 (her *u*-value of Outdoor–Sun prospect). When the posterior probability is 0.4, Kim chooses Indoors, and the *u*-value of this alternative is 0.63 (same as the *u*-value of the Indoor deal with no detector). When the posterior probability is 0, Kim goes Indoors, and the *u*-value of this alternative is 0.67 (her *u*-value of the Indoor–Rain prospect).

By multiplying the preposterior probability of a detector indication, $\{T|\&\}$ by the *e*-value of the *u*-value of the party given this indication, column (k), and summing over all possible values of T, we get

$$e\text{-value of the } u\text{-values of party deal with free detector} = 0.72886$$

$$\text{Value of party deal with free detector} = \$57.07$$

Since Kim is a deltaperson and values the party at \$45.87 with no detector, her value of the 100 degree detector is

$$\text{Value of the detector} = \$57.07 - \$45.87 = \$11.20$$

Note that this analysis is in principle similar to that of the simple detector analysis. The main difference is the complexity arising from the 100 degrees. The use of tables similar to Table 20.1 is convenient for handling these types of problems.

20.2.1 Valuing the Indication $\leq T$

Suppose the salesman now offers Kim a new alternative: The possibility of receiving an indication of the form stated here.

> *"The detector indication, T, is less than or equal to some value T_0."*

For example, he will tell her whether the detector indication is less than or equal to 50. How much is this indication worth?

Table 20.2 illustrates this calculation. The quantities in Columns (a), (b), (c), and (d) have previously been calculated in Table 20.1. To answer this new question, we first calculate the posterior probabilities of the weather given this indication from the salesman. For example, to calculate the posterior probability of Sun given this indication, we have

$$\{S|T \leq T_0,\&\} = \frac{\{T \leq T_0|S,\&\}\{S|\&\}}{\{T \leq T_0|\&\}} = \frac{\{T \leq T_0, S|\&\}}{\{T \leq T_0|\&\}}$$

The term $\{T \leq T_0|S,\&\}$ is the cumulative probability distribution of the detector indication given Sun. This is simple to obtain from Table 20.1, since it is the summation of the cells in the column up to a given value of T_0. Table 20.2, Column (e), represents this cumulative distribution given Sun. We have

$$\{T \leq 1|S,\&\} = \{T = 1|S,\&\} = 0.016,$$
$$\{T \leq 2|S,\&\} = \{T \leq 1|S,\&\} + \{T = 2|S,\&\} = 0.016 + 0.016 = 0.032 \ldots \text{etc.}$$

Column (f) in Table 20.2 shows the joint distribution $\{S, T \leq T_0|\&\}$ obtained by multiplying the cells in Column (e) by $\{S|\&\} = 0.4$. Column (g) is the cumulative probability distribution for an indication $\{T \leq T_0|\&\}$. This is obtained by summing the cells in Column (d) of $\{T|\&\}$. Column (h) is the posterior distribution of Sun given the detector report $T \leq T_0$. Thus,

$$\{S|T \leq T_0, \&\} = \frac{\{T \leq T_0|S, \&\}\{S|\&\}}{\{T \leq T_0|\&\}}$$

This result is produced by multiplying Column (e) by 0.4 and then dividing by Column (g).

Column (i) shows the indication $\{R|T \leq T_0, \&\} = 1 - \{S|T \leq T_0, \&\}$.

Column (j) shows the optimal party location decision for a given indication $T \leq T_0$ and given posterior probability $\{S|T \leq T_0, \&\}$. Column (k) shows the *u*-value of the optimal decision for a given indication.

Columns (l) through (r) show the analysis for an indication of the form stated here.

"The detector indication, T, is greater than or equal to some value T_0."

To calculate the posterior probability, we have

$$\{S|T > T_0, \&\} = \frac{\{T > T_0|S, \&\}\{S|\&\}}{\{T > T_0|\&\}}$$

Note that the indication $\{T > T_0|S, \&\} = 1 - \{T \leq T_0|S, \&\}$. Column (l) is obtained by taking 1 minus the cells of column (e). Column (m) is obtained by multiplying the cells in column (l) by 0.4, since $\{T > T_0, S|\&\} = \{T > T_0|S, \&\}\{S|\&\} = 0.4\{T > T_0|S, \&\}$. Column (n) is obtained by taking 1 minus the cells of Column (g). Column (o) is the posterior distribution of Sun given an indication, $T > T_0$, $\{S|T > T_0, \&\}$. Column (p) is obtained by taking 1 minus the cells of Column (o). Columns (q) and (r) show the optimal party location decisions and the *u*-value of the optimal decision for a given indication.

Let us now calculate the value of an indication $T \leq 50$ or $T > 50$ for this detector. From Table 20.2, the probability of receiving an indication is $\{T \leq 50|\&\} = 0.441$ and $\{T > 50|\&\} = 1 - \{T \leq 50|\&\} = 0.559$.

For the indication $T \leq 50$, the posterior probability of Sun is $\{S|T \leq 50, \&\} = 0.719$; the optimal party decision is the Porch with a *u*-value equal to 0.774. For the indication $T > 50$, the posterior probability of Sun is $\{S|T > 50, \&\} = 0.148$; the optimal decision is Indoors with a *u*-value equal to 0.652.

The *e*-value of *u*-values of this free detector indication is

$$0.441 * 0.774 + 0.559 * 0.652 = 0.706$$

Kim's certain equivalent of the deal with this free detector indication = \$54.39. For Kim, the value of this detector indication is therefore \$54.39 − \$45.83 = \$8.56.

This value is close to that of the Acme Rain Detector with the two degrees discussed earlier, which was \$8.82. In fact, we can think of the Acme Rain Detector as a detector with multiple degrees, but for which we know only that the indication is above or below a particular number: In this case, 50.

TABLE 20.2 Analysis of Detector Indication Above/Below Certain Value

(a)	(b)	(c)	(d)	(e)	(f)	(g)	(h)	(i)
x	$\{T = x \mid S,\&\}$	$\{T = x \mid R,\&\}$	$\{T \mid \&\}$	$\{T <= x \mid S,\&\}$	$\{T <= x, S \mid \&\}$	$\{T <= x \mid \&\}$	$\{S \mid T <= x,\&\}$	$\{R \mid T <= x,\&\}$
1	0.016	0.000	0.006	0.016	0.006	0.006	1.000	0.000
2	0.016	0.000	0.006	0.032	0.013	0.013	1.000	0.000
3	0.016	0.000	0.006	0.048	0.019	0.019	1.000	0.000
...	...	...	...	...	...	...	...	...
36	0.016	0.000	0.006	0.571	0.229	0.229	1.000	0.000
37	0.016	0.000	0.006	0.587	0.235	0.235	1.000	0.000
38	0.016	0.016	0.016	0.603	0.241	0.251	0.962	0.038
39	0.016	0.016	0.016	0.619	0.248	0.267	0.929	0.071
...	...	...	...	...	...	...	...	...
40	0.016	0.016	0.016	0.635	0.254	0.283	0.899	0.101
...	...	...	...	...	...	...	...	...
49	0.016	0.016	0.016	0.778	0.311	0.425	0.731	0.269
50	0.016	0.016	0.016	0.794	0.317	0.441	0.719	0.281
51	0.016	0.016	0.016	0.810	0.324	0.457	0.708	0.292
...	...	...	...	...	...	...	...	...
63	0.016	0.016	0.016	1.000	0.400	0.648	0.618	0.382
64	0.000	0.016	0.010	1.000	0.400	0.657	0.609	0.391
...	...	...	...	...	...	...	...	...
82	0.000	0.016	0.010	1.000	0.400	0.829	0.483	0.517
83	0.000	0.016	0.010	1.000	0.400	0.838	0.477	0.523
84	0.000	0.016	0.010	1.000	0.400	0.848	0.472	0.528
85	0.000	0.016	0.010	1.000	0.400	0.857	0.467	0.533
...	...	...	...	...	...	...	...	...
99	0.000	0.016	0.010	1.000	0.400	0.990	0.404	0.596
100	0.000	0.016	0.010	1.000	0.400	1.000	0.400	0.600

20.2.2 Continuous Uniform Detector

As the number of degrees of the detector indication increases, it will be convenient to switch to a continuous representation for the detector indications (even though the indication may be discrete with a very high resolution). In this representation, we assume that the detector will provide any number from a continuous domain. We then represent the likelihood functions of the detector accuracy by **continuous functions**.

Suppose, for example, that the detector provides any value between 0 and 63 with equal probability if the Weather is Sun and between 37 and 100 with equal probability if the Weather is Rain. If the detector indication is continuous, the likelihood distributions for the detector indication can be expressed as

$$\{T|S,\&\} = \begin{cases} \dfrac{1}{63}, & 0 \le T \le 63 \\ 0, & T > 63 \end{cases} \qquad \{T|R,\&\} = \begin{cases} 0, & 0 \le T \le 37 \\ \dfrac{1}{63}, & T > 37 \end{cases}$$

Figures 20.7 and 20.8 show these curves graphically. In the continuous case, we refer to these distributions as **probability density functions**. Just as in the discrete case, a probability distribution must sum to 1, in the continuous case, a probability "density" function must integrate to 1.

TABLE 20.2 (continued)

(k)	(l)	(m)	(n)	(o)	(p)	(q)	(r)
u-value	$\{T > x\|S,\&\}$	$\{T > x, S\|\&\}$	$\{T > x\|\&\}$	$\{S\|T > x,\&\}$	$\{R\|T < x,\&\}$	Location	*u*-value
1.000	0.984	0.394	0.994	0.396	0.604	I	0.628
1.000	0.968	0.387	0.987	0.392	0.608	I	0.628
1.000	0.952	0.381	0.981	0.388	0.612	I	0.629
...	...	...	...	...	...	...	...
1.000	0.429	0.171	0.771	0.222	0.778	I	0.645
1.000	0.413	0.165	0.765	0.216	0.784	I	0.646
0.962	0.397	0.159	0.749	0.212	0.788	I	0.646
0.929	0.381	0.152	0.733	0.208	0.792	I	0.646
...	...	...	...	...	...	...	...
0.899	0.365	0.146	0.717	0.204	0.796	I	0.647
...	...	...	...	...	...	...	...
0.782	0.222	0.089	0.575	0.155	0.845	I	0.652
0.774	0.206	0.083	0.559	0.148	0.852	I	0.652
0.767	0.190	0.076	0.543	0.140	0.860	I	0.653
...	...	...	...	...	...	...	...
0.710	0.000	0.000	0.352	0.000	1.000	I	0.667
0.705	0.000	0.000	0.343	0.000	1.000	I	0.667
...	...	...	...	...	...	...	...
0.626	0.000	0.000	0.171	0.000	1.000	I	0.667
0.622	0.000	0.000	0.162	0.000	1.000	I	0.667
0.619	0.000	0.000	0.152	0.000	1.000	I	0.667
0.621	0.000	0.000	0.143	0.000	1.000	I	0.667
...	...	...	...	...	...	...	...
0.627	0.000	0.000	0.010	0.000	1.000	I	0.667
0.627	0.000	0.000	0.000	0.000	1.000	I	0.667

For example, the area under the curve must equal 1. The reader can verify that this is indeed the case.

Since the probability density on detector indication is different for the two degrees of weather, the relevance diagram remains the same as that of the discrete case in Figure 20.3.

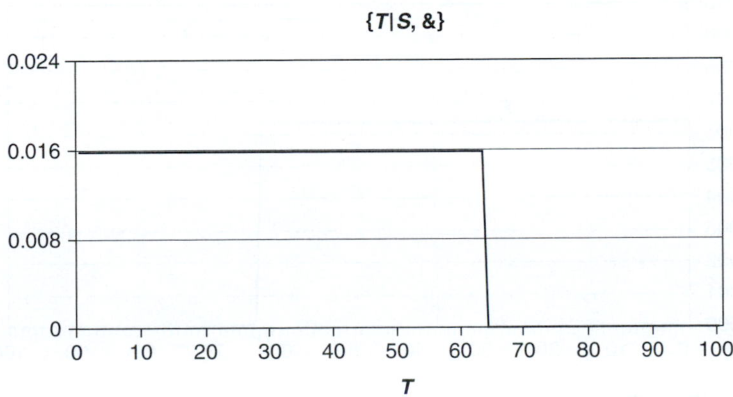

FIGURE 20.7 Probability Density Function for Detector Indication Given Sun

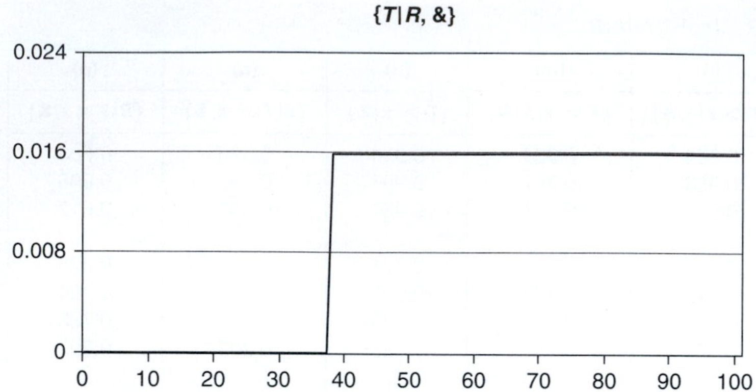

FIGURE 20.8 Probability Density Function for Detector Indication Given Rain

What is the value of this detector indication? To find it, we must again reverse the tree to get the posterior distributions—but this time in continuous form. Once more, we first find the joint distributions of detector indications and weather:

$\{T, S|\&\}$ is simply the multiplication of the curve $\{T|S, \&\}$ by 0.4, and $\{T, R|\&\}$ is the multiplication of the $\{T|R, \&\}$ curve by 0.6. These curves are shown in Figure 20.9. Hence,

$$\{S, T|\&\} = \begin{cases} \dfrac{4}{630}, & 0 \le T \le 63 \\ 0, & T > 63 \end{cases} \qquad \{R, T|\&\} = \begin{cases} 0, 0 \le T \le 37 \\ \dfrac{6}{630}, & T > 37 \end{cases}$$

The preposterior probability of receiving an indication $\{T|\&\}$ is the sum of the curves $\{S, T|\&\} + \{R, T|\&\}$, which we represent graphically as a simple addition of the two curves in Figure 20.9.

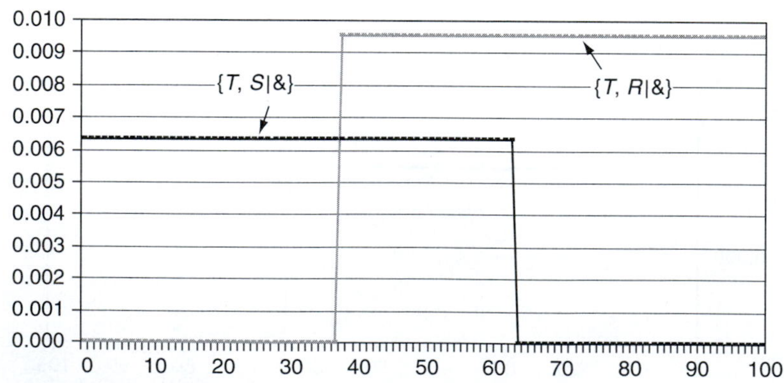

FIGURE 20.9 Density Functions for $\{S, T|\&\}$ and $\{R, T|\&\}$

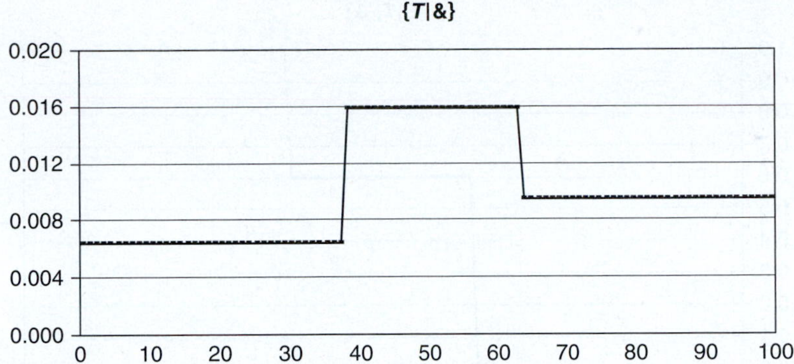

FIGURE 20.10 Preposterior Probability Density Function for the Indication

We can also represent this distribution analytically as

$$\{T|\&\} = \{S, T|\&\} + \{R, T|\&\} = \begin{cases} \dfrac{4}{630}, & 0 \leq T \leq 37 \\ \dfrac{1}{63}, & 37 < T \leq 63 \\ \dfrac{6}{630}, & 63 < T \leq 100 \end{cases}$$

The posterior probabilities of Sun and Rain for a given indication are

$$\{S|T, \&\} = \frac{\{S, T|\&\}}{\{T|\&\}}, \quad \{R|T, \&\} = \frac{\{R, T|\&\}}{\{T|\&\}}$$

Since we have already calculated the terms $\{S, T|\&\}, \{R, T|\&\}$, and $\{T|\&\}$, these posterior probabilities can be obtained by dividing the curves of Figure 20.9 by the curve in Figure 20.10, which then appear as Figures 20.11 and 20.12.

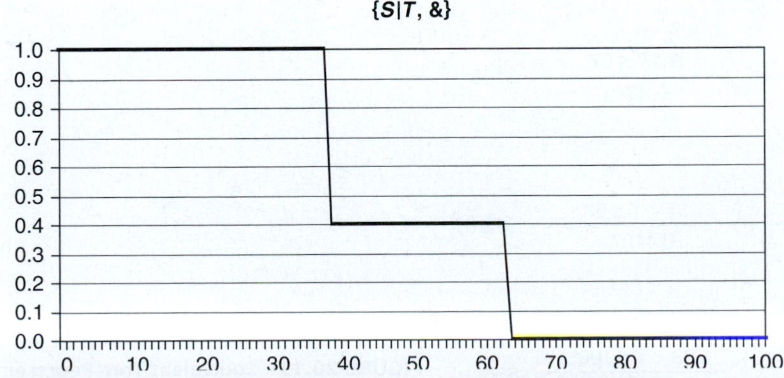

FIGURE 20.11 Posterior Probability for Sun Given Indication *T*

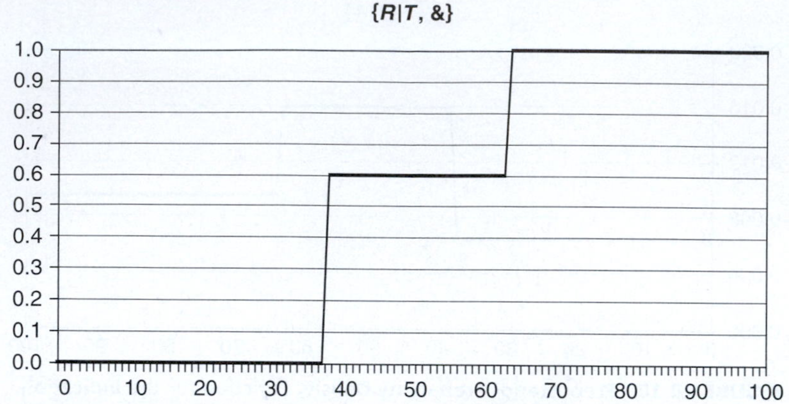

FIGURE 20.12 Posterior Probability for Rain Given Indication *T*

The area under the curve of Figure 20.10 between 0 and 37 is 0.23492, which is the probability the indication will lie in that range. For indications in this region, Kim's posterior probability for Sun will be 1. She therefore will have the party Outdoors, and her CE for the party will be \$100.

There is a 0.4127 probability that the detector will produce an indication between 37 and 63, in which case her posterior probability for Sun will be 0.4, which is the same as her prior probability. The detector will have provided no material information, and she will have the party Indoors with a CE of \$45.83.

Finally, there is a 0.3524 probability the detector will be between 63 and 100, in which case her posterior probability for Sun is 0. We know Kim will have the party Indoors with a CE \$50.

Therefore, Kim's deal with a free continuous detector can be represented by the probability tree in Figure 20.13.

The certain equivalent for this free detector deal is \$57.06. Her value of the detector is \$11.20, which is the same value we obtained with the 100° detector.

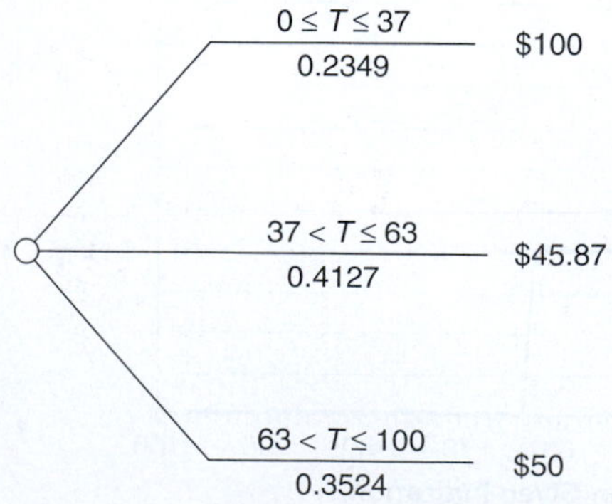

FIGURE 20.13 Equivalent Tree Representation for Continuous Detector

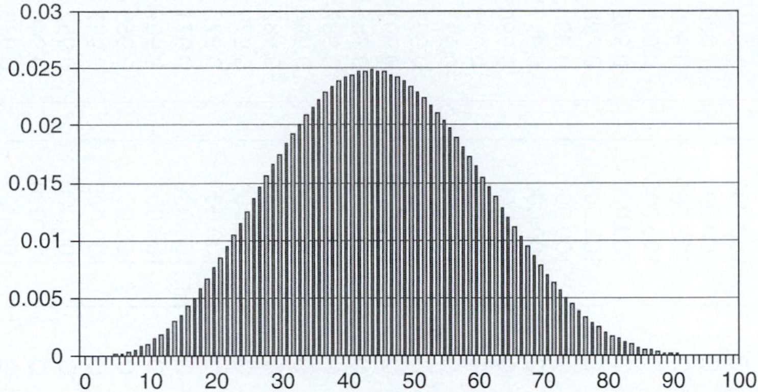

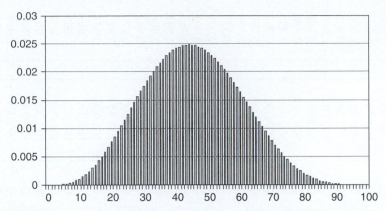

FIGURE 20.14 Detector Indication Given Sun with Specified Probabilities

20.2.3 Discrete Non-Uniform Detector

The detector indications for the weather do not need to be uniform. Suppose the detector salesman now presents Kim with another detector that also provides 100° but whose indications given Sun and Rain have probabilities represented in Figures 21.14 and 21.15, respectively. Note that the detector indications are again relevant to the weather—a necessary condition from our earlier discussion—but they no longer have equal probabilities for ranges of indications.

The analysis of a detector with multiple degrees and non-uniform probabilities is, in principle, similar to that of the uniform detector. Columns (b) and (c) of Table 20.3 show the exact values of probabilities of indications of this detector given the weather.

To determine the value of this detector, we again calculate the posterior probabilities of the weather given an indication. Column (d) shows the joint probability of an indication and Sun obtained by multiplying the cells in Column (b) by 0.4, where $\{S, T|\&\} = \{S|\&\}\{T|S, \&\} = 0.4\{T|S, \&\}$.

Column (e) shows the joint probability of an indication and Rain obtained by multiplying the cells in Column (c) by 0.6, where $\{R, T|\&\} = \{R|\&\}\{T|R, \&\} = 0.6\{T|R, \&\}$.

Column (f) shows the preposterior probability of an indication obtained by adding the cells in the same row of Columns (d) and (e); $\{T|\&\} = \{R, T|\&\} + \{S, T|\&\}$. Columns (g) and (h) show the posterior probabilities of Sun and Rain, respectively, that are obtained using the

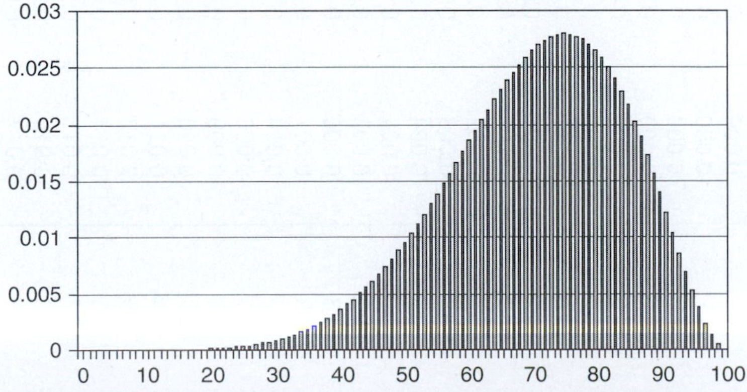

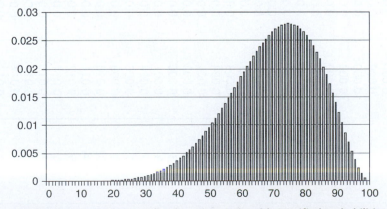

FIGURE 20.15 Detector Indication Given Rain with Specified Probabilities

TABLE 20.3 Analysis of 100° Non-Uniform Detector

(a)	(b)	(c)	(d)	(e)	(f)	(g)	(h)	(i)	(j)	(k)
Indication, T	$\{T\|S,\&\}$	$\{T\|R,\&\}$	$\{S, T\|\&\}$	$\{R, T\|\&\}$	$\{T\|\&\}$	$\{S\|T,\&\}$	$\{R\|T,\&\}$	Decision	CE of Party	u-Value of Party
1	0.000	0.000	0.000	0.000	0.000	1.000	0.000	O	100.000	1.000
2	0.000	0.000	0.000	0.000	0.000	1.000	0.000	O	100.000	1.000
3	0.000	0.000	0.000	0.000	0.000	1.000	0.000	O	100.000	1.000
4	0.000	0.000	0.000	0.000	0.000	1.000	0.000	O	100.000	1.000
5	0.000	0.000	0.000	0.000	0.000	1.000	0.000	O	100.000	1.000
6	0.000	0.000	0.000	0.000	0.000	0.999	0.001	O	100.000	0.999
7	0.000	0.000	0.000	0.000	0.000	0.999	0.001	O	100.000	0.999
8	0.001	0.000	0.000	0.000	0.000	0.999	0.001	O	100.000	0.999
9	0.001	0.000	0.000	0.000	0.000	0.998	0.002	O	100.000	0.998
10	0.001	0.000	0.000	0.000	0.000	0.998	0.002	O	100.000	0.998
11	0.001	0.000	0.001	0.000	0.001	0.997	0.003	O	100.000	0.997
12	0.002	0.000	0.001	0.000	0.001	0.996	0.004	O	100.000	0.996
13	0.002	0.000	0.001	0.000	0.001	0.995	0.005	O	100.000	0.995
14	0.003	0.000	0.001	0.000	0.001	0.994	0.006	O	100.000	0.994
15	0.004	0.000	0.001	0.000	0.001	0.992	0.008	O	100.000	0.992
16	0.004	0.000	0.002	0.000	0.002	0.991	0.009	O	100.000	0.991
17	0.005	0.000	0.002	0.000	0.002	0.989	0.011	O	100.000	0.989
18	0.006	0.000	0.002	0.000	0.002	0.987	0.013	O	100.000	0.987
19	0.007	0.000	0.003	0.000	0.003	0.984	0.016	O	100.000	0.984
20	0.008	0.000	0.003	0.000	0.003	0.982	0.018	O	100.000	0.982
21	0.008	0.000	0.003	0.000	0.003	0.979	0.021	O	100.000	0.979
22	0.009	0.000	0.004	0.000	0.004	0.975	0.025	O	100.000	0.975
23	0.010	0.000	0.004	0.000	0.004	0.971	0.029	O	100.000	0.971
24	0.011	0.000	0.005	0.000	0.005	0.967	0.033	O	100.000	0.967
25	0.012	0.000	0.005	0.000	0.005	0.963	0.037	O	100.000	0.963
26	0.014	0.000	0.005	0.000	0.006	0.957	0.043	O	100.000	0.957
27	0.015	0.000	0.006	0.000	0.006	0.952	0.048	O	100.000	0.952
28	0.016	0.001	0.006	0.000	0.007	0.946	0.054	O	100.000	0.946
29	0.017	0.001	0.007	0.000	0.007	0.939	0.061	O	100.000	0.939
30	0.017	0.001	0.007	0.001	0.008	0.932	0.068	O	100.000	0.932
31	0.018	0.001	0.007	0.001	0.008	0.924	0.076	O	100.000	0.924
32	0.019	0.001	0.008	0.001	0.008	0.915	0.085	O	46.000	0.915

33	0.020	0.001	0.008	0.001	0.009	0.905	0.095	O	46.000	0.905
34	0.021	0.002	0.008	0.001	0.009	0.895	0.105	O	46.000	0.895
35	0.022	0.002	0.009	0.001	0.010	0.884	0.116	O	46.000	0.884
36	0.022	0.002	0.009	0.001	0.010	0.873	0.127	O	46.000	0.873
37	0.023	0.002	0.009	0.001	0.011	0.860	0.140	P	46.000	0.862
38	0.023	0.003	0.009	0.002	0.011	0.847	0.153	P	46.000	0.854
39	0.024	0.003	0.010	0.002	0.011	0.832	0.168	P	46.000	0.845
40	0.024	0.004	0.010	0.002	0.012	0.817	0.183	P	46.000	0.835
41	0.024	0.004	0.010	0.002	0.012	0.801	0.199	P	46.000	0.825
42	0.025	0.005	0.010	0.003	0.013	0.784	0.216	P	46.000	0.815
43	0.025	0.005	0.010	0.003	0.013	0.766	0.234	P	46.000	0.803
44	0.025	0.006	0.010	0.003	0.013	0.747	0.253	P	46.000	0.792
45	0.025	0.006	0.010	0.004	0.014	0.728	0.272	P	46.000	0.779
46	0.025	0.007	0.010	0.004	0.014	0.707	0.293	P	46.000	0.767
47	0.024	0.007	0.010	0.004	0.014	0.686	0.314	P	46.000	0.753
48	0.024	0.008	0.010	0.005	0.015	0.664	0.336	P	46.000	0.739
49	0.024	0.009	0.010	0.005	0.015	0.641	0.359	P	46.000	0.725
50	0.023	0.010	0.009	0.006	0.015	0.618	0.382	P	46.000	0.710
51	0.023	0.010	0.009	0.006	0.015	0.594	0.406	P	46.000	0.696
52	0.022	0.011	0.009	0.007	0.016	0.570	0.430	P	46.000	0.680
53	0.022	0.012	0.009	0.007	0.016	0.545	0.455	P	46.000	0.665
54	0.021	0.013	0.008	0.008	0.016	0.520	0.480	P	46.000	0.649
55	0.020	0.014	0.008	0.008	0.016	0.495	0.505	P	46.000	0.633
56	0.020	0.015	0.008	0.009	0.017	0.470	0.530	I	46.000	0.621
57	0.019	0.016	0.008	0.009	0.017	0.444	0.556	I	46.000	0.623
58	0.018	0.017	0.007	0.010	0.017	0.419	0.581	I	46.000	0.625
59	0.017	0.018	0.007	0.011	0.017	0.395	0.605	I	46.000	0.628
60	0.016	0.019	0.007	0.011	0.018	0.370	0.630	I	46.000	0.630
61	0.015	0.019	0.006	0.012	0.018	0.346	0.654	I	46.000	0.633
62	0.015	0.020	0.006	0.012	0.018	0.323	0.677	I	46.000	0.635
63	0.014	0.021	0.005	0.013	0.018	0.300	0.700	I	46.000	0.637
64	0.013	0.022	0.005	0.013	0.018	0.278	0.722	I	46.000	0.639
65	0.012	0.023	0.005	0.014	0.019	0.257	0.743	I	46.000	0.642
66	0.011	0.024	0.004	0.014	0.019	0.236	0.764	I	46.000	0.644
67	0.010	0.025	0.004	0.015	0.019	0.217	0.783	I	46.000	0.646
68	0.009	0.025	0.004	0.015	0.019	0.198	0.802	I	46.000	0.647
69	0.009	0.026	0.003	0.016	0.019	0.181	0.819	I	46.000	0.649

(Continued)

TABLE 20.3 (Continued)

(a)	(b)	(c)	(d)	(e)	(f)	(g)	(h)	(i)	(j)	(k)
Indication, T	$\{T\|S,\&\}$	$\{T\|R,\&\}$	$\{S, T\|\&\}$	$\{R, T\|\&\}$	$\{T\|\&\}$	$\{S\|T,\&\}$	$\{R\|T,\&\}$	Decision	CE of Party	u-Value of Party
70	0.008	0.026	0.003	0.016	0.019	0.164	0.836	I	49.546	0.651
71	0.007	0.027	0.003	0.016	0.019	0.148	0.852	I	49.546	0.652
72	0.006	0.027	0.003	0.016	0.019	0.134	0.866	I	49.546	0.654
73	0.006	0.028	0.002	0.017	0.019	0.120	0.880	I	49.546	0.655
74	0.005	0.028	0.002	0.017	0.019	0.107	0.893	I	49.546	0.656
75	0.004	0.028	0.002	0.017	0.019	0.095	0.905	I	49.546	0.658
76	0.004	0.028	0.002	0.017	0.018	0.084	0.916	I	49.546	0.659
77	0.003	0.028	0.001	0.017	0.018	0.074	0.926	I	49.546	0.660
78	0.003	0.027	0.001	0.016	0.018	0.065	0.935	I	49.546	0.661
79	0.002	0.027	0.001	0.016	0.017	0.057	0.943	I	49.546	0.661
80	0.002	0.027	0.001	0.016	0.017	0.049	0.951	I	49.546	0.662
81	0.002	0.026	0.001	0.015	0.016	0.043	0.957	I	49.546	0.663
82	0.001	0.025	0.001	0.015	0.016	0.036	0.964	I	49.611	0.663
83	0.001	0.024	0.000	0.014	0.015	0.031	0.969	I	49.670	0.664
84	0.001	0.023	0.000	0.014	0.014	0.026	0.974	I	49.722	0.664
85	0.001	0.022	0.000	0.013	0.013	0.022	0.978	I	49.768	0.665
86	0.001	0.020	0.000	0.012	0.012	0.018	0.982	I	49.808	0.665
87	0.000	0.019	0.000	0.011	0.011	0.015	0.985	I	49.844	0.666
88	0.000	0.017	0.000	0.010	0.011	0.012	0.988	I	49.874	0.666
89	0.000	0.016	0.000	0.009	0.009	0.009	0.991	I	49.900	0.666
90	0.000	0.014	0.000	0.008	0.008	0.007	0.993	I	49.923	0.666
91	0.000	0.012	0.000	0.007	0.007	0.005	0.995	I	49.941	0.666
92	0.000	0.010	0.000	0.006	0.006	0.004	0.996	I	49.957	0.667
93	0.000	0.009	0.000	0.005	0.005	0.003	0.997	I	49.969	0.667
94	0.000	0.007	0.000	0.004	0.004	0.002	0.998	I	49.979	0.667
95	0.000	0.005	0.000	0.003	0.003	0.001	0.999	I	49.986	0.667
96	0.000	0.004	0.000	0.002	0.002	0.001	0.999	I	49.992	0.667
97	0.000	0.002	0.000	0.001	0.001	0.000	1.000	I	49.996	0.667
98	0.000	0.001	0.000	0.001	0.001	0.000	1.000	I	49.998	0.667
99	0.000	0.001	0.000	0.000	0.000	0.000	1.000	I	49.999	0.667
100	0.000	0.000	0.000	0.000	0.000	0.000	1.000	I	50.000	0.667

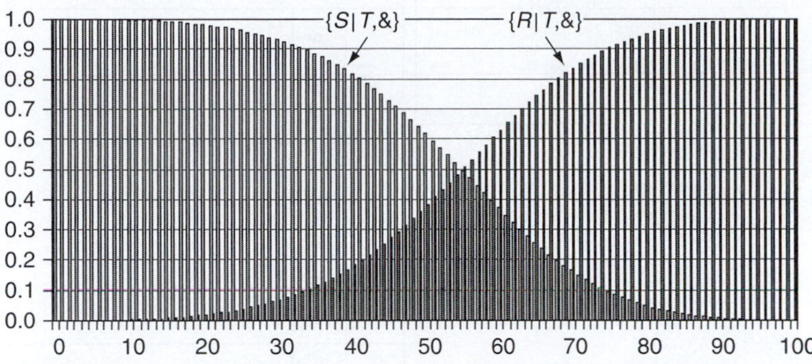

FIGURE 20.16 Posterior Probabilities of Weather Given Detector Indication

relations $\{S|T,\&\} = \dfrac{\{S,T|\&\}}{\{T|\&\}}, \{R|T,\&\} = \dfrac{\{R,T|\&\}}{\{T|\&\}}$. Figure 20.16 plots these posterior probabilities for any given indication. Note that $\{S|T,\&\} + \{R|T,\&\} = 1$.

For a given indication, *T,* Kim must decide where to have her party. Once again, we can use the results of the sensitivity analysis chapter to determine the optimal party decision for a given posterior probability of weather. Column (i) shows the optimal party location for a given indication. Note that Kim's choices include all three party locations, depending on the indication, in contrast to the case of the discrete uniform detector where she only went Outdoors or Indoors. Columns (j) and (k) show the certain equivalents and the *u*-value of the party location decision for a given indication.

By multiplying the corresponding cells in Columns (f) and (k) and summing, we get the *e*-value of the *u*-value of the party with the free detector. This is equal to the *e*-value of the *u*-values of the party with the free detector = 0.7127.

Therefore, the certain equivalent of the party with a free detector is \$55.15

Since the value of the party with no detector is \$45.83, and since Kim is a deltaperson, the value of this detector is \$9.33. (*Note:* The value of this detector is higher than that of the discrete uniform detector discussed.)

20.2.4 Valuing the Indication ≤ *T* for the Discrete Non-Uniform Dectector

The salesman also offers Kim the possibility of receiving an indication of the form (*T* is less than or equal to a certain number) for this detector. How much should Kim pay for this indication? Once again, we consider the possible indications in this situation. The detector can say either $T <= T_0$ or $T > T_0$.

Using analysis similar to that conducted for the uniform detector, we first calculate the posterior probability of an indication $\{S|T \le T_0,\&\} = \dfrac{\{T \le T_0|S,\&\}\{S|\&\}}{\{T \le T_0|\&\}}$ and $\{R|T \le T_0,\&\} = 1 - \{S|T \le T_0,\&\}$. The term $\{T \le T_0|S,\&\}$ is the cumulative probability of the detector indication given Sun. This is shown in Column (e) of Table 20.4 and is obtained by summing the cells of Column (b) up to a given indication. Column (f) shows the joint probability of an indication $\{T \le T_0, S|\&\} = \{T \le T_0|S,\&\}\{S|\&\}$, obtained by multiplying the cells in Column (e) by 0.4. Column (g) shows the cumulative preposterior distribution of an indication $\{T \le T_0|\&\}$ that is obtained by summing the corresponding cells in Column (d) for $\{T|\&\}$.

TABLE 20.4 Analysis for Indication T Less Than or Equal to a Number

(a)	(b)	(c)	(d)	(e)	(f)	(g)	(h)	(i)	(j)	(k)
Indication, T	$\{T \mid S, \&\}$	$\{T \mid R, \&\}$	$\{T \mid \&\}$	$\{T <= x \mid S, \&\}$	$\{T <= x, S \mid \&\}$	$\{T <= x \mid \&\}$	$\{S \mid T <= x, \&\}$	$\{R \mid T <= x, \&\}$	Location	u-Value
1	0.000	0.000	0.000	0.000	0.000	0.000	1.000	0.000	O	1.000
2	0.000	0.000	0.000	0.000	0.000	0.000	1.000	0.000	O	1.000
3	0.000	0.000	0.000	0.000	0.000	0.000	1.000	0.000	O	1.000
...	...	...	...	...	...	...	...	...	...	...
37	0.023	0.002	0.011	0.320	0.128	0.138	0.929	0.071	O	0.929
38	0.023	0.003	0.011	0.343	0.137	0.149	0.923	0.077	O	0.923
39	0.024	0.003	0.011	0.367	0.147	0.160	0.917	0.083	O	0.917
40	0.024	0.004	0.012	0.391	0.156	0.172	0.910	0.090	O	0.910
...	...	...	...	...	...	...	...	...	...	...
57	0.019	0.016	0.017	0.782	0.313	0.423	0.740	0.260	P	0.787
58	0.018	0.017	0.017	0.800	0.017	0.440	0.039	0.961	P	0.779
59	0.017	0.018	0.017	0.817	0.017	0.457	0.038	0.962	P	0.771
...	...	...	...	...	...	...	...	...	...	...
99	0.000	0.016	0.010	1.000	0.400	0.990	0.404	0.596	I	0.627
100	0.000	0.016	0.010	1.000	0.400	1.000	0.400	0.600	I	0.627

(l)	(m)	(n)	(o)	(p)	(q)	(r)
$\{T > x \mid S, \&\}$	$\{T > x, S \mid \&\}$	$\{T > x \mid \&\}$	$\{S \mid T > x, \&\}$	$\{R \mid T > x, \&\}$	Location	u-Value
1.000	0.400	1.000	0.400	0.600	I	0.627
1.000	0.400	1.000	0.400	0.600	I	0.627
1.000	0.400	1.000	0.400	0.600	I	0.627
...	...	...	...	...	...	...
0.680	0.272	0.862	0.316	0.684	I	0.636
0.657	0.263	0.851	0.309	0.691	I	0.636
0.633	0.253	0.840	0.302	0.698	I	0.637
0.609	0.244	0.828	0.294	0.706	I	0.638
...	...	...	...	...	...	...
0.218	0.087	0.577	0.151	0.849	I	0.652
0.200	0.080	0.560	0.143	0.857	I	0.653
0.183	0.073	0.543	0.135	0.865	I	0.654
...	...	...	...	...	...	...
0.000	0.000	0.010	0.000	1.000	I	0.667
0.000	0.000	0.000	0.000	1.000	I	0.667

Column (h) shows the posterior distribution $\{S|T \leq T_0, \&\}$ obtained by dividing the cells in Columns (f) by those in Column (g). Column (i) shows the posterior probability of Rain for a given indication, where $\{R|T \leq T_0, \&\} = 1 - \{S|T \leq T_0, \&\}$. Column (j) shows the optimal party location for a given indication, and Column (k) shows the u-value of the optimal location decision.

To calculate the posterior probability of an indication, $\{S|T > T_0, \&\}$, we use the formula $\{S|T > T_0, \&\} = \dfrac{\{T > T_0|S, \&\}\{S|\&\}}{\{T > T_0|\&\}}$. Column (l) shows the probability of an indication $\{T > T_0|S, \&\}$, which is the excess distribution of $\{T|S, \&\}$ and is obtained from $\{T > T_0|S, \&\} = 1 - \{T \leq T_0|S, \&\}$, and so was just 1 minus the cells in Column (e). Column (m) shows the probability of a joint indication $\{T > T_0,\ S|\&\}$, that is obtained by multiplying the cells of Column (l) by 0.4. Column (n) shows the probability $\{T > T_0|\&\} = 1 - \{T \leq T_0|\&\}$, obtained by 1 minus the cells in Column (g). Columns (o) and (p) calculate the posterior probability of weather given an indication $T > T_0$, and Columns (j) and (k) calculate the optimal party location and u-values of the location decision.

Note that the report that the indication is less than or equal to any particular value can lead to choosing any of the three party locations; however, the report that the indication is greater than or equal to any particular value always leads to an indoor party with various u-values.

From the results of Table 20.4, we can now value any indication T less than or equal to a certain indication. For example, consider the indication $T \leq 58$. We have $\{T \leq 58|\&\} = 0.44$ and $\{S|T \leq 58|\&\} = 0.727$. Therefore, Kim will have her party on the Porch, and the u-value of this party is equal to 0.779. Kim may also get an indication $T > 58$, with probability $\{T > 58|\&\} = 0.56$ and so $\{S|T > 58|\&\} = 0.143$. In this case, Kim will have her party Indoors with a u-value 0.653.

By calculating the e-value of u-values, we get the u-value of the optimal party location decision for an indication $T \leq 58$, as $0.44 * 0.779 + 0.56 * 0.653 = 0.7083$.

The certain equivalent of the party deal with this free detector indication = \$54.66.

The value of this free indication is, therefore, = \$54.66 − \$45.87 = \$8.79, which is close to the value of the original Acme Detector. This no surprise, since the indication of below or above 58 has the same sensitivity and specificity as the Acme Detector.

20.3 THE CONTINUOUS BETA DETECTOR

We can also consider a continuous form of the previous detector. The **Beta probability density function** is one very useful functional form. The formula for the Beta probability density function on the domain [0, 1] is

$$f(x, r, k) = cx^{r-1}(1 - x)^{k-1}$$

where r and k are parameters and c is a normalizing constant (to make the area under the curve integrate to one).[1] The Beta density has many possible shapes, depending on the choice of its parameters. For example, Figure 20.17 shows some examples of the Beta probability density function.

Beta densities can also be defined on a scale [a, b] using the equation

$$f(x, r, k, a, b) = c(x - a)^{r-1}(b - x)^{k-1}$$

[1]In some cases, the beta density is also written as $f(x, r, n) = cx^{r-1}(1 - x)^{n-r-1}$, where $n = r + k$.

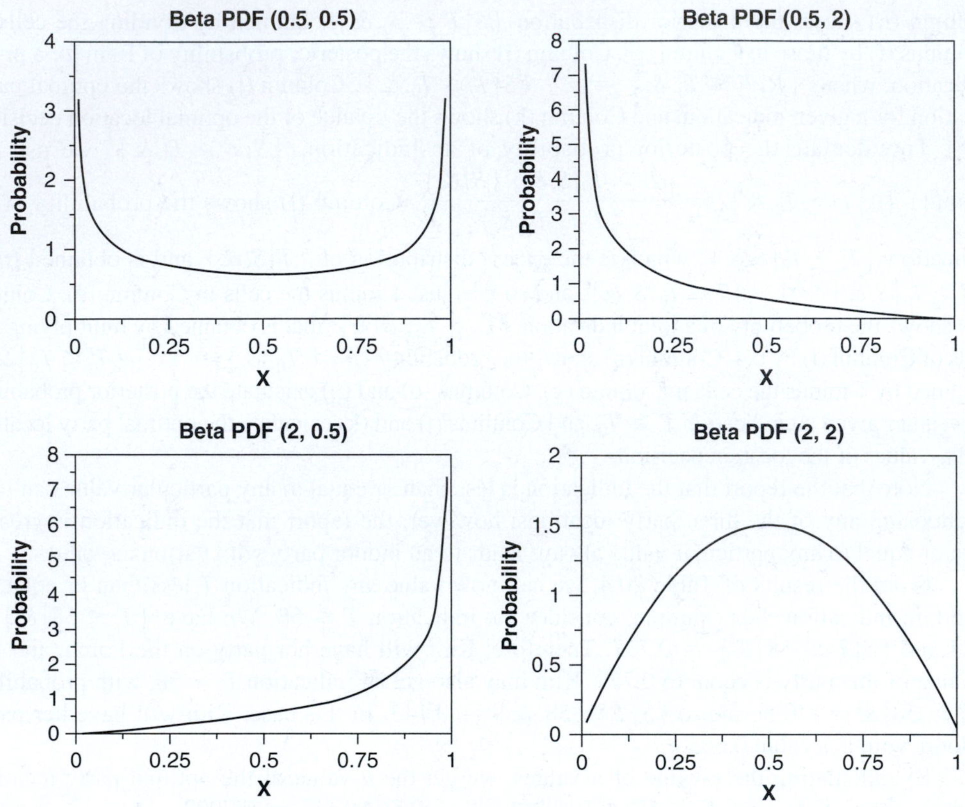

FIGURE 20.17 Examples of Beta Probability Density Functions

Now, suppose we have a continuous detector with probability density function for its indication given Sun by $\{T|S,\&\} = \text{Beta}(4.4,5.6,0,100)$ and an indication given Rain by $\{T|R,\&\} = \text{Beta}(7,3,0,100)$. This density function is represented by the top two curves of Figure 20.18. To determine the value of this detector, we first calculate

$$\{T,S|\&\} = \{S|\&\} * \{T|S,\&\} = 0.4 * \{T|S,\&\}$$

$$\{T,R|\&\} = \{R|\&\} * \{T|R,\&\} = 0.6 * \{T|R,\&\}$$

The preposterior probability density function of the detector indication is then $\{T|\&\} = \{T,S|\&\} + \{T,R|\&\}$

The posterior distributions for Sun and Rain given a detector indication are shown in Figure 20.19 and are obtained by

$$\{S|T,\&\} = \frac{\{T,S|\&\}}{\{T|\&\}}, \{R|T,\&\} = \frac{\{T,R|\&\}}{\{T|\&\}}$$

Note the posterior distributions have the whole range of values from 0 to 1. Kim will now go with Outdoors, Porch, or Indoors, depending on the detector indication. (Compare to the uniform detector, whose indications led to posterior distributions of Figures 20.11 and 20.12

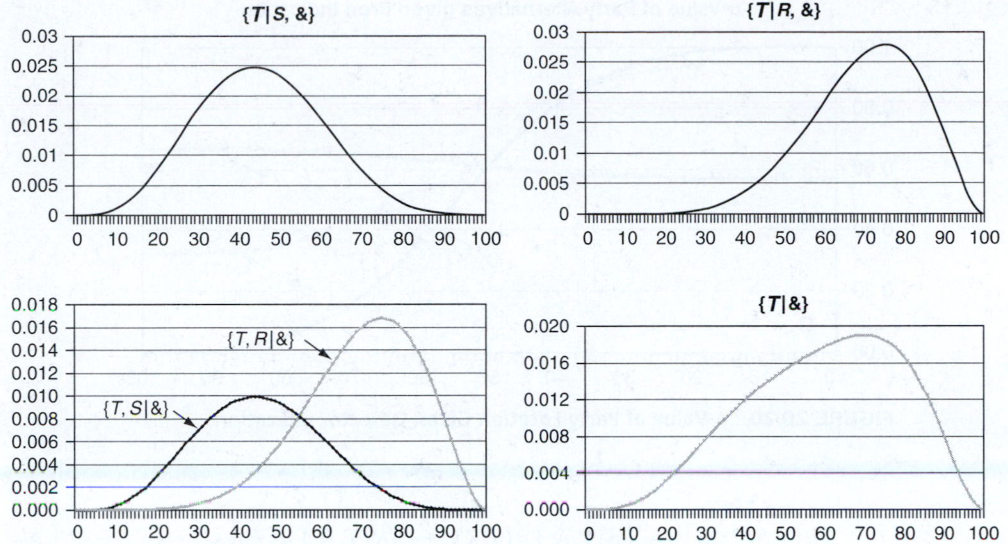

FIGURE 20.18 Likelihood, Joint, and Preposterior Probabilities

with values 0, 0.4, 0.6, and 1, which led her to go either Indoors or Outdoors.) The generality of the Beta function allows for a continuous range of posterior distributions.

At each indication, Kim should choose the alternative with the highest u-value. For every value of $\{S|T,\&\}$ and $\{R|T,\&\}$ we calculate the e-value of the u-values for each party location.

$$\begin{aligned} u(O,S|T,\&) &= \{S|T,\&\}u(O,S) + \{R|T,\&\}u(O,R) \\ &= \{S|T,\&\}.1 + \{R|T,\&\}.0 \\ &= \{S|T,\&\} \end{aligned}$$

$$\begin{aligned} u(P,S|T,\&) &= \{S|T,\&\}u(P,S) + \{R|T,\&\}u(P,R) \\ &= \{S|T,\&\}.(0.95) + \{R|T,\&\}.(0.32) \end{aligned}$$

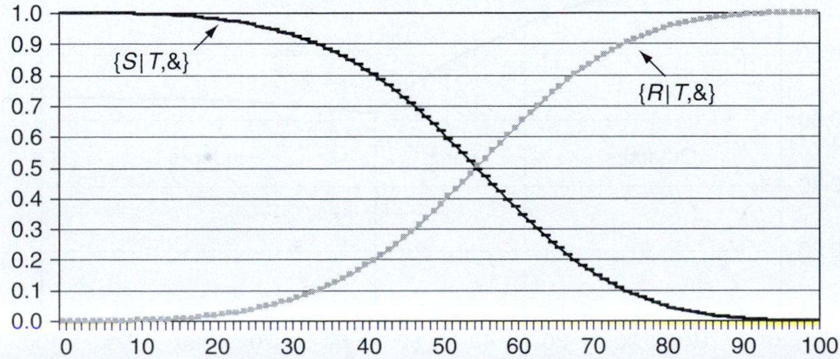

FIGURE 20.19 Posterior Distributions for Sun and Rain Given Detector Indication

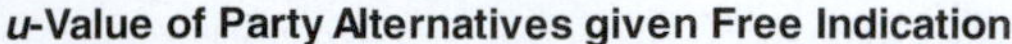

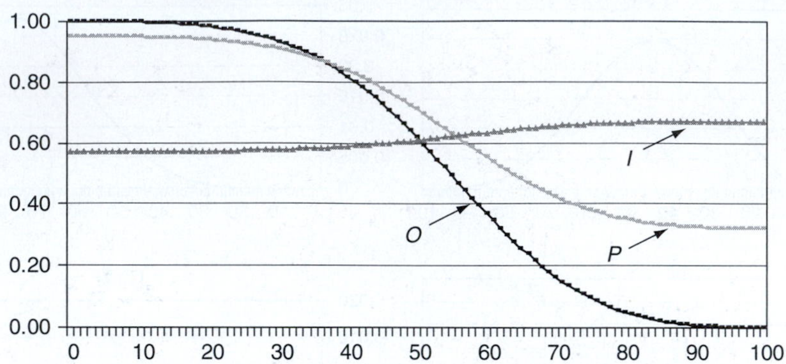

FIGURE 20.20 ***u*-Value of Party Location Given Detector Indication**

$$u(I,S|T,\&) = \{S|T,\&\}u(I,S) + \{R|T,\&\}u(I,R)$$
$$= \{S|T,\&\}.(0.57) + \{R|T,\&\}.(0.67)$$

In fact, these equations are just those of Kim's sensitivity in u-space to the probability of sunshine first shown in Figure 12.2. When the posterior probabilities on sunshine given each indication of Figure 20.19 are incorporated, we create the three curves in Figure 20.20, showing how the u-Value for each alternative changes with the detector indication.

The maximum u-value of the party decision for any detector indication is the envelope of the curves in Figure 20.20, and is shown in Figure 20.21.

As in the discrete case, the u-value of any detector indication is the summation of the product of the probability of that detector indication $\{T|\&\}$ and the maximum u-value given that indication. In other words, we multiply the preposterior distribution $\{T|\&\}$ from Figure 20.18 and a maximum u-Value from Figure 20.21 to obtain Figure 20.22, and then we find the area under this product curve.

Max u-Value of Party Alternatives with Free Indication

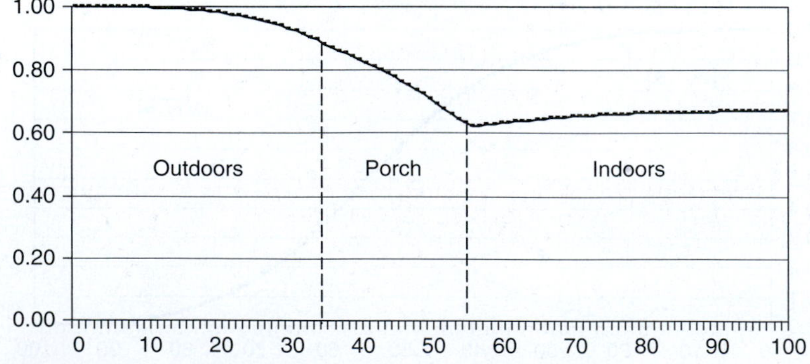

FIGURE 20.21 ***u*-Value of Best Party vs. Detector Indication**

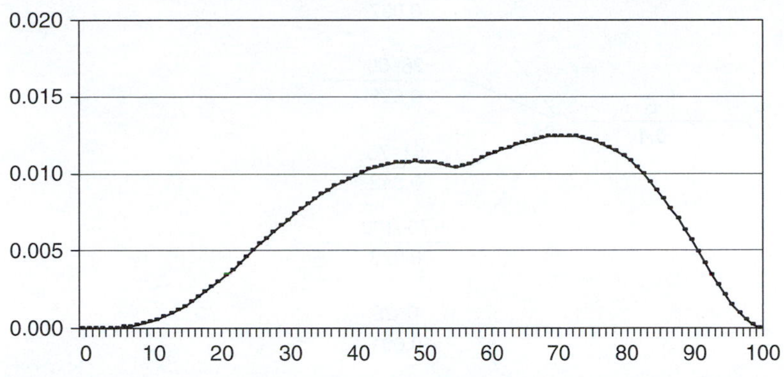

FIGURE 20.22 Product of $\{T|\&\}$ and Figure 20.21

The area under this curve is the e-value of the u-values of the deal with free detector indication and is about 0.714, corresponding to Kim's certain equivalent of approximately \$55.31, and a detector value of \$9.44.

20.3.1 Discretizing the Continuous Beta Detector Indication

Another (approximate) way of calculating the detector value is to discretize its indication into intervals. For example, we may use equal intervals. The clairvoyant can predict into which interval the indication will fall, and the probabilities can be calculated using the continuous distributions. For example, Table 20.5 shows four intervals and the corresponding probability of an indication falling within each interval for Sun and Rain.

The probabilities in the cells of Table 20.5 are calculated from the continuous distributions as:

$$\begin{aligned}\{0 \le T \le 25|S,\&\} &= \{T \le 25|S,\&\} - \{T \le 0|S,\&\} \\ &= \text{Beta}(25, 4.4, 5.6, 0, 100) - \text{Beta}(0, 4.4, 5.6, 0, 100) \\ &= 0.097 - 0 = 0.097\end{aligned}$$

Similarly,

$$\begin{aligned}\{25 < T \le 50|S,\&\} &= \{T \le 50|S,\&\} - \{T \le 25|S,\&\} \\ &= \text{Beta}(50, 4.4, 5.6, 0, 100) - \text{Beta}(25, 4.4, 5.6, 0, 100) \\ &= .6348 - 0.097 = 0.538\end{aligned}$$

TABLE 20.5 Analysis of Discretized Detector

Interval	$\{T \text{ in Interval}\|S,\&\}$	$\{T \text{ in Interval}\|R,\&\}$
0 − 25	0.097	0.001
>25 − 50	0.538	0.091
>50 − 75	0.342	0.514
>75 − 100	0.023	0.394

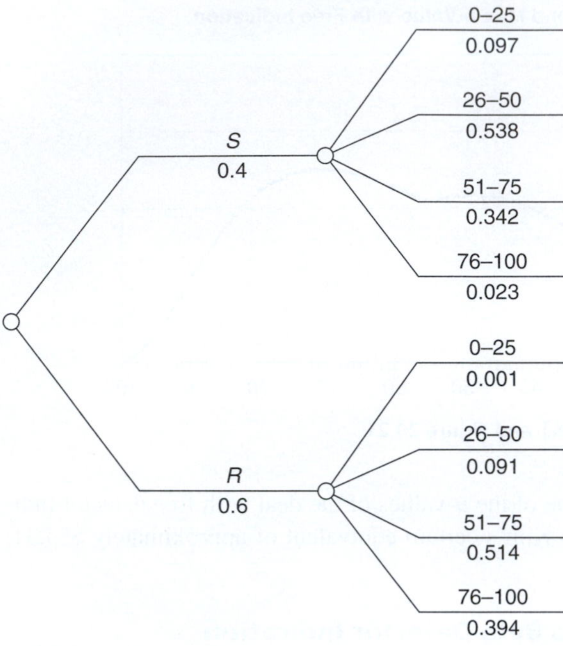

FIGURE 20.23 Assessed Probability Tree for Discretized Detector Analysis

The remaining cells in the table are obtained in the same way from the cumulative distribution. Figure 20.23 shows the probability tree for this discretized detector analysis.

Calculating the value of the detector is now straightforward using a simple tree analysis and is also shown in Table 20.6. From the table, the *e*-value of *u*-value of party with free detector is 0.707.

This corresponds to a value of the party with free detector is $54.50.

TABLE 20.6 Using Simple Tree Analysis to Calculate Detector Value

Interval	{*T* in Interval|*S*, &}	{*T* in Interval|*R*, &}
0 – 25	0.097	0.001
26 – 50	0.538	0.091
51 – 75	0.342	0.514
76 – 100	0.023	0.394

{*T*, *S*|&}	{*T*, *R*|&}	{*T*|&}	{*S*|*T*, &}	{*R*|*T*, &}	Location	*u*-Value of Alternative
0.039	0.001	0.040	0.978	0.022	O	0.039
0.215	0.054	0.270	0.798	0.202	P	0.222
0.137	0.308	0.445	0.307	0.693	I	0.283
0.009	0.236	0.246	0.037	0.963	I	0.163
		1.000			EU Free Detector	0.707

Since Kim is a deltaperson, her value of the detector is equal to the value of the party with a free detector less the value with no detector.

$$\text{Value of detector} = \$54.50 - \$45.87 = \$8.63$$

This value can be compared to the one obtained using a continuous detector analysis of $9.44. Therefore, discretizing the detector indication provides less accuracy, but at the same time, it has the benefits of simplicity, as we have seen.

20.4 SUMMARY

This chapter illustrates the level of complexity that may arise when detectors provide multiple indications. The basic steps and analysis for calculating the value of information and for updating belief are similar, but we now have the added complication of dealing with multiple degrees.

We have shown how to conduct the analysis for detectors with multiple discrete indications in table format. We have also shown several methods to compute the value of a continuous detector. One method uses direct integrals and calculates areas under the curve. A second method discretizes the detector indication into many degrees (such as the 100º indication). A third method discretizes the detector into an even smaller number of degrees, gaining simplicity, but at the same time, losing some accuracy. The art is to choose the appropriate framework for the problem at hand. In a future chapter, we will also discuss methods to solve problems with continuous detectors using simulation.

An important observation on what we have done in this chapter is that detectors with multiple indications, either discrete or continuous, require no further concepts beyond those we used in understanding basic information gathering through our analysis of the two-indication Acme Detector in Chapter 13. Another important concept is the fact that we can discretize detectors with a large number of degrees into ones with fewer degrees. Finally, we introduced beta distributions that will be useful in modeling uncertainty for a variety of decisions.

KEY TERMS

- Detector with multiple indications
- Continuous detector
- Beta distribution

PROBLEMS

1. Suppose the 100° uniform detector accuracy is given by

$$\{T|S,\&\} = \begin{cases} \frac{1}{80}, 1 \le T \le 80 \\ 0, \text{otherwise} \end{cases} \quad \text{and} \quad \{T|R,\&\} = \begin{cases} \frac{1}{80}, 21 \le T \le 100 \\ 0, \text{otherwise} \end{cases}$$

This implies more overlap than the example discussed in this chapter.

a. Do you think it will have a higher or lower detector value to Kim?

b. Using similar analysis as that presented in this chapter, calculate the value of this detector for Kim.

2. Suppose the 100° uniform detector accuracy is

$$\{T|S,\&\} = \frac{1}{100}, 1 \le T \le 100 \quad \text{and} \quad \{T|R,\&\} = \frac{1}{100}, 1 \le T \le 100$$

This implies 100% overlap in the detector indication with equal probabilities.

a. Do you think it will have a higher or lower value than the detector discussed in this chapter?

b. Without any further calculations, what is the value of this detector?

3. Suppose the 100° uniform detector accuracy is

$$\{T|S,\&\} = \begin{cases} \frac{1}{50}, 1 \le T \le 50 \\ 0, \text{otherwise} \end{cases} \quad \text{and} \quad \{T|R,\&\} = \begin{cases} \frac{1}{50}, 51 \le T \le 100 \\ 0, \text{otherwise} \end{cases}$$

This implies no overlap in the detector indications. Without any further calculations, how much is this detector worth to Kim?

4. On your own, repeat the analysis of the following tables presented in this chapter

(i) Table 20.1

(ii) Table 20.2

(iii) Table 20.3

(iv) Table 20.4.

CHAPTER

21

Decisions with Influences

CHAPTER CONCEPTS

After reading this chapter, you will be able to explain the following concepts:

- The role of influence arrows in decision diagrams
- The value of clairvoyance in decisions with influences
- The value of a detector in decisions with influences
- Canonical decision diagrams
- Economies of assessments in decisions with influences

21.1 INTRODUCTION

In many instances of decision analysis, the choice of an alternative will influence the actual distinction of interest. Depending upon the alternative chosen, the decision maker will assign different probabilities to the distinction. In decision diagrams, we indicate this influence with an arrow from a decision node to a chance node in Chapter 14. As we discussed earlier, we call this arrow an influence arrow.

Influences are not unusual. For example, a research and development executive might believe that the expenditure on research and development (R&D) will change the chance of technical success. In a decision diagram, we would represent this belief by drawing an influence arrow from the decision node representing R&D funding to the chance node representing technical success. As another example, a marketing manager would use an influence arrow to show that spending more money on a marketing campaign would increase the chance of marketing success. In both marketing and R&D departments, allocating a sufficient budget can influence the chances of both technical and marketing success. As we shall see, we must treat the use of influences in decision diagrams with care, particularly with value of information considerations.

21.2 SHIRLEY'S PROBLEM

To illustrate the use of influences, we again revisit the party problem introduced back in Chapter 9. In this chapter, we introduce Shirley, another party planner. She is similar to Kim in every respect, including risk preference, except that she believes that the weather is more likely to be better if

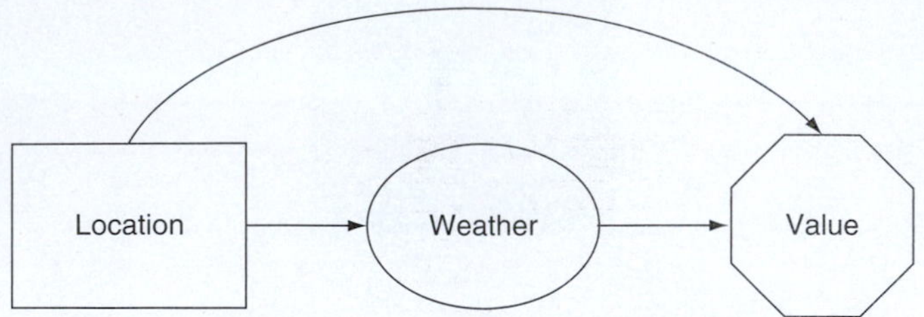

FIGURE 21.1 Shirley's Decision Diagram for the Party Problem

she exposes herself to nature.[1] For example, she believes that sunshine is more likely if she has an outdoor party than if she has an indoor party. Her belief is that if she opens herself to nature, nature will cooperate with her. This is contrary to the belief of many people, who think that washing their car makes it more likely to rain. Shirley's decision diagram appears in Figure 21.1.

This diagram looks just like Kim's, except in Shirley's diagram, there is an influence arrow from the Location node to the Weather node.

21.2.1 Best Alternative with Influences

The decision tree corresponding to Shirley's decision diagram appears in Figure 21.2. This looks just like Kim's decision tree, except for the probabilities that are assigned to the weather. Shirley assigns a 0.5 chance to Sunshine when she has the party Outdoors; 0.4 when she has the party on the Porch; and 0.3 when she has the party Indoors. Rolling back the decision tree in our usual way, we found that the Outdoor, Porch, and Indoor alternatives have *u*-values of 0.5, 0.57, and 0.64, respectively, and certain equivalents \$33.90, \$40.60, and \$46.85, respectively. Like Kim, Shirley will choose to have the party Indoors; however, her certain equivalent of \$46.85 is somewhat higher than Kim's certain equivalent of \$45.83.

Note that the existence of the influence and the corresponding difference in probabilities of Sunshine following each alternative did not affect our method for determining the best

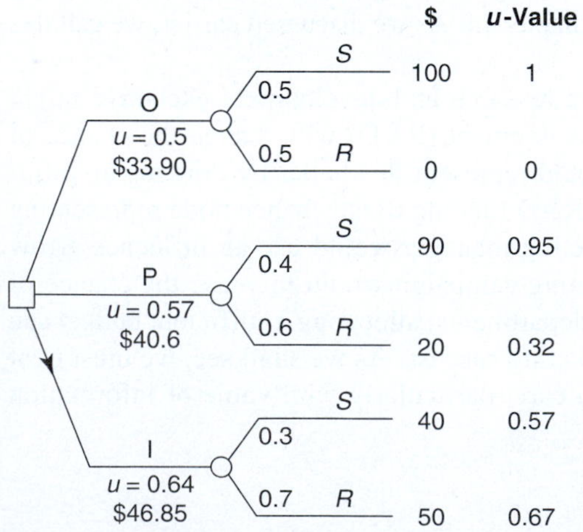

FIGURE 21.2 Shirley's Decision Tree for the Party Problem

[1]This problem was named Shirley's problem by a student who thought these beliefs were similar to those of a Hollywood star with this first name.

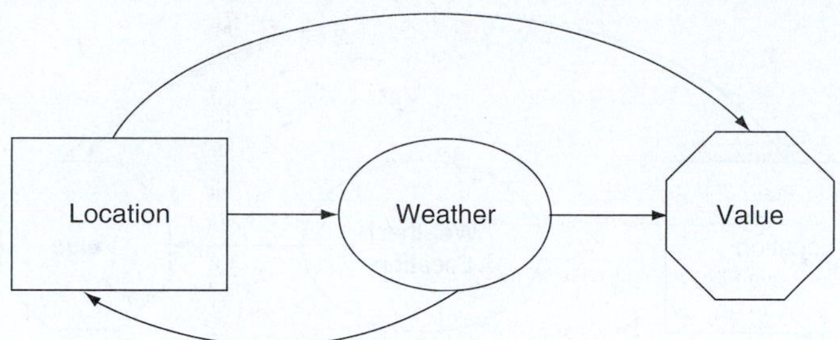

FIGURE 21.3 The Problem of Clairvoyance with Influence

alternative in any way other than the specific numerical calculations required. The existence of influences—whether in Shirley's party problem or in R&D funding decisions—creates no difficulty in determining the best course of action.

21.2.2 Value of Clairvoyance with Influences

The existence of influences, however, is not nearly so benign when we consider calculating the value of clairvoyance (VOC). For example, if we attempt to find the VOC on the weather in the usual way—by adding an information arrow from weather to location in Shirley's decision diagram of Figure 21.1—we obtain the diagram of Figure 21.3. This implies that we know the Weather before making the party Location decision.

The difficulty appears immediately when we see that there is a loop formed from Location to Weather and from Weather to Location. Recall that decision diagrams cannot have a cycle. It is clear that we will also never be able to place this diagram in decision tree form; therefore, we cannot carry out the VOC calculation.

To illustrate this point further, suppose we wish to calculate the value of information for Shirley using the normal method we demonstrated earlier, but using Kim's decision tree (see Figure 10.8). The question that immediately arises is "Which probability do we assign for her belief about the clairvoyant's telling her sunshine or rain?" Depending on the alternative she chooses, we now have three different probabilities.

The philosophical reason for this difficulty is that Shirley is requesting information from the clairvoyant that will depend partly on her own actions. To this extent, he is being asked to predict what she will do, and if he could, she would lose free will in making the decision as a result. By allowing influences, we have blurred the distinction between decisions (under our control) and uncertainties (not under our control). We pay the price for this imprecision when we seek to compute the VOC.

21.2.3 Placing Decision Diagrams in Canonical Form

We can avoid difficulties in computing the VOC by drawing our decision diagrams without influences. Then, there will be no possibility of creating a VOC computation that we cannot handle. We say that decision diagrams without influences are decision diagrams in **canonical form.**

Placing diagrams in canonical form requires some additional work on the part of the analyst and, should VOC questions arise, possibly more assessments on the part of the decision maker. The canonical form of Shirley's decision diagram it is shown in Figure 21.4.

In this diagram, the uncertainty is no longer simply the weather, but it is the weather conditioned on Shirley's choice of party location. In other words, the distinction Weather|Location is a probability assessment of the weather (Sun|Rain) for each possible party location (Outdoor|Porch|Indoor).

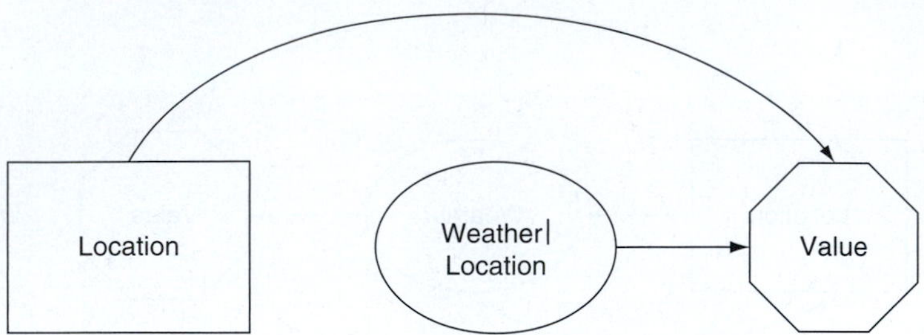

FIGURE 21.4 **General Canonical Form for Shirley's Problem**

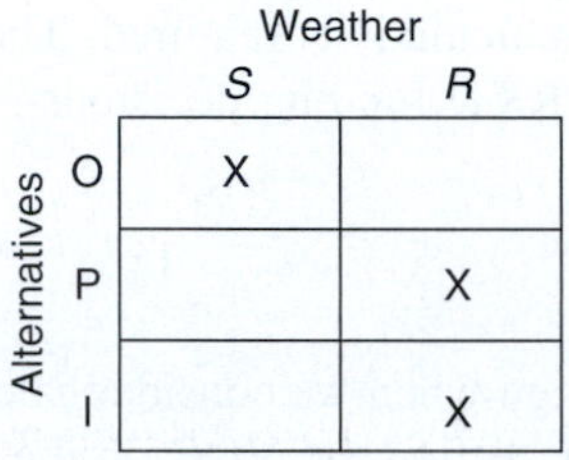

Alternatives	Weather: *S*	Weather: *R*
O	X	
P		X
I		X

FIGURE 21.5 **The Clairvoyant's Questionnaire**

21.2.4 The Clairvoyant's Questionnaire

To be specific, we imagine that we are going to prepare a questionnaire for the clairvoyant allowing him to record for each of Shirley's possible actions what the weather will be like. The questionnaire is shown in Figure 21.5. There are three rows, one for each of the alternatives—Outdoor (O), Porch (P), or Indoor (I)—that Shirley might follow and two columns for the two possibilities for weather—Sunshine, *S*, and Rain, *R*. Shirley's request to the clairvoyant is that he place one mark in each row showing what will happen to the weather if she makes a decision corresponding to that particular row.

As shown in Figure 21.5, the clairvoyant may report that, if Shirley has the party Outdoors, the weather will be Sunny, but if she has the party either on the Porch or Indoors, the weather will be Rainy. Note that there are eight possible responses that the clairvoyant could provide. In general, if there are n different alternatives and m different degrees of distinction for the uncertainty, there will be m^n possible responses from the clairvoyant.

SHIRLEY'S PROBABILITY ASSESSMENT ON THE CLAIRVOYANT'S QUESTIONNAIRE We can think of the Weather given Location node of Figure 21.4 as composed of the three chance nodes shown in Figure 21.6.

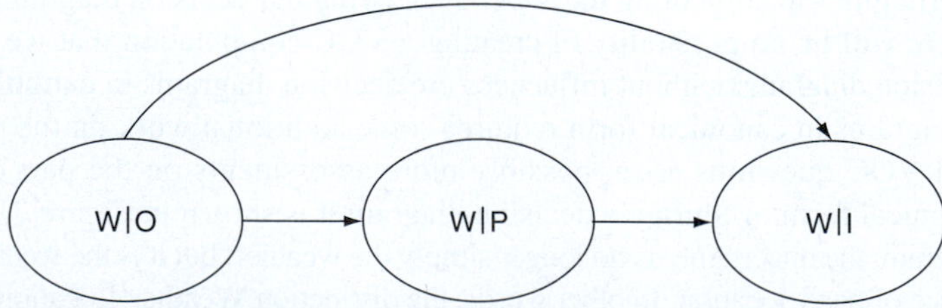

FIGURE 21.6 **Relevance Diagram for Clairvoyant's Questionnaire**

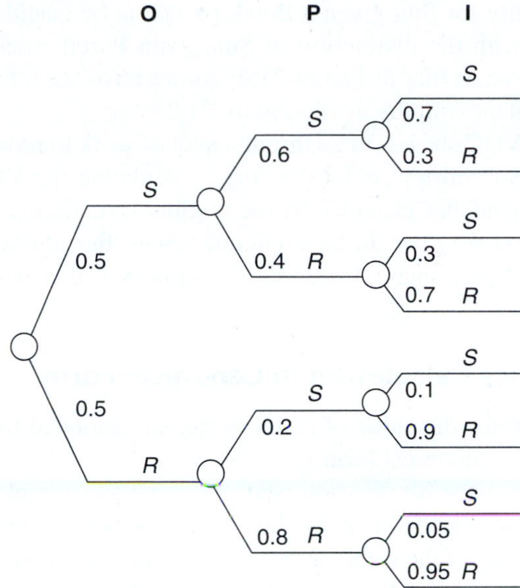

FIGURE 21.7 Shirley's Probability Tree

They correspond to Weather given Outdoor party (W|O), Weather given Porch party (W|P), and Weather given Indoor party (W|I). Each of these distinctions is a binary distinction for Sun or Rain. We have shown a particular assessment order in Figure 21.6; the order may be chosen to suit Shirley's convenience. The arrows allow the possibility that how the clairvoyant fills out one row of his questionnaire may tell Shirley something about how he will fill out other rows of the questionnaire.

Refer to Figure 21.7, and suppose that Shirley has provided us with this probability tree to represent her assessments for her relevance diagram. The three layers of the tree, as shown at the top, correspond to the three alternatives of Outdoor (O), Porch (P), and Indoor (I).

The first layer in the tree shows Shirley's belief that if she has a party Outdoors, then there a 50/50 chance the weather will be Sun, *S* or Rain, *R*. This corresponds to the tree in Figure 21.2 where the probability *S* given Outdoors is 0.5.

The second layer in the tree represents Shirley's belief about a Porch party given the Clairvoyant's indication for an Outdoors party. Shirley believes that there is a 60% chance the weather will be Sunshine on the Porch if the Clairvoyant says it will be Sunshine for an Outdoors party. She also believes there is a 20% chance it will be Sunshine on the Porch if the Clairvoyant says an Outdoor party will have Rain. Note that Kim's probability of Sunshine for a Porch party given Sunshine for an Outdoors party is 1, and the probability of Sunshine for a Porch party given an outdoors rainy party is 0.

The third layer of the tree shows, that if an Outdoor and Porch party would have taken place in the Sunshine there will be a 70% chance that an Indoor party would take place in the Sunshine *S*. However, if both the Outdoor and Porch parties would take place in the Rain *R*, there will be a 0.05 chance that an Indoor party would be in the Sun *S*. Finally, reports of an Outdoor party in the Sunshine *S* and a Porch party in the Rain *R* would lead to a 0.3 chance of Sunshine *S* for an Indoor party; an Outdoor Rainy party and Porch Sunny party would cause Shirley to assign a 0.1 chance to Sunshine.

It takes only a moment to check that the probability assignments of this tree are consistent with those of Figure 21.2. For example, the marginal probability of Sun given an Outdoor party is still 0.5, while the marginal probability for Sun given a Porch party can be calculated by flipping the tree in Figure 21.7 to start with the distinction of Sun given Porch equal to $0.5 * 0.6 + 0.5 * 0.2 = 0.4$, which is the same as that in Figure 21.2. As we have seen, these conditional assessments also should be consistent with the tree in Figure 21.2.

We see that if Shirley wishes to get the VOC she has to go through a lot of work to provide the necessary additional assessments. If instead Shirley decides to forgo calculating the VOC, she will have to be content with the decision guidance provided by the original formulation that allowed for an influence between Location and Weather. In most circumstances, the additional benefit of VOC comes at the cost of additional assessment. In some cases, however, this assessment may not be as onerous as it first appears.

21.2.5 Shirley's Value of Clairvoyance Calculation in Canonical Form

Figure 21.8 shows Shirley's decision diagram for the value of clairvoyance in canonical form. (***Note:*** This figure has no cycles because it is in canonical form.)

We have said that, once she makes the location decision, she will have the completed clairvoyant's questionnaire available to her. The probability that the clairvoyant's questionnaire will be filled out in each possible way is the probability of the eight end points of the probability tree in Figure 21.7. We compute Shirley's VOC by employing the decision tree shown in Figure 21.9.

The first three layers of the tree in Figure 21.9 are identical to the probability assessments of Figure 21.7 with each corresponding to the eight possible clairvoyance reports. Following each report, Shirley must make a decision about the party location.

For example, if the clairvoyant reports (as he does in the top trajectory) that for any party location the weather will be Sunny, we know that by following each of the three alternatives—Outdoor (O), Porch (P), and Indoor (I)—she will receive payoffs of \$100, \$90, and \$40. These three possible values are shown in parentheses adjacent to the decision node at the end point corresponding to this clairvoyant's report. It is clear that she will do best by having the party Outdoors and achieving the value of \$100. This dollar value is recorded in an adjacent column, as is the *u*-value of the amount, namely 1.

Proceeding downward to the second end point, we observe the report that an Outdoor and Porch party will be held in the Sunshine, but an Indoor party will be held in the Rain. This

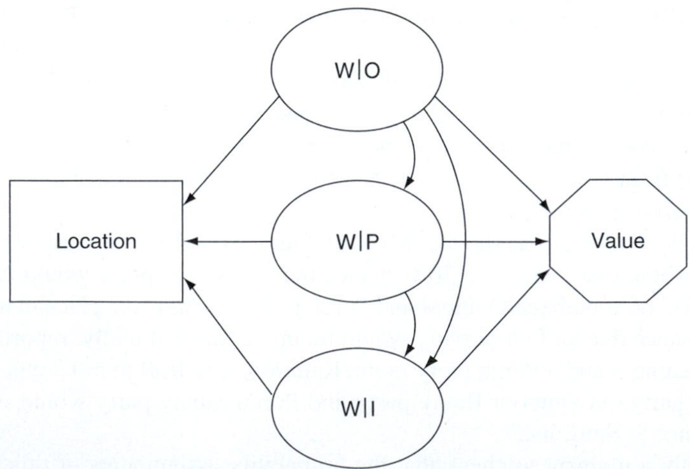

FIGURE 21.8 Shirley's Canonical Form Decision Diagram

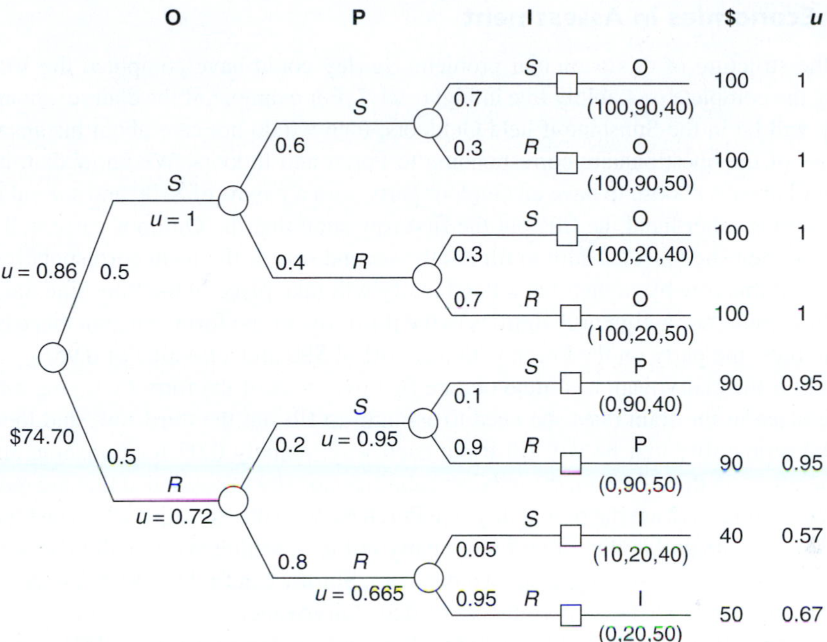

FIGURE 21.9 Computing Shirley's Value of Clairvoyance

produces payoffs for the three alternatives of $100, $90, and $50, so again she will choose an Outdoor party with a payoff of $100 and a *u*-value of 1. The next two reports also lead to an Outdoor party with the same payoff and *u*-value. Therefore, we see that if the clairvoyant reports that an Outdoor party will be Sunny, then, as we would expect, she will choose to have the party Outdoors with a value of $100 and a *u*-value of 1.

The next end point in Figure 21.9 corresponds to an Outdoor party in the Rain but otherwise Sunny with payoffs of 0, $90, and $40 for the three alternatives, so she will choose to have the party on the Porch with a payoff of $90 and a *u*-value of 0.95. The end point beneath differs only in that an Indoor party will take place in the Rain, so she stays with a Porch party and the same end point values.

If the clairvoyant reports that the Outdoor and Porch parties will experience Rain, but that an Indoor party will take place in the Sunshine, then she faces payoffs of 0, $20, and $40, so she will choose an Indoor party with a payoff of $40 and a *u*-value of 0.57. Lastly, a report that she will have a Rainy party no matter what the location produces payoffs of 0, $20, and $50, so she will choose an Indoor party with a payoff of $50 and a *u*-value of 0.67.

When we take the *e*-value of the eight end point *u*-values in this tree, we obtain at the beginning a *u*-value of 0.86, which is Shirley's certain equivalent of $74.70. Since, like Kim, Shirley satisfies the delta property and her certain equivalent without clairvoyance was $47.85, this means that her value of clairvoyance is about $27.81. This value of clairvoyance is considerably higher than the $20 value of clairvoyance computed for Kim. Note that if Shirley were not a deltaperson, we would have to use the general method of subtracting a price from all payoffs and iterating to find the price of clairvoyance that would make her indifferent to buying it.

21.2.6 Economies in Assessment

Due to the structure of this particular problem, Shirley could have computed the VOC without assigning the complete probability tree in Figure 21.7. For example, if the clairvoyant answers that the party will be in the Sunshine if held Outdoors, then we do not care about his answers on the other rows of the questionnaire corresponding to Porch and Indoors. We know that, in this case, Shirley will always choose to have an Outdoor party with a payoff of $100 and a *u*-value of 1.

If, on the other hand, he fills out the first row such that the Outdoor party will take place in the Rain, then she must ask him to fill out the second row of the form corresponding to Porch. If he fills out this row by saying that a Porch party will take place in the Sunshine, then she need not ask him about nor assign probabilities to the third row of the form, since in this case, she will choose to have the party on the Porch with a payoff of $90 and a *u*-value of 0.95.

Only if the clairvoyant has filled out the first two rows of the form by saying that the party will take place in the Rain does she need to ask him to fill out the third row, and then she must assign the probability that he will fill it out each way, namely 0.05 for Sunshine and 0.95 for Rain. Therefore, in this problem, the only probabilities not already assigned that are required to be assigned in Figure 21.7 are the probability of a Porch party in the Sunshine given an Outdoor party in the Rain, and the probability of an Indoor party in the Sunshine given both Outdoor and Porch parties in the Rain. By assigning these two numbers, Shirley can find the VOC on the weather.

We can summarize the consequences of free clairvoyance as shown in Figure 21.10. There is a 50% chance that Shirley will have an Outdoor party in the Sunshine, a 10% chance that she will have a Porch party in the Sunshine, a 0.02 chance that she will have an Indoor party in the Sunshine, and a 0.38 chance that she will have an Indoor party in the Rain. This deal is worth $74.70 to Shirley. Since Shirley is a deltaperson, her VOC can now be obtained as the value of the deal with free clairvoyance minus the value of the deal with no clairvoyance.

> ***Note:*** Once the diagram is in canonical form, each of the distinctions W|O, W|P, W|I can be treated as a separate uncertainty and the analysis is similar to that of calculating the value of clairvoyance on multiple distinctions that we discussed in Chapter 18 for Kim's modified party problem.

21.2.7 Partial Clairvoyance on Influenced Distinctions

From our discussion, it should now be clear that clairvoyance can sometimes be achieved without completely filling out the clairvoyant questionnaire. For example, it would be possible for Shirley to buy clairvoyance only on what the weather will be if she has an Outdoor party—information provided by the first row of the clairvoyance questionnaire. This would mean that in Figure 21.8, there would be an arrow from Weather given Outdoors to Location, but not from the other two chance nodes to Location. We can evaluate clairvoyance on any individual row of the questionnaire or on

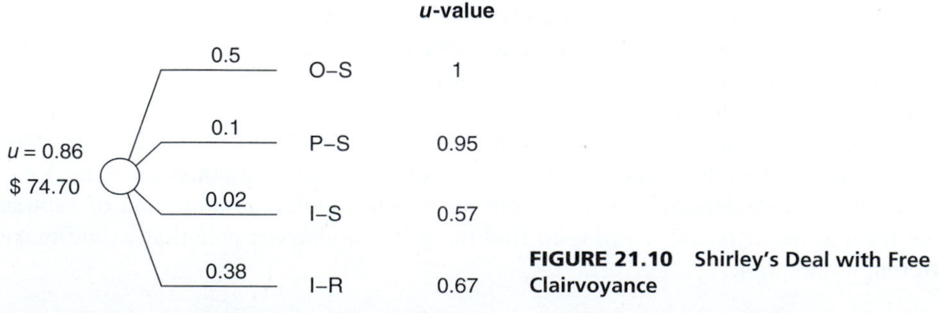

FIGURE 21.10 Shirley's Deal with Free Clairvoyance

any collection of rows on the questionnaire. We can also think of the value of filling out any part of the questionnaire given that the decision maker already knows how other parts of the questionnaire have been filled out. While we seldom need to investigate clairvoyance at this level of complexity, it is reassuring to know that our principles and procedures are adequate to the task.

By the way, you may wish to verify that Shirley's VOC on weather given an Outdoor party—the first row of the clairvoyance questionnaire—is almost \$23.47. This is 83.4% percent of the value of the fully completed questionnaire: \$27.81.

Another problem we may wish to consider for Shirley is her value of imperfect information on the weather. How much would Shirley value the Acme Detector if it was offered to her with an accuracy of 80%? We will address this problem in the following section.

21.2.8 Imperfect Information on Distinctions with Influence

Consider Shirley's value of imperfect information for a detector indication on weather. In Figures 21.8 and 21.9, we have seen that the three distinctions W|O, W|P, and W|I are mutually relevant given her state of information. For example, if the clairvoyant told her the outcome of W|O, this information would update her belief about both W|P and W|I.

Now we consider the Acme Rain Detector for Shirley. Similar to the case of valuing clairvoyance, the detector indication will also determine the indication given Shirley's choice of location. The detector will give three indications for W|O, W|P, and W|I. The detector indications must be relevant to the canonical distinctions of Weather|Location (W|L); otherwise the indication would be of no value.

Next, we need to consider whether the indications of W|O, W|P, and W|I are relevant or irrelevant given the canonical distinctions "W|O," "W|P," and "W|I." The answer to this question must come from Shirley based on her beliefs about the detectors and their operation. For example, one relevance diagram she might use to represent this situation is shown in Figure 21.11.

In Figure 21.11, the distinctions "W|O," "W|P," and "W|I" are irrelevant given the distinctions W|O, W|P, and W|I. For example, if Shirley receives the indication of W|O from a detector, she will not update her belief about the probabilities that the detector will give the indication for W|P except for what she has learned about the weather. This would correspond

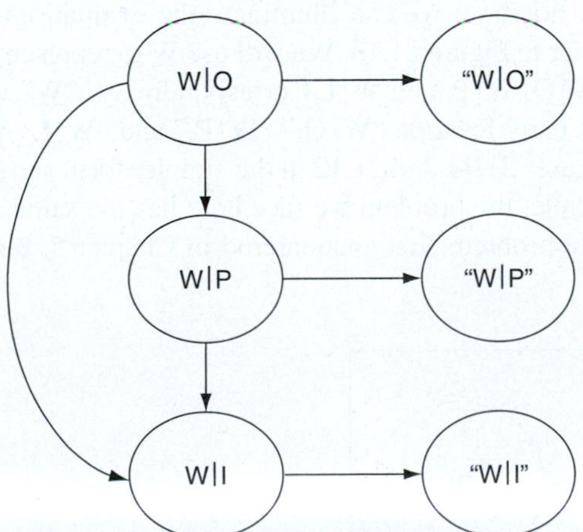

FIGURE 21.11 Possible Relevance Diagram for Shirley's Detector Indication

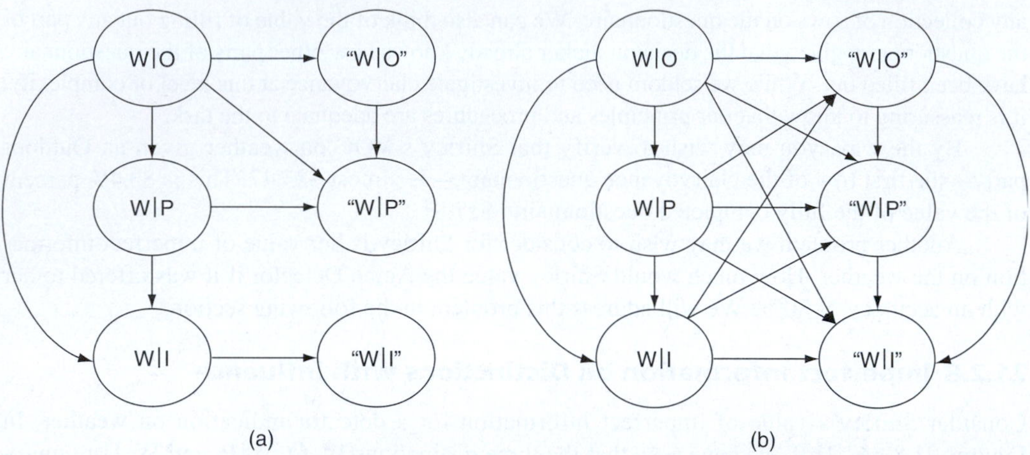

FIGURE 21.12 Examples of More Relevance Diagrams Representing Belief about Detector Indications

to Shirley's belief that although her choice of party location will likely affect the weather, it will not affect the workings of the detector. In Figure 21.11, for example, if the detectors are symmetric and have an accuracy of 0.8, the conditional probability that Shirley receives three Sun indications from the detectors—given W|O = Sun, W|P = Sun, and W|I = Rain—is $0.8 * 0.8 * 0.2 = 0.128$. This is the conditional probability that the two detectors W|O and W|P give correct indications and that the detector W|I gives an incorrect indication.

We note, however, that there can be other relevance diagrams for Shirley depending on her belief. For example, knowledge of the weather condition of W|O may update her belief about all detector indications. Furthermore, the accuracy of detector indication may not even be symmetric.

Figure 21.12a shows an example of a situation where Shirley believes the detector indication "W|O" will update her information about the indication "W|P" even if she has knowledge of W|O and W|P. Figure 21.12b shows an even more general decision diagram where knowledge of any node updates her belief about the other nodes. For each of these relevance diagrams, we would still need to assess the full probability tree in assessed form. The more arrows, the larger the number of conditional probability assessments needed.

By simplifying our notation, we can illuminate the evaluation of information gathering when influences exist. Refer to Figure 21.13. We will use $\underline{W}$ to represent the weather states conditional on party location W|O, W|P, and W|I. Correspondingly, "$\underline{W}$" will represent the possible detector indications given party location "W|O," "W|P," and "W|I." We can now represent the relevance diagrams of Figures 21.11 and 21.12 in the simpler form shown in Figure 21.13.

Notice that in principle, the problem we face here has the same form as the original Beer Drinker–College Graduate problem first encountered in Chapter 5, but with more details to be worked out.

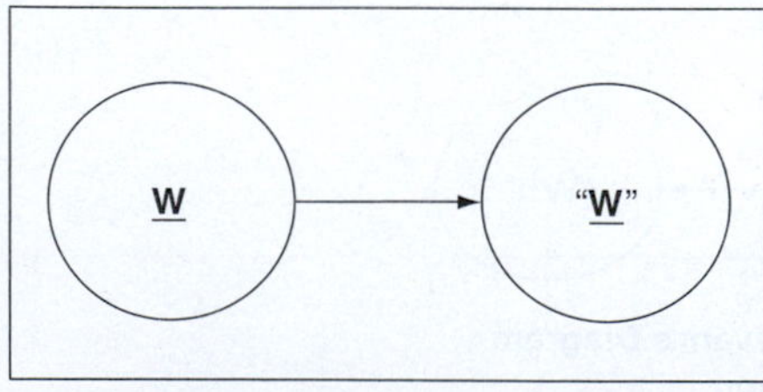

FIGURE 21.13 Simplified Relevance Diagram

In the following discussion, we will consider the situation shown in Figure 21.11 and assume symmetry in the detector indications with an accuracy of 0.8; they are Acme Rain Detectors. The value of these imperfect detectors is now the value of joint clairvoyance on the three distinctions of "W|O," "W|P," "W|I," or "W".

To simplify the notation, we will use three letters to refer to the three possible detector indications or weather conditions. For example, "SO"–"SP"–"SI" refers to the weather condition "Sunny Outside"–"Sunny Porch"–"Sunny Indoors," one of the eight possibilities for W. Of the eight possible detector reports for "W", "SO"–"SP"–"RI" refers to the detector indication of "Sunny Outside"–"Sunny Porch"–"Rainy Indoors."

Shirley has already assigned her prior distribution {W|&} on the weather conditions in Figure 21.7. To complete the assessed form of the relevance diagram in Figure 21.13, we must compute the likelihood function {"W"|W&}. Table 21.1 shows the conditional probabilities, or likelihood, of possible detector indications for the different weather conditions in assessed form. For example, the number 0.512 in the first cell in the table represents the probability 0.8 * 0.8 * 0.8 that all three of the detector indications will correctly indicate "S" in each location when it is sunny in all locations. The entry 0.128 to the right of the first cell refers to the probability that the detector indication will be "SO"–"SP"–"RI" given the weather is all Sunny, *SSS*. This probability is 0.8 * 0.8 * 0.2 = 0.128, where the indication was correct for Outdoors and Porch by reporting "SO"–"SP", and was incorrect for the indoor party "RI".

To determine the joint distribution {W, "W"|&} that represents the relevance diagram of Figure 21.11, we now need to multiply the likelihood values of Table 21.1 by Shirley's prior

TABLE 21.1 {"W"|W &}: Likelihood Values for Imperfect Detector Indication Using Relevance Diagram of Figure 21.11

	Outdoors-Porch-Indoors				
	"SO"–"SP"–"SI"	"SO"–"SP"–"RI"	"SO"–"RP"–"SI"	"SO"–"RP"–"RI"	
SSS	0.512	0.128	0.128	0.032	
SSR	0.128	0.512	0.032	0.128	
SRS	0.128	0.032	0.512	0.128	
SRR	0.032	0.128	0.128	0.512	
RSS	0.128	0.032	0.032	0.008	
RSR	0.032	0.128	0.008	0.032	
RRS	0.032	0.008	0.128	0.032	
RRR	0.008	0.032	0.032	0.128	
	Detector Indication|Location				
	"RO"–"SP"–"SI"	"RO"–"SP"–"RI"	"RO"–"RP"–"SI"	"RO"–"RP"–"RI"	
SSS	0.128	0.032	0.032	0.008	1
SSR	0.032	0.128	0.008	0.032	1
SRS	0.032	0.008	0.128	0.032	1
SRR	0.008	0.032	0.032	0.128	1
RSS	0.512	0.128	0.128	0.032	1
RSR	0.128	0.512	0.032	0.128	1
RRS	0.128	0.032	0.512	0.128	1
RRR	0.032	0.128	0.128	0.512	1

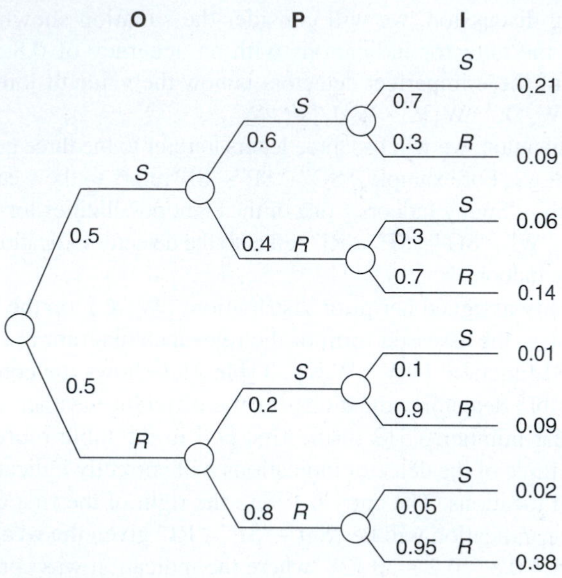

FIGURE 21.14 {W|&} Shirley's Prior on Weather Conditions

for each of the eight possible weather conditions. This assessment was already provided in Figure 21.7. We repeat it in Figure 21.14 for convenience. For example, we know that Shirley's belief of O|S, P|S, I|S is equal to 0.5 * 0.6 * 0.7 = 0.21.

If we multiply the marginal prior probability of each of the eight possible weather conditions by the detector likelihood values of Table 21.1, we obtain the joint distribution of weather condition and detector indication for each location, as shown below in Table 21.2. For example, the first cell (equal to 0.10752) is obtained by multiplying the prior probability of *SSS* (equal to 0.21) by the likelihood of the detector indication of "SO"–"SP"–"SI" from Table 21.1 (equal to 0.512). The cell to the right of this cell (0.0268) corresponds to the joint probability of weather being *SSS* and detector indication of "SO"–"SP"–"RI" obtained by multiplying the prior probability of 0.21 by the likelihood of the detector indication from Table 21.1 (0.128) to get 0.02688. Note that the sum of cells in this matrix is equal to one.

To make use of the detector indication, we need to flip the tree to change the diagram into inferential form. Flipping will produce the preposterior distribution on detector indication and then the posterior conditional probability of weather given detector indication and location.

To obtain the preposterior distribution {"W"|&} in inferential form for each detector indication, we first sum the joint probabilities for each detector indication. Summing each column of the joint distribution in Table 21.2 produces the preposterior probability distribution shown in the last row of the table. For example, the preposterior probability of the detector indication "SO"–"SP"–"SI" is equal to 0.139. The sum of the posterior probabilities is equal to 1 as show in the right column of the table.

The next step is to obtain the posterior probability distribution for weather for each detector indication {W| "W" &}: by flipping the tree. To obtain this posterior distribution, we divide each cell in the joint distribution of Figure Table 21.2 by the preposterior

TABLE 21.2 {W, "W"|&} Joint Distribution for Weather and Detector Indication

	"SO"–"SP"–"SI"	"SO"–"SP"–"RI"	"SO"–"RP"–"SI"	"SO"–"RP"–"RI"	"RO"–"SP"–"SI"	"RO"–"SP"–"RI"	"RO"–"RP"–"SI"	"RO"–"RP"–"RI"	
SSS	0.108	0.027	0.027	0.007	0.027	0.007	0.007	0.002	0.21
SSR	0.012	0.046	0.003	0.012	0.003	0.012	0.001	0.003	0.09
SRS	0.008	0.002	0.031	0.008	0.002	0.000	0.008	0.002	0.06
SRR	0.004	0.018	0.018	0.072	0.001	0.004	0.004	0.018	0.14
RSS	0.001	0.000	0.000	0.000	0.005	0.001	0.001	0.000	0.01
RSR	0.003	0.012	0.001	0.003	0.012	0.046	0.003	0.012	0.09
RRS	0.001	0.000	0.003	0.001	0.003	0.001	0.010	0.003	0.02
RRR	0.003	0.012	0.012	0.049	0.012	0.049	0.049	0.195	0.38
{"W"|&}	0.139	0.117	0.094	0.150	0.064	0.120	0.083	0.233	1

TABLE 21.3 {W | "W" &} Posterior Probability Given Detector Indications

	"SO"–"SP"–"SI"	"SO"–"SP"–"RI"	"SO"–"RP"–"SI"	"SO"–"RP"–"RI"
SSS	0.7733	0.2298	0.2855	0.0448
SSR	0.0829	0.3940	0.0306	0.0769
SRS	0.0552	0.0164	0.3263	0.0513
SRR	0.0322	0.1532	0.1903	0.4784
RSS	0.0092	0.0027	0.0034	0.0005
RSR	0.0207	0.0985	0.0076	0.0192
RRS	0.0046	0.0014	0.0272	0.0043
RRR	0.0219	0.1040	0.1291	0.3246
	1	1	1	1

	"RO"–"SP"–"SI"	"RO"–"SP"–"RI"	"RO"–"RP"–"SI"	"RO"–"RP"–"RI"
SSS	0.4190	0.0561	0.0813	0.0072
SSR	0.0449	0.0961	0.0087	0.0123
SRS	0.0299	0.0040	0.0929	0.0082
SRR	0.0175	0.0374	0.0542	0.0768
RSS	0.0798	0.0107	0.0155	0.0014
RSR	0.1796	0.3845	0.0348	0.0494
RRS	0.0399	0.0053	0.1239	0.0110
RRR	0.1895	0.4059	0.5886	0.8337
	1	1	1	1

probability of its detector indication with the result shown in Table 21.3. For example, the first cell in this table is $.10752/.13904 = 0.7733$ and is obtained by dividing the first cell in Table 21.2 (equal to 0.108) by the preposterior probability of "SO"–"SP"–"SI" (equal to 0.139). The cell to the right of this cell in Table 21.3 (equal to 0.2298) is obtained by dividing the corresponding cell (0.027) in Table 21.2 by the value of the preposterior probability of "SO"–"SP"–"RI" (0.117).

Each column in the matrix of Table 21.3 represents the posterior distribution for Weather given location for each possible detector indication. As shown in the table, each column sums to 1.

For each detector indication, Shirley now has a decision to make about where to have the party in view of her updated distributions of weather given the detector indications. For example, if the detector indication is "SO"–"SP"–"SI", Shirley's decision tree appears as shown in Figure 21.15.

Based on the this tree, Shirley will choose to go outdoors if she receives a detector indication "SO"–"SP"–"SI." Her certain equivalent for the party deal given this detector indication is $88.73.

Figure 21.16 shows the certain equivalent for each location decision alternative given each of the eight possible detector indications.

If we take the maximum cell for each column in Table 21.4 that are shown as the last row in the table, we can determine the best location of the party given each detector indication. By observing the first four columns, we see that Shirley will have an Outdoor party if the detector indicates that such a day will be Sunny. The next two columns imply a Porch party; the last two columns indicate an Indoor party.

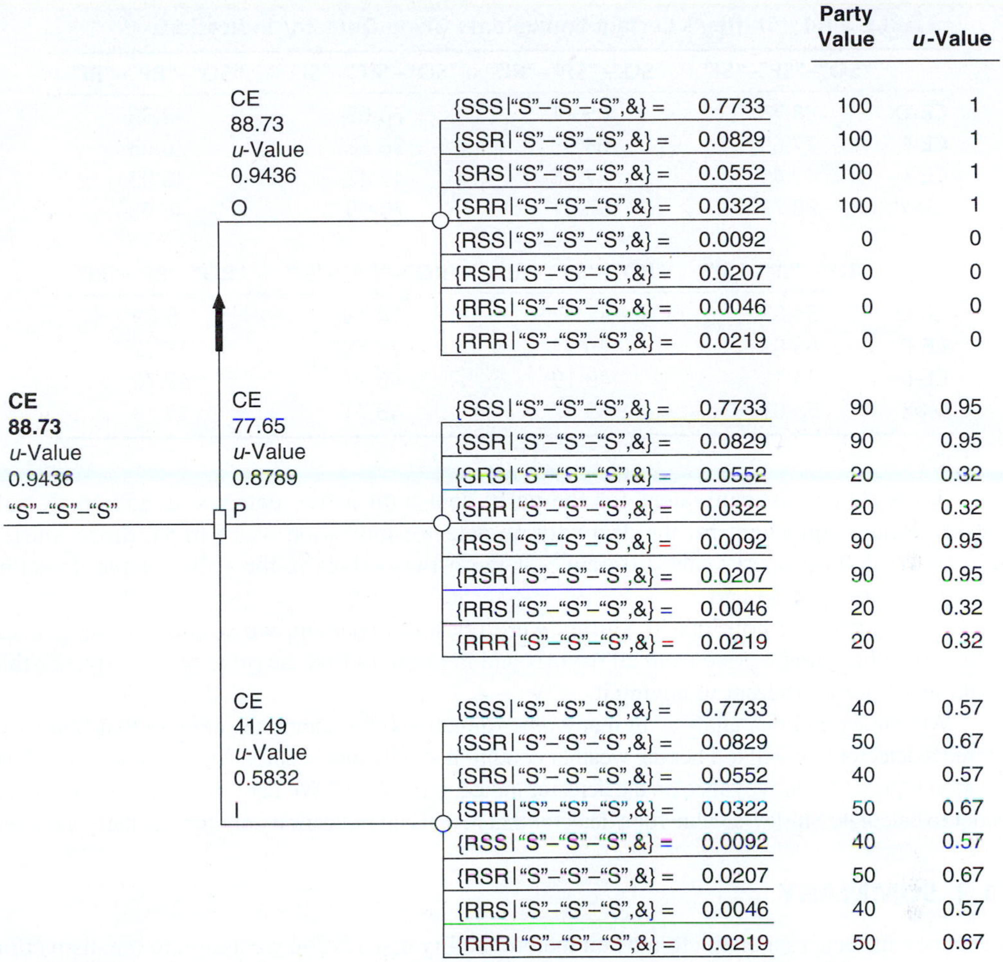

CE 88.73, u-Value 0.9436, "S"–"S"–"S"

Alternative	CE	u-Value	Outcome	Probability	Party Value	u-Value
O	88.73	0.9436	{SSS|"S"–"S"–"S",&} =	0.7733	100	1
			{SSR|"S"–"S"–"S",&} =	0.0829	100	1
			{SRS|"S"–"S"–"S",&} =	0.0552	100	1
			{SRR|"S"–"S"–"S",&} =	0.0322	100	1
			{RSS|"S"–"S"–"S",&} =	0.0092	0	0
			{RSR|"S"–"S"–"S",&} =	0.0207	0	0
			{RRS|"S"–"S"–"S",&} =	0.0046	0	0
			{RRR|"S"–"S"–"S",&} =	0.0219	0	0
P	77.65	0.8789	{SSS|"S"–"S"–"S",&} =	0.7733	90	0.95
			{SSR|"S"–"S"–"S",&} =	0.0829	90	0.95
			{SRS|"S"–"S"–"S",&} =	0.0552	20	0.32
			{SRR|"S"–"S"–"S",&} =	0.0322	20	0.32
			{RSS|"S"–"S"–"S",&} =	0.0092	90	0.95
			{RSR|"S"–"S"–"S",&} =	0.0207	90	0.95
			{RRS|"S"–"S"–"S",&} =	0.0046	20	0.32
			{RRR|"S"–"S"–"S",&} =	0.0219	20	0.32
I	41.49	0.5832	{SSS|"S"–"S"–"S",&} =	0.7733	40	0.57
			{SSR|"S"–"S"–"S",&} =	0.0829	50	0.67
			{SRS|"S"–"S"–"S",&} =	0.0552	40	0.57
			{SRR|"S"–"S"–"S",&} =	0.0322	50	0.67
			{RSS|"S"–"S"–"S",&} =	0.0092	40	0.57
			{RSR|"S"–"S"–"S",&} =	0.0207	50	0.67
			{RRS|"S"–"S"–"S",&} =	0.0046	40	0.57
			{RRR|"S"–"S"–"S",&} =	0.0219	50	0.67

FIGURE 21.15 Shirley's Decision Tree for Detector Indication of "SO"–"SP"–"SI"

Shirley now faces the following deal if she receives a free detector report for all possible party locations:

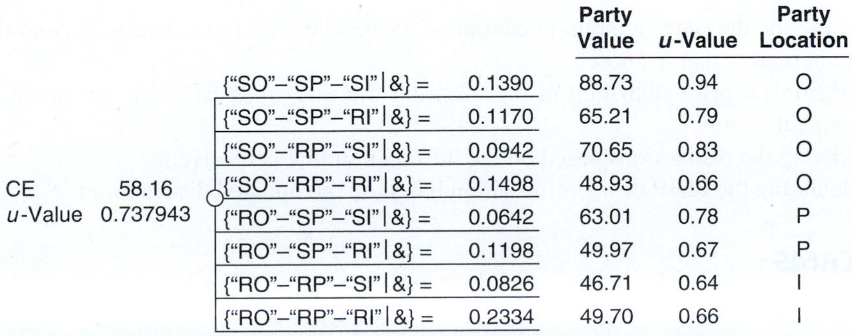

CE 58.16, u-Value 0.737943

Indication	Probability	Party Value	u-Value	Party Location
{"SO"–"SP"–"SI"|&} =	0.1390	88.73	0.94	O
{"SO"–"SP"–"RI"|&} =	0.1170	65.21	0.79	O
{"SO"–"RP"–"SI"|&} =	0.0942	70.65	0.83	O
{"SO"–"RP"–"RI"|&} =	0.1498	48.93	0.66	O
{"RO"–"SP"–"SI"|&} =	0.0642	63.01	0.78	P
{"RO"–"SP"–"RI"|&} =	0.1198	49.97	0.67	P
{"RO"–"RP"–"SI"|&} =	0.0826	46.71	0.64	I
{"RO"–"RP"–"RI"|&} =	0.2334	49.70	0.66	I

FIGURE 21.16 Value of the Party Deal for Shirley with Free Detector

TABLE 21.4 Shirley's Certain Equivalents Given Detector Indications

	"SO"–"SP"–"SI"	"SO"–"SP"–"RI"	"SO"–"RP"–"SI"	"SO"–"RP"–"RI"
CE-O	88.73	65.21	70.65	48.36
CE-P	77.65	63.16	36.38	26.63
CE-I	41.49	47.36	43.42	48.93
Max	88.73	65.21	70.65	48.93
	"RO"–"SP"–"SI"	**"RO"–"SP"–"RI"**	**"RO"–"RP"–"SI"**	**"RO"–"RP"–"RI"**
CE-O	34.88	11.32	14.13	5.89
CE-P	63.01	49.97	26.58	23.22
CE-I	44.15	49.19	46.71	49.70
Max	63.01	49.97	46.71	49.70

Shirley's certain equivalent for the party deal with a free detector is \$58.16. Recall that her certain equivalent for the deal with no detector indication was \$46.85. Since Shirley follows the delta property, the difference of these two values is the value of the detector: \$58.16 − \$46.85 = \$11.31.

Once again, we note that if Shirley were not a deltaperson, we would use the general method of subtracting a price from all payoffs and iterating to find the price of clairvoyance that would make her indifferent to buying it.

We can extend this analysis in many other directions. For example, suppose that Shirley is offered a detector that will tell her the weather condition at only one location (such as outdoors). This situation represents clairvoyance on the detector indication "W | O." We leave it as an exercise for the reader to calculate Shirley's value for a detector that reveals an indication only for one party location.

21.3 SUMMARY

A decision influences a distinction when the probability distribution we assign to the distinction depends on the alternative that is chosen. In decision diagrams, this is represented by an arrow from the decision node to the uncertainty node.

Calculating the best decision alternative for decisions with influences follows the same procedure used for decisions without influences.

Calculating the value of clairvoyance for decisions with influences requires:

- Placing the decision diagram in canonical form, where each distinction is conditioned on the alternative that is chosen.
- Assigning a prior distribution representing the relevance between the distinctions in canonical form.
- Updating the distinctions based on the information that is received.
- Calculating the PIBP of the information based on the updated distributions.

KEY TERMS

- Canonical form
- Influence arrow

PROBLEMS

Problems marked with an asterisk (*) are considered more challenging.

1. The clairvoyant chooses not to give Shirley full clairvoyance on the weather conditions given the three locations. However, the clairvoyant offers to sell her one line of the questionnaire, but he will not give her information for all three locations. For instance, if she buys the Porch report, he might say "If you have a Porch party, it will be Sunny." Shirley can buy any one of the three lines of the report (Outdoors, Indoors, or Porch).
   a. Draw a decision diagram in canonical form representing Shirley's decision to buy one line from the clairvoyant's report.
   b. Draw a decision tree representing Shirley's decision to buy one line from the clairvoyant's report. Using this tree, calculate the PIBP for the information given by the clairvoyant.
   *c. Which line of Clairvoyant's report has the highest value? (Outdoors, Indoors or Porch?)

*2. What would Figure 21.7 look like for Kim?

*3. On your own, redo the calculations of
   a. The value of clairvoyance for Shirley
   b. The value of an imperfect detector for Shirley as shown in Tables 21.1, 21.2, 21.3 and 21.4.

CHAPTER

22

The Logarithmic *u*-Curve

CHAPTER CONCEPTS

After reading this chapter, you will be able to explain the following concepts:

- Properties of the logarithmic *u*-curve:
  - Functional form
  - Certain equivalent calculations
  - Risk aversion function
- Deals with large monetary prospects:
  - Implications for the exponential *u*-curve
  - Implications for the logarithmic *u*-curve
- The St. Petersburg paradox

22.1 INTRODUCTION

In Chapter 11, we discussed two types of *u*-curves:

1. The exponential *u*-curve, which we expressed as either

$$u(x) = a + be^{-\gamma x}$$

or

$$u(x) = a + br^{-x}$$

2. The linear *u*-curve, which is a special case of the exponential,

$$u(x) = a + bx$$

For convenience, we shall refer to either of these forms as the **exponential *u*-curve,** and to a person who has one of those *u*-curves as a *deltaperson*. Recall that the exponential *u*-curves allow simple assessment of their parameters and the computational advantages offered by the delta property.

1. We do not need to consider the initial wealth of the decision maker in evaluating uncertain deals, because the valuation of the deal does not depend on the initial wealth.
2. The value of information on an uncertainty is the value of the deal with free information on this uncertainty less the value of the deal without the information.
3. The risk-aversion is characterized by a single parameter: either the risk tolerance or its reciprocal—the risk-aversion coefficient; or the risk odds.
4. The Personal Indifferent Buying Price (PIBP) of an uncertain deal you do not own is equal to the Personal Indifferent Selling Price (PISP) of this uncertain deal if you owned it. It is important to understand that this equality of PIBP and PISP does not depend on the notion of a cycle of ownership for exponential *u*-curves. The PIBP is equal to the PISP even if they are calculated at different wealth levels. Therefore, we can simplify the calculation of the PIBP by calculating the PISP, which is easier to calculate.

With the delta property, we have seen that if we increase all outcomes of a deal by the same amount, z, then the certain equivalent of the deal will also increase by the amount z. Of course, we can envision situations where the delta property may not hold: The certain equivalent may increase by an amount that is greater than or less than z.

As we shall see, despite some of the advantages of a u-curve possessing the delta property, many other u-curves also have appealing features and may be more representative of attitudes towards risk particularly when a large range of monetary values is considered. We now introduce some additional u-curves and characterize each of their properties. We begin our discussion by presenting the *logarithmic* u-curve. We will then examine the relative advantages of having a u-curve in either exponential or logarithmic form. Finally, we will end the chapter by presenting historical motivation for the logarithmic u-curve.

22.2 THE LOGARITHMIC u-CURVE

As an alternative to the exponential u-curve we might consider the **logarithmic u-curve,** which depends on a person's total wealth and also has several implications. To highlight one implication, notice that, when the same percentage wage increases are proposed for all people in an organization, there appears to be a sense of fairness—even though some people will be receiving much more money than others. This leads to the notion of a u-curve described by the property that small percentage changes in wealth lead to equal changes in the u-value regardless of the wealth level.

For example, consider a u-curve with the property that the u-value of any deal increases by the same amount if we modify all its outcomes by the same percentage or, equivalently, multiply all its outcomes by the same constant. This u-curve has the form

$$u(y) = \ln(y)$$

where y is the person's total wealth, which must be positive and 'ln' is the natural logarithmic function. Even a beggar will have positive total wealth: It is the least you would have to pay the beggar to forgo begging forever.

To see why this u-curve satisfies the multiplicative property mentioned here, we refer to the property of logarithms that

$$\ln(zy) = \ln(z) + \ln(y)$$

If we multiply all outcomes by an amount z (corresponding to a percentage increase or decrease), the resulting increase in u-value of the deal is $\ln(z)$ regardless of the actual deal and of the initial wealth.

When a person with initial wealth α considers a deal with payoff x, the logarithmic u-curve has the form

$$u(x) = \ln(x + \alpha)$$

Figure 22.1 shows the logarithmic u-curve plotted versus total wealth $(x + \alpha)$.

Note *When the total wealth approaches 0, the logarithmic u-curve becomes negative infinity, making such a prospect unacceptable.*

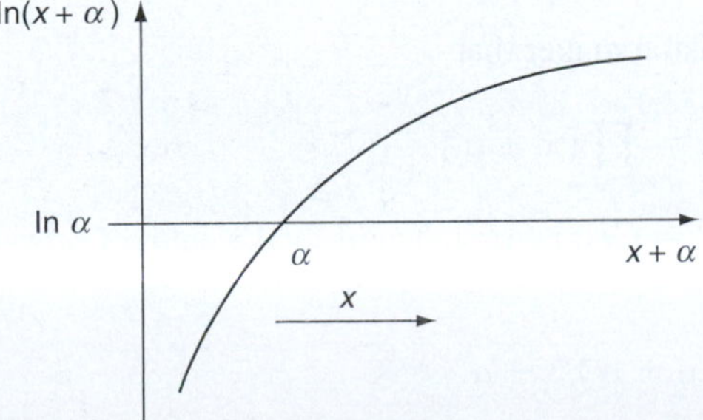

FIGURE 22.1 Logarithmic u-Curve over Total Wealth

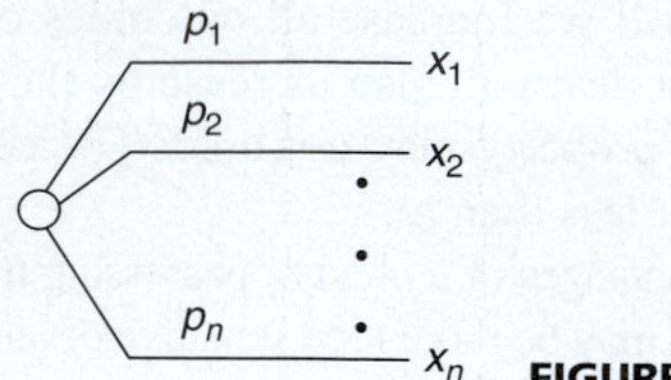

FIGURE 22.2 Uncertain Deal

22.2.1 Certain Equivalent of Uncertain Deals using the Logarithmic *u*-Curve

Suppose that a decision maker with a logarithmic *u*-curve faces the uncertain monetary deal shown in Figure 22.2. What is the certain equivalent?

Each prospect x_i has a *u*-value $u(x_i) = \ln(x_i + \alpha)$ defined over total wealth. We now calculate the *e*-value of the *u*-values for this deal and observe that, by definition, this *e*-value of *u*-values also must be equal to the *u*-values of the wealth certain equivalent $\tilde{x} + \alpha$. Therefore,

$$e\text{-value of } u\text{-values} = u(\tilde{x} + a) = \sum_{i=1}^{n} p_i \ln(x_i + \alpha)$$

Using properties of logarithms (see Appendix A), we can also write this as

$$e\text{-value of } u\text{-values} = \sum_{i=1}^{n} \ln(x_i + \alpha)^{p_i}$$

Therefore,

$$\ln(\tilde{x} + \alpha) = \sum_{i=1}^{n} \ln(x_i + \alpha)^{p_i} = \ln\left(\prod_{i=1}^{n} (x_i + \alpha)^{p_i}\right)$$

where the symbol $\prod_{i=1}^{n} x_i$ means the multiplication of the terms $x_1, x_2, \ldots, x_n$, and we again used the property of logarithms that $\ln(x) + \ln(y) = \ln(xy)$.

Raising both sides as powers of *e* and noting that $e^{\ln(x)} = x$, we can now write

$$\tilde{x} + \alpha = \prod_{i=1}^{n} (x_i + \alpha)^{p_i}$$

The geometric mean of a deal that provides total payoffs $y_1, y_2, \ldots, y_n$ with probabilities $p_1, p_2, \ldots, p_n$, respectively, is by definition equal to

$$\text{Geometric mean} = \prod_{i=1}^{n} y_i^{p_i}$$

Therefore, we can write for a logarithmic decision maker that

$$\text{Certain equivalent} + \text{Initial wealth} = \tilde{x} + \alpha = \prod_{i=1}^{n} (x_i + \alpha)^{p_i}$$
$$= \text{Geometric mean of total wealth of prospects}$$

Rearranging gives

$$\tilde{x} = \prod_{i} (x_i + \alpha)^{p_i} - \alpha$$

Or the certain equivalent for a logarithmic decision maker is given by

Certain equivalent = Geometric mean of total wealth of prospects − Initial wealth

For a logarithmic decision maker, two deals are valued equally if their geometric means are equal. On the other hand, a risk-neutral decision maker will value deals if their arithmetic means (e-values) are equal. Since the geometric mean is always less than or equal to the arithmetic mean, the certain equivalent of an uncertain deal for a decision maker with a logarithmic u-curve must be less than the arithmetic mean of the deal. This tells us right away that the decision maker is risk-averse.

The following examples demonstrate applications and calculations with the logarithmic u-curve.

EXAMPLE 22.1 Logarithmic Decision Maker Buying Insurance

Suppose an individual has a property worth of \$10,000 that will be lost with probability 0.05. This individual can buy insurance at a premium of \$800. How much must his initial wealth, α, be in order for him to be indifferent to whether or not he purchases insurance?

SOLUTION

The prospects for insurance and no insurance are shown in Figure 22.3.

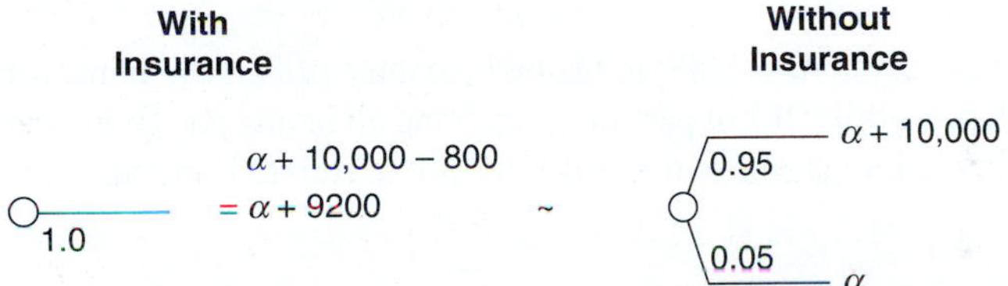

FIGURE 22.3 Choosing between Insurance vs. No Insurance

The left-hand side of Figure 22.3 shows the deal after buying insurance. The right-hand side shows the deal without insurance.

For indifference, we equate the certain equivalents for both sides of Figure 22.3.

$$\alpha + 9200 = (\alpha + 10{,}000)^{0.95}\alpha^{0.05}$$

Solving gives the initial wealth that makes him indifferent as

$$\alpha = \$5043$$

EXAMPLE 22.2 Logarithmic Decision Maker Selling Insurance

Now, we consider the insurer's problem, assuming he has a logarithmic u-curve. What must the initial wealth of the insurer be for him to be indifferent between selling and not selling insurance at an \$800 premium?

SOLUTION

Let the insurer's initial wealth be β. The prospects for selling and not selling insurance are shown in Figure 22.4.

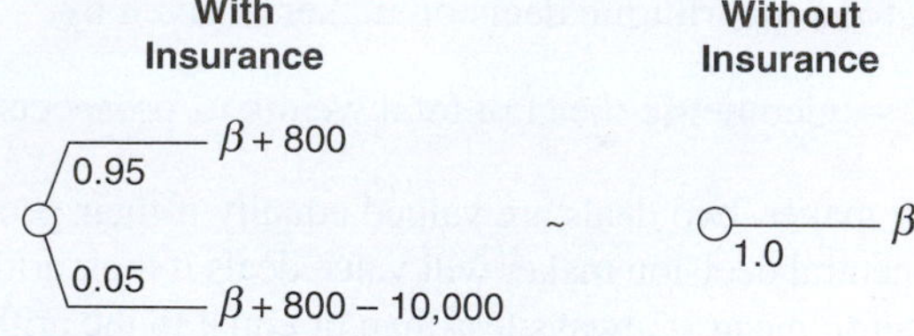

FIGURE 22.4 Selling Insurance

Equating the certain equivalent of both sides gives

$$(\beta + 800)^{0.95}(\beta + 800 - 10{,}000)^{0.05} = \beta$$

Solving gives the initial wealth as

$$\beta = \$14{,}243$$

(Comparing the equations for α and β shows by inspection that β is just $\alpha + 9200$.)

Note *We could not answer these questions for a deltaperson buyer and deltaperson seller of insurance, because neither the buying nor selling prices of insurance depend on the the initial wealth.*

EXAMPLE 22.3 Transporting Goods

An individual has goods worth \$4,000 in his own country (this is his initial wealth) and \$8,000 abroad. Ships with probability 0.1 of perishing can bring all or any part of his foreign goods home. What are his foreign holdings worth to him if he uses one ship to bring his foreign goods home?

SOLUTION

For one ship, the deal is represented in Figure 22.5.

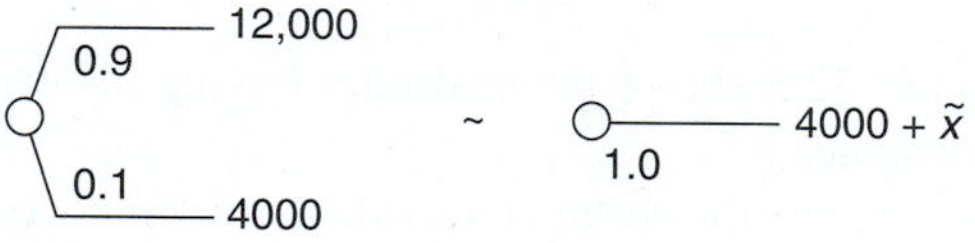

0.9 — 12,000
0.1 — 4000
~ 1.0 — $4000 + \tilde{x}$

FIGURE 22.5 Deal with One Ship

The certain equivalent for this deal is calculated as

$$(12{,}000)^{0.9}(4000)^{0.1} = 4000 + \tilde{x}$$

which implies

$$\tilde{x} = 6751$$

On the other hand, if he uses two ships (each carrying half the shipment), his deal is now the new one shown in Figure 22.6.

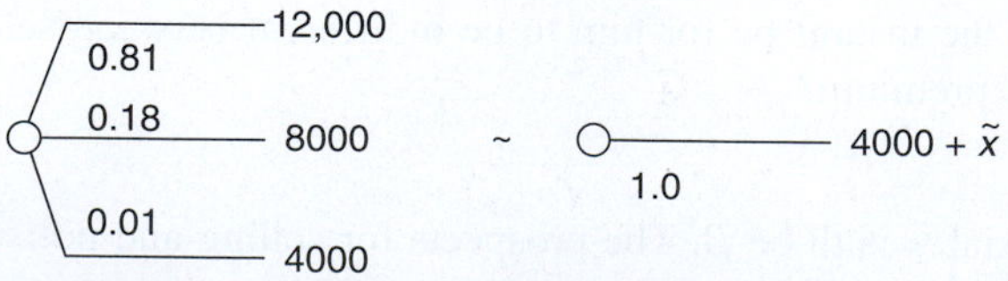

0.81 — 12,000
0.18 — 8000
0.01 — 4000
~ 1.0 — $4000 + \tilde{x}$

FIGURE 22.6 Deal with Two Ships

To calculate the certain equivalent, we write

$$(12{,}000)^{0.81}(8000)^{0.18}(4000)^{0.01} = 4000 + \tilde{x}$$

which implies

$$\tilde{x} = 7033$$

As the number of ships among which he distributes his foreign goods increases, their certain equivalent increases. However, it can never exceed the *e*-value of the goods, $0.9 \times 8000 + 0.1 \times 0 = \$7{,}200$.

22.2.2 Source for the Logarithmic *u*-curve

All of the results in this section on the logarithmic *u*-curve, including the insurance and shipping examples, were published by Daniel Bernoulli in 1738.

"Specimen Theoriae Novae de Mensura Sortis"
Commentarii Academiae Scientiarum Imperialis Petropolitanae, Tomus V
[Papers of the Imperial Academy of Sciences in Petersburg, Vol. V], 1738: 175–192.

The title when translated is "Exposition of a New Theory on the Measurement of Risk." Though many are still unfamiliar with the concept and usage of *u*-curves, the idea is over 250 years old. In this chapter, we shall later present the motivation for Bernoulli's work.

22.3 DEALS WITH LARGE MONETARY PROSPECTS FOR A DELTAPERSON

To understand why other *u*-curves besides the exponential might be appealing, let us see how a deltaperson deals with monetary prospects that are large compared to the risk tolerance. From the party problem in Chapter 9, recall Kim's exponential *u*-curve was

$$u(x) = \frac{4}{3}\left(1 - \left(\frac{1}{2}\right)^{\frac{x}{50}}\right) = \frac{1 - e^{\frac{x}{\rho}}}{1 - e^{\frac{100}{\rho}}}$$

with a risk tolerance $\rho = \dfrac{1}{\gamma} = 100/\ln 4 = \$72.13$.

Suppose Kim faces the deal in Figure 22.7, where she has a 0.5 probability of receiving a monetary amount, *X*, and a 0.5 probability of receiving an amount of 0.

To calculate Kim's certain equivalent for this deal, we calculate the *u*-value of *X* and 0 dollars as *u*(*X*) and *u*(0), respectively. We then calculate the *e*-value of the *u*-values of the deal to get the calculated *u*-value of the deal, as

$$u(\tilde{x}) = \frac{1}{2}u(X) + \frac{1}{2}u(0)$$

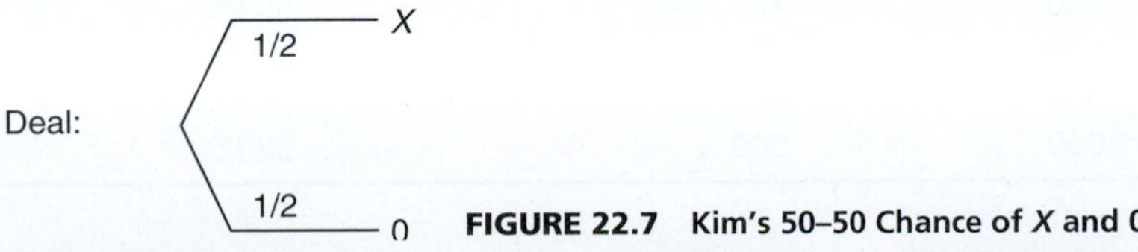

FIGURE 22.7 Kim's 50–50 Chance of *X* and 0

Finally, we obtain the certain equivalent of the deal by projecting the *u*-value of the deal on the *u*-curve to get

$$\tilde{x} = u^{-1}\left(\frac{1}{2}u(X) + \frac{1}{2}u(0)\right)$$

The *e*-value of this deal (which is also the certain equivalent for a risk neutral decision maker) is equal to the *e*-value of the payoffs. Thus,

$$e\text{-value} = \frac{1}{2}X + \frac{1}{2}0 = \frac{X}{2}$$

Table 22.1 shows Kim's certain equivalent and the *e*-value for this deal for different values of *X*.

Since Kim is risk-averse, she values the deal less than its *e*-value for any value of *X*. From Table 22.1, we see that when $X = \$100$, the *e*-value of the deal is \$50, and Kim's certain equivalent is \$33.90. The difference is \$16.10, which is known as the risk premium.

What happens when the value of *X* increases? When $X = \$500$, the *e*-value of the deal is \$250, and Kim's certain equivalent is \$49.93. Kim's certain equivalent increases by only \$16.03 as *X* changes from \$100 to \$500. When $X = \$1000$, the *e*-value of the deal is \$500, and Kim's certain equivalent is \$50 (only a 7-cent increase in certain equivalent from the previous value of $X = \$500$, while the *e*-value changes by \$250).

TABLE 22.1 e-Value and Certain Equivalent of Figure 22.7 Deal

X	e-Value of Deal	Certain Equivalent of Deal
0	0	0
50	25	20.75
100	50	33.90
150	75	41.50
200	100	45.62
250	125	47.78
300	150	48.88
350	175	49.44
400	200	49.72
450	225	49.86
500	250	49.93
550	275	49.96
600	300	49.98
650	325	49.99
700	350	49.99
750	375	49.99
800	400	50.00
850	425	50.00
900	450	50.00
950	475	50.00
1000	500	50.00

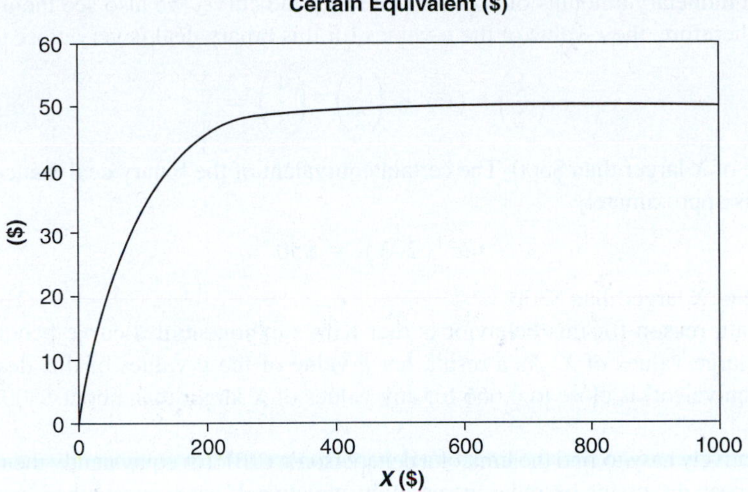

FIGURE 22.8 Sensitivity Analysis: Kim's Certain Equivalent vs. *X*

We can extend this calculation further and note that when $X = \$1{,}000{,}000$, the *e*-value of the deal is \$500,000, but Kim's certain equivalent is still only \$50. Furthermore, the certain equivalent of the deal never exceeds \$50, no matter how large we make *X* (see Figure 22.8).

Does it make sense that Kim would value a 50–50 chance of \$500 and 0 for \$49.93, while her certain equivalent would increase by only 7 cents for a 50–50 chance of \$1,000,000 and 0? To verify the mathematical correctness of this phenomenon, we plot Kim's *u*-curve over a range from 0 to \$1000 in Figure 22.9.

Notice that the *u*-curve has a maximum value of $4/3 = 1.333$ as *X* approaches infinite values. However, the *u*-curve reaches approximately 98% of its maximum value after only four times the risk tolerance (\$288). As a result, the *u*-value of *X* is very close to 4/3 for any value of

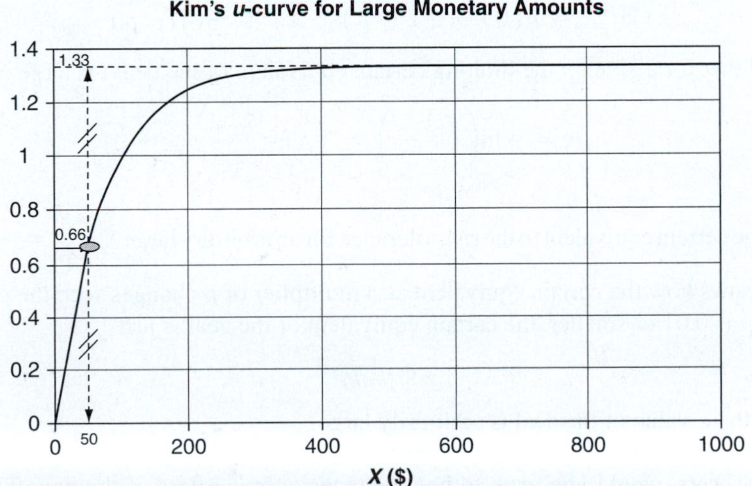

FIGURE 22.9 Kim's *u*-Curve Over the Domain [0,1000]

X larger than monetary amounts of about \$300. From the curve, we also see the u-value of 0 is equal to 0. Therefore, the e-value of the u-values for this binary deal is very close to

$$\left(\frac{1}{2}\right) * (0) + \left(\frac{1}{2}\right) * \left(\frac{4}{3}\right) = \frac{2}{3}$$

for any value of X larger than \$300. The certain equivalent of the binary deal that corresponds to this u-value is approximately

$$u^{-1}(2/3) = \$50$$

for any value of X larger than \$300.

The main reason for this behavior is that Kim's exponential u-curve becomes virtually constant for large values of X. As a result, her e-value of the u-values of this deal (and hence her certain equivalent) is close to 0.666 for any values of X larger than about \$300, as shown in Figure 22.9.

It is relatively easy to find the limit of a deltaperson's PIBP (or equivalently their PISP), as the deal we have been discussing becomes increasingly attractive. If we represent the u-curve simply by

$$u(x) = -e^{-x/\rho}$$

then when X is arbitrarily large, $u(X) \approx 0$ and so

$$u(\tilde{x}) = -e^{-\tilde{x}/\rho} = \frac{1}{2}u(0) + \frac{1}{2}u(X) \approx \frac{1}{2}u(0) = -\frac{1}{2}$$

Solving for $\tilde{x}$ shows that the certain equivalent of the deal will be no greater than

$$\rho \ln 2 = .693\rho$$

regardless of the size of X.

Using the same argument, we find the deltaperson's limiting PISP (or PIBP) for a deal where with probability p the prize will either be an arbitrarily large X or 0. We write

$$u(\tilde{x}) = -e^{-\tilde{x}/\rho} = p\,u(X) + (1-p)\,u(0)$$
$$= p(0) + (1-p)(-1) = -(1-p)$$

We find that for a given p the limiting certain equivalent of the deal for large X will be

$$\tilde{X} = \rho \ln\left(\frac{1}{1-p}\right) = -\rho \ln(1-p)$$

The ratio of the certain equivalent to the risk tolerance for an arbitrary large X is $\frac{\tilde{X}}{\rho} = -\ln(1-p)$.

Table 22.2 shows how the certain equivalent as a multiplier of ρ changes with the probability p.

When p is 0.01 or smaller, the certain equivalent of the deal is just

$$\tilde{x} \approx \rho p$$

even though the e-value of the deal is arbitrarily large.

With these observations about the behavior of the certain equivalent with large positive monetary prospects, would you want to be a deltaperson regardless of the payoffs? If so, then you would exhibit similar behavior for large monetary amounts.

TABLE 22.2 Ratio of Certain Equivalent to Risk Tolerance for Large Values of X

ρ	$\tilde{X}/\rho = -\ln(1 - p)$
0.5	0.693147
0.1	0.105361
0.01	0.01005
0.001	0.001001
0.0001	0.0001

Note *This is not a problem created by accepting the five Rules of Actional Thought, since they do not require you to be a deltaperson. However, as we know, being a deltaperson over a certain domain simplifies many of the mathematical calculations.*

22.4 PROPERTIES OF THE LOGARITHMIC u-CURVE

As Bernoulli postulated, we can think of the logarithmic u-curve as a u-curve defined on total wealth. Notice that when the total wealth approaches zero, the logarithmic u-curve becomes negative infinity, making such a prospect extremely undesirable. Unlike the exponential, this u-curve continually increases as wealth increases. Since this u-curve has the property that any line connecting two points on it will always lie below the curve, it represents a risk averse attitude. As Bernoulli showed, above the certain equivalent of a logarithmic person always will be less than that of a risk-neutral person for the same deal. Also note that, unlike the exponential, the logarithmic cannot represent risk-preferring behavior.

As we discussed, the logarithmic u-curve has the form

$$u(x) = \log(x + \alpha)$$

where x is the outcome of the deal and α is the inital wealth. As we might expect, a decision maker with a logarithmic u-curve will need to determine his initial wealth when valuing uncertain deals and, furthermore, such a person will not satisfy the delta property. Consequently, the value of information calculation is more involved. There will be a difference in the value of the PIBP and PISP of an uncertain deal—a difference that may be desirable.

22.4.1 Risk-Aversion Function for the Logarithmic u-Curve

Recall that the risk-aversion coefficient for a decision maker with an exponential u-curve was a constant, γ. This means that the risk aversion does not change with the decision maker's initial wealth. If we add any fixed amount of money to all outcomes of a deal, the certain equivalent increases by that same fixed amount.

On the other hand, we have seen that the valuation of uncertain deals, and hence the risk aversion, changes with initial wealth for a logarithmic decision maker. Therefore, it is not just a constant, but is also a function of wealth. The expression for the risk aversion function for a logarithmic u-curve (see Appendix B) is

$$\gamma(x) = \frac{1}{x + \alpha}$$

and the risk-tolerance function of a logarithmic decision maker is

$$\rho(x) = \frac{1}{\gamma(x)} = x + \alpha$$

The risk tolerance increases and the risk-aversion coefficient decreases as the monetary prospects increase. We will see an implication of this phenomenon in the following section. In Chapter 24, we will refer back to the risk-aversion function and illustrate its role in providing approximate expressions for the certain equivalent.

22.4.2 Effects of Initial Wealth on the Valuation of an Uncertain Deal with the Logarithmic *u*-Curve

As we have discussed, the certain equivalent calculation will depend on the initial wealth level for a decision maker with a logarithmic *u*-curve. To illustrate this idea, consider the deal shown below in Figure 22.10 that provides a 50–50 chance of receiving \$1,000 and \$0.1 (one cent). Suppose the initial wealth was \$100.

The *e*-value of *u*-values of this deal is

$$\begin{aligned} 0.5\ln(1000 + \alpha) + 0.5\ln(0.1 + \alpha) &= 0.5\ln(1000 + 100) + 0.5\ln(0.1 + 100) \\ &= 0.5 * 7 + 0.5 * 4.61 \\ &= 5.8 \end{aligned}$$

This also must be equal to the *u*-value of the certain equivalent. Thus,

$$u(\tilde{x}) = \ln(\tilde{x} + 100) = 5.8$$

The certain equivalent is then

$$\tilde{x} = u^{-1}(5.8) = e^{5.8} - 100 = \$231.82$$

The *e*-value of this deal is

$$e\text{-value} = 0.5 * 1000 + 0.5 * 0.1 = \$500.05$$

Since the decision maker values the deal less than its *e*-value, he is risk-averse.

Let us now investigate the effects of the initial wealth on the certain equivalent of this deal. Figure 22.11 plots a sensitivity analysis of the certain equivalent versus initial wealth, α. As we can see, the certain equivalent increases as the initial wealth increases. This means that the decision maker will value the deal more highly when he becomes "richer."

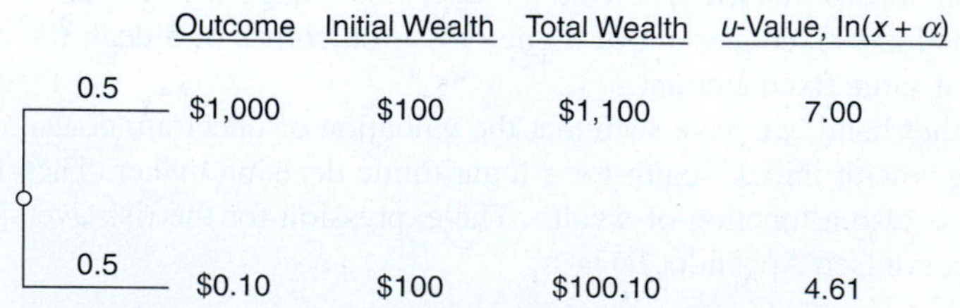

Outcome
Initial Wealth
Total Wealth
u-Value, ln(x + α)
0.5
$1,000
$100
$1,100
7.00
0.5
$0.10
$100
$100.10
4.61

FIGURE 22.10 Uncertain Deal with Payoffs \$1000 and \$0.1

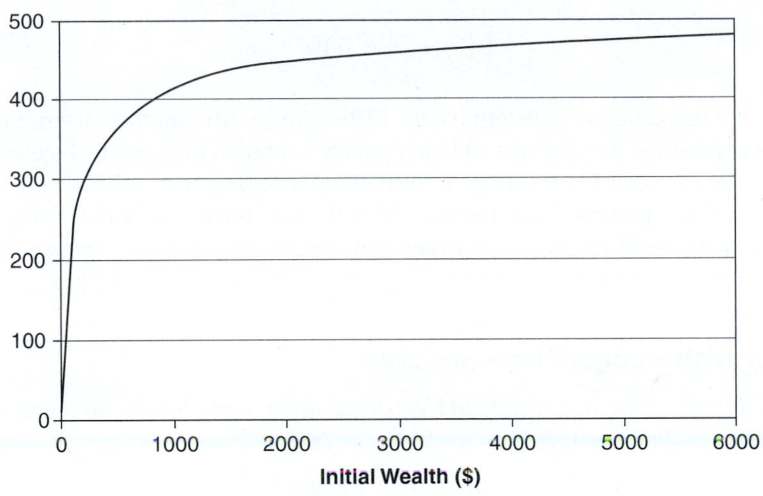

FIGURE 22.11 Sensitivity Analysis: Certain Equivalent vs. Initial Wealth

Note that as the initial wealth increases, the certain equivalent becomes closer to the *e*-value of the deal. This means that as his initial wealth increases, the decision maker is becoming closer to risk-neutral. We refer to this phenomenon as decreasing risk aversion with wealth. This also implies that if we add any fixed amount to all outcomes of the deal we face, the certain equivalent increases by an amount that is greater than this fixed amount. We value deals more as our initial wealth increases.

22.4.3 Selling and Buying Prices of Deals Using the Logarithmic *u*-Curve

As Bernoulli computed, the certain equivalent or PISP of the uncertain deal of Figure 22.2 is given by

$$\text{Certain equivalent} = \text{Geometric mean of prospects} - \text{Initial wealth}$$

To calculate the PIBP, *b*, of the uncertain deal of Figure 22.2 with probabilities p_i of receiving x_i, we equate the *u*-values

$$\sum_{i=1}^{n} p_i \ln(x_i + \alpha - b) = u(\alpha) = \ln(\alpha)$$

Once again, we use properties of the logarithm and write the left-hand side as

$$\sum_{i=1}^{n} (\ln(x_i + \alpha - b))^{p_i} = \ln\left(\prod_{i=1}^{n} (x_i + \alpha - b)^{p_i}\right)$$

Equating this expression with the right-hand side gives

$$\ln\left(\prod_{i=1}^{n} (x_i + \alpha - b)^{p_i}\right) = \ln(\alpha)$$

Therefore the PIBP, b, must satisfy the equation

$$\prod_{i=1}^{n}(x_i + \alpha - b)^{p_i} = \alpha$$

Note that solving this equation to determine the PIBP is more difficult than determining the PISP using a direct expression. Solving this equation requires iteration or numerical methods to find b. It is generally the case that PIBPs are more difficult to calculate than PISPs, which emphasizes an advantage of the exponential u-curve for which the two prices are equal. For a deltaperson, we simply calculate the PISP (since it is easier) but for any other u-curve, the two values will be different.

22.4.4 Kim with a Logarithmic *u*-Curve

We will now discuss a case in which Kim has a logarithmic u-curve with an initial wealth equal to her risk tolerance. For example,

$$\alpha = \rho = \$72.13$$

We can write this u-curve in the form

$$u(x) = \frac{\ln\left(1 + \frac{x}{\rho}\right)}{\ln\left(1 + \frac{100}{\rho}\right)}$$

This form ensures that, just like her exponential u-curve, Kim's logarithmic u-curve will be 0 at 0 dollars and 1 for \$100. Figure 22.12 plots Kim's exponential u-curve and her logarithmic u-curve. Note how close they are over the region 0 to \$100, and that for large monetary prospects, the logarithmic curve continues to grow, whereas the exponential approaches the value 4/3.

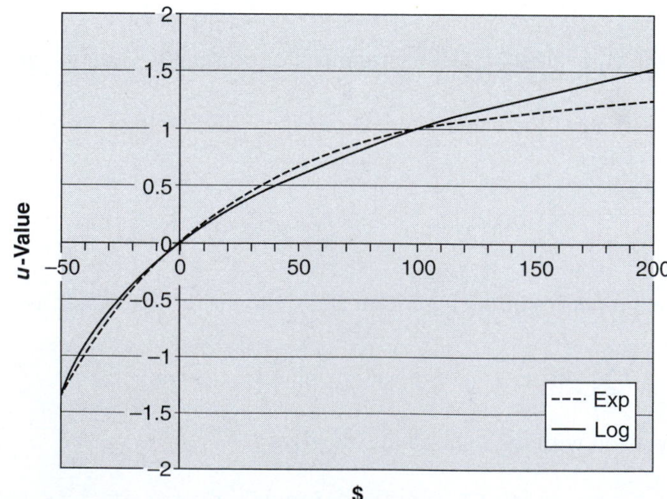

FIGURE 22.12 Comparison of Exponential and Logarithmic *u*-Curves

22.4.5 The Logarithmic Kim Facing Large Monetary Prospects

We now calculate Kim's certain equivalent for the deal, where she has a 50–50 chance of receiving \$*X* or 0 using her logarithmic *u*-curve. For any value of *X*, her certain equivalent or PISP for the deal—if she owned it—is

$$\tilde{x} = (X + 72.13)^{0.5}(0 + 72.13)^{0.5} - 72.13$$

For comparison, we also calculate Kim's PIBP, *b*, for the same deal that is determined by the equation

$$(\$100 + \$72.13 - b)^{0.5}(\$0 + \$72.13 - b)^{0.5} = \$72.13$$

Note

It is easier to calculate the PISP of the deal than it is to calculate the PIBP, because the PISP, $\tilde{X}$, is computed directly without having to solve for the calculation of the PIBP, *b*.

When Kim has a logarithmic *u*-curve with $\alpha = 72.13$ and faces a 50–50 chance of 0 or \$100, her PISP $= \tilde{X} = 39.3$ and her PIBP $= b = 34.36$. Recall that Kim with an exponential *u*-curve of risk tolerance of 72.13 had a PIBP = PISP = 33.9.

Table 22.3 shows a comparison of Kim's PIBP and PISP for the deal if she has a logarithmic *u*-curve. From Table 22.3, we see that the certain equivalent increases substantially as *X*

TABLE 22.3 PIBP and PISP for Uncertain Deal of Figure 22.7

X	PISP (Logarithmic)	PIBP (Logarithmic)
1	0.50	0.50
2	0.99	0.99
4	1.97	1.97
8	3.89	3.89
16	7.60	7.56
32	14.54	14.25
64	26.96	25.22
128	48.02	39.70
256	81.71	53.21
512	133.13	62.16
1024	209.05	67.07
2048	318.93	69.59
4096	476.18	70.86
8192	699.94	71.49
16384	1017.36	71.81
32768	1466.95	71.97
65536	2103.26	72.05
131072	3003.49	72.09
262144	4276.85	72.11
524288	6077.84	72.12
1048576	8624.94	72.13

increases, while the PIBP increases at a much slower rate and eventually saturates at a value of 72.13, which is her initial wealth. This happens because the certain equivalent is a selling price for something you own. As X increases, owning the deal makes you wealthier. The increased wealth creates a larger risk tolerance, and a larger PISP.

On the other hand, the PIBP is a buying price for a deal you do not own. As X increases, you do not become wealthier. As a result, the risk tolerance does not increase with \$$X$, and you value the deal with a lower risk tolerance. Therefore, Kim would never pay more than her wealth (\$72.13) for any deal—no matter how attractive X is. Therefore,

> *While the exponential u-curve leads to identical PIBP and PISP values for uncertain deals, the logarithmic u-curve creates differences in these values. The PIBP of a deal you do not own (at a given initial wealth) is not equal to the PISP of the same deal if you owned it at the same initial wealth.*

We have already noted that, for the deal of Figure 22.7, someone with an exponential *u*-curve like Kim will have a PISP (and PIBP) that approaches a fixed value. In her case, $\rho \ln 2 = 0.693\rho = \$50$—no matter how large X becomes. Switching to the logarithmic *u*-curve with a wealth of $\alpha = \rho = \frac{1}{\gamma} = 100/\ln 4 = \$72.13$ will now make her PISP increase without limit as X increases. However, with this logarithmic *u*-curve, as X increases, her PIBP for the deal will approach a limit. This limit is Kim's wealth, $\rho = \frac{1}{\gamma} = 100/\ln 4 = \$72.13$, since no matter how attractive the deal, she can never pay more than her wealth for it. This is the value approached by the PIBP in Table 22.3 as X increases without limit.

The observation that a logarithmic person can never allow a decrease in wealth below zero—no matter how improbable—contrasts with the limitless negative payoffs that can be considered by a deltaperson.

Recall that in Chapter 2 we had to introduce the concept of escrow to make sure that people facing uncertain deals understood that they could not accept a deal believing that they would receive positive payoffs while avoiding any associated negative consequences. There could be no attitude of "Give me my winnings, but try to catch me if I lose." The escrow principle should apply regardless of the type of *u*-curve. *u*-curves, such as the logarithmic, ensure that even in the case of extreme negative payoffs the principle will still hold.

22.5 CERTAIN EQUIVALENT OF TWO MUTUALLY IRRELEVANT DEALS

We now highlight some important points in the valuation of mutually irrelevant deals for exponential and logarithmic decision makers. Recall from Section 22.4 that Kim's certain equivalent for a 50–50 chance of 0 and \$100 is equal to \$33.90 when she has an exponential *u*-curve with a risk tolerance of \$72.13. When we used a logarithmic *u*-curve of initial wealth $\alpha = \$72.13$, her PIBP was \$34.36, and her certain equivalent was \$39.30. Now, consider the PIBP and PISP for two deals, each of which gives a 50–50 chance of 0 and \$100. The two deals are mutually irrelevant as shown by the probabilities in the tree of Figure 22.13.

Is the certain equivalent (or PIBP) of the sum of two mutually irrelevant deals equal to the sum of their certain equivalents (or PIBPs)? In the next section, we answer this question.

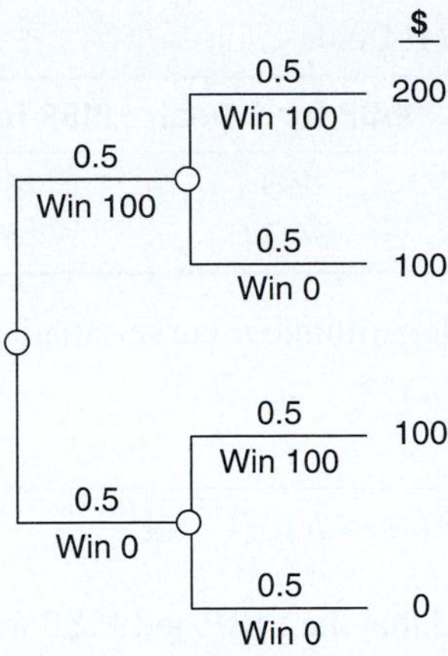

FIGURE 22.13 Two Mutually Irrelevant Deals

22.5.1 PIBP and PISP Calculations for a Deltaperson

The PIBP and PISP are equal for a deltaperson and can be found by calculating either the PIBP or the PISP of the deal. The certain equivalent for a single deal which has a 50–50 chance of 0 and 100 satisfies

$$e^{-\gamma\tilde{x}} = 0.5e^{-\gamma 100} + 0.5e^{-\gamma(0)}$$

Rearranging gives

$$\text{PIBP for a single deal} = \text{PISP for a single deal} = \tilde{x} = \frac{-1}{\gamma}\ln(0.5e^{-\gamma 100} + 0.5)$$

Similarly, for two mutually irrelevant deals as shown in Figure 22.13, we have

$$e^{-\gamma\tilde{x}} = 0.25e^{-\gamma 200} + 0.5e^{-\gamma 100} + 0.25e^{-\gamma(0)}$$

Rearranging gives

$$\tilde{X} = \frac{-1}{\gamma}\ln(0.25e^{-\gamma 200} + 0.5e^{-\gamma 100} + 0.25e^{-\gamma(0)}$$

PIBP for two mutually irrelevant deals = PISP for two mutually irrelevant deals

$$= \frac{-1}{\gamma}\ln(0.25e^{-\gamma 200} + 0.5e^{-\gamma 100} + 0.25)$$

$$= \frac{-1}{\gamma}\ln(0.5e^{-\gamma 100} + 0.5)(0.5e^{-\gamma 100} + 0.5)$$

$$= \frac{-1}{\gamma}\ln(0.5e^{-\gamma 100} + 0.5) + \frac{-1}{\gamma}\ln(0.5e^{-\gamma 100} + 0.5)$$

which is just the sum of the PISPs or PIBPs for each deal.

22.5.2 PIBP and PISP Calculations for a Logarithmic Decision Maker

The PISP for one deal which has a 50–50 chance of 0 and 100 using a logarithmic *u*-curve can be calculated using the equation

$$\text{PISP} = \tilde{x} = (100 + \alpha)^{0.5}(0 + \alpha)^{0.5} - \alpha$$

and the PISP for the two deals of Figure 22.13 can be calculated using

$$\text{PISP} = \tilde{x} = (200 + \alpha)^{0.25}(100 + \alpha)^{0.5}\alpha^{0.25} - \alpha$$

TABLE 22.4 PIBP and PISP for Two Mutually Irrelevant Deals

	PISP for 1 Deal	PISP for 2 Deals	PIBP for 1 Deal	PIBP for 2 Deals
Exponential *u*-curve	33.9	67.8	33.9	67.8
Logarithmic *u*-curve	39.3	83.16	34.37	61.61

On the other hand, the PIBP for a single deal using a logarithmic *u*-curve satisfies

$$(100 + \alpha - b)^{0.5}(0 + \alpha - b)^{0.5} = \alpha$$

and the PIBP of the two deals satisfies

$$(200 + \alpha - b)^{0.25}(100 + \alpha - b)^{0.5}(\alpha - b)^{0.25} = \alpha$$

By direct calculation using $\alpha = \frac{1}{\gamma} = \$72.13$, we find that the PIBP and PISP for the deals are as shown in Table 22.4.

From Table 22.4, we see that for an exponential *u*-curve the certain equivalent (PISP) for two mutually irrelevant deals is equal to 67.81: The sum of the PISP of each deal separately. Similarly, the PIBP for the two mutually irrelevant deals, 67.81, is equal to the sum of their PIBPs.

For a logarithmic *u*-curve, however, the PISP for the two deals is 83.16, which is more than twice the PISP for the individual deals. Furthermore, the PIBP for the two mutually irrelevant deals is less than the sum of the PIBPs for the two individual deals. For a logarithmic *u*-curve, neither the PIBP nor the PISP for two mutually irrelevant deals is equal to the sum of those quantities for the individual deals.

Here is an example of mutually irrelevant deals with a negative outcome.

EXAMPLE 22.4 Mutually Irrelevant Deals with a Negative Outcome

Consider the deal shown in Figure 22.14.

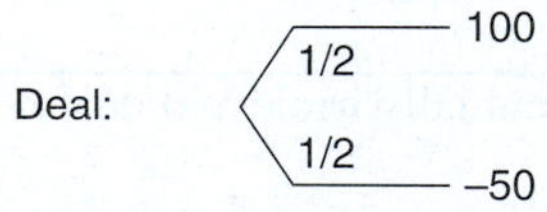

Deal: 1/2 → 100; 1/2 → –50

FIGURE 22.14 Deal with Negative Outcome

Using similar analysis, we calculate the PIBP and PISP for this individual deal for an exponential and logarithmic Kim with $\rho = \alpha = 72.13$. We also calculate the PIBP and PISP in Table 22.5 for two of these deals when they are mutually irrelevant.

TABLE 22.5 PIBP and PISP for Two Mutually Irrelevant Deals

	PISP for 1 Deal	PISP for 2 Deals	PIBP for 1 Deal	PIBP for 2 Deals
Exponential *u*-curve	−8.5	−17	−8.5	−17
Logarithmic *u*-curve	−10.41	Could not own	−6.92	−31.63

Again, we find for an exponential *u*-curve that the PISP and PIBP of two mutually irrelevant deals are equal to the sum of PISP and PIBP of the individual deals respectively. Note that Kim's PISP for one deal is negative, −8.5: She would pay that much to get rid of it. Her PIBP for one deal is the same: She would pay 8.5 not to own it. These amounts double for two such deals.

For a logarithmic u-curve, her PISP for one deal is -10.41: She would pay that much to get rid of it. Her PIBP for one deal is -6.92, and she would have to be paid at least 6.92 to take it on.

The case of two mutually irrelevant deals shows first that she could never have owned such deals because her wealth would be negative, so there is no question of her selling them. Likewise, she is in no position to buy them: Paying her whole wealth for them, \$72.13, the most she could pay, would leave her in a state of negative wealth if both of the deals turned out badly for her. As we discussed, a person with a logarithmic u-curve cannot take on any deal that has the possibility of making his total wealth negative. However, like Kim, she could be paid to take on both, and she would do if she were paid more than 31.63. If paid this amount, the lowest her wealth could be even if she lost in both would be $72.13 + 31.63 - 100 = 3.73$, which is a positive wealth.

22.5.3 Assessment Difficulty for Logarithmic *u*-Curve

The previous sections provided some insights into the valuation of uncertain deals using the exponential and logarithmic u-curves and demonstrated some desirable features for each of them. We now highlight some difficulties that arise when using logarithmic u-curves.

1. Assessing the value of initial wealth may be a challenge to determine in practice.
2. Using a logarithmic u-curve, we cannot value even an uncertain deal that is irrelevant to the deals we own without considering all other deals in our possession. Therefore, while the logarithmic u-curve resolves the issue of saturation for the PISP of large monetary prospects, it comes with an added difficulty in calculating the certain equivalent of uncertain deals.
3. Recall that, since a logarithmic decision maker does not follow the delta property, the calculation of the value of information is also complicated, since we cannot just subtract the value of a deal with no information from its value with information.

22.6 THE ST. PETERSBURG PARADOX

Now, we will discuss Bernoulli's motivation for introducing the logarithmic u-curve in 1738. He based it on the notion that preference for wealth rested on percentage changes in wealth rather than on the wealth itself. His original paper contained the discussion of computing certain equivalents using the geometric mean and the numerical examples of buying and selling insurance presented in Section 22.2. His intention was to resolve the **St. Petersburg Paradox,** which is a conundrum based on the following game.

Suppose a game is played by tossing a coin until it lands heads. The total number of tosses, n, that is needed for the first show of heads determines the value of the reward, and is equal to $\$2^n$. The coin is tossed in such a way that we believe it is equally likely to land either heads or tails and that the results of successive tosses are mutually irrelevant.

For example, if the result of the first coin toss is heads, then the game ends and the reward will be $\$2^1 = \$2$. If it is not heads, then the coin is tossed again. If it lands heads on the second toss, then the reward will be $\$2^2 = \$4$, and the game ends. Otherwise, it is tossed again, and so on.

The probability of getting a prize of $\$2^n$ (first head on the nth toss) is equal to $\frac{1}{2^n}$. The game has an infinite number of possible prizes. The e-value of the St. Petersburg game is therefore equal to

$$e\text{-value} = \sum_{i=1}^{\infty} \frac{1}{2^i}\left(2^i\right) = \infty$$

However, few people would be willing to pay more than a few dollars to play the game: That is the paradox. Bernoulli proposed the logarithmic *u*-curve to explain why.

As we found in Section 22.3.3, a person with a logarithmic *u*-curve and initial wealth α will have a PIBP b for a deal with payoffs x_i received with probability p_i computed from

$$\ln\left(\prod_{i=1}^{n}(x_i + \alpha - b)^{p_i}\right) = \ln(\alpha)$$

or equivalently,

$$\sum_{i=1}^{n} p^i \ln(x_i + \alpha - b) = \ln(\alpha)$$

Since for the St. Petersburg game, $x_i = 2^i$ and $p_i = (1/2^i)$, then

$$\sum_{i=1}^{\infty} \frac{1}{2^i} \ln(2^i + \alpha - \text{PIBP}) = \ln(\alpha)$$

This calculation leads to a finite value for the PIBP that depends on the value of initial wealth, as shown in Figure 22.15.

Table 22.6 shows the PIBP of the St. Petersburg game for a logarithmic decision maker for different values of the maximum *n* and different amounts of initial wealth. As we can see, a decision maker with an initial wealth of \$1 million would value an $n = 50$ version of this game for only \$20.87.

For comparison, Table 22.7 shows the PIBP for the deal for an exponential decision maker for different values of the maximum *n* and different risk tolerances. The values of the PIBP are quite small—even though the payoffs can be in the order of 10^{15}.

Some have suggested that Bernoulli's solution would not be valid if the rate of increase of the prizes was sufficient to overcome the diminution provided by the logarithmic function.

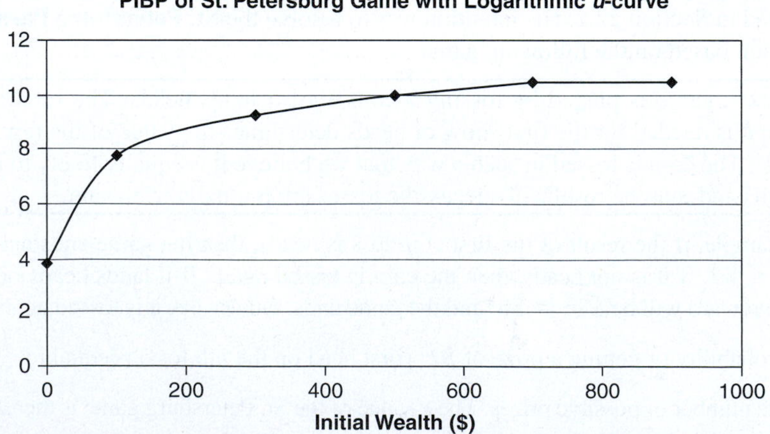

FIGURE 22.15 PIBP vs. Initial Wealth for Game Using Logarithmic *u*-Curve

TABLE 22.6 Two-way Sensitivity Analysis of PIBP for Logarithmic Decision Maker vs. *n* and Initial Wealth

n	**Initial Wealth**					
	100	1,000	10,000	100,000	1,000,000	10,000,000
5	5.74	5.97	6.00	6.00	6.00	6.00
10	7.66	10.00	10.86	10.99	11.00	11.00
15	7.79	10.86	13.87	15.58	15.95	16.00
20	7.79	10.95	14.23	17.43	19.94	20.85
25	7.79	10.95	14.24	17.55	20.83	23.83
30	7.79	10.95	14.24	17.55	20.87	24.18
35	7.79	10.95	14.24	17.55	20.87	24.20
40	7.79	10.95	14.24	17.55	20.87	24.20
45	7.79	10.95	14.24	17.55	20.87	24.20
50	7.79	10.95	14.24	17.55	20.87	24.20

TABLE 22.7 PIBP for Exponential Decision Maker

n	ρ					
	100	1,000	10,000	100,000	1,000,000	10,000,000
5	5.73	5.97	6.00	6.00	6.00	6.00
10	7.00	9.84	10.86	10.99	11.00	11.00
15	7.00	10.13	13.40	15.55	15.95	16.00
20	7.00	10.13	13.41	16.72	19.77	20.85
25	7.00	10.13	13.41	16.72	20.04	23.35
30	7.00	10.13	13.41	16.72	20.04	23.36
35	7.00	10.13	13.41	16.72	20.04	23.36
40	7.00	10.13	13.41	16.72	20.04	23.36
45	7.00	10.13	13.41	16.72	20.04	23.36
50	7.00	10.13	13.41	16.72	20.04	23.36

For example, if the first head is obtained on the *n*th toss, the prize received could be changed from $\$2^n$ to $\$2^{2^n}$. Instead of payments on getting the first head on successive tosses being \$ 2, 4, 8, 16, . . . they would be \$ 4, 16, 256, 65536, . . . Presumably, one would be willing to pay more for this modified game, since the prizes are much larger. But how much more? Recall that the logarithmic risk attitude requires that the person can never participate in any deal that has the possibility of resulting in negative wealth. Suppose a person with a wealth of \$10 pays \$10 for this deal. If the buyer has the bad luck of obtaining heads on the first toss, the resulting wealth would be \$4, not negative. Obtaining heads for the first time on any future toss would result in an even higher wealth state. Therefore, a logarithmic person regardless of wealth would forego his total wealth in exchange for this enhanced deal. Thus, observers are correct that enhanced deals could require a logarithmic person to give up everything for them. However, this is a theoretical result of no practical guidance to a logarithmic person, as we shall see.

Would a person with a logarithmic *u*-curve be willing to pay almost everything he had for this enhanced opportunity? No. This conclusion is incorrect—not because of any fault in the mathematical computation, but because of inappropriate modeling of prospects in the game.

To see why, consider the original St. Petersburg game where payments grow by 2^n. If you were actually going to play this game in our world today, how much could you possibly win? You could play with someone whose maximum ability to pay you was \$1000, \$1 million, or even many billions of dollars. But let us go even further. Suppose you were able to play for the entire financial value of our planet. To estimate this, the GDP of the entire world is about \$75 trillion. If we multiply this by a factor of 15 to allow for the accumulation of wealth, the wealth of the planet is about \$1000 trillion or $\$10^{15}$. Since 10^{15} equals approximately 2^{50}, the expression for the *e*-value of the deal to a risk neutral given above as a sum that goes to infinity would, in fact, only go to the number 50. A risk-neutral person playing for the wealth of the world would have an *e*-value of \$50 for the deal. For any risk-averse person, the certain equivalent would be even less. There are no infinite PIBP values here.

Let us return to the enhanced payment game playing for the wealth of the world. Table 22.8 shows the PIBP for the game of logarithmic people with wealth ranging from \$10 to \$100 billion. As wealth decreases, the percentage of wealth to pay for the game increases until it is all wealth at a wealth of \$10.

TABLE 22.8 PIBP of Modified Game vs. Wealth

Wealth, \$	10	100	1,000	10,000	100,000	1,000,000	10,000,000	100,000,000	1 Billion	10 Billion	100 Billion
PIBP, \$	10.0	9.96E+01	8.70E+02	7.82E+03	6.99E+04	6.25E+05	5.58E+06	4.85E+07	4.01E+08	3.21E+09	2.58E+10
% of Wealth	100%	99.55%	86.95%	78.20%	69.93%	62.49%	55.83%	48.47%	40.16%	32.10%	25.84%

Correctly representing the situation faced by a person playing the St. Petersburg paradox game or the enhanced game shows that any attempt to make PIBP infinite is doomed. To make a more general point, while the notion of infinity is very useful in mathematics, there is no place for infinity in the world we experience, or in any decision we face in that world. An infinite result in any practical problem is an artifact of an inappropriate model.

22.7 SUMMARY

The exponential *u*-curve saturates at monetary amounts of about three times the risk tolerance. When used to value deals whose prospects are significantly higher than the risk tolerance, we observe a saturation effect in the valuation of the certain equivalent. This is not an issue with the five Rules of Actional Thought, but is instead, it is a consequence of choosing to use the exponential *u*-curve over the entire monetary domain. Consequently, we would not want to follow the delta property over the entire domain of monetary prospects.

For a logarithmic decision maker, the following are valid:

- The PIBP and PISP of a deal depend on the initial wealth.
- The value of clairvoyance is NOT necessarily equal to the value of a deal with free clairvoyance less the value of the deal with no clairvoyance.

- The PIBP of a deal you do not own is not necessarily equal to the PISP of the same deal if you owned it at the same initial wealth.
- If we multiply all outcomes of a deal by a fixed amount, the *u*-value of the deal increases by the same amount—regardless of the deal and regardless of the initial wealth.

The risk-aversion function is constant for a deltaperson, but it is a decreasing function for a logarithmic person.

In general, calculating the PISP of an uncertain deal is easier than calculating the PIBP.

The PISP and PIBP for a deltaperson are equal and approach a limit as the deal spans a large range of monetary outcomes. The PISP for a logarithmic person continually increases with the attractiveness of the deal, but the PIBP can never exceed the person's wealth.

To assess a logarithmic person's *u*-curve requires knowing all of the mutually irrelevant and relevant unresolved deals owned by the person.

Be careful when assuming infinite resources in a model. It might lead to unrealistic results.

KEY TERMS

- Exponential *u*-curve
- Logarithmic *u*-curve

APPENDIX A The Logarithmic Function and Its Properties

The log function has the properties

$$\begin{aligned} \log(1) &= 0, \log(0) = -\infty, \log(\infty) = \infty \\ \log_a(a) &= 1 \\ \log(xy) &= \log(x) + \log(y) \\ \log(x^y) &= y\log(x) \\ a^{\log_a(x)} &= x \end{aligned}$$

The logarithmic u-curve has the form

$$u(x) = \log(x + \alpha)$$

where α is the inital wealth. We can use any base for the log operation, since moving from one base to the other can be obtained by a simple multiplication operation. To move from base a to base b, we have

$$\log_a(x + \alpha) = \frac{\log_b(x + \alpha)}{\log_b(a)}$$

For the special case of base $e \approx 2.7182818$ (natural log), we use the notation $\ln(X + \alpha)$ for the u-curve.

APPENDIX B The Risk-Aversion Function

The risk-aversion function for any given u-curve is equal to the negative the ratio of the second to the first derivative of the u-curve. Thus,

$$\gamma(x) = -\frac{u''(x)}{u'(x)}$$

The sign of the risk-aversion function determines whether a person is risk-averse or risk preferring over a certain range. A decision maker is risk-averse if the risk-aversion function is positive and is risk preferring if it is negative. As we illustrate in Chapter 24, the risk-aversion function plays an important role in approximating the certain equivalent of a deal.

The risk-tolerance function is the reciprocal of the risk-aversion function and is given as

$$\rho(x) = \frac{1}{\gamma(x)} = -\frac{u'(x)}{u''(x)}$$

RISK-AVERSION FUNCTION FOR THE EXPONENTIAL *u*-CURVE

Consider an exponential u-curve, $u(x) = -e^{-\gamma x}$. We have $u'(x) = \gamma e^{-\gamma x}$, $u''(x) = -\gamma^2 e^{-\gamma x}$. Therefore

$$\gamma(x) = -\frac{u''(x)}{u'(x)} = -\frac{-\gamma^2 e^{-\gamma x}}{\gamma e^{-\gamma x}} = \gamma = \text{Constant}$$

$$\rho(x) = \frac{1}{\gamma(x)} = \frac{1}{\gamma} = \rho = \text{Constant}$$

Both the risk-aversion function and the risk tolerance are constant for an exponential u-curve.

RISK-AVERSION FUNCTION FOR THE LOGARITHMIC *u*-CURVE

Consider a logarithmic u-curve, $u(x) = \ln(x + \alpha)$. For a logarithmic u-curve,

$$u'(x) = \frac{1}{x + \alpha} \text{ and } u''(x) = \frac{-1}{(x + \alpha)^2}$$

which leads to

$$\gamma(x) = \frac{1}{x + \alpha}$$

$$\rho(x) = \frac{1}{\gamma(x)} = x + \alpha$$

The risk tolerance increases linearly with an increase in x, and the risk aversion decreases linearly with an increase in x.

APPENDIX C A Student's Question Following an Economist Article

After a class seminar covering the five Rules of Actional Thought, one of the participants presented the following article, which appeared in the *Economist* magazine. The article is directly related to our previous discussions of the five Rules of Actional Thought and refers to *Expected Utility Theory* (which, in our terminology, is the *e*-value of the *u*-values) as a wrong theory. Having read the article, the student wondered whether using these rules for decision making was even appropriate. Read the article for yourself and see what you think.

"Averse to Reality,"[1] *The Economist* 9, Aug. 2001.
Based on the Article: Thaler, Richard and Rabin, Matthew. "Anomalies: Risk Aversion." *Journal of Economic Perspectives*, Volume 15, Number 1, Pages 219–232, Year 2001.

Economics relies on a standard theory of how people deal with risk. This theory is obviously wrong.

A GREAT deal of economic theory turns on how people cope with risk—one of the least escapable facts of economic life. The model that most economists rely on when they need to take account of risk in their pure or applied research is expected-utility theory. The trouble is, this theory has implications so absurd that it cannot be true.

People are commonly observed to be "risk averse" in everyday life—that is, they reject better-than-fair gambles. Suppose you were offered a 50−50 bet that paid you $11 if you won and cost you $10 if you lost. Given the odds and the pay-offs, the expected return for accepting this bet is 50 cents (50% of $11 less 50% of $10); and since this is a positive number, the gamble looks attractive. The fact that many people do turn down such bets does not trouble expected-utility theory. It has an explanation: diminishing marginal utility. As your wealth rises, each extra dollar is worth less to you than the previous one. Because the utility of extra wealth declines, it is not necessarily illogical to attach a lower subjective value to the upside of the gamble (50% of $11) than to the downside (50% of $10). All seems well: the facts and the theory sit comfortably together.

Unfortunately, if you think about it, they do not, as Matthew Rabin of the University of California, Berkeley, and Richard Thaler of the University of Chicago point out in a recent article*[1]. Consider the bet described in the previous paragraph, and imagine some unremarkably risk-averse person who turns it down. Now ask yourself this: knowing nothing else about the person, and assuming expected-utility theory to be true, how big a prize would you need to offer in a 50−50 bet to persuade him to risk losing $100?

Knowing he turned down the $11 prize, you might guess it would have to be more than $110. Would $220 be enough? The expected payoff of that bet would be $60 (50% of $220 less 50% of $100). Looks good—yet our putative risk-avoider would still turn it down. Things get worse. What about $2,000? He would turn that down as well. All right, $20,000. No, still too risky. Very well, $2 million; wait, what the heck, $2 billion. Still no. Given only what you know about this risk-avoiding person, plus the truth of expected-utility theory, *you are forced to conclude that he will reject ANY 50–50 bet costing $100, regardless of the prize.*

Risk-aversion of this degree is literally insane—yet rational, according to the theory. What is going on? To understand what the theory is doing, as Mr. Rabin and Mr. Thaler explain, you need to follow along with some arithmetic. Suppose that the person's initial wealth is W.

[1]Matthew Rabin, "Averse To Reality," in Anomalies: Risk Aversion, Journal of Economic Perspectives, Volume 15, Number 1 (Richard Thaler, ed.).

Then rejecting the original lose \$10 gain \$11 bet implies that on average he values each of the dollars between W and $(W + 11)$ by at most ten-elevenths of the average value he puts on dollars between W and $(W - 10)$. This implies that the value he puts on the $W + 11$th dollar is at most ten-elevenths of the value he puts on the $W - 10$th dollar. In effect, then, our subject's marginal utility of wealth falls as his wealth rises, and rises as his wealth falls, at a rate of around 10% for every change of \$21. This phenomenally powerful multiplier so inflates the value he attaches to a loss of \$100, and so deflates the value he attaches to any gains, that no gain can be big enough to make the bet seem attractive.

The absurdity, as Mr. Rabin and Mr. Thaler emphasize, is not a trick reflecting particular assumptions, but is wired into the standard theory. "Expected-utility theory says risk attitudes derive solely from changes in marginal utility associated with fluctuations in lifetime wealth. Hence, the theory says that people will not be averse to risks involving monetary gains and losses that do not alter lifetime wealth enough to affect significantly the marginal utility one derives from that lifetime wealth." The theory, in other words, implies that people should be risk-neutral towards gambles involving small stakes—but they aren't.

ODDS AND ENDS

The question is how to make sense of the fact that people will reject small-stake gambles and yet accept, as they are wont to, moderate-stake gambles provided the terms are good. The authors call for an approach based on two ideas (both mentioned on earlier occasions in this space): loss aversion and mental accounts.

Loss aversion is the idea that people feel the pain of a loss more acutely than the pleasure of a gain of equal size: changes in wealth, and their direction, are what count, regardless of levels. This directly explains why people turn down even very small gambles with positive expected gains. Mental accounting plays a complementary role. It is the idea that people judge financial risks in isolation, rather than alongside overall wealth and other risks. Small, better-than-fair gambles may look irresistible in relation to total wealth, because any losses will be negligible in that context; judged in isolation, especially given loss aversion, such gambles are much easier to turn down.

The authors argue that such decision isolation is pervasive, and explains many otherwise perplexing features of economic life—from the "equity-premium puzzle" to the otherwise contradictory facts that (a) lotteries are popular and (b) people are willing to pay outlandish prices to insure themselves against easily affordable losses (as with, for instance, optional extended warranties on consumer durables). The evidence against the expected-utility approach seems overwhelming, and the broad shape of an alternative, thanks to the earlier work of Mr. Thaler and others, is reasonably clear. The greatest puzzle, perhaps, is that the old theory has not yet been discarded.

*[1]"Anomalies: Risk Aversion" By Matthew Rabin and Richard Thaler. *Journal of Economic Perspectives*, Volume 15, Number 1."

To answer the student's question, we examined the arguments made in this article in a bit more detail—particularly the statement that confused him that if a person turned down a 50–50 bet that would give either a prize of \$11 or a cost of \$10, then:

"You are forced to conclude that he will reject ANY 50–50 bet costing \$100, regardless of the prize."

Suppose that a person with an exponential u-curve faces a deal with 0.5/0.5 probability of \$11 and \$$-10$ (Figure 22.16).

Suppose he rejects this deal, what must his risk aversion be? Figure 22.17 plots a sensitivity analysis of the certain equivalent of this deal versus the risk-aversion coefficient, γ, for a deltaperson. As we have discussed, the certain equivalent is equal to the PIBP for a deltaperson.

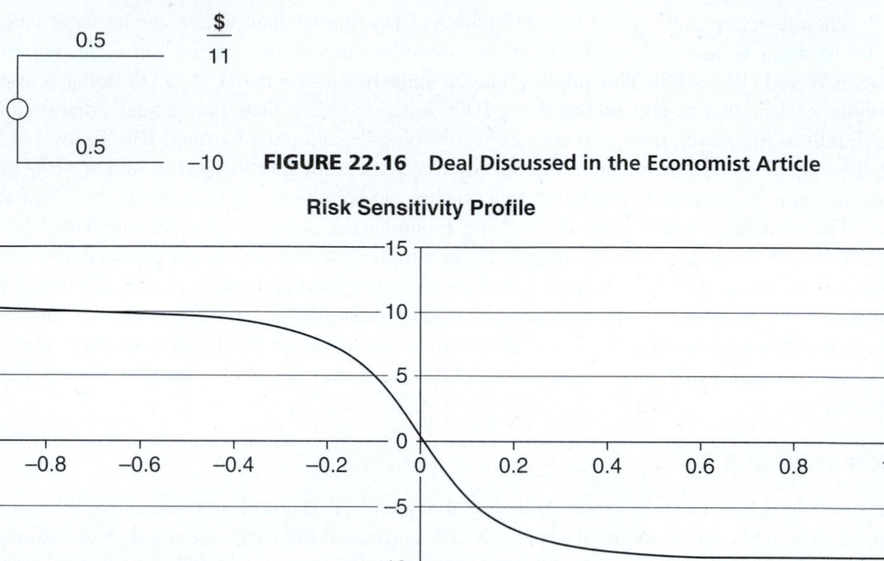

FIGURE 22.16 Deal Discussed in the Economist Article

FIGURE 22.17 Certain Equivalent vs. Risk-Aversion Coefficient for a Deltaperson

If the person is risk preferring, he will value the deal more highly than its e-value and, therefore, will accept it. If he is risk-averse, then he will value it less than its e-value. By direct calculation, a person with a risk-aversion coefficient equal to 0.009 (risk tolerance of \$110.08) will value this deal at 0 dollars. A person with a higher risk-aversion coefficient (lower risk tolerance) will find this deal has a negative certain equivalent.

Consider now such a deltaperson with a risk attitude described by $\gamma = 0.01$ and $\rho = 100$ who would reject this deal. Suppose this person faces the second deal presented in the article where he has a 0.5/0.5 probability of receiving -100 and $+X$. We use the delta property to investigate, as shown in Figure 22.18. To see whether this person would see this deal as attractive for a given X, we add \$100 to both prizes and see whether he would have a certain equivalent for the transformed deal greater than \$100.

Indeed, we find that the certain equivalent of the deal on the right-hand side never exceeds \$69.32 for any value of X. Why is this happening? We have discussed this phenomenon earlier in Table 22.1 and Figure 22.8. For large values of X, the exponential u-curve saturates and therefore the certain equivalent of the deal also saturates.

However, as we discussed earlier, this is not a problem with the five Rules of Actional Thought. Rather, it is a feature of the exponential u-curve when monetary values span a range that is much larger than the order of the risk tolerance.

$ 0.5 X | $0 ? | 0.5 −100 → $100 ? | $ 0.5 X + 100 | 0.5 0

FIGURE 22.18 Using the "Delta Property" for the Modified Deal

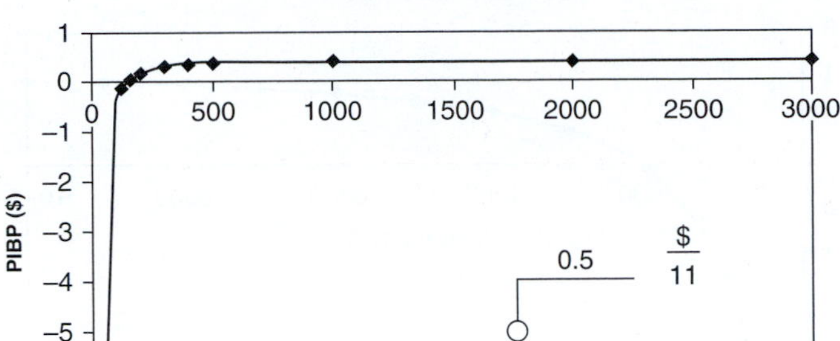

FIGURE 22.19 PIBP vs. Initial Wealth for a Logarithmic Decision Maker

Suppose, in the previous example, the decision maker has a logarithmic u-curve, $U(x) = \ln(x + \alpha)$ and rejects a deal with a 0.5/0.5 probability of getting -10 and 11 (Figure 22.16). This means that his PIBP for the deal is negative. We can plot a sensitivity analysis for the PIBP and certain equivalent of this deal versus initial wealth, α.

Figure 22.19 shows the PIBP of the deal versus the initial wealth. By direct calculation, his initial wealth must be less than $109.99 to reject this deal, as it will have a negative PIBP and also a negative certain equivalent if he owned it. If his initial wealth is larger than $109.99, then the logarithmic decision maker will value this deal at a positive amount.

Let us now provide a counter example for the argument made in the *Economist* article. Consider a decision maker with a logarithmic u-curve and initial wealth level of $105, as shown in Figure 22.20. As we discussed, this person will reject the .5/.5 deal of -10 and 11. However, suppose this decision maker faces the second deal of 0.5/0.5 of -100 and *X*. Is it true that this decision maker will not accept this second deal for any value of X? The answer is no (see Figure. 22.21).

Figure 22.21 plots a sensitivity analysis of the PIBP versus *X*, for a logarithmic decision maker with an initial wealth of $105 facing a deal of 0.5/0.5 of -100 and $*X*. Indeed, the PIBP

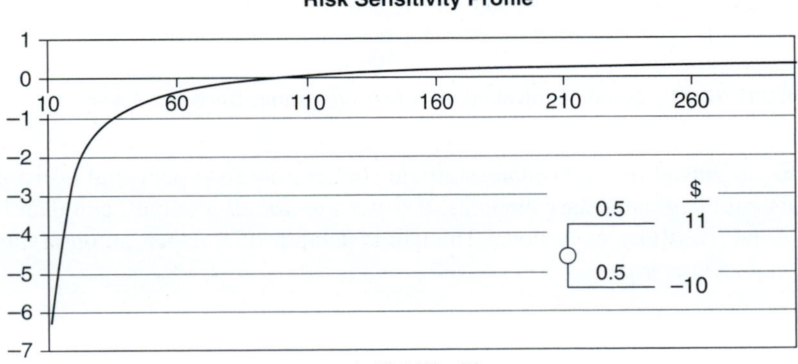

FIGURE 22.20 Certain Equivalent vs. Initial Wealth

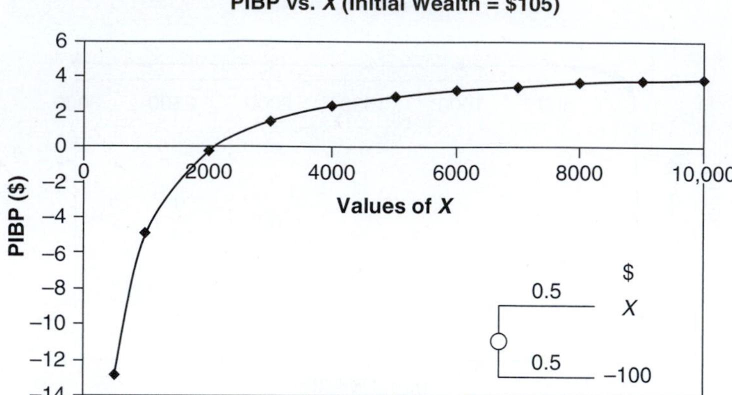

FIGURE 22.21 PIBP vs. *X* for Logarithmic Decision Maker

is negative when $X < 2100$, but it is positive when X is above this value. Therefore, the decision maker would readily accept this deal for values of X larger than 2100.

Figure 22.22 shows a sensitivity analysis of the certain equivalent of this deal for a logarithmic decision maker (if he owned the deal). As we can see from Figure 22.22, the certain equivalent of a logarithmic decision maker with initial wealth \$105 becomes positive when the value of X is greater than \$2100.

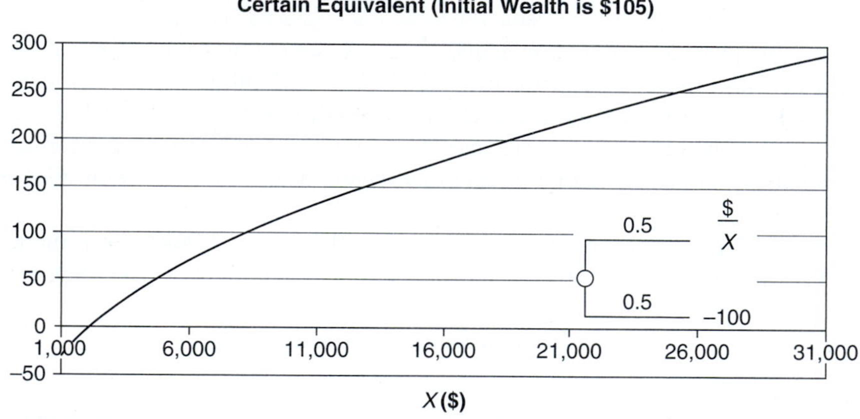

FIGURE 22.22 Certain Equivalent vs. *X* for Logarithmic Decision Maker

The authors' argument in the *Economist* article applies only to exponential decision makers facing deals with large monetary amounts. It is not true for all *u*-curves, and therefore, their criticism of the five Rules of Actional Thought is invalid. It is, rather, an observation about the use of exponential *u*-curves.

PROBLEMS

Problems marked with a dagger (†) are considered quantitative.

1. How many of the following are true for a decision maker with a logarithmic u-curve?
   I. The value of clairvoyance on a distinction of interest is equal to the value of the deal with free clairvoyance on the distinction less the value with no clairvoyance.
   II. The decision maker follows the delta property.
   III. A logarithmic decision maker exhibits more risk aversion than an exponential decision maker.
   a. 0
   b. 1
   c. 2
   d. 3
2. Given an initial wealth, w, and an uncertain deal, how many of the following must be true if you are a logarithmic decision maker?
   I. The PIBP of the deal (if you do not own it) is equal to the PISP of the deal if you owned it and have (in addition) a wealth w.
   II. If you buy the deal at your PIBP, your PISP for the deal will be equal to the PIBP at which you bought it.
   III. The PIBP of two deals is equal to the sum of PIBPs of the individual deals.
   a. 0
   b. 1
   c. 2
   d. 3

†3. A decision maker with a logarithmic u-curve owns a deal that provides a 50/50 chance of receiving either \$100 or 0. His certain equivalent for this deal is \$30.
   a. What is the value of his initial wealth, w?
   b. If a decision maker with an exponential u-curve had the same certain equivalent for this deal, what is the value of his risk tolerance?
   c. Plot a sensitivity analysis of the certain equivalent to the initial wealth from $w = \$5$ to $w = \$10{,}000$.
   d. What is the certain equivalent of the deal if the initial wealth is \$1,000; \$2,000; \$4,000; \$10,000?
   e. Does the certain equivalent increase or decrease with wealth? Compare to an exponential decision maker.

†4. A decision maker with a logarithmic u-curve is interested in purchasing a deal that provides a 50/50 chance of receiving either \$100 or 0. His PIBP for this deal is \$30.
   a. What is the value of his initial wealth, w?
   b. If a decision maker with an exponential u-curve had the same certain equivalent for this deal, what is the value of his risk tolerance?
   c. Plot a sensitivity analysis of the buying price to the initial wealth from $w = \$5$ to $w = \$10{,}000$.
   d. What is the buying price of the deal if the initial wealth is \$1,000, \$2,000, \$4,000, \$10,000?
   e. Does the buying increase or decrease with wealth? Compare to an exponential decision maker.

†5. A decision maker with a logarithmic u-curve and initial wealth \$20 faces two mutually irrelevant deals.
   I. Deal 1: Pays $\$-10$ and \$50 with equal chances.
   II. Deal 2: Pays $\$-5$ and \$100 with equal chances.
   a. What is his PIBP of Deal 1?
   b. What is his PIBP of Deal 2?
   c. What is his PIBP for the two deals?
   d. If he buys Deal 1 at his PIBP, what is his PIBP for Deal 2?
   e. If he buys Deal 1 at his PIBP, what is his PISP for Deal 1?

†6. A decision maker with a logarithmic *u*-curve owns a deal that pays $100 and $1,000 with equal probabilities. For simplicity, assume the initial wealth is zero.

a. What is his certain equivalent of this deal? Due to unforeseen circumstances, the decision maker will now receive only 70% of the deal (i.e., the decision maker will now receive either $70 or $700).

b. What is his certain equivalent of this modified deal? How does it compare to the certain equivalent of the original deal?

†7. If the decision maker did not own the deals in problem 6, but had a wealth of $1000, what is his PIBP for each deal? What is his PIBP for both deals?

†8. Calculate the value of clairvoyance for Kim if she has a logarithmic *u*-curve and has an initial wealth of $72.

CHAPTER

23

The Linear Risk Tolerance *u*-Curve

CHAPTER CONCEPTS

After reading this chapter, you will be able to explain the following concepts:

- Linear risk tolerance
- Constructing *u*-curves exhibiting linear risk tolerance

23.1 INTRODUCTION

We have now discussed three types of *u*-curves: the linear and the exponential—both of which satisfy the delta property and have constant risk aversion—and the logarithmic *u*-curve, whose risk-aversion function decreases with wealth.

In this chapter, we will present the **linear risk-tolerance *u*-curve,** which is a generalization of these *u*-curves, and some methods to assess its parameters.

23.2 LINEAR RISK TOLERANCE

Consider a *u*-curve with the property that the risk tolerance used to evaluate any monetary prospect is linear in the value of the prospect. The risk-tolerance function would be

$$\rho(x) = \rho + \eta x \tag{1}$$

where ρ and η are constants and η is positive when risk tolerance increases with wealth. This linear risk attitude is also known as the **hyperbolic absolute risk aversion (HARA),** since the risk-aversion function is hyperbolic, as

$$\gamma(x) = \frac{1}{\rho + \eta x} = \frac{1}{\rho(x)}$$

and is decreasing with wealth when η is positive. Since the risk tolerance in Equation (1) is increasing with wealth when $\eta > 0$, the risk aversion is decreasing for such positive values of η.

If we wish to have a risk tolerance linear in each prospect value as per Equation (1), what functional form must the *u*-curve have? In Appendix A, we show that the *u*-curve must have the form

$$u(x) = (\rho + \eta x)^{1-\frac{1}{\eta}}, \qquad \eta \neq 0 \text{ or } 1,\ x > -\frac{\rho}{\eta}$$

Note

The curve is only defined when $x > -\frac{\rho}{\eta}$. Any deal with a payoff less than this amount cannot be accepted by this person. It is important to remember this restriction whenever using a curve in this family.

23.2.1 Special Cases

The idea of linear risk tolerance generalizes the previously discussed *u*-curves. To illustrate:

- $\eta = 0$: This implies $\rho(x) = \rho$, and is the case of an exponential *u*-curve with constant risk tolerance.
- $\eta = 1$: This implies $\rho(x) = \rho + x$, which is the case of a logarithmic *u*-curve, as we discussed, and where ρ is the initial wealth.

As a further illustration of the generality of this *u*-curve, consider the case where $\eta = \frac{1}{2}$. This case provides the reciprocal *u*-curve $u(x) \propto -\left(\rho + \frac{1}{2}x\right)^{-1}$

23.2.2 Assessment of Parameters

Since the linear risk tolerance *u*-curve has two parameters, ρ and η, we need two indifference assessments from two uncertain deals. Then, we could solve the corresponding equations to determine the value of the parameters.

To illustrate, if a decision maker with a linear risk-tolerance *u*-curve is indifferent between owning the deal in Figure 23.1 or not, this indifference implies

$$u(0) = 0.5\,u(a) + 0.5\,u\left(-\frac{a}{2}\right)$$

This implies (by direct substitution) that

$$\rho^{1-\frac{1}{\eta}} = 0.5(\rho + \eta a)^{1-\frac{1}{\eta}} + 0.5\left(\rho - \frac{\eta a}{2}\right)^{1-\frac{1}{\eta}}$$

Note *Recall that for exponential u-curves, this value of a is approximately equal to the risk tolerance, ρ.*

Next, we determine the value of b that makes the decision maker indifferent to the deal in Figure 23.2 . This implies that

$$u(b) = 0.5\,u(2b) + 0.5\,u\left(\frac{b}{2}\right)$$

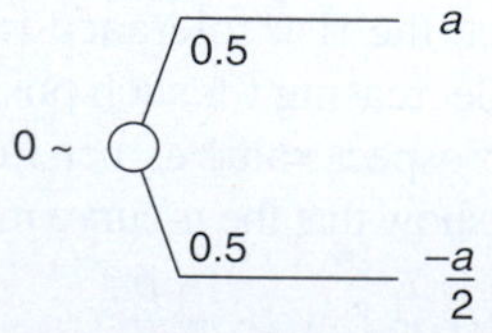

FIGURE 23.1 Indifference Assessment

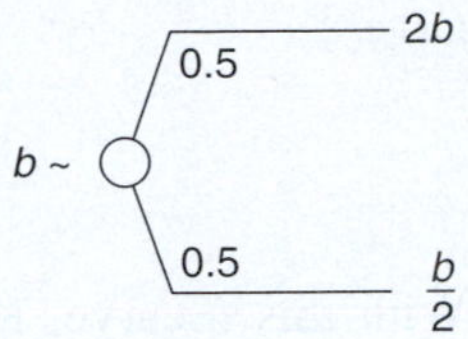

FIGURE 23.2 Indifference Assessment

By direct substitution, we get

$$(\rho + b\eta)^{1-\frac{1}{\eta}} = 0.5(\rho + \eta(2b))^{1-\frac{1}{\eta}} + 0.5\left(\rho + \eta\left(\frac{b}{2}\right)\right)^{1-\frac{1}{\eta}}$$

Solving the previous equations corresponding to the two indifference assessments gives the values of ρ and η.

We also have an approximate method to assess the parameters of the linear risk-tolerance u-curve directly (of course, with this case, we still need two assessments).

First, we offer the decision maker the deal shown in Figure 23.1 and find the value of a makes him indifferent to receiving this deal and getting 0.

This value of a makes the risk tolerance at $x = 0$ equal to

$$\rho(0) = \rho$$

The first equation we have is

$$\rho \approx a \tag{2}$$

Next, we determine the value of b that makes the decision maker indifferent to receiving the deal of Figure 23.2 or b.

This value of b is equal to the risk tolerance at $x = b$ and can be matched to the expression of the linear risk tolerance valuated at b. The second equation is

$$\rho(b) = b = \rho + \eta b \tag{3}$$

Solving gives the value of η as

$$\eta \approx 1 - \frac{a}{b} \tag{4}$$

These approximate expressions for ρ and η work well when $\eta \approx 0$, because it approaches an exponential u-curve.

23.2.3 Kim's Linear Risk-Tolerance *u*-Curve

Now we will find a linear risk-tolerance u-curve for Kim. First, we ask her to make the assessment in Figure 23.1. She says that she would be just indifferent if a were \$72. This means that she would be indifferent to a deal equally likely to pay her that amount or lose her half that, \$36. Note that \$72 is very close to the risk tolerance of her exponential curve, \$72.13. Thus, from her first assessment, we find $\rho = \$72$.

We now proceed to the second assessment of Figure 23.2. Here, she is indifferent when $b = \$144$. She finds that amount as attractive as a deal were she is equally likely to gain twice that amount, \$288, or half that amount, \$72. Her parameter η is therefore

$$\eta = 1 - \frac{a}{b} = 1 - \frac{72}{144} = 0.5$$

meaning that her linear risk-tolerance u-curve is the reciprocal u-curve

$$u(x) \propto -\left(\rho + \frac{1}{2}x\right)^{-1}$$

discussed previously. It is defined when $x > -\frac{\rho}{\eta} = -144$. With this u-curve, Kim will consider taking on any deal that does not have a larger possibility of loss than \$144.

If we want to make $u(0) = 0$, and $u(100) = 1$ as we did before, then we first write

$$u(x) = a\left(\rho + \frac{1}{2}x\right)^{-1} + b$$

and substitute to determine the values of a, b. We have

$u(0) = 0 \Rightarrow 0 = \frac{a}{\rho} + b$, and so $u(x) = a\left(\frac{1}{\rho + \frac{1}{2}x} - \frac{1}{\rho}\right)$. Then we use

$u(100) = 1 \Rightarrow a\left(\frac{\frac{1}{2}100}{\rho\left(\rho + \frac{1}{2}100\right)}\right) = 1 \Rightarrow a = \frac{\rho(\rho + 50)}{50}$. The expression for the reciprocal u-curve becomes

$$\boxed{u(x) = \left(\frac{\rho + 50}{50}\right)\left(\frac{\frac{1}{2}x}{\rho + \frac{1}{2}x}\right)}$$

Figure 23.3 plots Kim's exponential and log u-curves from Figure 22.12, as well as her linear risk tolerance u-curve for $\eta = 0.5$.

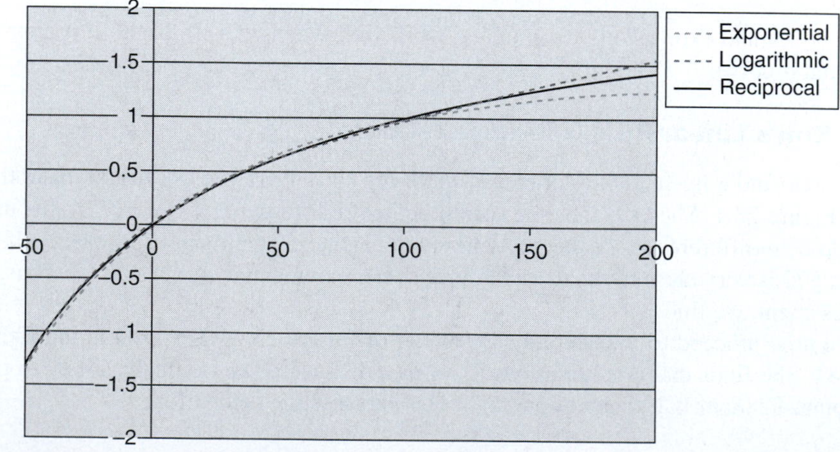

FIGURE 23.3 **Kim's Exponential, Reciprocal, and Logarithmic *u*-Curves**

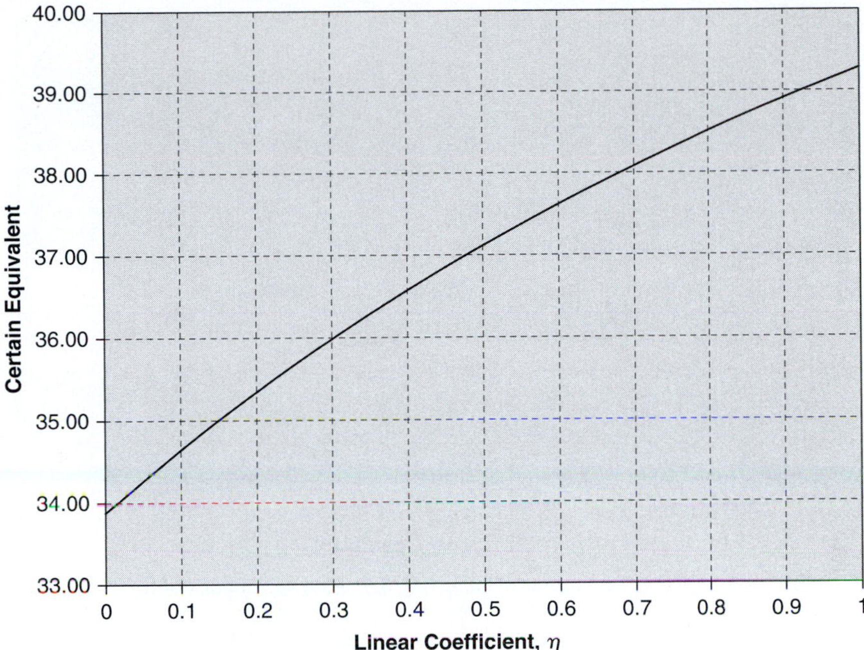

FIGURE 23.4 Certain Equivalent of [0.5, 100; 0.5, 0] vs. Linear Coefficient, η

The reciprocal curve lies between the exponential and logarithmic curves everywhere except at the points $x = 0$ and $x = 100$. Recall that Kim's exponential and logarithmic certain equivalents for the 50–50 chance of winning \$100 or 0 (which we write as [0.5, 100; 0.5, 0]), as developed in the last chapter, were \$33.90 and \$39.30. For the reciprocal u-curve, the certain equivalent is \$37.11.

As the η parameter increases from above 0 to almost 1, the LRT curve changes smoothly from the exponential to the logarithmic, and the certain equivalent of this deal does the same. Figure 23.4 shows how the certain equivalent for the [0.5, 100; 0.5, 0] deal changes as η varies from 0 to 1.

You might be tempted to think that the certain equivalent for any deal derived from the reciprocal curve will lie between those for the exponential and logarithmic curves, provided that all have the same value of ρ. Recall the deal we evaluated for Kim in Chapter 22: $[0.5, 100; 0.5, -50]$, which is a 50–50 chance of winning 100 and of losing 50. We found that exponential Kim ($\eta = 0$) had a certain equivalent of -8.54, whereas logarithmic Kim's ($\eta = 1$) was -10.51, which is an amount lower than the exponential.

Is it true that, as η changes from 0 to 1, the certain equivalent will keep decreasing? Figure 23.5 shows how the certain equivalent for this deal changes with the linear coefficient η when $\rho = \$72$.

We see that the certain equivalent at first increases with η and then decreases. The first effect is the result of increasing risk tolerance; the second is a result of increasing concern about approaching zero wealth. A peak occurs for a value of η approximately 0.33. For the reciprocal u-curve with $\eta = 0.5$ the certain equivalent is -8.28 and does not lie between the certain equivalents of the exponential and logarithmic curves.

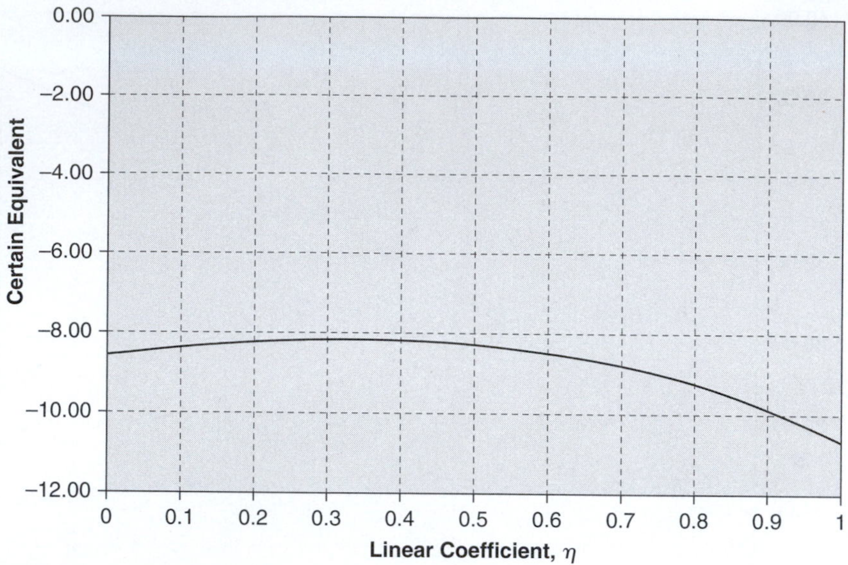

FIGURE 23.5 Certain Equivalent of [0.5, 100; 0.5, −50] vs. Linear Coefficient, η

Now, we will return to the question of finding the certain equivalent of mutually irrelevant deals. We know that the linear risk-tolerance *u*-curve reduces to the exponential only when $\eta = 0$; only in this case will it demonstrate the delta property.

In the case of reciprocal Kim ($\eta = 0.5$), let us begin by examining the case of two mutually irrelevant deals, each of the form [0.5, 100; 0.5, 0] with a resulting deal [0.25, 200; 0.5, 100; 0.5, 0], by extending what we saw in Table 22.4 (See Table 23.1).

We see that the results for the reciprocal fall between those for the other two *u*-curves.

Continuing to the case of two mutually irrelevant deals of the form [0.5,100;0.5, −50] where each deal and its composite [0.25,200;0.5,100;0.25, −100] have the possibility of a loss, we extend Table 22.5.

In Table 23.2, we see that reciprocal Kim values owning one of the deals negatively (−8.28)—but not as negatively as does either exponential or logarithmic Kim. Unlike logarithmic Kim, she could own two of them, but her certain equivalent of owning both of them (−32.71) is much more negative than exponential Kim's. As for buying such deals, reciprocal Kim would require slightly more incentive (6.97). Unlike logarithmic Kim, who would have to be paid more than 31.63 to take on both deals, reciprocal Kim would do so if paid more than 19.8.

TABLE 23.1 PIBP and PISP for Two Mutually Irrelevant Deals [0.5, 100; 0.5, 0]

	PISP for 1 Deal	PISP for 2 Deals	PIBP for 1 Deal	PIBP for 2 Deals
Exponential *u*-curve	33.9	67.8	33.9	67.8
Reciprocal *u*-curve	37.11	77.63	34.34	66.51
Logarithmic *u*-curve	39.3	83.16	34.37	61.61

TABLE 23.2 PIBP and PISP for Two Mutually Irrelevant Deals [0.5, 100; 0.5, −50]

	PISP for 1 Deal	PISP for 2 Deals	PIBP for 1 Deal	PIBP for 2 Deals
Exponential *u*-curve	−8.5	−17	−8.5	−17
Reciprocal *u*-curve	−8.28	−32.71	−6.97	−19.8
Logarithmic *u*-curve	−10.41	Could not own	−6.92	−31.63

Now, consider the square root LRT curve with $\eta = 2$, as

$$u(x) = a(\rho + 2x)^{\frac{1}{2}} + b, \quad x > -\frac{\rho}{2}$$

If Kim used it, she could not own even one [0.5, 100; 0.5, −50] deal. This is because Kim with a wealth of \$72 and an η of 2 cannot take on any deal that could lose her half of her wealth \$36. Figure 23.6 shows the variation of the certain equivalent with η. Once again, we observe that the variation is not monotonic when the deal has negative prospects. People with the same ρ and values of linear coefficient η above 1.4 cannot own this deal. This is because the *u*-curve is undefined for the (outcome \$ − 50) for $\rho = \$72$ and values of η above 1.44 since $u(x) = (\rho + \eta x)^{1-\frac{1}{\eta}} = (72 + \eta(-50))^{1-\frac{1}{\eta}}$ is zero when $\eta = 1.4$ and undefined above this value of η. The curve only makes sense when $-50 > -\rho/\eta$ or $\eta < \rho/50 = 1.44$.

Figure 23.7 plots the variation of the certain equivalent with η for the no-loss deal [0.5, 100, 0.5, 0]. Here the curve is monotonic, and there are no surprises.

The linear risk-tolerance *u*-curves give us a readily access and compute way to the increased choice in specifying risk preference. They both moderate and share the extreme behavior of the exponential for very high payoffs and the assessment difficulties of the logarithmic because of the need to consider mutually irrelevant owned deals in assessment.

23.2.4 St. Petersburg Problem for Linear Risk Tolerance

Table 23.3 shows the PIBP for the St. Petersburg game for the reciprocal *u*-curve $\eta = \frac{1}{2}$ with different values of the maximum n and different ρ. The PIBP increases as ρ increases and, of course, increases with the number of games, n.

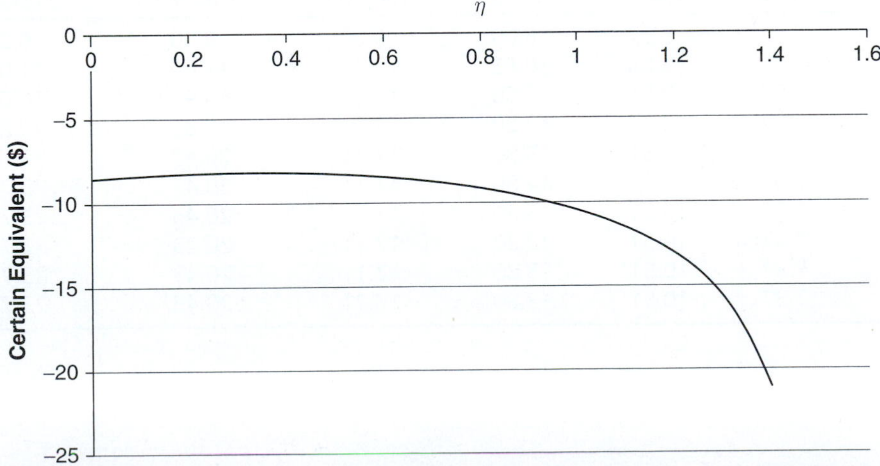

FIGURE 23.6 Sensitivity of Certain Equivalent of Deal [0.5, 100; 0.5, −50] to Value of η

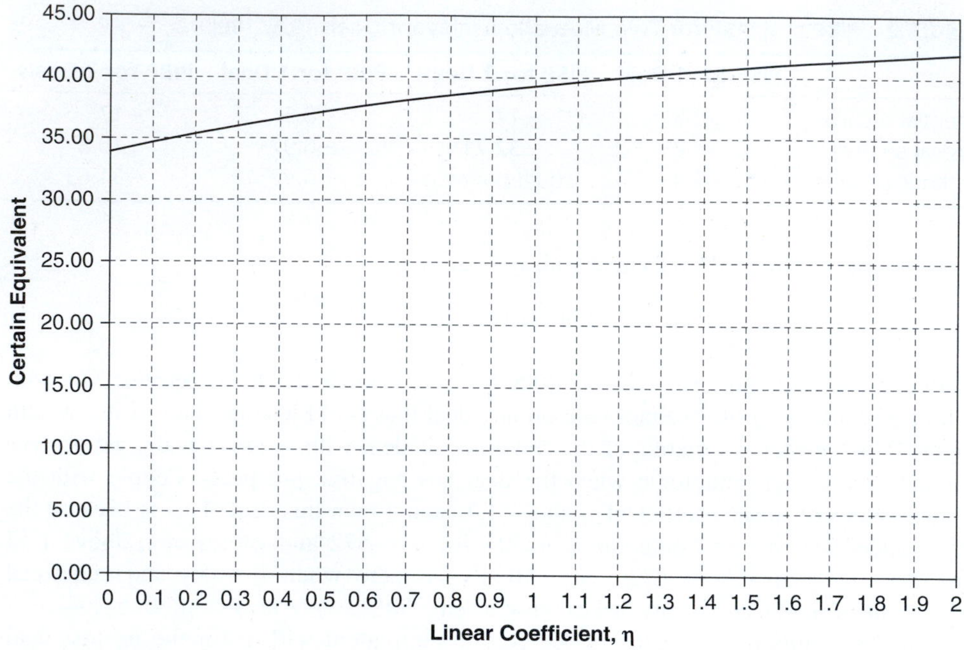

FIGURE 23.7 Sensitivity of Certain Equivalent of Deal [0.5, 100; 0.5, 0] to Value of η

Note that these values lie between the corresponding values for logarithmic and exponential *u*-curves discussed in the previous chapter. This should not come as a surprise, since there are no losses in the St. Petersburg payoffs.

TABLE 23.3 PIBP for St. Petersburg Problem for LRT *u*-Curve with $\eta = \frac{1}{2}$

n	*ρ*					
	100	1,000	10,000	100,000	1,000,000	10,000,000
5	5.74	5.97	6.00	6.00	6.00	6.00
10	7.35	9.88	10.86	10.99	11.00	11.00
15	7.37	10.51	13.66	15.56	15.95	16.00
20	7.37	10.51	13.80	17.09	19.86	20.85
25	7.37	10.51	13.80	17.11	20.43	23.63
30	7.37	10.51	13.80	17.11	20.43	23.75
35	7.37	10.51	13.80	17.11	20.43	23.75
40	7.37	10.51	13.80	17.11	20.43	23.75
45	7.37	10.51	13.80	17.11	20.43	23.75
50	7.37	10.51	13.80	17.11	20.43	23.75

TABLE 23.4 PIBP for St. Petersburg Problem for LRT u-Curve with $\eta = 2$

n	ρ					
	100	1,000	10,000	100,000	1,000,000	10,000,000
5	5.75	5.97	6.00	6.00	6.00	6.00
10	8.13	10.10	10.86	10.99	11.00	11.00
15	8.63	11.60	14.16	15.60	15.95	16.00
20	8.72	11.89	15.04	17.94	20.05	20.85
25	8.74	11.94	15.20	18.44	21.53	24.13
30	8.74	11.95	15.23	18.53	21.81	25.00
35	8.74	11.95	15.24	18.55	21.86	25.16
40	8.74	11.95	15.24	18.55	21.87	25.19
45	8.74	11.95	15.24	18.55	21.87	25.20
50	8.74	11.95	15.24	18.55	21.87	25.20

For comparison, Table 23.4 shows the PIBP for the St. Petersburg game for $\eta = 2$ and various values of the maximum n and ρ.

***Note** The PIBP in both tables increases as η increases.*

Notice also that these values lie above the corresponding values for logarithmic u-curve discussed in the previous chapter. When there is no possibility of loss, these square-root u-curve people are more willing than logarithmic people to pay a higher fraction of their wealth for the game.

23.3 SUMMARY

Linear risk-tolerance u-curves can overcome the saturation effects of the exponential for very large payoffs. This has particular practical application to pharmaceutical drug development. Each new drug tested may require tens to hundreds of millions of dollars of investment for a relatively low probability of creating a blockbuster drug that will produce tens of billions in sales.

The linear risk tolerance family of u-curves models a wide range of preferences that include the exponential and logarithmic u-curves. The curve has two parameters, ρ and η, that define a risk tolerance as $\rho(x) = \rho + \eta x$. The case $\eta = 0$ implies $\rho(x) = \rho$ and represents the exponential u-curve with constant risk tolerance. The case $\eta = 1$ implies $\rho(x) = \rho + x$, which represents the logarithmic u-curve.

The linear risk-tolerance u-curve exhibits decreasing risk aversion with wealth when $\eta > 0$.

We have an approximate method to assess both parameters.

If a deal has losses, the certain equivalent does not necessarily change monotonically with the value of the linear coefficient, η.

KEY TERMS

- Hyperbolic absolute risk aversion (HARA)
- Linear risk-tolerance u-curve

APPENDIX A Derivation of Linear Risk Tolerance *u*-Curve

Note that risk tolerance can be expressed in terms of the derivative of a *u*-curve as

$$\rho(x) = -\frac{u'(x)}{u''(x)} = \left(-\frac{d}{dx}\ \ln(u'(x))\right)^{-1}$$

The equation for the *u*-curve is

$$-\frac{d}{dx}\ \ln(u'(x)) = \frac{1}{\rho + \eta x}$$

Taking the integral of both sides with respect to x gives

$$-\ln(u'(x)) = \frac{1}{\eta}\ \ln(\rho + \eta x) + c, \qquad \eta \neq 0,\ x > -\frac{\rho}{\eta}$$

where c is a constant of integration. Raising both sides as exponents to e gives

$$u'(x) = (\rho + \eta x)^{-\frac{1}{\eta}} k, \qquad \eta \neq 0, x > -\frac{\rho}{\eta}$$

where k is a constant equal to e^c. Integrating both sides again with respect to x gives

$$u(x) = k\frac{1}{\eta}\frac{\eta}{\eta - 1}\ (\rho + \eta x)^{1-\frac{1}{\eta}} + d, \quad \eta \neq 0 \text{ or } 1,\ x > -\frac{\rho}{\eta}$$

where d is a constant of integration.

We can now write the equation of the *u*-curve that satisfies linear risk tolerance as

$$u(x) \infty (\rho + \eta x)^{1-\frac{1}{\eta}}, \quad \eta \neq 0 \text{ or } 1,\ x > -\frac{\rho}{\eta}$$

Therefore,

$$u(x) = a\,(\rho + \eta x)^{1-\frac{1}{h}} + b, \quad \eta \neq 0 \text{ or } 1, x > -\frac{\rho}{\eta}$$

and a, b are constants.

APPENDIX B Student's Problem Using Linear Risk Tolerance *u*-Curve

Let us return to the *Economist* article discussed in Appendix C of Chapter 22. There we showed that the apparent absurdity in preference discussed was the result of assuming that the u-curve was exponential. We showed that, if the person had a logarithmic u-curve, the issue disappeared. Moreover, it will disappear with many other u-curves, including the linear risk-tolerance u-curves discussed in this chapter.

We shall illustrate with the reciprocal u-curve characterized by a linear risk-tolerance coefficient of $^1/_2$. First, we show someone a deal with a 50–50 chance of winning \$11 and losing \$10 is [0.5, 11; 0.5, -10]. The person will not accept the deal. By calculation, we find that this will be true if the ρ parameter of the u-curve is 109.99 or less. We fix this parameter at $\rho = 105$ (as we did in the previous chapter) and then evaluate the other question raised in the article: How large would X have to be for the person to accept the deal [0.5, X; 0.5, -100]? We find that this deal will have a positive certain equivalent for X larger than \$2,100. Recall that, for a logarithmic person who made the same initial choice about [0.5, 11; 0.5, -10], X also would have to be greater than \$2,100.

However, as revealed by Table 23.5, the analysis has more complexity. For the linear coefficients shown, not only is it impossible to find an X for $\eta = 0$—as we expect since it is the exponential case—but it is also impossible to find one for coefficients 0.1, 0.2, and 0.3. Furthermore, the values of X required decrease from about 0.45 to 0.8 and thereafter increase until they are the same for coefficients 0.5 and 1.

TABLE 23.5 Payment X Required to Accept [0.5. X; 0.5, -100] after Refusing [0.5,11; 0.5, -10] for $\rho = 105$ and Various Linear Coefficients η (* = no possible X)

Linear Coefficient	X Required
0	*
0.1	*
0.2	*
0.3	*
0.4	*
0.45	5937.38
0.5	2100
0.6	1306.09
0.7	1132.32
0.8	1125.91
0.9	1285.03
1	2100

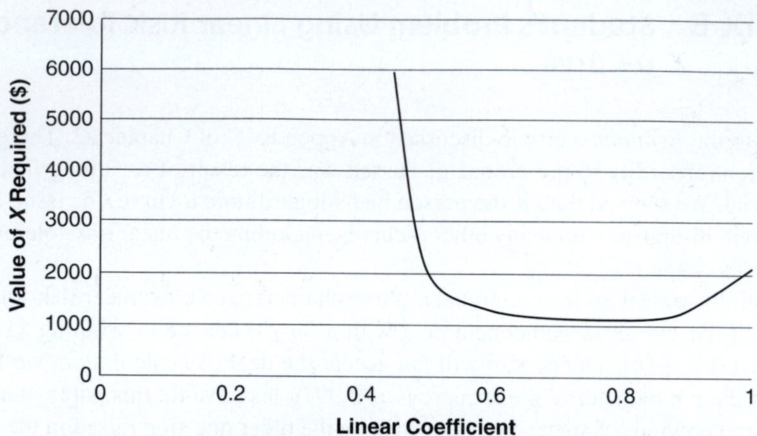

FIGURE 23.8 Payment X Required to Accept [0.5, X; 0.5, −100] after Refusing [0.5,11; 0.5, −10] for Various Linear Coefficients η

This behavior appears graphically in Figure 23.8.

The rapid increase of X as the coefficient falls below 0.45 suggests that, as was the case with exponential Kim, there is a minimum coefficient below which increases in X will not make the deal more desirable. We find that an X of \$1,000,000 would not make the deal desirable to a person with a linear coefficient of 0.4 or below, and that even an X of \$1,000,000,000 would have a negative certain equivalent for someone whose coefficient was 0.4 or below.

PROBLEMS

Problems marked with an asterisk (*) are considered more challenging.

1. Consider the two assessments in difference shown in Figures 23.1 and 23.2 that we repeat for convenience.

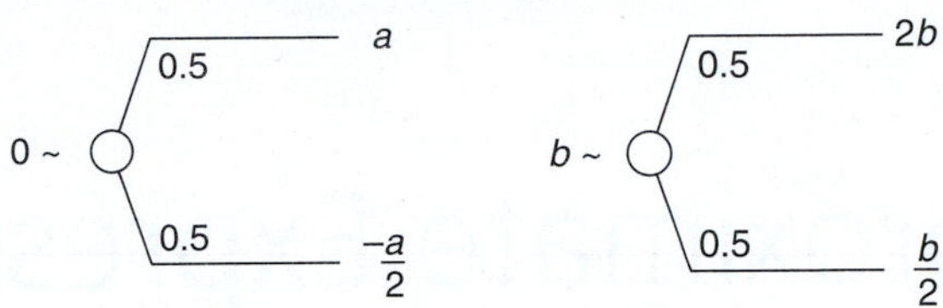

For a decision maker with a linear risk-tolerance u-curve, we have

$$u(X) = (\rho + \eta X)^{1-\frac{1}{\eta}}$$

Calculate the exact and approximate values of ρ and η when

I. $a = 100$; $b = 200$
II. $a = 100$; $b = 500$
III. $a = 100$; $b = 150$

***2.** Calculate the PIBP and PISP for a deal [0.5, 0; 0.5, X] for a linear risk tolerance decision maker with $\rho = 72$ and $\eta = 0.5$

X	PISP	PIBP
10		
100		
1000		
10,000		
100,000		
1,000,000		

CHAPTER

24

Approximate Expressions for the Certain Equivalent

CHAPTER CONCEPTS

After reading this chapter, you will be able to explain the following concepts:

- Moments of a measure
- Central moments of a measure
- Approximating the certain equivalents using moments and central moments
- Risk premium
- Cumulants of a measure

24.1 INTRODUCTION

Sometimes, we find it convenient to compute the approximate certain equivalent of a deal instead of calculating its exact value. This approach provides computational simplicity and also shows the role of the risk-aversion function and other properties of the distribution in determining the certain equivalent.

In this chapter, we develop approximate expressions for the certain equivalent of an uncertain deal in terms of the risk-aversion function and properties called moments of the probability distribution for the deal. We will define moments and illustrate their calculation. Later, in Chapter 35, we will show how to use the moments of a distribution to convert a continuous distribution into a form that takes on only discrete values.

24.2 MOMENTS OF A MEASURE

24.2.1 The First Moment

The **first moment** of a measure is simply its *e-value* or *arithmetic mean*. The first moment is equal to the sum of all the products of each value of the measure with its corresponding probability. To illustrate, consider the probability tree in Figure 24.1.

For this tree, we can calculate

$$\text{First moment } (e\text{-value}) = p_1x_1 + p_2x_2 + p_3x_3 + p_4x_4 = \sum_{i=1}^{4} p_i x_i.$$

To illustrate the calculation of the first moment numerically, consider the probability tree and corresponding probability distribution of Figure 24.2.

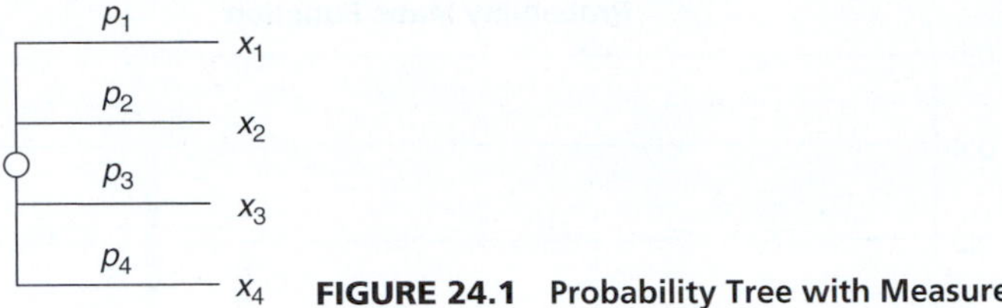

FIGURE 24.1 Probability Tree with Measure

The first moment of the value measure shown in the tree is

$$\text{First Moment } (e\text{-value}) = 0.1(-1) + 0.3(2) + 0.4(3) + 0.2(4) = 2.5$$

From here on, we use the notation $<x|\&>$ to refer to the first moment of the measure x (its e-value), where the symbol $<>$ specifies that this is a moment calculation. We use it to differentiate this notation from the notation for a probability distribution, $\{x|\&\}$. The & in the $<x|\&>$ expression reminds us that this calculation is based on our state of information.

The quantity to the left of the vertical bar is the measure whose first moment we are calculating. Since the first moment has special importance and is widely used, it is often given a special symbol, $\bar{x}$, for compactness.

We begin our approximation of the certain equivalent by computing the e-value of the deal. As we have seen, a risk-neutral decision maker's certain equivalent of an uncertain deal is equal to the e-value of the value measure. We have also shown that the certain equivalent of a deal is less than its e-value for a risk-averse decision maker and more than its e-value for a risk-preferring decision maker. Therefore, we must make further adjustments to the e-value to provide a more accurate expression for the certain equivalent for those who are not risk-neutral.

The first moment has two geometric interpretations.

1. The first interpretation uses the **probability distribution**, or **probability mass function**. Think of the columns in Figure 24.3 as if they had weights equal to their heights and were sitting on a weightless bar. Suppose you had to place a pivot that balanced the bar so it would not tilt in either direction. Where would you place it? The first moment is the position of this pivot. If we shift the distribution to the right or left by any amount, it is clear that the pivot also will be shifted by the same amount. This means that, if we add a fixed amount to all values of a measure, the first moment of the new distribution will be the original first moment plus that fixed amount.

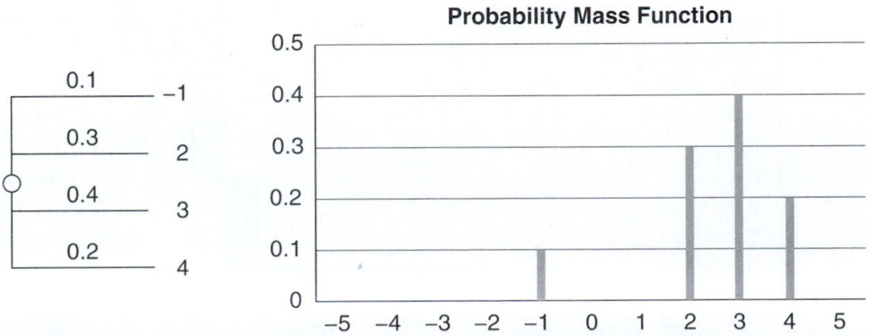

FIGURE 24.2 Probability Tree and Corresponding Probability Distribution

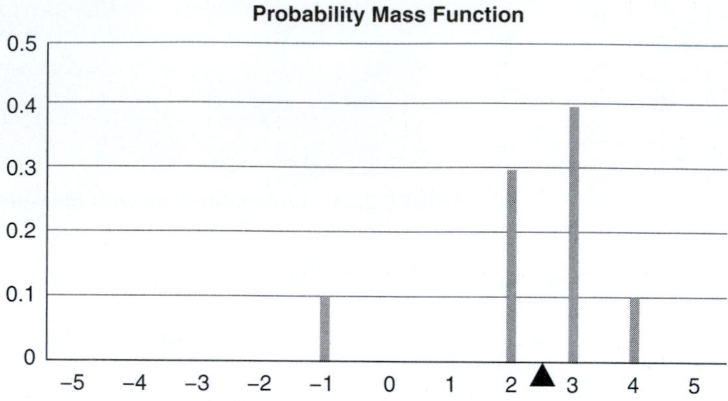

FIGURE 24.3 The First Moment is the Pivot

2. The second geometric interpretation of the first moment uses the cumulative probability distribution of the deal shown in Figure 24.4. Suppose you drew a vertical line in the plane of the cumulative distribution such that the area to the left of the line that lies under the cumulative distribution curve and above zero (Area A_1) is equal to the area to the right of the line that lies above the cumulative distribution curve and below 1 (area A_2). The value of

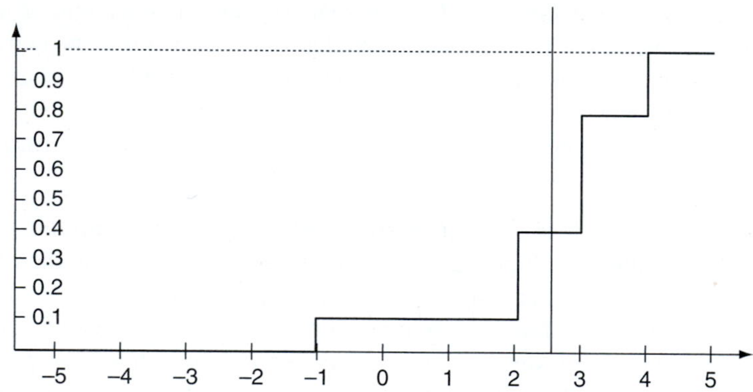

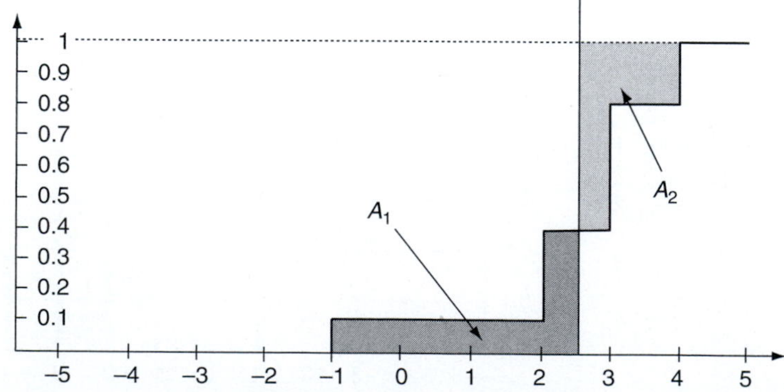

FIGURE 24.4 Interpreting the First Moment Using the Cumulative Distribution

the first moment is the equal to the location of the line that makes the two areas equal. You may want to confirm that this method will always yield the same first moment produced by the pivot method.

We have discussed what happens to the first moment when we add a fixed amount to all outcomes of the measure. But what happens to the first moment if we multiply all values of the measure by a fixed amount m? The new first moment will be m times the previous first moment, since

$$< mx|\&> = p_1 mx_1 + p_2 mx_2 + p_3 mx_3 + p_4 mx_4 = m\sum_{i=1}^{4} p_i x_i = m < x|\&> = m\bar{x}$$

24.2.2 The Second (and Higher Order) Moments

The **second moment** of a measure is the sum of all the products of the measure squared and their corresponding probabilities. For consistency, we use the notation $<x^2|\&>$ for the second moment, where for the tree in Figure 24.1, we have

$$<x^2|\&> = p_1 x_1^2 + p_2 x_2^2 + p_3 x_3^2 + p_4 x_4^2 = \sum_{i=1}^{n} p_i x_i^2$$

For the tree in Figure 24.2, the second moment is

$$<x^2|\&> = 0.1(-1)^2 + 0.3(2)^2 + 0.4(3)^2 + 0.2(4)^2 = 8.1$$

Note *If there is no uncertainty, and the pivot is at 0, then the second moment is 0, since* $<x^2|\&> = 1(0)^2 = 0$.

Similarly, we can calculate the kth moment of a measure and refer to it as

$$<x^k|\&> = p_1 x_1^k + p_2 x_2^k + p_3 x_3^k + p_4 x_4^k = \sum_{i=1}^{n} p_i x_i^k$$

For the tree in Figure 24.2, the third moment is

$$<x^3|\&> = 0.1(-1)^3 + 0.3(2)^3 + 0.4(3)^3 + 0.2(4)^3 = 25.9$$

If a distribution is symmetric and its first moment, or **pivot**, is at the origin, then its third moment must be 0. The reason is that for every value of x_i that occurs with a probability p_i there will be a value of $-x_i$ that occurs with a probability p_i.

$$<x^3|\&> = p_1(-x_1)^3 + p_1(x_1)^3 + p_2(-x_2)^3 + p_2(x_2)^3 + \ldots + p_n(-x_n)^3 + p_n(x_n)^3 = 0$$

What happens to the second moment if we multiply all measure values by any number m? The new second moment will be m^2 times the previous second moment, since

$$\text{Second moment of new deal} = p_1(mx_1)^2 + p_2(mx_2)^2 + p_3(mx_3)^2 + p_4(mx_4)^2 = m^2\sum_{i=1}^{4} p_i x_i^2$$

Similarly, if we multiply all measure values by m, the new kth moment will be m^k times the previous kth moment, since

$$\text{New } k\text{th moment} = \sum_{i=1}^{4} p_i (m x_i)^k = m^k \sum_{i=1}^{4} p_i x_i^k$$

24.3 CENTRAL MOMENTS OF A MEASURE

We calculate **central moments** of a measure by first subtracting the first moment from every measure value and then computing the moments of the new measure values. Equivalently, we shift the probability distribution to the left by an amount equal to the first moment and then calculate the moments of this new probability distribution. Of course, the pivot of the new distribution will be at zero, since the first moment follows the same shift amount, and so the first central moment of a distribution must be zero.

To numerically illustrate the calculation of the central moments, refer back to Figure 24.2. If we subtract 2.5, which is the first moment, from every prospect in the tree, we get the new tree of Figure 24.5, which represents a deal with payoffs $-3.5, -0.5$, 0.5, and 1.5 having the same probabilities as the first deal. The central moments of the tree in Figure 24.2 are the moments of the tree in Figure 24.5.

Suppose, we also shifted the probability distribution of Figure 24.2 to the right by an amount equal to 2.5, which is its first moment. We create a new probability distribution whose pivot is—of course—now at zero. The central moments determined by the probability distribution of Figure 24.2 and the left side of Figure 24.6 are the moments of the probability distribution on the right side of Figure 24.6.

We can easily verify that the first central moment, which is the *e*-value of this tree, is zero. The first moment of the shifted tree is

$$0.1(-3.5) + 0.3(-0.5) + 0.4(0.5) + 0.2(1.5) = 0 \text{ (as expected)}$$

The second central moment of the tree in Figure 24.2 is

$$0.1(-3.5)^2 + 0.3(-0.5)^2 + 0.4(0.5)^2 + 0.2(1.5)^2 = 1.85$$

Note *The second central moment cannot be negative, nor can any other even central moment.*

The third central moment is

$$0.1(-3.5)^3 + 0.3(-0.5)^3 + 0.4(0.5)^3 + 0.2(1.5)^3 = -3.6$$

The second central moment has an interesting interpretation. As we discussed, if there is no uncertainty, the probability distribution must be concentrated at the pivot, or the mean value. Since there is no positive probability assigned to any measure value besides zero and when there is no uncertainty, the second central moment is zero. Therefore, the second central moment is referred to as the **variance of the deal**. We also give it a special symbol, x^v. In accordance with the pivot analogy of Figure 24.3, the variance corresponds to the torque required to rotate the figure about a vertical axis through the pivot point: the larger the variance, the larger the torque.

If two probability distributions are similar in every aspect except that one is shifted to the right or left with respect to the other, they must have the same central moments. Their position has no effect on their central moments.

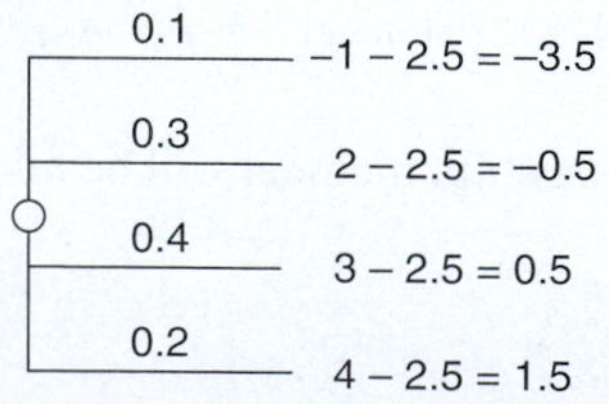

0.1 —— $-1 - 2.5 = -3.5$
0.3 —— $2 - 2.5 = -0.5$
0.4 —— $3 - 2.5 = 0.5$
0.2 —— $4 - 2.5 = 1.5$

FIGURE 24.5 Subtracting Mean Value from Every Prospect in Figure 24.2

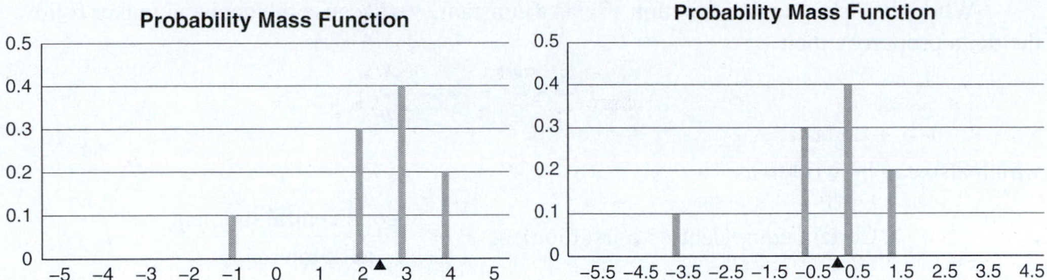

FIGURE 24.6 Shifting Probability Distribution to the Right by Amount Equal to First Moment

It is also easy to verify that the second central moment is equal to the second moment minus the first moment squared.

If there is no uncertainty, then the third central moment must also be 0. Furthermore, if a distribution is symmetric about its mean, then its third central moment—and indeed all odd central moments—must be zero.

However, if a distribution is not symmetric—for example, by being **skewed**, or stretched out more in one direction than the other—the third central moment will not be 0. For example, in the tree of Figure 24.6, the distribution is skewed to the left. As a result, the third central moment is negative. If it were skewed to the right, the third central moment would be positive.

If we multiply all outcomes at the end of the tree by a fixed amount m, the new kth central moment will be m^k times the previous kth central moment, since

$$k\text{th central moment} = \sum_{i=1}^{4} p_i(mx_i - m <x|\&>)^k = m^k \sum_{i=1}^{4} p_i(x_i - <x|\&>)^k$$

For example, the variance of a scaled deal is m^2 times the variance of the unscaled deal, as

$$(mx)^v = m^2 x^v$$

24.4 APPROXIMATING THE CERTAIN EQUIVALENT USING FIRST AND SECOND CENTRAL MOMENTS

We now provide an approximation for the certain equivalent, $\tilde{x}$, in terms of the first moment and second central moment as

$$\tilde{x} \approx \bar{x} - \gamma(\bar{x})\frac{x^v}{2}$$

where $\gamma(\bar{x})$ is the risk-aversion function evaluated at the first moment.

The certain equivalent is approximately the first moment of the deal less an amount equal to one-half the second central moment (variance) times the risk-aversion function evaluated at the first moment of the deal. The difference between the certain equivalent and the first moment is known as the **risk premium**. Since the second central moment cannot be negative, this expression implies that the certain equivalent is less than or equal to the first moment when the risk-aversion function is positive. When the risk-aversion function is negative (risk preferring), the certain equivalent is larger than the first moment.

When the risk-aversion function $\gamma(x)$ is a constant, γ, (because the decision maker follows the delta property), then

$$\tilde{x} \approx \bar{x} - \gamma \frac{x^v}{2} = \bar{x} - \frac{x^v}{2\rho}$$

which also can be written as

$$\text{Certain equivalent} \approx \text{First moment} - \frac{1}{2}\,\frac{\text{Second central moment}}{\text{Risk tolerance}}$$

We have illustrated how to determine the risk tolerance by direct assessment. All we must do to find the approximate certain equivalent is to calculate the first moment and second central moment of the uncertain deal.

24.4.1 Accuracy of the Certain Equivalent Approximation

To examine the accuracy of the certain equivalent approximation, we offer the example of a deltaperson facing a deal with a 0.50 probability of paying \$100 and a 0.5 probability of paying 0.

The first moment of this deal, $\bar{x}$, is

$$\sum_{i=1}^{2} p_i(x_i) = 0.5 * 100 + 0.5 * 0 = 50$$

The second moment, $\overline{x^2}$, is

$$\sum_{i=1}^{2} p_i(x_i)^2 = 0.5 * (100)^2 + 0.5 * (0)^2 = 5000$$

The second central moment is the second moment less the first moment squared, so $5000 - 2500 = 2500$.

The approximate certain equivalent for this deal is

$$\tilde{x} \approx 50 - \gamma \frac{2500}{2}$$

To derive the exact certain equivalent for a deltaperson with risk-aversion coefficient, γ, and u-curve $U(x) = -e^{-\gamma x}$, we calculate the e-value of the u-values as

$$U(\tilde{x}) = 0.5 * \left(-e^{-\gamma 100}\right) + 0.5\left(-e^{-\gamma 0}\right) = -0.5\left(1 + e^{-\gamma 100}\right) = -e^{-\gamma \tilde{x}}$$

which gives an exact value of the certain equivalent as

$$\tilde{x} = -\frac{1}{\gamma} \ln\left[0.5\left(1 + e^{-\gamma 100}\right)\right]$$

Figure 24.7 shows the risk sensitivity profile comparing the exact and approximate certain equivalents for different values of the risk-aversion coefficient γ. The horizontal axis also shows the risk tolerance ρ and the reciprocal of γ. The straight dashed line on the plot is the certain equivalent approximation.

Note that, at a risk-aversion coefficient of 0.02 or a risk tolerance of 50, the certain equivalent of the deal is 28.31, whereas the approximation gives a certain equivalent of 25. For a risk-aversion coefficient of 0.01 or less, the approximation is quite close to the actual certain

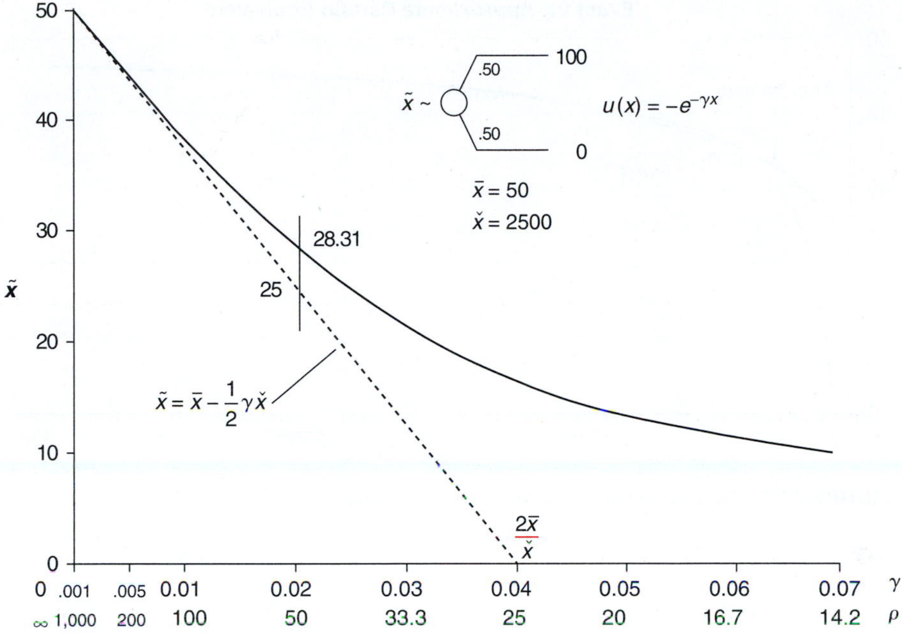

FIGURE 24.7 Sensitivity Analysis–Certain Equivalent vs. Risk Aversion

equivalent and is typically excellent when the spread of the distribution (or the difference between maximum and minimum values) is less than the risk tolerance. However, it still may be excellent in other cases, depending on the shape of the *u*-curve.

We can also repeat the analysis for a logarithmic *u*-curve as $u(x) = \ln(x + \alpha)$. As we discussed in Chapter 22, the risk-aversion function for a logarithmic *u*-curve is $\gamma(x) = \frac{1}{x + \alpha}$, so $\gamma(\bar{x}) = \frac{1}{\bar{x} + \alpha}$. The approximate certain equivalent is then

$$\tilde{x} \approx 50 - \frac{1}{50 + \alpha}\frac{2500}{2}$$

Figure 24.8 plots the exact versus the approximate certain equivalent for different values of the wealth, α. The figure shows that the approximation is better as the wealth α increases.

For this example, note that the approximate certain equivalent is higher than the exact. For an initial wealth of $\alpha = 50$, the approximate certain equivalent is 37.50, whereas the exact is 36.60. The difference becomes even smaller as the initial wealth α increases.

24.5 APPROXIMATING THE CERTAIN EQUIVALENT USING HIGHER ORDER MOMENTS

We can also provide a better approximation for the certain equivalent by including the third central moment as

$$\tilde{x} \approx \bar{x} - \gamma(\bar{x})\frac{x^{v}}{2} - \eta(\bar{x})\frac{< (x - \bar{x})^3 | \& >}{6}$$

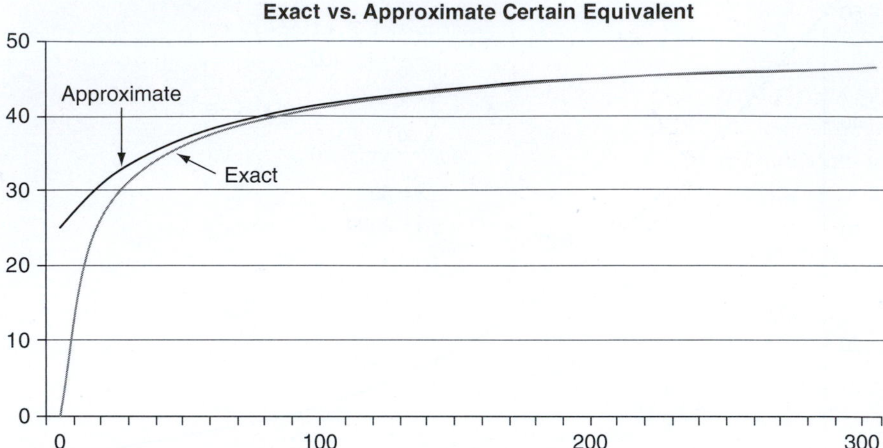

FIGURE 24.8 Exact vs. Approximate Certain Equivalent

where $\eta(x) = -\dfrac{U'''(x)}{U'(x)}$ is the negative ratio of the third derivative to the first derivative of the utility function and $< (x - \bar{x})^3 | \& >$ is the third central moment of the deal.

For an exponential u-curve, $u(x) = -e^{-\gamma x}$, $u'(x) = \gamma e^{-\gamma x}$, $u''(x) = -\gamma^2 e^{-\gamma x}$, and $u'''(x) = \gamma^3 e^{-\gamma x}$.

Hence, $\gamma(x) = \gamma$, $\eta(x) = -\gamma^2$ and

$$\tilde{x} \approx \text{First moment} - \gamma \frac{\text{Second central moment}}{2} + \gamma^2 \frac{\text{Third central moment}}{6}$$

The first central moment is equal to

$$0.25 * 5.875 + 0.25 * 0.875 + 0.125 * -1.125 + 0.375 * -4.125 = 0 \text{ (as expected)}$$

The second central moment is equal to

$$0.25 * (5.875)^2 + 0.25 * (.875)^2 + 0.125 * (-1.125)^2 + 0.375 * (-4.125)^2$$
$$= 0.25 * (34.52) + 0.25 * (0.77) + 0.125 * (1.27) + 0.375 * (17.02) = 15.359$$

We can also calculate the second central moment using the first and second moments calculated previously as

$$<x|\&>^v = <x^2|\&> - <x|\&>^2 = 32.375 - (4.125)^2 = 15.359 \text{ (as expected)}$$

The third central moment is

$$0.25 * (5.875)^3 + 0.25 * (.875)^3 + 0.125 * (-1.125)^3 + 0.375 * (-4.125)^3$$
$$= 0.25 * (202.77) + 0.25 * (.6699) + 0.125 * (-1.42) + 0.375 * (-70.1895) = 24.363$$

EXAMPLE 24.1 Certain Equivalent Calculations

Consider the probability tree shown in Figure 24.9. We now calculate the certain equivalent for an exponential decision maker using the exact versus the approximate expression using the first three moments.

The *e*-value (first moment) is equal to

$$<x|\&> = 0.25 * 10 + 0.25 * 5 + 0.125 * 3 + 0.375 * 0 = 4.125$$

To calculate the central moments, we first subtract the *e*-value ($4.125) to get the new tree shown in Figure 24.10, which is equivalent to calculating the moments of a tree that subtracts the *e*-value from every prospect value.

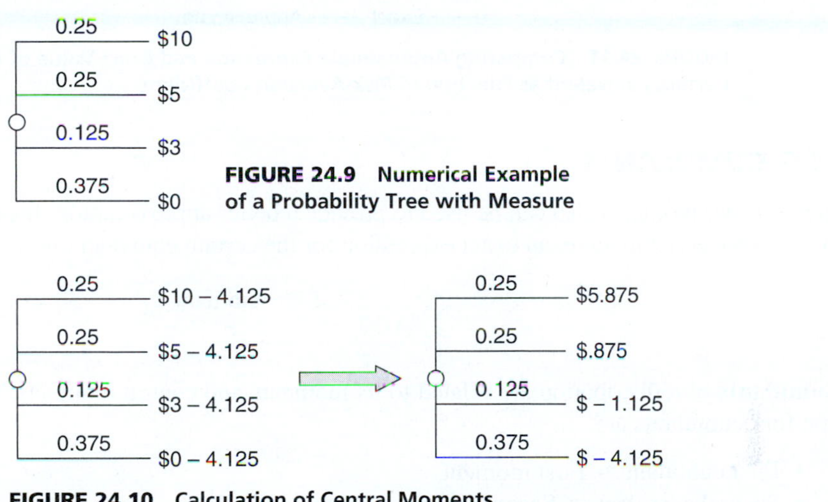

FIGURE 24.9 **Numerical Example of a Probability Tree with Measure**

FIGURE 24.10 **Calculation of Central Moments**

The approximate certain equivalent is

$$\tilde{x} \approx \text{First moment} - \gamma \frac{\text{Second central moment}}{2} + \gamma^2 \frac{\text{Third central moment}}{6}$$

$$= 4.125 - \gamma \frac{15.359}{2} + \gamma^2 \frac{24.363}{6}$$

We notice from this approximate expression that when the risk aversion is positive (risk-averse decision maker), the increase in variance decreases the certain equivalent. On the other hand, a positive skewness of the deal—when skewed to the right—increases the certain equivalent. Figure 24.11 plots the certain equivalent versus the risk-aversion coefficient for an exponential decision maker using the approximate formula and the exact calculation.

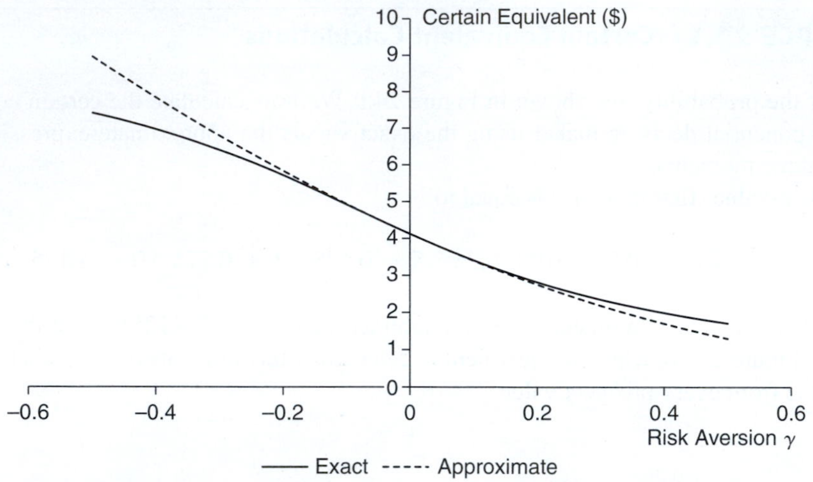

FIGURE 24.11 Comparing Approximate Expression and Exact Value of the Certain Equivalent as Function of Risk-Aversion Coefficient

24.6 CUMULANTS

Higher-order moments also can be used to provide a better approximation. It can be shown that, for an exponential *u*-curve, an exact expression for the certain equivalent is

$$\tilde{x} = \sum_{i=1}^{\infty} (-\gamma)^{i-1} \frac{(i\text{th cumulant})}{i!}$$

Cumulants of a distribution are related to its moments and central moments. For example, the first four cumulants are

- First cumulant = First moment
- Second cumulant = Second central moment
- Third cumulant = Third central moment
- Fourth cumulant = Fourth central moment $-3 \times$ (Second central moment)^2

24.7 SUMMARY

Calculating moments of a deal: The *k*th moment of a deal is denoted $<x^k | \& >$ and is

$$<x^k | \& > = p_1 x_1^k + p_2 x_2^k + p_3 x_3^k + p_4 x_4^k = \sum_{i=1}^{n} p_i x_i^k$$

The first moment is often given the special symbol $\bar{x}$. The first moment of a distribution has two graphical interpretations.

- In the probability distribution (mass function), it is the pivot.
- In the cumulative probability distribution, it is the point that would be chosen such that the area to the left of a vertical bar passing through the point and the under the cumulative distribution curve is equal to the area to the right of the vertical bar above the cumulative distribution curve.

Scaling expression for the moments: If we receive a fraction (or multiple), m, of the deal, the new moments are

$$k\text{th moment of scaled deal} = < (mx)^k | \& > = \sum_{i=1}^{n} p_i (mx_i)^k = m^k \sum_{i=1}^{n} p_i (x_i)^k = m^k < x^k | \& >$$

For example, the first moment of the scaled deal is m times the first moment of the unscaled deal.

Calculating central moments of a deal:

$$k\text{th central moment} = \sum_{i=1}^{n} p_i (x_i - < x | \& >)^k = \sum_{i=1}^{n} p_i (x_i - < x | \& >)^k$$

The variance is the second central moment. It is a measure of spread of the distribution around the pivot. It is given the special symbol x^v.

The skewness is the third central moment. It is a measure of asymmetry.

Scaling expression for the central moments: If we receive a fraction (or multiple), m, of the deal, the new central moments are

$$k\text{th central moment} = \sum_{i=1}^{n} p_i (mx_i - m < x | \& >)^k = m^k \sum_{i=1}^{n} p_i (x_i - < x | \& >)^k$$

The approximate expression for certain equivalent of a deal using three central moments is

$$\tilde{x} \approx \bar{x} - \gamma(\bar{x}) \frac{x^v}{2} - \eta(\bar{x}) \frac{< (x - \bar{x})^3 | \& >}{6}$$

KEY TERMS

- First moment
- Central moment
- Probability distribution
- Probability mass function
- Risk premium
- Second moment
- Skewed
- Variance of the deal

PROBLEMS

Problems marked with an asterisk (*) are considered more challenging.

1. Calculate the first four moments, central moments, and cumulants of the following deals:
   I. 0.5 probability of paying $100 and a 0.5 probability of paying zero; i.e. [0.5, 100; 0.5, 0].
   II. The deal [0.1, 0; 0.2, 100; 0.6, 500; 0.1, 1000].

***2.** Plot the approximate certain equivalent using the first moment, second central moment, and third central moment approximation vs. the risk aversion coefficient for an exponential decision maker for the deal [0.5, 100; 0.5, 0].

***3.** Plot the approximate certain equivalent using the first moment, second central moment, third central moment approximation vs. the initial wealth for a logarithmic decision maker for the deal [0.5, 100; 0.5, 0].

4. Plot the cumulative probability distribution and probability distribution of the deal [0.1, 0; 0.2, 100; 0.6, 500; 0.1, 1000]. Determine the first moment graphically.

CHAPTER

25

Deterministic and Probabilistic Dominance

CHAPTER CONCEPTS

After reading this chapter, you will be able to explain the following concepts:

- Deterministic dominance
- First-order probabilistic dominance
- Second-order probabilistic dominance

25.1 INTRODUCTION

Dominance conditions are possible properties of the cumulative probability distributions. When they exist, they can simplify the analysis of calculating the best decision significantly. For example, if the sole aim is to find the best alternative, they may obviate the need to specify the risk preference of a decision maker who follows the Five Rules of Actional Thought.

25.2 DETERMINISTIC DOMINANCE

In choosing between two uncertain deals you face, *A* and *B*, your criterion is to choose the one with the higher *e*-value of *u*-values (or higher *e*-value of preference probabilities). This would typically require eliciting the probability distributions of the alternatives; the preference probabilities of their prospects; and then calculating the *e*-value of the preference probabilities for each alternative. However, suppose that the deals were constructed in such a way that you would always receive a better outcome with one than with the other. Regardless of your preference probabilities or your beliefs about the probabilities of the outcomes themselves, you already would have clarity of action. This would facilitate the analysis of the decision significantly, since we would not need to assess preference probabilities or even probabilities of the outcomes. If you are guaranteed a better outcome with Deal *B* than with Deal *A*, we say that Deal *B* has **deterministic dominance** over Deal *A*.

To illustrate, let us refer back to the decision tree of Hamlet's dilemma, covered in a previous chapter, which we reprint in Figure 25.1 for convenience. Suppose that Hamlet prefers his current life (To Be), which is characterized by Prospect *A*, over either of the Prospects *B* or *C*. In this case, he would be guaranteed a better outcome by choosing To Be than any outcome he would get by choosing Not to Be. If Hamlet's preferences were indeed that way, $A > B$ and $A > C$, then the probability, p, would not matter. This would be an example of deterministic dominance for the To Be alternative.

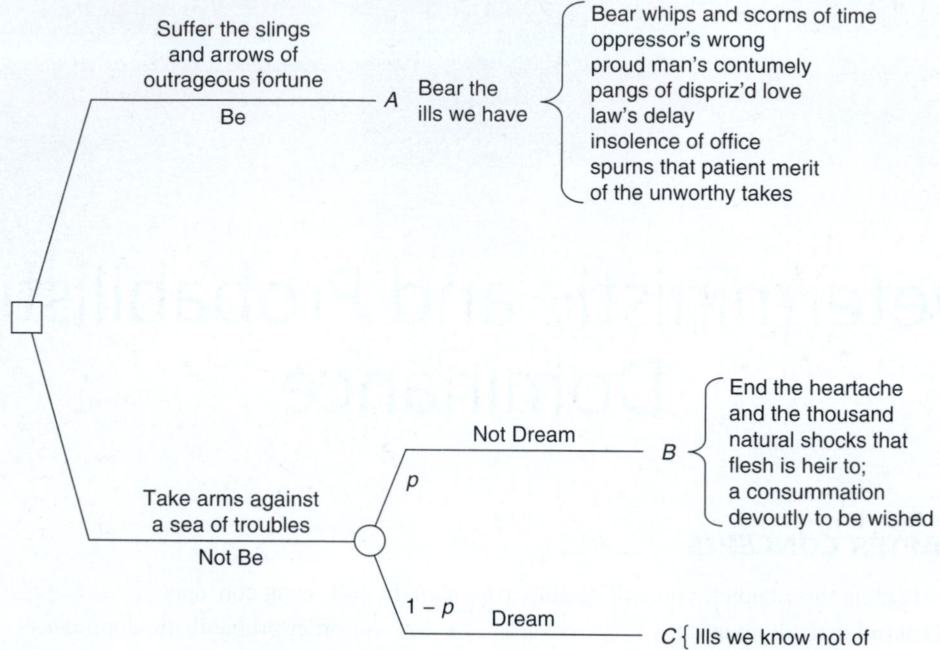

FIGURE 25.1 Hamlet's Decision Tree

On the other hand, if Hamlet preferred the alternative Not to Be, characterized by either of the prospects *B* or *C*, over the prospect *A*, he always would be better off with Not Be—an example of deterministic dominance for the Not to Be alternative.

We see that even when the results of the deals are uncertain and when the prospects of the deal do not involve money, one deal may deterministically dominate another.

We will now present several examples of deterministic dominance for deals with prospects completely described by monetary value measures.

EXAMPLE 25.1 Deals without Uncertainty

Suppose you face a choice between the following two deals:

Deal *A*: Receive $5

Deal *B*: Receive $10

You have no uncertainty: Both deals are deterministic. You will receive a better outcome with Deal *B* than you will with Deal *A*. Deal *B* dominates Deal *A* deterministically. You do not need to assess any *u*-values to make the choice. Anyone who prefers more money to less will prefer Deal *B* to Deal *A*.

EXAMPLE 25.2 Deterministic Dominance with Uncertain Deals

Suppose you faced a choice between the two deals of Figure 25.2. Which would you choose?

In this example, the worst possible outcome of Deal *B*, $15, is better than the highest possible outcome of Deal *A*, $10. Even though the deals are uncertain, you will receive a

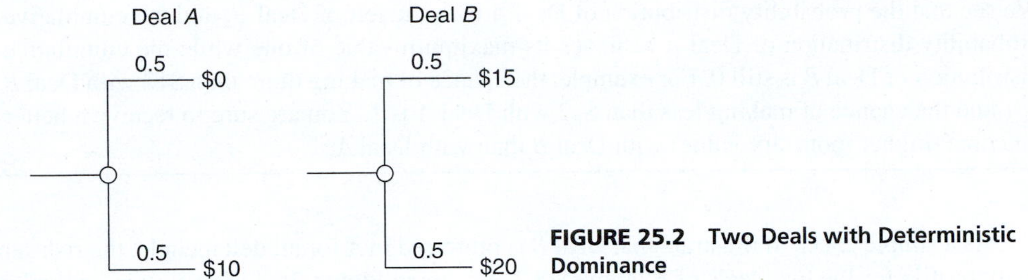

FIGURE 25.2 Two Deals with Deterministic Dominance

higher value with Deal *B* than with Deal *A*. Deal *B* dominates Deal *A* deterministically. Once again, we do not need to elicit any *u*-values to determine the best choice for these deals, and we do not even need to elicit the probabilities of the deals. Regardless of outcome probabilities, any decision maker—even the risk seeker—who follows the five Rules of Actional Thought and prefers more money to less, will prefer Deal *B* to Deal *A*.

However, it is still useful to look at the probability distributions of the two deals of Figure 25.2 to recognize deterministic dominance. The probability distributions and cumulative probability distributions, for the two deals of Example 25.2 appear in Figure 25.3.

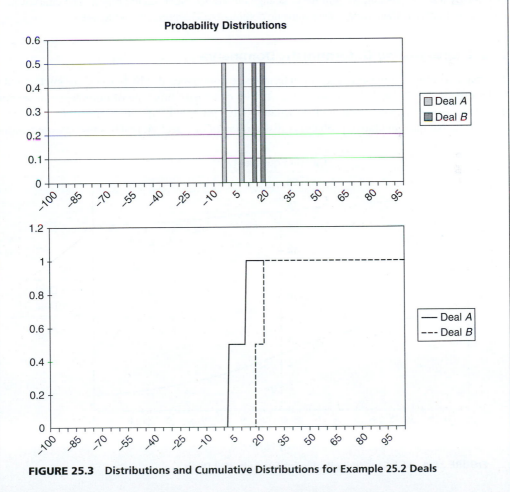

FIGURE 25.3 Distributions and Cumulative Distributions for Example 25.2 Deals

We see that the probability distribution of Deal *A* is to the left of Deal *B*, and the cumulative probability distribution of Deal *A* achieves its maximum value of one while the cumulative distribution of Deal *B* is still 0. For example, the chance of making more than \$12 with Deal *B* is 1 and the chance of making less than \$12 with Deal *A* is 1. You are sure to receive a better outcome (higher monetary value) with Deal *B* than with Deal *A*.

As a simple check to illustrate that Deal *B* is preferred to *A* for all deltapeople, the risk sensitivity profile for the two deals of Example 25.2 appears in Figure 25.4. The certain equivalent of Deal *A* ranges from 0 (for extreme risk aversion) to \$10 (for extreme risk seeking), while the certain equivalent for Deal *B* ranges from \$15 (for extreme risk aversion) to \$20 (for extreme risk seeking). Therefore, Deal *B* has a higher certain equivalent for any risk-aversion coefficient.

However, as we discussed, this result does not depend on the actual *u*-curve we use. The certain equivalent of Deal *B* is higher than that of Deal *A* with any *u*-curve provided only that more value is preferred to less.

Again with regard to Example 25.2, we will now determine the value of clairvoyance (VOC). To begin, note that the clairvoyant's report will always specify a higher value outcome for Deal *B* than for Deal *A*. As a result, we will never change our choice based on the report. The clairvoyant's report is immaterial when there is deterministic dominance. Therefore, the VOC is 0.

25.2.1 Recognizing Deterministic Dominance

The main criterion for the existence of deterministic dominance is whether you are guaranteed a better outcome with Deal *B* than with Deal *A*. You can verify this type of dominance in several different ways. First, you can examine the cumulative probability distributions of the two deals. If you find that the cumulative probability distribution of one deal, *A*, reaches its maximum value

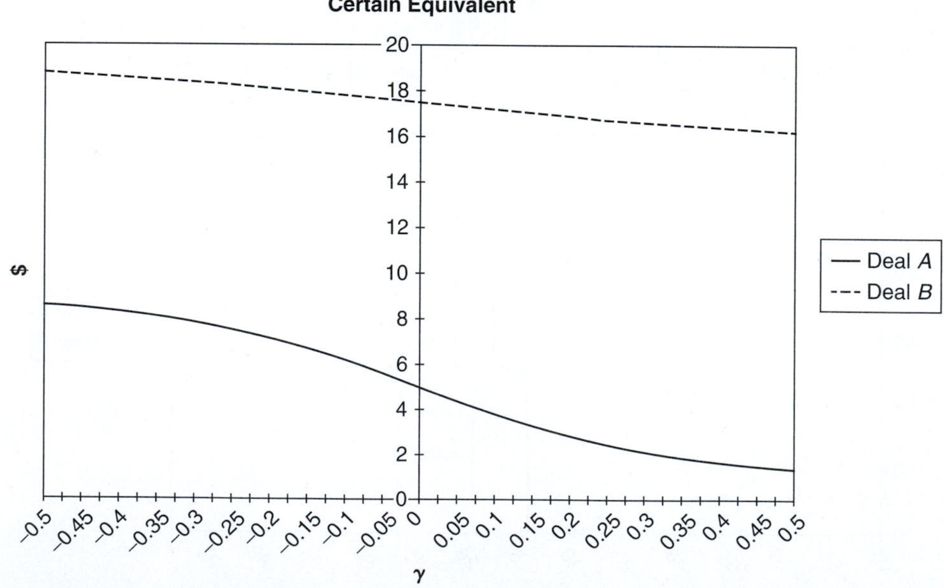

FIGURE 25.4 Risk Sensitivity Profile for Example 25.2 Deals

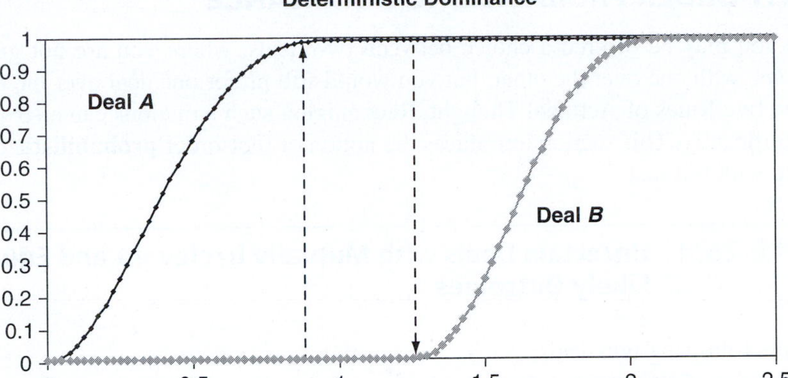

FIGURE 25.5 Two Deals with Deterministic Dominance. The Cumulative Probability of Deal *A* is 1 before that of Deal *B* is Positive

of 1 before that of the other deal, *B*, achieves a positive value, as shown in Figure 25.5, then Deal *B* deterministically dominates Deal *A*. This is because the highest value you can achieve with Deal *A* is less than the lowest value you can achieve with Deal *B*.

Deterministic dominance also can have other forms that are not determined by comparing the cumulative probability distributions. We illustrate this below.

EXAMPLE 25.3 Deterministic Dominance with Uncertain Deals Generated by Same Uncertainty

Suppose you faced two deals:

Deal *A* pays $10 if the outcome of a coin toss is heads and 0 if its outcome is tails.

Deal *B* pays $15 if the outcome of the *same* coin toss is heads and $5 if its outcome is tails.

Since the outcomes are generated by the same uncertainty, heads will yield a payoff of $10 for Deal *A* and $15 for *B*, while tails will yield 0 for *A* and $5 for *B*. So we are guaranteed a better outcome with Deal *B* than with Deal *A*, and *B* deterministically dominates *A*. In this example, however, the cumulative probability distributions are the same as those of Figure 25.7 for Example 25.4 do not exhibit the same type of separation as do those of Figure 25.5, since the plots do not contain information about the relevance between the outcomes of the two deals.

We summarize the properties of deals with deterministic dominance as follows:

1. We always receive a better outcome with one deal than with the other.
2. The value of clairvoyance is 0.
3. We do not need to assess *u*-curves to choose the better decision alternative.

As we have seen, the existence of deterministic dominance simplifies the analysis. In practice, deterministic dominance conditions may seldom occur. However, other types of dominance can also trim the analysis task.

25.3 FIRST-ORDER PROBABILISTIC DOMINANCE

Sometimes you may be offered a choice between two deals, where you are not guaranteed a better outcome with one over the other, but you would still prefer one deal over the other if you followed the five Rules of Actional Thought. Recognizing such situations can also simplify the analysis significantly. This section introduces the notion of first-order **probabilistic dominance** that can help with this task.

EXAMPLE 25.4 Uncertain Deals with Mutually Irrelevant and Equally Likely Outcomes

Consider the following two deals:

Deal *A* pays $10 if the outcome of a coin toss is heads and $0 if its outcome is tails.

Deal *B* pays $15 if the outcome of a *different* coin toss is heads and $5 if its outcome is Tails.

The deals appear in Figure 25.6. Unlike Example 25.3, different coin tosses determine the outcomes, and we believe the outcome of each deal is irrelevant to the outcome of the other. Which deal would you prefer?

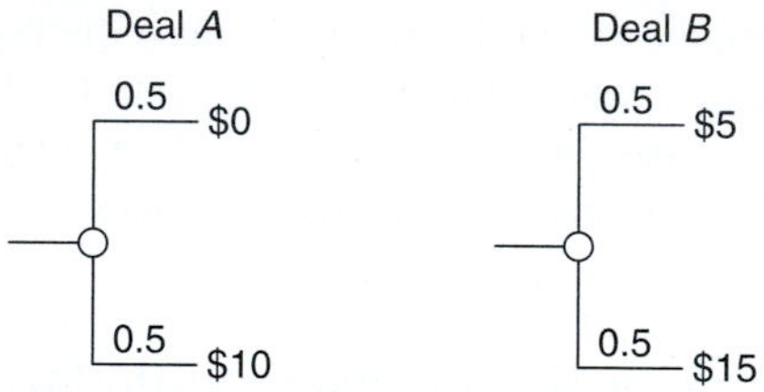

Deal *A*: 0.5 → $0; 0.5 → $10

Deal *B*: 0.5 → $5; 0.5 → $15

FIGURE 25.6 Two Mutually Irrelevant Deals with First-Order Probabilistic Dominance

The first thing we notice about these two deals in comparison to the previous example is that we are not guaranteed a better outcome with Deal *B* than with Deal *A*. For example, it is possible to win $5 with Deal *B* and $10 with Deal *A*. As a result, the two deals do not have deterministic dominance.

On close examination of the deals, however, most people would answer that they prefer Deal *B* to Deal *A*. They see that for any monetary amount, v, they have a probability of winning at least v that is as high or higher with Deal *B* than with Deal *A*. This makes Deal *B* more attractive—even though we are not guaranteed a higher monetary value with Deal *B*. This brings us back to our very first distinction of the difference between a decision and its outcome.

We say that Deal *B* dominates Deal *A* with first-order probabilistic dominance if the probability of exceeding any value is at least as high with Deal *B* as it is with Deal *A* and is higher for at least one value. Using the choice rule, any decision maker who prefers more money to less would prefer Deal *B* to Deal *A*. As a result, we do not need to assess risk preference to determine the better decision alternative when first-order probabilistic dominance exists.

The probability distributions and cumulative probability distributions for both deals of Example 25.4 appear in Figure 25.7. Note that the two cumulative distributions in Figure 25.7 are the same as those for Example 25.3 because they do not contain information about the relevance between the outcomes of both deals. Note further, the curves do not cross, although they do meet at a few points. This is the test for first-order probabilistic dominance using the cumulative distributions. We say that Deal *B* dominates Deal *A* with first-order probabilistic dominance if its cumulative probability distribution curve is below that of Deal A somewhere and is never above it.

Deal B dominates Deal A with first-order probabilistic dominance when

$$\{X_B \leq x|\&\} \leq \{X_A \leq x|\&\}$$

for all values of x with inequality for at least one value of x. Equivalently, the difference between the cumulative probability distributions, $d(x)$, will be non-negative, as

$$d(x) = \{X_A \leq x|\&\} - \{X_B \leq x|\&\} \geq 0$$

for all values of x, and greater for at least one value of x.

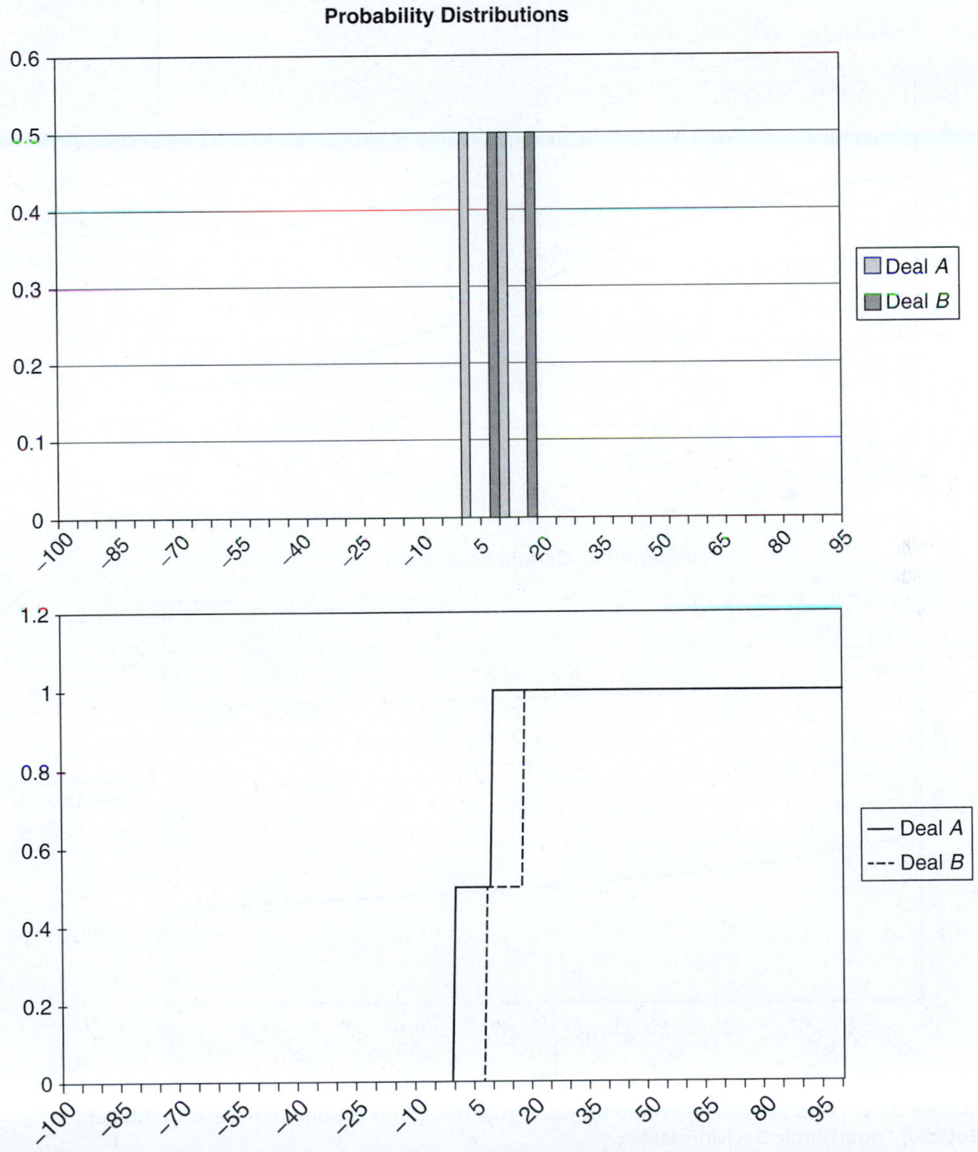

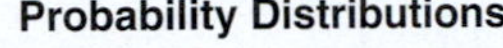

FIGURE 25.7 Example of First-Order Probabilistic Dominance

***Note** If Deal B deterministically dominates Deal A, then it will exhibit first-order probabilistic dominance of Deal A.*

This was manifested by Example 25.3 where deterministic dominance existed, yet the plot of cumulative distributions was the same as that for Example 25.4 where first-order dominance existed.

Examine the risk sensitivity profile for the two deals shown in Figure 25.8 for both a deltaperson and a logarithmic decision maker. We see that Deal *B* has a higher certain

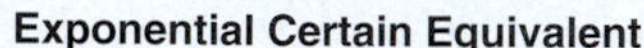

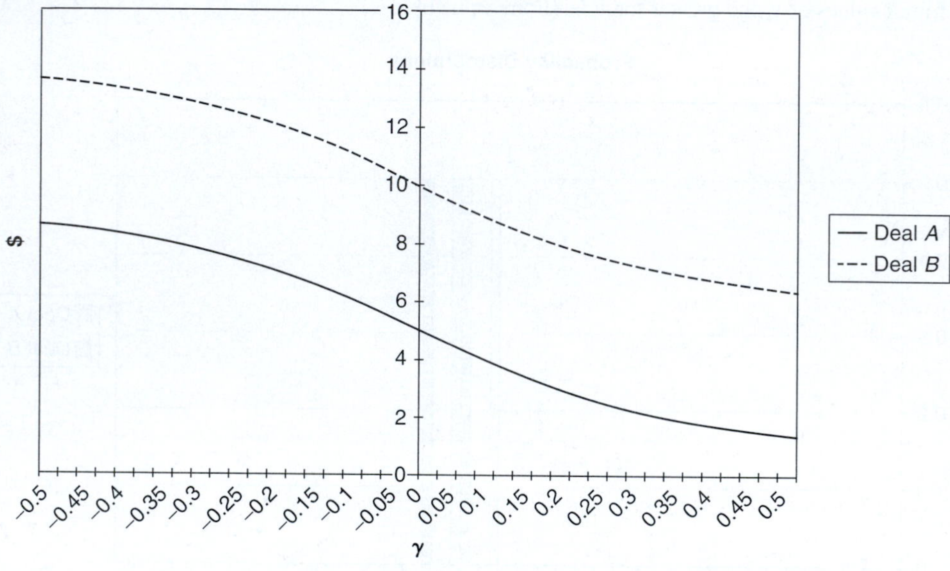

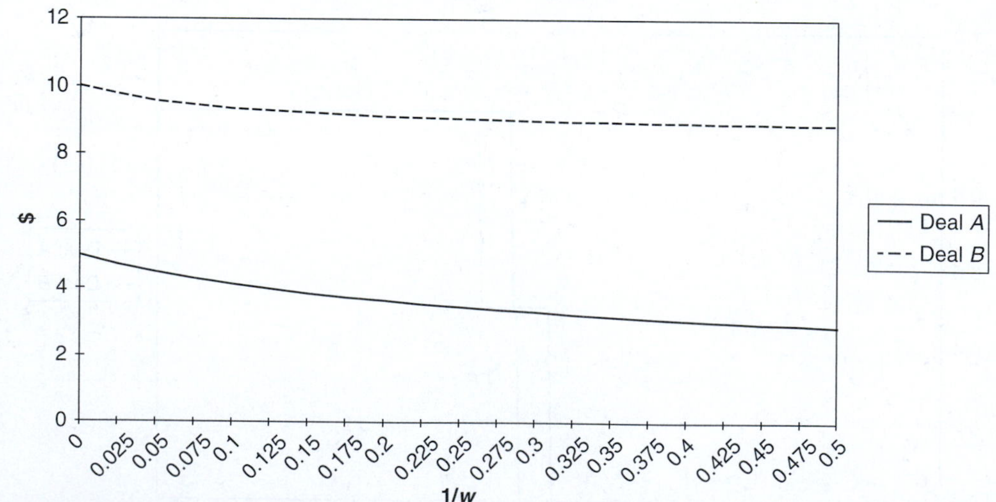

FIGURE 25.8 Risk Sensitivity Profile for Deals A and B: (Top) Exponential Decision Maker; (Bottom) Logarithmic Decision Maker

equivalent for any risk-aversion coefficient for a deltaperson—even one who is risk preferring. This is also true for any initial wealth, w, for a logarithmic decision maker who is necessarily risk-averse. In general, if Deal B probabilistically dominates Deal A in the first-order, then it has a higher certain equivalent than Deal A for any u-curve. As a result, we do not need to assess u-curves to determine the better decision alternative if first-order probabilistic dominance conditions exist.

Note *We are not guaranteed a better outcome with Deal B than with Deal A.*

To drive home this point, suppose that you and a friend who has not studied decision analysis confront the choice of Deals A and B. You say, "We learned about this in class, so I know that I must choose the probabilistically dominant Deal B." Your less-learned friend chooses Deal A. When the coins are tossed, you win \$5 and he wins \$10. He says, "Ignorance is bliss."

We will now consider the VOC for Example 25.4. Since we are not guaranteed a better outcome with Deal B than with Deal A, we may change our choice if the clairvoyant tells us we will get a better outcome with Deal A than with Deal B. Therefore, the VOC need not be 0 when first-order probabilistic dominance exists.

For a risk-neutral person, we can illustrate this easily. Without further information, the best choice is Deal B with a certain equivalent of \$10 for the deal. You can confirm that the VOC on either the result of Deal A or Deal B alone would be 0: The choice would remain Deal B. However, if the clairvoyant provides results of both deals, then only in the case where Deal A would pay \$10 and Deal B would pay \$5 does the choice change to Deal A. The probability of this report that produces \$5 of value increase is 0.25; the VOC is \$1.25.

25.3.1 First-Order Dominance Implies Mean Dominance

If Deal B has first-order dominance with respect to Deal A, then it also has a higher mean and a higher geometric mean than Deal A. This immediately implies that a risk-neutral decision maker will prefer Deal B to Deal A and (from the discussion of logarithmic u-curves) any logarithmic decision maker must prefer Deal B to Deal A. Of course, as we have discussed, the results of first-order dominance apply to any other u-curve as well.

Here, we summarize the properties of first-order probabilistic dominance of Deal B over Deal A.

1. The cumulative probability distribution of Deal A never exceeds that of Deal B and must be below it somewhere.
2. The certain equivalent of Deal B is higher than the certain equivalent of Deal A for any u-curve even for risk seekers.
3. We may not receive a better outcome with Deal B.
4. The mean of Deal B is higher than that of Deal A.
5. The geometric mean of Deal B is higher than that of Deal A.
6. We do not need u-values to determine the better decision alternative.
7. If choosing between the deals, the value of clairvoyance need not be 0 and depends on the risk preference.
8. While first-order dominance determines the deal with the higher certain equivalent, it determines neither the actual value of the certain equivalents nor your indifferent buying prices (if you do not own them).

25.4 SECOND-ORDER PROBABILISTIC DOMINANCE

If first-order probabilistic dominance conditions do not exist, we may still simplify the analysis of the problem if second-order dominance exists. We have seen that first-order dominance implies mean dominance. Suppose we have two deals with the same mean but with different variances, as shown in the following example.

EXAMPLE 25.5 Two Deals with the Same Mean and Different Variances

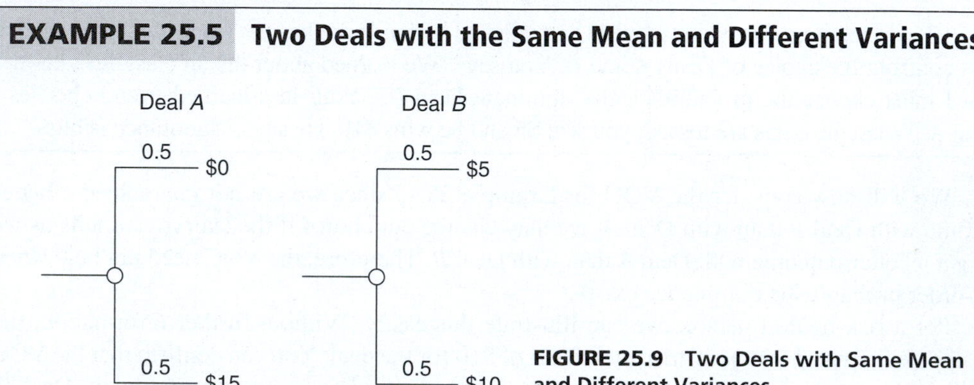

FIGURE 25.9 Two Deals with Same Mean and Different Variances

Consider the mutually irrelevant deals of Figure 25.9 and consider which you would prefer. The probability distributions and cumulative probability distributions of the two deals appear in Figure 25.10. Note that the two deals have the same mean of \$7.50. You have a higher chance of receiving at least \$12 with Deal A $(\{X_A \geq 12|\&\} = 0.5)$ than with Deal B $(\{X_B \geq 12|\&\} = 0)$, but you have a higher chance of receiving at least \$10 with Deal B $(\{X_B \geq 10|\&\} = 1)$ than with Deal A $(\{X_A \geq 10|\&\} = 0.5)$. Therefore, determining which alternative is better is not as clear as it was in the previous examples.

Since the two cumulative probability distributions cross, there is no first-order dominance. The difference between the two cumulative probability distribution curves, $d(x)$, changes from positive to negative and appears in Figure 25.11. For some values of x,

$$d(x) = \{X_A \leq x|\&\} - \{X_B \leq x|\&\} \geq 0$$

and for others,

$$d(x) = \{X_A \leq x|\&\} - \{X_B \leq x|\&\} \leq 0$$

Now we sum (integrate) the area under the curve in Figure 25.10 from left to right. We can do this in table format by plotting the equation of the curve

$$I(x) = \sum_{-\infty}^{x} d(x) \times (\text{increment})$$

The increment determines the level of discretization. Table 25.1 shows how the calculation of $I(x)$ is carried out for discrete values of x for the two deals. Here, we have discretized the values of x into increments of 5. Columns i and ii show the probabilities of achieving a given value for each deal. Columns iii and iv present the cumulative probability values obtained by summing the probabilities in Columns i and ii, respectively. Column v is the difference

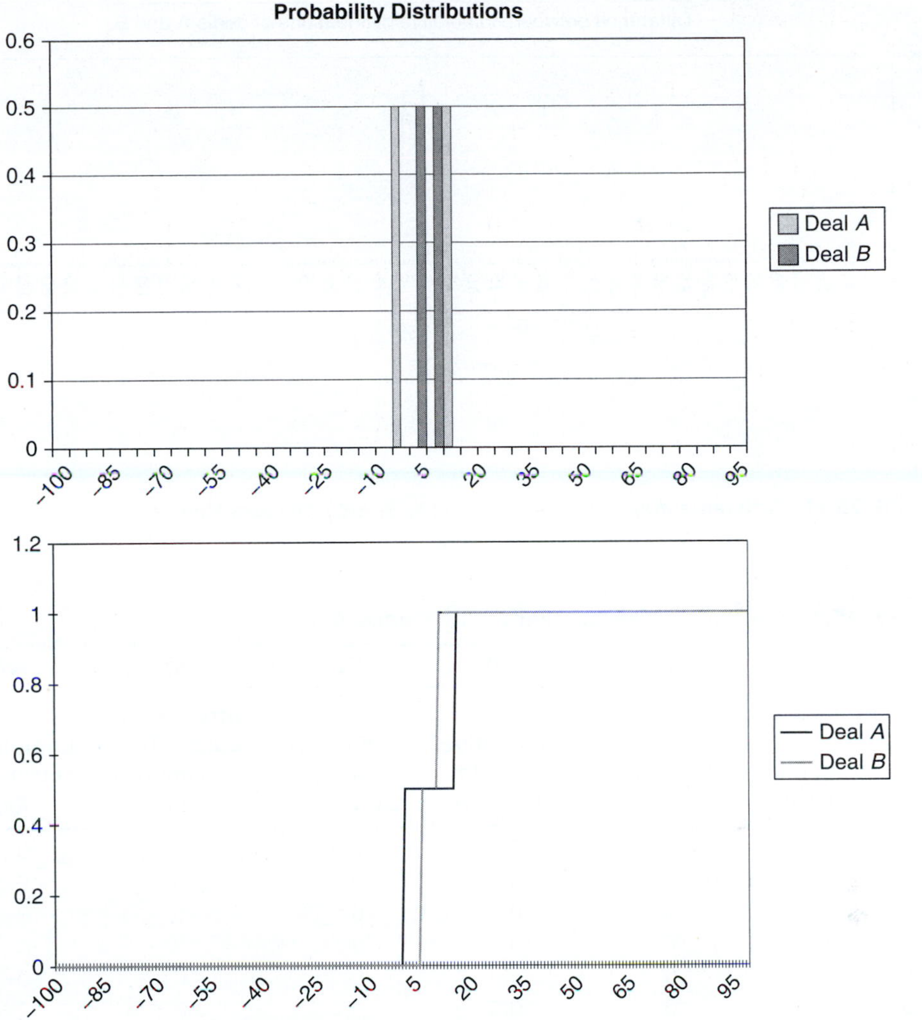

FIGURE 25.10 Example 25.5 Distributions and Cumulative Distributions

between Columns iii and iv, namely $d(x)$. Column vi represents the cumulative area under the curve of Column v using the formula:

$$I(x_{\min}) = d(x_{\min})$$
$$I(x + 5) = I(x) + d(x) \times 5$$

For example, for an increment of 5, the value of $I(x)$ that corresponds to $x = 5$ is obtained by summing the value of $I(x)$ for $x = 0$ and the cumulative difference at $x = 5$ (namely, $d(x) = 0.5$) multiplied by 5. We multiply by 5, because the table is discretized into units of increments equal to five, and we are interested in the area under the curve.

The plot of $I(x)$ versus x appears in Figure 25.12.

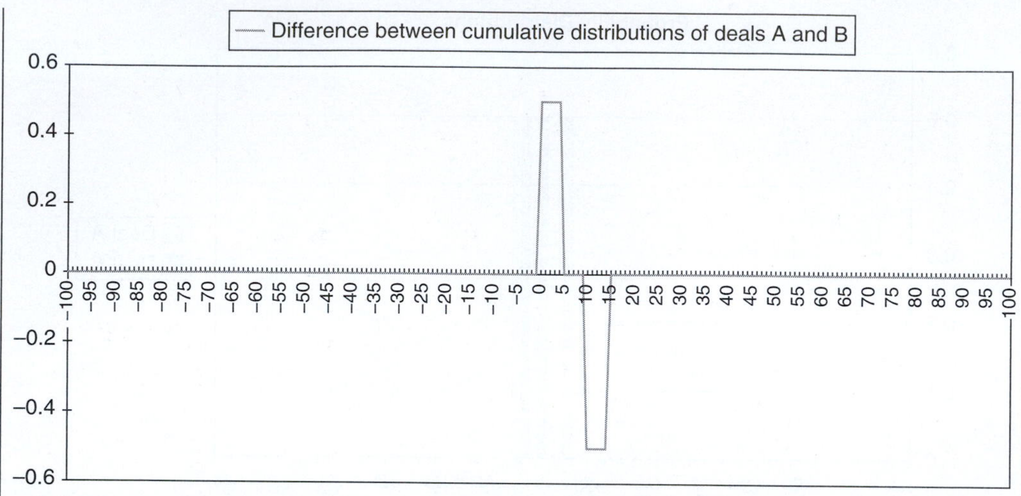

FIGURE 25.11 Difference $d(x) = \{X_A \leq x|\&\} - \{X_B \leq x|\&\}$ Changes Sign

TABLE 25.1 Calculating the Cumulative Difference, $I(x)$

	(i)	(ii)	(iii)	(iv)	(v)	(vi)
x	Probability of Deal A	Probability of Deal B	Cumulative Probability of Deal A	Cumulative Probability of Deal B	Difference in Cumulative Probabilities, $d(x)$	Cumulative Difference, $I(x)$
−25	0	0	0	0	0	0
−20	0	0	0	0	0	0
−15	0	0	0	0	0	0
−10	0	0	0	0	0	0
−5	0	0	0	0	0	0
0	0.5	0	0.5	0	0.5	2.5
5	0	0.5	0.5	0.5	0	2.5
10	0	0.5	0.5	1	−0.5	0
15	0.5	0	1	1	0	0
20	0	0	1	1	0	0
25	0	0	1	1	0	0

Note: *$I(x)$ in this example does not change sign and is non-negative.*

In general, we say that Deal B dominates Deal A with second-order probabilistic dominance if the curve $I(x)$ is never negative and is positive for at least one value of x.

$$I(x) \geq 0$$

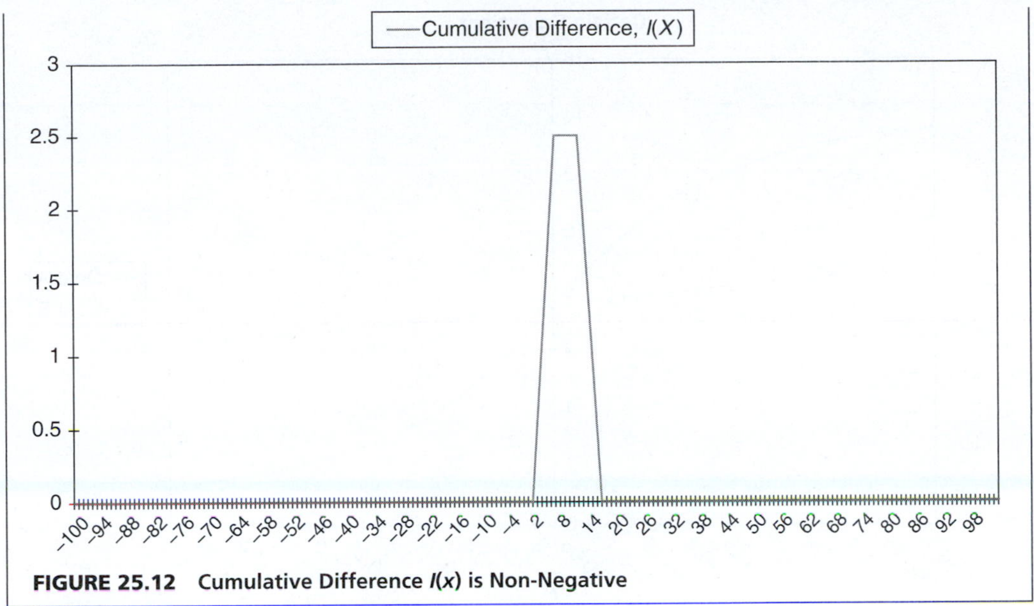

FIGURE 25.12 Cumulative Difference *I*(*x*) is Non-Negative

25.4.1 Implications of Second-Order Probabilistic Dominance

If Deal *B* has second-order probabilistic dominance over Deal *A*, then any risk-averse decision maker will prefer Deal *B* to Deal *A*, regardless of any other property of his *u*-curve. As a result, we need only verify that the decision maker is risk-averse to determine the better decision alternative.[1]

Second-order probabilistic dominance determines the deal with the higher certain equivalent for a risk-averse decision maker. It does not determine the value of the certain equivalent, the indifferent buying price if not owned, or the additional amount of money that would make the decision maker indifferent to receiving the least preferred deal in exchange for the better one. Determining these values requires additional calculation.

For a deltaperson, the risk sensitivity profile for the two deals of Figure 25.13 appears below. Note that Deal *B* has a higher certain equivalent for positive values of the risk-aversion coefficient (risk-averse), while Deal *A* has a higher certain equivalent for negative values of the risk-aversion coefficient (risk seeking). The figure also shows the risk sensitivity profile for a logarithmic decision maker (who is always risk-averse). Here, for all wealth levels, we see that Deal *B* dominates Deal *A*.

Notice that, if Deal *B* has first-order probabilistic dominance with respect to Deal *A*, the difference between the cumulative distributions, $d(x) = \{X_A \leq x|\&\} - \{X_B \leq x|\&\} \geq 0$ does not change sign for all values of x and is positive for some value. As a result, $I(x) \geq 0$ for all values of x and $I(x) > 0$ somewhere. Therefore, Deal *B* also has second-order probabilistic dominance with respect to Deal *A*. Therefore, first-order probabilistic dominance implies second-order probabilistic dominance.

[1]Though of much less interest, any risk-preferring decision maker will always prefer Deal *A* over Deal *B* if $\sum_{x}^{\infty} d(x) \times$ (increment) does not change sign. This is known as risk-preferring second-order probabilistic dominance.

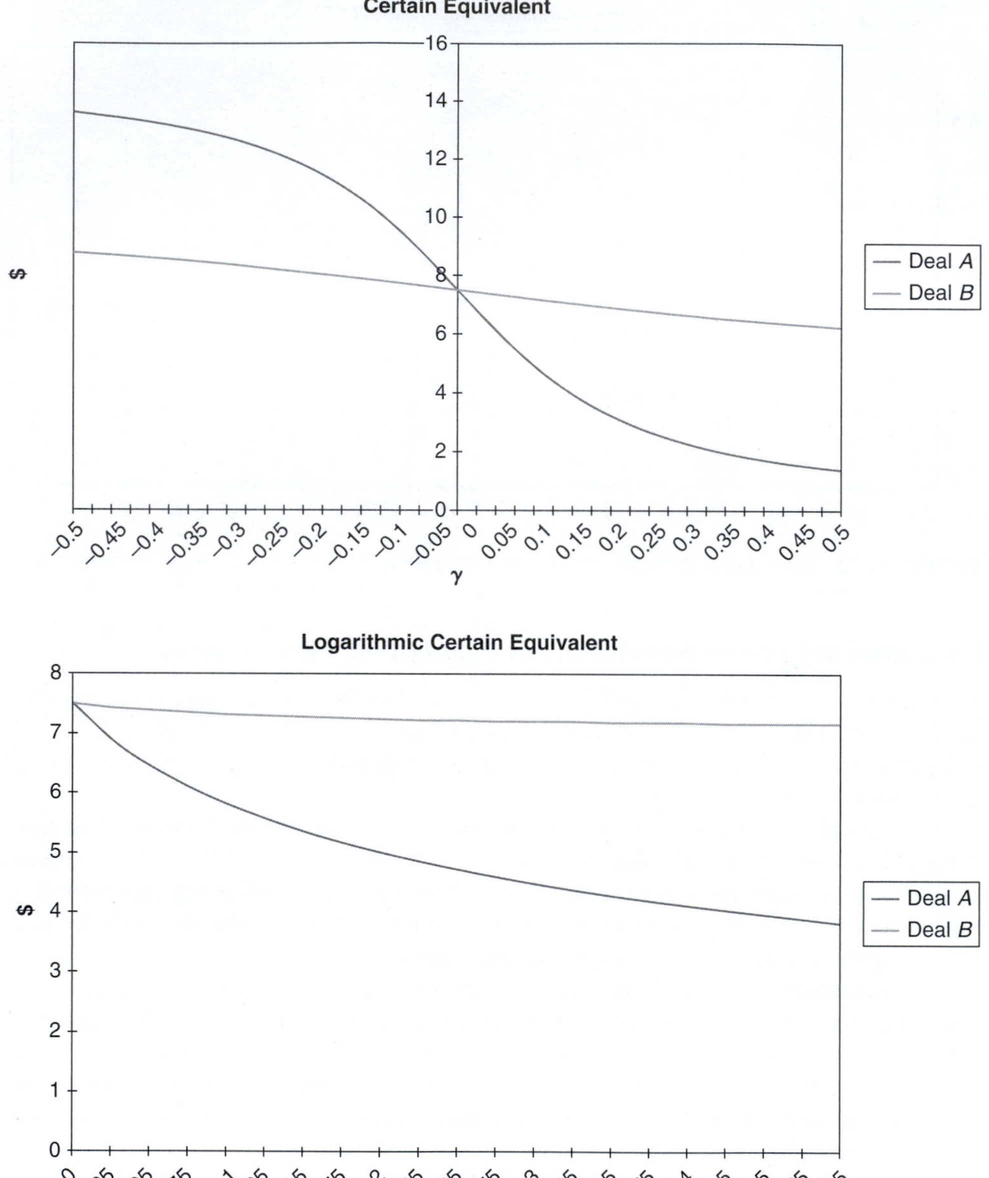

FIGURE 25.13 Risk Sensitivity Profile for Second-Order Dominance

25.5 DOMINANCE FOR ALTERNATIVES IN THE PARTY PROBLEM

Figure 25.14 revisits the alternatives in the party problem from Chapter 9, their corresponding payoffs and their cumulative probability distributions.

From the figure, we can see right away that neither deterministic nor first-order probabilistic dominance conditions exist between any of the alternatives for the specified values and probability of Sun. Therefore, the choice will depend on the *u*-curve that is chosen.

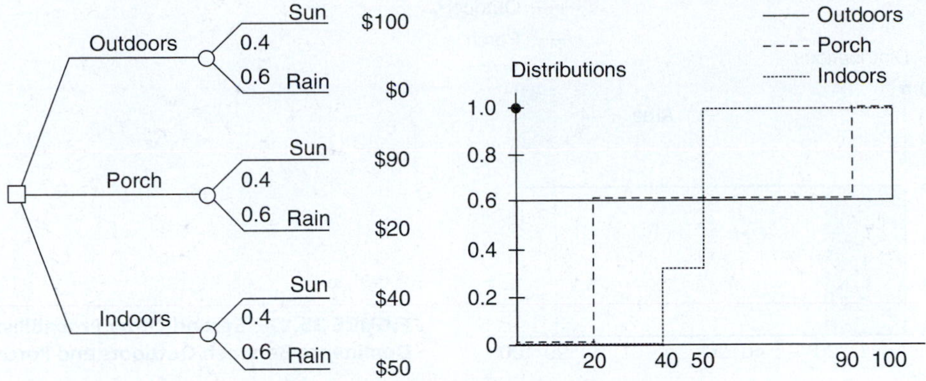

FIGURE 25.14 Party Problem Alternatives and Cumulative Distributions

Figure 25.15 examines the second-order probabilistic dominance relations between the Porch alternative and the Indoors alternative by calculating the area between the curves Porch–Indoors. From Figure 25.15, we see that area is first +14, and then an additional −16 is added, which leads it to change sign. Therefore, there is no second-order probabilistic dominance between these two alternatives.

Figure 25.16 examines the second-order dominance probabilistic between Outdoors and Indoors alternatives calculating the area between the curves of Outdoors–Indoors. From the figure, we see a +26 area followed by a −20 area, so the difference will not change sign. Therefore, Indoors has second-order probabilistic dominance with respect to Outdoors, and any risk-averse

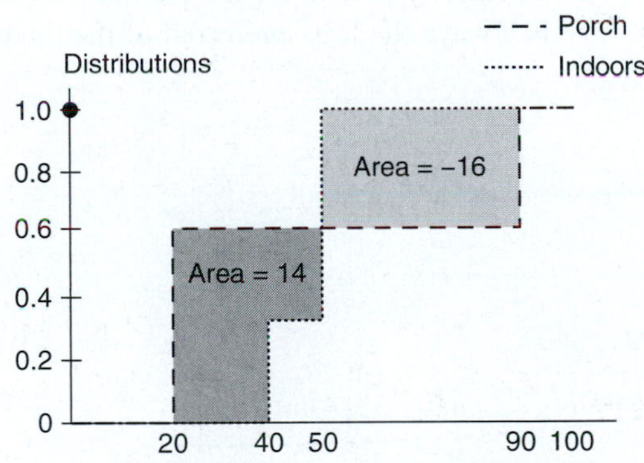

FIGURE 25.15 No Second-Order Probabilistic Dominance Between Porch and Indoors

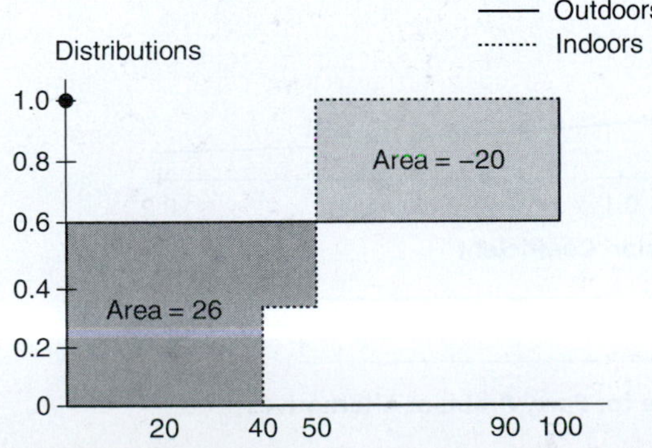

FIGURE 25.16 Second-Order Probabilistic Dominance Between Outdoors and Indoors

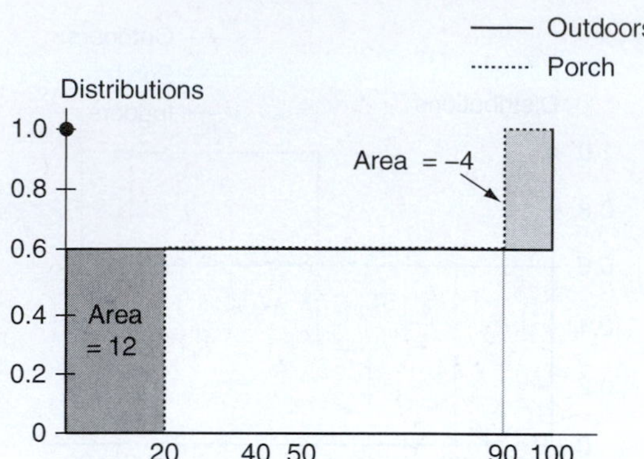

FIGURE 25.17 Second-Order Probabilistic Dominance Between Outdoors and Porch

person will prefer Indoors over Outdoors. Once we know only that Kim is risk-averse, we know that based on her present information and values, Kim will not choose the Outdoors alternative. Finally, Figure 25.17 examines the second-order probabilistic dominance relations between the Outdoors and Porch alternatives. In a similar way, we see that Porch has second-order probabilistic dominance with respect to Outdoors, since the sum of areas does not change sign.

So long as Kim is risk-averse, has the current probability of 0.4 for Sunshine, and has the same values for party locations, she will never choose the Outdoor alternative—regardless of the actual *u*-curve she decides to use. Kim could even convert from being a deltaperson to a logarithmic person without changing her refusal to use the Outdoor alternative when the other two are available.

If Kim remains a deltaperson, revisiting the risk sensitivity profile for the alternatives of the party problem shows that the Outdoors alternative is always the least preferred of the three

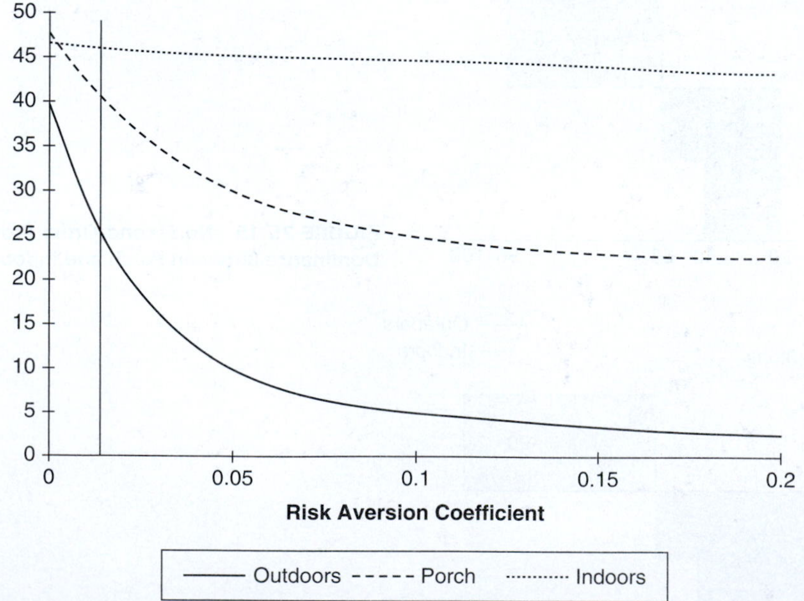

FIGURE 25.18 Risk Sensitivity Profile for Party Problem Alternatives

for any positive value of the risk-aversion coefficient (a manifestation of second-order dominance), as shown in Figure 25.18. Her preference for the Indoors over the Porch alternative changes when her risk-aversion coefficient reaches 0.00354, corresponding to a risk tolerance of \$282 and not the \$72.13 she currently has.

25.6 SUMMARY

Deal B dominates Deal A deterministically if you are guaranteed a better outcome with Deal B than with Deal A. The value of clairvoyance (VOC) in choosing between the two deals is 0. Deterministic dominance does not require the deals to be represented monetarily. Deterministic dominance does not require assessing any preference probabilities or probabilities for the outcomes.

First-order probabilistic dominance does not guarantee a better outcome, but it does guarantee that the certain equivalent is higher for any decision maker who follows the five Rules of Actional Thought. The condition is that the two cumulative distributions do not cross or $d(x) = \{X_A \leq x|\&\} - \{X_B \leq x|\&\} \geq 0$ with strict inequality for at least one value of x.

Second-order probabilistic dominance does not guarantee a better outcome but does guarantee that the certain equivalent is higher for any risk-averse decision maker who follows the five Rules of Actional Thought. The condition is that the accumulated difference between the two cumulative distributions does not change sign. First-order dominance is a sufficient condition for second-order dominance to hold.

Dominance arguments determine the best deal (with the highest certain equivalent) for certain types of u-functions, but they do not determine the value of the certain equivalent or the additional payment that would make the decision maker indifferent to receiving the less preferred deal.

KEY TERMS

- Deterministic dominance
- Dominance conditions
- Probabilistic dominance

PROBLEMS

Problems marked with an asterisk (*) are considered more challenging.

1. Consider the following two deals.

I. Roll a die you believe has equal probability of 1 through 6. If it lands on 1, you receive $100; otherwise you receive 0.

II. Toss a coin with equal probability of heads or tails. If it lands as heads, you receive $100; otherwise you receive 0.

a. Plot the probability distribution and cumulative probability distribution of each deal.
b. Test for deterministic, first-order, and second-order dominance.
c. Are you guaranteed a better outcome with one deal over the other?

2. In this chapter, using Examples 25.2, 25.3, and 25.4:

a. Plot the probability distribution of each deal.
b. Plot the cumulative probability distribution of each deal.
c. Plot the risk profile for each deal assuming an exponential *u*-curve (certain equivalent vs. risk aversion coefficient).
d. Test for deterministic, first-order, and second-order dominance.

3. Considering the party problem, plot the distributions and cumulative probability distributions for the Outdoors–Indoors–Porch alternatives when the probability of Sun = 0.5. Test for deterministic, first-order, or second-order dominance.

***4.** Consider the decision tree shown in the Figure. Plot the probability distribution and cumulative probability distribution for each alternative. Test to see if there is deterministic, first-order, or second-order dominance.

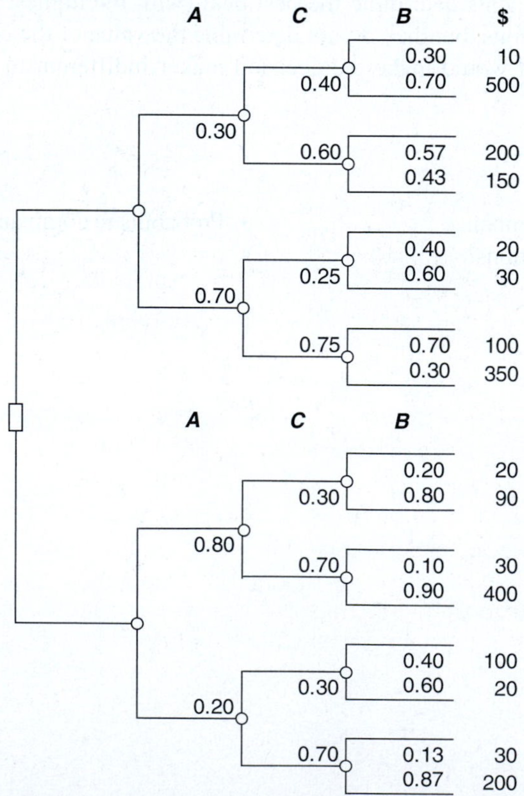

CHAPTER

26

Decisions with Multiple Attributes (1)–Ordering Prospects with Preference and Value Functions

CHAPTER CONCEPTS

After reading this chapter, you will be able to explain the following concepts:

- Direct and indirect value attributes
- Preference functions
- Value functions

26.1 INTRODUCTION

In many of the previous chapters, we discussed decisions with prospects characterized by a single measure, such as money. We developed a *u*-curve over this measure and used the *e*-value of the *u*-values of the prospects to determine the best decision alternative. The use of money as a measure also enabled the calculation of the PIBP and PISP for an uncertain deal and the value of clairvoyance (VOC) on a distinction of interest.

Sometimes, it may be necessary to include additional attributes in the characterization of the prospects. An **attribute** is simply another distinction about the prospect that contributes to its characterization. For example, when thinking about the purchase of a new vehicle, we may encounter attributes such as color, trunk space, fuel mileage, and acceleration. All of these attributes can change our preference for the vehicle we purchase. Recalling the party problem introduced in Chapter 9, we considered two attributes: Location and Weather. Both of these attributes contributed to Kim's ordering of the prospects and her valuation.

In some situations, an attribute can be expressed as a numeric measure. For example, an attribute such as trunk space can be expressed in cubic feet. We can also characterize trunk space by naming ranges of cubic feet, such as Large, Medium, and Small. Other attributes, like the type of car transmission, may be categorical—for example, Automatic or Manual.

As we shall see, the presence of additional attributes (or additional measures associated with them) does not require a new set of Rules of Actional Thought, nor does it require different criteria for selecting the best decision alternative. We still need to characterize the prospects now described by multiple attributes, and we still need to assign a probability to them in accordance with the probability rule. We also need to order the prospects according to the order rule, and when uncertainty is present, we must assign a preference probability to the prospects in accordance with the equivalence rule.

In Chapter 6, we saw an example in which prospects were characterized by two attributes with measures: dollar earnings and number of car washings. We characterized our belief about the occurrence of these two attributes using a joint probability for the measures car washes and dollars earned (Figure 6.36). The same analysis applies to any other decision with multiple measures. In this chapter, we build on Chapter 6 by examining how to order and assign value measures to the prospects with multiple measures once probability has been assigned. Later, in Chapter 28, we discuss assigning preference probabilities and *u*-values to multiple attribute prospects.

26.2 STEP 1: DIRECT VS. INDIRECT VALUES

Before we begin discussing prospect ordering, we first make an important distinction between direct and indirect values. To do this, we use the following example.

Consider the case of an African resort hotel owner who is vitally interested in tourism—because of the revenues associated with tourists—but who is indifferent to the presence of African wildlife. However, the owner knows that the wild animals attract tourists: therefore, the owner would favor policies that promote wildlife. We would say that the owner places a **direct value** on tourism and an **indirect value** on wildlife.

Figure 26.1 shows the assignments of value to the distinctions in a relevance diagram and a corresponding value measure at the end of a probability tree. The arrow from tourism to value is the direct value of tourism. The arrow from wildlife to tourism is the possible relevance of wildlife to tourism and is the source of the indirect value on wildlife.

The tree representation on the right side of Figure 26.1 shows that the decision maker is indifferent between the two prospects Wildlife–High Tourism and No Wildlife–High Tourism, as he values them both at V_1. He is also indifferent between Wildlife–Low Tourism and No Wildlife–Low Tourism, as he values them both at V_2. Consequently, wildlife does not play a role in the valuation of the prospects. It is an indirect value attribute. Therefore, this decision maker is indifferent between any two prospects having identical values of the direct value attributes, even if they have different values of the indirect value attributes. The characterization of prospects by the second attribute, Wildlife in this example, does not affect the preference ordering of the prospects.

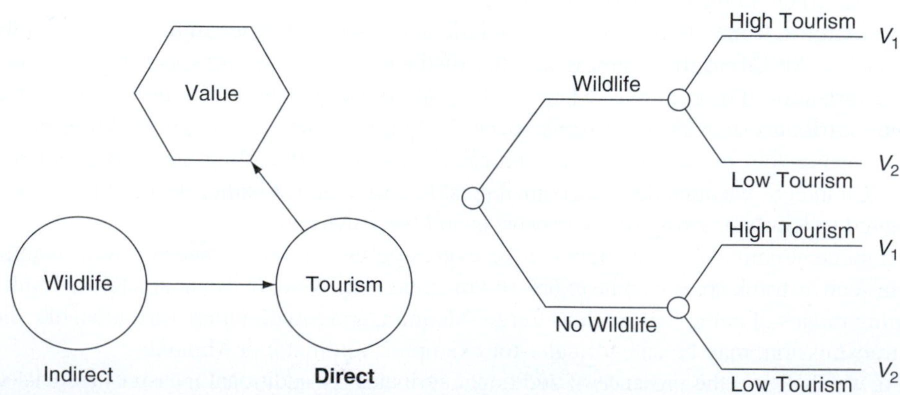

FIGURE 26.1 Direct vs. Indirect Values for Hotel Owner

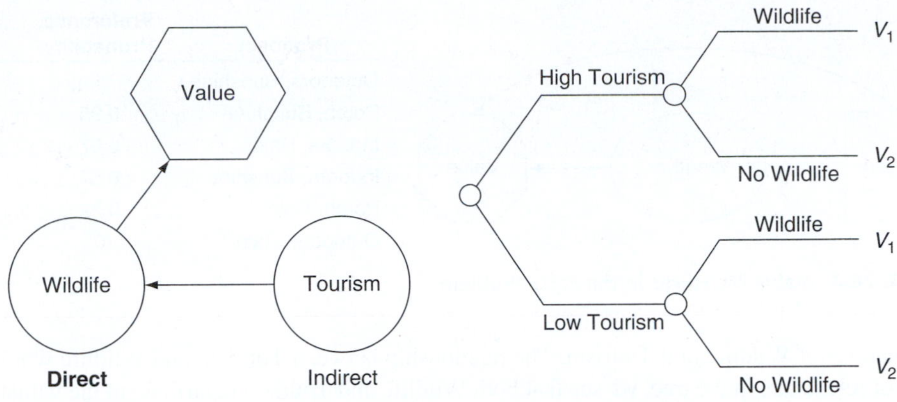

FIGURE 26.2 Direct vs. Indirect Values for Environmentalist

The relevance of wildlife to tourism reflects the knowledge of the owner. If additional information gained by the owner revealed that, in actuality, wildlife was not relevant to tourism, then the owner would no longer have an indirect value of Wildlife. Furthermore, if the owner should learn that wildlife actually decreased tourism, the relevance arrow would reappear, but the indirect value would be negative. Direct values derive from preference; indirect values come from information.

In contrast, consider an environmentalist who cares only about the welfare of African wildlife and is indifferent to the presence of tourists. Figure 26.2 shows the assignments of value to the distinctions in a relevance diagram. The environmentalist would place a direct value on Wildlife but not on Tourism.

However, if the environmentalist believed that tourism would provide funding to preserve animal habitat and defend animals against poachers, then the environmentalist would place an indirect value on Tourism. Once again, if new information showed that there was no relevance of Tourism to Wildlife, then Tourism would no longer be an indirect value to the environmentalist. As before, if the information showed that tourism was deleterious to wildlife, then the indirect value of Tourism would be negative. Once again, the tree representation shows that prospects having the same level of direct value attributes are valued equally—regardless of the level of the indirect value attribute.

Someone who placed a direct value on both Tourism and Wildlife would have value connections like those in Figure 26.3. The Value node would show how overall value depends on

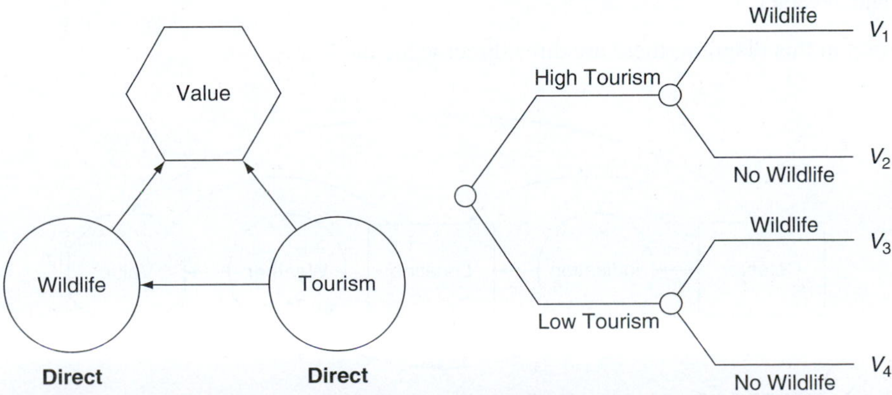

FIGURE 26.3 Direct Values for Both Tourism and Wildlife

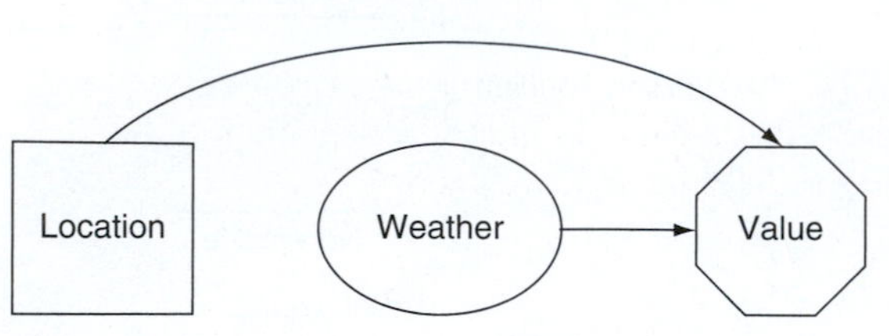

Prospect	Preference Probability	Dollar Value
Outdoors, Sunshine	1	$100
Porch, Sunshine	0.95	$90
Indoors, Rain	0.67	$50
Indoors, Sunshine	0.57	$40
Porch, Rain	0.32	$20
Outdoors, Rain	0	$0

FIGURE 26.4 Value Measures in the Party Problem

combinations of Wildlife and Tourism. The relationship between Tourism and Wildlife would be a matter of relevance. In the tree, we see that both Wildlife and Tourism contribute to the valuation of the prospect. If he believed that there was no relevance between the two attributes of Tourism and Wildlife, the hotel owner would still care about both attributes and place a direct value upon them.

The concepts of direct and indirect value can considerably simplify the formulation and solution of decision problems. While there are many features of a decision problem that a decision maker may care about, we can often identify a few direct values and recognize other values as indirect. To order prospects, we need only consider their direct value attributes. Two prospects having different indirect values but the same direct values would be preferentially equivalent.

26.2.1 Direct and Indirect Values in the Party Problem

To further illustrate this direct and indirect value distinction, refer back to the decision diagram of Figure 14.4 (repeated in Figure 26.4 for convenience) concerning the party problem. Recall that Kim's value for each prospect of the party depends on both party location and weather. Neither the weather alone nor the location alone is sufficient to determine her value of any prospect. However, given both of these attributes, we can determine how much Kim values any of the prospects she faces. Looking at the diagram, we say both Weather and Location are direct values for Kim in her party decision. As a result, both of these nodes connect to the Value node in Figure 26.4.

Now, consider Kim's decision diagram for the detector purchase decision in Figure 26.5. The diagram has three nodes with arrows into the Value Node:

1. The Detector purchase decision (since she has to pay for the detector if she uses it)
2. The party Location decision.
3. The Weather.

Therefore, in this diagram, there are three direct value measures.

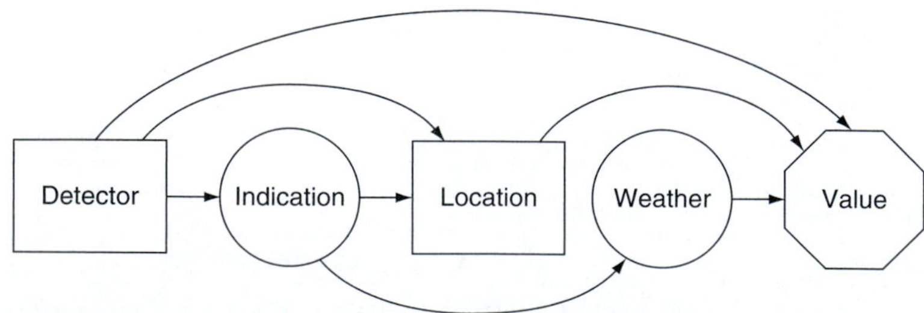

FIGURE 26.5 Direct and Indirect Values in Detector Decision

Observe, however, that the detector indication does not connect to the Value node, because the actual detector indication does not affect Kim's preferences for the party. The detector indication is relevant to another distinction, Weather, upon which Kim places a direct value. However, if the clairvoyant told Kim that the weather would be either Sun or Rain, she would no longer care about the actual detector indication in thinking about her party prospects. The detector indication is therefore not a direct value.

26.2.2 Direct and Indirect Values for an Automobile

Consider a person purchasing an automobile. We examine two automotive attributes: cost of ownership and fuel mileage. For many people, there would be a direct value on cost of ownership and an indirect value on fuel mileage. These people would not care whether a particular cost of ownership was the result of high fuel mileage and low maintenance cost, for example, or the reverse.

Suppose, however, that the person buying the car is an environmentalist. Owning a fuel-efficient car is essential to this person's public image. Even if the environmentalist rarely used the car and would consume little fuel in any car owned, the "bragging rights" of owning an extremely efficient car would require a direct value on fuel mileage.

When analyzing any decision, we must distinguish between the direct and indirect value measures. To help you think more clearly about the distinction between direct and indirect values, suppose the clairvoyant revealed the exact level of one of your direct value attributes. You cannot change it—no matter what you do. Would you still care about the other value measures available to you? If your answer is yes, then the others are direct values. If not, then they are indirect values.

26.2.3 Direct and Indirect Values for a Company

We once conducted a session with managers from a large oil company involved in everything from office expansion to space exploration. We began by asking each of them about their most valued attributes, and they suggested about twenty or thirty in total. Next, we discussed direct and indirect values, using the African environmentalist example described previously. After agreeing that the company would take only legal and ethical actions, we examined which of the many suggested distinctions had direct—rather than indirect—values.

As you might expect, one direct value was the profit of the company. We asked them to imagine the company profit for each year out into the indefinite future. Next, we discussed which of the other distinctions they would be willing to change—even to the detriment of profit.

One distinction we discussed was market share. Specifically, we addressed the questions, "Why would you want more market share if it had no impact on profit? If the clairvoyant told you the profit level you would receive from selling the product over its whole lifetime, would you still care about your market share?" The answer was no. Therefore, market share was recognized as an indirect value.

Another distinction we considered was the goodwill of the community. The executive in charge of real estate management said that when land was no longer needed for company operations, it was often donated to the local government entity. When we asked why the company did this, he said that it encouraged good relations that eased permitting and regulation in that community. When we asked why they would make such donations if they knew the donations would have no effect on profit, it was clear that we were discussing an indirect value: a means to an end rather than an end in itself.

As we went around the room, we discovered that virtually all of the value topics originally suggested were sources of indirect value. These included such things as treating employees well, paying high wages for dangerous work, and many others. You can see how these conclusions evolved. In the case of treating employees well, for example, managers believed that doing so would attract higher-quality employees who would stay with the company for longer times and ultimately would help to create higher profits.

Other than profit, the only measure of company operations that received a direct value was the number of people not associated with the company who were killed or injured as the result of company actions. This value was in addition to any insurance payments or legal judgments that the victims might receive. In order to avoid such casualties, the executives were even willing to receive lower profits.

Although this particular set of value judgments might not apply to all companies, every company and every decision maker can benefit by separating preferences from information by using the concepts of direct and indirect values.

Reflection

If you were managing a manufacturing company, which of the following values would you consider direct or indirect? Explain. Mention how you would test if they are indeed direct or indirect values.

Market share	Employee motivation
Profit	Machining costs
Technical success of a product	Man months required for manufacturing
Demand	Direct material costs
Quality of a product	Speed to market
Ease of use of a product	Ease of recycling
Environmental impact of a product	Aesthetics

26.3 STEP 2: ORDERING PROSPECTS CHARACTERIZED BY MULTIPLE "DIRECT VALUE" ATTRIBUTES

When a decision is characterized by multiple attributes, we first identify the direct value and indirect value attributes. Next, we refer to a prospect by placing the direct value attributes in parenthesis. The prospects in the party problem can be represented by multiple direct value attributes, such as Indoors, Sun as (I, S), Outdoors, Rain as (O, R), etc.

When the attributes can be represented by measures, we place the amounts of the measures in parentheses. For example, (x_1, y_1) represents a prospect with two measures having values x_1 and y_1.

As we have seen in the party problem, if the number of prospects is small, ordering the prospects is a straightforward task even when multiple attributes are present. When we have a large number of prospects, however, the ordering may be more involved. To facilitate this task, it is useful to assign a measure to each of the direct value attributes and then define a preference function.

26.3.1 Preference Functions

Preference functions associate a number to any prospect. The larger the number assigned by the preference function, the more preferred is the prospect. Recall that two prospects having the same amount of direct value attributes but different levels of indirect value attributes must have the same preference ordering. Therefore, the preference function should assign equal values to equally preferred prospects. Thus, ordering the prospects reduces the problem of assigning an appropriate preference function for the measures with direct value to the decision.

The qualitative features of the problem provide insights into the type of preference function we should use. The art of mathematical modeling is used to properly represent the qualitative features present without making the formulation too difficult to solve. Preference functions are simplified when the attributes can be characterized by measures.

To illustrate, we present several examples on ordering prospects using a preference function and direct value measures.

EXAMPLE 26.1 More of a Measure is Preferred to Less: A Multiplicative Form of Preference Functions

Consider a decision with two measures: (1) length of remaining healthy life and (2) wealth measured in dollars. Both these measures can be represented numerically after meeting the clarity test to define "healthy life" and "wealth."

To order the prospects of a decision involving these two direct value measures using a preference function, we first capture the qualitative properties. We know that in the case of these measures more of each measure is preferred to less—that is, a healthy person will prefer a longer life, and most everyone will prefer more money to less. Many preference functions could satisfy this property; one of them is

$$P(x,y) = xy^{\eta}$$

where x is the remaining healthy lifetime, y is wealth in millions of dollars, and η is a constant. Note that when any measure is 0, the preference function is 0—in this case, regardless of the level of the other measure. In some cases, this feature might be desirable, but we might want to modify this feature such that the preference is 0 when an attribute takes on a particular value. For example, if x_0 is the minimum wealth needed to survive (for food or shelter), we can modify the function as

$$P(x,y) = (x - x_0)y^{\eta}, \quad x \geq x_0, y \geq 0$$

where the preference value is 0 if either the remaining life time is 0 or wealth is below the minimum consumption needed to survive.

We refer to the set of prospects (x,y) that provide the same numeric value for the preference function as an isopreference contour. An example of such contours appears in Figure 26.6 for $P(x,y) = xy^{\eta}$ where $\eta = 1$.

In Figure 26.6, Prospects A and E lie on the same isopreference contour and therefore are equally preferred. The same applies to Prospects D and F. However, Prospects A and B do not lie on the same contour. Both A and B have the same health state, but B has a higher wealth level. Since we always prefer more money to less, B is preferred to A. Similarly,

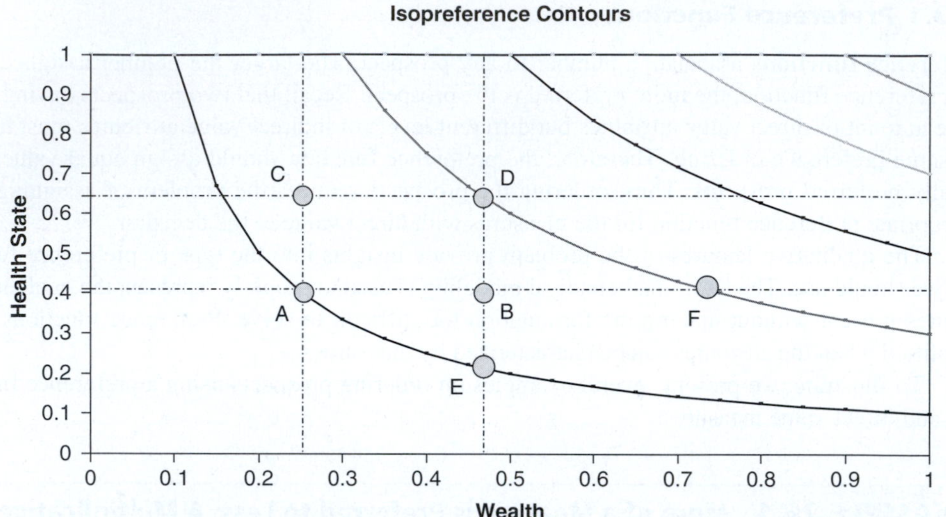

FIGURE 26.6 Isopreference Contours for the Given Preference Function where $\eta = 1$

prospects *C* and *A* have the same wealth level, but *C* has a better health state. Since we prefer a better health state to a worse one, *C* is preferred to *A*. We also know that *D* is preferred to *A*, *B*, or *C*, since it provides better health states and more wealth.

However, the difficulty arises when we compare *B* and *C*. Prospect *B* has a higher wealth and *C* has a better health state. Here, a trade-off between having better health and more wealth needs to be determined. The trade-off function implicitly incorporates the preference, since it provides the complete ordering of the prospects. The higher the preference value obtained with *B* or *C*, the more preferred is the prospect. In this example, this trade-off is determined by the value of η.

To illustrate the idea behind trade-offs among the measures, suppose we start with a prospect (x_1, y_1) and then reduce x_1 by a small amount Δx to yield $(x_1 - \Delta x, y_1)$. Naturally, we would not like this new prospect as much as the previous one. But suppose we then increase *y* by a small amount Δy to produce $(x_1 - \Delta x, y_1 + \Delta y)$. What is the amount of increase Δy needed to compensate for the decrease Δx so that we would be indifferent between (x_1, y_1) and $(x_1 - \Delta x, y_1 + \Delta y)$? In Appendix A, we show that these small increments must be related by

$$(\Delta y) = -(\Delta x)\frac{y}{x\eta}$$

In other words, if we decrease measure *x* by an amount Δx, the amount of increase in measure *y* and Δy needed to make us just indifferent is larger when we have smaller amounts of *x* and larger amounts of measure *y*. This makes intuitive sense. We might be willing to exchange measure *x* for *y* if we have more of it, but we would be reluctant to exchange *x* for *y* if we have less of it.

Consider, for example, two measures representing levels of food consumption in two consecutive years. We may be willing to trade some consumption in the first year for an increased consumption in the second, but not if we have 0 levels of consumption in the first year. Even an infinite amount of consumption in the second year would not be sufficient, since zero consumption in the first year would guarantee we would not be alive to enjoy the increase in the second year. The preference function under consideration captures this qualitative feature.

Rearranging also gives an interpretation for the parameter, η, as

$$\eta = -\frac{\left(\frac{\Delta x}{x}\right)}{\left(\frac{\Delta y}{y}\right)}$$

The parameter η is the ratio of the (small) fractional decrease in measure x to the (small) fractional increase in measure y needed to make us just indifferent between the two prospects. For example, suppose the decision maker states that a 2% decrease in wealth and a 1% increase in health state would make him indifferent to his current health and wealth states. Then his value is $\eta = 2$.

The parameter η determines the trade-offs we would be willing to make to substitute one measure for another and have equally preferred prospects. Now, we return to the prospect preferences of Figure 26.6. Consider the Prospects $A = (0.26, 0.4)$, $B = (0.46, 0.4)$, $C = (0.26, 0.63)$, $D = (0.46, 0.63)$, $E = (0.46, 0.23)$, and $F = (0.75, 0.4.)$ shown in Table 26.1. With the preference function shown previously and $\eta = 1$, we have the preference order $F \sim D > B > C > A \sim E$. We see that B is preferred to C, which is a question not settled in the figure. We confirm that the decision maker is indifferent between A and E and between F and D, as shown in the figure. In each case, an increase in one attribute is offset by a decrease in the other.

If the decision maker sets $\eta = 2$, Table 26.1 reveals the preference order $D > F > C > B > A > E$, where E and A no longer lie on the same contour and neither do F and D. For both $\eta = 1$ and $\eta = 2$, however, $F > A, F > B, F > E$, $D > B, D > C, D > A, D > E, C > A$, and $B > A$, and $B > E$, as expected.

Recall that, if there is no uncertainty in the decision, the best alternative is the one that yields the most preferred prospect, and the preference function is sufficient to achieve this ordering task.

TABLE 26.1 Different Orders for Prospects with Different Values of η

			$\eta = 1$	$\eta = 2$
	X	Y	V	V
A	0.26	0.4	0.104	0.042
B	0.46	0.4	0.184	0.074
C	0.26	0.63	0.164	0.103
D	0.46	0.63	0.290	0.183
E	0.46	0.23	0.106	0.024
F	0.725	0.4	0.290	0.116

EXAMPLE 26.2 More of a Measure is Not Preferred to Less: The Peanut Butter and Jelly Sandwich

Suppose you were choosing among several peanut butter and jelly sandwiches. After thinking about the sandwich in Figure 26.7 for a while, you declare that the direct value measures are (1) thickness of both slices of bread, (2) thickness of peanut butter, (3) thickness of jelly, and (4) the ratio of the thickness of peanut butter to jelly.

Here, our preference for the sandwich is not necessarily increasing with the thickness of peanut butter, jelly, or bread. For example, we would not want a sandwich that has no bread (0 thickness of bread), but we would also not want a sandwich whose thickness of bread is 4 inches (it would be too difficult to eat). Hence, the preference function of the previous example would not be valid for this situation.

There is some value for the thickness of bread intermediate to the extremes that would make the sandwich most enjoyable. There are also optimal thicknesses of peanut butter and of jelly that would most suit us. However, the optimal thickness of peanut butter may change with thickness of jelly. For example, we may prefer more peanut butter if there is more jelly. Therefore, a fourth value measure would be the ratio of thickness of peanut butter to jelly.

We can capture these qualitative features in a preference function for the sandwich given any values of the four measures. An example of such a preference function is

$$P(b, j, p, f) = \frac{bjpf}{(b^* j^* p^* f^*)^2}(2b^* - b)(2j^* - j)(2p^* - p)(2f^* - f)$$

where b, j, p, and f are the thicknesses of bread, jelly, peanut butter, and fraction of peanut butter to jelly, respectively, and b^*, j^*, p^*, and f^* are their optimal values: all measurements are in inches.

Suppose a decision maker's preference function is given by the equation for $b^* = 1, j^* = .1, p^* = .05$, and $f^* = .5$. Suppose he faces two sandwiches, as shown in Table 26.2. Note that the second sandwich has more bread, more jelly, and more peanut butter, but it is less preferred to the first, since the decision maker does not want too much of any of these measures.

FIGURE 26.7 Peanut Butter and Jelly Sandwich
(2C Squared Studios/Ocean/Corbis)

TABLE 26.2 Preference Function Applied to Two Sandwiches

	b	*j*	*p*	*f*	Preference
Sandwich 1	1	0.09	0.05	0.56	0.98
Sandwich 2	1.2	0.15	0.08	0.53	0.46

This example does not include any uncertainty about the sandwich that the decision maker will get. We can determine the desired sandwich using the preference function because it is a choice between deterministic alternatives.

Note *This preference function does not require a monetary measure.*

To the reader: If you like peanut butter and jelly sandwiches, how would you set the parameters in this preference function? Can you construct another one that would better describe your preferences?

Note *Why the peanut butter and jelly Sandwich?*

We have chosen the peanut, butter and jelly sandwich example because, while simple to reason about, it involves a complicated form of preferences that would otherwise be difficult to reason about. We did not require that more of any attribute be preferred to less when constructing the preference function. We also used levels of attributes that could be clearly communicated and that the clairvoyant could answer. It is easy to ask the clairvoyant about the thickness of peanut butter, jelly, or bread. We did not use attributes like convenience, taste, or hunger level using constructed scales that could be difficult to for the clairvoyant to understand.

26.3.2 Preference Functions with a Monetary Value Measure: Value Functions

In principle, we do not need a monetary measure to determine the best decision alternative. In the party problem, Kim used two attributes of interest, location and weather, to characterize the prospects. We asked her to order the prospects and then state her preference probabilities for them. We were able to determine her best decision alternative without the use of a monetary measure. Likewise, in the peanut butter and jelly sandwich, we are able to order any set of sandwiches using the preference function.

In many cases, one of the measures in the preference function will be a **value measure**, which we interpret as the amount for which we would be just willing to sell a prospect possessing given levels of the other attributes if we owned it. In other words, this is our PISP for the prospect described by these attributes. In these cases, we can then obtain a monetary equivalent for any prospect when the remaining attributes are set at fixed values. As a special case, the magnitude of the preference function itself may be a monetary value measure describing our PISP for the prospect. We will call such preference functions **value functions**.

If the ownership of the prospect does not materially affect our wealth (think of the Rolls-Royce example in Chapter 3) then this same value function will give us our Personal Indifferent Buying Price (PIBP) for a prospect that we do not own.

Using a value function provides both an ordering of all prospects and the convenience of having a value measure associated with each prospect. Furthermore, as was the case when we originally introduced a value measure, having a value measure permits calculating the certain equivalent of an uncertain deal with multiple measures as well as computing the value of clairvoyance in decisions with such prospects.

Consider a value function that returns a dollar amount (our PISP and PIBP) of the peanut butter and jelly sandwich is

$$V(b, j, p, f) = V_{\max} \frac{bjpf}{(b^* j^* p^* f^*)^2} (2b^* - b)(2j^* - j)(2p^* - p)(2f^* - f)$$

where $V_{\max}$ is the PIBP dollar value for the best sandwich.

With this value function, we can determine our PIBP and PISP for any sandwich with different combinations of bread, jelly, and peanut butter. An agent can even determine our preferences on our behalf if he has access to this value function.

Suppose $V_{\max} = \$3$, $b^* = 1$, $j^* = .1$, $p^* = .05$, and $f^* = 0.5$. Now we can determine the decision maker's PIBP for the two sandwiches previously specified in Table 26.2, as shown in Table 26.3.

TABLE 26.3 Value Function Applied to Determine the PIBPs for Two Sandwiches

	b	*j*	*p*	*f*	Value
Sandwich 1	1	0.09	0.05	0.56	\$2.93
Sandwich 2	1.2	0.15	0.08	0.53	\$1.38

Not only do we have a preference ordering for the two sandwiches, but we can also tell how much we value each sandwich and how much we would need to be paid to give up Sandwich 1 for Sandwich 2. In this case, it is \$2.93 − \$1.38 = \$1.55.

Having a value function also gives us the ability to determine the value of changes in the sandwich, which is a sensitivity analysis that may be very useful. For example, we can see for this example that increasing the thickness of jelly in Sandwich 1 from 0.09 to 0.1 would increase the value of the sandwich by seven cents and make it the optimal sandwich.

EXAMPLE 26.3 Value as a Linear Additive Weighting of Attribute Measures

Would your PIBP for having two peanut butter and jelly sandwiches for lunch equal the sum of your PIBPs for having each sandwich individually? Usually, it would not, because you would probably be satisfied after eating the first sandwich. We have previously discussed that the PIBP of the sum of two items is not necessarily equal to the sum of their PIBPs.

Now, consider a different situation with a different set of measures, such as salary and vacation time. Would your PIBP for getting two prospects of salary and vacation time be equal to the sum of PIBPs of getting each prospect separately?

Consider two prospects (x_1, y_1), (x_2, y_2) and a value function that provides the PIBP for each prospect. If the PIBP of the sum is equal to the sum of PIBPs for all such prospects, then

$$V(x_1 + x_2, y_1 + y_2) = V(x_1, y_1) + V(x_2, y_2)$$

This equality implies that the value function must be of the form

$$V(x, y) = w_x x + w_y y$$

with an additive value function, where w_x, w_y are the weights assigned to the different measures and x, y determine the level of each attribute for any prospect.

This additive value function also implies constant trade-offs between the levels of each of the two measures. For example, suppose $x_1 < x_2$ and $y_1 > y_2$, starting from (x_1, y_1). By moving to (x_2, y_2), you get more of measure x but less of measure y. You would be indifferent between these two prospects if

$$w_x x_1 + w_y y_1 = w_x x_2 + w_y y_2$$
$$w_x(x_2 - x_1) = w_y(y_1 - y_2)$$

Therefore,

$$\frac{\text{Decrease in } x}{\text{Increase in } y} = \frac{(x_2 - x_1)}{(y_1 - y_2)} = \frac{w_y}{w_x} = \text{Constant}$$

Herein lies the potential problem with this value function. It might be that your trade-offs are constant indeed over a small interval but may not be constant across the whole domain. For example, you might be willing to trade-off a certain number of vacation days for a fixed increase in salary provided that you still have enough vacation left, but you would ask for a much higher exchange price if you had no vacation left. (Compare this to the case of the preference function $V(x, y) = xy^{\eta}$, where the trade-off depended on the amount you have for each measure and the trade-off value increases as that amount of the measure you have decreases).

Consider again the example of the two measures of food consumption in Year 1 and food consumption in Year 2. We can envision getting less consumption in Year 1 if we are guaranteed a higher consumption in Year 2, but how much less consumption in Year 1 are we able to accommodate?

Suppose we got 0 consumption in Year 1. Is there any level of consumption in Year 2 that would make us just indifferent? Presumably not, if we prefer to live than to die. Yet this value function would imply that our trade-offs for any decrease in consumption of Year 1 versus Year 2 will be constant. This example illustrates how this additive value function fails to capture preferences for consumption.

Additive value functions may be constructed for more than two attributes. They imply that our trade-off between any two attributes does not depend on the level of the remaining attributes.

We can envision situations where this additive value function may apply, such as for measures having monetary values that represent investments over time. The trade-off parameters simply represent our time preference. We may be willing to have constant investment trade-offs for a certain number of years. We discuss this topic in more detail in the next chapter.

26.4 SUMMARY

When a decision involves multiple attributes, or measures of importance, our first step is to identify the direct versus indirect value measures. Two prospects having the same direct value measures but different indirect value measures would have equal preference in the order rule.

Direct value attributes need not have a numeric measure, but we can still order the prospects. For example, in the party problem, we had attributes of Location and Weather, and Kim was able to order her prospects.

We construct preference and value functions only for direct value measures. Indirect value measures represent probabilistic relevance to direct values. For example, we might care about market share only because the higher the market share, the more likely it is that the profit will be higher. However, if the clairvoyant told us the amount of profit we will get, then we will no longer care about market share. In this case, market share is an indirect value, and profit is a direct value. We construct the value function for profit.

If the problem is deterministic, a preference function is sufficient to rank order the prospects and determine the best alternative. This preference function does not require a monetary measure.

When one of the measures can be represented in monetary terms, we can construct a value function to determine our monetary value for each prospect. This lets us determine the amount of money that would make us indifferent between receiving one prospect and another.

Weight and rate (additive) value functions are simple to construct and use, but we must make sure they correctly characterize our preferences before we use them. Weight and rate value functions imply constant trade-offs between any two subsets of the attributes—regardless of the level of the remaining attributes.

Value functions of the form $V(x, y) = xy^{\eta}$ imply that the fractional increase in an attribute needed to compensate for a fractional decrease in the other is constant. Thus,

$$\eta = -\frac{\left(\dfrac{\Delta x}{x}\right)}{\left(\dfrac{\Delta y}{y}\right)}$$

is a constant value.

Value functions need not be increasing as we increase each of the attributes. An example is the value function for the peanut butter and jelly sandwich.

KEY TERMS

- Attribute
- Direct value
- Indirect value
- Preference functions
- Value functions
- Value measure

APPENDIX A Deriving the Relation Between Increments in x and y as a Function of η in the Preference Function

Consider

$$V(x, y) = xy^{\eta}$$

Across an isopreference contour, we observe the increment in preference must be 0 as we increase x and decrease y. Here, we use calculus and let Δ represent a small increment in the level of each measure. Hence, it must be that

$$\Delta V = 0 = y^{\eta}(\Delta x) + x\eta y^{\eta-1}(\Delta y)$$

Rearranging gives

$$(\Delta y) = -(\Delta x)\frac{y}{\eta x}$$

PROBLEMS

Problems marked with an asterisk (*) are considered more challenging.

1. Construct a value function to determine your PIBP and PISP for a peanut butter and jelly sandwich. Provide any required parameters, such as best thickness of peanut butter, jelly, and bread.
   a. How much is your best sandwich worth to you?
   b. If you were going to get 90% of your optimal thickness of bread, how much would this new sandwich be worth to you?

*2. Construct a value function that returns your value of an automobile. Think of the direct versus indirect value attributes. Do you think the automobile manufacturer will have a different value function?

*3. In the preference function of Example 26.1, plot the isopreference contours for values of $\eta = 1, 2, 3$. What is the effect of changing η on the trade-off between the two measures?

*4. Calculate the ratio $\eta = -\dfrac{\left(\dfrac{\Delta x}{x}\right)}{\left(\dfrac{\Delta y}{y}\right)}$ for an additive value function.

CHAPTER

27

Decisions with Multiple Attributes (2)–Value Functions for Investment Cash Flows: Time Preference

CHAPTER CONCEPTS

After reading this chapter, you will be able to explain the following concepts:

- Investment cash flows
- Present equivalent of a cash flow
- Time preference
- Valuing an annuity

27.1 INTRODUCTION

In this chapter, we will discuss how to construct a value function to value a pattern of deterministic resource amounts that you receive over time. The resource can be any fungible resource, but we shall consider it to be money. The pattern then represents a sequence of cash flows and in this discussion; we shall assume that the cash flows represent an investment. In this chapter, we shall not consider the case where you are consuming a resource (consumption), although this topic is important and will be covered in Chapter 34. Here, we shall address how to value a pattern of deterministic cash flows for investment purposes.

27.1.1 Representing Cash Flows

Let us suppose that the payments that you receive or make will be made at instants of time numbered 0, 1, 2,.... The time period between payments is constant and is usually, but is not necessarily, a time of year. We shall describe the cash flow by the sequence $x_0, x_1, x_2, \ldots, x_N$, where x_j denotes the payment at time j and N is the time of the last payment. Each payment may be positive or negative; we shall consider payments we receive as positive and payments we disburse as negative. We can describe this pattern of cash flows succinctly by a cash flow vector X with the components we have defined. We can also represent this cash flow as a prospect where the payoff in each year represents a direct value measure. For example, the vector

$$X = (x_0, x_1, x_2, x_3) = (-2, 4, 8, -1)$$

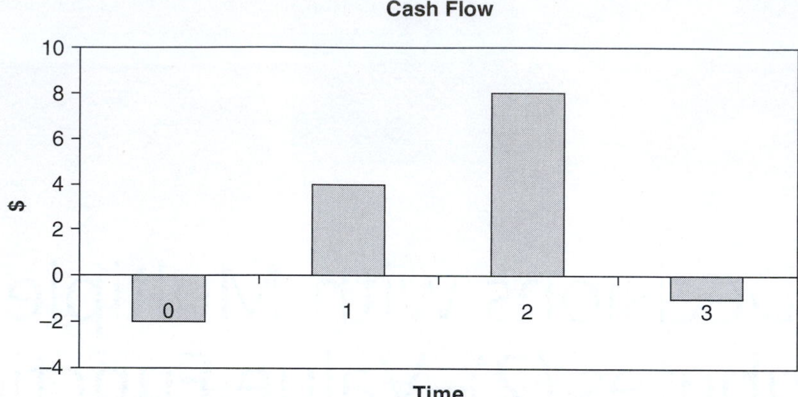

FIGURE 27.1 Graphical Representation of a Cash Flow

represents a payment of \$2 at time 0, receipt of \$4 and \$8 at times 1 and 2, respectively, and a payment of \$1 at time 3. Figure 27.1 represents this prospect graphically.

Note that in this formulation there is no uncertainty about the actual amount you will receive or pay at any time step. Our purpose is to find a single payment you could either receive or pay now that would make you indifferent to incurring the sequence of payments. Since this is a deterministic cash flow, we do not need u-values to solve this problem.

27.1.2 The Present Equivalent of a Cash Flow, $pe(X)$

We shall call the payment that makes you indifferent between receiving the payment now and receiving the cash flow your **present equivalent** of the cash flow, and we denote it by $pe(X)$. The present equivalent may be either positive or negative. We use the term present equivalent for this type of value function to emphasize that it involves cash flows over time and, consequently, trade-offs between present and future.

> **Note**
>
> The initial payment in the cash flow x_0 is, in fact, paid now and could be simply ignored in our calculation and then added at the end to the present equivalent we otherwise calculate. However, we shall include it in the cash-flow pattern for convenience.

27.2 RULES FOR EVALUATING INVESTMENT CASH FLOWS

The way we shall determine the present equivalent of a cash-flow pattern is to prescribe a few rules that govern the process. If you choose to abide by them, you can reduce the present equivalent calculation to a very simple form.

RULE 1: THE SCALING RULE The first rule we shall invoke is the **scaling rule**. This rule states that if we multiply each of the cash flow payments by any constant k, then to maintain equivalence, we must multiply the present equivalent of the cash flow vector by the same constant k. In equation form, this means

$$pe(kX) = k\,pe(X)$$

The interpretation of the scaling rule is simple. If, for example, we doubled all the cash flows, we would have to double the present equivalent. This rule also applies for $k = 0$, which means that the present equivalent of receiving no payments is 0. Remember that this rule applies to a deterministic cash flow where there is no uncertainty. This rule need not be true for uncertain cash flows.

RULE 2: THE ADDITION RULE The second rule is the **addition rule**, which we also discussed in the previous chapter. This states that if the cash flow can be divided into two parts, cash flow A and cash flow B, then the present equivalent of A plus the present equivalent of B must equal the present equivalent of the original cash flow. To be precise, let

$$X = X_A + X_B$$

where every cash flow $x_j = x_{jA} + x_{jB}$. In this decomposition of x_j, there is no restriction on the sign of x_{jA} or x_{jB}. Then we must have

$$pe(X) = pe(X_A) + pe(X_B)$$

As a particular case of the addition rule, we can choose to decompose the cash flow X in time. For example, we can let

$$X_A = (x_0, 0, 0, 0,. . . . .)$$
$$X_B = (0, x_1, x_2, x_3, . . .)$$

In other words, X_A is composed only of the first cash flow, whereas X_B is composed of the remainder. Then, in accordance with the addition rule, we have

$$pe(X) = pe(x_0, 0, 0, 0, ...) + pe(0, x_1, x_2, x_3, . . .)$$

We can repeat this procedure and split X_B into x_1 and all other components. Then, we would have

$$pe(X) = pe(x_0, 0, 0, 0, ...) + pe(0, x_1, 0, 0, . . .) + pe(0, 0, x_2, x_3, x_4 . . .)$$

By continuing this process, we can write the present equivalent as the sum of the present equivalents of cash flows that are potentially nonzero in only one component:

$$pe(X) = pe(x_0, 0, 0, 0, ...) + pe(0, x_1, 0, 0, . . .) + pe(0, 0, x_2, 0, 0, . . .)$$
$$+ pe(0, 0, 0, x_3, 0, . . .) + \ . . .$$

If we let

$$p_j(x_j) = pe(0, 0, . . ., 0, x_j, 0, 0, . . .)$$

then we can write the present equivalent of any cash flow vector X as

$$pe(X) = \sum_{j=1}^{N} p_j(x_j)$$

Let us further examine the quantity $p_j(x_j)$. It represents the present equivalent of a cash flow pattern that is 0 everywhere except at time point j where it is equal to x_j. As a result of the scaling rule, the present equivalent of such a pattern must be just x_j times the present equivalent of a unit payment at the same time. In other words,

$$p_j(X_j) = x_j p_j(1)$$

It then follows that the present equivalent of the cash-flow pattern X must be given by

$$pe(X) = \sum_{j=1}^{N} x_j p_j(1)$$

This means that to evaluate a cash-flow pattern, all we must do is multiply the magnitude of the cash flow at each time point by the present equivalent of a unit payment at that time point, and then sum over all time points in the cash flow.

We can write this result in even simpler form by letting

$$p_j(1) = \beta_j$$

The quantity β_j is the present equivalent of a unit payment at time point j, which is a quantity we shall call the time point j **discount factor**. Then the present equivalent of the cash flow pattern X is given by

$$pe(X) = \sum_{j=1}^{N} x_j \beta_j$$

This form for the present equivalent is as far as the scaling and addition rules can take us. In fact, it is the proper form if it is necessary to associate the value of the payments with specific calendar times.

INVESTMENT VERSUS CONSUMPTION CASH FLOWS We can see why the present equivalent of this cash flow would be appropriate for an investment cash flow but not for consumption. If an investor is willing to abide by the scaling and addition rules, then he will have an additive value function. This implies that the trade-offs do not depend on the actual amount received or paid in any year. However, if this were a consumption cash flow, then it is quite plausible that the trade-offs will depend on the actual amount consumed in each year. A consumer might not be willing to decrease consumption in a given year in exchange for any increase in consumption in a future year if it falls below his consumption level needed for survival that year.

So far, the scaling and addition rules did not specify anything about the trade-off amount in each year. The investor may assign specific trade-offs for each calendar year. However, if there is nothing special about particular time periods, we can invoke another rule.

RULE 3: THE CALENDAR TIME INVARIANCE RULE The **calendar time invariance rule** states that our time preference is concerned only with the delays in receiving or dispersing payments and not with the actual calendar times at which the payments are made. For example, one unit of payment one year from now will be worth β_1 now. One unit of payment two years from now will be worth β_2 now. If we say that only the time differences, and not the calendar times, are important, then one unit of payment two years from now will be worth $\beta_1 x \beta_1$ one year from now and consequently worth β_1^2 now; thus, $\beta_2 = \beta_1^2$. (See Figure 27.2). Since a unit payment three periods from now will be worth β_2 two periods from now and has $\beta_1 x \beta_2$ value now, we have $\beta_3 = \beta_1 \beta_2 = \beta_1^3$. Continuing this process, we have $\beta_j = \beta_1^j$.

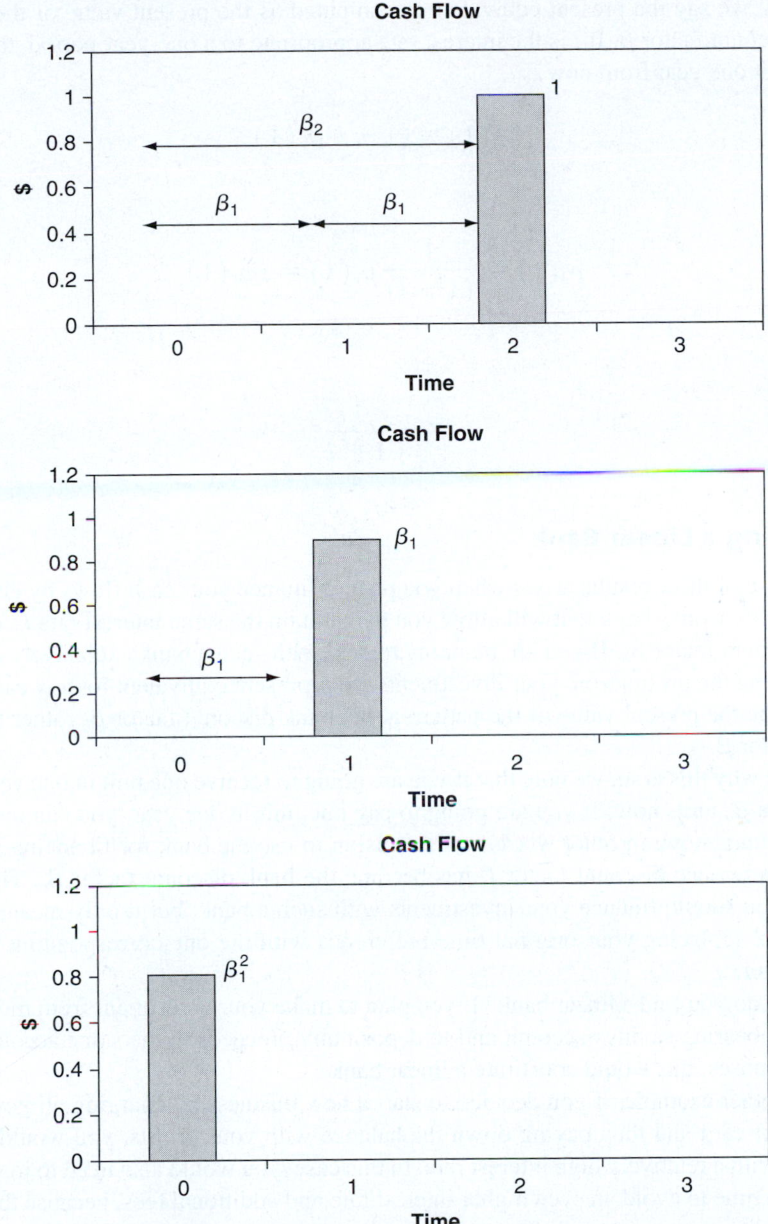

FIGURE 27.2 Valuing Cash Flows with Calendar Time Invariance, $\beta_2 = \beta_1^2$

Calendar time invariance means that the present equivalent of unit payments at different times in the future must be just powers of a unit payment one period from now. In this case, we can write β_1 more simply as β and place our present equivalent rule in the form

$$pe(X) = \sum_{j=1}^{N} x_j \beta^j$$

In this form, we say the present equivalent is computed as the present value of the cash flows with the discount factor β. If i is the interest rate appropriate to a one-year period, the value of a unit payment one year from now is

$$p_1(1) = (1 + i)p_0(1)$$

or

$$p_0(1) = \frac{1}{(1 + i)}p_1(1) = \beta p_1(1)$$

Therefore,

$$\beta = \frac{1}{1 + i}$$

27.2.1 Using a Linear Bank

A special case of these results arises when you plan to finance your cash flows by either lending to or borrowing from a bank that will allow you to maintain the same interest rate i_b, corresponding to a discount factor β_b. If you are planning to deal with such a bank exclusively in financing and reinvesting the profits from your investments, your present equivalent for any cash-flow pattern should be the present value of the pattern at the bank discount factor β_b rather than at your discount factor β.

To see why this is so, we note that if you are going to receive one unit in one year, the bank will give you β_b units now. If you are going to pay one unit in one year, you can instead pay to the bank β_b units now. In other words, your decision to use the bank for financing implies that your time preference discount factor β has become the bank discount factor β_b. This does not imply that you should finance your investments with such a bank, but it only means that if you do so you are replacing your original time preference with the one corresponding to the bank discount factor.

Where do you find a linear bank? If you plan to make your investment from money in your own interest-bearing savings account and to deposit any proceeds in the same account, then for practical purposes, this would constitute a linear bank.

As another example, if you decided to start a new business by charging all your expenses to your credit card and then paying down the balance with your profits, you would be using a linear bank with a relatively high interest rate. In this case, you would also need to make required payments on time to avoid an even higher interest rate and additional fees, because then it would no longer be linear.

VALUING AN ANNUITY FOR *m* YEARS Following our discussion, the present equivalent for receiving \$1.00 every year for m years, which we denote as $pe[a(m)]$, is then

$$pe\left[a(m)\right] = 1 + \beta + \beta^2 + \cdots + \beta^m$$
$$= \frac{1 - \beta^{m+1}}{1 - \beta}$$

where β is the present value of a unit payment one year in the future at the prevailing interest rate i. Thus,

$$\beta = \frac{1}{1 + i}$$

An annuity that pays $X every year for m years has a present equivalent of

$$Xpe\left[a(m)\right] = \$X\frac{1 - \beta^{m+1}}{1 - \beta}$$

In Chapter 28, we shall calculate the value of an annuity that pays $X every year for an uncertain number of years.

27.2.2 Determining Future Value

The rules for present equivalent that we have discussed also can be used to determine the future value of any cash flow pattern. Suppose, for example, we have agreed to follow the scaling and addition rules and we have computed the present equivalent of a cash flow pattern X in terms of the quantities $\beta_1, \beta_2, \ldots$.

Now, we ask what payment received at some future time moment k would have this present equivalent. Since one unit of payment at time moment k would be worth β_k now, it follows that we would have to divide the present equivalent of the cash flow pattern by β_k to obtain the future value in period k. Of course, if the calendar time-invariance rule is also satisfied, then β_k is simply β^k.

27.2.3 Illustration of Time Preference Concepts

To illustrate the use of these time preference concepts for investments, consider three $1,000 investments of *A*, *B*, and *C*—each of which will yield cash returns at the end of each of the next ten periods. Their cash flows appear in Table 27.1.

Investment *A* is like putting the $1,000 in a bank that pays 20% ($200) interest per period on your investment. In this case, your principal is returned at the conclusion of Period 10. Investment *B* pays 25% ($250) per period on your $1,000 investment for ten periods with no return of principal. Investment *C* pays $500 for three periods, then $300, $200, $150, $100, and finally $50 for the last three periods.

Now, we will examine in detail each of these investments in the situation where you will finance the investment by withdrawing the $1,000 from a bank and then depositing the proceeds

TABLE 27.1 Ten Period Cash Flows for Three $1,000 Investments

	Period										
Investment	**0**	**1**	**2**	**3**	**4**	**5**	**6**	**7**	**8**	**9**	**10**
A	($1,000)	$200	$200	$200	$200	$200	$200	$200	$200	$200	$1,200
B	($1,000)	$250	$250	$250	$250	$250	$250	$250	$250	$250	$250
C	($1,000)	$500	$500	$500	$300	$200	$150	$100	$50	$50	$50

in the same bank. We will consider bank interest rates per period of 5%, 10%, and 20%. All interest payments or charges are made at the end of each period.

PRESENT VALUE WITH A 5% INTEREST RATE Table 27.2 illustrates the results of the three investments when the financing will be done in a 5% bank. The first columns show the results for Investment *A*. The receipts (payments) column shows the $1,000 withdrawn from the bank at time 0, then nine $200 receipts and finally the $1,200 receipt in Period 10 composed of the return of the $1,000 principal and the final $200 interest payment.

The second column shows the interest that would be received (paid) at the conclusion of each period. For example, in Period 1, you would owe a $50 interest payment (5% of $1,000).

The next column shows the amount in your bank account at the end of the period. Therefore, at the end of Period 1, you have a negative balance of $850 in the bank, composed of the original withdrawal of $1,000, the first $200 receipt from the investment, and the interest of $50 owed to the bank. The row corresponding to Period 2 shows a receipt of $200 and an interest payment of $42.50 (5% of $850) for a net bank account of –$692.50.

We see that as the periods pass, the bank account grows less negative until at Period 6 it becomes a positive amount of $20.29. In future periods, you will receive interest on your positive bank account in addition to your receipts from the investment. Finally, in Period 10 you receive the principal of $1,000, the receipt of $200, and interest of $32.70 to produce a final bank account of $1,886.68. If you invest in Investment *A* and finance it in a bank with a 5% interest rate, the amount that you will have in your bank account after ten periods will be $1,886.68 more than you started with.

The final row of the table shows the amount that would have to be in your bank account now to grow to the same $1,886.68 if you did not make the investment. This amount is the present value of Investment *A* at 5% or $1,158.26, which is computed as $\frac{1886.68}{(1.05)^{10}}$. Consequently, we can say that investing in Investment *A* when you are going to finance it in a 5%-interest bank is like placing $1,158.26 in that same bank today. Clearly, the higher the future value of the bank account, the higher the present value of that same investment today.

The central columns of Table 27.2 show the same calculations for Investment *B*. The bank account for Investment *B* turns positive in Period 5; however, it reaches only $1,515.58 by Period 10. The present value of this amount is $930.43. Therefore, Investment *A* will produce a higher future value and, consequently, a higher present value than will Investment *B*.

The final columns of Table 27.2 show the results for Investment *C*. Here, the bank account turns positive in Period 3, but the decreasing payments result in a bank account at the end of Period 10 of $1,702.05 with a present value of $1,044.91. We see that Investment *C* produces a higher future value and present value than does Investment *B*, but this is less than the corresponding quantities produced by Investment *A*. Thus, of the three investments, Investment *A* would be best if all investments were to be financed in a 5%-per-period bank.

Notice that we have compared the investments using the common-sense notion of which investment will produce the largest bank account at the end of its ten-period life. We know that the investment with the highest future value also will be the investment with the highest present value. Since investments may have different lifetimes, the simplest procedure is to compare them in terms of present value.

PRESENT VALUE WITH A 10% INTEREST RATE Table 27.3 compares the three investments in the situation where they all will be financed in a 10%-per-period bank. We see that the investment with the highest present value (and future value) is Investment *C*; its present value is $772.31.

TABLE 27.2 Results of Three Investments Using a 5% Bank

	A			*B*			C		
Period	Receipts (Payments)	Interest Received (Paid)	Bank Account	Receipts (Payments)	Interest Received (Paid)	Bank Account	Receipts (Payments)	Interest Received (Paid)	Bank Account
0	($1,000)	$0.00	($1,000)	($1,000.00)	$0.00	($1,000)	($1,000)	$0.00	($1,000)
1	$200	($50.00)	($850)	$250.00	($50.00)	($800)	$500	($50.00)	($550)
2	$200	($42.50)	($693)	$250.00	($40.00)	($590)	$500	($27.50)	($78)
3	$200	($34.63)	($527)	$250.00	($29.50)	($370)	$500	($3.88)	$419
4	$200	($26.36)	($353)	$250.00	($18.48)	($138)	$300	$20.93	$740
5	$200	($17.67)	($171)	$250.00	($6.90)	$105	$200	$36.98	$977
6	$200	($8.56)	$20	$250.00	$5.26	$360	$150	$48.83	$1,175
7	$200	$1.01	$221	$250.00	$18.02	$628	$100	$58.77	$1,334
8	$200	$11.07	$432	$250.00	$31.42	$910	$50	$66.71	$1,451
9	$200	$21.62	$654	$250.00	$45.49	$1,205	$50	$72.54	$1,573
10	$1,200	$32.70	$1,887	$250.00	$60.27	$1,516	$50	$78.67	$1,702
Present Value:			$1,158.26			$930.43			$1,044.91

TABLE 27.3 Results of Three Investments Using a 10% Bank

	A			*B*			*C*		
Period	**Receipts (Payments)**	**Interest Received (Paid)**	**Bank Account**	**Receipts (Payments)**	**Interest Received (Paid)**	**Bank Account**	**Receipts (Payments)**	**Interest Received (Paid)**	**Bank Account**
0	($1,000)	$0.00	($1,000)	($1,000.00)	$0.00	($1,000)	($1,000)	$0.00	($1,000)
1	$200	($100.00)	($900)	$250.00	($100.00)	($850)	$500	($100.00)	($600)
2	$200	($90.00)	($790)	$250.00	($85.00)	($685)	$500	($60.00)	($160)
3	$200	($79.00)	($669)	$250.00	($68.50)	($504)	$500	($16.00)	$324
4	$200	($66.90)	($536)	$250.00	($50.35)	($304)	$300	$32.40	$656
5	$200	($53.59)	($389)	$250.00	($30.39)	($84)	$200	$65.64	$922
6	$200	($38.95)	($228)	$250.00	($8.42)	$157	$150	$92.20	$1,164
7	$200	($22.84)	($51)	$250.00	$15.73	$423	$100	$116.42	$1,381
8	$200	($5.13)	$144	$250.00	$42.31	$715	$50	$138.07	$1,569
9	$200	$14.36	$358	$250.00	$71.54	$1,037	$50	$156.87	$1,776
10	$1,200	$35.79	$1,594	$250.00	$103.69	$1,391	$50	$177.56	$2, 003
Present Value:			$614.46			$536.14			$772.31

The next best investment is Investment *A* with a present value $614.46. The worst investment is *B* with a present value of $536.14.

> ***Note*** *The use of the higher interest rate bank for financing has reduced the present value of all investments from those for the 5% bank.*

We can say simply that investing in Investment *C* is like putting $772.31 in a 10% bank today.

PRESENT VALUE WITH 20% INTEREST RATE Table 27.4 shows the results for the three investments when they are financed with a 20% bank. The calculations for Investment *A* are particularly simple, since in this case, the interest owed in each period is exactly equal to the receipt from the investment. The net bank balance at the end of the ten periods is 0 as is, of course, the present value. Investment *A* is therefore like putting nothing in a bank account that pays 20% per period. Investment *B* has a present value of $48.12; Investment *C*, the winner, has a present value of $385.83. Note that this present value is associated with a very impressive future value in Period 10 of $2,388.95. Also, notice that this is the first of the interest rates we have considered that makes Investment *B* better than Investment *A*.

All of these results stand regardless of your personal time preference rate, as we have proved. If a person is going to finance these investments with a 5%-per-period bank, then he will have the highest future value at the end of ten periods by investing in Investment *A*. No matter what personal time preference rate he uses to discount the future value, Investment *A* will always produce the highest present value. Thus, there is a complete decoupling from personal time preference rate in the situation where the investment will be financed using a bank with a particular interest rate. The bank interest rate always will be the appropriate one to use for discounting purposes.

PRESENT VALUE PLOTS We can gain a deeper understanding into these results by plotting the present value of the three investments as a function of the interest rate, as shown in Figure 27.3. In accordance with our detailed calculations, Investment *A* has the highest present value for an interest rate of 5%, whereas Investment *C* has the highest present value for interest rates of 10% and 20%. The crossover points of the present values are shown in the figure. The present value of *C* rises above the present value of *B* at 2%. The present value of *C* rises above the present value of *A* at 6.4%. Finally, the present value of *B* rises above the present value of *A* at 14.5%. We can see from the figure that one of Investments *A* and *C* will always be preferred to Investment *B*: as long as both *A* and *C* are available, Investment *B* would never be chosen regardless of interest rate.

SUM OF CASH FLOWS A particularly interesting feature of Figure 27.3 is the intercept of the present value curve on the present value axis. This intercept is simply the sum of the cash flows throughout the whole investment without discounting. For Investment *A*, it is $2,000; for Investment *B*, it is $1,500; and for Investment *C*, it is $1,400. The present value as a function of the interest rate plot is analogous to the certain equivalent as a function of the risk-aversion coefficient plot for the exponential risk-preference person. The certain equivalent plot starts at $\gamma = 0$ at the risk-neutral evaluation of the deal, namely, the *e*-value of the monetary measure. As the risk-aversion coefficient increases, the certain equivalent falls.

In the present value plot, we can think of the present value at a 0 interest rate—the sum of the cash flows—as the present equivalent of each investment to a time-indifferent person. Time-indifferent people do not discount their investments; risk-neutral people compute certain equivalents as *e*-values.

TABLE 27.4 Results of Three Investments using a 20% Bank

	A			*B*			*C*		
Period	**Receipts (Payments)**	**Interest Received (Paid)**	**Bank Account**	**Receipts (Payments)**	**Interest Received (Paid)**	**Bank Account**	**Receipts (Payments)**	**Interest Received (Paid)**	**Bank Account**
0	($1,000)	$0.00	($1,000)	($1,000.00)	$0.00	($1,000)	($1,000)	$0.00	($1,000)
1	$200	($200.00)	($1,000)	$250.00	($200.00)	($950)	$500	($200.00)	($700)
2	$200	($200.00)	($1,000)	$250.00	($190.00)	($890)	$500	($140.00)	($340)
3	$200	($200.00)	($1,000)	$250.00	($178.00)	($818)	$500	($68.00)	$92
4	$200	($200.00)	($1,000)	$250.00	($163.60)	($732)	$300	$18.40	$410
5	$200	($200.00)	($1,000)	$250.00	($146.32)	($628)	$200	$82.08	$692
6	$200	($200.00)	($1,000)	$250.00	($125.58)	($504)	$150	$138.50	$981
7	$200	($200.00)	($1,000)	$250.00	($100.70)	($354)	$100	$196.20	$1,277
8	$200	($200.00)	($1,000)	$250.00	($70.84)	($175)	$50	$255.43	$1,583
9	$200	($200.00)	($1,000)	$250.00	($35.01)	$40	$50	$316.52	$1,949
10	$1,200	($200.00)	$0	$250.00	$7.99	$298	$50	$389.83	$2,389
Present Value:			$0.00			$48.12			$385.83

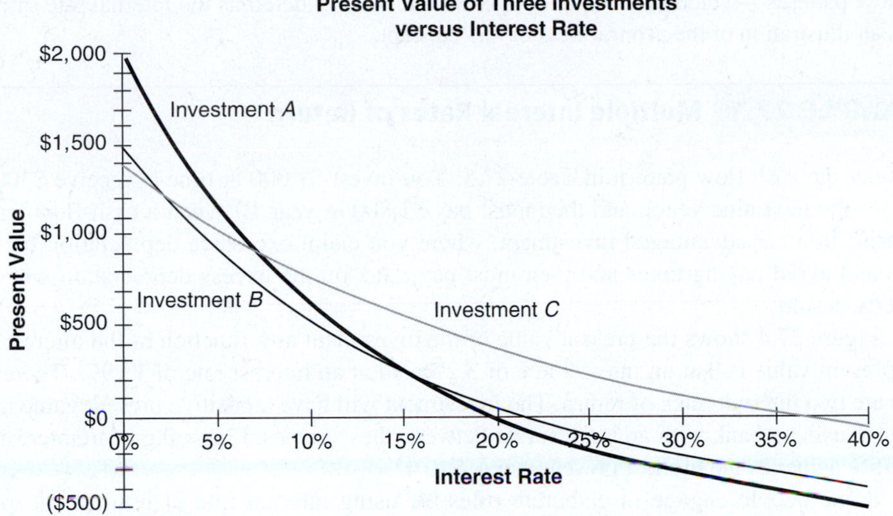

FIGURE 27.3 Present Value of Three Investments as Function of Interest Rate

27.3 METHODS NOT EQUIVALENT TO THE PRESENT EQUIVALENT

Sometimes people choose to compare investments using other measures that do not satisfy the specified rules for time preference. This section discusses three such measures: (1) the internal rate of rate return, (2) the payback period, and (3) the benefit–cost ratio. We shall discuss the issues that may arise with their use.

27.3.1 Internal Rate of Return

The internal rate of return of an investment is the interest rate that makes its present value equal to 0. This is found by determining the intercept of each present value curve of Figure 27.3 with the interest rate axis.

We see that Investment *A* has an internal rate of return of 21%, Investment *B* has an internal rate of return of 22%, and Investment *C* has an internal rate of return of 36.5%. Many banks calculate the *internal rate of return* (*IRR*) of their projects and choose the project with the highest IRR. This might imply that we should prefer the investments in the order *C*, *B*, *A* as determined by their internal rate of return. Yet it is clear from Table 27.2 that preferring the investments in the order *C*, *B*, *A* will not produce the highest future value of the investment if it is to be financed by a 5% bank.

If internal rate of return does not produce a correct result, why is it used? Could it be because it is simpler to calculate? The answer is no. Computing the present value of an investment is a very simple numerical task; however, computing the internal rate of return requires solving an algebraic equation that can have many roots. If there are *N* periods in the investment, it will be necessary to solve an *N*th-order polynomial equation to find the interest rate. The following example illustrates this idea.

This observation reveals not only the difficulty of the numerical calculation, but it also uncovers an additional problem that can be of extreme practical importance in certain situations. Using the fundamental theorem of algebra, such an equation will have *N* roots. If there is more than one real root—and there can be more than one real root even in only moderately complicated cash-flow patterns—which of these real roots should be considered as the internal rate of return? This is an illustration of the arbitrariness of this concept.

EXAMPLE 27.1 Multiple Internal Rates of Return

Consider the cash flow pattern in Table 27.5. You invest $1,000 at time 0, receive $300 per year for the next nine years, and then must pay $1,800 in year 10. Such a cash-flow pattern can arise in a tax advantaged investment, where you claim excessive depreciation for nine years and avoid paying taxes and then must pay a tax on the excess depreciation when the property is sold.

Figure 27.4 shows the present value of the investment as a function of the interest rate. The present value is 0 at an interest rate of 3.2% and at an interest rate of 13.9%. Therefore, there are two internal rates of return. The investment will have a positive present value if it is financed using a bank with an interest rate between these values. Financing at an interest rate of 7.75% achieves the highest present value, $40.41.

Some people engage in elaborate rules for using internal rate of return concepts to choose the proper investment. However, the gold standard by which the correct investment

TABLE 27.5 A Cash Flow with Two Internal Rates of Return

Period:	0	1	2	3	4	5	6	7	8	9	10
Payment:	−1000	300	300	300	300	300	300	300	300	300	−1800

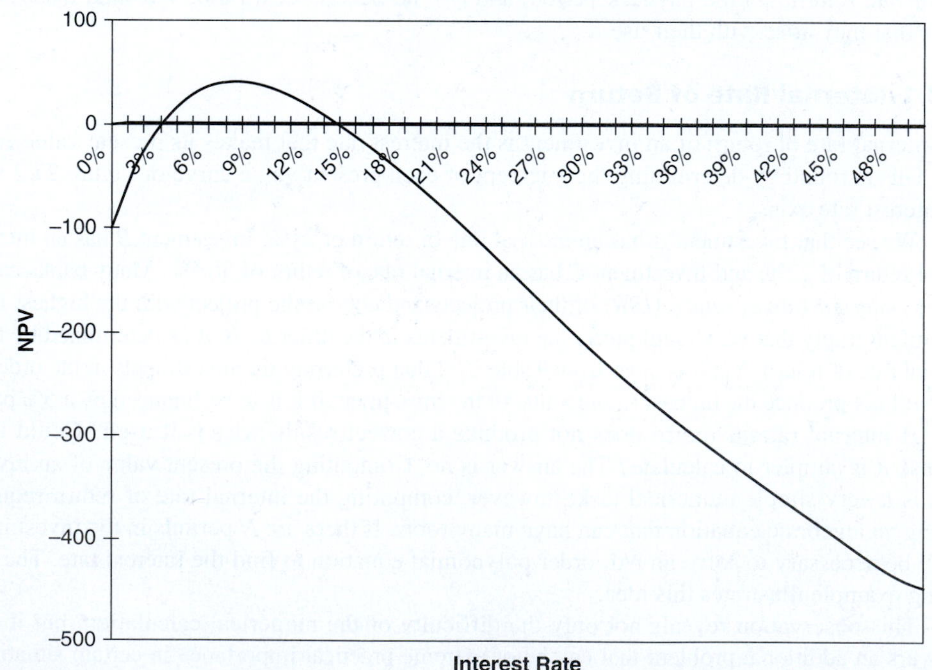

FIGURE 27.4 Present Value Curve Showing Two Internal Rates of Return

is determined is always the present value. Since present value calculation is theoretically correct and practically much simpler to use, why do some people still bother with internal rate of return?

Why does internal rate of return seem like a better idea than it is? Perhaps the answer is this. In most early banking transactions, people dealt with simple accumulations of interest and simple constant payment mortgages. In these cases, as we found out when considering Investment *A* in the 20% bank, the internal rate of return calculation will produce exactly the bank interest rate. Suppose that now the world becomes more complex and that people are considering investments with much more complicated cash flows. They know that they could compare banks by simply looking at the interest rates, and they have seen that in a world of simple investments the internal rate of return reproduces the bank interest rate. Why not, then, compute internal rate of return for possible investments and simply select the one with the highest internal rate of return? The logic is reasonable; unfortunately, it is also wrong.

A vice president of an investment firm revealed how they took advantage of a bank; call it Bank *B*. The two set of bankers were meeting in a hotel to design a corporate bond.

They would meet and construct a possible deal and then return to their individual hotel rooms to evaluate the proposed deal and then get back together again and so on. The vice president soon realized that Bank *B* was maximizing IRR, so they proposed a deal where Bank *B* got an immense IRR on a very small investment. The vice president took practically all the profit on a larger investment with a much lower IRR. Would you rather have the IRR or the money? Nobody ever puts IRR in their pocket.

27.3.2 Payback Period

To complete this historical aspect of time preference, we might mention an early measure of time preference that is still sometimes used today: the **payback period**. In investments that start with a negative cash flow and then produce positive returns, there may come a period when the sum of all the cash flows first turns positive. This is commonly called the payback period. For example, when we look at the cash flows of our investments in Table 27.1, we see that Investment *A* will have returned its original $1,000 payment by Period 5. This means that the payback period for Investment *A* is five periods. Investment *B* returns its $1,000 investment by the end of Period 4; its payback period is 4. Finally, Investment *C* has a payback Period of 2. People like investments where they will "get their money back" as soon as possible. In this case, they would prefer Investment *C* over Investment *B* over Investment *A*, which is the same ranking produced by the internal rate of return measure. As we have seen, it is certainly reasonable. Unfortunately, it is not necessarily right.

27.3.3 Benefit–Cost Ratios

Another flawed method of choice involves calculating the total benefit (sum of positive cash flows), dividing that by the total cost (sum of negative cash flows), and then choosing the cash flow that has the highest benefit–cost ratio. For the three cash flows considered, *A* has a benefit–cost ratio of $3000/1000 = 3$, *B* has a ratio of 2.5, and *C* has a ratio of 2.4. This criterion would favor Investment *A*, but as we have seen, it is not always the case.

In general, ratios are not good criteria for choosing between alternatives. Think of a cash flow where you pay $1 in advance and receive $10 a week later. This has a benefit–cost ratio of 10. Now, think of another cash flow where you pay $100 and you receive $500 a week later. This has a benefit–cost ratio of 5, but many people would prefer the second cash flow (with lower benefit–cost ratio) to the first. Ratios do not take into account the time in which you will receive the payments and do not take into account your personal time preference.

Refer back to Chapter 1 where we discussed the story of Arno Penzias who selected projects based on their benefit–cost ratio. We have seen that this is not necessarily the best decision criterion as prescribed by the five Rules of Actional Thought.

27.4 CASH FLOWS: A SINGLE MEASURE

The present equivalent value function shows that we can think of the direct value measures of a cash flow as a single measure; the present equivalent. Therefore, we can reduce decisions with multiple direct-value cash flows over several time periods into decisions with a single direct-value measure.

As we will see in the next chapter, this idea can be generalized to other direct-value measures by appropriately choosing the value function.

27.5 SUMMARY

The additive rule guarantees our present equivalent value function for cash flows is, in fact, additive (with constant trade-offs between flows of any two different years). This calculation of the present equivalent can be appropriate for investment—but not consumptive—cash flows.

Adding the scaling rule to the additive rule guarantees that we can use the discount factor of each time period as weights.

Adding the calendar time-invariance rule to the additive and scaling rules guarantees that we can represent our time preference by a single parameter: the discount factor. We can also calculate our discounting for n time periods in terms of our discounting for one time period.

The internal rate of return is not a good criterion for selecting cash flows. Moreover, it is not necessarily unique.

The payback period is not a good criterion for evaluating cash flows.

Benefit–cost ratios are not good criteria for evaluating cash flows. They do not take into account the actual monetary values considered (if we multiply all benefit and all cost by 100, we get the same benefit cost ratio but very different present equivalent).

When considering decisions with multiple direct value cash flows over n years, we simply convert the cash flows into a single measure: the present equivalent. For deterministic cash flows, the present equivalent is sufficient to determine the best decision alternative. The higher the present equivalent, the better.

KEY TERMS

- Additive rule
- Calendar time-invariance rule
- Discount factor
- Payback period
- Present equivalent
- Scaling rule

PROBLEMS

Problems marked with an asterisk (*) are considered more challenging.

***1.** Consider two cash flows, where one has a higher internal rate of return (IRR) but lower present equivalent. Which would you prefer?

***2.** Consider two cash flows, where one has a higher benefit–cost ratio but lower present equivalent. Which would you prefer?

3. Calculate the present equivalent of an annuity that pays $1000 every year for

i. 5 years
ii. 10 years
iii. 15 years

Assume that $\beta = 0.9$.

CHAPTER

28

Decisions with Multiple Attributes (3) – Preference Probabilities Over Value

CHAPTER CONCEPTS

After reading this chapter, you will be able to explain the following concepts:

- Constructing multiattribute *u*-functions by first constructing a preference or a value function and then assigning a one-dimensional *u*-curve over a value measure.

28.1 INTRODUCTION

In Chapters 26 and 27, we discussed how to order multiattribute prospects using preference and value functions. As we discussed in the five Rules of Actional thought, if the decision situation is deterministic (there is no uncertainty about which prospect you will get), the order rule is sufficient to determine the best decision alternative. This concept applies whether or not we have multiple attributes. For example, if we were offered a choice between two peanut butter and jelly sandwiches, then either the preference function or the value function would be sufficient to determine our preferred sandwich as well as the preference order for any given sandwich.

If a value function is used to provide a monetary value measure, then we can also determine the willingness to pay to receive one peanut butter and jelly sandwich over another. Similarly, if we were offered a choice between two deterministic cash flows, the net present value function that we discussed in Chapter 27 would be sufficient to determine the best cash flow, as well as the present equivalents for any given cash flow and the willingness to pay to receive one cash flow instead of another.

When uncertainty is present, we need to state our preference probabilities for the ordered prospects in correspondence with the **equivalence rule**. If we have assigned a value measure for the prospects—as determined by a value function—then all that is needed to determine the *u*-values of the multiple attribute prospects is a *u*-curve over dollars. In this chapter, we shall discuss how to assign preference probabilities and *u*-values for prospects characterized by multiple attributes. We shall see how the equivalence rule for multiple attributes is significantly simplified when a value function is constructed.

28.2 STATING PREFERENCE PROBABILITIES WITH TWO ATTRIBUTES

As we discussed in Chapter 26, multiattribute prospects have further characterizations, but the main idea of the equivalence rule is still the same. Given three prospects *A*, *B*, and *C* with strict preference $A > B > C$, the equivalence rule requires us to state a preference probability of receiving *B* for certain or receiving a deal which gives *A* with that stated preference probability and gives *C* with one minus that preference probability.

We have already seen examples of applying the equivalence rule when multiple attributes are present. Recall that in the party problem, Kim stated her preference probabilities for prospects characterized by two attributes. For example, Kim stated a preference probability of 0.57 for the prospect (Indoors, Sun): she was indifferent between receiving the prospect for certain or receiving a deal that gives (Outdoors, Sun) with a probability 0.57 and (Outdoors, Rain) with a probability 0.43. In its basic form, the party problem had only six prospects. Therefore it was easy to verify whether a more preferred prospect on the ordered list had a higher preference probability.

When the attributes are defined on a continuous domain, ordering the prospects is simplified by having a preference or a value function. In some cases, the qualitative properties of the problem may also help provide some partial order. For example, suppose we have a decision with two attributes, *X* and *Y*, such as health and wealth. In this decision, we may assume that more of any attribute is preferred to less. As we have discussed in Chapter 26, this feature alone is not sufficient to provide a complete preference order for prospects with multiple attributes. But it does provide partial order. Let (x, y) be a prospect of this decision, where x, y represent certain values of *X* and *Y*, respectively. Also, let x^* be the highest wealth level under consideration and y^* be the highest value of health state. Therefore, (x^*, y^*) is the most preferred prospect in this decision. If x^0 is the lowest wealth level and y^0 is the lowest health state, then (x^0, y^0) is the least preferred prospect in this decision.

Because more of any attribute is preferred to less in this example, any prospect (x, y) with $x^0 < x < x^*$ and $y^0 < y < y^*$ must satisfy the preference relation

$$(x^0, y^0) < (x, y) < (x^*, y^*)$$

Therefore, in correspondence with the equivalence rule, we can state a preference probability p_{xy} of receiving this prospect (x, y) for certain or a binary deal that gives the best prospect (x^*, y^*) with a probability p_{xy} and the worst prospect (x^0, y^0) with a probability $1 - p_{xy}$ (as shown in Figure 28.1).

We can assert in this decision that a prospect with any higher value of x or y will have a higher preference probability. This observation may help us with the preference probability assignment when more than one attribute is present. However, we cannot assert without a preference or value function that a prospect with a higher value of *X* and a lower value of *Y* would have a higher preference probability than a given prospect. It is generally difficult to reason about the preference probabilities for the prospects of the decision and to verify their relative magnitude if an ordering relation for the prospects is not specified.

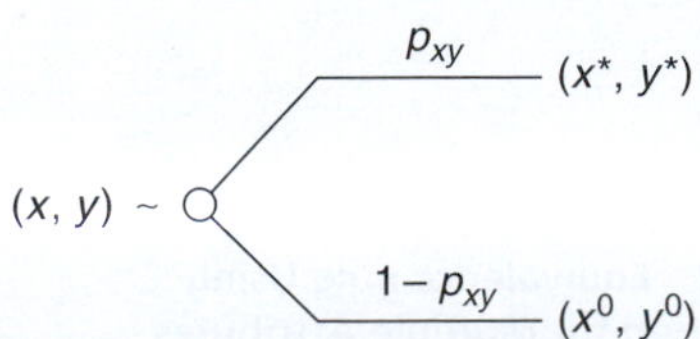

FIGURE 28.1 Equivalence Rule is the Same for Multiple Attributes: (x^*, y^*) is Preferred to (x, y) which is Preferred to (x^0, y^0)

It is much easier to reason about the preference probabilities for multiple attributes by first reasoning about the deterministic trade-offs among the attributes using a value function and then making the preference probability assignment. For example, if we were to assign preference probabilities for the prospects encountered in a peanut butter and jelly sandwich in terms of the best and the worst sandwich, it would not be the case that more of an attribute is preferred to less and it would not even be the case that increasing the value of an attribute would result in a higher preference probability. If a value function for the peanut butter and jelly sandwich is constructed, however, then we can simply substitute a prospect for its value measure. The problem of stating multiattribute preference probabilities would then reduce to that of stating a preference probability over a one-dimensional value measure. With this value function approach, we could also verify whether a more preferred prospect has a higher preference probability. We discuss this method of assigning preference probabilities over value in more detail in the next section.

28.3 STATING PREFERENCE PROBABILITIES WITH A VALUE FUNCTION

Having a value measure simplifies the preference probability assignment. As we discussed earlier, the value measure need not be money, but money is a convenient measure as it both fungible and divisible. Refer back to Figure 28.1. If we have a dollar value associated with each prospect through a value function of $V(x, y)$, we can define the maximum value as $\$V^* = V(x^*, y^*)$; the minimum value as $\$V^0 = V(x^0, y^0)$, and the value as $\$V_{xy} = V(x, y)$ corresponding to a prospect (x, y). Because of the previous preference relation $(x^0, y^0) < (x, y) < (x^*, y^*)$ that we specified in this example, we must also have $V^0 < V_{xy} < V^*$. The preference probability for (x, y) now can be expressed in terms of a preference probability over a single value measure using indifference between the deals shown in Figure 28.2.

We have now reduced the analysis into the same type of analysis we used earlier in the party problem when we stated preference probabilities over monetary prospects. Recall, for example, that Kim's value for (Indoors, Sun) is $40, for (Outdoors, Sun) is $100, and for (Outdoors, Rain) is 0. The value function converts the prospects into a dollar value measure. Therefore, Kim states a preference probability of getting $40 for certain in terms of a binary deal that gives $100 with probability 0.57 and 0 with probability 0.43. We can now assess a *u*-curve over the dollar values associated with each prospect, as we did in the party problem.

28.4 STATING A *u*-CURVE OVER THE VALUE FUNCTION

Because the value function assigns a dollar equivalent for each prospect, we can represent any multiple attribute prospect with its dollar equivalent. In the peanut butter and jelly sandwich example of Chapter 27, the value function provides a dollar value measure for each prospect, so

$$V(b, j, p, f) = V_{\max} \frac{bjpf}{(b^*j^*p^*f^*)^2}(2b^* - b)(2j^* - j)(2p^* - p)(2f^* - f)$$

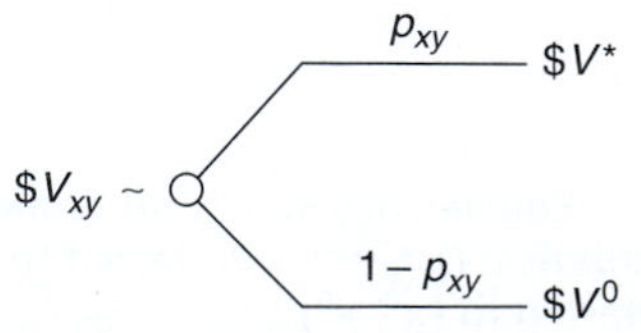

$$\$V_{xy} \sim \begin{cases} p_{xy} & \$V^* \\ 1 - p_{xy} & \$V^0 \end{cases}$$

FIGURE 28.2 Equivalence Rule Using a Value Function for Multiple Attributes

where each attribute can take values from 0 to twice the optimal values. Here, the best prospect is $(p^*j^*b^*f^*)$, which corresponds to a dollar value of $V_{\max}$. There are many less preferred prospects. For example, $(0, 0, 0, 0)$ and $(2p^*, 2j^*, 2b^*, 2f^*)$ have a dollar value of 0.

The preference probability for any sandwich in terms of the best and the worst sandwiches can be determined using the equivalence rule probabilities over the multiple attribute prospects. But it is much simpler to use the value function to convert any sandwich into its dollar equivalent and then assess the u-curve over money. By doing this, we decompose the assessment into two assessments: deterministic trade-offs among the attributes and risk-aversion over a monetary measure. The following example illustrates this idea.

EXAMPLE 28.1 *u*-Values for Peanut Butter and Jelly Sandwich

Having assessed a value function for the sandwich, if we now assess a u-curve over this value measure, we can write the u-value of any sandwich as

$$U(p, j, b, f) = U_V(V(p, j, b, f))$$

where U_V is the u-curve over dollars. For example, if the u-curve over dollars is exponential, we need to assess the decision maker's risk tolerance over value, as

$$U(p, j, b, f) = \frac{1 - e^{-\gamma_V V(p, j, b, f)}}{1 - e^{-\gamma_V V_{max}}} = \frac{1 - e^{-\gamma_V V_{\max} \frac{bjpf}{(b^* j^* p^* f^*)^2}(2b^* - b)(2j^* - j)(2p^* - p)(2f^* - f)}}{1 - e^{-\gamma_V V_{max}}}$$

where γ_V is the risk-aversion coefficient on dollars. This risk-aversion coefficient is exactly the same that would be assessed on money even if we were not dealing with a peanut butter and jelly sandwich. Therefore, the whole problem of assigning a u-value for multiple measures reduces to the same old one-dimensional u-value assessment over one attribute once the value function is determined. Your standard u-curve works equally well for prospects with multiple attributes. Suppose the decision maker has stated that $V_{\max} = \$3$, $b^* = 1$, $j^* = .1$, $p^* = .05$, and $f^* = .5$. We now assign a one-dimensional exponential utility function over value with a risk-aversion coefficient of $\gamma_V = 0.01$. Figure 28.3a plots the u-values for different values of b and p.

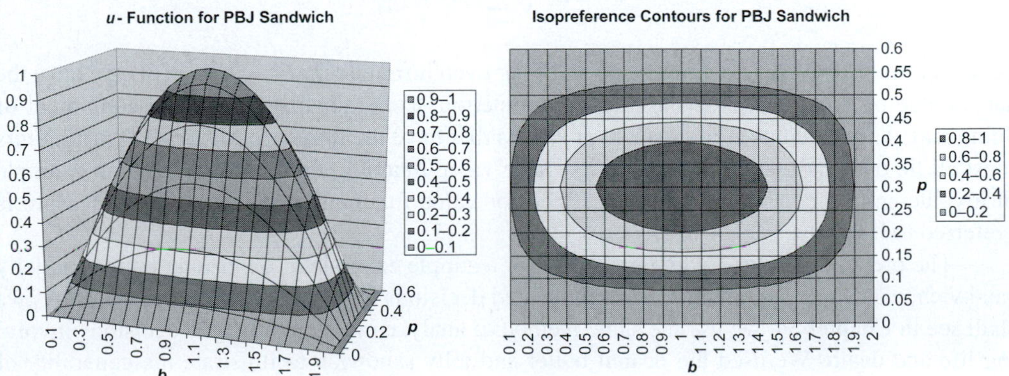

FIGURE 28.3 (a) *u*-Values for Peanut Butter and Jelly Sandwich Plotted for Different Values of *b* and *p* (b) Isopreference Contours for Peanut Butter and Jelly Sandwich

Figure 28.3b plots the isopreference contours of constant value. In this situation, as we can see, more of any attribute is not necessarily preferred to less.

The approach of assigning a u-curve over value is not specific to the exponential u-curve.

You also can have a logarithmic u-curve or a linear risk tolerance u-curve over value. For a logarithmic decision maker having the same value function for the peanut butter and jelly sandwich, the u-function would be

$$U(b,j,p,f) = \ln\left(V_{\max}\frac{bjpf}{(b^*j^*p^*f^*)^2}(2b^* - b)(2j^* - j)(2p^* - p)(2f^* - f) + w\right)$$

28.5 THE VALUE CERTAIN EQUIVALENT

Having a value measure for the multiple attribute prospects enables the calculation of a certain equivalent for each (multiple attribute) alternative as well as the willingness to pay to receive one multiple attribute deal over another. To illustrate, if the value associated with each multiple attribute prospect of i is V_i, we can determine the e-value of the u-values for each alternative as

$$e\text{-value of } u\text{-values} = \sum_{i=1}^{n} p_i U(V_i)$$

where V_i is the value assigned to each prospect using the value function.

To determine the certain equivalent of any alternative, we find the prospect(s) whose u-value is equal to the e-values of the u-values, so

$$U(\widetilde{V}) = \sum_{i=1}^{n} p_i U(V_i)$$

where $\widetilde{V}$ is the certain equivalent of value for the uncertain deal with multiple attributes. Rearranging gives

$$\widetilde{V} = u^{-1}\left(\sum_{i=1}^{n} p_i u(V_i)\right)$$

When multiple attributes are present, several—or even infinitely many—prospects may have the same value of $\widetilde{V}$ as determined by the value function. Consequently, the certain equivalent of value may define a contour of possible prospects that have the same preference. These contours must all lie on the isopreference contours of the value function. Figure 28.4 shows an example of a value-certain equivalent contour defined for value functions where more of a measure is preferred to less.

The previous analysis, while presented for a simple example of the peanut butter and jelly sandwich, also applies to much more complicated decisions involving multiple attributes. As we shall see in Chapter 34, we use the same method to analyze more complicated decisions involving life and death. We used the peanut butter and jelly sandwich to illustrate the generality of the approach and the generality of u-values that can be modeled by assigning a u-curve over the value function. We did not assume that more of an attribute is preferred to less in this example and, we were able to reason about our risk aversion and deterministic trade-offs separately.

Isopreference Contours
1
0.9
0.8
0.7
0.6
0.5
0.4
0.3
0.2
0.1
0
Attribute y
Value Certain Equivalent
0
0.1
0.2
0.3
0.4
0.5
0.6
0.7
0.8
0.9
1
Attribute x

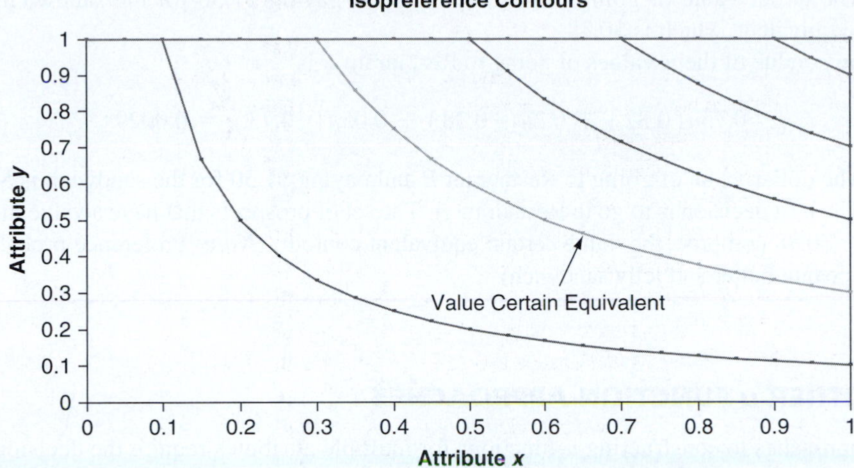

FIGURE 28.4 Value Certain Equivalent Contour: More of a Measure is Preferred to Less

EXAMPLE 28.2 The Peanut Butter and Jelly Sandwich (with Uncertainty)

Suppose that a decision maker with an exponential *u*-curve and a risk tolerance of \$50 has a choice of two outlets where he can get a peanut butter and jelly sandwich for \$1.50. Both restaurants use the same ingredients for their sandwiches but use different quantities of the ingredients in each sandwich. Out of experience from previous visits, he draws the decision tree in Figure 28.5 for the sandwich he will get at each restaurant.

In this example, we needed to have the *u*-function over value because there was uncertainty in the decision problem. The value function alone would not suffice.

Let *u*-function for dollars be $u(x) = 1 - e^{-\frac{x}{50}}$.

The *e*-value of the *u*-values of going to Restaurant *A* is

$$0.2u(-0.47) + 0.6u(1.5) + 0.2u(0.42) = 0.0175$$

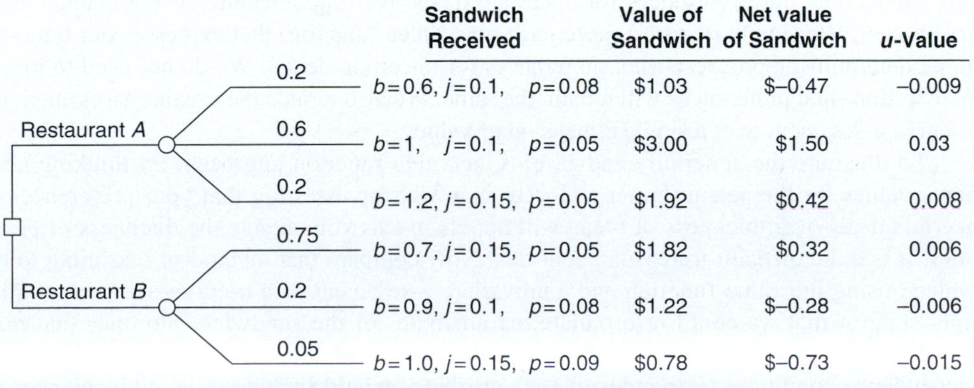

Sandwich Received
Value of Sandwich
Net value of Sandwich
u-Value
Restaurant A
0.2
b=0.6, j=0.1, p=0.08 $1.03 $-0.47 -0.009
0.6
b=1, j=0.1, p=0.05 $3.00 $1.50 0.03
0.2
b=1.2, j=0.15, p=0.05 $1.92 $0.42 0.008
Restaurant B
0.75
b=0.7, j=0.15, p=0.05 $1.82 $0.32 0.006
0.2
b=0.9, j=0.1, p=0.08 $1.22 $-0.28 -0.006
0.05
b=1.0, j=0.15, p=0.09 $0.78 $-0.73 -0.015

FIGURE 28.5 Decision Tree for Two Restaurants

The dollar value of going to Restaurant A and paying \$1.50 for the sandwich is the certain equivalent, which is \$0.88.

The e-value of the u-values of going to Restaurant B is

$$0.75u(0.32) + 0.2u(-0.28) + 0.05u(-0.73) = 0.0029$$

The dollar value of going to Restaurant B and paying \$1.50 for the sandwich is \$0.15.

The best decision is to go to Restaurant A. The set of prospects that have a value measure equal to \$0.88 comprise the value certain equivalent contour. (***Note:*** Preference probabilities for the peanut butter and jelly sandwich)

28.6 OTHER *u*-FUNCTION APPROACHES

Some approaches to constructing u-functions for multiple attributes require the decision maker to make certain assertions that preferences between uncertain deals over an attribute do not change when levels of another attribute vary. For example, in a medical decision with two attributes, such as health state and wealth, they would require assertions like preferences for uncertain deals (or investments) over money would not change with our health state. These types of assertions are referred to as "utility independence" conditions and are used to determine the functional form of the u-function. Common assumptions that are also made include the idea of "mutual utility independence," where preferences between uncertain deals over any subset of the attributes do not change as we vary the levels of the remaining subset. Such assumptions result in additive or multiplicative combinations of single attribute u-curves over each of the individual attributes.

We have several issues with this type of "utility independence" reasoning. First, we find such assumptions very limiting for modeling preferences of decision makers in most practical multiattribute problems. It is very unlikely that preferences for investments would not vary with our health state. Note that these independence conditions do not even apply in simpler problems, such as the peanut butter and jelly sandwich. Enforcing these "utility independence" assumptions results in functional forms that are simple, but quite frequently they will not represent the preferences of the decision maker.

If we use the value function approach, we do not need to think about (or make any assumptions about) how our preferences for uncertain deals over one attribute do not change as we vary another. We simply require a preference (or a value function) that expresses our trade-offs among deterministic prospects (not preferences for uncertain deals). We do not need to impose any assertions that preferences will remain the same. We also reduce the u-value assessment into a u-curve assessment over a single dimension of value.

To illustrate the generality and ease of the value function approach, try thinking about your u-values for the peanut butter and jelly sandwich by asserting that your preferences for uncertain deals over thickness of bread will not change as you change the thickness of peanut butter. It is quite difficult to reason about this. Now compare this method of reasoning to that obtained using the value function and a univariate assessment of a u-curve over value. Some might suggest that we could reformulate the attributes of the sandwich into ones that make the value of the sandwich strictly increasing with each attribute and that would satisfy utility independence conditions. Examples of such attributes would include taste, convenience, and comfort while eating it. The problem with such attributes is that they require constructed scales that are difficult to define and reason about. What does taste really mean? The attributes we have chosen in the peanut butter and jelly sandwich example are clear. The clairvoyant can tell you the thickness of each attribute in the sandwich you will get. Furthermore, how can we really assert (and reason about the fact) that preferences for each of these attributes do not change as we change another.

By making such limiting assumptions of utility independence in multiattribute problems, we can easily end up with u-values that do not represent the actual preferences of the decision maker.

To help provide a reality check into our reasoning about a problem, it is often useful to consider the following sequence:

(i) Think about the decision situation we start with;

(ii) Think about the assumptions that are made to formulate our model;

(iii) Solve the model to arrive at a solution but then think about the solution we obtain for the model we have imposed, and finally,

(iv) Verify whether the solution to our model actually is an answer to the original decision situation we have started with.

Keeping in mind this useful distinction between a solution to a model and an answer to the original decision situation we started with can help us clarify our reasoning in multiattribute problems.

28.7 STATING A u-CURVE OVER AN INDIVIDUAL ATTRIBUTE WITHIN THE VALUE FUNCTION

Even if the output of the preference (or value) function is not a monetary measure but one of the attributes in the decision is money, we can assess the u-curve over that monetary attribute to construct the u-values. Consider, for example, the isopreference contours described by a value function of consumption (measured in monetary units) and health state (measured in remaining lifetime). Figure 28.6 plots isopreference contours for prospects of this decision. The horizontal axis is consumption and the vertical axis is health state (normalized from 0 to 1).

In Figure 28.6, the u-value of E must be equal to the u-value of A and also the u-value of B, since they all lie on the same isopreference contour. Therefore, if we assess only the u-curve for consumption at health state $= 1$ (the top horizontal axis) for example, then we can trace any point E back along its isopreference contour to get an equivalent point A for the u-value we have assessed. Alternatively, we may assess the u-curve for health state at a particular consumption level (such as \$1 M) and trace any point back to that axis. Now, we can determine the u-value of any point by the u-function over either health state or wealth. Therefore, once we know the value function, a single u-curve is sufficient to determine the u-values over all possible consequences.

Furthermore, if we have assessed our trade-off over these deterministic prospects using a preference function, then we know that the isopreference contours we provided for the u-values are consistent with our deterministic preferences. If we assess individual u-curves for consumption and health state and then make the arbitrary assumption that no change in preference will occur for uncertain lotteries over any attribute as we vary either attribute, we will end up with

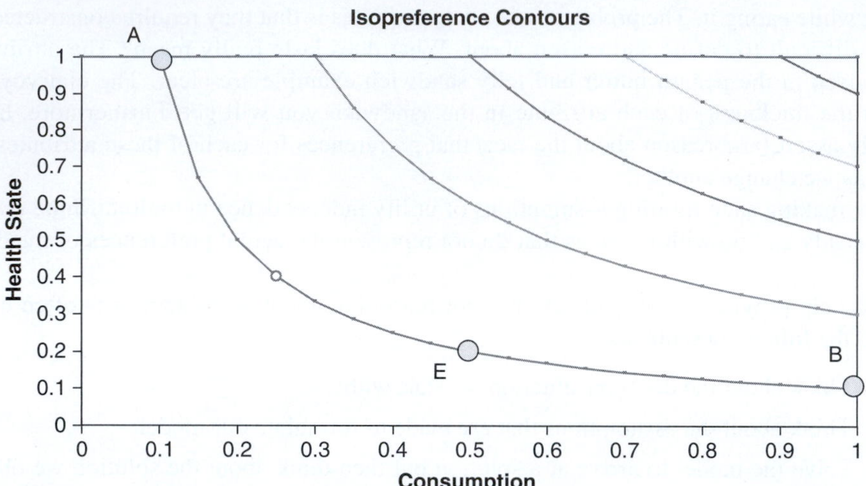

FIGURE 28.6 Many Dimensions for Assessing a *u*-Curve in Decisions with Multiple Attributes

u-values that are either additive or multiplicative combinations of the individual assessments. The isopreference contours resulting from these combinations need not match the contours of our value function. If we start with the value function specification and then think about our deterministic trade-offs (which is simpler than thinking about uncertain deals), we need only one dimension of a *u*-curve assessment. We are guaranteed to produce multiple attribute *u*-values that are consistent with our deterministic preferences.

EXAMPLE 28.3 Linear Value Function with Exponential and Logarithmic *u*-Curves

Suppose the value function is additive over two attributes as

$$V(x, y) = ax + by$$

If the decision maker has (1) an exponential or (2) a logarithmic *u*-curve over value, what would be his *u*-function for the two measures *x* and *y*?

1. Exponential *u*-Function Over Value If the *u*-curve over value is $u(V) = -e^{-\gamma V}$, we derive the *u*-function for the two measures by substitution as

$$u(V(x, y)) = -e^{-\gamma(ax+by)}$$

From properties of exponential functions, we can also write

$$u(x, y) = -e^{-\gamma ax}e^{-\gamma by}$$

This function therefore can be written as a product of the two functions $e^{-\gamma ax}$ and $e^{-\gamma by}$. Suppose we had assessed a u-function directly over the measures x and y and, furthermore, had assigned exponential u-curves for each measure with a risk-aversion coefficient γa for x and γb for y. Suppose further that we stated that our u-value for both measures jointly is the product of the u-values of the individual measures. We would have then obtained the same result as that obtained using the value function.

Our approach using the value function relies only on assessing deterministic trade-offs among the attributes and then assigning a u-curve over value without making assumptions about the form of the u-function.

2. Logarithmic *u*-Function Over Value

If we construct the u-values by make assumptions about the functional form, such as a product or an additive function of u-curves over the individual attributes, we end up with a very special case. To illustrate, we now consider what happens when a logarithmic u-curve is assigned over the additive value function for net present value as

$$u(V) = \log(V + \alpha) = \log(ax + by + \alpha)$$

Assume the product form of u-values as individual u-curves over each attribute results in an inaccurate representation of the decision maker's preferences, since the u-curve over the net present value measure cannot be expressed as a product of two functions. It is, of course, possible that a decision maker can have a logarithmic u-curve over an additive value function, particularly when the value function represents the present equivalent of a cash flow.

EXAMPLE 28.4 Multiplicative Value Function with Exponential and Logarithmic *u*-Curves

Suppose the value function is multiplicative over two measures. Thus,

$$V(x, y) = axy$$

If the decision maker has (1) an exponential or (2) a logarithmic u-function over value, what is his u-function for the two measures x and y?

1. Exponential By substitution, we have

$$u(V(x, y)) = -e^{-\gamma V(x,y)} = -e^{-\gamma axy}$$

Note that in this case, the u-function over the measures cannot be expressed as a product of two functions for x and y.

2. Logarithmic *u*-Curve Over Value By substitution, if $u(V) = \log(V)$ with initial wealth equal to 0, we have

$$u(V(x, y)) = \log(axy) = \log(a) + \log(x) + \log(y)$$

and we now have an additive u-function over the two measures. Once again, the assumption of an additive u-function for the individual measures is a special case that will not generally hold.

EXAMPLE 28.5 Risk Aversion and Time Preference

The use of a value function and a u-curve over value enables us to determine how our risk aversion should change for money received today versus money received in the future. To illustrate, consider a decision situation with two measures: x = money received today and y = money received a year from today. The present equivalent value function for this situation is the additive function

$$V(x, y) = x + \beta y$$

where β represents our trade-off between money received today and a year from today.

Suppose that a decision maker has an exponential u-curve for money received today with a risk aversion coefficient of γ. We have

$$u(V) = -e^{-\gamma V}$$

His multiple attribute u-values are

$$u(x, y) = -e^{-\gamma(x + \beta y)}$$

This is a product of two functions—both of which are exponential—that have different risk-aversion coefficients. The risk-aversion coefficient for x (money received today) is γ, while the risk-aversion coefficient for y (money received in a year) is $\gamma\beta$.

We now have a relation between the risk aversion today and the risk aversion in one year. The risk aversion for money a year from today is multiplied by the time preference discount factor. Equivalently, the risk tolerance for money received a year from today compounds at the same rate as our time-preference interest rate:

Risk-aversion coefficient for money received today = Risk-aversion coefficient for money received in the future multiplied by the discount factor

By doing this, we have u-values that are consistent with our deterministic trade-offs as determined by the net present value function. If we had assessed our u-curves for each year separately and then multiplied the u-curves, we can get u-values that have arbitrary risk preferences and that do not correspond to the time preference discount factor.

28.8 VALUING UNCERTAIN CASH FLOWS

28.8.1 Joint Risk and Time Preference

Thinking in terms of a value function and a u-curve over value helps us achieve clarity in many problems. In this section, we illustrate how we can value uncertain deals involving future cash flows using a net present value function and a u-curve over net present value.

Suppose you were offered a deal that would pay \$100 in 15 years with a probability of 0.2 and –\$10 in two years with a probability of 0.8. Figure 28.7 illustrates this deal.

Note

If we multiply the dollar amounts by hundreds of millions of dollars, this example resembles the decision problems faced by pharmaceutical companies in drug development. Each new chemical entity has a relatively small chance of producing a huge return after many years and a large probability of loss in the next few years if the proposed drug fails its initial safety and efficacy tests.

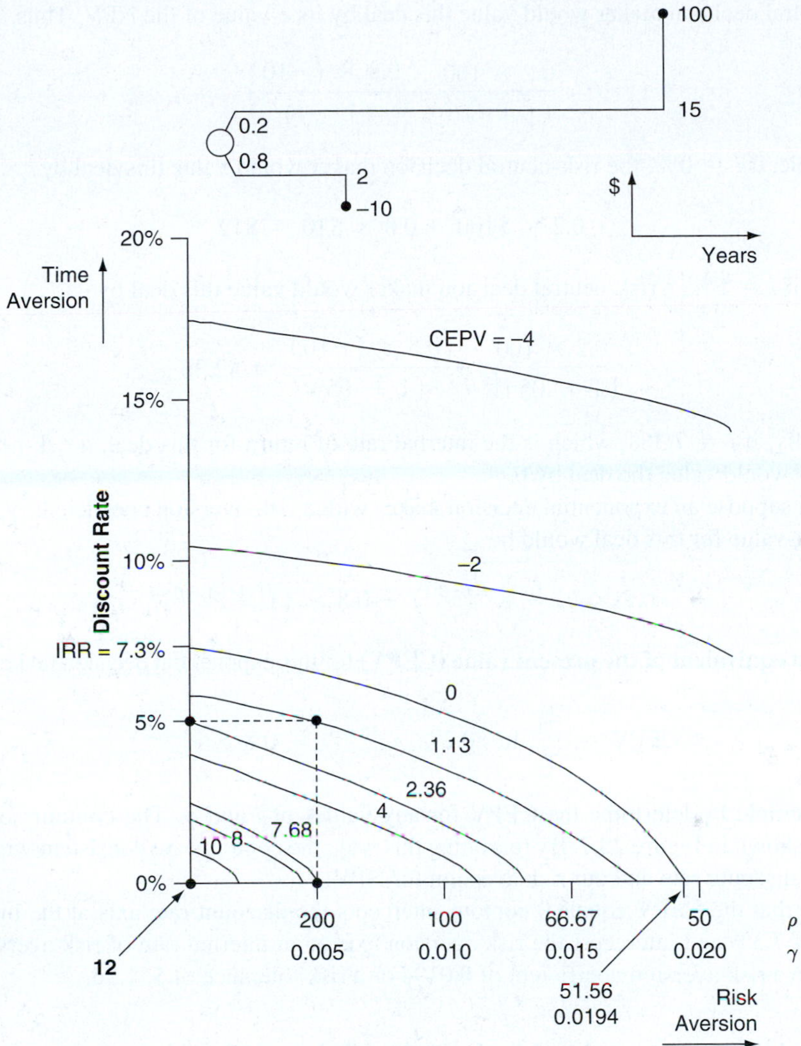

FIGURE 28.7 Joint Risk and Time Preference

To think about our certain equivalent for this deal, we first determine our present equivalent for each possible outcome. Then we think about our risk aversion over present dollars. To determine the present equivalent, we consider the discount factor, $\beta = \dfrac{1}{1+i}$, where i is the time preference discount rate.

The present equivalent of \$100 in 15 years is $100\beta^{15} = \dfrac{100}{(1+i)^{15}}$.

The present equivalent of $-\$10$ in 2 years is $-10\beta^{2} = \dfrac{-10}{(1+i)^{2}}$.

A risk neutral decision maker would value this deal by its *e*-value of the NPV. Thus,

$$\frac{0.2 \times 100}{(1+i)^{15}} + \frac{0.8 \times (-10)}{(1+i)^2}$$

For example, if $i = 0\%$, the risk-neutral decision maker would value this deal by

$$0.2 \times \$100 - 0.8 \times \$10 = \$12$$

Similarly, if $i = 5\%$, a risk-neutral decision maker would value this deal by

$$\frac{0.2 \times 100}{(1+.05)^{15}} + \frac{0.8 \times (-10)}{(1+.05)^2} = \$2.36$$

Finally, if $i = 7.3\%$, which is the internal rate of return for this deal, a risk-neutral decision maker would value the deal by 0.

Now suppose an exponential decision maker with a risk-aversion coefficient, γ, faced this deal. The *u*-value for this deal would be

$$0.2 \times (1 - e^{-100\gamma\beta^{15}}) + 0.8 \times (1 - e^{10\gamma\beta^{2}})$$

The **certain equivalent of the present value (CEPV)** for this exponential decision maker would be

$$\text{CEPV} = \frac{-1}{\gamma} \ln (0.2 \times e^{-100\gamma\beta^{15}} + 0.8 \times e^{10\gamma\beta^{2}})$$

It is now simple to determine the CEPV for any values of i and γ. The contours of constant CEPV are shown in Figure 28.7. By reasoning this way, the contours are consistent with our time preference discount rate and our risk aversion for NPV.

Note that the CEPV equals 0 contour intercepts the discount rate axis at the internal rate of return of 7.3% and intercepts the risk aversion axis at an internal rate of risk aversion corresponding to a risk aversion coefficient of 0.0194 or a risk tolerance of $51.56.

28.8.2 Valuing an Annuity for the Rest of your Life

In Chapter 27, we determined the value of an annuity that pays $*X* every year for exactly *m* years. Now, we will show how to value uncertain cash flows, such as an annuity that pays a constant amount for the rest of your uncertain lifetime.

Suppose you are offered an annuity that will pay you $*X* every year for the rest of your life. Suppose that you are risk-neutral. How would you value this annuity? Since a person's lifetime, *m*, is uncertain, the present equivalent represented by the term $pe[a(m)]$ is also uncertain. Figure 28.8 shows an example of the probability tree for a dollar payment every year for *m* years where *m* is uncertain. The first payment is made immediately.

To value this annuity, we first calculate the present equivalent of a cash flow that pays $*X* up to a specific (fixed) year *i*. We have seen in the previous chapter that this is equal to

$$Xpe[a(i)] = \$X \frac{1 - \beta^{i+1}}{1 - \beta}$$

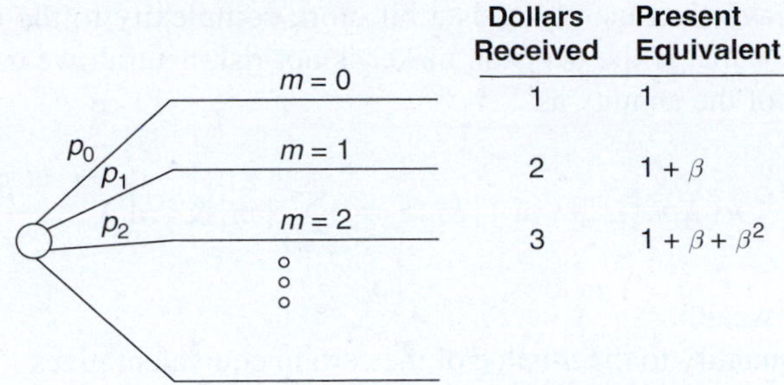

FIGURE 28.8 Annuity for Uncertain Number of Years

This expression represents the prospect of receiving payments up to precisely *i* years. Now, we need $\{m|\&\}$, which is the probability distribution for the remaining number of years for which we will receive the payment. This is important, because we will need to get the *e*-value of the *u*-values of the different prospects. The distribution $\{m|\&\}$ depends again on the decision maker's belief, but it is sometimes helpful to look at demographics. Insurance companies keep track of these statistics. For example, Figure 28.9 shows a histogram for the remaining lifetime distribution for a male at the age of 30. This histogram was obtained from historic records.

The *e*-value of the annuity is simply the *e*-value of present possible present equivalents it provides, so

$$< Xpe[a]|\& > = \sum_{m=0}^{n} p_m X \frac{1-\beta^{m+1}}{1-\beta} = X\frac{1-<\beta^{m+1}|\&>}{1-\beta}$$

where

$$< \beta^{m+1}|\& > = \sum_{m=0}^{n} p_m \beta^{m+1} = \sum_{m=0}^{n} \{m|\&\}\beta^{m+1}$$

and p_m is the probability of a payment at year *m*, which is obtained from the distribution $\{m|\&\}$.

We refer back to this annuity calculation in Chapter 34.

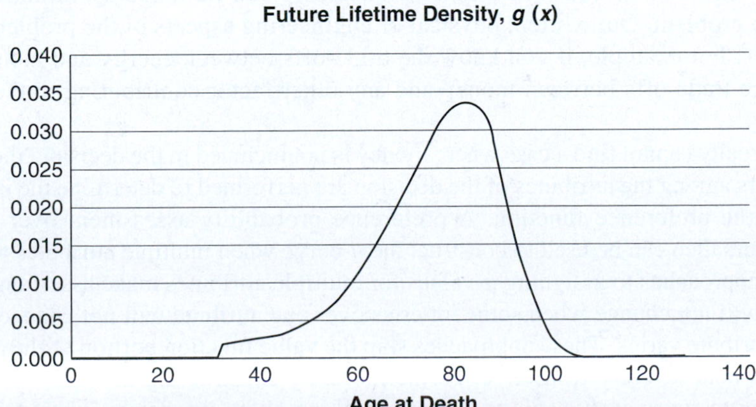

FIGURE 28.9 Remaining Lifetime Distribution

Including risk aversion merely adds a bit more complexity to the calculations, but the concepts are still the same. If the decision maker is not risk-neutral, we can use the u-curve to calculate the u-value of the annuity as

$$u\text{-value} = \langle u(Xpe[a] + w) | \& \rangle = \sum_{m=0}^{n} \{m|\&\} u\left(X\frac{1-\beta^{m+1}}{1-\beta} + w\right)$$

where w is the initial wealth.

Equating this quantity to the u-value of the certain equivalent gives

$$\sum_{m=0}^{n} \{m|\&\} u\left(X\frac{1-\beta^{m+1}}{1-\beta} + w\right) = u(\tilde{x} + w)$$

from which we can calculate the certain equivalent, $\tilde{x}$.

28.9 DISCUSSION

The five Rules of Actional Thought still apply when prospects are characterized by multiple attributes. You still need to characterize the alternatives you face in terms of prospects and you still need to assign a probability for each prospect as required by the probability rule. You also need to order the prospects in accordance with the order rule. Then you need to state a preference probability for each prospect as per the equivalence rule. You also need to follow the substitution rule, and then make your choice according to the choice rule.

When prospects are characterized by multiple attributes, the order rule can be facilitated by constructing a deterministic preference (or value) function as we have seen in Chapter 26. The equivalence rule can be applied to the multiple attribute prospects directly, but as we have seen, it is easier to apply the equivalence rule to a value measure that characterizes each prospect.

In most applications, you can construct a value function with monetary outputs or at least have a monetary measure as one of the attributes. The monetary scale is meaningful because it is fungible and divisible. Constructed scales such as taste, convenience, or comfort are challenging to quantify and difficult to reason about. When a monetary attribute is present, the problem of assigning u-values or preference probabilities then can be reduced to that of assigning a single-attribute u-curve over money.

It is often the case that deterministic trade-offs can be made by thinking about the nature of the problem. Quite often, physical or engineering aspects of the problem determine the trade-offs. For example, if you know the trade-offs between energy and money, you can determine the trade-offs between money and any other factor contributing to the generation of the energy.

If you really cannot find a case where money is not included in the decision, then deterministic trade-offs among the attributes of the decision are performed to determine the isopreference contours of the preference function. A preference probability assessment over the isopreference contours then can be used to construct the u-curve when multiple attributes are present.

Other approaches to assigning u-values for multiple attributes make assertions about how preferences will not change when some lotteries over one attribute will not change as the level of another attribute varies. These approaches skip the value function portion of the analysis. We find these models severely restrictive. Furthermore, by constructing the u-curve using single attribute u-curves over each attribute without thinking about the deterministic trade-offs and then combining them into some additive or multiplicative combination, you may end up with arbitrary shapes of isopreference contours. You also miss out on finding the structure of the problem that may facilitate constructing the trade-offs among the attributes.

28.10 SUMMARY

When uncertainty is present, we need to construct a multiattribute *u*-function.

A multiple attribute *u*-function can be constructed by assigning a *u*-curve over the value function output or any of the attributes within the value function. This approach enables us to reduce the *u*-curve assignment into a single dimension and also enables us to place a value on a deal with multiple measures.

Once the value function is specified and a one-dimensional *u*-curve is assigned over the value function, we do not have any further choice in assigning individual *u*-curves for each of the individual attributes: They are determined.

A multiple attribute *u*-function need not be additive or multiplicative. It does not need to be increasing with each of the attributes. It simply takes the shape of the surface defined by the *u*-curve over value. The peanut butter and jelly sandwich is an example where this increasing behavior was not necessarily satisfied. Starting with an inappropriate objective function may result in an inappropriate answer to the problem.

To value uncertain cash flows, we first determine the present equivalent for each possible prospect and then get the *e*-values of the *u*-values of the prospects. From that, we can determine the certain equivalent of the uncertain cash flow.

KEY TERMS

- Certain equivalent of the present value (CEPV)
- Equivalence rule
- Joint time-risk preference

PROBLEMS

Problems marked with an asterisk (*) are considered more challenging.

***1.** Calculate the value of an annuity with uncertain duration: it will pay $1000 every year for either 5 years, 10 years; or 15 years with equal probability. Assume the discount factor is $\beta = 0.9$ and the decision maker is a deltaperson with risk tolerance equal to
   i. $\rho = 5000$
   ii. $\rho = 10{,}000$

***2.** Repeat Problem 1 for a logarithmic decision maker with initial wealth
   i. $w = 5000$
   ii. $w = 10{,}000$

   Calculate both the PIBP and also the PISP if he owned it at this initial wealth.

CHAPTER

29

Betting on Disparate Belief

CHAPTER CONCEPTS

After reading this chapter, you will be able to explain the following concepts:

- How to construct a deal that two parties would find attractive if they have a difference in belief
- How to make money by constructing a deal that two parties would find attractive

29.1 INTRODUCTION

This chapter presents an important concept related to probability assignment. This concept can be applied whenever two people have disparate beliefs and, as a result, assign different probabilities to the same observable distinction. The difference in probabilities permits constructing a deal between them that will have a positive certain equivalent for both of them. Furthermore, the person constructing the deal may even be able to charge them for doing so.

29.2 BETTING ON DISPARATE PROBABILITIES

29.2.1 Demonstration: The Rain Bet

In the following situation, an instructor is speaking to his students about decision making.

> ***Note*** *"I" refers to instructor and "C" refers to one or more individuals in the class. Class participants may also be referred to by name.*

I: We now create a simple distinction about rain tomorrow, November 1, in Palo Alto; with two degrees: Either it will rain or it will not rain. Does this distinction pass the clarity test?

C: No. We still need to define what you mean by "rain" and what time you mean when you say "tomorrow." And which part of Palo Alto do you mean?

I: By tomorrow, I mean any time from midnight tonight until midnight tomorrow. I also mean the Stanford Campus in Palo Alto. We must realize that people may differ on the amount of moisture that must be falling to deserve the description "rain." They may also see strong evidence of rain when they leave a building, such as puddles and wet sidewalks, but not actually see rainfall; perhaps it is a residue of lawn sprinkling. To achieve clarity on the rain distinction, we shall define it as the result of a class vote at our next class. In two days' time, in our next session, I will ask the class "did it rain yesterday?" I will then count the number of positive and negative responses. A majority vote will determine the resolution of the distinction. Should there be a tie, we will flip a coin to settle the matter. We expect everyone to vote "best knowledge and conscience." Is the distinction "*R*: Rain in Palo Alto tomorrow" now clear?

C: Yes, that means it does not matter what any weather report will say; what matters is just the class vote.

I: Correct. Now, I would like each of you to write down your probability on *R*, that the class will vote "it rained" in the next class.

C: (*Writing their probabilities.*)

I: Who has a probability larger than 0.5?

John: I have a probability of 0.7.

I: Ok, anyone higher than 0.7?

Bill: I have a probability of 0.9.

I: Ok, anyone higher than 0.9? (*Waits*) No. OK, Bill you have the highest probability of rain.

I: Does anyone have a probability lower than 0.5?

Lisa: I have a probability of 0.2.

Mary: I have a probability of 0.08.

I: Anyone have a probability lower than 0.08? (*Waits*) No. OK, Mary, you have the lowest probability of rain.

Bill assesses the probability of rain tomorrow as $\{R|\&_{\text{Bill}}\} = 0.9$

Mary assesses the probability of rain tomorrow as $\{R|\&_{\text{Mary}}\} = 0.08$

Now, suppose we give Bill the deal on the left hand side of Figure 29.1 and give Mary the deal on the right-hand side. (*Shows Figure 29.1 to class*)

I: If both players were risk-neutral within this range of monetary prospects—and they should be for such small amounts—they would value the deals at their *e*-value as

$$\text{Bill would value his deal at } 0.9 \times 5.1 + 0.1 \times (-4.9) = \$4.1$$

$$\text{Mary would value her deal at } 0.08 \times (-5.1) + 0.92 \times (4.9) = \$4.1$$

Now, we customarily ask both participants if they actually want to commit to these deals in the upcoming class. If they do and they are free to say no (though this is rare), we actually collect from them to create the pot: Bill would contribute $4.90, and Mary would contribute $5.10. If the class votes for "rain" in two days, Bill will

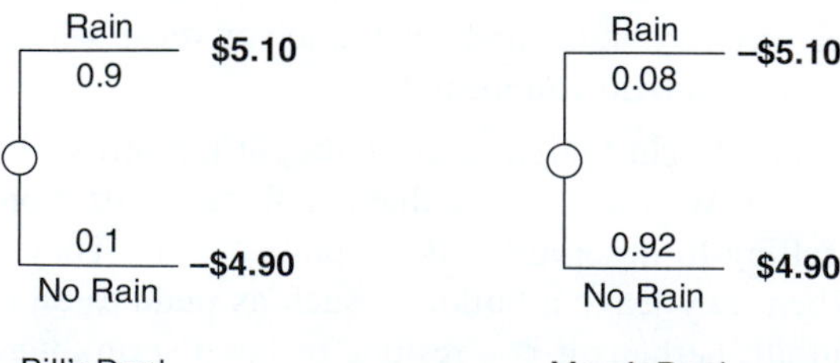

FIGURE 29.1 Deals Presented to Players Bill and Mary According to Their Beliefs

collect the pot; otherwise, it will be Mary. Should either one not wish to participate, that opportunity would be given to the person with the second highest or second lowest probability of rain, as appropriate. The discussion may then continue.

John: If both Bill and Mary have assigned their true beliefs about the probability of rain, they would have the same *e*-value for these deals.

I: Correct. They would also happily accept their deals if they were offered to them, since they have a positive certain equivalent. In fact, we can even charge each of them up to \$4.10 for getting this deal; at that point, they would be indifferent as to whether or not they got their deals.

John: Can we always construct a deal like this when two people assign different probabilities to the same distinction?

I: Yes. Furthermore, we can even construct the deals such that they will have the same positive certain equivalents. In our construction procedure, we will make the *e*-values the same for both participants.

First, I want you all to notice that the sum of the absolute values of the total payoffs in these deals is set equal to \$10, that is, \$5.10 + \$4.90 = \$10. This is why we created a pot of \$10: \$5.10, contributed by Mary, and \$4.90, contributed by Bill. If the result is Rain, Bill collects the pot; if not Rain, Mary collects the pot. The pot can be any size agreeable to the participants. If we had decided on a pot of \$100, then Mary would contribute \$51, Bill \$49, and each would have a certain equivalent of \$41.

29.2.2 Setting Up the Deal

To generalize this setup for any two participants A and B, suppose that

$$\{R|\&_A\} = p \text{ and } \{R|\&_B\} = q, \text{ and that } p > q.$$

Now, we construct the two deals such that Player B pays Player A an amount equal to $\$a$ in the event of rain, and Player A pays Player B an amount equal to $\$b$ in the event of no rain (Figure 29.2). The total pot will be

$$m = a + b$$

To calculate the values of a and b, we need two equations relating the *e*-values of the two deals. We have

$$e\text{-value of Deal } A = pa - (1 - p)b$$
$$e\text{-value of Deal } B = -qa + (1 - q)b$$

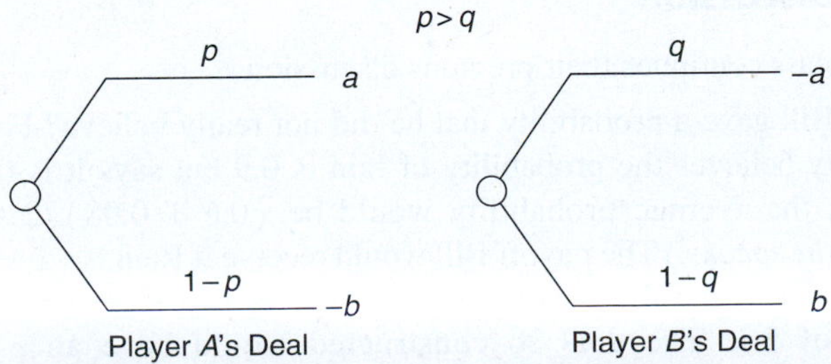

FIGURE 29.2 Constructing the Deals Having the Same Mean

Equating the two e-values gives

$$pa - (1-p)b = -qa + (1-q)b$$
$$\Rightarrow p(a+b) - b = b - q(a+b) \quad (\text{Divide by } (a+b))$$
$$\Rightarrow p - \frac{b}{a+b} = \frac{b}{a+b} - q$$
$$\Rightarrow \frac{b}{a+b} = \frac{1}{2}(p+q)$$

Now we define $k = \dfrac{b}{a+b}$. Substituting into the previous equation gives

$$k = \frac{1}{2}(p+q)$$

which is the average probability of rain for the two participants, or $(0.9 + 0.08)/2 = 0.49$ in our case. The total pot m the sum of $a + b$. So we have

$$a = (1-k)m \quad \text{and} \quad b = km$$

For any value of m and any two probabilities of rain, we can calculate the average probability, k, and the values of a and b. Furthermore, the e-value of the deal for each player would be

$$e\text{-value Deal } A = e\text{-value Deal } B = pa - (1-p)b$$
$$= p(1-k)m - (1-p)km$$
$$= (p-k)m = \left(\frac{p-q}{2}\right)m$$

The e-value of payoff to each participant is thus half the difference in their probabilities times the size of the pot. For our example, this was

$$\left(\frac{p-q}{2}\right)m = \left(\frac{0.9-0.08}{2}\right)10 = 4.1$$

The larger is the difference in probabilities, the larger is the e-value. If one person assigns a probability 1 and the other a probability 0, the e-value is one-half the size of the pot, which is the maximum size it can be.

29.2.3 Further Discussion

In this section, the class continues their previous discussion.

I: What if Bill gave a probability that he did not really believe? For example, suppose Bill really believes the probability of rain is 0.9 but says it is 0.6. Then, using our analysis, the average probability would be $(0.6 + 0.08)/2 = 0.34$. (*Writes on board as he speaks.*) The payoff Bill would receive if Rain is $(1 - 0.34)(10) = 6.6$ instead of 4.9, and the payoff Mary would receive if no Rain is $0.34(10) = 3.4$ instead of 5.1. The deal so constructed would have an e-value to Mary of $0.08(-6.6) + 0.92(3.4) = 2.6$, as predicted by the difference in probabilities

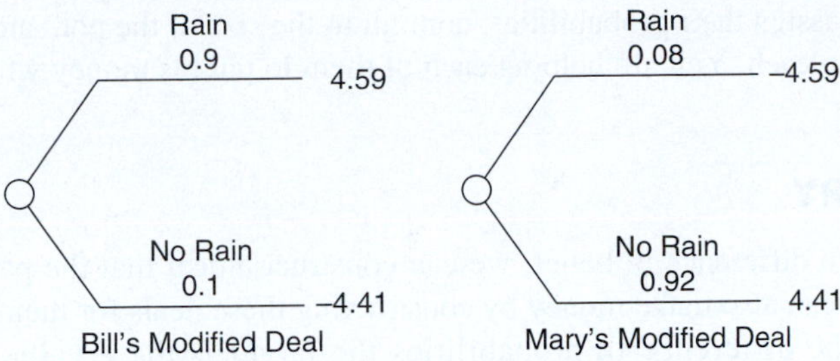

FIGURE 29.3 Bill's and Mary's Modified Deals

$\left(\frac{p-q}{2}\right)m = \left(\frac{0.6 - 0.08}{2}\right)*10 = \$2.6$ and less than the value when Bill assigned a 0.9 probability at \$4.1. So Mary is worse off by \$1.50 than she was before. What about Bill? Since Bill really has a 0.9 probability of Rain, his deal is 0.9 probability of receiving 6.6 and 0.1 probability of losing 3.4: An *e*-value of 5.6. By setting up a deal based on an asserted probability of 0.9 rather than his assigned probability of 0.6, Bill has increased his *e*-value to 5.6, an increase of $5.6 - 4.1 = \$1.50$. Bill's gain is Mary's loss.

This might be an issue if Bill and Mary were negotiating to create the deal. In the class, however, those stating their probabilities do not know how they will be used. Note that if Bill had claimed a probability of rain of 0.6, he would have been succeeded by John at 0.7 and maybe by someone with an even higher probability.

John: Is it possible for the person constructing the deal to profit from the disparate probabilities?

I: Yes. Suppose I just take a dollar out of the \$10 pot for my troubles in constructing the game, leaving \$9 for them to share. Since their average probability is still 0.49, the payoffs are $0.49(9) = 4.41$ instead of 4.9, and $0.51(9) = 4.59$ instead of 5.1. Consider the modified deals of Bill and Mary in Figure 29.3.

I: Each of these modified deals has a positive *e*-value of half the difference in probabilities times the remaining pot, $0.41(9) = 3.69$, rather than $0.41(10) = 4.1$.

John: What if the players were risk averse? Can we still design such deals?

I: Yes. If we had wished to do so, we would have collected the *u*-curves of everyone in the class before beginning the demonstration. Then we could have also constructed similar deals this time equating the certain equivalents for both players. Furthermore, as noted in our discussion of the pot, there is no limit on the amount of money we can set for the payoffs. We can scale the payoffs by any factor, *m*, to make these deals more appealing to the players.

29.3 PRACTICAL USE

If you hear two colleagues arguing about anything that will be observable, you can usually abbreviate the argument by applying what we have learned in this chapter. Whether they are arguing about who will win the game, whether the contract will be awarded, or whether the mission will be successful, all you have to do is create the clarity test definition of the observable distinction, have each of them assign their probabilities, commit to the size of the pot, and collect the appropriate amounts from each. You are helping each of them to put his money where his mouth is.

29.4 SUMMARY

Whenever there is a difference in belief, we can construct a deal that the parties involved will find attractive. We can also make money by constructing these deals for them.

The larger the difference in probabilities the larger is the *e*-value of the deal for a given pot.

KEY TERMS

- Disparate beliefs

PROBLEMS

1. Sean and Luce are asked to assign their probability of rain this coming Thursday night based on their true belief. After meeting the clarity test on a binary distinction Rain and Sun, they assign their probabilities as

$$\{\text{Rain}|\&_{\text{Sean}}\} = 0.11 \text{ and } \{\text{Rain}|\&_{\text{Luce}}\} = 0.65$$

and are given the deals shown in the figure.

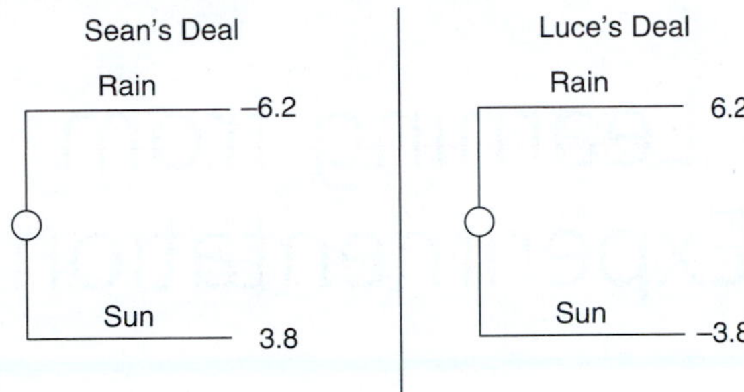

Given the low range of monetary prospects involved in these deals, we can assume that both Sean and Luce are risk neutral. Circle the statements that are true:

I. Both Sean and Luce value their deals at \$2.70.
II. Even if Sean's probability of rain were 0.33 (more than three times the probability he stated), he would still prefer the deal he got to 0.
III. If Luce had stated a lower probability than his true belief, he would have gotten a deal with a higher certain equivalent.
IV. Uncertain deals should not be valued based on their outcome, but rather on their certain equivalent.

2. A friend of yours issues a pair of coupons on a binary uncertainty. Each coupon pays an amount equal to \$10 if a certain outcome occurs. For example, one coupon pays \$10 if the stock market goes up, and the other pays \$10 if the stock market goes down. In other words, a coupon pair will always pay its holder an amount of \$10, since we know that exactly one outcome will occur and the bearer will always get paid \$10. (Ignore the case that the stock market index remains the same.)

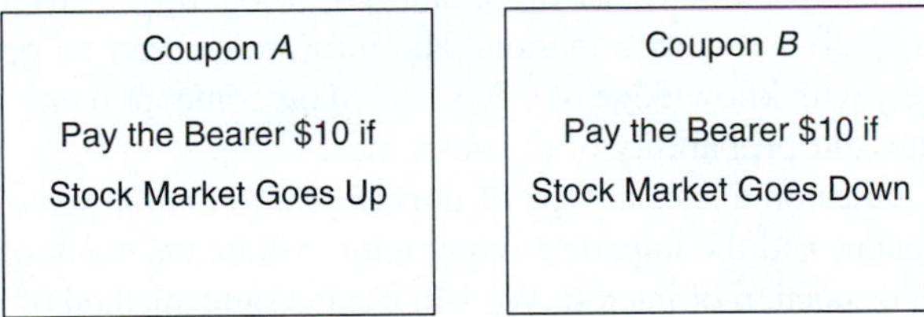

Coupon *A*
Pay the Bearer \$10 if
Stock Market Goes Up
Coupon *B*
Pay the Bearer \$10 if
Stock Market Goes Down

I. Decision maker 1 is risk-averse (risk tolerance equals \$20) and believes the stock market will go up with a probability of 0.8.
II. Decision maker 2 is risk-neutral and believes the stock market will go up with a probability of 0.7.
   a. What is decision maker 1's PIBP for the coupon pair?
   b. After some thought, decision maker 1 says he is risk-neutral. You are called to design the pricing and selling schemes for the two coupons. Which decision maker will you sell each coupon to? What is the maximum profit you can generate for your friend?

3. Bill assesses the probability of rain tomorrow as $\{R|\&_{\text{Bill}}\} = 0.9$.
Mary assesses the probability of rain tomorrow as $\{R|\&_{\text{Mary}}\} = 0.08$.
Mary is risk-neutral and Bill has a risk tolerance of \$100. Construct a rain bet such that the certain equivalent of the deal for both Bill and Mary will be equal.

CHAPTER

30

Learning from Experimentation

CHAPTER CONCEPTS

After reading this chapter, you will be able to explain the following concepts:

- Assigning probability of head and tail to thumbtack
- Updating probability of head and tail after observing results of a toss (or multiple tosses)
- Conjugate prior distributions
- Does observing a head make the probability of a head on the next toss more likely?

30.1 INTRODUCTION

In previous chapters, we discussed how to update our probability distribution for a distinction of interest when we receive new information. The information can be perfect, eliminating our uncertainty completely with knowledge of a guaranteed outcome, or it can be imperfect, in which case, it merely updates our probability of the outcomes.

This chapter focuses on a special type of imperfect information for which the distinction of interest will be continuous and the imperfect information will be the result of a discrete experiment that is relevant to that distinction of interest. We will use the same method of reasoning that we used to update belief with an imperfect detector, but in this case, we will also show how to update our belief graphically.

An important application of this formulation is how to update our probability of the degrees of an unfamiliar distinction. In Chapter 2, for example, the thumbtack demonstration showed that people were uncomfortable choosing the thumbtack deal over the medallion deal—even though they could not be worse off by choosing the thumbtack. Their unfamiliarity with thumbtacks was what led them to this decision—even though, as we showed in Chapter 6, the thumbtack deal can be converted into a medallion deal by a simple toss of a coin.

If people are uncertain about the way a thumbtack will land, they may be tempted to toss it a few times first. Tossing the tack a few times does not tell us how it will land when the person actually calls. But if we toss it and it lands "Pin Up" or "Pin Down," we can use this information to update our probability about the way the tack will land in the future. This chapter shows how.

30.2 ASSIGNING PROBABILITY OF HEAD AND TAIL FOR THE THUMBTACK

To begin, we must review what we know about the coin tossing process. We may believe that successive tosses of the coin are *physically independent*. That is, the way the coin has landed on preceding trials has no physical effect on how it will land on the next toss. This would not be true if we were tossing something that might wear or melt, like dice made of ice. It will also not be true if the tossing mechanism changes or the surfaces involved vary, like sometimes tossing on a hard floor and sometimes tossing on a carpet. If the successive tosses are physically independent, we believe the behavior of the coin on each toss has no effect on any future tosses. This means that the probability of any sequence of toss results would be the same regardless of the order in which those results were achieved.

Let us denote "Pin Up" as Tail and "Pin Down" as Head, as we did in Chapter 2. With this belief, all that we need to know to describe the process is the number ϕ that we would assign as the probability of Head on any toss in the sequence.

At this point, we have exploited all our basic *a priori* knowledge of the process. We do not know the number ϕ, only that it is between 0 and 1. We now recall the process we used to assign a probability distribution to the weight of the chair in our exercise in Chapter 15. The person making the probability assignment does not know the weight of the chair, but he does have information about it based on factors such as size, construction, materials, and dynamic movement.

If we want to assign a probability distribution to ϕ, we must consider what we believe about tossing dynamics that we discussed as the end of Chapter 2. Some students believe that the minimum energy position (Tail) is more likely. Others favor the nail–coin concept. They say first to imagine that the thumbtack has its pin cut off and so is very close to a disk like the coin with a probability of Head equal to 0.5. They then imagine that the thumbtack has a very long pin similar to a nail. In this case, the probability of Head would approach 1. Since the actual pin length is somewhere between these extremes, they believe that the thumbtack is more likely to fall Head. However you might reason, you would have to assign a probability distribution on ϕ between 0 and 1.

As we have said, if we knew ϕ, we would assign it as the probability of Head (H). In our notation

$$\{H|\phi, \&\} = \phi$$

But given our uncertainty about ϕ, what number would we assign as our probability for Head on the next toss—for example, what is $\{H|\&\}$?

We can think of this probability using the relevance diagrams in Figure 30.1. The first distinction is ϕ, and the second Head is H.

The top figure in Figure 30.1 shows the assessed form. Here we assign our prior as $\{\phi|\&\}$, and our likelihood (conditional probability of Head given the revelation of ϕ) as $\{H|\phi, \&\}$.

To compute $\{H|\&\}$, which is the preposterior, we need to place the diagram in inferential form. As we did with the case of trees, we multiply each probability of ϕ by the conditional probability $\{H|\phi, \&\}$ and then sum over all possible values of ϕ. If ϕ takes on a certain number of discrete values, we have the summation

$$\{H|\&\} = \sum_{\phi}\{H|\phi, \&\}\{\phi|\&\}$$

If ϕ is represented as a continuous distribution, we have the integral

$$\{H|\&\} = \int_{\phi}\{H|\phi, \&\}\{\phi|\&\}$$

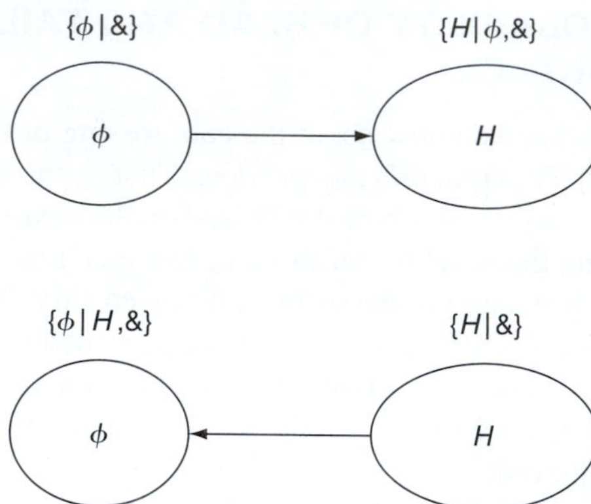

FIGURE 30.1 Assigning Probability of Next Toss, $\{H|\&\}$, by Thinking About ϕ.

As we discussed, if ϕ were known to this individual, he would assign it as the probability of Head. Thus,

$$\{H|\phi, \&\} = \phi$$

Therefore,

$$\{H|\&\} = \int_{\phi} \phi\{\phi|\&\} = <\phi|\&>$$

which is the mean of the distribution of ϕ. Someone who is uncertain about ϕ must assign the mean of the ϕ distribution as the probability of a Head on the next toss. This is not some type of approximation but a logical requirement of our formulation.

Similarly, we can calculate the probability of Tail (T) on the next toss as

$$\{T|\&\} = \int_{\phi} \{T|\phi, \&\}\{\phi|\&\}$$

Of course,

$$\{T|\&\} = 1 - \{H|\&\} = 1 - <\phi|\&>$$

Two observations on these computations are

1. Even though we are uncertain about ϕ and we assign a distribution to it, the actual probability of Head we assign to the next toss is just one number: the mean of the distribution of ϕ. There is no uncertainty about this probability.
2. No matter what distribution we assign to ϕ, namely $\{\phi|\&\}$, we would still choose the thumbtack deal over the medallion. For example, if the mean were 0.2, we would choose the tack and simply call Tail, and if the mean were 0.8, we would call Head. And if the mean were 0.5 and we believed the Medallion was equally likely to land either side, we would be indifferent.

30.3 PROBABILITY OF HEADS ON NEXT TWO TOSSES

What about the probability of getting two Heads in a row in the next two tosses given this uncertainty about ϕ? To calculate this probability, we assign our prior of $\{\phi|\&\}$ and our likelihood of $\{H, H|\phi, \&\}$. We then obtain the preposterior, $\{H, H|\&\}$, by multiplying the prior and likelihood and summing over each value of ϕ.

We have

$$\{H, H|\&\} = \int_\phi \{H, H|\phi, \&\}\{\phi|\&\}$$

As we discussed, if ϕ were known to the individual, he would assign it as the probability of Head. Hence

$$\{H, H|\phi, \&\} = \phi^2$$

Hence the probability of getting two Heads would be

$$\{H, H|\&\} = \int_\phi \phi^2 \{\phi|\&\} = <\phi^2|\&>$$

which is the second moment of the distribution of ϕ. Note that the second moment of a distribution is not equal to the first moment squared. The difference arises from the uncertainty about ϕ.

30.4 PROBABILITY OF ANY NUMBER OF HEADS AND TAILS

Similarly, we can calculate the probability of getting m Heads. We assign our prior of $\{\phi|\&\}$ and our likelihood of $\{H_1, H_2, \ldots, H_m|\phi, \&\}$, where H_1 means Head on the first toss, H_2 means Head on the second toss, etc. We have

$$\{H_1, H_2, \ldots, H_m|\&\} = \int_\phi \{H_1, H_2, \ldots, H_m|\phi, \&\}\{\phi|\&\}$$

$$= \int_\phi \phi^m \{\phi|\&\} = <\phi^m|\&>$$

which is the mth moment of the distribution.

How about the probability of observing a Head on the first toss, then a Tail on the second toss? We have

$$\{H_1, T_2|\&\} = \int_\phi \{H_1, T_2|\phi, \&\}\{\phi|\&\}$$

If ϕ were known to the individual, he would assign

$$\{H_1, T_2|\phi, \&\} = \phi(1 - \phi)$$

The probability of getting a Head on the first toss, then a Tail on the second toss would be

$$\{H, T|\&\} = \int_\phi \phi(1 - \phi)\{\phi|\&\} = <\phi(1 - \phi)|\&> = <\phi|\&> - <\phi^2|\&>$$

On the other hand, the probability of getting a Head and a Tail in two tosses without specifying a particular order would be either Head on the first toss, H_1, and Tail on the second, T_2; or Tail on the first toss, T_1, and Head on the second, H_2. Therefore,

$$\{H_1, T_2|\&\} + \{T_1, H_2|\&\} = 2<\phi(1 - \phi)|\&> = 2(<\phi|\&> - <\phi^2|\&>)$$

Similarly, we can calculate the probability of any number or sequence of Heads and Tails for the thumbtack given the prior distribution as $\{\phi|\&\}$.

30.5 LEARNING FROM OBSERVATION

We now know how to calculate the probability of any number or sequence of Heads and Tails for the thumbtack given the prior distribution of $\{\phi|\&\}$. But we have not discussed how this probability might change if we observe the results of tossing the thumbtack a few times. This is an experiment, and its results will change the distribution of ϕ.

Now, we discuss the question of how our knowledge of ϕ is changed by observing the outcomes of tosses (the experiment). Then, we show how to update our probability of Head as $\{H|\&\}$, given the updated distribution of ϕ.

30.5.1 Observing a Head or a Tail

Suppose that the individual has observed one Head in addition to his information &. What has he learned? To answer this, we need to calculate the posterior probability $\{\phi|H, \&\}$. We have

$$\{\phi|H, \&\} = \frac{\{H|\phi, \&\}\{\phi|\&\}}{\{H|\&\}}$$

However, we know that $\{H|\phi, \&\} = \phi$ and $\{H|\&\} = <\phi|\&>$, so this expression can be written as

$$\{\phi|H, \&\} = \frac{\phi\{\phi|\&\}}{<\phi|\&>}$$

The effect of observing a Head is to multiply the ϕ distribution (density function) by ϕ and divide it by a normalization factor that is equal to $<\phi|\&>$, which is its mean.

Similarly, if a tail T is observed, the effect on the ϕ distribution is given by

$$\{\phi|T, \&\} = \frac{\{T|\phi, \&\}\{\phi|\&\}}{\{T|\&\}}$$

We have $\{T|\phi, \&\} = 1 - \{H|\phi, \&\} = 1 - \phi$ and $\{T|\&\} = 1 - \{H|\&\} = 1 - <\phi|\&>$.

This expression now can be written as

$$\{\phi|T, \&\} = \frac{(1-\phi)\{\phi|\&\}}{1 - <\phi|\&>}$$

The distribution is multiplied by $1 - \phi$ and divided by a normalization factor equal to one minus its mean.

The normalization factor is always the integral of the numerator over all values of ϕ.

30.5.2 Updating the Distribution of ϕ Graphically

We could now design a computer to perform the updating job. It would start with the prior density function of $\{\phi|\&\}$. If we observe a Head, we would multiply this density function by ϕ and renormalize it to have total area one. Observing a Tail would cause multiplication by $1 - \phi$ and the same renormalization.

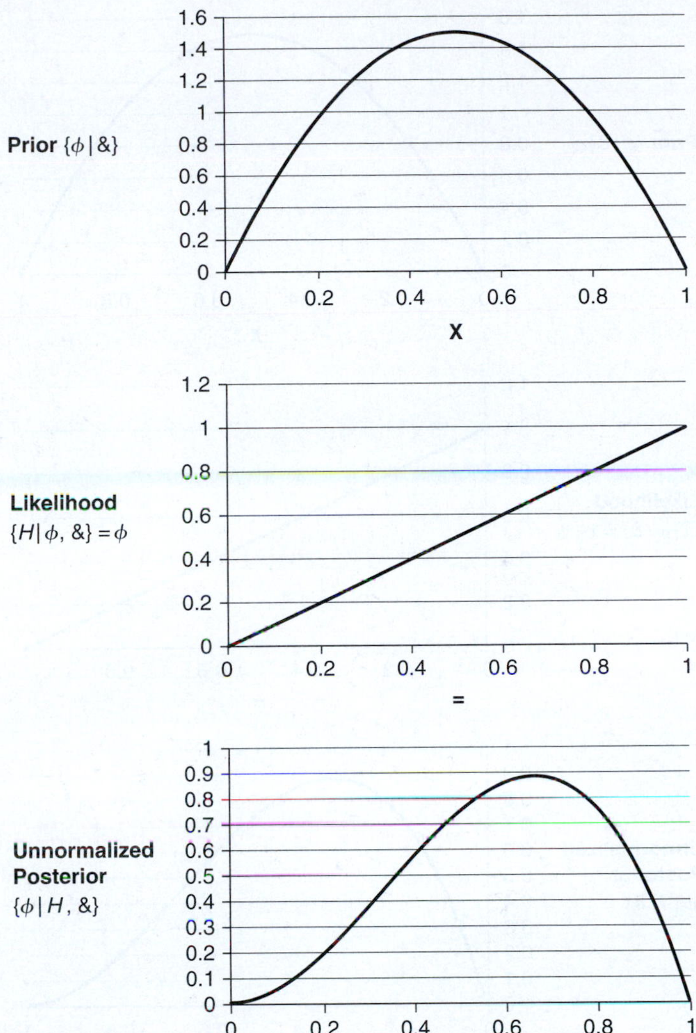

FIGURE 30.2 Graphical Updating of Distribution of ϕ by Observation of a Head

Once we have the new distribution for ϕ, we can calculate the new probability of Head on the next toss by calculating its mean. Figure 30.2 shows that the effect of observing a Head on the ϕ density function.

To get the normalized posterior, we simply divide the probability density by the mean of $\{\phi|\&\}$, so that $<\phi|\&>$. The mean of the normalized posterior is the probability of a Head given we have observed a Head.

Figure 30.3 shows that the effect of observing a Tail on the ϕ density function. To get the normalized posterior, we simply divide the probability density by the one minus the mean of $\{\phi|\&\}$, so that $1 - <\phi|\&>$. The mean of the normalized posterior distribution is the probability of Head given we have observed a Tail.

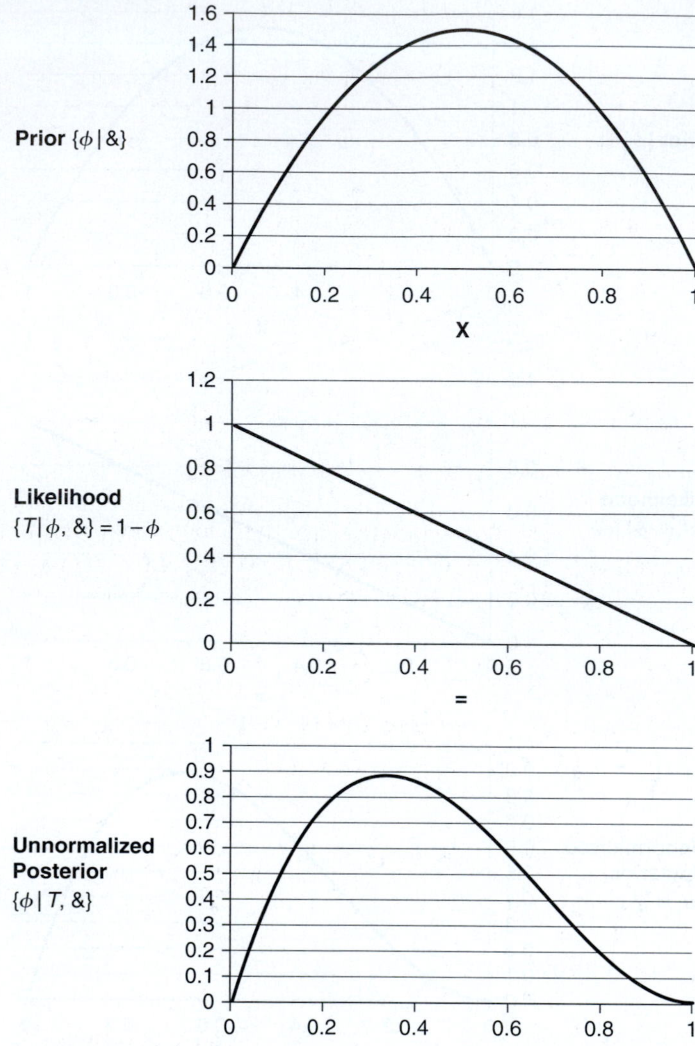

FIGURE 30.3 Graphical Updating of Distribution of ϕ by Observation of Tail To Get Normalized Posterior $\{\phi|T, \&\}$, Divide Product by $1 - \langle\phi|\&\rangle$

The graphical procedure used to obtain $\{\phi|H, \&\}$ or $\{\phi|T, \&\}$ can be used regardless of the prior distribution $\{\phi|\&\}$. Observe that for any prior on ϕ, the result of multiplying it by the likelihood function in Figure 30.2 for every Head you observe and by the likelihood function in Figure 30.3 for every Tail you observe will be a continual narrowing of the posterior. If sufficiently large numbers of Heads and Tails are observed, the mean of the distribution will have virtually all its mass concentrated at the ratio equal to the number of Heads divided by the number of tosses, which is the fraction of Heads observed in this very extensive experiment.

30.5.3 Effect of Observing *r* Heads and *k* Tails

To make this behavior clear, suppose that r Heads are observed in n tosses where $n = r + k$. Then, in view of the effect of observing Heads and Tails that we have discussed, the posterior distribution on ϕ, $\{\phi | r, n, \&\}$ will be given by

$$\{\phi | r, n, \&\} = a\phi^r(1 - \phi)^{n-r}\{\phi | \&\}$$

where a is whatever normalizing constant is required to make the posterior a density function.

The posterior distribution of ϕ can also be determined graphically by multiplying the prior $\{\phi | \&\}$ by the curve $\phi^r(1 - \phi)^{n-r}$ and normalizing the resulting curve.

Once we have obtained the posterior distribution $\{\phi | r, n, \&\}$, the probability of Head given this observation is the mean of this new distribution, i.e., $<\phi | r, n, \&>$.

30.6 CONJUGATE DISTRIBUTIONS

While the previous results constitute a complete procedure for updating information, they do not have the analytical convenience necessary if we are to demonstrate decision making and experimentation. Our work would be easier if the prior and posterior density functions in the previous equations could be members of the same family. This would be so if their dependence on ϕ were of the form $\phi^l(1 - \phi)^m$ where l and m are constants; then both sides of the equation would have that form.

If we choose a prior density function of the form

$$\{\phi | r', k', \&\} = c\phi^{r'-1}(1 - \phi)^{k'-1}$$

where r' and k' are constants and c is a normalizing constant, the posterior density function following observation of r Heads and k Tails will depend on ϕ as

$$\begin{aligned}\{\phi | r, r', k, k', \&\} &= ac\phi^r(1 - \phi)^k\phi^{r'-1}(1 - \phi)^{k'-1} \\ &= ac\phi^{r+r'-1}(1 - \phi)^{k+k'-1} \\ &= ac\phi^{r''-1}(1 - \phi)^{k''-1}\end{aligned}$$

where we have defined posterior parameters $r'' = r + r'$ and $k'' = k + k'$ to show that the posterior is, in fact, in the same family as the prior.

Indeed, this prior is the Beta density function, which we defined in Chapter 20 as

$$\text{Beta}(\phi | r', k') = c\phi^{r'-1}(1 - \phi)^{k'-1}, \quad 0 \le \phi \le 1$$

with mean equal to

$$\text{Mean of Beta}(r', k') = \frac{r'}{r' + k'}$$

and variance equal to

$$\text{Variance of Beta}(r', k') = \frac{\text{Mean} \times (1 - \text{Mean})}{r' + k' + 1}$$

30.6.1 What Does All This Math Mean?

It means very simply that if you started with a prior for $\{\phi | \&\}$ that is a Beta distribution with parameters r', k', and if you observed r Heads and k Tails of the thumbtack, then your posterior distribution for ϕ would be a Beta distribution with parameters r'', k'', where $r'' = r + r'$ and $k'' = k + k'$.

Moreover, if you were to assign a probability for Heads on the next toss after observing these tosses, it would be the mean of the new Beta distribution, namely, $\dfrac{r''}{r'' + k''}$.

When we can find a family of distributions such that the prior and posterior belong to the family for some experimental process, we say that the family is a conjugate family with respect to the experimental process. We have found such a family for the binary toss process and, furthermore, have shown that the posterior parameters are obtained simply by adding the numbers that describe the experimental outcome to the prior parameters, a very simple procedure.

30.6.2 Numeric Example

Your prior belief about the ϕ distribution for a thumbtack landing Head is a Beta distribution Beta (r, k). You observe two Heads and two Tails in four tosses of the tack. What is your prior probability and posterior probability of Heads on the next toss if

(a) $r = 1, k = 2$?
(b) $r = 20, k = 21$?

(a) As we discussed, Prior probability of Head is $\{H|\&\} = <\phi|\&>$. If $r = 1, k = 2$, then

$$\{H|\&\} = 1/(1 + 2) = 1/3$$

which is the mean of a Beta $(1, 2)$ distribution
The posterior distribution for ϕ is

$$\{\phi|H, H, T, T, \&\} = \text{Beta}(r + 2, k + 2) = \text{Beta}(3, 4)$$

Hence, the posterior probability of Head is

$$\{H|H, H, T, T, \&\} = <\{\phi\,|H, H, T, T, \&\}\,|\&> = 3/(3 + 4) = 3/7$$

which is the mean of a Beta $(3, 4)$ distribution
(b) The prior probability of Head is $\{H|\&\} = <\phi|\&>$. If $r = 20, k = 21$, then

$$\{H|\&\} = 20/41$$

The posterior distribution of ϕ is $\{\phi|H, H, T, T, \&\} = \text{Beta}(r + 2, k + 2) = \text{Beta}(22, 23)$
Hence, the posterior probability of Head is

$$\{H|H, H, T, T, \&\} = 22/(22 + 23) = 22/45$$

30.7 DOES OBSERVING A HEAD MAKE THE PROBABILITY OF A HEAD ON THE NEXT TOSS MORE LIKELY?

The framework we have developed can be used to determine the effect of new information on the probability that an event will occur. We find the posterior distribution of ϕ given the information and then calculate its mean. Another way is to incorporate new information is based only on the prior ϕ distribution. Consider, for example, the probability of observing a second Head given we have observed a Head on the first toss and we know the distribution of ϕ. Therefore, we would like the probability $\{H_2|H_1, \phi, \&\}$.

We know that the probability of getting two Heads in a row is

$$\{H_2, H_1|\phi, \&\} = <\phi^2|\&>$$

and we know that the probability of getting a Head on the next toss is

$$\{H_1|\phi, \&\} = <\phi|\&>$$

Therefore, we simply calculate the required probability as

$$\{H_2|H_1, \phi, \&\} = \frac{\{H_2, H_1|\phi, \&\}}{\{H_1|\phi, \&\}} = \frac{<\phi^2|\&>}{<\phi|\&>}$$

Furthermore, since the variance of the ϕ distribution ${}^{v}\!<\phi|\&>$ is just the difference between the second moment and the first moment squared, we have

$${}^{v}\!<\phi|\&> = <\phi^2|\&> - <\phi|\&>^2$$

we can also write the probability $\{H_2|H_1, \phi, \&\}$ as

$$\{H_2|H_1, \phi, \&\} = \frac{<\phi|\&>^2 + {}^{v}\!<\phi|\&>}{<\phi|\&>} = <\phi|\&> + \frac{{}^{v}\!<\phi|\&>}{<\phi|\&>}$$

This result shows that the probability of getting a Head on the second toss after observing a Head on the first toss and given the uncertainty about ϕ is equal to the mean of the distribution ϕ plus another term that is the ratio of the variance of the distribution of ϕ to its mean. Since the variance is non-negative, the occurrence of the first Head raises the probability of a second Head by the ratio of the variance to the mean of the ϕ density function. Only if there is no variance in the ϕ distribution—no uncertainty about ϕ—will the occurrence of the event not increase its probability.

Observe the evolution of our thinking about this process over the course of our education. Children often believe that seeing Heads occur in this type of situation makes the occurrence of a Head more likely next time. When, later, they study probability in school, they learn that tossing a two-state device is a Bernoulli process and that the chance of observing a success or Head next time depends in no way on the history of previous results. However, as college or graduate students reading this book, you are now learning that if you are at all uncertain about the parameter of the process, then seeing a success will increase the probability of future success.

30.8 ANOTHER THUMBTACK DEMONSTRATION

Let us refer back to the thumbtack demonstration. Suppose we have two thumbtacks in the jar instead of one, and consider the following dialogue between an instructor (I) and selected members of his class (C):

I: How many ways can the thumbtacks land?

C: The thumbtacks can land either (1) both pins up. (2) both pins down, or (3) they can land one pin facing up and the other facing down.

I: Correct. Let us call "Same" when we have either both pins up or both pins down and call "Different" when we have one pin facing up and the other facing down, as shown in Figure 30.4. Is this clear?

C: Yes, now we know what we can call.

I: Would you call "Same" or "Different" if you acquired the certificate?

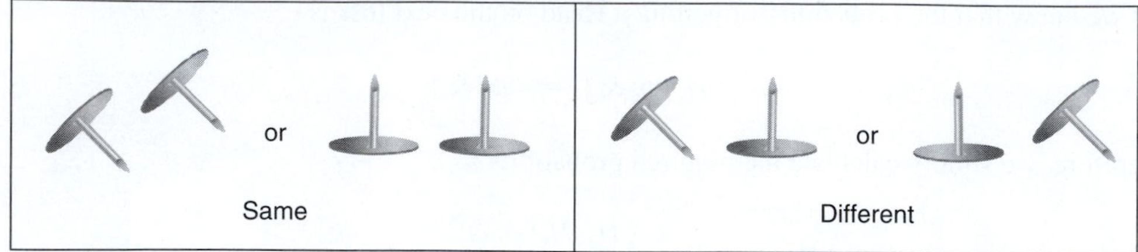

FIGURE 30.4 Calling Alternatives for the Two Thumbtacks

C: If we assign a probability, p, for each thumbtack landing pin up and $1 - p$ for it landing pin down, and if we assume the way one thumbtack lands is irrelevant to the way the other lands, then the probability of both thumbtacks landing the same way is

$$\{\text{Same} \mid \&\} = \{\text{Two pins up} \mid p, \&\} + \{\text{Two pins down} \mid p, \&\}$$
$$= p^2 + (1 - p)^2$$

for any value of p.

I: Well, given your response, I know you have assumed the way one thumbtack lands is irrelevant to the way the other lands. That means if you observed the first thumbtack landing, you would not learn anything about the way the second thumbtack would land.

C: Correct.

I: Suppose I told you that the first thumbtack landed pin up, does that update your belief about the probability of the second thumbtack landing pin up?

C: (*Thinking*)

I: OK. This is what we will be talking about today. Let us first plot the probability $\{\text{Same} \mid p, \&\}$ for different values of p as you mentioned. (*Shows them Figure 30.5.*)

I: Note that $\{\text{Same} \mid p, \&\}$ is equal to 0.5 only when $p = 0.5$. Therefore, the probability of the thumbtack landing the "Same" way is always greater than or equal to their landing

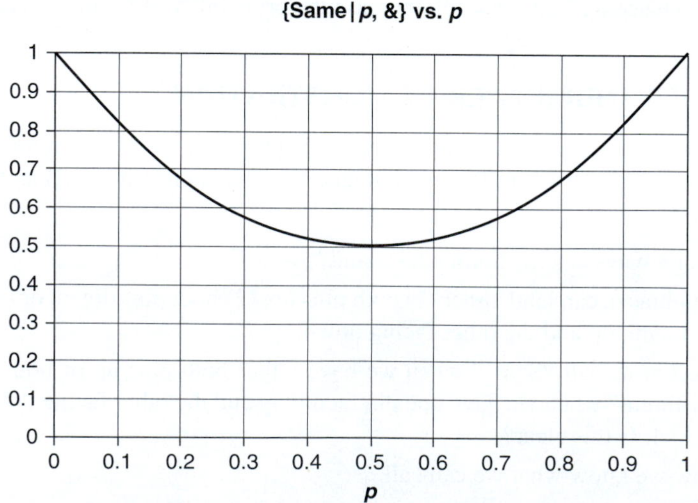

FIGURE 30.5 Sensitivity Analysis {Same | *p*, &} vs. *p*

"Different." The Choice Rule states that we should call "Same," since it will provide us with a higher probability of getting the \$100, unless $p = 0.5$, in which case we would be indifferent.

I: Now, suppose that observing a thumbtack lands in one direction updates my probability of the second thumbtack landing that way.

C: In this case, the two are not irrelevant.

I: Correct. How then do I update my probability?

C: This is easy. We can incorporate this using the previous discussion on learning from experimentation. Suppose we have an uncertainty about ϕ. The probability of $\{\text{Same}|\&\}$ is

$$\begin{aligned}\{\text{Same}|\&\} &= \{H_1, H_2|\&\} + \{T_1, T_2|\&\} = <\phi^2|\&> + <(1-\phi)^2|\&> \\ &= 1 - 2<\phi|\&> + 2<\phi^2|\&>\end{aligned}$$

I: What about $\{\text{Different}|\&\}$?

C: Here it is:

$$\begin{aligned}\{\text{Different}|\&\} &= \{H_1, T_2|\&\} + \{T_1, H_2|\&\} = 2<\phi(1-\phi)|\&> \\ &= 2<\phi|\&> - 2<\phi^2|\&>\end{aligned}$$

I: Correct.

30.8.1 Is Uncertainty About ϕ Good or Bad?

There is another point to make in this discussion. As we have extensively discussed, when the probability of Heads is p, the probability of "Same", $\{\text{Same}|p, \&\}$ is equal to

$$p^2 + (1-p)^2$$

and is plotted in Figure 30.5. As the plot shows, this will be greater than 0.5 except when p equals 0.5. This expression also can be written as

$$1 - 2p(1-p)$$

Now let us turn our attention to the expression above using the ϕ distribution:

$$\begin{aligned}\{\text{Same}|\&\} &= \{H_1, H_2|\&\} + \{T_1, T_2|\&\} = <\phi^2|\&> + <(1-\phi)^2|\&> \\ &= 1 - 2<\phi|\&> + 2<\phi^2|\&>\end{aligned}$$

or since

$$^v<\phi|\&> = <\phi^2|\&> - <\phi|\&>^2$$

as

$$\begin{aligned}\{\text{Same}|\&\} &= 1 - 2<\phi|\&> + 2(^v<\phi|\&> + <\phi|\&>^2) \\ &= 1 - 2<\phi|\&>(1 - <\phi|\&>) + 2^v<\phi|\&>\end{aligned}$$

Suppose that we now have a prior distribution of ϕ with a mean of one half but also a positive variance. Even when we substitute the mean of 0.5 in the expression, the probability of "Same" will be greater than one half because of the positive contribution of the variance. In fact, for all distributions of ϕ that have a mean of one half, the higher their variance, the higher your chance of winning by calling "Same." An extreme example is when you are tossing a coin that is equally likely to be two-headed or two-tailed. Your chance of calling a toss correctly is one half, but you will always win by saying that two tosses will come up the same.

30.8.2 Insight: Modeling and the Clairvoyant

Now that we have discussed the question of updating information using the notion of uncertainty about ϕ, we can deal with it at a more fundamental level. After reflecting on what we have done in this chapter, students sometimes ask whether the clairvoyant could tell them the value of ϕ.

You might recall that the clairvoyant can only tell us the value of quantities that are defined using the clarity test; that is, with no use of judgment. The notion of ϕ exists only in our minds as the parameter of a representation of the process: a model. The clairvoyant has no knowledge of models, which are our mental creations. Any story you are telling yourself about the future has no effect on how the clairvoyant will respond. Thinking that today is your lucky day will have no effect on whether the clairvoyant will tell you that the next toss of the thumbtack will be a Head. Your notion of ϕ as the number you would assign, if you only knew it, as the probability of a Head on the next loss of the thumbtack does not exist in the clairvoyant's world—only in yours. The clairvoyant can only tell you whether the thumbtack will land on Head or Tail.

What you are doing when you use a model with a parameter to describe your knowledge is to extend the notion of assigning a probability to a quantity meeting the clarity test to a quantity that exists only in your representation. Note that if you and another person agree on the thumbtack model but disagree on the probability distribution of its parameter ϕ, you would not be able to bet on ϕ but only on the result of tosses that are within the realm of the clairvoyant. This extension of probability to model parameters is not a major step, since all probabilities you assign are in your head and not in the province of the clairvoyant. As long as you know what you are thinking about, the clairvoyant's inability to share your thoughts is irrelevant.

The observation that the clairvoyant has no role in modeling extends to any degree of complexity. An investigator may think of several competing models to represent a process and may assign a prior probability to each of them being correct. As the process is observed, the probability of each will update just as it does in the case of thumbtack tossing. However, the clairvoyant would never be able to tell the investigator which (if any) of the models is correct.

30.9 SUMMARY

If we are uncertain about ϕ, we assign the mean of that distribution as the probability of Head on the next toss. The probability is just one number. There can be no uncertainty about it.

If you are uncertain about ϕ, then observing a Head makes the probability of a future Head more likely.

To get the posterior distribution of ϕ after observing a Head as $\{\phi|H, \&\}$, we multiply the prior density by a straight line and normalize the product to get the posterior distribution.

If the prior is a Beta distribution (r', k') and you observe r Heads and k Tails, the posterior distribution for ϕ is Beta $(r' + r, k' + k)$.

The prior probability of a future Head, $\{H_1|\&\} = <\phi|\&>$, becomes $\{H_2|H_1, \&\} = <\phi|\&> + {}^{v}<\phi|\&>/<\phi|\&>$ after observing a Head. This means that the probability of Head increases after observing a Head.

It is always a good decision to call "Same" instead of "Different" in the two thumbtack game, even though you may lose.

KEY TERMS

- Experimentation
- Updating belief graphically
- Conjugate distributions

PROBLEMS

1. Suppose your prior distribution ϕ about how a thumbtack would land Heads is
   a. $\{\phi|\&\} = \text{Beta}(2, 2)$
   b. $\{\phi|\&\} = \text{Beta}(8, 8)$

   For each case, determine:
   i. The probability $\{H|\&\}$ on the next toss.
   ii. The probability of Heads after observing 5 Heads and 2 Tails. What do you notice?
2. For the prior distributions of problem 1, determine the shape of the posteriority distribution graphically after observing 1 Toss.

   Hint: It might be helpful to first construct a table whose first column has the degrees of ϕ discretized; whose second column has the corresponding probability $\{\phi|\&\}$, and the third column has $1 - \phi$. Plot the results. Note the sum in the fourth column need not be 1 but the column can be normalized if we want the actual distribution and not just its shape.

ϕ	$\{\phi\|\&\}$	$1 - \phi$	Product $\{\phi\|\&\}(1 - \phi)$
0			
0.01			
0.02			
.			
.			
.			
.			
.			
.			
1			
			Σ Sum

3. Repeat Problem 2 if you observe 2 Heads and 1 Tail.

CHAPTER

31

Auctions and Bidding

CHAPTER CONCEPTS

After reading this chapter, you will be able to explain the following concepts:

- The optimal bid in:
  - Sealed auctions
  - Second price auctions
  - Descending auctions
  - Ascending open auctions
- The value of a bidding opportunity
- The winner's curse effect

31.1 INTRODUCTION

In Chapter 2, we presented a bidding demonstration for a certificate using the thumbtack example. We also discussed the notion of the Personal Indifferent Buying Price (PIBP) for the certificate. However, we did not discuss what the optimal bid should be for this bidding situation given your PIBP.

In this chapter, we present the analysis of that bidding situation as well as several other auction types using an in-class demonstration. We show that many auctions, while different, have similar elements that require us to think about our PIBP for the bidding item and our belief about the competitive bids.

While most of our discussion in this chapter is based on an individual who is making a bidding decision in an auction, the reasoning and methods apply equally well to businesses engaged in any type of competitive bidding. In fact, the only change necessary is to replace the PIBP with the **Corporate Indifferent Buying Price (CIBP)**.

A word on terminology: When auction bidders are successful in obtaining the object of their bidding, they often call this "winning." Of course, if the person has bid too much for what they have "won," then in actuality, they have lost. Here, we will use the term "acquirer" to represent the person whose bid completed the auction and who acquired the bidding item. Keep in mind that in the auction process, our ultimate goal should not be just to *acquire*, but to *gain*.

31.2 ANOTHER THUMBTACK DEMONSTRATION

In the following situation, an instructor is speaking to students.

> ***Note*** *"I" refers to instructor and "C" refers to one or more individuals in the class. Class participants may also be referred to by name.*

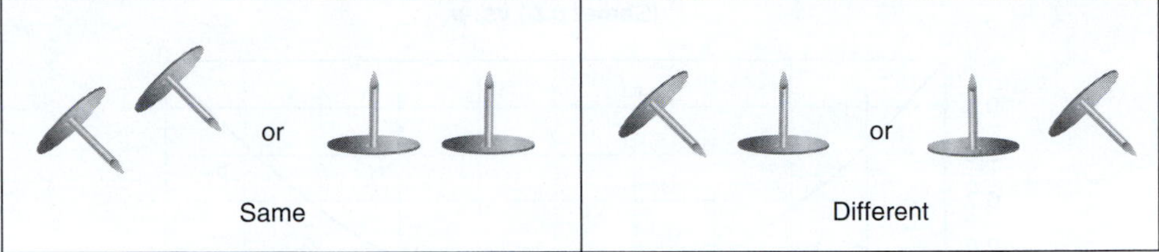

FIGURE 31.1 Calling Alternatives for the Two Thumbtacks

I: Today we shall bid for a certificate that pays \$100 if you call correctly or 0 if you call incorrectly the outcome of an uncertain device. The device will be an opaque plastic jar containing two thumbtacks. (*Instructor shows the jar.*) We will shake the jar upside down, place it on its cap, carefully unscrew the jar from the cap, and see how the thumbtacks have fallen.

C: Is this the same certificate we used in the last class (Chapter 30)?

I: Yes. We have two thumbtacks again.

C: And the thumbtacks can land (1) with both pins up, (2) with both pins down, or (3) with one pin up and one pin down.

I: Correct. Again, we call "Same" when we have either both pins up or both pins down, and we call "Different" when we have one pin facing up and the other facing down. (*Shows them Figure 31.1.*) Is this clear?

I: OK, now the acquirer of the certificate will still be the highest bidder and will call either "Same" or "Different." If he calls correctly, he gets \$100, and if he calls incorrectly, he gets 0. Clear?

C: Yes. And do the usual rules apply about no collusion, no syndicates, no ability to resell an owned certificate, and resolution of ties by coin tosses?

I: Yes, but in this case, we will be holding several types of auctions. Before we analyze these different auctions, let me first ask a question: If you actually acquired the certificate, would you call "same" or "different?"

C: From our earlier discussions, if we assign a probability, p, for any thumbtack landing pin up and $1 - p$ for it landing pin down, and if we assume the way one thumbtack lands is irrelevant to the way the other lands, then the probability of both thumbtacks landing the same way is (*writes equation on board*)

$$\begin{aligned}\{\text{Same}|\&\} &= \{\text{Two pins up}|p, \&\} + \{\text{Two pins down}|p, \&\} \\ &= p^2 + (1 - p)^2\end{aligned}$$

for any value of p.

I: True. Let me again plot the probability (*writes on board*) $\{\text{Same}|p, \&\}$ for different values of p to remind you (*shows Figure 31.2*).

I: Note that (*points to board*) $\{\text{Same}|p, \&\}$ is equal to 0.5 only when $p = 0.5$. Therefore, the probability of the thumbtacks landing the same way is always greater than or equal

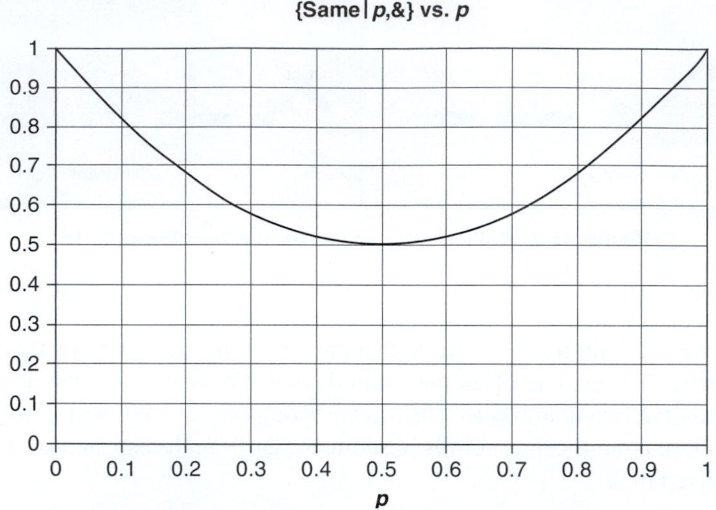

FIGURE 31.2 Sensitivity Analysis {Same| *p*, &} vs. *p*

to the probability of their landing differently. For $p = 0.5$, we have the same probability of their landing either way.

I: Suppose reflection leads you to think of different possible values of p. What would you choose?

C: In this case, we can use the results of the previous chapter to determine the probability of $\{\text{Same} | \&\}$. We would still choose to call "same," and we can calculate our probability of $\{\text{Same} | \&\}$ as we discussed.

I: Correct. Nothing new here. Now, we have determined what we should call if we acquired the certificate, but we have not discussed how much to bid. We will now present four types of auctions and determine how much to bid for each one. You will see that we will not learn anything about the bids of other bidders as we consider the different auctions, so your bidding will not be affected by the order. Then, we will execute one of the auctions using a coin toss (flipped twice), so we are equally likely to deal with any one of them.

31.2.1 Auction 1: Closed Bid Auction

I: In this auction, each of you will submit a closed bid for the certificate. The highest bidder will acquire the certificate and pay his bid amount. This is the same type of auction we discussed in the Chapter 2 thumbtack demonstration. What is the best bidding strategy?

C: Since the PIBP for any item is the amount that the person would pay for the certificate such that they would be indifferent to receiving it, anyone participating in a closed bid auction for the certificate should bid an amount less than their PIBP to obtain any possibility of a benefit from this bidding situation.

I: Right. If there is uncertainty about the value of the item, we can use our u-curve to determine the PIBP, as we discussed in Chapter 10. For example, in the case of two thumbtacks, we can calculate our PIBP for a deal that pays \$100 with a probability $\{\text{Same} | p, \&\}$ and 0 with a probability $\{\text{Different} | p, \&\}$.

C: It makes sense that increasing the bid amount will increase our chances of acquiring the certificate. However, acquiring at a larger bid reduces the net value of the bidding situation, since we will pay more for it.

I: Correct. There is a trade-off between the chance of acquiring the item and the value of the item if acquired. We have yet to discuss what the optimal bid ought to be in this situation.

Now submit your cards with your name and bid for Auction 1.

31.2.2 Auction 2: Second Price Closed Bid Auction

I: Next, let's discuss another closed bid auction. In this auction, the highest bidder acquires the certificate but pays an amount equal to the second highest bid.

C: So the difference between this and the first auction is that the highest bidder does not actually pay his bid amount, but pays the second highest bid amount?

I: That's right. Given this rule, what is the optimal bidding strategy?

C: (*Thinking*)

I: As we will see, in this auction, the optimal bidding strategy is simple. You bid your PIBP for the certificate. To understand why, suppose you do so and then the clairvoyant offers to tell you the amount of the highest bid submitted by any of the other bidders, thereby removing all material uncertainty about what the other bidders will do. We will refer to this highest bid amount as the **maximum competitive bid (MCB)** and discuss three possibilities for the clairvoyant's report. In each case, would you change your bid of PIBP after receiving the report?

The clairvoyant's first possible report is that the MCB from all of the other players is an amount greater than your PIBP: MCB > PIBP. In this case, you will not acquire the certificate; the net value is 0. There is no way to benefit by changing your bid of PIBP. If you bid more than MCB, you would acquire it at the MCB, since you pay the second highest price, and at a value that is higher than your PIBP and negative to you.

C: We got it. If the MCB is greater than the PIBP, there is no point bidding higher than our PIBP.

I: *The clairvoyant's second possible report* is that the MCB from all of the other players is an amount less than your PIBP: MCB < PIBP. In this case, if your bid is your PIBP, you will acquire the item and pay only the MCB (which is less than your PIBP). As a result, you acquire the item with a net positive value of (PIBP − MCB). Bidding any value between MCB and PIBP will still result in acquiring the certificate at the MCB. There is again no reason to change your bid of PIBP.

C: OK, if the MCB is less than the PIBP, bidding the PIBP will result in acquiring the certificate and a positive gain.

I: *The clairvoyant's third possible report* is that the MCB from all of the other players is an amount equal to your PIBP: MCB = PIBP. There is no reason to change your bid since the best you can do has a value of 0. If you bid more, you will acquire the item at a price equal to your PIBP. As a result, its value is 0. If you bid less, you do not acquire the item, and your value is still 0. Finally, if you bid equal, you will acquire and pay the MCB, which is equal to your PIBP; the value is still 0.

C: For any of the possibilities, there is no better choice than to bid the PIBP.

I: Correct, and it is the best you can do even if you have complete information about the maximum competitive bid. As a result, you can do no better in this auction than bidding your PIBP.

(*Instructor writes on board*: Optimal bid = PIBP)

I: Class, note that this optimal bid (your PIBP) gives the auction an interesting property: You would not change your bid even if you somehow knew all the competing bids—unless your goal was to have someone else pay as much as possible for an item—which would be a different frame.

Now submit your cards with your name and bid for Auction 2.

31.2.3 Auction 3: Descending Closed Bid Auction

I: Now, we will consider a third type of auction. In this auction, the auctioneer announces possible bid amounts verbally and in descending order. The auctioneer can see all participants, but they cannot see each other. When a participant hears his bid amount, he raises his hand. In commercial settings, the auction would now be over. In educational settings, however, all bids can be collected by continuing. In this "blindfolded" design, all participants close their eyes, lower their heads, and silently raise a hand for a count of three when their bid amount is reached: No participant learns the bids of others. As the hands are raised, the names of the participants and their bid amounts are recorded by the auctioneer. When the countdown is finished, the auction is over, and the highest bidder acquires the item for his highest bid amount.

C: OK, so you will count down from 100 to 0, we close our eyes, and when we hear our bid amount, we raise our hands?

I: Yes. Now let us conduct Auction 3 as we have specified.

31.2.4 Auction 4: Ascending Open Bid Auction

I: Now, we will discuss the open bidding auction, which is one of the most widely used types of auctions. In this type of auction, people call out their bids, as they do when bidding for antiques. The auctioneer will take higher and higher bids until no one wishes to bid any higher. Should this auction procedure be selected, the person making the last bid will be declared the highest bidder and will pay that amount to acquire the certificate.

C: Yes, we have seen many auctions like this.

I: Quite often, we observe some "macho" competitive behavior between participants, who end up bidding even higher than their PIBP values. Such people must be deriving additional value by thwarting the intentions of others. For those who attach value only to acquisition of the item, the optimal strategy is to bid incrementally up to the PIBP. Now we will conduct Auction 4. Is it true that you have not learned anything of the bids of others in the first three auctions?

C: Correct, we have not.

The open bid auction is then conducted.

A coin is tossed twice to determine which of the four auctions will be executed.

Heads, Heads ⇒ Auction 1
Heads, Tails ⇒ Auction 2
Tails, Heads ⇒ Auction 3
Tails, Tails ⇒ Auction 4

Money changes hands.

31.2.5 Bidding Strategies for the Four Auctions

I: We spent some time finding the best bidding strategy for Auction 2, the second price auction. It is to bid the PIBP. What is the best strategy for Auction 3, the descending closed-bid auction?

C: Upon close examination, we see that it is the same as the closed-bid auction. The only difference is the form in which bids are made.

I: Right. Auctions 1 and 3 should have the same bidding strategy, which we have yet to discuss.

Now, we will summarize the optimal bidding strategies for each auction.

1. **Auction 1:** Bid = Some amount less than PIBP
2. **Auction 2:** Bid = PIBP
3. **Auction 3:** Bid = Same as Auction 1
4. **Auction 4:** Bid up to your PIBP

> ***Note*** *If we asked an agent to go to the auction and bid for us, then it would be sufficient to give him our PIBP for Auctions 2 and 4 so that he can participate effectively in the auction on our behalf. For Auctions 1 and 3, however, we can give him our PIBP and tell him to bid less, but exactly how much less remains in question. Our next section will address this topic.*

31.3 AUCTIONS 1 AND 3 FOR A DELTAPERSON

Since the bidding strategies for Auctions 2 and 4 merely require the PIBP, we can determine the optimal bidding strategy by calculating the PIBP for an uncertain deal as we discussed earlier. In the last chapter, we specifically discussed how to calculate the PIBP for a certificate involving two thumbtacks. This same analysis is all that is required for Auctions 2 and 4.

We will now analyze Auctions 1 and 3 in more detail. The decision we are facing is exactly how much to bid. Each bid amount, b, represents a decision alternative. There is uncertainty about whether or not we will acquire the certificate at this bid and about whether or not we will call correctly. The generic decision tree for this situation appears in Figure 31.3.

If the person is a deltaperson, we can again simplify the analysis. We can replace the portion of the tree following "Acquire" with the PIBP of the certificate—less the bid—to get the simpler tree in Figure 31.4. A deltaperson considers acquiring the deal as a result of his bid to be equivalent to receiving a payment equal to his PIBP less his bid. We can use this result to simplify our analysis. Recall that, for a deltaperson, we can calculate the PIBP by calculating the certain equivalent of a deal, since in this case, the PIBP and PISP are equal.

There is now only one uncertainty in the tree of Figure 31.4, which is the probability of acquiring the item for a given bid. If we acquire the certificate at a given bid amount, we would

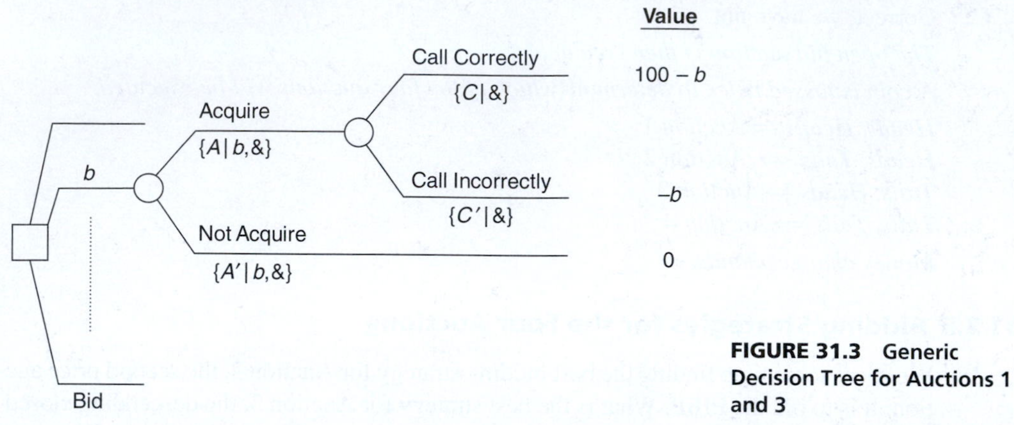

FIGURE 31.3 Generic Decision Tree for Auctions 1 and 3

have a net value of the PIBP − bid. If we do not acquire, then we have 0 value, since we neither pay nor receive anything.

> ***Note*** *The probability of acquiring the certificate in Figure 31.4 is influenced by the bid amount. As the bid amount increases, the probability of acquiring increases.*

Now, we expand our reasoning about this probability of acquiring. We acquire only if our bid is higher than any of the competitors' bids. We referred to the highest competitive bid earlier as the maximum competitive bid (MCB). Therefore, the probability of acquiring at any given bid is equal to the probability that the MCB is less than our bid amount, so

$$\{\text{Acquire} \mid b, \&\} = \{\text{MCB} < b \mid \&\}$$

Therefore, the main uncertainty is the probability distribution of the MCB, $\{\text{MCB} \mid \&\}$. We can use the methods discussed in Chapter 15 to encode our probability. Alternatively, we can use a specified form of the probability distribution function and assess its parameters, as we discussed earlier.

The value we get if we acquire the auction is equal to the PIBP bid, and the decision we have to make is the actual bid amount. The decision diagram for a deltaperson facing this bidding situation appears in Figure 31.5.

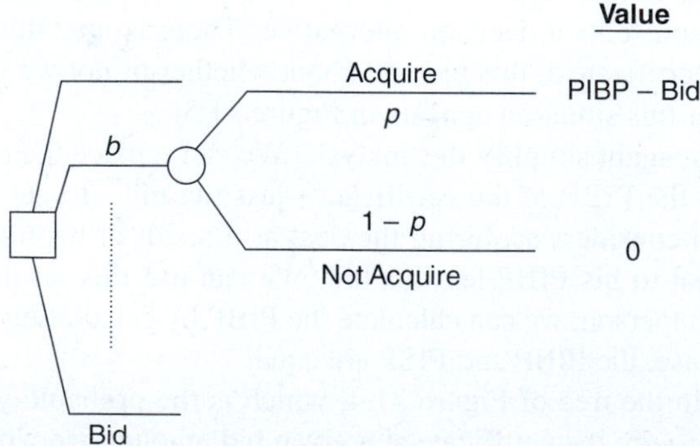

FIGURE 31.4 Generic Decision Tree for Auctions 1 and 3 for a Deltaperson

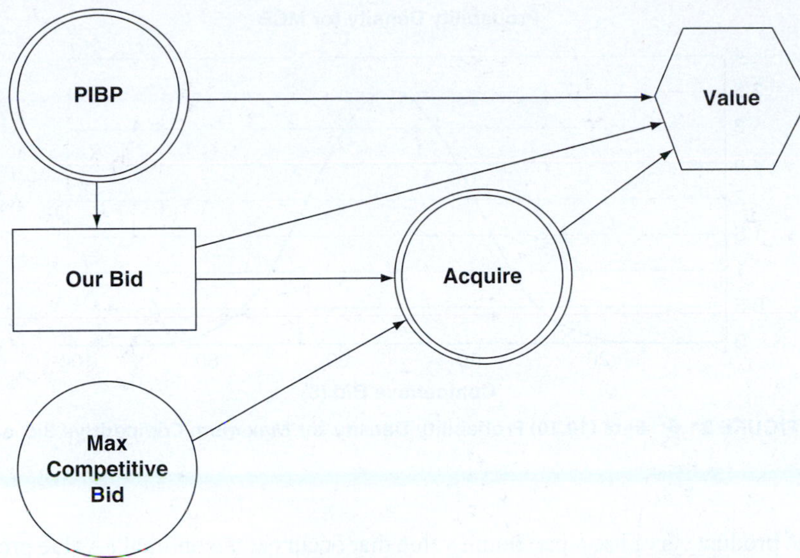

FIGURE 31.5 Decision Diagram for a Deltaperson Facing Auctions 1 and 3

The deterministic node "Acquire" determines whether or not we acquire the item given the bid amount and the MCB.

Now, we analyze this bidding problem for a risk neutral decision maker—a risk-neutral deltaperson. The optimal bid amount is the one with the largest *e*-value of monetary prospects. From the tree in Figure 31.4, we can see that the *e*-value of each bid alternative is

$$e\text{-value} = p(\text{PIBP} - \text{Bid})$$

where p is the probability of acquiring at this bid amount read directly from the cumulative distribution for the MCB.

31.3.1 Determining the Optimal Bid for a Risk-Neutral Decision Maker

Suppose a risk-neutral decision maker is facing Auction 1 or 3. He believes that the probability of calling correctly is 0.7, so his PIBP for the certificate is \$70. The decision maker assigns a scaled Beta (10,10) probability density function for the MCB, as shown in Figure 31.6. The horizontal axis represents the range of monetary bids from 0 to \$100.

Since the probability of acquiring is equal to the cumulative distribution function of the MCB distribution, we now plot the cumulative distribution function of the Beta (10,10), together with the straight line $(\text{PIBP} - \text{bid})$.

We call the two curves, as shown in Figure 31.7, the **opposing forces of bidding**, since one of them increases as we increase the bid amount (probability of acquiring) and the other one decreases (value of the item at that particular bid amount). From the *e*-value equation, we take the product of these two curves (as shown in Figure 31.8) to calculate the *e*-value for each bidding alternative. The product is 0 at the origin and 0 at the value of bid = PIBP. Above the amount bid = PIBP, the product is negative, and below the amount bid = PIBP, the product is

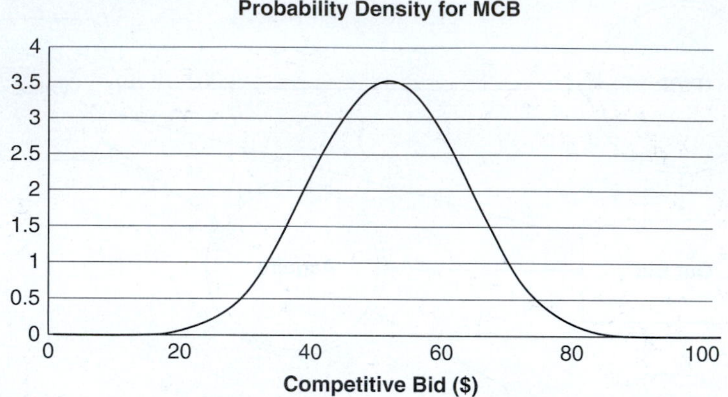

FIGURE 31.6 Beta (10,10) Probability Density for Maximum Competitive Bid

positive. The product curve has a maximum value that occurs at the optimal *e*-value produced by the optimal bid amount.

Notice, in Figure 31.8, that the optimal bid is $52.50, which is less than the PIBP of $70. The difference between the optimal bid and our PIBP depends on our belief about the MCB distribution and our PIBP for the bid item.

The optimal bid amount has an *e*-value determined by the magnitude of the peak, $11.28, in Figure 31.8. Since this peak is positive, the auction provides the bidder with a positive *e*-value, which in turn is the minimum amount the bidder would have to receive to forgo the bidding opportunity.

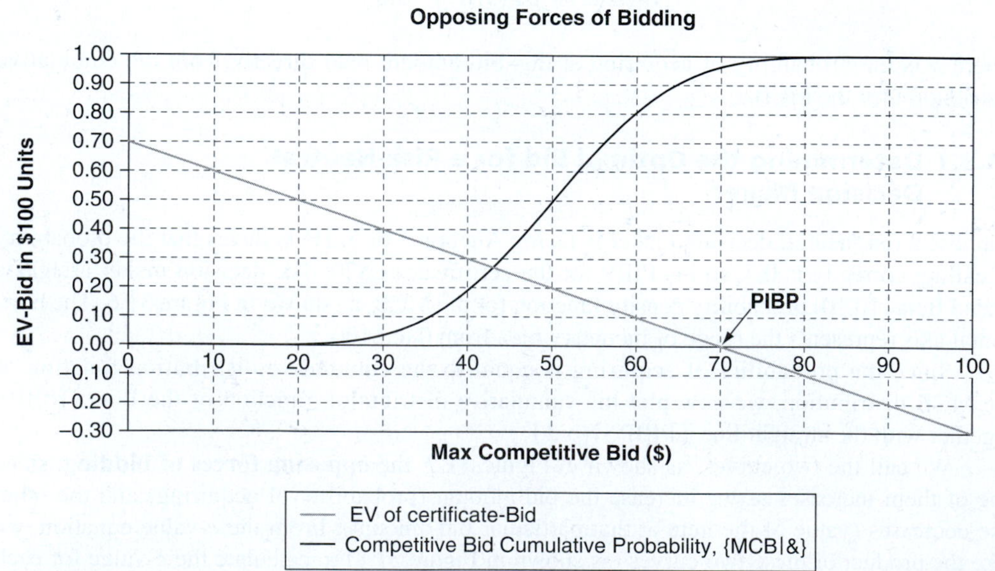

FIGURE 31.7 Cumulative Distribution of MCB, and (PIBP-Bid) [Vertical Scale for EV-Bid in $100 Units]

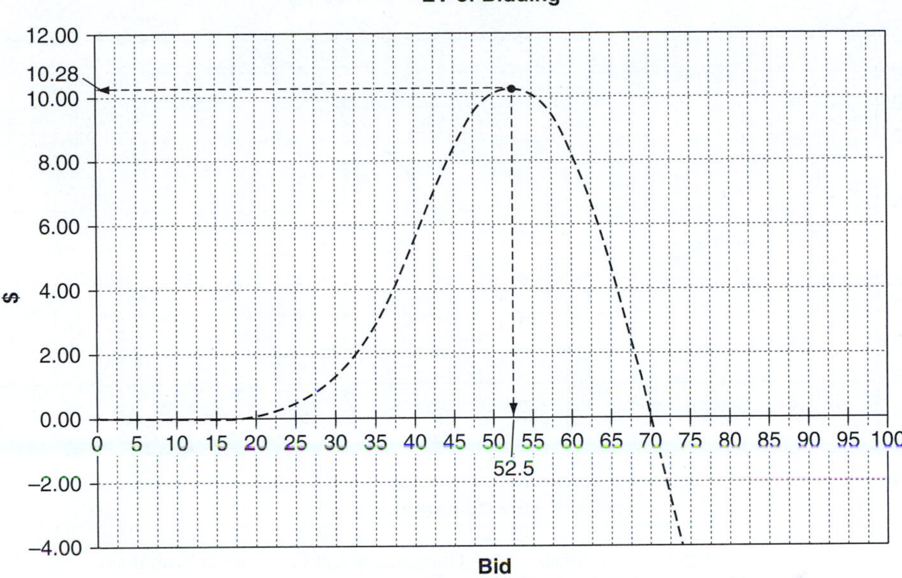

FIGURE 31.8 *E*-Value of Bid vs. Bid Amount. Optimal bid is $52.50.

31.3.2 Incorporating Risk Aversion for Auctions 1 and 3

In the previous section, we assumed that the decision maker is risk-neutral. We can easily extend the analysis to all deltapeople. In this case, the optimal bid is the one with the highest *e*-value of the *u*-values. In other words, it maximizes

$$e\text{-value of } u\text{-value of bid} = p.u(\text{PIBP} - \text{Bid})$$

As we discussed, if the decision maker follows the delta property, then the PIBP of the certificate is equal to the certain equivalent, which is simple to calculate. We illustrate this idea through the following example.

EXAMPLE 31.1 Bidding for a Risk-Averse Deltaperson

Consider Auctions 1 and 3 again. The decision maker provides the same distribution for the MCB as Beta (10,10) on a scale 0 to 100 and the same probability of calling correctly at 0.7.

The PIBP for the certificate for an exponential decision maker with a risk tolerance of $100 is $58.40, which is his certain equivalent of a deal providing $100 with probability 0.7 and 0 otherwise.

Figure 31.9 shows the opposing forces of bidding for both a risk-neutral and an exponential decision maker (risk tolerance = $100), which is expressed using the cumulative distribution for MCB and the value PIBP − Bid.

The next step is to determine the certain equivalent of each bid. This is the certain equivalent of a deal that provides $(58.4 − bid) with probability $\{\text{MCB} <= \text{bid} | \&\}$ and 0 otherwise.

Figure 31.10 plots the certain equivalent of each bid for both the risk-neutral and the exponential decision maker. The risk-averse person should bid about $47.50 and will have

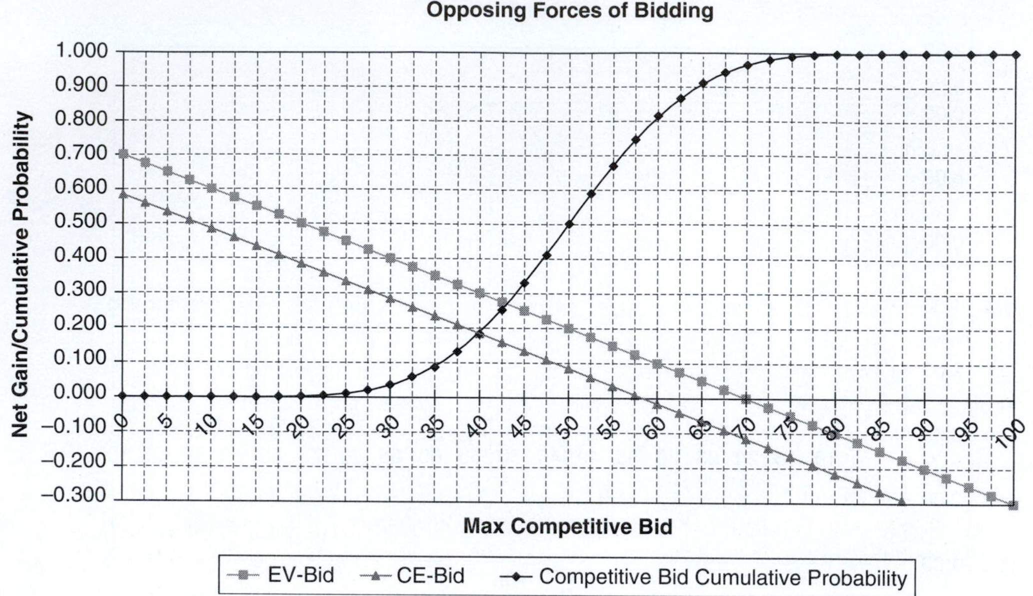

FIGURE 31.9 Opposing Forces of Bidding (PIBP-Bid) vs. Bid for a Risk-Neutral and Risk-Averse Bidder, Plus Cumulative Probability of Maximum Competitive Bid

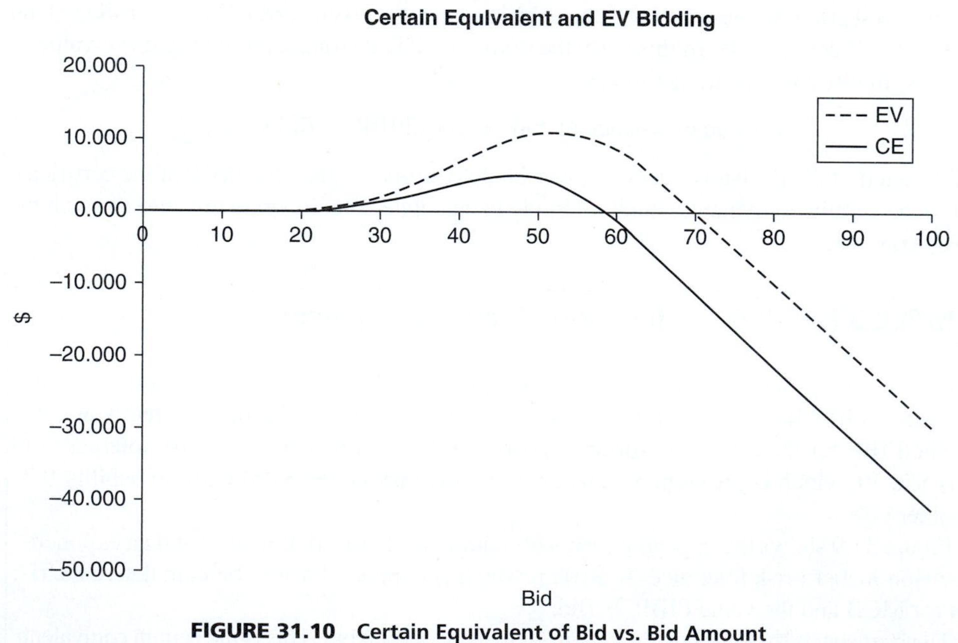

FIGURE 31.10 Certain Equivalent of Bid vs. Bid Amount

a certain equivalent of \$4.36 for this bidding situation. The risk-neutral person, as we have seen, should bid about \$52.50; the value of the deal will be \$10.28. Note that the optimal bid amount is less for the risk-averse decision maker than that for the risk-neutral decision maker. Furthermore, the certain equivalent of the optimal bid is less for a risk-averse decision maker than for a risk-neutral decision maker; the risk-averse decision maker would value the bidding opportunity less than the risk-neutral decision maker.

31.4 NON-DELTAPERSON ANALYSIS

Figure 31.5 presented the decision diagram of how much to bid in the two-thumbtack example. However, we should note that this diagram only applies if the decision maker is a deltaperson. Recall that a person's PIBP for an attractive deal can only be determined by increasing the price of the deal until the person is indifferent to buying it, and it is not necessarily equal to the certain equivalent. Therefore, in general, we cannot replace acquiring the certificate at a given bid with a simple PIBP $- b$.

Suppose that a non-deltaperson bids b. Let A be "Acquiring" the deal, and A' be "Not Acquiring" the deal. Let C be calling "Same" or "Different" on correctly calling the tossing of the thumbtacks, and C' be not doing so. The decision tree would appear as shown in Figure 31.3, which we repeat for convenience in Figure 31.11.

Suppose that the bidder is not a deltaperson but is someone with a logarithmic u-curve and with total wealth of \$100. The bidder assigns a probability 0.7 of "Same" and therefore one of calling correctly. By calling "Same," he believes he would receive \$100 with probability 0.7 and nothing with probability 0.3.

This logarithmic bidder would have a certain equivalent for the certificate giving a 0.7 chance of \$100 and a 0.3 chance of 0 equal to

$$(200)^{0.7}(100)^{0.3} - 100 = \$62.45$$

However, this bidder would have a PIBP obtained by the equation

$$\log(100) = 0.7\log(200 - \text{PIBP}) + 0.3\log(100 - \text{PIBP})$$

which implies that his PIBP $=$ \$56.75.

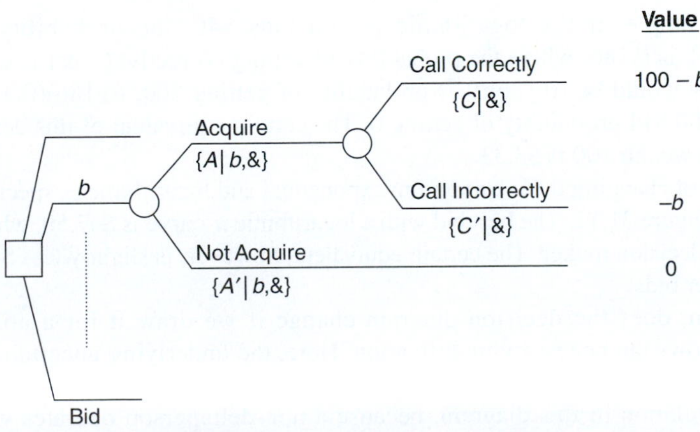

FIGURE 31.11 Non-Deltaperson Analysis

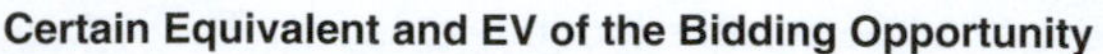

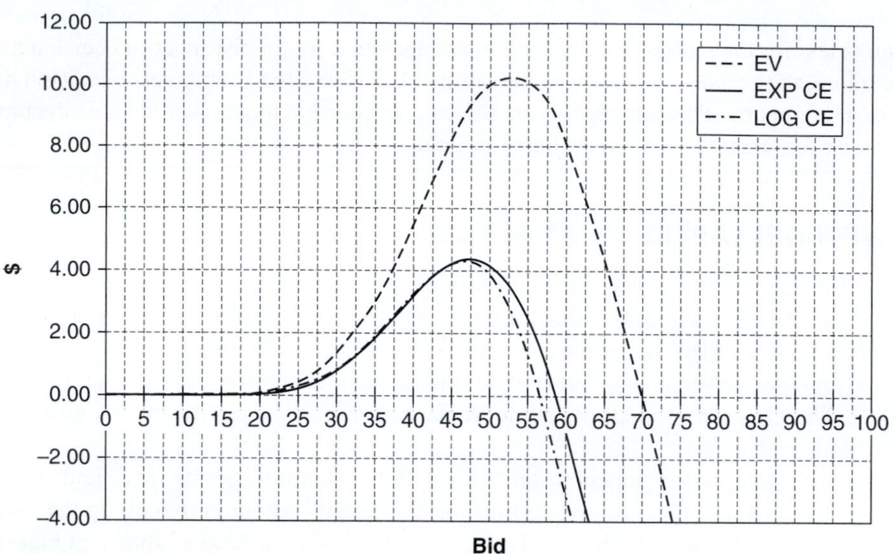

FIGURE 31.12 Comparing the Certain Equivalents of the Bidding Opportunity for Logarithmic and Delta Bidders

For a person with a logarithmic *u*-curve, the PIBP for the deal establishes only the maximum that this person would ever bid—in this case, \$56.75.

The bidding analysis for a non-deltaperson is more complicated, because we cannot replace the certificate deal with its PIBP as we did in Figure 31.4. The bidding opportunity at a bid, *b*, would then be described by Figure 31.3 with the certain equivalent of a deal having

Probability $\{A \mid b, \&\}\{C \mid \&\}$ of receiving \$100 $- b$
Probability $\{A \mid b, \&\}\{C' \mid \&\}$ of receiving $-\$b$
Probability $\{A' \mid b, \&\}$ of receiving 0.

In our example, if the logarithmic person bids \$40, the probability of acquiring. $\{A \mid b = 40, \&\}$ is 0.186, where the probability of calling correctly$\{C \mid \&\}$ is 0.7, and the bidding opportunity would be (0.186)(0.7) probability of getting \$60; (0.186)(0.3) probability of getting \$40; and 0.814 probability of getting 0. The certain equivalent of this deal to a logarithmic person with wealth 100 is \$3.33.

The result of changing the *u*-curve from exponential and logarithmic as specified for varying bids appears in Figure 31.12. The best bid with a logarithmic *u*-curve is \$47.50, which is similar to the exponential decision maker. The certain equivalent, however, is slightly less \$4.30 and is also different for other bids.

How, then, does the decision diagram change if we draw it for a non-deltaperson? Figure 31.13 shows the necessary modification. Here, the underlying uncertainty is the actual amount that the bidder will receive by making the better call of “Same” or “Different.” There is no PIBP calculation in this diagram, because a non-deltaperson operates with the tree of Figure 31.3.

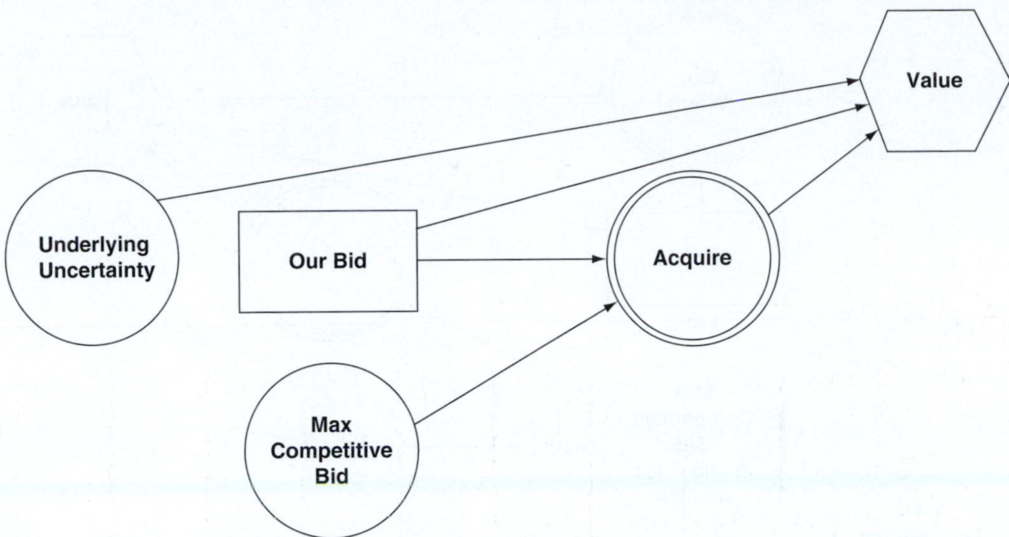

FIGURE 31.13 Bidding Decision Diagram for Non-Deltaperson

31.5 THE VALUE OF THE BIDDING OPPORTUNITY FOR AUCTION 2

Up to this point, we have focused primarily on how much to bid. We have also addressed the value to the bidder—the PISP of selling the right to bid to another party—should such a transaction be possible. In the example based on Auctions 1 and 3, we found that this would be \$10.28 for a risk-neutral bidder and \$4.35 for a deltaperson with a \$100 risk tolerance. When we examined the results of the same example for a logarithmic non-deltaperson with a wealth of \$100, we found that the value of the bidding opportunity became \$4.30.

Suppose, however, we are dealing with the second price Auction 2. We know that the best strategy for this auction is to bid the PIBP of what is on auction. For the risk-neutral person, this would be \$70; for the deltaperson with a risk tolerance of \$100, it would be \$58.40. But what is the value of the bidding opportunity to each of them? What is the PISP for each of them? To answer these questions, we need the decision diagram shown in Figure 31.14.

This diagram shows a new uncertainty: the MCB in a second price auction in which people are bidding their PIBPs. We know that in a first price auction, no one will bid as high as their PIBP, since they would have nothing to gain. If we have already assessed the distribution of the MCB for the first price auction of some deal, this may guide us in assessing the generally higher probability distribution of bids for a second price auction. The figure as drawn is correct for deltapeople. They will bid their PIBP and acquire if their bid is greater than the MCB in the second price auction. If they acquire, their net value will be their PIBP less the amount they had to pay: the highest competitive bid just below their bid.

To illustrate, let us specify the maximum competitive distribution in this second price auction as a Beta (35, 15) scaled to a \$100 maximum; the *e*-value of this distribution is (35/50) times 100 or \$70. The distribution of the MCB in this second price auction is the bidder's belief about the PIBP of the highest competitive bidder. Figure 31.15 compares the distributions of MCB for Auction 1 and the second price auctions. The mean of this distribution is a bit higher than that for the first price auction, because the bidder believes people will bid higher in this second price auction.

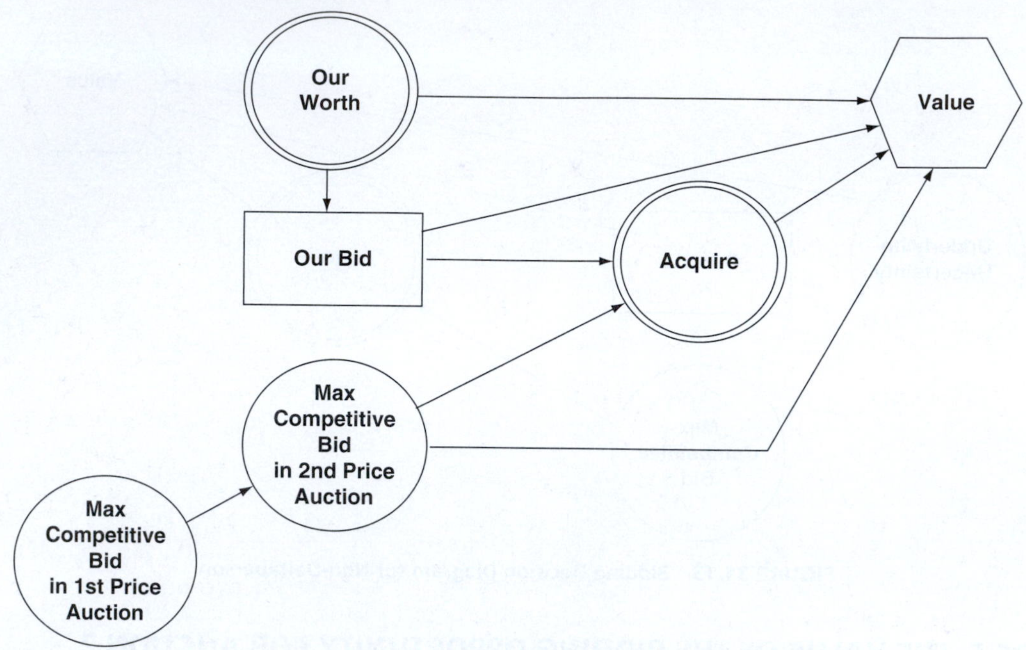

FIGURE 31.14 Decision Diagram for Second Price Auction for a Deltaperson.

The probability you will acquire is still the probability that this MCB is less than your bid. The amount you will have to pay is described by the MCB distribution conditional on that MCB being less than your bid.

What is unusual about this analysis is that what you will have to pay if you acquire is uncertain because you do not know what the second highest bid is. What you will pay if you acquire is determined by the distribution of the MCB for this second price auction.

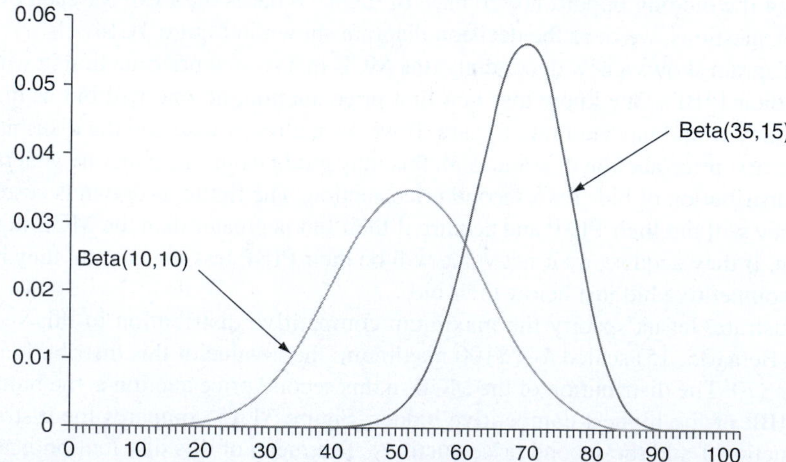

FIGURE 31.15 Comparison of MCB Distributions for the First Price Auction, Beta(10,10) and the Second Price Auction Beta (35,15) on a Scale of 0 to 100.

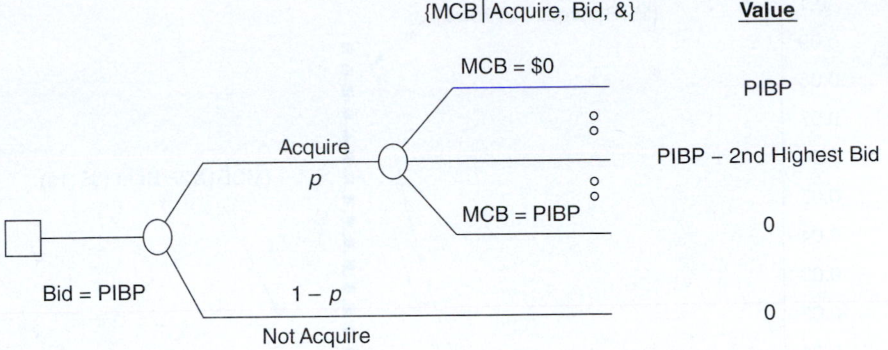

FIGURE 31.16 Generic Decision Tree for Value of the Second Price Auction for a Deltaperson.

We now illustrate the analysis for a deltaperson. The value of the bidding situation for a deltaperson starts by computing the PIBP (or equivalently, PISP) for what is being auctioned. As we have discussed, this amount will be your bid.

There are now two possibilities following the bid: Either you will acquire it, or you will not. If you do not acquire it, you will receive nothing, 0. If you do acquire it, your certain equivalent is determined by your PIBP less the amount you have to pay for it. What you have to pay given you acquired is an uncertain quantity determined by the distribution {MCB | Acquire, Bid, &}. Figure 31.16 illustrates the decision tree for this situation.

The distribution of {MCB | Acquire, &} when you bid your PIBP can be determined from the distribution {MCB | &} using the formula

$$\{\text{MCB}|\text{Acquire, Bid, }\&\} = \frac{\{\text{MCB}|\&\}\{\text{Acquire}|\text{MCB, Bid, }\&\}}{\{\text{Acquire}|\text{Bid, }\&\}}$$

To determine $\{\text{MCB}|\text{Acquire, }\&\}$, we distinguish two cases (and assume you acquire if $\text{PIBP} = \text{MCB}$).

Case 1: $\text{MCB} > \text{PIBP}$

In this case, you are guaranteed not to acquire by bidding your PIBP and so $\{\text{Acquire}|\text{MCB, Bid, }\&\} = 0$. Therefore, $\{\text{MCB}|\text{Acquire, Bid, }\&\} = 0$.

Case 2: $\text{MCB} \leq \text{PIBP}$

If $\text{MCB} \leq \text{PIBP}$, then you are guaranteed to acquire by bidding your PIBP. Therefore $\{\text{Acquire}|\text{MCB, Bid, }\&\} = 1$.

Furthermore, $\{\text{Acquire}|\&\} = \{\text{MCB} \leq \text{PIBP}|\&\}$ and so

$$\{\text{MCB}|\text{Acquire, Bid, }\&\} = \frac{\{\text{MCB}|\&\}}{\{\text{MCB} \leq \text{PIBP}|\&\}}$$

What this means graphically is that to calculate $\{\text{MCB}|\text{Acquire, Bid, }\&\}$ you can start with the distribution of {MCB | &}, truncate it at the value of the PIBP, and then normalize the truncated distribution. Figure 31.17 shows an example of {MCB | Acquire, Bid, &} calculated from the distribution {MCB | &} when you bid a PIBP of $70.

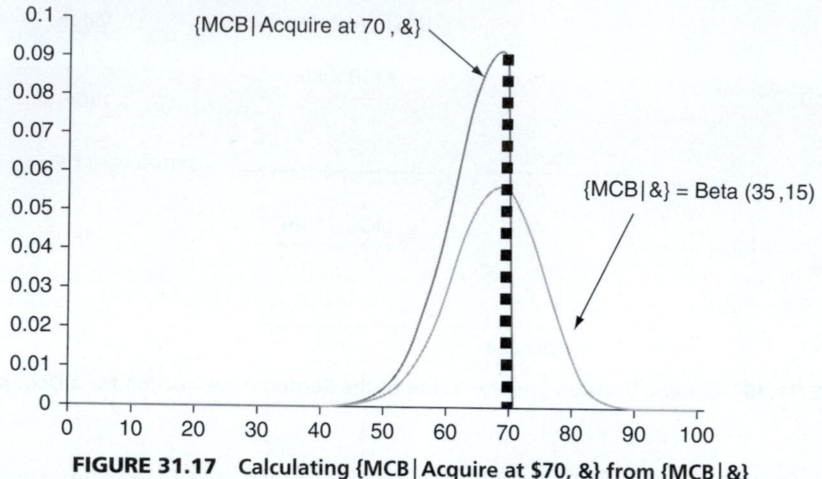

FIGURE 31.17 Calculating {MCB | Acquire at \$70, &} from {MCB | &}

For a deltaperson, we can simplify the analysis by first calculating the certain equivalent of what you have to pay if you acquire: the **CEcost**. The CEcost, in turn, is the certain equivalent of paying each of the bids lower than yours, which is the certain equivalent of the distribution {MCB | Acquire, Bid, &} and is illustrated by the tree in Figure 31.18. The probabilities in the tree are obtained by dividing the probabilities of {MCB | &} by the quantity $\{\text{MCB} <= \text{PIBP} | \&\}$. For example, if $\{\text{MCB} = 0 | \&\} = p_0$ and you bid your PIBP, then $\{\text{MCB} = 0 | \text{Acquire}, \&\} = \frac{p_0}{\{\text{MCB} \leq \text{PIBP} | \&\}}$.

For a deltaperson, the analysis now simplifies to the decision tree in Figure 31.19.

Therefore, the certain equivalent of the bidding opportunity is the certain equivalent of the deal of getting \$(PIBP − CEcost) with a probability, $p = \{\text{Acquire} | \text{PIBP}, \&\}$ and a chance of getting 0 with a probability $1 - p$.

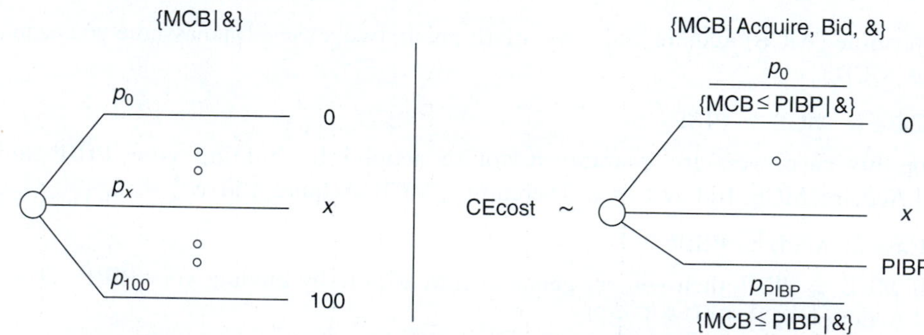

FIGURE 31.18 CEcost is the Certain Equivalent of What You Will Pay If You Acquire

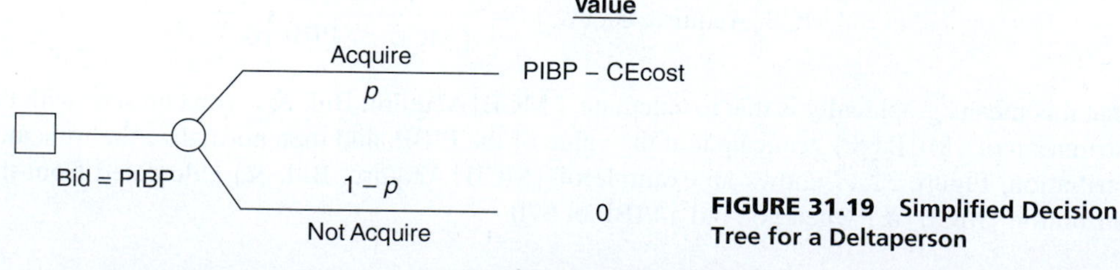

FIGURE 31.19 Simplified Decision Tree for a Deltaperson

To illustrate, let us begin with the risk-neutral person having a PIBP of \$70 for what is on auction and who bids this amount. The probability of acquiring with this bid, given the Beta (35, 15) distribution, is $p = 0.484$.

The CEcost of lower bids is just their *e*-value, or \$65.17. This is the mean of the distribution $\{\text{MCB} \mid \text{Acquire at } 70, \&\}$ in Figure 31.17. Therefore, the certain equivalent of the bidding opportunity is the certain equivalent of 0.484 chance of $\$(70 - 65.17)$, and 0.516 chance of getting 0, which is \$2.33. This would be the risk-neutral person's PISP for the opportunity to bid in the second price auction.

We turn now to the deltaperson with a risk tolerance of \$100, and a certain equivalent of \$58.42 for what is to be auctioned (\$100 with probability 0.7 and \$0 with probability 0.3). Based on our discussion, this person will bid his PIBP. The probability of acquiring with this bid given the MCB distribution is 0.04 and the CEcost of lower bids is \$55.57. The risk-averse decision maker therefore has a much lower probability of acquiring at 0.04 than the risk-neutral decision maker, which was 0.484. The certain equivalent of the bidding opportunity is that of a 0.04 chance of $\$58.42 - \$55.547 = \$2.83$ and a 0.96 chance of 0 or \$0.11. This risk-averse decision maker has a lower value for bidding in a second price auction than the risk-neutral decision maker.

We leave it as an exercise to the reader to calculate the value of this bidding opportunity for a non-deltaperson.

31.6 THE WINNER'S CURSE

A phenomenon called "**the winner's curse**" can arise in bidding situations when the participants in the auction have different levels of knowledge about the value of what is on offer. For example, suppose the auction is for an oil lease on a particular property. Some bidders may have more knowledge about the amount of oil because of their operation on adjacent properties. If you are one of the more ignorant bidders and you acquire the deal, acquiring it may bear the curse that the reason for your success was that the more knowledgeable bidders knew how little oil the land contained. The good news is that you have acquired the lease; the bad news is that you paid too much for it.

31.6.1 Decision Diagram

We will now show how to take the winner's curse phenomenon into account when determining the best bidding strategy for a deltaperson. Refer to Figure 31.20 for the decision diagram. The idea will be to (1) replace the PIBP node with an Underlying Uncertainty node—since we are no longer sure of the value of the item—and (2) add a relevance arrow from the Underlying Uncertainty to the Maximum Competitive Bid node, since knowing the value of the item should update our belief about the MCB distribution.

The underlying uncertainty here is the amount of oil in the ground. The bidder can assess a probability distribution on this quantity, but also knows that competitors will have different distributions. In high-value auctions, considerable effort can be devoted to representing the knowledge of the various parties. The essence is to characterize the relationship between the information available to the bidder and to the competing bidders. As usual, this diagram can be transformed into the equivalent inferential decision diagram of Figure 31.21.

This diagram will allow our bidder to make proper decisions in a situation susceptible to the winner's curse phenomenon. Our bidder believes generally that the more oil there is in the ground, the higher will be the competitive bids. He can model this belief using the relevance diagrams of Figure 31.22. If he knew the other bidders were bidding high, this would update his information about the amount of oil in the ground, and vice versa.

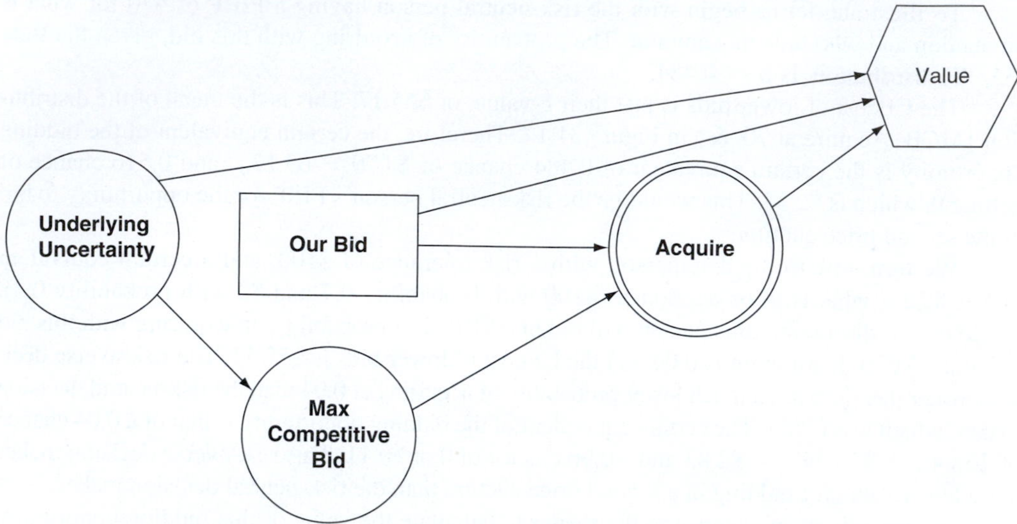

FIGURE 31.20 Assessed Decision Diagram for Winner's Curse

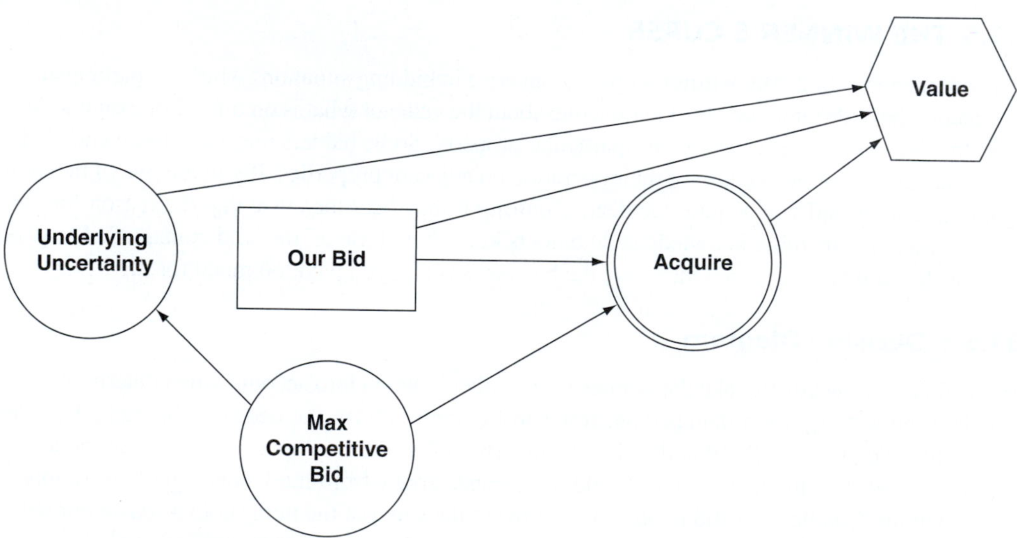

FIGURE 31.21 Inferential Decision Diagram for Winner's Curse

31.6.2 Relevance between Underlying Uncertainty and Maximum Competitive Bid

The underlying uncertainty, v, is the value of the actual amount of oil in the ground. We begin by assigning the probability distribution on v. For simplicity, we let v take on only five possible values: 0.5, 0.6, 0.7, 0.8, and 0.9. These numbers could be in units of tens of millions of barrels, for example. The probability assigned by the bidder to each of these values appears in Table 31.1. The most likely value is 0.7; higher or lower values of v decrease in probability as they depart from 0.7. The mean of v is 0.7.

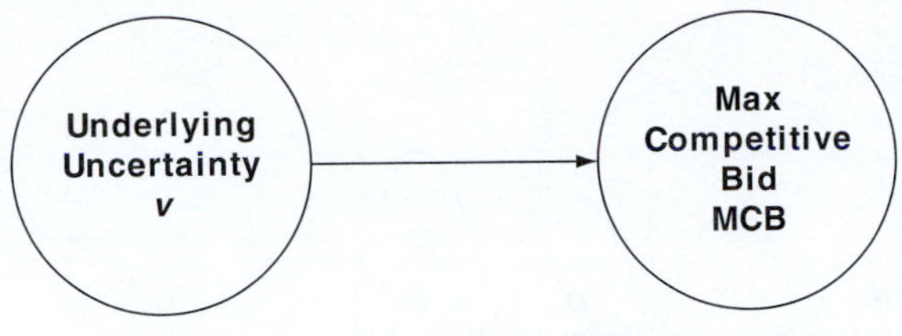

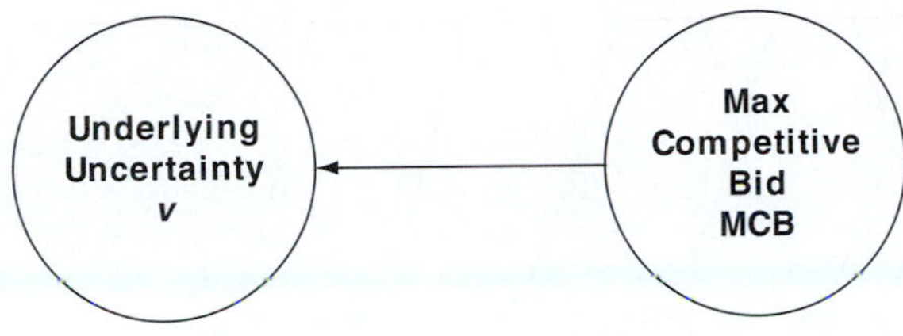

FIGURE 31.22 Relevance Diagrams to Represent Information

TABLE 31.1 Prior on v, $\{v|\&\}$

v	0.5	0.6	0.7	0.8	0.9
$\{v \mid \&\}$	0.025	0.25	0.45	0.25	0.025

Next, we consider our belief about the maximum competitive bid (MCB). To capture the relevance between v and MCB, the bidder must assign $\{\text{MCB} \mid v, \&\}$, which is the probability distribution of the maximum competitive bid given the actual value of v. Each of these five distributions labeled *A*, *B*, *C*, *D*, and *E* is modeled as a Beta distribution with parameters and mean shown in Table 31.2. Note that the mean of each distribution is increasing with the value of v.

These five distributions are plotted in Figure 31.23 on a normalized scale of [0,1].

Figure 31.24 shows a generic tree showing the conditional distributions given the distribution of v.

From Figure 31.24, we observe the trend that higher levels of v imply higher maximum competitive bids: the driver of the winner's curse. The distributions $\{v \mid \&\}$ and $\{\text{MCB} \mid v, \&\}$ complete the specification of the upper relevance diagram in Figure 31.22. Table 31.3 shows the conditional probability distributions of $\{\text{MCB} \mid v, \&\}$ for each value of v. Note that the most probable value of MCB—its mode—increases as v increases.

TABLE 31.2 Probability of MCB Given v, $\{\text{MCB}|v, \&\}$

Label	*A*	*B*	*C*	*D*	*E*
v	0.5	0.6	0.7	0.8	0.9
Beta r	15	20	25	30	35
Beta $n = r + k$	50	50	50	50	50
Mean: r/n	0.3	0.4	0.5	0.6	0.7

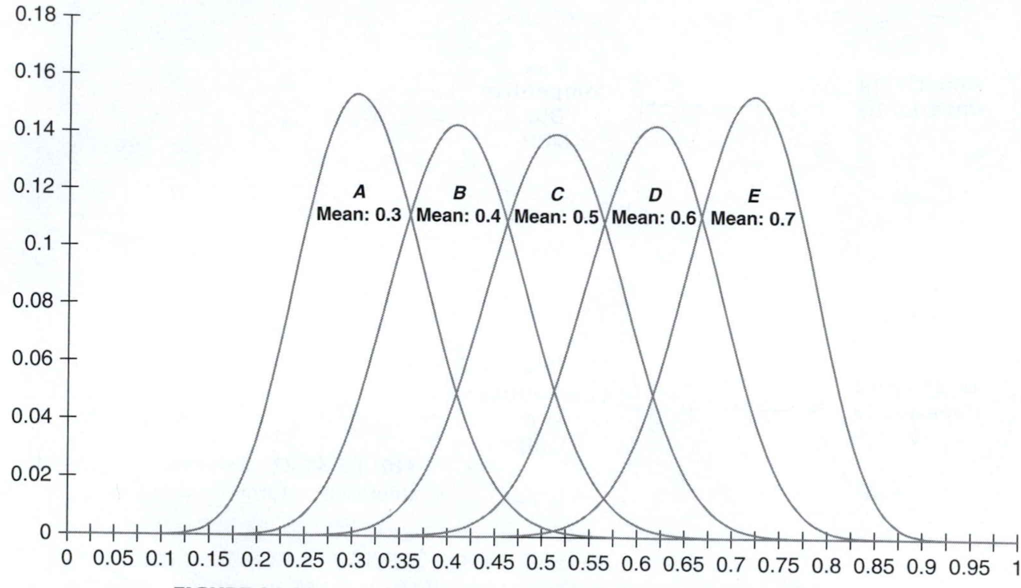

FIGURE 31.23 Conditional Probability Distributions of {MCB | *v*, &}

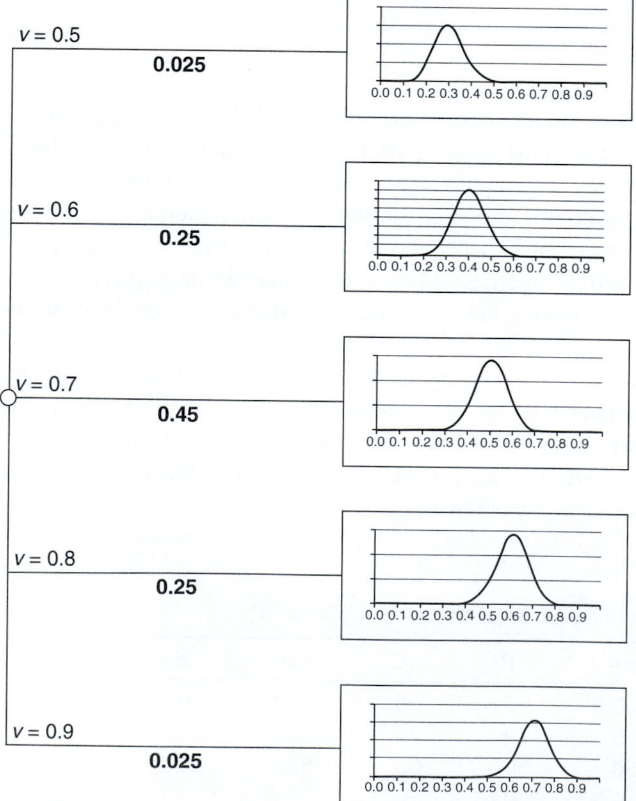

FIGURE 31.24 Tree showing Conditional Probability of Maximum Competitive Bid Given *v*

TABLE 31.3 Conditional Probability Distributions for MCB Given *v*

{*v* | &}	0.025	0.25	0.45	0.25	0.025
	0.5	0.6	0.7	0.8	0.9
MCB			**{MCB | *v*, &}**		
0.000	0	0	0	0	0
0.025	6.6E-13	1.3E-19	3.1E-27	1E-35	4.05E-45
0.050	9.5E-09	6.6E-14	5.8E-20	6.8E-27	9.78E-35
0.075	1.8E-06	1.1E-10	7.9E-16	8.1E-22	9.92E-29
0.100	5.5E-05	1.6E-08	5.6E-13	2.7E-18	1.61E-24
0.125	0.00062	6.2E-07	7.7E-11	1.3E-15	2.72E-21
0.150	0.00361	1E-05	3.8E-09	1.8E-13	1.08E-18
0.175	0.01323	9.7E-05	8.8E-08	1.1E-11	1.59E-16
0.200	0.03423	0.00058	1.2E-06	3.3E-10	1.11E-14
0.225	0.06741	0.00244	1.1E-05	6.3E-09	4.43E-13
0.250	0.10612	0.00773	6.8E-05	8E-08	1.12E-11
0.275	0.13829	0.01939	0.00033	7.3E-07	1.97E-10
0.300	0.15291	0.03975	0.00124	5.1E-06	2.53E-09
0.325	0.14603	0.06831	0.00384	2.8E-05	2.5E-08
0.350	0.12203	0.10026	0.00987	0.00013	1.97E-07
0.375	0.09003	0.12744	0.02158	0.00048	1.27E-06
0.400	0.05902	0.1417	0.04064	0.00153	6.85E-06
0.425	0.03452	0.13887	0.06668	0.00419	3.14E-05
0.450	0.01805	0.12054	0.09604	0.01001	0.000124
0.475	0.00843	0.09295	0.12212	0.02099	0.00043
0.500	0.00352	0.06374	0.13758	0.03884	0.001307
0.525	0.00131	0.03884	0.13758	0.06374	0.003519
0.550	0.00043	0.02099	0.12212	0.09295	0.008434
0.575	0.00012	0.01001	0.09604	0.12054	0.018046
0.600	3.1E-05	0.00419	0.06668	0.13887	0.03452
0.625	6.8E-06	0.00153	0.04064	0.1417	0.059024
0.650	1.3E-06	0.00048	0.02158	0.12744	0.090032
0.675	2E-07	0.00013	0.00987	0.10026	0.122026
0.700	2.5E-08	2.8E-05	0.00384	0.06831	0.146035
0.725	2.5E-09	5.1E-06	0.00124	0.03975	0.152908
0.750	2E-10	7.3E-07	0.00033	0.01939	0.138288
0.775	1.1E-11	8E-08	6.8E-05	0.00773	0.106116
0.800	4.4E-13	6.3E-09	1.1E-05	0.00244	0.067406
0.825	1.1E-14	3.3E-10	1.2E-06	0.00058	0.034235
0.850	0	1.1E-11	8.8E-08	9.7E-05	0.013226
0.875	0	1.8E-13	3.8E-09	1E-05	0.003609
0.900	0	0	7.7E-11	6.2E-07	0.00062
0.925	0	0	5.6E-13	1.6E-08	5.51E-05
0.950	0	0	0	1.1E-10	1.78E-06
0.975	0	0	0	6.6E-14	9.45E-09
1.000	0	0	0	0	6.56E-13
Sum	1	1	1	1	1

31.6.3 Joint Probability Distribution {MCB, *v*|& }

As usual, we reverse the direction of the arrow in Figure 31.22 by first calculating the joint probability distribution. Table 31.4 represents the joint probability distribution as {MCB, *v*|&}, which

TABLE 31.4 Joint Probability Distribution, {MCB, *v*|&}

MCB	0.5	0.6	0.7	0.8	0.9	{MCB, *v*|&}
0.000	0	0	0	0	0	
0.025	1.64E-14	3.2E-20	1.41E-27	2.5671E-36	1.01E-46	
0.050	2.36E-10	1.64E-14	2.6E-20	1.7114E-27	2.45E-36	
0.075	4.45E-08	2.63E-11	3.56E-16	2.0131E-22	2.48E-30	
0.100	1.38E-06	3.9E-09	2.52E-13	6.8267E-19	4.04E-26	
0.125	1.55E-05	1.54E-07	3.49E-11	3.2909E-16	6.79E-23	
0.150	9.02E-05	2.61E-06	1.7E-09	4.5822E-14	2.7E-20	
0.175	0.000331	2.43E-05	3.97E-08	2.6877E-12	3.96E-18	
0.200	0.000856	0.000145	5.41E-07	8.317E-11	2.78E-16	
0.225	0.001685	0.000609	4.83E-06	1.5699E-09	1.11E-14	
0.250	0.002653	0.001933	3.07E-05	1.9956E-08	2.81E-13	
0.275	0.003457	0.004846	0.000148	1.833E-07	4.92E-12	
0.300	0.003823	0.009936	0.000559	1.281E-06	6.33E-11	
0.325	0.003651	0.017077	0.001726	7.0812E-06	6.26E-10	
0.350	0.003051	0.025066	0.004441	3.1895E-05	4.93E-09	
0.375	0.002251	0.031859	0.009709	0.00011978	3.18E-08	
0.400	0.001476	0.035426	0.01829	0.00038186	1.71E-07	
0.425	0.000863	0.034717	0.030006	0.001048	7.86E-07	
0.450	0.000451	0.030135	0.043218	0.0025032	3.11E-06	
0.475	0.000211	0.023237	0.054954	0.00524696	1.07E-05	
0.500	8.8E-05	0.015934	0.061913	0.00971071	3.27E-05	
0.525	3.27E-05	0.009711	0.061913	0.01593424	8.8E-05	
0.550	1.07E-05	0.005247	0.054954	0.02323694	0.000211	
0.575	3.11E-06	0.002503	0.043218	0.03013532	0.000451	
0.600	7.86E-07	0.001048	0.030006	0.03471748	0.000863	
0.625	1.71E-07	0.000382	0.01829	0.03542597	0.001476	
0.650	3.18E-08	0.00012	0.009709	0.03185904	0.002251	
0.675	4.93E-09	3.19E-05	0.004441	0.02506586	0.003051	
0.700	6.26E-10	7.08E-06	0.001726	0.01707741	0.003651	
0.725	6.33E-11	1.28E-06	0.000559	0.00993637	0.003823	
0.750	4.92E-12	1.83E-07	0.000148	0.00484648	0.003457	
0.775	2.81E-13	2E-08	3.07E-05	0.00193277	0.002653	
0.800	1.11E-14	1.57E-09	4.83E-06	0.00060923	0.001685	
0.825	2.8E-16	8.32E-11	5.41E-07	0.00014485	0.000856	
0.850	0	2.69E-12	3.97E-08	2.4309E-05	0.000331	
0.875	0	4.58E-14	1.7E-09	2.6114E-06	9.02E-05	
0.900	0	0	3.49E-11	1.5432E-07	1.55E-05	
0.925	0	0	2.52E-13	3.8973E-09	1.38E-06	
0.950	0	0	0	2.6272E-11	4.45E-08	
0.975	0	0	0	1.6431E-14	2.36E-10	
1.000	0	0	0	0	1.64E-14	

is obtained by multiplying each cell in Table 31.3 with the corresponding probability of $\{v \mid \&\}$. For example, the cell corresponding to MCB $= 0.5$ and $v = 0.5$ in Table 31.4 (with a joint probability of 8.8×10^{-5}) is obtained by multiplying the cell corresponding to the conditional probability of MCB $= 0.5$ given $v = 0.5$ (namely, the conditional probability 0.00352) with the probability of $v = 0.5$ (which is 0.025).

31.6.4 Maximum Competitive Bid Distribution

The marginal distribution of $\{\text{MCB} \mid \&\}$ is obtained by summing the joint probability $\{\text{MCB}, v \mid \&\}$ over all possible values of v. This is obtained by summing horizontally all the rows in Table 31.4 to get Table 31.5.

TABLE 31.5 {MCB|&} Obtained by Summing Rows in Table 31.4 Horizontally

MCB	{MCB|&}	MCB	{MCB|&}
0.000	0	0.525	0.087679
0.025	1.64E-14	0.550	0.083659
0.050	2.36E-10	0.575	0.076311
0.075	4.45E-08	0.600	0.066636
0.100	1.38E-06	0.625	0.055573
0.125	1.56E-05	0.650	0.043939
0.150	9.28E-05	0.675	0.03259
0.175	0.000355	0.700	0.022461
0.200	0.001001	0.725	0.01432
0.225	0.002299	0.750	0.008451
0.250	0.004616	0.775	0.004616
0.275	0.008451	0.800	0.002299
0.300	0.01432	0.825	0.001001
0.325	0.022461	0.850	0.000355
0.350	0.03259	0.875	9.28E-05
0.375	0.043939	0.900	1.56E-05
0.400	0.055573	0.925	1.38E-06
0.425	0.066636	0.950	4.45E-08
0.450	0.076311	0.975	2.36E-10
0.475	0.083659	1.000	1.64E-14
0.500	0.087679		

Figure 31.25 shows a plot of $\{\text{MCB} \mid \&\}$.

Figure 31.26 compares $\{\text{MCB} \mid \&\}$ obtained in Table 31.4 to a Beta (10, 10) distribution. This Beta (10, 10) distribution was the maximum competitive bid distribution for Auction 1—a problem equivalent to this except for the winner's curse effect.

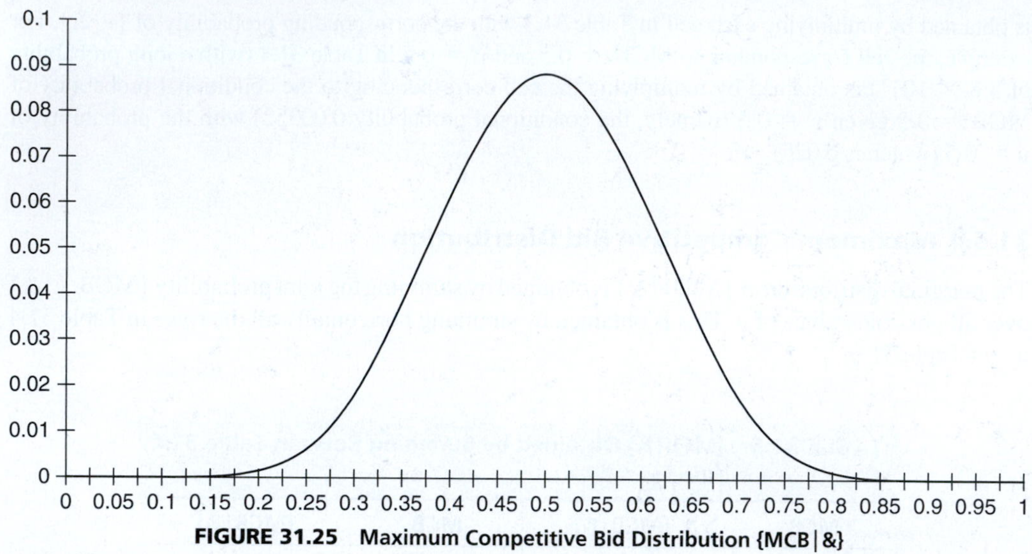

FIGURE 31.25 Maximum Competitive Bid Distribution {MCB | &}

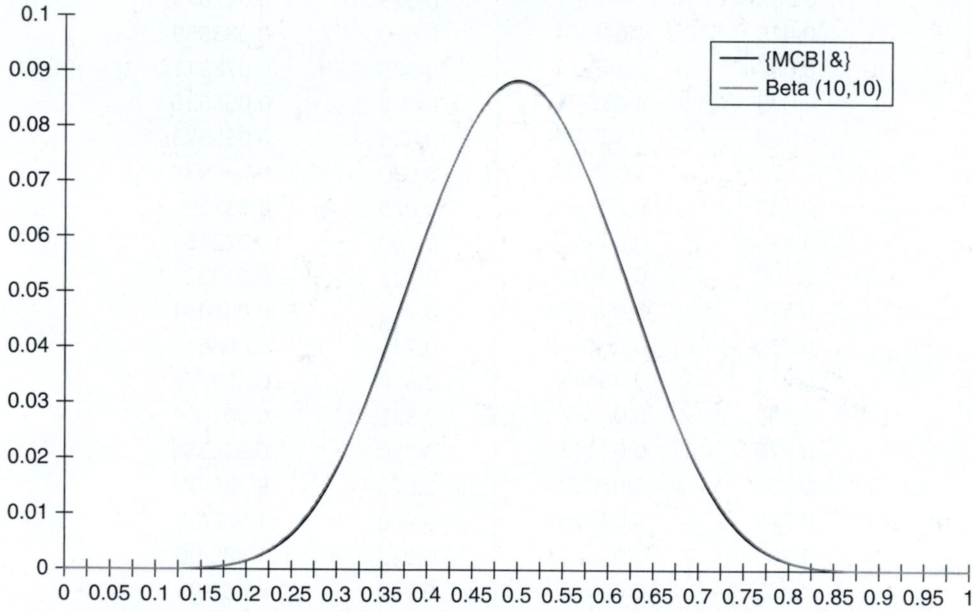

FIGURE 31.26 Comparison of Maximum Competitive Bid Distributions

31.6.5 Updating Distribution of Underlying Uncertainty Given Maximum Competitive Bid

Proceeding with the winner's curse analysis, for the lower relevance diagram in Figure 31.23, we can compute the conditional probability $\{v|\text{MCB}, \&\}$, since

$$\{v|\&\}\,\{\text{MCB}|v, \&\} = \{v, \text{MCB}|\&\} = \{\text{MCB}|\&\}\,\{v|\text{MCB}, \&\}$$

Therefore,

$$\{v|\text{MCB}, \&\} = \frac{\{\text{MCB}, v|\&\}}{\{\text{MCB}|\&\}}$$

The conditional probability of $\{v \mid \text{MCB}, \&\}$, $\{\text{MCB}, v \mid \&\}$ is obtained by dividing each cell in Table 31.4 by its corresponding probability $\{\text{MCB} \mid \&\}$ in Table 31.5. The conditional probabilities are shown in Table 31.6. Each row represents the conditional probability distribution $\{v \mid \text{MCB}, \&\}$.

For example, the row corresponding to MCB = 0.5 shows that when MCB = 0.5, $\{v = 0.5 \mid \text{MCB} = 0.5, \&\} = 0.001003372$, $\{v = 0.6 \mid \text{MCB} = 0.5, \&\} = 0.181734244$, etc.

TABLE 31.6 Rows Representing Conditional Probability Distributions {v|MCB, &}

MCB	0.5	0.6	0.7	0.8	0.9	
	{v|MCB}					
0.000						
0.025	0.99999805	1.9502E-06	8.59235E-14	1.56531E-22	6.18E-33	1
0.050	0.999930519	6.94805E-05	1.10067E-10	7.24221E-18	1.04E-26	1
0.075	0.999410031	0.000589961	7.99406E-09	4.52051E-15	5.57E-23	1
0.100	0.997180881	0.002818937	1.82415E-07	4.93775E-13	2.92E-20	1
0.125	0.99013274	0.009865031	2.22907E-06	2.10376E-11	4.34E-18	1
0.150	0.971854192	0.02812754	1.82672E-05	4.93558E-10	2.91E-16	1
0.175	0.931411963	0.068476132	0.000111897	7.57082E-09	1.12E-14	1
0.200	0.854789263	0.144670655	0.000539999	8.30655E-08	2.78E-13	1
0.225	0.732923343	0.264976339	0.002099635	6.82801E-07	4.82E-12	1
0.250	0.574669266	0.418673879	0.006652532	4.32293E-06	6.08E-11	1
0.275	0.409065316	0.573448235	0.017464759	2.16888E-05	5.82E-10	1
0.300	0.266953473	0.693892778	0.039064286	8.94586E-05	4.42E-09	1
0.325	0.162540762	0.76030483	0.076839115	0.000315265	2.79E-08	1
0.350	0.093607782	0.769137026	0.136276355	0.000978686	1.51E-07	1
0.375	0.051225757	0.725076534	0.220970905	0.002726081	7.23E-07	1
0.400	0.02655243	0.637465134	0.329108067	0.006871288	3.08E-06	1
0.425	0.012950955	0.521005273	0.450304582	0.0157274	1.18E-05	1
0.450	0.005911885	0.394903911	0.566340626	0.032802814	4.08E-05	1
0.475	0.002520314	0.277756997	0.656876011	0.062718224	0.000128	1
0.500	0.001003372	0.181734244	0.706136579	0.110753201	0.000373	1
0.525	0.000372603	0.110753193	0.706136601	0.18173423	0.001003	1
0.550	0.000128454	0.062718224	0.656876011	0.277756997	0.00252	1
0.575	4.07642E-05	0.032802814	0.566340626	0.394903911	0.005912	1
0.600	1.17898E-05	0.0157274	0.450304582	0.521005273	0.012951	1
0.625	3.08094E-06	0.006871288	0.329108067	0.637465134	0.026552	1
0.650	7.22798E-07	0.002726081	0.220970905	0.725076534	0.051226	1
0.675	1.51203E-07	0.000978686	0.136276355	0.769137026	0.093608	1
0.700	2.78603E-08	0.000315265	0.076839115	0.76030483	0.162541	1
0.725	4.41887E-09	8.94586E-05	0.039064286	0.693892778	0.266953	1
0.750	5.81974E-10	2.16888E-05	0.017464759	0.573448235	0.409065	1
0.775	6.08183E-11	4.32293E-06	0.006652532	0.418673879	0.574669	1
0.800	4.81786E-12	6.82801E-07	0.002099635	0.264976339	0.732923	1
0.825	2.79979E-13	8.30655E-08	0.000539999	0.144670655	0.854789	1
0.850	0	7.57081E-09	0.000111897	0.068476132	0.931412	1
0.875	0	4.93583E-10	1.82672E-05	0.02812754	0.971854	1
0.900	0	0	2.22907E-06	0.009865031	0.990133	1
0.925	0	0	1.82414E-07	0.002818936	0.997181	1
0.950	0	0	0	0.000589961	0.99941	1
0.975	0	0	0	6.9535E-05	0.99993	1
1.000	0	0	0	0	1	1

31.6.6 Updating Distribution of Underlying Uncertainty Given Knowledge of Acquiring at a Bid, *b*

The distribution $\{v \mid \text{MCB}, \&\}$ shows how the bidder's belief on the value of the deal changes as a result of knowing the MCB. Of course, in many bidding situations, he may not know the MCB, but only that he did or did not acquire it with his bid. If the bidder acquires, then he does not know the MCB, but only that the MCB was less than his bid. Then, the bidder can derive a new distribution on the value of the deal, not given the maximum competitive bid, but given that it was less than his bid. Therefore, the logical sequence of thought for bidder is as follows: What is the chance of acquiring for a given bid, what do I learn about the value of the deal by acquiring at this bid, and finally, how well did I do in the presence of the winner's curse phenomenon?

To see how this works, we first compute the cumulative probability distribution on $\{\text{MCB} \mid \&\}$ from the probability distribution in Table 31.5 and plot it as Figure 31.27.

Now, we will determine the effect of bidding 0.5. From this distribution, the probability of acquiring is $\{\text{MCB} \le 0.5 \mid \&\} = 0.544$.

If the deal is acquired at this bid, then the MCB must have been less than 0.5. Given that the MCB was less than 0.5, the probability distribution on value given this information, $\{v \mid \text{MCB} < 0.5, \&\}$, needs to be determined. This is shown in Table 31.7.

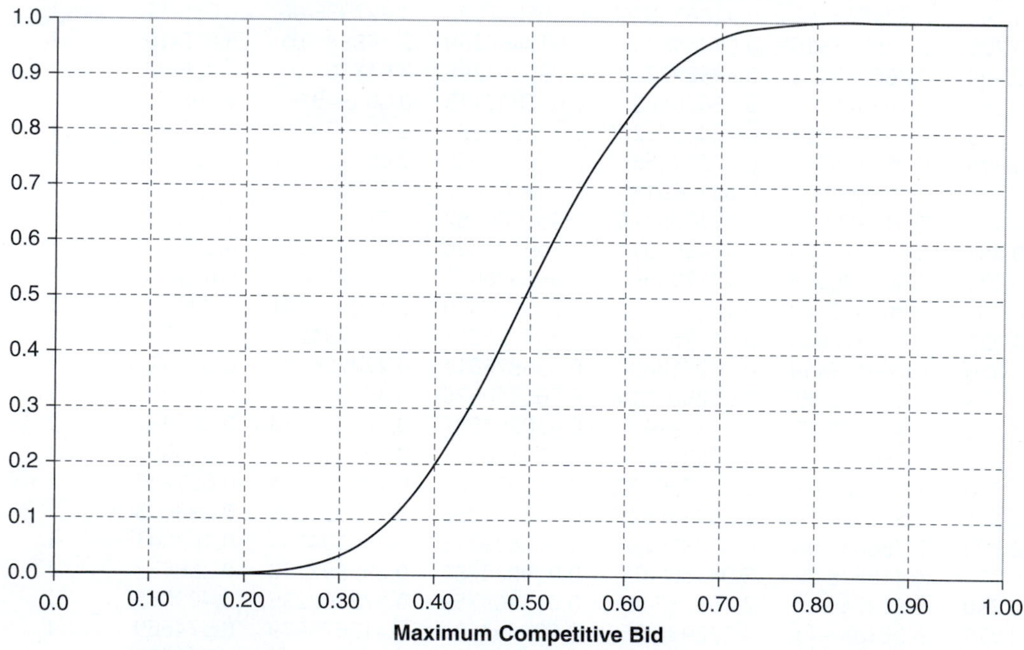

FIGURE 31.27 Maximum Competitive Bid Cumulative Probability

TABLE 31.7 Probability Distribution on Value Given Maximum Competitive Bid < 0.5

v	0.5	0.6	0.7	0.8	0.9	1
$\{v \mid \text{MCB} < 0.5, \&\}$	0.049905	0.461898	0.45	0.038102	9.5E-05	1

This table is computed from

$$\{v|\text{MCB} \le c, \&\} = \frac{\sum_{\text{MCB}=0}^{c} \{v, \text{MCB}|\&\}}{\{\text{MCB} \le c|\&\}} = \frac{\sum_{\text{MCB}=0}^{c} \{v, \text{MCB}|\&\}}{\sum_{\text{MCB}=0}^{c} \{\text{MCB}|\&\}}$$

For example, the probability

$$\{v = 0.6|\text{MCB} \le 0.5, \&\} = \frac{\sum_{i=0}^{0.5} \{v = 0.6, \text{MCB} = i|\&\}}{\{\text{MCB} \le 0.5|\&\}}$$

The numerator is obtained by summing the cells of the column corresponding to $v = 0.6$ of the joint probability table (Table 31.4) from MCB $= 0$ to MCB $= 0.5$. The denominator sums the marginal probability of Table 31.5 from MCB $= 0$ to MCB $= 0.5$. Table 31.8 determines the conditional probability $\{v|\text{MCB} \le c, \&\}$ for all possible values of v and c. Each row represents the conditional distribution on value, v, given we know MCB $\le c$ (i.e., that we acquired at a bid amount c).

TABLE 31.8 $\{v|\text{MCB} \le c, \&\}$

Bid	0.5	0.6	0.7	0.8	0.9	Sum
0.000						
0.025	0.999998	1.95E-06	8.59E-14	1.57E-22	6.18E-33	1
0.050	0.999931	6.95E-05	1.1E-10	7.24E-18	1.04E-26	1
0.075	0.999413	0.000587	7.95E-09	4.5E-15	5.54E-23	1
0.100	0.997251	0.002749	1.77E-07	4.78E-13	2.83E-20	1
0.125	0.990728	0.00927	2.06E-06	1.93E-11	3.98E-18	1
0.150	0.974785	0.025199	1.57E-05	4.2E-10	2.47E-16	1
0.175	0.941666	0.058245	8.92E-05	5.88E-09	8.59E-15	1
0.200	0.882337	0.117266	0.000397	5.86E-08	1.93E-13	1
0.225	0.791103	0.20746	0.001437	4.4E-07	3.02E-12	1
0.250	0.671899	0.32379	0.004309	2.58E-06	3.49E-11	1
0.275	0.539938	0.449136	0.010914	1.22E-05	3.1E-10	1
0.300	0.414458	0.56164	0.023854	4.77E-05	2.2E-09	1
0.325	0.308919	0.644869	0.046052	0.00016	1.29E-08	1
0.350	0.22752	0.691849	0.080161	0.000469	6.52E-08	1
0.375	0.168	0.703067	0.127701	0.001231	2.87E-07	1
0.400	0.125673	0.683437	0.18797	0.002919	1.12E-06	1
0.425	0.095908	0.640545	0.257242	0.006301	3.94E-06	1
0.450	0.075012	0.583511	0.32901	0.012454	1.25E-05	1
0.475	0.060304	0.521474	0.395533	0.022653	3.6E-05	1
0.500	0.049905	0.461898	0.45	0.038102	9.5E-05	1
0.525	0.042515	0.409509	0.488214	0.059531	0.000231	1
0.550	0.037233	0.366293	0.509232	0.086726	0.000516	1
0.575	0.033437	0.332255	0.515061	0.11818	0.001067	1
0.600	0.030702	0.306352	0.509762	0.151145	0.002039	1
0.625	0.02874	0.287219	0.49822	0.182215	0.003605	1
0.650	0.027358	0.27354	0.484889	0.208318	0.005895	1
0.675	0.026416	0.264154	0.472884	0.22763	0.008916	1
0.700	0.025804	0.258037	0.463703	0.239979	0.012477	1
0.725	0.025428	0.25428	0.457518	0.246591	0.016184	1

TABLE 31.8 $\{v|MCB \le c, \&\}$ **(continued)**

Bid	0.5	0.6	0.7	0.8	0.9	Sum
0.750	0.025211	0.252113	0.453767	0.249376	0.019532	1
0.775	0.025094	0.250945	0.451695	0.250161	0.022104	1
0.800	0.025037	0.250367	0.45066	0.250195	0.023741	1
0.825	0.025012	0.250116	0.450209	0.250089	0.024574	1
0.850	0.025003	0.250027	0.450049	0.250025	0.024896	1
0.875	0.025	0.250004	0.450008	0.250004	0.024984	1
0.900	0.025	0.25	0.450001	0.25	0.024999	1
0.925	0.025	0.25	0.45	0.25	0.025	1
0.950	0.025	0.25	0.45	0.25	0.025	1
0.975	0.025	0.25	0.45	0.25	0.025	1
1.000	0.025	0.25	0.45	0.25	0.025	1

31.6.7 The Value of the Underlying Uncertainty Given You Acquired

The PIBP of the deal then can be calculated for any decision maker using the tools we presented earlier. For a deltaperson, we can simply calculate the certain equivalent of the deal instead of the PIBP (which is more complicated numerically).

The PIBP can now be calculated from the distribution of v given we acquire at any bid amount. For example, the e-value of the bidding item given we acquire at a bid of 0.5 is equal to

$$0.5 \times 0.049905 + 0.6 \times 0.461898 + 0.7 \times 0.45 + 0.8 \times 0.038102 + 0.9 \times 9.5E - 05 = 0.647$$

which corresponds to the e-value of v from Table 31.8. Similarly, the PIBP for an exponential decision maker with a normalized risk tolerance of 1 is 0.645.

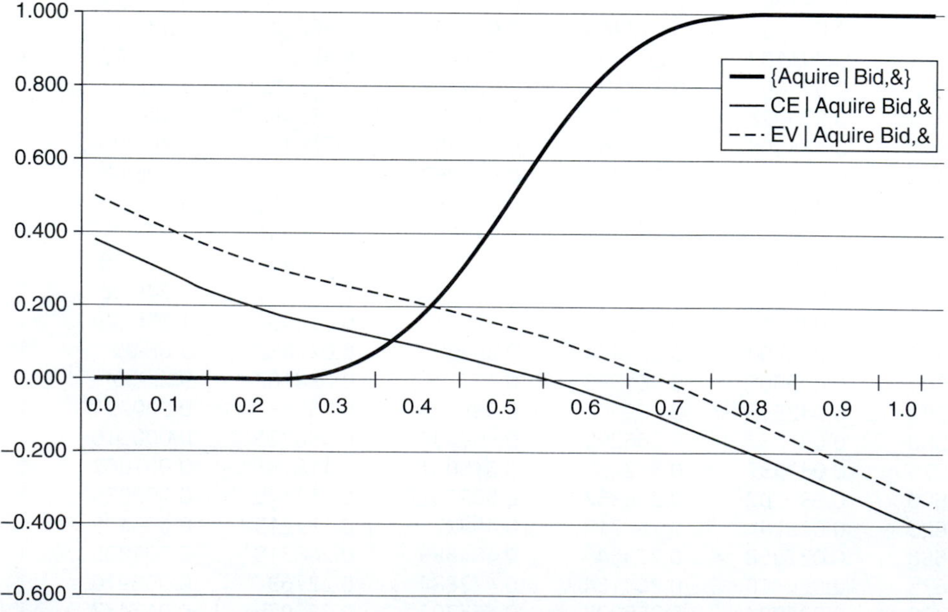

FIGURE 31.28 The Opposing Forces of Bidding, {MCB | &} and PIBP Given Acquire at a Given Bid

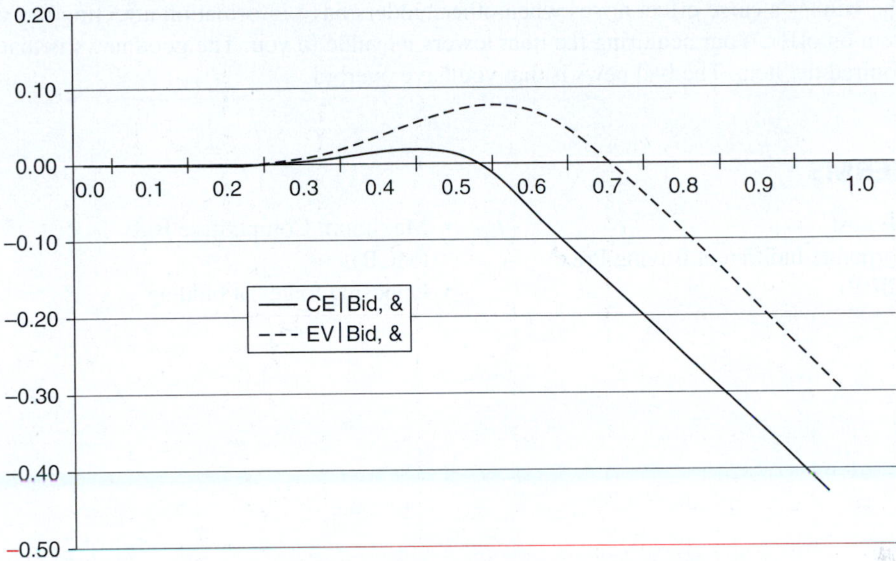

FIGURE 31.29 Certain Equivalent and EV of Bids

The implications of the existence of the curse become clear when we view the opposing forces of bidding in Figure 31.28. The curves of EV | Acquire, Bid and CE | Acquire, Bid are obtained from the distribution of v for any bid.

As before, increasing the bid increases the chance of acquiring, but it decreases the value of acquiring. However, the decrease in value is no longer linear, as it was in Figure 31.7. At the bid value of 0.5, the probability of acquiring is 0.544 (obtained from {MCB | &}). The EV of what would be acquired after paying the bid of 0.5 is now 0.147.

Figure 31.29 shows how the value of this bidding opportunity depends on the amount bid for both the risk-averse and risk-neutral decision makers.

We find that the best bid for the risk-averse person is 0.45 with a corresponding certain equivalent of the opportunity of 0.018. The risk-neutral person would bid 0.52 and have a certain equivalent of 0.077.

31.7 SUMMARY

You should never bid higher than your PIBP in any auction. The PIBP plays an important role.

- In first price auctions like Auctions 1 and 3, bidding your PIBP is pointless because you cannot gain by doing so. The difference between what you should bid and your PIBP for the item is determined by your risk aversion and your belief about the competitive bids.
- For Auction 2, the second price auction, you should bid your PIBP.
- For Auction 4, you bid up to your PIBP.

If knowing the other people's bids updates your PIBP for the item on offer, then acquiring the item will update your PIBP, because you know that the maximum competitive bid must be less than your own bid.

The winner's curse effect arises when other bidders have information affecting the value of the item on offer. Your acquiring the item lowers its value to you. The good news is that you have acquired the item. The bad news is that you have overbid.

KEY TERMS

- CE cost
- Corporate Indifferent Buying Price (CIBP)
- Maximum Competitive Bid (MCB)
- Opposing forces of bidding

PROBLEMS

Problems marked with an asterisk (*) are considered more challenging.

Read the following description of a sealed bid auction and then answer the related Problems 1 through 6.

Baragon considers placing a sealed bid in an on-line auction for a rare Gamera Action Card. The graphs represent Baragon's beliefs about this auction. Baragon is risk-neutral over the range of values below.

1. What is Baragon's PIBP for the card?
2. Baragon considers a bid of $25. Evaluate this bid from the perspective of each of the graphs above. Is it a reasonable bid?

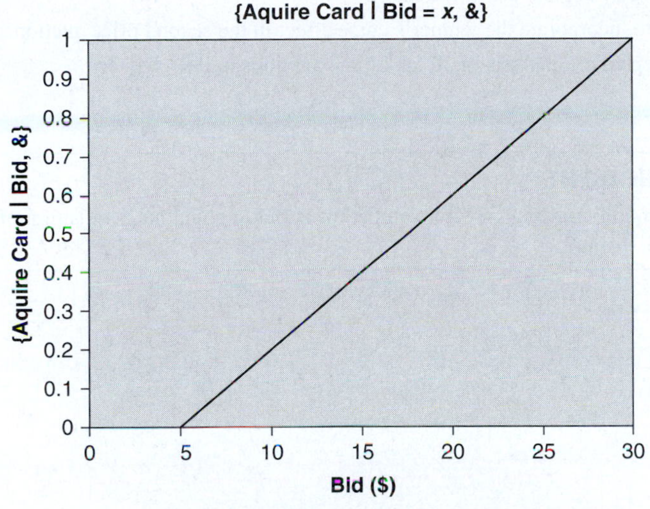

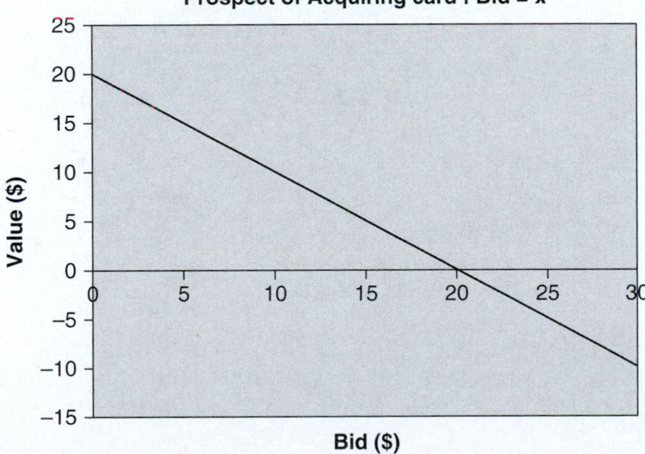

3. Baragon considers a bid of $35. Evaluate this bid from the perspective of each of the graphs above. Is it a reasonable bid?
4. Baragon considers a bid of $1. Evaluate this bid from the perspective of each of the graphs above. Is it a reasonable bid?

5. Sketch a graph of the certain of equivalent of each bid vs. the bid amount. Label any significant values on the graph.

6. Estimate Baragon's optimal bid for the card. Explain.

***7.** Value of a Second Price Auction: Determine the value of a second price auction for a logarithmic decision maker with initial wealth equal to $100. Use the same parameters and distributions as those in Section 31.6 of this chapter. Is the value higher or lower than that for a deltaperson with risk tolerance equal to $100?

***8.** Bidding with a fixed PIBP: A deltaperson is bidding on a $100 bill. His maximum competitive bid distribution, is Beta (10, 10) on a scale 0 to 100. Plot the optimal bid versus the risk-aversion coefficient. Does the optimal bid increase or decrease with risk aversion?

***9.** Repeat Problem 8 if the item on offer is a certificate offering $100 with a probability of 0.7 and 0 with a probability of 0.3. Explain any differences.

***10.** Build a model to incorporate the winner's curse effect in the second price auction.

***11.** On your own repeat the analysis of all tables and auctions in this chapter.

FOOD FOR THOUGHT

How would you modify the analysis if a competitor is bidding and his acquisition of the item on offer has negative value to you?

CHAPTER

32

Evaluating, Scaling, and Sharing Uncertain Deals

CHAPTER CONCEPTS

After reading this chapter, you will be able to explain the following concepts:

- Optimal scaling of uncertain deals
- Optimal share in an investment
- Optimal risk aversion for a partnership
- The surety of a deal
- Covariance and correlation between two deals
- Optimal share in a portfolio

32.1 INTRODUCTION

Sometimes, an investment opportunity is so large in its financial requirements that owning a piece of it is the only way an investor would invest. For example, investing in a major real estate project might be more appealing if multiple shareholders were involved. In this case, the investor needs to ask what fraction of the investment would be most desirable. The investor might also ask which partner to add to a partnership to make the investment more appealing to everybody involved. This chapter discusses the optimal share in an investment and the optimal risk aversion of a partnership for a given investment.

32.2 SCALING AND SHARING RISK

Often, people or businesses have the opportunity to invest in an uncertain financial deal, such as the purchase of a stock with a positive *e*-value of net present value at the given purchase price. As the number of shares purchased increases, the *e*-value of the deal will increase, but the uncertainty in the resulting payoff will also increase. The is due to the larger variance resulting from owning many shares.

A risk-neutral investor will always find owning more shares of a stock with positive *e*-value more desirable than owning less because of the larger *e*-value of the investment. But a risk-averse investor will need to consider the certain equivalent of the deal and the increase in its variance. Even a risk-neutral decision maker may become risk-averse as the payoffs under consideration increase. The risk-averse investor's most important consideration will be how many shares to buy in order to get the best deal; the highest certain equivalent.

The procedure for finding how many shares of stock to purchase, or identifying the best partnership fraction, is known as the process of **scaling** the risk to suit the investor. A related problem is that of a partnership. If the investment is so large, like the real estate investment, that no single individual in a partnership could possibly take it on, then how should the partnership divide the investment among the partners so that each one will be comfortable with their investment? We call this **sharing** the risk among partners.

This chapter discusses both scaling and sharing of risk. We illustrate the main results by referring back to the *approximate expression* for the certain equivalent of an uncertain deal that we first discussed back in Chapter 24. Related results also can be derived using the exact expression for the certain equivalent if needed. The approximate expression provides several insights into optimal scaling and sharing for a given investment and partnership. We first use the approximate expression to determine the best fraction or multiple of the deal that can be obtained by scaling its monetary payoffs. We then find the most desirable percentages of deals when multiple people form a syndicate to share the same deal. Finally, we determine the optimal scaling percentages for a portfolio of uncertain deals.

32.3 SCALING AN UNCERTAIN DEAL

As we have discussed in Chapter 24, the certain equivalent, $\tilde{x}$, of the monetary deal with mean, $\bar{x}$, and second central moment (variance), x^v, to a deltaperson with risk tolerance, ρ, is approximately given by

$$\text{Certain equivalent} \approx \text{First moment} - \frac{1}{2}\frac{\text{Variance}}{\text{Risk tolerance}}$$

or equivalently,

$$\tilde{x} \approx \bar{x} - \frac{x^v}{2\rho}$$

Suppose that a deltaperson can choose any fraction or multiple of this deal, m. For example, instead of receiving any prospect, x, he will receive mx.

If the deltaperson chooses to own mx, then the mean of what he receives will be $m\bar{x}$ and the variance will be m^2x^v. This is so because, as we discussed in Chapter 24, a deal with mx_i for each prospect has an e-value equal to

$$\sum_{i=1}^{n} p_i(mx_i) = m\sum_{i=1}^{n} p_i x_i = m\bar{x}$$

The second moment of this modified deal is

$$\sum_{i=1}^{n} p_i(mx_i)^2 = m^2\sum_{i=1}^{n} p_i(x_i)^2$$

The variance of such a deal is

$$m^2\sum_{i=1}^{n} p_i(x_i)^2 - (m\bar{x})^2 = m^2\Big[\sum_{i=1}^{n} p_i(x_i)^2 - \bar{x}^2\Big] = m^2x^v$$

The certain equivalent of what he receives, $\tilde{x}(m)$, therefore will be approximately

$$\tilde{x}(m) \approx m\bar{x} - \frac{m^2x^v}{2\rho}$$

Figure 32.1 shows the approximate certain equivalent as a function of m.

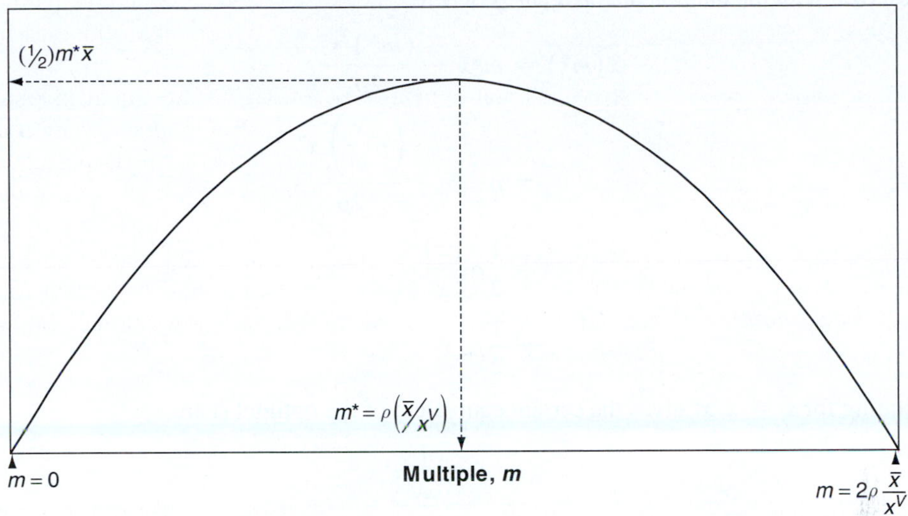

FIGURE 32.1 Sensitivity of Approximate Certain Equivalent to Multiple, *m*

Note *The certain equivalent will be 0 when m equals 0 (since he owns 0% of the deal) and is 0 again when*

$$\bar{x} - (1/2)(1/\rho)\,m x^{v} = 0$$

or

$$m = 2\rho\frac{\bar{x}}{x^{v}}$$

By inspection of the parabola, the maximum occurs when $m = m^*$ given by

$$m^* = \rho\frac{\bar{x}}{x^{v}}$$

We call the ratio $\bar{x}/x^{v}$, which is the ratio of the mean of the deal to its variance, *the surety* s_x of the deal i.e.,

$$s_x = \bar{x}/x^{v}$$

We see that, given the approximation, the optimum share of the deal can be expressed in terms of the surety as

$$m^* = \rho s_x$$

We also see that the certain equivalent becomes 0 when the multiple $m = 2m^* = 2\rho s_x$

Having calculated the optimal share, m^*, we now substitute to get the certain equivalent of this best deal with the multiple m^* as $\tilde{x}(m^*)$, where

$$\tilde{x}(m^*) = m^*\bar{x} - \frac{(m^*)^2 x^v}{2\rho}$$

$$= \rho\frac{\bar{x}}{x^v}\bar{x} - \frac{\left(\rho\frac{\bar{x}}{x^v}\right)^2 x^v}{2\rho}$$

$$= \frac{1}{2}\rho\frac{(\bar{x})^2}{x^v}$$

$$= \frac{1}{2}\rho s_x \bar{x}$$

Substituting for $s_x = \bar{x}/x^v$ gives the certain equivalent of the optimal share,

$$\tilde{x}(m^*) \approx \frac{1}{2}m^*\bar{x}$$

Result: *The certain equivalent of the deal with the optimal share is approximately one-half the optimal share times the mean of the original deal.*

The concept of scaling uncertain deals allows us to take some fraction (or multiple) of these deals that we may otherwise find unattractive if we had to accept the deal as a whole. We demonstrate this result through the following example.

EXAMPLE 32.1 Risk Scaling

Suppose that a deltaperson with a risk tolerance, $\rho = 2000$ faces a deal x with probability one-quarter of getting \$2,000, one-half of getting nothing, and one-quarter of losing \$1,000, as shown in Figure 32.2; and we write it as [114,2000; 112,0; 114 −1000].

By direct calculation, the certain equivalent of this deal to the person is −\$8.28, so it is undesirable. However, a portion of the deal will be attractive, and we will now find what that share should be.

The mean of the deal is equal to

$$\bar{x} = \frac{1}{4}\cdot(2000) + \frac{1}{2}\cdot 0 + \frac{1}{4}\cdot(-1000) = \$250$$

The second moment $\bar{x^2}$ is

$$\bar{x^2} = \frac{1}{4}\cdot(2000)^2 + \frac{1}{2}\cdot 0^2 + \frac{1}{4}\cdot(-1000)^2 = 1{,}250{,}000$$

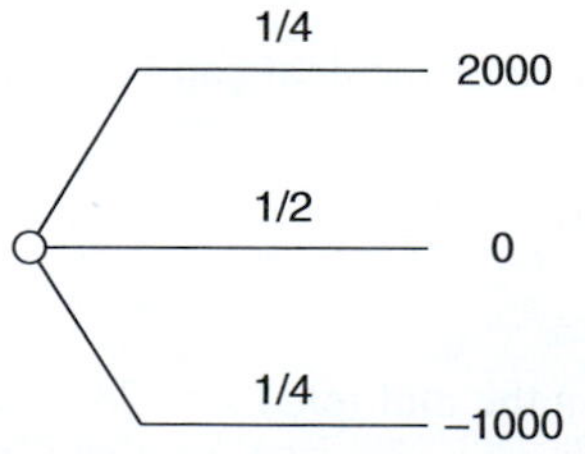

FIGURE 32.2 A Scaled Deal

The variance of the deal is

$$x^v = \bar{x^2} - \bar{x}^2 = 1{,}250{,}000 - (250)^2 = 1{,}187{,}500$$

The surety, s_x, is

$$s_x = \bar{x}/x^v = 2.105 \times 10^{-4}$$

and the optimum share of the deal, given the approximation, is

$$m^* = \rho s_x = 0.421$$

or about 42%.

The approximate certain equivalent the deltaperson will receive is

$$\tilde{x}(m^*) \approx (1/2)m^*\bar{x} = \$52.63$$

The certain equivalent of this share of the deal with no approximation is \$55.42.

Notice that while the original deal had a negative certain equivalent for the decision maker and was undesirable, receiving a 42% fraction of the deal is desirable, and the decision maker would even be willing to pay \$50 to receive this fraction. Therefore, scaling a deal can improve the value of a deal that a decision maker may be facing.

32.4 RISK SHARING OF UNCERTAIN DEALS

Suppose there is a deal described by a probability distribution on payoff x that has a positive mean $\bar{x}$, but whose variance x^v is so large that it has a negative certain equivalent for you. In this situation, we know that by scaling the deal you may increase your certain equivalent. But when you scale the deal, would another decision maker (or group of decision makers) be willing to accept the remaining portion?

Suppose you could form a syndicate to take the deal and share it in such a way that all members of the syndicate would have a positive certain equivalent for their share. If this can be achieved, it would be a nice way to participate in deals that you would not want to assume alone. We show how to do this and discuss the characteristics of the syndicate that would be needed next.

Let us suppose that you have a group of size n willing to form a syndicate to share the deal and that all in the group are deltapeople with individual positive risk tolerances ρ_i, $i = 1, 2, \ldots, N$. We shall now show when and how to form a syndicate that all of you will find attractive.

To determine the optimal share of each group member, you first compute a "group" risk tolerance ρ_g that is defined as the sum of the individual risk tolerances for members of the group. Thus,

$$\rho_g = \sum_{i=1}^{N} \rho_i$$

where ρ_i is the risk tolerance of member i.

If the deal has a positive certain equivalent $\tilde{x}_g$ using this group risk tolerance, then a profitable syndicate is possible. Furthermore, it can be shown that the sum of the certain equivalents of the syndicate's members will be maximized if each member receives a fraction of the deal f_i that is the ratio of i's risk tolerance ρ_i to the group risk tolerance ρ_g

$$f_i = \rho_i/\rho_g$$

Finally, the ith member's certain equivalent $\tilde{x}_i$ under this sharing rule will be just a fraction f_i of the groups certain equivalent,

$$\tilde{x}_i = f_i\tilde{x}_g$$

EXAMPLE 32.2 Risk Sharing

Consider again the deal we discussed in Figure 32.2 under risk scaling. A one-quarter probability of winning $2,000, one-half of winning 0, and one-quarter of losing $1,000, is now offered to a group of three deltapeople with risk tolerances

$$\rho_1 = 200, \rho_2 = 1000, \rho_3 = 2000$$

If each of these people were offered the deal individually, their exact certain equivalents would be

$\tilde{x}_1 = -725.42$, $\tilde{x}_2 = -193.43$, and $\tilde{x}_3 = -8.28$, respectively. None of them would want it.

Now suppose they form a syndicate with a group risk tolerance equal to the sum

$$\rho_g = \rho_1 + \rho_2 + \rho_3 = 3200$$

Using the approximate expression for the certain equivalent, the group certain equivalent is

$$\tilde{x}_g \approx 250 - \frac{1{,}187{,}500}{2(3200)} = 64.45$$

which is positive. The fraction of this deal that each member would receive would be

$$f_1 = {}^{\rho_1}/_{\rho_g} = \frac{200}{3200} = \frac{1}{16}$$

$$f_2 = {}^{\rho_2}/_{\rho_g} = \frac{1000}{3200} = \frac{5}{16}$$

$$f_3 = {}^{\rho_3}/_{\rho_g} = \frac{2000}{3200} = \frac{10}{16}$$

and their certain equivalents would be

$$\tilde{x}_1 = f_1\tilde{x}_g = 4.96$$
$$\tilde{x}_2 = f_2\tilde{x}_g = 24.78$$
$$\tilde{x}_3 = f_3\tilde{x}_g = 49.56$$

Each person receives a share of the deal that is scaled to that person's risk-taking preferences. Furthermore, the total wealth of the group is maximized.

Suppose, for example, that Person 1 simply gives his share to Person 2, so that Person 2's share is now $6/16$ or $3/8$ of the deal. Now, Person 2 faces the deal

$$[1/4, 750; 1/2, 0; 1/4, -375]$$

which has a certain equivalent of 18.33 to a person with a risk tolerance of 1000 like that of Person 2. We see that Person 1 is worse off after this transaction because he now has no share with a certain equivalent of zero rather than a share with a certain equivalent of 4.96. However, Person 2 is also worse off. His certain equivalent has fallen from 24.78 to 18.33. Person 3 still has a certain equivalent of 49.56. The total worth of all three has decreased to $49.56 + 18.33 = 67.89$, rather than the 79.29 provided by the original optimal shares.

32.4.1 Finding the Best Partner

Is there any way the syndicate of Example 32.2 could do better? To answer this question, let's return to risk scaling. What fraction of the deal would each person want if he could name his fraction?

Recall that the fraction of the deal anyone with risk tolerance ρ would want is $m^* = \rho s_x$, where $s_x = \bar{x}/x^v$. As we calculated in Example 32.2, $s_x = \bar{x}/x^v = 2.105 \times 10^{-4}$.

Taking the whole deal would be the best action for the whole group if the group risk tolerance ρ_g were such that $m^* = 1$, so

$$m^* = 1 = \rho_g s_x$$

or

$$\rho_g = \frac{1}{s_x} = \frac{1}{2.105 \times 10^{-4}} = 4751$$

A group with a risk tolerance of 4751 would take the whole deal. The group has three members with a total risk tolerance of 3200, which is 1551 shy. Suppose they find a fourth syndicate member with a risk tolerance of 1551. Then $\rho_g = 4751$, and the certain equivalent of the deal for the group is 118.45. The members would receive fractions of

$$f_1 = {}^{\rho_1}/_{\rho_g} = {}^{200}/_{4751} = 0.04$$
$$f_2 = {}^{\rho_2}/_{\rho_g} = {}^{1000}/_{4751} = 0.21$$
$$f_3 = {}^{\rho_3}/_{\rho_g} = {}^{2000}/_{4751} = 0.42$$
$$f_4 = {}^{\rho_4}/_{\rho_g} = {}^{1551}/_{4751} = 0.33$$

Their certain equivalents would be

$$x_1 = f_1\tilde{x}_g = 4.74$$
$$x_2 = f_2\tilde{x}_g = 24.87$$
$$x_3 = f_3\tilde{x}_g = 49.75$$
$$x_4 = f_4\tilde{x}_g = 39.08$$

Everyone is better off than he would have been without the new syndicate member. Note, however, that if the risk tolerance of the new member exceeds 1128, the certain equivalents of the original members of the group will decrease. If in the extreme they add a person who is risk-neutral, that person will take the whole deal and receive a certain equivalent of 250; the original members will have none of the deal and, hence, certain equivalents of 0.

When forming a syndicate, you should think carefully about your members. Ideally, the sum of risk tolerances of the group should be $\rho_g = {}^1/_{s_x}$.

32.5 OPTIMAL INVESTMENT IN A PORTFOLIO

Now, we discuss the optimal choices and percentages that a deltaperson should make when facing n uncertain deals. For example, the person may be facing n stocks or n different investments, and he is interested in the portfolio of the deals and the percentage of each deal that would maximize his certain equivalent. We will use the approximation for the certain equivalent in terms of the mean and variance of the deals. We will therefore calculate the mean and variance of the portfolio.

32.5.1 Mean Value of Portfolio

Consider choosing multiples $m_1, m_2, \ldots, m_n$ of n possible deals $x_1, x_2, \ldots, x_n$. For any given multiples $m_1, m_2, \ldots, m_n$, the total payoff, $X(m)$, of the portfolio is equal to that provided by the additive value function, so

$$X(m) = m_1 x_1 + m_2 x_2 + \cdots + m_n x_n$$

The mean payoff is equal to $X(m)$. Thus,

$$X(m) = m_1 \bar{x}_1 + m_2 \bar{x}_2 + \cdots + m_n \bar{x}_n$$

where $\bar{x}_1, \bar{x}_2, \ldots, \bar{x}_n$ are the mean values of deals $x_1, x_2, \ldots, x_n$.

32.5.2 Variance of Portfolio

The **variance** of $X(m)$ can be considered for two cases: when the deals are mutually irrelevant (the simple case) and when there is relevance between them.

1. Mutually Irrelevant Deals In this case, the variance expression is

$$\text{Variance of } X(m) = X(m)^v = m_1^2 x_1^v + m_2^2 x_2^v + \cdots + m_n^2 x_n^v = \sum_{i=1}^{n} m_i^2 x_i^v$$

For a deltaperson, if the deals are mutually irrelevant, then the certain equivalent of the portfolio is approximately

$$\begin{aligned}
\tilde{x}(m) &\approx \bar{X}(m) - \frac{1}{2\rho} X(m)^v \\
&= m_1 \bar{x}_1 + m_2 \bar{x}_2 + \cdots + m_n \bar{x}_n - \frac{1}{2\rho}\left[m_1^2 x_1^v + m_2^2 x_2^v + \cdots + m_n^2 x_n^v\right] \\
&= \left(m_1 \bar{x}_1 - \frac{m_1^2 x_1^v}{2\rho}\right) + \left(m_2 \bar{x}_2 - \frac{m_2^2 x_2^v}{2\rho}\right) + \cdots + \left(m_n \bar{x}_n - \frac{m_n^2 x_n^v}{2\rho}\right) \\
&= m_1 \tilde{x}_1 + m_2 \tilde{x}_2 + \cdots + m_n \tilde{x}_n
\end{aligned}$$

where $\tilde{x}_1, \tilde{x}_2, \ldots, \tilde{x}_n$ are the certain equivalents of each deal.

2. Deals that are Not Mutually Irrelevant In this case, the variance depends on the correlation between the deals. See Appendix A for more information about the covariance and correlation coefficient of uncertain deals. The general definition of **covariance** between two deals x and y is

$$\text{Cov}(x, y) = \sum_{i=1}^{n} p_i (x_i - \bar{x})(y_i - \bar{y})$$

where p_i is the elemental probability at the end of the tree and x_i, y_i are the corresponding elemental values of x, y at each end node.

Note

The covariance of a deal with itself is the variance of the deal, as

$$\text{Cov}(x_i, x_i) = \sum_{i=1}^{n} p_i(x_i - \bar{x})(x_i - \bar{x}) = \sum_{i=1}^{n} p_i(x_i - \bar{x})^2 = x_i^v$$

The variance of the sum of multiples of uncertain deals can be expressed in terms of the covariance between them as

$$\text{Variance of } X(m) = X(m)^v = \sum_{i=1}^{n}\sum_{j=1}^{m} m_i m_j \text{Cov}(x_i, x_j)$$

where $\text{Cov}(x_i, x_j)$ is the covariance between deals x_i and x_j.

We show in Appendix A that when the deals are mutually irrelevant

$$\text{Cov}(x_i, x_j) = 0$$

Therefore, the expression for the variance $\sum_{i=1}^{n}\sum_{j=1}^{m} m_i m_j \text{Cov}(x_i, x_j)$ of two mutually irrrelevant deals reduces to $\sum_{i=1}^{n} m_i^2 x_i^v$.

Note

If two deals x_i and x_j are irrelevant, then the covariance between them is equal to 0. However, if the covariance is equal to 0, it does not mean that the two deals are irrelevant.

32.5.3 Calculating Multiples $m_1, m_2, \ldots, m_n$ and Certain Equivalent of a Portfolio

It will be convenient to use vector and matrix notation for our derivations here. If you are not familiar with matrices, Appendix C presents a brief tutorial on matrix multiplication and inversion. We will also illustrate the final results of the section numerically, using several examples. You will notice a similarity between the results in matrix form and those for a single deal.

The deals and multiples of the deals will be represented by column vectors $\underline{X}$ and $\underline{m}$, respectively, as

$$\underline{X} = \begin{pmatrix} x_1 \\ x_2 \\ \cdot \\ \cdot \\ x_n \end{pmatrix}, \underline{m} = \begin{pmatrix} m_1 \\ m_2 \\ \cdot \\ \cdot \\ m_n \end{pmatrix}$$

As we discussed, the mean is given by

$$\bar{X}(m) = m_1\bar{x}_1 + m_2\bar{x}_2 + \cdots + m_n\bar{x}_n$$

but this is also equal to the dot product of the two vectors

$$\bar{X}(m) = \underline{m}^T\underline{X}$$

Here $\underline{\bar{X}}$ is a column vector whose elements are $\bar{x}_1, \bar{x}_2, \ldots, \bar{x}_n$, and $\underline{m}^T$ is the transpose of the column vector $\underline{m}$. The transpose of a column vector is a row vector, so

$$\underline{m}^T = (m_1, m_2, \ldots, m_n)$$

The variance also can be expressed using matrix notation as

$$\operatorname{var}(X(m)) = \underline{m}^T \underline{V} \underline{m}$$

where $\underline{V}$ is a covariance matrix whose elements are equal to the covariance between the assets. Thus,

$$\underline{V} = \begin{pmatrix} \operatorname{Cov}(x_1, x_1) & \dots & \operatorname{Cov}(x_1, x_n) \\ \vdots & \ddots & \vdots \\ \operatorname{Cov}(x_n, x_1) & \dots & \operatorname{Cov}(x_n, x_n) \end{pmatrix}$$

The certain equivalent approximation for the portfolio of deal can be written in matrix notation as

$$\tilde{x}(m) \approx \overline{X}(m) - \frac{1}{2\rho} \operatorname{var}(X(m))$$

$$= \underline{m}^T \underline{\overline{X}} - \frac{1}{2\rho} \underline{m}^T \underline{V} \underline{m}$$

Note that this expression also applies for the case of mutually irrelevant deals, where the covariance matrix has a special structure: All elements that are not on the diagonal are 0, and elements on the diagonal correspond to the variance of the deal. Therefore,

$$\underline{V} = \begin{pmatrix} \operatorname{Var}(x_1) & \dots & 0 \\ \vdots & \ddots & \vdots \\ 0 & \dots & \operatorname{Var}(x_n) \end{pmatrix}$$

To find the optimal share, we maximize the certain equivalent of the portfolio using derivatives to get

$$\underline{m}^* = \rho \underline{V}^{-1} \underline{\overline{X}}$$

where $\underline{V}^{-1}$ is the inverse of the covariance matrix (see Appendix C for matrix inverses).

The **surety vector** is defined as

$$\text{Surety vector} = \underline{V}^{-1} \underline{\overline{X}}$$

The certain equivalent of the optimal portfolio is

$$\tilde{x}(m^*) \approx \underline{m}^{*T} \underline{\overline{X}} - \frac{1}{2} \gamma \underline{m}^{*T} \underline{V} \underline{m}^*$$

$$= \rho \underline{\overline{X}}^T \underline{V}^{-1} \underline{\overline{X}} - \frac{1}{2} \gamma \rho \underline{\overline{X}}^T \underline{V}^{-1} \underline{V} \rho \underline{V}^{-1} \underline{\overline{X}}$$

$$= \frac{1}{2} \rho \underline{\overline{X}}^T \underline{V}^{-1} \underline{\overline{X}}$$

$$= \frac{1}{2} \underline{m}^{*T} \underline{\overline{X}}$$

Note the similarity between this expression and the case of a single deal.

32.5.4 Using Correlation Coefficients

It is often helpful to express the relevance relations in terms of the correlation coefficients between them rather than the covariance. The correlation coefficient is

$$r_{ij} = \frac{\text{Cov}(x_i, x_j)}{\sigma_{x_i}\sigma_{x_j}}$$

where $\sigma_{x_i} = \sqrt{\text{Var}(x_i)}$ = standard deviation of deal x_i.

If the two deals x_i and x_j are irrelevant, then the correlation coefficient is equal to zero. However, if the correlation coefficient is equal to 0, it does not mean that the two measures are irrelevant.

A diagonal matrix of standard deviations of the deals is defined as

$$\underline{S} = \begin{pmatrix} \sigma_{x_1} & \cdots & 0 \\ \vdots & \ddots & \vdots \\ 0 & \cdots & \sigma_{x_n} \end{pmatrix}$$

and the correlation coefficient matrix, $\underline{R}$, as

$$\underline{R} = \underline{S}^{-1}\underline{V}\underline{S}^{-1}$$

Alternatively, we can write

$$\underline{V} = \underline{S}\underline{R}\underline{S} \text{ and } \underline{V}^{-1} = \underline{S}^{-1}\underline{R}^{-1}\underline{S}^{-1}$$

EXAMPLE 32.3 Optimal Share in a Portfolio with Relevant and Irrelevant Deals

An exponential decision maker with a risk tolerance of \$1,000 faces three deals of x_A, x_B, and x_C with means and standard deviations (square root of the variances) given by the matrices as

$$\underline{\bar{X}} = \begin{pmatrix} \bar{x}_A \\ \bar{x}_B \\ \bar{x}_C \end{pmatrix} = \begin{pmatrix} 10 \\ 15 \\ 0 \end{pmatrix}, \underline{S} = \begin{pmatrix} \sigma_A & 0 & 0 \\ 0 & \sigma_B & 0 \\ 0 & 0 & \sigma_C \end{pmatrix} = \begin{pmatrix} 15 & 0 & 0 \\ 0 & 20 & 0 \\ 0 & 0 & 1 \end{pmatrix}$$

This means that $\bar{x}_A = 10$, $\bar{x}_B = 15$, and $\bar{x}_C = 0$, and the standard deviations of these deals are $\sigma_A = 15$, $\sigma_B = 20$ and $\sigma_C = 1$ respectively.

Determine the approximate optimal shares that the decision maker should have in each of the deals and the approximate certain equivalent for this optimal share for the following.

a. The deals are all mutually irrelevant

b. The deals are relevant, having the same standard deviations but with a correlation matrix given by

$$\underline{R} = \begin{pmatrix} 1 & 0.8 & 0 \\ 0.8 & 1 & 0 \\ 0 & 0 & 1 \end{pmatrix}$$

c. The deals are relevant, having the same standard deviations but with a correlation matrix given by

$$\underline{R} = \begin{pmatrix} 1 & -0.8 & 0 \\ -0.8 & 1 & 0 \\ 0 & 0 & 1 \end{pmatrix}$$

SOLUTIONS

a. When the deals are all mutually irrelevant,

$$\underline{R} = \begin{pmatrix} 1 & 0 & 0 \\ 0 & 1 & 0 \\ 0 & 0 & 1 \end{pmatrix}$$

$$\underline{V} = \underline{S}\,\underline{R}\,\underline{S} = \underline{S}\,\underline{S} = \begin{pmatrix} 225 & 0 & 0 \\ 0 & 400 & 0 \\ 0 & 0 & 1 \end{pmatrix}$$

$$\underline{V}^{-1} = \begin{pmatrix} \frac{1}{225} & 0 & 0 \\ 0 & \frac{1}{400} & 0 \\ 0 & 0 & 1 \end{pmatrix}$$

$$\underline{m}^* = \rho \underline{V}^{-1}\underline{\overline{X}} = 1000 \cdot \begin{pmatrix} \frac{1}{225} & 0 & 0 \\ 0 & \frac{1}{400} & 0 \\ 0 & 0 & 1 \end{pmatrix} \cdot \begin{pmatrix} 10 \\ 15 \\ 0 \end{pmatrix} = \begin{pmatrix} 44.44 \\ 37.50 \\ 0 \end{pmatrix}$$

The optimal portfolio would constitute 44.44 shares of Deal A, 37.5 of Deal B and 0 of Deal C. The certain equivalent of this optimal portfolio is

$$\tilde{x}(m^*) \approx \frac{1}{2}\underline{m}^{*T}\underline{\overline{X}} = \$503.45$$

b. The deals are relevant with a correlation matrix $\underline{R} = \begin{pmatrix} 1 & 0.8 & 0 \\ 0.8 & 1 & 0 \\ 0 & 0 & 1 \end{pmatrix}$. Hence,

$$\underline{V} = \underline{S}\,\underline{R}\,\underline{S} = \begin{pmatrix} 225 & 240 & 0 \\ 240 & 400 & 0 \\ 0 & 0 & 1 \end{pmatrix}$$

$$\underline{V}^{-1} = \begin{pmatrix} 0.0123 & -0.0074 & 0 \\ -0.0074 & 0.0069 & 0 \\ 0 & 0 & 1 \end{pmatrix}$$

$$\underline{m}^* = \rho \underline{V}^{-1}\underline{\overline{X}} = 1000 \cdot \begin{pmatrix} 0.0123 & -0.0074 & 0 \\ -0.0074 & 0.0069 & 0 \\ 0 & 0 & 1 \end{pmatrix} \cdot \begin{pmatrix} 10 \\ 15 \\ 0 \end{pmatrix} = \begin{pmatrix} 12.3 \\ 30 \\ 0 \end{pmatrix}$$

$$\tilde{x}(\underline{m}^*) \approx \frac{1}{2}\underline{m}^{*T}\underline{\overline{X}} = \$287.40$$

***Note** When the deals are positively correlated, the overall variance increases with the same mean, so the deal is less desirable than the case where the deals are irrelevant.*

c. The deals are relevant with a correlation matrix $\underline{R} = \begin{pmatrix} 1 & -0.8 & 0 \\ -0.8 & 1 & 0 \\ 0 & 0 & 1 \end{pmatrix}$

$$\underline{V} = \underline{S}\underline{R}\underline{S} = \begin{pmatrix} 225 & -240 & 0 \\ -240 & 400 & 0 \\ 0 & 0 & 1 \end{pmatrix}$$

$$\underline{V}^{-1} = \begin{pmatrix} 0.0123 & 0.0074 & 0 \\ 0.0074 & 0.0069 & 0 \\ 0 & 0 & 1 \end{pmatrix}$$

$$\underline{m}^* = \rho \underline{V}^{-1}\underline{\overline{X}} = 1000 \cdot \begin{pmatrix} 0.0123 & -0.0074 & 0 \\ -0.0074 & 0.0069 & 0 \\ 0 & 0 & 1 \end{pmatrix} \cdot \begin{pmatrix} 10 \\ 15 \\ 0 \end{pmatrix} = \begin{pmatrix} 234.56 \\ 178.24 \\ 0 \end{pmatrix}$$

$$\tilde{x}(\underline{m}^*) \approx \frac{1}{2}\underline{m}^{*T}\underline{\overline{X}} = \$2{,}509.60$$

***Note** When the deals are negatively correlated, the overall variance decreases with the same mean, so the deal is more desirable than the case where the deals are irrelevant.*

EXAMPLE 32.4 Shorting a Deal

An exponential decision maker with a risk tolerance of \$1,000 faces three deals: x_A, x_B, and x_C, with means and standard deviations given by the matrices

$$\underline{\overline{X}} = \begin{pmatrix} \overline{x}_A \\ \overline{x}_B \\ \overline{x}_C \end{pmatrix} = \begin{pmatrix} 10 \\ 15 \\ 5 \end{pmatrix}, \underline{S} = \begin{pmatrix} \sigma_A & 0 & 0 \\ 0 & \sigma_B & 0 \\ 0 & 0 & \sigma_C \end{pmatrix} = \begin{pmatrix} 15 & 0 & 0 \\ 0 & 20 & 0 \\ 0 & 0 & 25 \end{pmatrix}$$

The deals are relevant with a correlation matrix $\underline{R} = \begin{pmatrix} 1 & 0.8 & 0.4 \\ 0.8 & 1 & 0.3 \\ 0.4 & 0.3 & 1 \end{pmatrix}$.

Find the approximate optimal portfolio the decision maker should have in these deals.

SOLUTION

Once again, we calculate the covariance matrix as

$$\underline{V} = \underline{S}\underline{R}\underline{S}$$

The inverse of the matrix is

$$\underline{V}^{-1} = \begin{pmatrix} 0.0133 & -0.0075 & -0.0014 \\ -0.0075 & 0.0069 & 0.0001 \\ -0.0014 & 0.00015 & 0.0019 \end{pmatrix}$$

$$\underline{m}^* = \rho \underline{V}^{-1}\underline{\overline{X}} = 1000 \cdot \begin{pmatrix} 0.0133 & -0.0075 & -0.0014 \\ -0.0075 & 0.0069 & 0.0001 \\ -0.0014 & 0.00015 & 0.0019 \end{pmatrix} \cdot \begin{pmatrix} 10 \\ 15 \\ 5 \end{pmatrix} = \begin{pmatrix} 14.27 \\ 29.91 \\ -2.60 \end{pmatrix}$$

$$\tilde{x}(m^*) \approx \frac{1}{2}\underline{m}^{*T}\underline{\overline{X}} = \$289.20$$

***Note** The best portfolio has negative holdings of Deal C. This is called* **shorting** *the deal, which means you need to sell it. To short a deal, you offer someone the deal and are responsible for paying him its outcomes.*

EXAMPLE 32.5 Valuing an Uncertain Deal Relevant to Other Deals

An exponential decision maker with a risk tolerance of \$500,000 faces two deals, x_A and x_B, with means of \$10,000 and \$15,000, respectively. The variances of the deals are $4 * 10^{11}$ and $6.25 * 10^{11}$, respectively: the correlation coefficient between them is 0.8. First, we calculate the optimal portfolio of these two deals and the certain equivalent of the optimal portfolio.

***Note** The certain equivalent for each of the two deals individually is negative.*

$$\tilde{x}_A \approx 10{,}000 - \frac{4 * 10^{11}}{500{,}000} = -\$390{,}000$$

$$\tilde{x}_B \approx 15{,}000 - \frac{6.25 * 10^{11}}{500{,}000} = -\$610{,}000$$

The covariance between the two deals is

$$\mathrm{Cov}(x_A, x_B) = 0.8 * \sqrt{4 * 10^{11}} * \sqrt{6.25 * 10^{11}} = 4 * 10^{11}$$

Now, we write out the mean vector and variance matrix as

$$\bar{x} = \begin{pmatrix} 10{,}000 \\ 15{,}000 \end{pmatrix}, \quad V = \begin{pmatrix} 4 * 10^{11} & 4 * 10^{11} \\ 4 * 10^{11} & 6.25 * 10^{11} \end{pmatrix}$$

Taking the inverse of the covariance matrix gives

$$V^{-1} = 10^{-11} * \begin{pmatrix} .6944 & -.444 \\ -.444 & -.444 \end{pmatrix}$$

Now, we calculate the optimal share as

$$m^* = \rho V^{-1}\bar{x} = \begin{bmatrix} 0.0014 \\ 0.0111 \end{bmatrix}$$

In other words, the decision maker would prefer approximately 1/1000 of Deal *A* and approximately 1/100 of Deal *B*.

The certain equivalent for this optimal portfolio is

$$\tilde{x}(m^*) \approx \frac{1}{2} m^{*T}\bar{x}$$

$$= 0.5 * [0.0014 \ \ 0.0111] * \begin{pmatrix} 10{,}000 \\ 15{,}000 \end{pmatrix} = \$90.28$$

Therefore, taking certain fractions of these deals can result in a portfolio with a positive certain equivalent for the decision maker.

If the decision maker has obtained his optimal share, m^*, and is now offered a deal, x_C, with a mean of \$10,000, variance of $3.24 * 10^9$, and has correlations with the previous two deals $r_{AC} = 0.4$, $r_{BC} = 0.3$, what is the decision maker's Personal Indifferent Buying Price (PIBP) for x_C?

Since the decision maker is an exponential decision maker, his PIBP is equal to his Personal Indifferent Selling Price (PISP). Therefore, we will calculate his certain equivalent for x_C. First, we calculate the approximate certain equivalent for the portfolio, including x_C, using a new covariance matrix and mean vector.

We have

$$\text{Cov}(x_A, x_C) = 0.4 * \sqrt{4 * 10^{11}} * \sqrt{3.24 * 10^9} = 1.44 * 10^{10}$$

$$\text{Cov}(x_B, x_C) = 0.3 * \sqrt{6.25 * 10^{11}} * \sqrt{3.24 * 10^9} = 1.35 * 10^{10}$$

The new covariance matrix is

$$V = \begin{pmatrix} 4 * 10^{11} & 4 * 10^{11} & 1.44 * 10^{10} \\ 4 * 10^{11} & 6.25 * 10^{11} & 1.35 * 10^{10} \\ 1.44 * 10^{10} & 1.35 * 10^{10} & 3.24 * 10^9 \end{pmatrix}$$

and the new mean vector is

$$\bar{x} = \begin{pmatrix} 10{,}000 \\ 15{,}000 \\ 10{,}000 \end{pmatrix}$$

Since the decision maker will not change his share in Deals A or B and he will be offered 100% of Deal C, his new share vector, m, is

$$m = \begin{bmatrix} 0.0014 \\ 0.0111 \\ 1.0000 \end{bmatrix}$$

The approximate certain equivalent for this new portfolio is

$$\tilde{x}(m) \approx m^T\bar{x} - \frac{1}{2\rho} m^T V m = \$6{,}510$$

If we subtract from this value the value of the portfolio without x_C, we find the certain equivalent of x_C to be

$$\tilde{x}_C = \$6{,}510 - \$90.28 \approx \$6{,}419$$

Note that the approximate certain equivalent for deal x_C alone is

$$\tilde{x}_C \approx 10{,}000 - \frac{3.24 * 10^9}{2 * 500{,}000} = \$6{,}760$$

which is not equal to the $6,419. As a result, even a deltaperson cannot value a deal that is relevant to other uncertain deals he is holding, individually, without considering the relevance relations between the deals.

32.6 SUMMARY

Risk scaling means getting a scaled amount of the deal. It has benefits because, if facing a deal with a positive e-value, you can create a positive certain equivalent by getting a fraction, or multiple, of the deal. The approximate optimal scale of a deal that a decision maker with a risk tolerance, ρ, would like to have is $m^* = \rho\left(\bar{x}/x^v\right)$. The approximate certain equivalent of this scaled deal is $\frac{1}{2}m^*\bar{x}$.

When you scale a deal, you might be interested in considering who would accept the remaining portion. If the deal is shared among a group of people, we can determine the best share for each member as follows:

- Define the group risk tolerance as the sum of risk tolerances of the individual members of the group, $\rho_g = \sum_{i=1}^{N} \rho_i$.
- The group certain equivalent of the deal using the approximation is $\tilde{x}_g = \bar{x} - \frac{x^v}{2\rho_g}$.
- Every individual will get a fraction equal to the ratio of his individual risk tolerance to the group risk tolerance, $f_i = \rho_i/\rho_g$.
- The certain equivalent for each individual's share is equal to $\tilde{x}_i = f_i\tilde{x}_g$.
- The best group (syndicate) is the one whose group risk tolerance is equal to the reciprocal of the surety of the deal, so $\rho_g = \frac{x^V}{\bar{x}}$.

If the two deals x_i and x_j are irrelevant their correlation coefficient is equal to zero. However, if the correlation coefficient is equal to zero, it does not mean that the two deals are irrelevant.

Consider a portfolio where you are choosing multiples $m_1, m_2, \ldots, m_n$ of n possible deals $x_1, x_2, \ldots, x_n$. For any given multiples $m_1, m_2, \ldots, m_n$, the total payoff of the portfolio is equal to that provided by the additive value function

$$X(m) = m_1x_1 + m_2x_2 + \cdots + m_nx_n$$

The mean payoff is equal to $X(m)$, where

$$X(m) = m_1\bar{x}_1 + m_2\bar{x}_2 + \cdots + m_n\bar{x}_n$$

where $\bar{x}_1, \bar{x}_2, \ldots, \bar{x}_n$ are the mean values of deals $x_1, x_2, \ldots, x_n$, and the variance of the portfolio is

$$\text{Var}(X(m)) = X(m)^v = \sum_{i=1}^{n}\sum_{j=1}^{m} m_i m_j \text{Cov}(x_i, x_j)$$

where $\text{Cov}(x_i, x_j)$ is the covariance between deals x_i and x_j.

In matrix notation, $X(m) = \underline{m}^T\underline{\bar{X}}$ here $\underline{\bar{X}}$ is a column vector whose elements are $\bar{x}_1, \bar{x}_2, \ldots, \bar{x}_n$, and m^T is the transpose of the column vector m. The transpose of a column vector is a row vector, as

$$\underline{m}^T = (m_1, m_2, \ldots, m_n)$$

The variance can also be expressed using matrix notation as

$$\text{Var}(X(m)) = \underline{m}^T \underline{V}\underline{m}$$

where $\underline{V}$ is a covariance matrix whose elements are equal to the covariance between the assets. Thus,

$$\underline{V} = \begin{pmatrix} \text{Cov}(x_1, x_1) & \cdots & \text{Cov}(x_1, x_n) \\ \vdots & \ddots & \vdots \\ \text{Cov}(x_n, x_1) & \cdots & \text{Cov}(x_n, x_n) \end{pmatrix}$$

The best multiples are given by

$$\underline{m}^* = \rho\underline{V}^{-1}\underline{\bar{X}}$$

KEY TERMS

- Covariance
- Scaling
- Sharing
- Surety vector
- Variance

APPENDIX A Covariance and Correlation

If we have a probability tree with two measures, we can calculate an additional quantity for the two measures known as the **cross moment**. The cross moment is equal to the first moment of the product of the two values of the measure for each prospect in the probability tree. Hence,

$$<xy|\&>= \sum_{i=1}^{n} p_i x_i y_i$$

For example, refer back to Chapter 8 and the two measures in Figure 8.9 of dollar winnings and car washes, which we repeat for convenience here.
The cross moment is

$$\begin{aligned}<\text{Dollar winning, Car wash}|\&> &= 0.07(-100)(-1) + 0.14(0)(0) + 0.28(0)(0) \\ &\quad + 0.21(100)(1) + 0.04(0)(0) + 0.06(0)(0) \\ &\quad + 0.06(0)(0) + 0.04(0)(0) + 0.03(300)(1) \\ &\quad + 0.04(200)(1) + 0.02(-200)(-1) \\ &\quad + 0.01(-300)(-1) \\ &= 52\end{aligned}$$

The covariance (Cov) between A and B is the cross central moment. Therefore,

$$\text{Cov} = \sum_{i=1}^{n} p_i(x_i - \bar{x})(y_i - \bar{y})$$

The covariance is also the difference between the cross moment and the product of the *e*-values of each measure. Thus,

$$\text{Cov} = \sum_{i=1}^{n} p_i x_i y_i - \bar{x}.\bar{y}$$

The mean values of dollar earnings and car wash winnings in the car wash example can be determined by first determining their distributions or directly from the tree of Figure 32.3.

$$\begin{aligned}<\text{Dollar winnings}|\&> &= 0.07(-100) + 0.14(0) + 0.28(0) + 0.21(100) \\ &\quad + 0.04(0) + 0.06(0) + 0.069(0) + 0.04(0) + 0.03(300) \\ &\quad + 0.04(200) + 0.02(-200) + 0.01(-300) \\ &= 24\end{aligned}$$

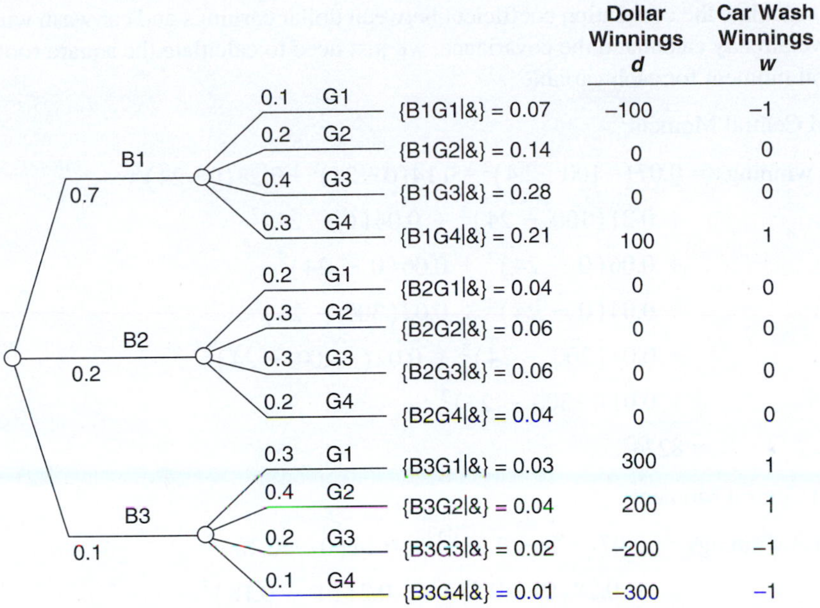

					Dollar Winnings d	Car Wash Winnings w
B1	0.7	0.1	G1	{B1G1|&} = 0.07	–100	–1
		0.2	G2	{B1G2|&} = 0.14	0	0
		0.4	G3	{B1G3|&} = 0.28	0	0
		0.3	G4	{B1G4|&} = 0.21	100	1
B2	0.2	0.2	G1	{B2G1|&} = 0.04	0	0
		0.3	G2	{B2G2|&} = 0.06	0	0
		0.3	G3	{B2G3|&} = 0.06	0	0
		0.2	G4	{B2G4|&} = 0.04	0	0
B3	0.1	0.3	G1	{B3G1|&} = 0.03	300	1
		0.4	G2	{B3G2|&} = 0.04	200	1
		0.2	G3	{B3G3|&} = 0.02	–200	–1
		0.1	G4	{B3G4|&} = 0.01	–300	–1

FIGURE 32.3 Probability Tree for Dollar Winnings and Car Washes

$$\begin{aligned}\langle \text{Car wash winnings} | \& \rangle &= 0.07(-1) + 0.14(0) + 0.28(0) + 0.21(1) \\ &\quad + 0.04(0) + 0.06(0) + 0.06(0) + 0.04(0) \\ &\quad + 0.03(1) + 0.04(1) + 0.02(-1) + 0.01(-1) \\ &= 0.18\end{aligned}$$

The covariance between dollar earnings and car was winnings is then

$$\text{Cov}(\text{Dollar winnings, Car wash winnings}) = 52 - (24)(0.18) = 47.68$$

Finally, the correlation coefficient between two measures is equal to the ratio of their covariance to the product of the square root of their second central moments. Thus,

$$\text{Correlation coefficient} = \frac{\text{Cov}}{\sqrt{\langle x | \& \rangle^{v}}\sqrt{\langle y | \& \rangle^{v}}}$$

The correlation coefficient ranges from -1 to 1. If the two measures X and Y are irrelevant, then the correlation coefficient is equal to 0. This is because the cross moment is equal to the product of the moments for irrelevant measures. However, if the correlation coefficient is equal to 0, it does not mean that the two measures are irrelevant.

Let us calculate the correlation coefficient between dollar earnings and car wash winnings. Since we have already calculated the covariance, we just need to calculate the square root of the second central moment for each variable.

Second Central Moment:

$$\begin{aligned}\text{Dollar winnings} &= 0.07(-100-24)^2+0.14(0-24)^2+0.28(0-24)^2\\&+0.21(100-24)^2+0.04(0-24)^2\\&+0.06(0-24)^2+0.06(0-24)^2\\&+0.04(0-24)^2+0.03(300-24)^2\\&+0.04(200-24)^2+0.02(-200-24)^2\\&+0.01(-300-24)^2\\&=8224\end{aligned}$$

Second Central Moment:

$$\begin{aligned}\text{Car wash winnings} &= 0.07(-1-0.18)^2+0.14(0-0.18)^2\\&+0.28(0-0.18)^2+0.21(1-0.18)^2\\&+0.04(0-0.18)^2+0.06(0-0.18)^2\\&+0.06(0-0.18)^2+0.04(0-0.18)^2\\&+0.03(1-0.18)^2+0.04(1-0.18)^2\\&+0.02(-1-0.18)^2+0.01(-1-0.18)^2\\&=0.3476\end{aligned}$$

The correlation coefficient is

$$\frac{47.68}{\sqrt{8224}\sqrt{0.3476}} = 0.892$$

Additional Observations About Covariance

1. When two deals are mutually irrelevant, $p_{ij} = p_i p_j$, and

$$\begin{aligned}\text{Cov}(x_i, x_j) &= \sum_{i=1}^{n}\sum_{j=1}^{n} p_{ij}(x-\bar{x}_i)(x-\bar{x}_j)\\&= \sum_{i=1}^{n} p_i(x-\bar{x}_i)\sum_{j=1}^{n} p_j(x-\bar{x}_j)\\&= (\bar{x}_j-\bar{x}_j)(\bar{x}_i-\bar{x}_i)=0\end{aligned}$$

2. When the two deals are identical,

$$\begin{aligned}\text{Cov}(x_i, x_i) &= \sum_{i=1}^{n}\sum_{j=1}^{n} p_{ji}(x-\bar{x}_i)(x-\bar{x}_j)\\&= \text{Var}(x_i)\end{aligned}$$

EXAMPLE A1

Consider the two deals shown in Figure 32.4.

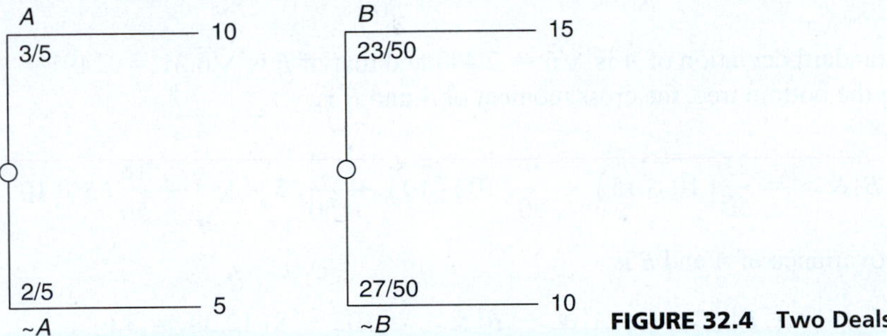

FIGURE 32.4 Two Deals

Suppose their joint probability tree is given as shown in Figure 32.5.

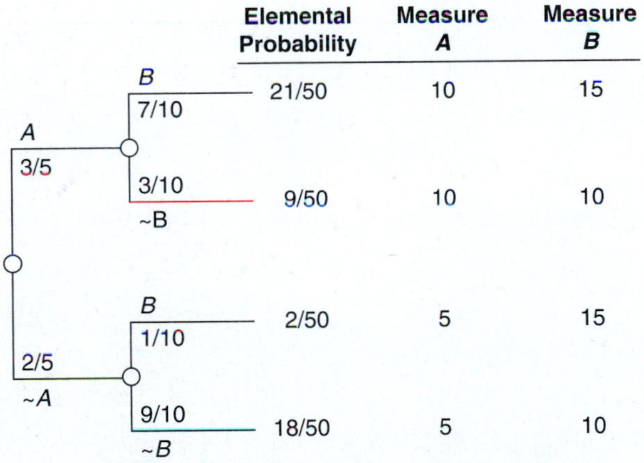

Branch	Elemental Probability	Measure A	Measure B
A 3/5, B 7/10	21/50	10	15
A 3/5, ~B 3/10	9/50	10	10
~A 2/5, B 1/10	2/50	5	15
~A 2/5, ~B 9/10	18/50	5	10

FIGURE 32.5 Probability Tree for Two Deals

1. Calculate the e-value, standard deviation, and variance of each deal.
2. Calculate the cross moment, covariance, and correlation coefficient between the two deals.

SOLUTION

1. Using the top two trees:

   The e-values of the two measures are given by

$$\langle A|\&\rangle = \frac{3}{5}10 + \frac{2}{5}5 = 8$$

$$\langle B|\&\rangle = \frac{23}{50}15 + \frac{27}{50}10 = 12.3$$

The variances of the measures are the second central moments

$$<A|\&>^{v} = \frac{3}{5}(10 - 8)^2 + \frac{2}{5}(5 - 8)^2 = 6$$

$$<B|\&>^{v} = \frac{23}{50}(15 - 12.3)^2 + \frac{27}{50}(10 - 12.3)^2 = 6.21$$

The standard deviation of A is $\sqrt{6} = 2.449$ and that of B is $\sqrt{6.21} = 2.491$.

2. Using the bottom tree, the cross moment of A and B is

$$<A, B|\&> = \frac{21}{50}(10)(15) + \frac{9}{50}(10)(10) + \frac{2}{50}(5)(15) + \frac{18}{50}(5)(10) = 102$$

The covariance of A and B is

$$<A, B|\&> - <A|\&><B|\&> = 102 - 8(12.3) = 3.6$$

The correlation coefficient between A and B is

$$\frac{3.6}{(2.449)(2.491)} = .589$$

APPENDIX B Scalar (Dot) Product of Vectors

A vector is a collection of elements in a compact (and ordered) form. For example, a column vector has the form $\begin{pmatrix} a \\ b \\ c \\ d \end{pmatrix}$, which represents four elements and has dimensions 4×1. A row vector has the form (a, b, c, d), which also represents four elements and has dimensions 1×4. The transpose of a column vector is row vector with the same elements and the transpose of a row vector is a column vector. A symbol T denotes the transpose operation. For example,

$$\begin{pmatrix} a \\ b \\ c \\ d \end{pmatrix}^T = (a, b, c, d)$$

The dot product of a column vector and a row vector results in a single number, which is the sum of the product of the elements in the vectors. For example,

$$(a\, b\, c\, d) \cdot \begin{pmatrix} e \\ f \\ g \\ h \end{pmatrix} = ae + bf + cg + dh$$

APPENDIX C 2 × 2 and 3 × 3 Matrix Multiplications and Matrix Inversion

A matrix is a compact representation of data. It has many uses, particularly in solving a set of linear equations. For example, if we have the two equations

$$2x + 3y = 6$$

$$3x + 4y = 8$$

and we are interested in determining the values of x, y, we can write them in matrix form as

$$\begin{pmatrix} 2 & 3 \\ 3 & 4 \end{pmatrix}\begin{pmatrix} x \\ y \end{pmatrix} = \begin{pmatrix} 6 \\ 8 \end{pmatrix}$$

This leads us to the idea of matrix multiplication. The product of two matrices $\begin{pmatrix} 2 & 3 \\ 3 & 4 \end{pmatrix}, \begin{pmatrix} x \\ y \end{pmatrix}$ is

$$\begin{pmatrix} 2 & 3 \\ 3 & 4 \end{pmatrix}\begin{pmatrix} x \\ y \end{pmatrix} = \begin{pmatrix} 2x + 3y \\ 3x + 4y \end{pmatrix}$$

The product of two 2 × 2 matrices is another 2 × 2 matrix given by

$$\begin{pmatrix} a & b \\ c & d \end{pmatrix}\begin{pmatrix} e & f \\ g & h \end{pmatrix} = \begin{pmatrix} ae + bg & af + bh \\ ce + dg & cf + dh \end{pmatrix}$$

Matrices can be used to solve the equations to get the values of x and y.
If

$$\begin{pmatrix} 2 & 3 \\ 3 & 4 \end{pmatrix}\begin{pmatrix} x \\ y \end{pmatrix} = \begin{pmatrix} 6 \\ 8 \end{pmatrix}$$

then

$$\begin{pmatrix} x \\ y \end{pmatrix} = \begin{pmatrix} 2 & 3 \\ 3 & 4 \end{pmatrix}^{-1}\begin{pmatrix} 6 \\ 8 \end{pmatrix}$$

where $\begin{pmatrix} 2 & 3 \\ 3 & 4 \end{pmatrix}^{-1}$ is known as the inverse matrix of $\begin{pmatrix} 2 & 3 \\ 3 & 4 \end{pmatrix}$.

The inverse of a 2 × 2 matrix $\begin{pmatrix} a & b \\ c & d \end{pmatrix}$ is

$$\begin{pmatrix} a & b \\ c & d \end{pmatrix}^{-1} = \frac{1}{ad - bc}\begin{pmatrix} d & -b \\ -c & a \end{pmatrix}$$

Therefore,

$$\begin{pmatrix} 2 & 3 \\ 3 & 4 \end{pmatrix}^{-1} = \frac{1}{8 - 9}\begin{pmatrix} 4 & -3 \\ -3 & 2 \end{pmatrix} = -\begin{pmatrix} 4 & -3 \\ -3 & 2 \end{pmatrix} = \begin{pmatrix} -4 & 3 \\ 3 & -2 \end{pmatrix}$$

and

$$\begin{pmatrix} x \\ y \end{pmatrix} = \begin{pmatrix} 2 & 3 \\ 3 & 4 \end{pmatrix}^{-1} \begin{pmatrix} 6 \\ 8 \end{pmatrix} = \begin{pmatrix} -4 & 3 \\ 3 & -2 \end{pmatrix} \begin{pmatrix} 6 \\ 8 \end{pmatrix} = \begin{pmatrix} 0 \\ 2 \end{pmatrix}$$

So the solution of the equations is obtained by

$$\begin{aligned} 2x + 3y &= 6 \\ 3x + 4y &= 8 \end{aligned}$$

for $x = 0, y = 2$.

Matrices also can have higher dimensions. A 3 × 3 matrix has nine elements. It can be used to summarize and solve three linear equations in three variables. For example,

$$\begin{aligned} 2x + 3y + 4z &= 20 \\ 5x - 2y + 2z &= 5 \\ x + y + 2z &= 9 \end{aligned}$$

In matrix form, this can be written as

$$\begin{pmatrix} 2 & 3 & 4 \\ 5 & -2 & 2 \\ 1 & 1 & 2 \end{pmatrix} \begin{pmatrix} x \\ y \\ z \end{pmatrix} = \begin{pmatrix} 20 \\ 5 \\ 9 \end{pmatrix}$$

To solve for x, y, z, we can write

$$\begin{pmatrix} x \\ y \\ z \end{pmatrix} = \begin{pmatrix} 2 & 3 & 4 \\ 5 & -2 & 2 \\ 1 & 1 & 2 \end{pmatrix}^{-1} \begin{pmatrix} 20 \\ 5 \\ 9 \end{pmatrix}$$

where $\begin{pmatrix} 2 & 3 & 4 \\ 5 & -2 & 2 \\ 1 & 1 & 2 \end{pmatrix}^{-1}$ is the inverse matrix of $\begin{pmatrix} 2 & 3 & 4 \\ 5 & -2 & 2 \\ 1 & 1 & 2 \end{pmatrix}$.

The inverse of a 3 × 3 matrix is a bit more complicated than the 2 × 2 matrix. Here is the formula

$$\begin{pmatrix} a_{11} & a_{12} & a_{13} \\ a_{21} & a_{22} & a_{23} \\ a_{31} & a_{32} & a_{33} \end{pmatrix}^{-1} = \frac{1}{\text{Det}} \begin{pmatrix} (a_{33}a_{22} - a_{32}a_{23}) & -(a_{33}a_{12} - a_{32}a_{13}) & (a_{23}a_{12} - a_{22}a_{13}) \\ -(a_{33}a_{21} - a_{31}a_{23}) & (a_{33}a_{11} - a_{31}a_{13}) & -(a_{23}a_{11} - a_{21}a_{13}) \\ (a_{32}a_{21} - a_{31}a_{22}) & -(a_{32}a_{11} - a_{31}a_{12}) & (a_{22}a_{11} - a_{21}a_{12}) \end{pmatrix}$$

where $\text{Det} = a_{11}(a_{33}a_{22} - a_{32}a_{23}) - a_{21}(a_{33}a_{12} - a_{32}a_{13}) + a_{31}(a_{23}a_{12} - a_{22}a_{13})$.

By direct substitution, we have

$$\begin{pmatrix} 2 & 3 & 4 \\ 5 & -2 & 2 \\ 1 & 1 & 2 \end{pmatrix}^{-1} = \begin{pmatrix} .75 & .25 & -1.75 \\ 1 & 0 & -2 \\ -.875 & -.125 & 2.375 \end{pmatrix}$$

Hence,

$$\begin{pmatrix} x \\ y \\ z \end{pmatrix} = \begin{pmatrix} 2 & 3 & 4 \\ 5 & -2 & 2 \\ 1 & 1 & 2 \end{pmatrix}^{-1} \begin{pmatrix} 20 \\ 5 \\ 9 \end{pmatrix}$$

$$= \begin{pmatrix} .75 & .25 & -1.75 \\ 1 & 0 & -2 \\ -.875 & -.125 & 2.375 \end{pmatrix} \begin{pmatrix} 20 \\ 5 \\ 9 \end{pmatrix} = \begin{pmatrix} .75 \times 20 + .25 \times 5 - 1.75 \times 9 \\ 1 \times 20 + 0 \times 5 - 2 \times 9 \\ -.875 \times 20 - .125 \times 5 + 2.375 \times 9 \end{pmatrix}$$

$$= \begin{pmatrix} 0.5 \\ 2 \\ 3.25 \end{pmatrix}$$

Therefore, $x = 0.5$, $y = 2$, and $z = 3.25$.

PROBLEMS

Problems marked with an asterisk (*) are considered more challenging.

1. Risk Scaling: Approximate Expression

Consider an exponential decision maker with a risk tolerance equal to $3000 who is offered the following deal

$$[0.25, -5000; 0.5, 1000, 0.25, 5000]$$

a. Calculate his certain equivalent for the deal using the approximate formula for the certain equivalent. Should the decision maker accept this deal?
b. Calculate the surety of this deal.
c. Using the approximate expression for the certain equivalent calculate the optimal fraction of this deal for this decision maker
d. Calculate the certain equivalent if the decision maker takes only the optimal fraction of the deal. Should the decision maker accepts this deal if he can take the optimal fraction?

***2.** Repeat Problem 1 using the exact expression for the certain equivalent.

3. Risk Sharing: Approximate Expression

Three investors with risk tolerance equal to $5000, $10,000, and $15000 face the following deal.

$$[0.25, -20000; 0.5, 0; 0.25, 30000]$$

a. What is the risk tolerance of the group?
b. Using the approximate expression, calculate the group certain equivalent.
c. What is the optimal fraction that each group member should take? What is the certain equivalent of each group member's optimal fraction?
d. The group has the option to include an additional investor. What is the group's preferred risk tolerance for this new investor?
e. What is the group certain equivalent and optimal share of each member if the preferred investor joined the group?

***4.** Repeat Problem 3 using the exact expression for the certain equivalent.

5. Certain Equivalent of a Portfolio: Approximate Expression

An exponential investor with a risk tolerance equal to $20000 faces the following two deals

$$[0.25, -20000; 0.5, 0; 0.25, 30000] \text{ and } [0.25, -5000; 0.5, 0; 0.25, 5000]$$

a. Using the approximate expression, calculate his certain equivalent for the deals if they are
   i. Mutually Irrelevant
   ii. Positively correlated with a correlation coefficient equal to 0.8
   iii. Negatively correlated with a correlation coefficient of −0.6
b. Calculate his optimal share in each deal for the three case of partial.
c. Calculate the certain equivalent of the optimal share in each case

CHAPTER

33

Making Risky Decisions

CHAPTER CONCEPTS

After reading this chapter, you will be able to explain the following concepts:

- The micromort value
- Micromorts in our daily lives
- Using micromorts in decision making

33.1 INTRODUCTION

At some point in your life, you will face the challenge of making a **risky decision**. By a *risky decision*, we mean a decision that has at least one alternative with at least one risk. By *risk,* we mean a prospect that has both a small probability of occurring and the promise of large loss if it does occur. The loss could be a loss of money, life, or other serious consequence.

In the English language, risk has connotations of being both unlikely to happen and bad if it does. We do not appear to have a corresponding term for a small probability of a very good outcome. We do not speak of the risk of receiving a large sum of money from a distant and unknown relative. We also do not speak of the risk of dying if we jump from a tall building: The chance of death is so high that using the word risk is almost comically inappropriate.

In making risky decisions, we often have to face several important issues. First, we often need to consider the value of "intangibles," such as pain and suffering—even though these are really the most tangible of consequences. But such consequences appear to pose greater valuation challenges than, for example, the Personal Indifferent Buying Price (PIBP) for a shirt. Second, we have to deal with the measurement and assessment of small probabilities. And finally, we may have to confront the prospect of death—as uncomfortable as that may be.

33.2 A PAINFUL DILEMMA

We begin by thinking about how we might value pain. Suppose that you went to the doctor and, after examining you and reviewing your test results, the doctor approaches with a serious face.

You ask, "What's the matter, doctor?"

The doctor says, "Well, I have some bad news and some good news. The bad news is that you have a fatal disease that will kill you painlessly within the next month."

If you are on the brink of suicide, this may seem to be good news, but for most of us, it will be bad. So you ask the doctor, "What is the good news?"

The doctor says, "There is a treatment for the disease that will cure you just by spending 24 hours in the hospital. The treatment is covered by your insurance, and you will be completely free of the disease at the end of your 24-hour hospital stay. Furthermore, your company will pay you your usual salary while you are in the hospital."

You say, "That's great!"

However, the doctor continues, "There is more bad news. My patients tell me that the treatment is excruciatingly painful."

You ask, "What do you mean 'excruciatingly painful'?"

The doctor says, "They tell me that it feels as if they were having a wisdom tooth, one of the last big molars in your mouth, pulled continuously for 24 hours. Notice that I say *continuously*. It's not as if the tooth is pulled and then you can recover from that experience."

You ask, "Isn't there something you can give me for the pain?"

The doctor says, "No, I am afraid that the experience of the pain is essential to the successful treatment of your disease."

Noticing the dismay on your face, the doctor says there is one more possibility. "I have a friend who has developed an experimental drug that may treat your disease successfully without pain. You still have to spend the 24 hours in the hospital, but you can just watch television and you will leave cured, and without any side effects."

You say, "That's wonderful, I'll take the drug!"

The doctor says, "There is one problem. The drug, as I said, is experimental, and consequently, your insurance company will not pay for it. I don't remember what it costs, but if you would just write down on a piece of paper the most you would pay for it (your PIBP) then I will call my friend and see what it costs. If it costs less than you are willing to pay, you can put that amount in escrow for payment, and I will sign you up for your painless stay in the hospital. If it costs more than you are willing to pay, then I will sign you up for your painful, but lifesaving, hospital visit."

Reflection

Please write down on a piece of paper your personal indifferent buying price, x, for the drug that will provide a painless cure.

When we asked this question of two colleagues, one said that he would pay $5,000 and the other $200,000. They looked at each other in disbelief. The one who said he would pay $200,000 said, "I hate pain. I want a pill before they give me the needle in the dentist's office. I would sell my vacation home to avoid an experience like this." The other said, "I was a colonel in the Marines, and I was wounded three times. I'm not going to spend a lot of money to avoid a little pain." Each one sounded very convincing to the other; neither questioned the appropriateness of his own or the other's answer.

33.2.1 Another Alternative

The doctor then says, "I have another alternative for you. There is a drug that would also cure you completely and without pain, and that drug is not experimental: Your insurance would cover its costs. You would still have to spend the same 24 hours in the hospital as you would before but again you can watch television."

You say, "Why didn't you tell me that before?"

The doctor says, "There is one potentially grave side effect of this drug: It may kill you. At the end of the 24 hours, you might just painlessly pass away."

You ask, "What is the probability that I would die?"

The doctor says, "I don't remember, but if you just write down on a piece of paper the probability of dying that would just make you indifferent between taking this drug and enduring the pain, then, as before, I will call to find out the fraction of people who die in the course of the treatment. If the probability that you write down is greater than this fraction, then I will order the drug for your hospital stay; otherwise, I will enroll you for your stay without the drug and wish you well in bearing the treatment."

Reflection

Please write down on a piece of paper the probability of death, *p*, by the drug that would make you just indifferent to enduring the pain.

33.2.2 The Equivalence of Choices

Presuming that you do not find suicide attractive, you are going to choose either the painful treatment or one of the two drugs. Your three alternatives are shown in the Figure 33.1.

The first alternative is to endure the pain. The second alternative is to pay some amount x for the drug that will provide a painless cure with no side effects. You have determined the amount x that would make you just indifferent between this and the first alternative. The third alternative is to take a free drug that will provide a painless cure but with some probability p of killing you. You have determined the value of p that will make you indifferent to taking the drug and enduring the pain.

Note *This means that you have established equivalence between two ways of avoiding the pain: paying x dollars and accepting an incremental probability of death, p.*

You can pay in the currency of dollars or in the currency of probability of death. What is the exchange rate? If you simply divide x by p, you will probably find a very large number that may be difficult to interpret.

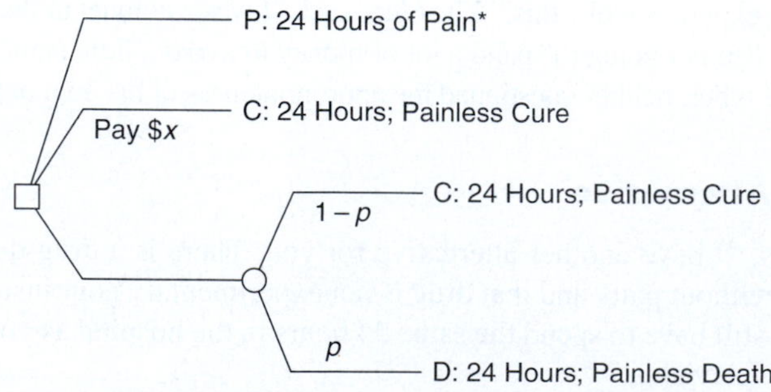

FIGURE 33.1 Three Equivalent Choices

33.3 SMALL PROBABILITIES

At this point, we turn to a discussion of how to deal with small probabilities. People often have trouble discussing and interpreting "small" probabilities. When doctors discuss the possibility of side effects with the intention that their patients will be able to provide "informed consent" for a medical procedure, they often consider probabilities less than 1/100 to be "too small to worry about." Yet we often take expensive and burdensome precautions, like wearing seat belts and helmets, to guard against a probability of death that is less than one in 100 per year. One reason for this difficulty is that we do not have an appropriate unit of probability for discussing "small" probabilities.

Having an inappropriate unit for any quantity can lead to inconvenience, if not confusion. If the focal length of common camera lenses were commonly measured in kilometers rather than millimeters, you would have to ask the clerk in a camera store about a 0.000050 km focal length lens, rather than a 50 mm focal length lens.

Yet it may take some time for most of us to become familiar with a new unit of measurement. Consider, for example, miles per hour (or kilometers per hour in the metric system). Such a unit of speed would make no sense to someone living at the turn of the 18th century. Conversely, if you asked someone how fast his horse could run, he might tell you that he could get to the next town in half an hour, but I doubt that any mention of miles per hour would enter into the discussion. Not until we had relatively constant velocity vehicles, such as trains, did the unit of velocity become widely used and understood.

Returning now to our discussion of small probabilities, we define a unit of probability of 1/1,000,000, as a **microprobability (μp)**. To get a feeling for the unit, the chance of getting 10 heads in 10 tosses of a coin with probability 1/2 of heads is about 1000 microprobabilities (actually, 976.6). The probability of getting 20 heads in 20 tosses of such a coin is about one microprobability (actually, 0.9537). The chance of dealing a royal flush—ace, king, queen, jack, ten in the same suit—is about 1.5 microprobabilities (actually, 1.539) if you are dealing from the apocryphally well-shuffled deck.

We can use our discussion of small probabilities to obtain more insight into the mortality process. We define a microprobability of death, which is a one in one million chance of dying, as a **micromort (μmt)**.

Now, we return to the exchange rate between money and probability of death, x and p, respectively, which we established earlier in our discussion of pain. If we divide this ratio by one million, forming

$$\frac{x}{1{,}000{,}000p}$$

we determine the exchange rate in the unit of dollars per micromort: our **micromort value**.

Reflection

Calculate your micromort value using your values of x and p.

33.4 USING MICROMORT VALUES

How do you decide whether an activity is in your best interests? You have to evaluate the benefits it provides, the impositions it creates, and the death or injury risks it entails. The benefits can include profits and enjoyments; the impositions can include costs and the willingness to avoid inconveniences.

As an example, suppose you are considering going on a ski trip. First, consider the benefits. You can think of how many minutes you will spend curving downhill in the clear and frosty air, attach a value to each, and multiply. You could also add the benefits of enjoying the scenery or a comfortable evening in the lodge. If you are spending time with co-workers or clients and advancing your position on the job, you may want to assign a dollar benefit to this feature.

Next, we come to the impositions, beginning with the costs. First, there are the costs of things such as transportation, lodging, lift tickets, and equipment. Then, there is what we might call cost of *hassle*—the minutes of waiting in lift lines or the hours spent in traffic jams on the way home.

Finally, we come to the risks of injury and death. For injuries, you would have to include your evaluation of the related pain and your willingness to pay to avoid the potential for days or weeks of restricted activity. We don't usually think of death risks in connection with skiing, but they do exist—both in the travel to and from the skiing destination and when on the slopes. We will now examine these death risks in more detail.

For the average American, motor vehicle accidents provide about one-half micromort per day. The typical skiing trip by automobile easily provides one micromort. But what about the death risk from downhill skiing itself?

Figure 33.2, shows that downhill skiing provides a risk of from 1 to 10 micromorts per day.

Ski Resorts	**Miccromarts Per Skiing Day**
Breckinredge, Colo.	1.5 μmt
Keystone, Colo.	0.87 μmt ←
Mammoth Lakes, Calif.	3.9 μmt
Vali, Colo.	9.4 μmt ←
Winter Park, Colo.	2.9 μmt

FIGURE 33.2 Microrisks in Skiing Resorts

If we choose three micromorts as a typical number and say that the trip lasts two days, we would have six micromorts from the skiing plus another from the driving for a total of seven micromorts. Using a nominal number like $10 per micromort would yield a death risk valuation of $70 for the entire trip. As with the costs, you need to subtract this amount from the benefits of the trip to see whether the trip is in your interest by being "profitable."

33.4.1 Skydiving Illustration

Once, after listening to this material, a student wanted to see whether he should continue skydiving. We followed the procedure we have outlined, assessing the benefits, costs, hassle costs, and, finally, assessing the death risk. It turned out that the activity was, in fact, profitable for him: He was getting a net benefit from skydiving.

However, when we assessed the higher risk he had faced as a beginner in the sport, we determined that he should never have started skydiving because of the much greater accident rate of beginners. Since he had survived the novice period though, the activity was now actually prudent for him.

He mentioned that he had just bought a device that would automatically deploy his parachute at a preset altitude should he, for any reason, be unable to release it himself. This could happen, for example, if he were somehow knocked unconscious exiting the airplane. We asked whether he would be interested in our new skydiving safety device (compliments of the wizard introduced in Chapter 9) that functions like a giant automobile airbag. If, for any reason, he approached the ground at a dangerous velocity, it would automatically lower him gently to earth. He was tempted for a moment, but then decided he wouldn't use it. He said to have perfect safety in an activity like skydiving would reduce it to the level of a theme park ride, where great danger is nothing but an illusion. Apparently, an essential element of the enjoyment of activities such as these is the knowledge that to some extent one is placing one's life on the line.

33.5 APPLICATIONS

The valuation and decision procedures described can be applied in several different settings. An indication of the kind of risks people typically face in life is shown in Figure 33.3.

Each of these statistics would have to be modified by the individual to account for personal behavior. For example, even though the table shows an annual death risk from airplane accidents of 0.9 micromorts per year, someone who never flew would not have to face this risk. However, they would still have to contend with the risk of falling airplanes, 0.06 micromorts per year, or more appropriately, 60 **nanomorts** per year. A nanomort is a one in one billion chance of death.

33.5.1 Driving

The risk of being the driver of an automobile has been extensively studied. One study estimated the risk of death to be about 12.5 micromorts per thousand miles driven. However, if the driver was a sober 40-year-old wearing seat belts and driving a car that weighed 700 lbs. more than average on a rural interstate highway, the risk went down to 0.8 micromorts per thousand miles. If the driver was an unbelted 18-year-old male with a blood alcohol

RISK SCALE

Odds are based on the number of deaths in the United States each year.

Decreasing Risk

Cause	Odds	μmt	nmt
All causes, age 80	8 in 100	80000	
Heart disease, age 65 and older	2 in 100	20000	
Stroke, age 65 and older	4 in 1000	4000	
Cancer, ages 45 to 64	25 in 1000	2500	
Lung cancer	6 in 10,000	600	
Unintentional injury	3.4 in 10,000	340	
Car accident	1.6 in 10,000	160	
Homicide	1 in 10,000	100	
Leukernia	7.5 in 100,000	76	
Accidental poisoning	3.5 in 100,000	35	
Fires/drowning	1.5 in 100,000	15	
Killed by a co-worker	9 in 1 million	9	
Tuberculosis	5 in 1 million	5	
Train accident	2 in 1 million	2	
Airplane accident	9 in 10 million	0.9	
Floods	4 in 10 million	0.4	
Lightning/insect sting	2 in 10 million	0.2	
Struck by fallen airplane	6 in 100 million	0.06	60
Hurricane	4 in 100 million	0.04	40

FIGURE 33.3 Common Risks of Living (Reprinted by permission from Risk Scale chart from Low Risk of Flying is Little Comfort After Crash, San Francisco Chronicle, Feb 5, 2000. pg. A1 by Carl T. Hall.)

level over 1.0 driving a car waiting 700 lbs. less than average on unspecified roads, the risk climbed to 931 micromorts per thousand miles or almost one micromort per mile. The factor of more than 1000 in risk shows the necessity of adjusting statistics to correspond to your own behavior.

33.5.2 Medical Error

Sooner or later, everyone will face the need for major or minor medical attention. In addition to the risks posed by the disease or injury, there is a risk of death from medical error. In the United States, there are about 32,000 deaths per year in hospitals due to medical error. Dividing this number by a population of about 280 million people would yield 114 micromorts per year per person from this cause. To find the death risk from medical error when a person enters the hospital, we must divide the number of deaths by the number of hospital admissions per year, about 35.6 million, and we learn that the death risk is 899 micromorts per hospital admission. Someone with a micromort value of $10 is incurring $9,000 of death risk just by being admitted to the hospital, regardless of complaint.

In the entire medical system, including hospitals, doctors offices, outpatient clinics, and elsewhere, there are 98,000 deaths per year from medical error in the population of 280 million, or 350 micromorts per person. For many of the people worried about the cost of medical care, a greater cost is their death risk from medical error.

Reflection: Calibrating your Micromort Value

The previous illustrations of micromorts in our daily lives serve as good mechanisms for calibrating our micromort values. Think about the amount of money you would be willing to pay to remove a particular risk (of low micromorts) from your daily life. If you observe a trend that the dollar values you would be willing to pay are higher than your micromort value multiplied by the probabilities of these small risks, then your micromort value might be too low and you should increase it. Conversely, if the dollar amount you would be willing to pay to avoid these risks is lower, than your micromort value multiplied by the probabilities of these risks, then you should decrease your micromort value.

33.6 FACING LARGER PROBABILITIES OF DEATH

The one consequence of life of which we can be sure is death; the only uncertainty is not whether, but when. Yet in the first several decades of life, the prospect of death seems remote and is seldom considered. We can understand this behavior by referring to Figure 33.4.

The graph shows the probability that a person will die in any year of life if that person is alive at the beginning of the year. It is called a mortality curve and is shown here for both males and females in the United States.

> ***Note*** *The male mortality curve lies above the female mortality curve consistent with the greater average lifetime of females.*

For the first several decades of life, the yearly probability of death is very small, but then it rises inexorably, until by age 95 when there is about a one-third chance of dying each year—even

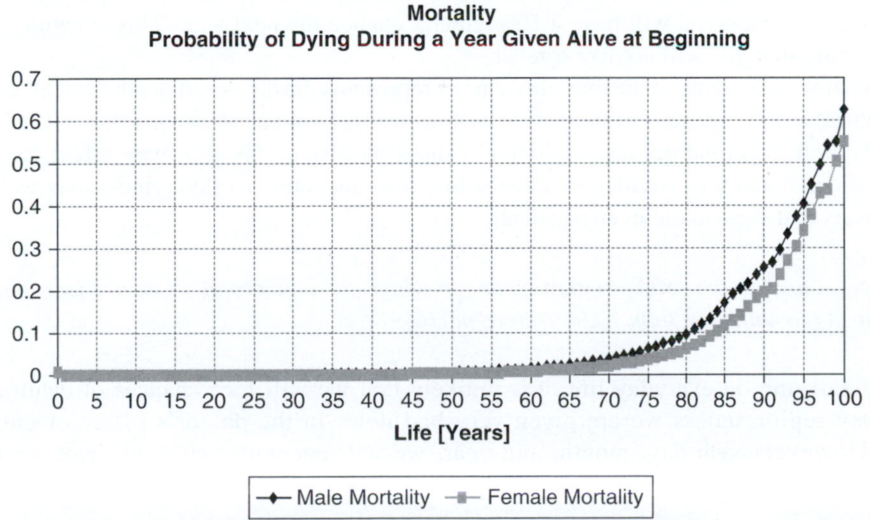

FIGURE 33.4 Probability of Dying in a Given Year

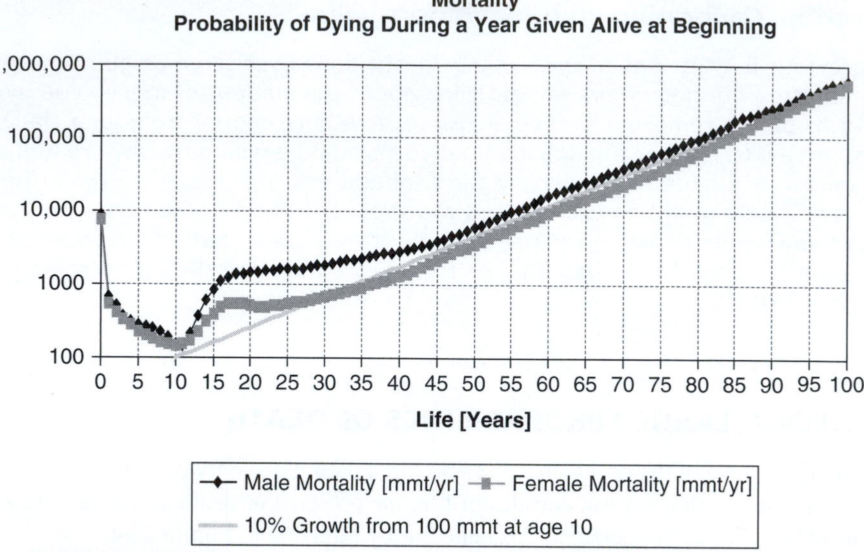

FIGURE 33.5 Probability of Dying During a Year Given Alive at Beginning

for women. We now plot the same mortality curves in the micromort unit and in a logarithmic scale, as shown in Figure 33.5.

We observe that the first year of life has a death risk close to 10,000 micromorts, which is a chance of death not reached again until people are in their sixties. However, after the first year, the risk of death falls until it is a minimum about age 10 and then increases through the teenage years, particularly for males. Note the straight line that begins at 100 micromorts at age 10 and increases at a 10% per year rate. This line closely parallels the male and female mortality curves after the forties. Therefore, in this period of life—no matter what chance of dying you have in any particular year—you will have a 10% greater chance the next year. This is nature's gift to you to ensure that you will not live forever.

What we are seeing in the mortality curves represents the chance of death from all causes—both natural and accidental. In the course of life, we make many decisions that will affect our mortality both from natural and accidental causes. Health habits may well affect the natural causes of death and the activities we engage in—from bicycling to skydiving—may well affect the chances that we will die in an accident.

It is useful to define a **safety region** *as one in which the chances of death are less than 1 in 100, or in our unit, less than 10,000 micromorts.*

In carrying on everyday life, it is unlikely that we will face chances of death outside the safety region unless we are given very bad news in the doctor's office or engage in battle. However, each day, month, and year, we will encounter risks of death within the safety region. For example, the chance of dying in an accident in a year in the United States is about 340 micromorts, or almost one micromort per day. The chance of dying in a motor vehicle accident is about 160 micromorts per year, or one half micromort per day. How can we make decisions about everyday activities that properly balance the prospect of death with other considerations?

Within the safety region—that is, for risks of death less than 10,000 micromorts—the micromort value is the amount you would pay to avoid incurring an additional micromort or that you would have to be paid to incur an additional micromort. If the number of micromorts under discussion falls outside the safety region, the micromort value we have computed would no longer apply. We will discuss this case again later.

Your micromort value should increase with income and decrease with age. In some sense, it is a measure of the desirability of your future life deal. Typical values might range from $1 to $20. Your micromort value should fit you like a suit of clothes. If you make your decisions using a micromort value that is too low, you will not feel safe. If you make your decisions using a micromort value that is too high, you will find yourself avoiding activities that seem perfectly acceptable to you.

As we will see in the next chapter, we can develop models to aid a person in determining his or her micromort value. Typical results of these models are shown in Figure 33.6.

The graph shows the micromort value per $10,000 of consumption (above the minimum level to sustain life) for males and females of various ages. For example, a 50-year-old male would have about a $1.00 micromort value for every $10,000 of annual consumption. If his consumption were $50,000 per year, then his micromort value would be $5.00. However, a woman of the same age and income would have a micromort value of about $1.30 per $10,000 annual consumption, or a micromort value of $6.50.

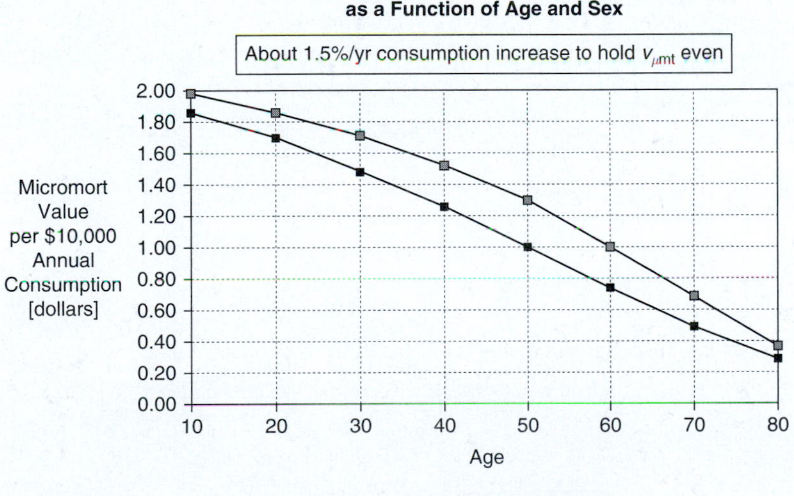

FIGURE 33.6 Micromort Dollar Values

33.7 SUMMARY

A microprobability (μp) is a unit of probability of 1 / 1,000,000.

A micromort is a microprobability of death.

A micromort value is the amount you would have to be paid (or pay) to be indifferent to an exposure (or be released from an exposure) of one micromort.

KEY TERMS

- Micromort value
- Microprobability
- Nanomorts
- Risky decision
- Safety region

PROBLEMS

1. Rita assesses the following curve representing her willingness to pay to avoid risks of death.

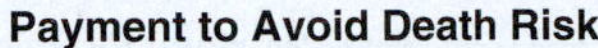

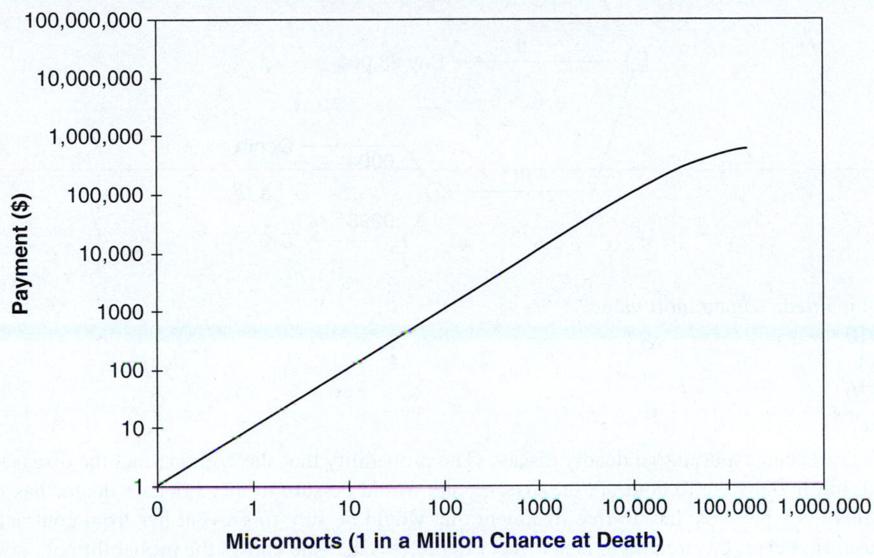

When given the following deal, Rita assigns a value of $p = 0.001$.

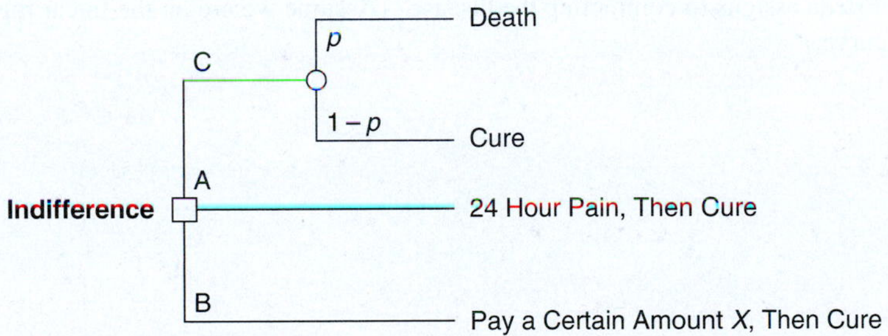

What value of X makes Rita just indifferent in the decision tree shown above?

a. $ 10
b. $ 100
c. Not enough information to tell
d. None of the above

2. The following situation applied to Q2 and Q3. Frieda is indifferent between alternatives I, II and III shown here.

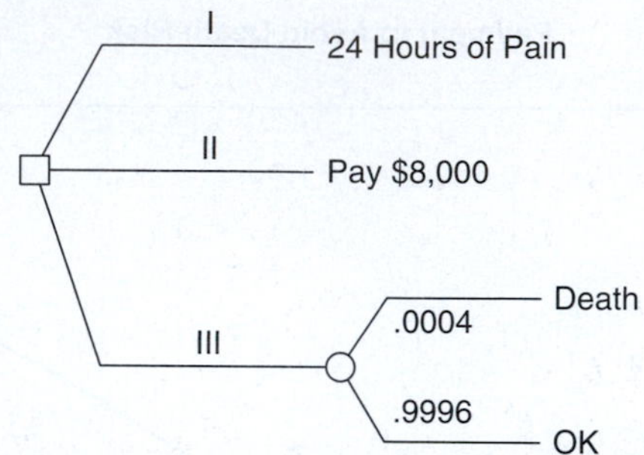

What is Frieda's micromort value?

a. $10
b. $4
c. $20
d. $8

3. Frieda has been exposed to a deadly disease. The probability that she will contract the disease is very small, but if she were to contract the disease, she would be sure to die. Frieda's doctor has told her that there is a painless, hassle-free treatment that would be sure to prevent her from contracting the disease. However, this treatment is not free. Frieda feels that she knows the probability of contracting the disease, and she has told her doctor that she is indifferent between paying $1,000 for the treatment and taking her chances with no treatment. Now, assume that her micromort value is $5. What is the probability Frieda assigns to contracting the disease? (Assume we are on the linear region of Frieda's micromort curve.)

a. .0002
b. 0.0001
c. 0.005
d. 0.002

CHAPTER

34

Decisions with a High Probability of Death

CHAPTER CONCEPTS

After reading this chapter, you will be able to explain the following concepts:

- Value function for:
  - Remaining life years
  - Consumption
- Determining micromort values:
  - The black pill
  - The white pill
- Equivalent perfect life probability

34.1 INTRODUCTION

In the last chapter, we discussed micromorts and decisions with a low probability of death. This chapter illustrates how to think about decisions involving larger probabilities of death. These include such decisions as whether to engage in combat or how to handle a life-threatening medical ailment.

The five Rules of Actional Thought also apply to decisions involving a high probability of death. In most cases, these decisions will involve more than direct value attribute. To handle such decisions, we first characterize them by creating necessary distinctions and associated probabilities. Next, we represent ordinal preferences and trade-offs among the attributes using a value function. Then, we assign a u-curve over value to represent preferences over uncertain deals. The next sections illustrate how to construct a value function and u-function for decisions involving a large probability of death.

34.2 VALUE FUNCTION FOR REMAINING LIFE YEARS AND CONSUMPTION

Two attributes that appear quite often in decisions involving life and death are length of life and consumption. We will now illustrate how to construct a value function for these situations and how to think about our trade-offs between these attributes.

Referring to our discussion of multiple attributes in Chapter 28, suppose that a person has two attributes of interest: constant annual consumption, c, for each year of life and remaining lifetime, l. **Constant annual consumption** is the fixed amount of money that you are indifferent to receiving every year for the rest of your life in exchange for all of your paychecks, commissions, and other sources of income that you would otherwise receive. We consider constant annual consumption to be that above the level of survival—an amount below which the person would rather be dead.

To represent preferences over constant annual consumption and lifetime, we need to think about the qualitivative properties required for this decision. We have seen that an additive value function of the form

$$V(c, l) = ac + bl$$

where a and b are constants with several undesirable qualitiative features when it comes to health and wealth. For example, it would imply that trade-offs between consumption and remaining lifetime are constant regardless of the wealth level or the remaining life time, since

$$\frac{\Delta c}{\Delta l} = -\frac{b}{a}$$

It is reasonable to assume that such trade-offs are not constant and that the increase in constant annual consumpotion that would make us indifferent to losing a certain amount of remaining lifetime would actually depend on the remaining lifetime.

Another type of functions that would avoid this problem is a value function of the form

$$V(c, l) = c\left(\frac{l}{\bar{l}}\right)^{\eta}$$

where $\bar{l}$ is the mean of the remaining lifetime distribution and η is the trade-off parameter discussed in Chapter 28. This value functions implies that the ratio of small fractional changes in c and l are constant, i.e., for small Δc and Δl, we have

$$\frac{\Delta c/c}{\Delta l/\bar{l}} = -\eta$$

The mean $\bar{l}$ can be determined from mortality tables as we illustrate in Figures 34.1 and 34.2.

> ***Note*** *If the consumption is* 0, *then the value is* 0, *since the decision maker is just at survival level. Furthermore, if the remaining lifetime is* 0, *then the value is also* 0.

34.2.1 Determining $\bar{l}$

The value of $\bar{l}$ depends on the particular person making that decision. Here, we will examine the case of a 30-year-old male rulesperson who uses a standard mortality table to describe his probability distribution on the length of his remaining life, $\{l|\&\}$. Of course, he should modify it according to any health or genetic issues that might affect his longevity. As an example, we will use the distribution for lifetime, l, shown in Figure 34.1, and the remaining lifetime distribution of Figure 34.2. The mean of the distribution, $\bar{l}$, in Figure 34.2 is 43.6 years. Appendix A at the end of this chapter provides the values for this lifetime distribution given you are alive at the age of 30.

To complete the value function, the man would have to assess both c and η.

34.2.2 Determining c

To simplify assessing c, we ask the 30-year-old man to specify:

> *The amount of annual consumption above survival in constant dollars for the rest of his life that would make him indifferent to his current economic future.*

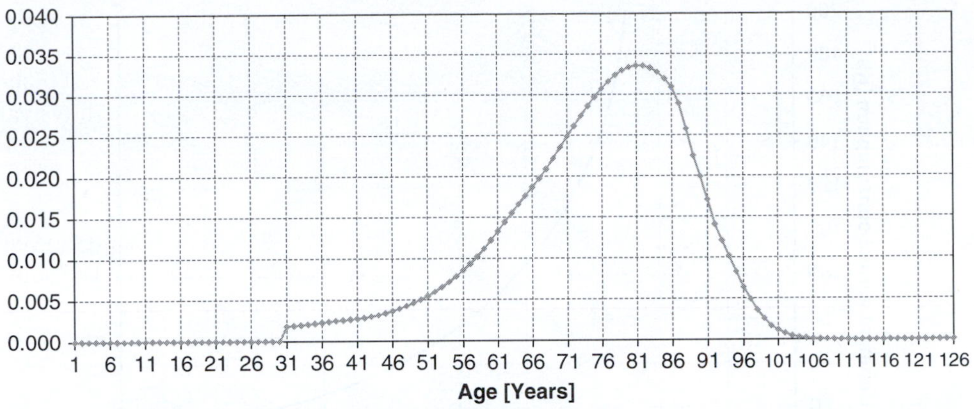

FIGURE 34.1 Probability Distribution on Year of Death for Male Age 30

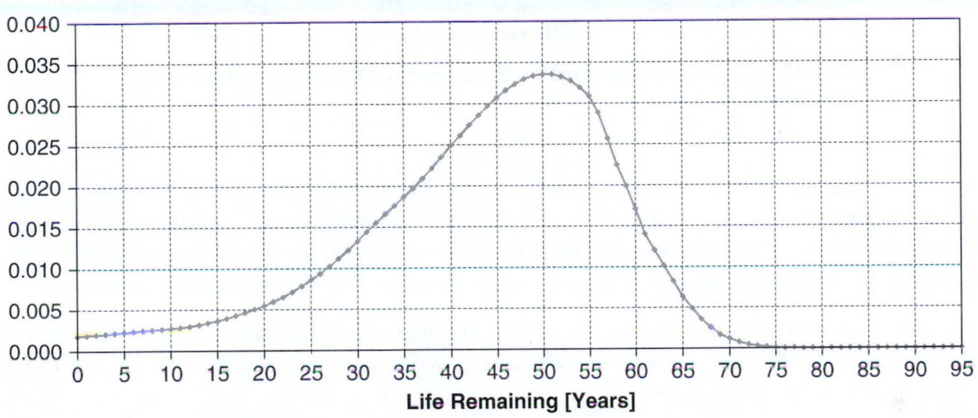

FIGURE 34.2 Remaining Life Distribution for Male Age 30

To simplify this assessment, it would be helpful for him to consider different scenarios. For example, the man can think about possible future jobs and associated potential earnings, or he can consider possibilities of entrepreneurial ventures. He then needs to think about the fixed amount that he would need to receive each year to be indifferent. The man also needs to specify the minimal survival consumption level: The value he needs to just survive. This includes shelter and minimum consumption. After some thought, he specifies $c = \$65{,}000$.

34.2.3 Determining η

To assess η, we note that a small percentage change in length of life will require a percentage change in constant annual consumption η times as large in the opposite direction for the person to remain indifferent, i.e., for small Δc and Δl, we have

$$\eta = -\frac{\Delta c/c}{\Delta l/\bar{l}}$$

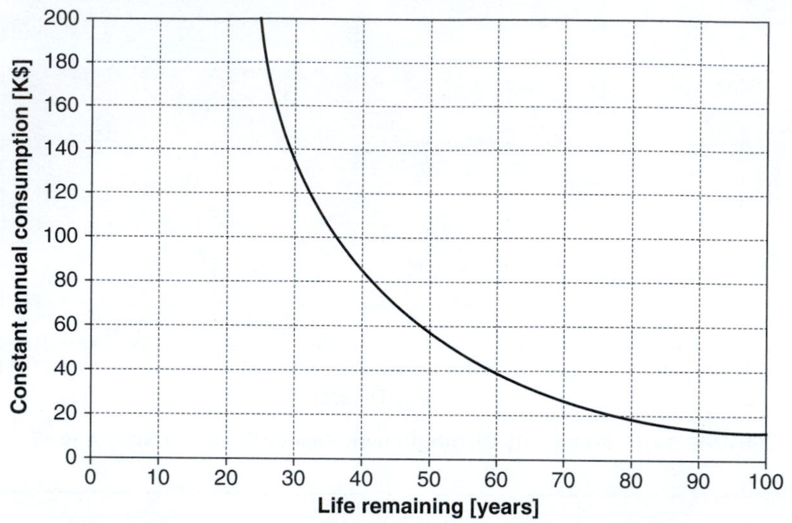

FIGURE 34.3 The Value Function for Constant Annual Consumption and Remaining Life Years

Therefore, if η equals 2, a 1% reduction in remaining life would have to be compensated by a 2% increase in constant annual consumption. Similarly, a 1% increase in remaining lifetime would just be balanced by a 2% decrease in constant annual consumption. Our 30-year-old male specifies that η is 2. With these specifications, the value function $V(c, l)$ is shown in Figure 34.3.

> ***Note*** *This man would require an increase in constant annual consumption from* \$65,000 *per year to* \$140,000 *per year to compensate for a decrease in mean remaining lifetime from* 43.6 *years to* 30 *years.*

34.3 ASSIGNING A *u*-CURVE OVER THE VALUE FUNCTION

We have determined the parameters of the value function. We now add the dimension of risk preference. To construct a multiple attribute u-function, we assign a u-function over the value function as we discussed. If the 30-year-old man is a deltaperson, this will be exponential, as

$$u_V(V) = -e^{-\gamma V}$$

Hence, by assigning this u-function over value, we get

$$u(c, l) = u_V(V(c, l)) = -e^{-\gamma c\left(\frac{l}{\bar{l}}\right)^{\eta}}$$

Here γ is the risk aversion coefficient for value.

34.3.1 Determining γ

To assess γ, we simply give the person the following deal in Figure 34.4. In this deal, he either keeps his constant annual consumption, c, at $\bar{l}$, or he gets a binary deal where he can double c or halve it with a probability, p, at $\bar{l}$. Perhaps it will be easier for him to think of this as a deal where he is equally likely to have all income amounts he receives in his life be either twice as big or half as big.

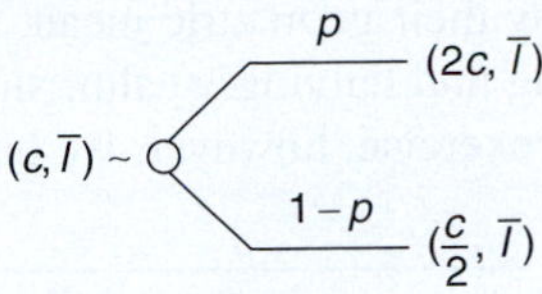

FIGURE 34.4 Assessing the Risk-Aversion Coefficient

Following the equivalence rule, the decision maker specifies the indifference probability, p, or the **doubling probability**, that would make him indifferent to the deal in Figure 34.4. Within this deal, the man is asked to think about the following.

1. A deterministic life deal where he lives exactly $\bar{l}$ years and receives the constant annual consumption, c.
2. A deal where he lives exactly $\bar{l}$ years, but he receives either $2c$ for the rest of his life with probability p or receives $c/2$ for the rest of his life with probability $1 - p$.

From his response to this deal, we can determine the risk-aversion coefficient as

$$u(c, \bar{l}) = pu(2c, \bar{l}) + (1 - p)u\left(\frac{c}{2}, \bar{l}\right)$$

Hence,

$$-e^{-\gamma c\left(\frac{\bar{l}}{\bar{l}}\right)^{\eta}} = -pe^{-\gamma 2c\left(\frac{\bar{l}}{\bar{l}}\right)^{\eta}} - (1 - p)e^{-\gamma\frac{c}{2}\left(\frac{\bar{l}}{\bar{l}}\right)^{\eta}}$$

which implies

$$-e^{-\gamma c} = -pe^{-\gamma 2c} - (1 - p)e^{-\gamma\frac{c}{2}}$$

If we know c and p, we can solve for γ.

Note *The resulting exponential u-curve will be on the value measure: Constant annual consumption at mean remaining lifetime.*

EXAMPLE 34.1 Calculating the Value of γ

To help our 30-year-old male deltaperson determine his γ, we ask him the following question. Suppose you were going to live exactly $\bar{l} = 43.6$ years and you had the opportunity to exchange your current constant annual consumption c for a deal that could either double it or halve it. What chance of doubling versus halving would make you just indifferent to the exchange?

A rulesperson must be able to answer this question, since for the equivalence rule, $2c$ is preferred to c, and c is preferred to $c/2$. This is his doubling probability.

The 30-year-old male in our example states a doubling probability of 0.8. Since he has already stated that $c = \$65,000$, this preference statement implies a risk-aversion coefficient γ of 4.84×10^{-5} and a risk tolerance $\rho = 1/\gamma$ of \$20,641.

People usually assign a doubling probability considerably above 0.5. Most people view losing half their lifetime income as much less desirable than doubling it. However, as we discussed in Chapter 22, a logarithmic person values deals by their geometric mean. Such a person would be just indifferent to equal chances of doubling and halving wealth, since the geometric mean of the deal is $(2c)^{0.5}(0.5c)^{0.5} = c$. In this exercise, however, we have not encountered any logarithmic people.

Note: We have now determined all of the parameters needed for the *u*-function for this example. Given any decision, medical or personal, we simply choose by calculating the alternative with the highest *e*-value of *u*-values of its prospects. In using this life model, we must realize that all assessments are dependent on your particular life situation. For example, if you should learn that you had contracted a disease that would make your life expectancy much shorter, you would create a life model with new parameters before making any decisions that rested upon it.

Once the *u*-curve has been established, we can use it to make decisions involving attributes of remaining life and consumption. Of particular interest, we shall present two decisions; the black pill and the white pill, that will relate this *u*-curve assessment to the micromort values discussed in Chapter 33. Before embarking on this undertaking, it will be helpful to consider—in the case of our 30-year-old man—the calculation of an annuity that pays a fixed amount for the rest of his life. This will be useful in calculating the amount of money that the decision maker would need to be paid each year in exchange for being exposed to a certain probability of death. In Chapter 28, we derived an expression for valuing this annuity payment. The following example illustrates this value for our 30-year-old man.

EXAMPLE 34.2 Value of an Annuity for 30-Year-Old Man

Following our discussion about time preference and multiple attribute problems in Chapter 28, a company selling an annuity determines its present equivalent by using an interest rate i to discount future payments. Suppose the company promises to pay the man one dollar per year of his life, starting immediately. If he lives m future years, the company will have paid out $m + 1$ dollars; we designate this cash flow of a dollar every year by $a(m)$.

The present equivalent to the person for 1 dollar every year, $pe[a(m)]$, is

$$pe[a(m)] = 1 + \beta + \beta^2 + \cdots + \beta^m$$
$$= \frac{1 - \beta^{m+1}}{1 - \beta}, \quad m = 0, 1, 2 \ldots$$

where m is the number of remaining life years and β is the present value of a unit payment one year in the future at the prevailing interest rate i, as

$$\beta = \frac{1}{1 + i}$$

Since the lifetime, m, is uncertain, the term $pe[a(m)]$ is also uncertain. The *e*-value of the present equivalent of the annuity to the company is therefore

$$<pe[a]|\&> = \sum_{m=0}^{n} p_m \frac{1 - \beta^{m+1}}{1 - \beta} = \frac{1 - <\beta^{m+1}|\&>}{1 - \beta}$$

where

$$<\beta^{m+1}|\&> = \sum_{m=0}^{n} p_m \beta^{m+1} = \sum_{m=0}^{n} \{m|\&\}\beta^{m+1}$$

A risk-neutral annuity company would have to receive an amount equal to the *e*-value of the present equivalent of a unit annuity, $<pe[a]|\&>$, to be able to fund it. If the company has an interest rate i of 3% for annuities paid in constant (inflation protected) dollars, β is $1/1.03 = 0.9709$. For the remaining life distribution of the 30-year-old male shown in Figure 34.2, which we presume is shared by the annuity company, we compute

$$<pe[a]|\&> = 24.04$$

Appendix A illustrates the calculation of this value.

The term $<pe[a]|\&>$ is the amount of money the annuity company would need to be paid to pay \$1 to the man every year for the rest of his life. Equivalently, if the company were to receive any amount x to fund an annuity, it would pay the man an annual amount ζx, where

$$\zeta = 1/<pe[a]|\&>$$

From the results of Appendix A,

$$\zeta = \frac{1}{<pe[a]|\&>} = \frac{1}{24.04} = 4.16\%$$

If the man gave the annuity company x, he would receive ζx immediately and for each year in the future that he is alive. For example, if he paid \$10, 000 to the annuity company, he would receive \$416 immediately and in each year in the future that he is alive.

34.4 DETERMINING MICROMORT VALUES

In this section, we will show how the *u*-function we have developed can help determine several characteristics of the decision maker, such as the micromort value, the maximum probability of death he would be willing to experience for any payment, and the maximum amount he would pay to avoid death.

34.4.1 The Black Pill Proposition

Suppose we offered our 30-year-old man an opportunity to take a special pill that is one of 10,000 indistinguishable pills in an urn, as shown in Figure 34.5.

The pill is a *black pill*, one that will kill him instantly and painlessly should he choose it. How much would he have to be paid to take the chance of drawing the black pill? What is his Personal Indifferent Selling Price (PISP) for the act of swallowing a pill from the urn? In other words, how much would we have to pay him to take on 100 micromorts or to sell 100 micromorts of safety? We can draw a decision tree for this decision, as shown in Figure 34.6.

Suppose that he is offered an amount of money x in exchange for a probability p of dying. If he refuses to take the pill, he continues with his present life and faces his future life deal. Note that this future life deal is full of *uncertainty* in such matters as wealth, health, and length of life,

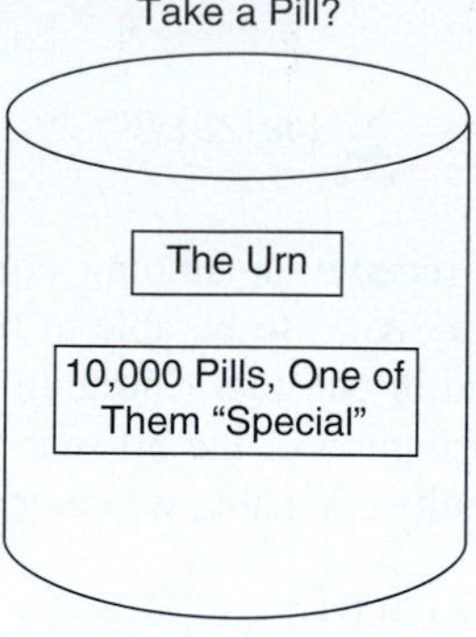

FIGURE 34.5 The "Black" Pill

but it is what it is. If he accepts the deal, he will be paid x, but with probability p, he will die an instant painless death, and he cannot bequest the money to anyone—not even his favorite university or charity. If he accepts the deal and survives, which he will with probability $1 - p$, he will face his future life deal augmented by the amount x that he was paid.

He prefers the prospects in this order, from most to least preferred: future life deal augmented by x, then the future life deal, and finally, m—instant painless death. As a rulesperson using the equivalence rule, for any x, he will be able to specify a p that will make him indifferent between the central prospect and the deal involving the best and worst prospects. Conversely, for any p, he will be able to find an x that will make him indifferent to the central prospect.

We will now build a model that will not only let you compute an appropriate micromort value but also will help you make decisions in situations that pose a major risk to life outside the region where a micromort value is all that is needed.

The black pill proposition offers a sum of money for taking the pill now. If he survives, he receives this money in equal annual installments for the rest of his life. In other words, he will receive his new income in the form of an *annuity*. We have previously determined how large an annuity the man will receive in exchange for a payment x.

The decision maker has now assessed:

- c, his constant annual consumption above survival for his remaining life. This reflects his current wealth.
- $\{l|\&\}$, his probability distribution on remaining lifetime, and, therefore, his mean remaining lifetime $\bar{l}$.
- η, to show his trade-off between consumption and length of life.

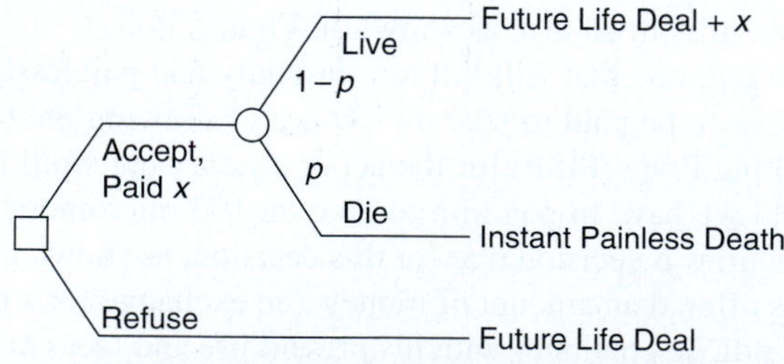

FIGURE 34.6 Decision Tree for Black Pill

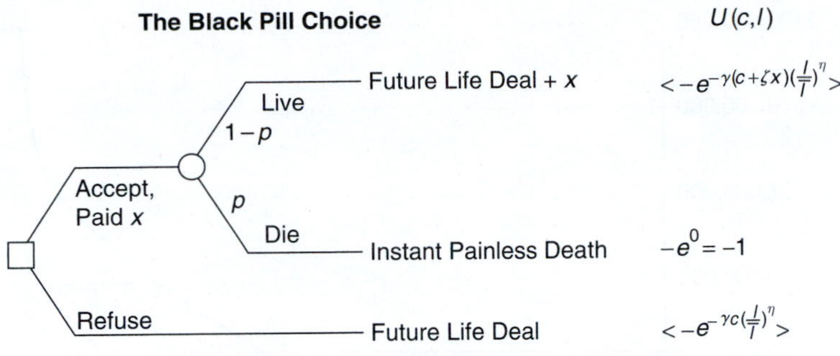

FIGURE 34.7 The Black Pill Decision Problem

- His doubling probability and, therefore, his exponential u-curve on constant annual consumption described by the risk-aversion coefficient γ.
- i, which is the prevailing interest rate that permits calculating β, and the annuital factor ζ that allows any cash payment now to be turned into annual income for as long as he lives.

We can now modify the decision tree for the black pill by presenting the u-values for the the prospects, as shown in Figure 34.7.

***Note** The prospects "Future Life Deal + x" and "Future Life Deal" involve uncertain life years and so we have taken the e-value of the u-values.*

At the point of indifference, we equate the e-values of the u-values in Figure 34.7 to get

$$(1-p) < -e^{-\gamma(c+\zeta x)\left(\frac{l}{l}\right)^{\eta}} | \& > + p(-1) = < -e^{-\gamma c\left(\frac{l}{l}\right)^{\eta}} | \& >$$

Rearranging gives

$$(1-p) < e^{-\gamma(c+\zeta x)\left(\frac{l}{l}\right)^{\eta}} | \& > + p = < e^{-\gamma c\left(\frac{l}{l}\right)^{\eta}} | \& >$$

Rearranging again gives

$$p = \frac{< e^{-\gamma c\left(\frac{l}{l}\right)^{\eta}} | \& > - < e^{-\gamma(c+\zeta x)\left(\frac{l}{l}\right)^{\eta}} | \& >}{< 1 - e^{-\gamma(c+\zeta x)\left(\frac{l}{l}\right)^{\eta}} | \& >}$$

This equation has two variables, x and p. For a given value of x, we can determine the corresponding value of p. As an example, Appendix B shows the calculation of p for $x = 10{,}000$ as 0.001. The plot of x versus p on logarithmic scales for the 30-year-old male appears in Figure 34.8. This shows what he would have to be paid to take on an additional risk of death or, in other words, to sell some of his safety.

Note that the curve is approximately linear for values of p smaller than 0.1, which we call the **safety region**. The slope of this linear curve is, in fact, the micromort dollar value. This 30-year-old male has a micromort value of about \$10, as estimated in Figure 34.8. As the number of micromorts increases by a factor of 10, so does the required payment, until the probability of death exceeds 10,000 micromorts and is, therefore, beyond the safety region. The 100-micromort black pill proposition using the urn is well within the safety region and would require a payment of about \$1,000. The value of a micromort to this person is, therefore, \$10 within the safety region, but it increases rapidly outside it until a death probability just above 100,000 micromorts—or one chance in 10—when the required payment becomes infinite.

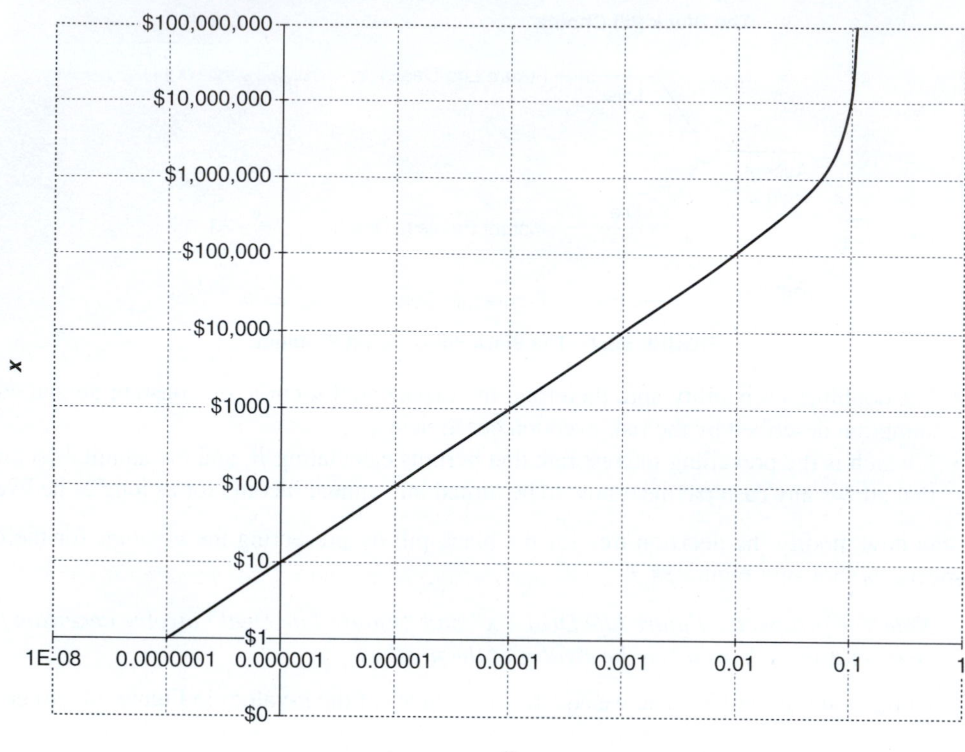

FIGURE 34.8 ***x* vs. *p* for Black Pill Problem**

This means that, for any probability of death in excess of this probability, there is no amount of money that could entice him to take on such a risk. We assign the name p_{max} to this maximum probability of death that would be acceptable for any amount of money. By setting x to be very large, direct substitution shows that

$$p_{\max} = < e^{-\gamma c\left(\frac{l}{i}\right)^{\eta}} | \& >$$

From the calculations of Appendix B, our 30-year-old male has a p_{max} equal to 0.115. Indeed, not many people would be willing to play the Russian roulette game[1]—no matter how much money they are paid.

34.4.2 The White Pill Proposition

The black pill proposition offers money to a person willing to take on an additional chance of death. Suppose the person already faces such a risk and is offered an opportunity to pay to avoid it. We call this the white pill proposition. You can think of the white pill as an antidote to a black pill that you may already have taken. If you were already facing a chance p of death, how much x would you pay to avoid it? The decision tree appears in Figure 34.9.

---

[1] Russian roulette is a game where the player puts a single bullet in a six-chambered revolver, spins the cylinder, puts the gun to his head, and pulls the trigger. He has a one in six chance of death.

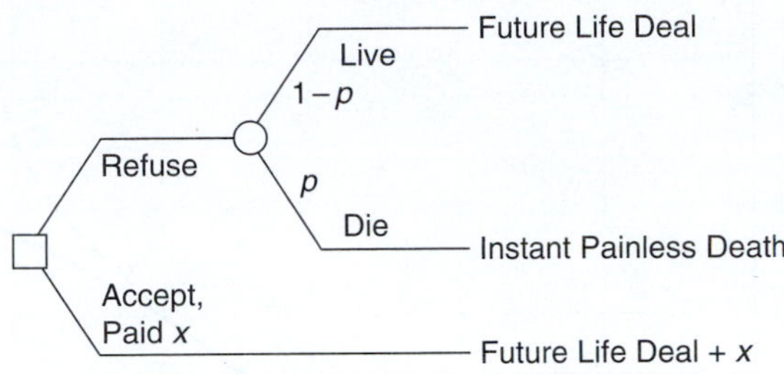

FIGURE 34.9 Decision Tree for White Pill

If our man refuses the white pill, he dies an instant painless death with probability p. With probability $1 - p$, he will live and face his future life deal. If he pays x for the white pill, he will face his future life deal with his bank account decreased by x. The future life deal less x is now the central prospect between the better one of the future life deal without payment and the worse one of an instant painless death.

Following the procedure of using an annuity to adjust lifetime consumption used in analyzing the black pill problem, we presume that he finances the payment x by decreasing yearly annual consumption by ζx.

We can modify the decision tree using the same value measures shown in Figure 34.10.

Equating the e-values of the u-values at indifference gives

$$(1 - p) < e^{-\gamma c \left(\frac{l}{l}\right)^{\eta}} | \& > + p = < e^{-\gamma (c - \zeta x)\left(\frac{l}{l}\right)^{\eta}} | \& >$$

Rearranging gives

$$p = \frac{< e^{-\gamma (c - \zeta x)\left(\frac{l}{l}\right)^{\eta}} | \& > - < e^{-\gamma c \left(\frac{l}{l}\right)^{\eta}} | \& >}{1 - < e^{-\gamma c \left(\frac{l}{l}\right)^{\eta}} | \& >}$$

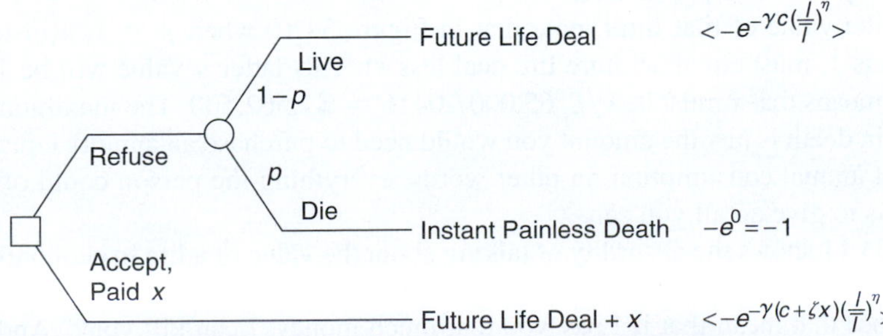

FIGURE 34.10 The White Pill Decision Problem

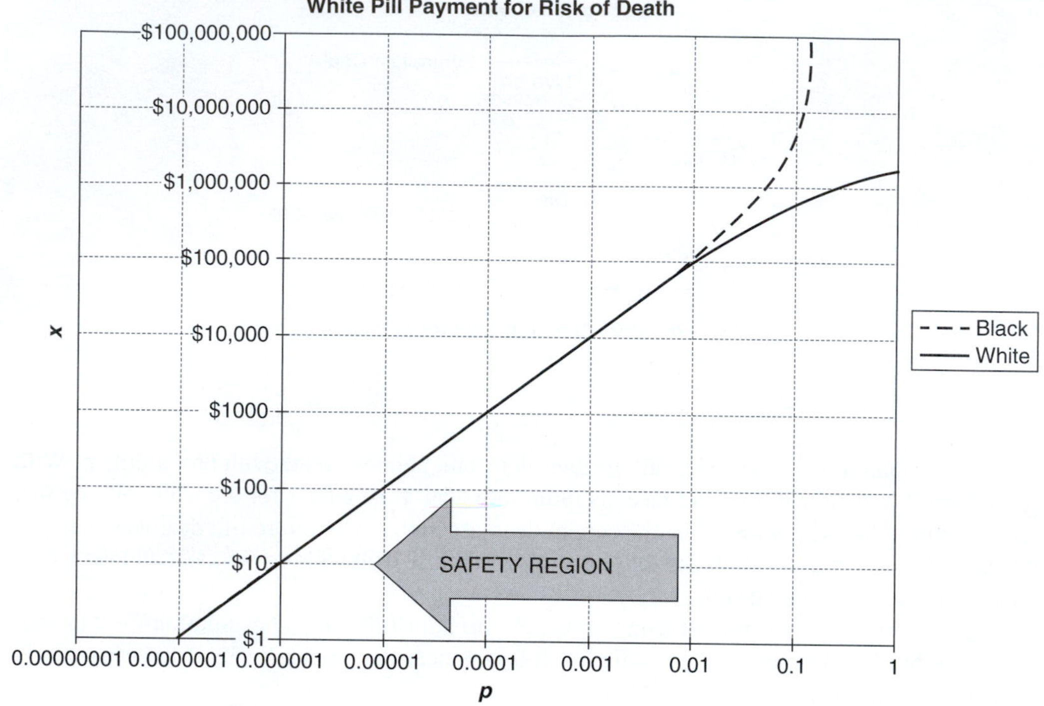

FIGURE 34.11 *x* **vs.** *p* **for Both Black and White Pill Problems**

Appendix C illustrates this calculation for $x = 10{,}000$ corresponding to a probability of $p = 0.001$.

Figure 34.11 plots x versus p for the 30-year-old male on the same logarithmic scales used for the black pill.

> ***Note*** *The curve for the white pill is virtually identical to that for the black pill within the safety region.*

The micromort value of $10 is appropriate for buying safety as well as for selling it within this region. However, as the probability of death rises beyond the safety region, the white pill curve bends down, showing that the value per micromort is decreasing. In fact, there is a finite payment x that he could make to avoid a certain death—namely, his total wealth beyond the level of survival. While there is no amount that might induce a person to face certain death, there is a limit to what anyone could pay to avoid it.

To see the value of that limit, note that in Figure 34.10 when $p = 1$, u(instant painless death), which is 1, must equal u(future life deal less x). This latter u-value will be 1 only when $c = \zeta x$. This means that x must be c/ζ, $65{,}000/.0416 = \$1{,}562{,}500$. The maximum payment x to avoid certain death is just the amount you would need to purchase an annuity equal to the person's constant annual consumption, in other words, everything the person could offer to avoid certain death is to give up all you can.

Figure 34.11 shows the absurdity of talking about the value of a life in economic terms. On occasion, a person will try to assign a monetary value to the value of a life. Someone might then ask them, "Does that mean that if I pay you that much money, I can kill you?" And, of course, their answer will be "No."

Yet, as the figure shows, it makes excellent sense to buy and sell risks to life in the safety region at your micromort value and to use the curves in Figure 34.11 to value probabilities of death even above the safety region.

34.4.3 Implications for Immediately Resolved Monetary Deals

This model has major implications for a person using it to evaluate decisions involving life and wealth. Suppose our 30-year-old male deltaperson faced a deal where he is equally likely to win \$100,000 or lose \$50,000. The risk tolerance of \$20,641 makes it appear that he would have no interest.

However, remember that this u-curve is on constant annual consumption—not wealth. If he decides to finance the deal by buying an annuity for the \$100,000 if he wins and by reducing his consumption sufficiently to pay for a \$50,000 annuity if he loses, he would convert this deal into one on constant annual consumption. With a probability of one-half, his constant annual consumption will increase by 100,000 ζ; with a probability of one-half, his constant annual consumption will decrease by 50,000 ζ. With ζ equal to .0416, he has an equal chance of increasing his annual consumption by \$4,160 and of decreasing it by \$2,080. Because these are well within his current annual consumption of \$65,000, he can afford the deal. Since he is a deltaperson, the certain equivalent change in his constant annual consumption is just the certain equivalent of [0.5, 4160; 0.5, −2080] using the risk tolerance 20,641, or \$805, which is an attractive increase (the e-value is \$1,040). Of course, this assumes that the annuity company is indifferent to buying and selling annuities. In addition, it is likely that he would have to pledge some property to ensure that he can make the payments on the \$50,000 annuity for as long as he lives—a condition that is beyond our model. Nevertheless, his risk tolerance on immediately resolved deals would be not \$20,641 but \$20,641/ζ or \$496,178. This person may be much less risk-averse than he believed by thinking of his whole lifetime prospects.

34.5 EQUIVALENT PERFECT LIFE PROBABILITY (EPLP)

The model we have developed thus far allows us to compute another quantity for an individual, the **Equivalent Perfect Life Probability (EPLP)**. Our reasoning is: If you accept the five Rules of Actional Thought, you are thereby a rulesperson. If you agree that if you face three prospects A, B, and C such that you prefer A to B and B to C, when you face a deal that could yield either A or C, and you will be able to state a preference probability of obtaining A rather than C that would make the deal equally preferred to receiving Prospect B for sure.

To continue, let Prospect A represent a prospect offered by our wizard of Chapter 9, which is perfect life in the terms we have been discussing. For example, you would live for 120 healthy years, enjoying an annual income of millions of dollars per year. The wizard will assure that if you receive Prospect A, you will lose all memory of how you came to receive it. You will just experience a very long, healthy, and wealthy life. Your doctor will be amazed that all of your old ailments have disappeared and that no new ones ever develop. The sudden arrival of millions of dollars each year will not surprise you, because you knew that eventually your valuable contributions would be appreciated.

Prospect C, also arranged by the wizard, is an instant, painless death.

Prospect B is your present life with all its vagaries in health and wealth attributes.

Unless you are on the brink of suicide, you will prefer *A* to *B* to *C*, and therefore, there is some probability of receiving *A* (the perfect life) versus *C* (instant painless death), which would make you just indifferent to the prospect *B* (continuing to live your present life without this deal). This is your **Equivalent Perfect Life Probability (EPLP)**.

A person with an EPLP of 0.93 is saying that his or her current life makes them just indifferent to exchanging it for a 0.93 chance of the perfect life we have defined and a 0.07 chance of instant painless death. I would not interpret the person's judgment as saying that the person's current life is 93% perfect.

The EPLP could become an alternative answer to the question "How's it going?" Instead of just saying "Fine," you could say, for example, "My EPLP is 0.92." Unfortunately, you are likely to be misunderstood and well may be considered rather bizarre by those unfamiliar with decision analysis.

The more promising the life and the future, the closer to 1 the EPLP will be; the worse, the closer to 0. We will not dwell on the philosophical implications of this quantity. However, it is worth mentioning how few of life's seemingly important events have any significant effect on our EPLP. Does seeing that your new car has received a ding in the parking lot change your EPLP? What about losing your job? Few financial reverses or windfalls would typically have the impact on EPLP occasioned by a major change in health, such as a diagnosis of cancer.

The model we have described for the individual allows us to compute the EPLP. We note that Prospect *A* with long life and great wealth will make the worth function $v(c, l)$ very large, and so $u(c, l) = -e^{-v(c, l)}$ is 0.

At the opposite extreme, Prospect *C*, which is instant painless death at $v(c, l)$, becomes equal to 0. So $u(c, l) = -e^{-v(c, l)} = -1$. The *u*-value of the present life situation $< u(c, l) | \& >$ must therefore equal

$$< u(c, l) | \& > = \text{EPLP}(0) + (1 - \text{EPLP})(-1)$$

or

$$\text{EPLP} = 1 - < e^{-\gamma c \left(\frac{l}{\bar{l}}\right)^{\eta}} | \& >$$

The equivalent perfect life probability in this model is just the *u*-value of the current life situation. By referring to the decision tree, we also see that for the black pill proposition when $p = p_{\max}$ we have

$$(1 - p_{\max})(0) + p_{\max}(-1) = < u(c, l) | \& >$$

Therefore,

$$\boxed{\text{EPLP} = 1 - p_{\max}}$$

The equivalent perfect life probability is just one minus the maximum probability of death that a person will accept—regardless of how much payment is offered. The 30-year-old man in our example therefore has an EPLP of 0.885.

Reflection

What is your EPLP?

34.6 SUMMARY

Decisions with large probabilities of death require the construction of a *u*-curve over a value function. Working with micromort values alone will not be sufficient. The value function used in these decisions often has the two attributes: remaining life and consumption.

Once the *u*-values have been constructed, we can use those values for any decision involving lifetime and consumption.

We can calculate the micromort value using both the black pill and white pill decisions presented in this chapter.

KEY TERMS

- Constant annual consumption
- Doubling probability
- Equivalent Perfect Life Probability (EPLP)
- Safety region

APPENDIX A Mortality Table for 30-Year-Old Male

We calculate the term

$$<pe[a]|\&> = \sum_{m=0}^{n} p_m \frac{1-\beta^{m+1}}{1-\beta} = \frac{1-<\beta^{m+1}|\&>}{1-\beta}$$

for β is $1/1.03 = 0.9709$, where $<\beta^{m+1}|\&> = \sum_{m=0}^{n} p_m\beta^{m+1} = \sum_{m=0}^{n} \{m|\&\}\beta^{m+1}$

The first column of Table 34.1 presents the age of the person. For ages greater than or equal to 30, we can define $m = \text{Age} - 30$. The probability of dying at any given year given the person was alive at age 30 is $\{m|\&\}$ and is given in the second column. The third column calculates the quantity β^{m+1} for a given year. The *e*-value of β^{m+1}, namely $<\beta^{m+1}|\&>$, is obtained by multiplying each cell in Column 2 with its corresponding cell in Column 3 and summing the result. The results show that $<\beta^{m+1}|\&> = 0.3$ and that $<l|\&> = 73.6$. Since the man is currently 30 years old, the *e*-value of remaining life time is 43.6. Substituting for

$$<pe[a]|\&> = \frac{1-<\beta^{m+1}|\&>}{1-\beta} = \frac{1-0.3}{1-0.9709} = 24.04$$

implies that $\zeta = \dfrac{1}{<pe[a]|\&>} = \dfrac{1}{24.04} = 4.16\%$.

TABLE 34.1 Mortality Table and Annuity Value for 30-Year Old Male

Age	{Death Probability |Alive at 30, &}	β^{m+1}
0	0	0.00
1	0	0.00
2	0	0.00
3	0	0.00
4	0	0.00
5	0	0.00
6	0	0.00
7	0	0.00
8	0	0.00
9	0	0.00
10	0	0.00
11	0	0.00
12	0	0.00
13	0	0.00
14	0	0.00
15	0	0.00
16	0	0.00
17	0	0.00
18	0	0.00

(Continued)

TABLE 34.1 (Continued)

Age	{Death Probability | Alive at 30, &}	β^{m+1}
19	0	0.00
20	0	0.00
21	0	0.00
22	0	0.00
23	0	0.00
24	0	0.00
25	0	0.00
26	0	0.00
27	0	0.00
28	0	0.00
29	0	0.00
30	0	0.00
31	0.002	0.97
32	0.002	0.94
33	0.002	0.92
34	0.002	0.89
35	0.002	0.86
36	0.003	0.84
37	0.003	0.81
38	0.003	0.79
39	0.003	0.77
40	0.003	0.74
41	0.003	0.72
42	0.003	0.70
43	0.003	0.68
44	0.003	0.66
45	0.004	0.64
46	0.004	0.62
47	0.004	0.61
48	0.005	0.59
49	0.005	0.57
50	0.005	0.55
51	0.006	0.54
52	0.007	0.52
53	0.007	0.51
54	0.008	0.49
55	0.009	0.48
56	0.009	0.46
57	0.01	0.45
58	0.011	0.44
59	0.012	0.42
60	0.013	0.41
61	0.014	0.40
62	0.016	0.39
63	0.017	0.38

(Continued)

TABLE 34.1 (Continued)

Age	{Death Probability |Alive at 30, &}	β^{m+1}
64	0.018	0.37
65	0.019	0.36
66	0.02	0.35
67	0.021	0.34
68	0.022	0.33
69	0.024	0.32
70	0.025	0.31
71	0.026	0.30
72	0.027	0.29
73	0.029	0.28
74	0.03	0.27
75	0.031	0.26
76	0.032	0.26
77	0.032	0.25
78	0.033	0.24
79	0.033	0.24
80	0.034	0.23
81	0.034	0.22
82	0.033	0.22
83	0.033	0.21
84	0.032	0.20
85	0.031	0.20
86	0.029	0.19
87	0.026	0.19
88	0.022	0.18
89	0.02	0.18
90	0.017	0.17
91	0.014	0.17
92	0.012	0.16
93	0.01	0.16
94	0.008	0.15
95	0.006	0.15
96	0.005	0.14
97	0.004	0.14
98	0.002	0.13
99	0.001	0.13
100	0.001	0.13
101	0.001	0.12
102	0	0.12
103	0	0.12
104	0	0.11
105	0	0.11
	1	
	$\langle l \| \& \rangle$	$\langle \beta^{m+1} \| \& \rangle$
	73.6	0.30

APPENDIX B Example of a Black Pill Calculation, $x = 10{,}000$

We now calculate the probability, p, for a payment of \$10,000 in the black pill problem for our 30-year-old man. We have

$$p = \frac{< e^{-\gamma c\left(\frac{l}{\bar{l}}\right)^{\eta}} | \& > - < e^{-\gamma (c+\zeta x)\left(\frac{l}{\bar{l}}\right)^{\eta}} | \& >}{< 1 - e^{-\gamma (c+\zeta x)\left(\frac{l}{\bar{l}}\right)^{\eta}} | \& >}$$

We are given the parameters: $\zeta = 0.0416, \bar{l} = 43.6$, and $\eta = 2, \gamma = 4.84 \times 10^{-5}$.

In Table 34.2, the first column indicates the remaining life time, l. The second column provides the probability of death at a given year given the man is alive at age 30. The third and forth columns compute the quantities $e^{-\gamma c\left(\frac{l}{\bar{l}}\right)^{\eta}}$ and $e^{-\gamma (c+\zeta x)\left(\frac{l}{\bar{l}}\right)^{\eta}}$ for a given year, l.

To calculate the e-value of these quantities, we multiply each probability in Column 2 with its corresponding value in Columns 3 and 4 and then sum over all possible values. Table 34.2 shows that

$$e^{-\gamma c\left(\frac{l}{\bar{l}}\right)^{\eta}} | \& > = 0.1152 \text{ and } e^{-\gamma (c+\zeta x)\left(\frac{l}{\bar{l}}\right)^{\eta}} | \& > = 0.1143$$

Substituting gives

$$p = \frac{0.1152 - 0.1143}{1 - 0.1143} = 0.001$$

TABLE 34.2 Black Pill Calculations for 30 Year Old Male

Age at Start of Year	Death Probability {l| Alive at 30, &}	$e^{-\gamma c\left(\frac{l}{\bar{l}}\right)^{\eta}}$	$e^{-\gamma (c+\zeta x)\left(\frac{l}{\bar{l}}\right)^{\eta}}$
30	0	0.0000	0.0000
31	0.002	0.9983	0.9983
32	0.002	0.9934	0.9934
33	0.002	0.9852	0.9851
34	0.002	0.9738	0.9737
35	0.002	0.9594	0.9592
36	0.003	0.9421	0.9417
37	0.003	0.9220	0.9216
38	0.003	0.8994	0.8988
39	0.003	0.8744	0.8737
40	0.003	0.8473	0.8464
41	0.003	0.8184	0.8173
42	0.003	0.7878	0.7866
43	0.003	0.7558	0.7545
44	0.003	0.7228	0.7213
45	0.004	0.6889	0.6872
46	0.004	0.6544	0.6526
47	0.004	0.6196	0.6177
48	0.005	0.5847	0.5827

(Continued)

TABLE 34.2 (Continued)

Age at Start of Year	Death Probability {*I*| Alive at 30, &}	$e^{-\gamma c\left(\frac{I}{i}\right)^{\eta}}$	$e^{-\gamma(c+\zeta x)\left(\frac{I}{i}\right)^{\eta}}$
49	0.005	0.5499	0.5478
50	0.005	0.5155	0.5133
51	0.006	0.4816	0.4794
52	0.007	0.4485	0.4462
53	0.007	0.4163	0.4140
54	0.008	0.3851	0.3828
55	0.009	0.3551	0.3528
56	0.009	0.3263	0.3240
57	0.01	0.2989	0.2966
58	0.011	0.2729	0.2706
59	0.012	0.2483	0.2461
60	0.013	0.2252	0.2230
61	0.014	0.2035	0.2015
62	0.016	0.1834	0.1814
63	0.017	0.1646	0.1627
64	0.018	0.1473	0.1455
65	0.019	0.1314	0.1297
66	0.02	0.1168	0.1152
67	0.021	0.1035	0.1020
68	0.022	0.0914	0.0900
69	0.024	0.0805	0.0792
70	0.025	0.0706	0.0694
71	0.026	0.0617	0.0607
72	0.027	0.0538	0.0528
73	0.029	0.0467	0.0458
74	0.03	0.0405	0.0397
75	0.031	0.0349	0.0342
76	0.032	0.0300	0.0294
77	0.032	0.0257	0.0252
78	0.033	0.0220	0.0215
79	0.033	0.0187	0.0183
80	0.034	0.0159	0.0155
81	0.034	0.0135	0.0131
82	0.033	0.0113	0.0110
83	0.033	0.0095	0.0093
84	0.032	0.0080	0.0077
85	0.031	0.0067	0.0065
86	0.029	0.0055	0.0054
87	0.026	0.0046	0.0044
88	0.022	0.0038	0.0037
89	0.02	0.0031	0.0030
90	0.017	0.0026	0.0025
91	0.014	0.0021	0.0020
92	0.012	0.0017	0.0016

(Continued)

TABLE 34.2 (Continued)

Age at Start of Year	Death Probability {*l* | Alive at 30, &}	$e^{-\gamma c\left(\frac{l}{l}\right)^{\eta}}$	$e^{-\gamma(c+\zeta x)\left(\frac{l}{l}\right)^{\eta}}$
93	0.01	0.0014	0.0013
94	0.008	0.0011	0.0011
95	0.006	0.0009	0.0009
96	0.005	0.0007	0.0007
97	0.004	0.0006	0.0006
98	0.002	0.0005	0.0004
99	0.001	0.0004	0.0004
100	0.001	0.0003	0.0003
101	0.001	0.0002	0.0002
102	0	0.0002	0.0002
103	0	0.0001	0.0001
104	0	0.0001	0.0001
105	0	0.0001	0.0001
	1	$< e^{-\gamma c\left(\frac{l}{l}\right)^{\eta}}$	$< e^{-\gamma(c+\zeta x)\left(\frac{l}{l}\right)^{\eta}}$
		0.1152	0.1143

APPENDIX C Example of a White Pill Calculation, $x = 10{,}000$

We now calculate the probability, p, for a monetary amount of \$10,000 in the white pill problem for our 30-year-old man. We have $p = \dfrac{< e^{-\gamma(c-\zeta x)\left(\frac{l}{\bar{l}}\right)^{\eta}} | \& > - < e^{-\gamma c\left(\frac{l}{\bar{l}}\right)^{\eta}} | \& >}{1 - < e^{-\gamma c\left(\frac{l}{\bar{l}}\right)^{\eta}} | \& >}$. Once again, we have the following parameters: $\zeta = 0.0416$, $\bar{l} = 43.6$, $\eta = 2$, and $\gamma = 4.84 \times 10^{-5}$.

In Table 34.3, the first column indicates the remaining life time, l. The second column provides the probability of death at a given year given the man is alive at age 30. The third and forth columns compute the quantities $e^{-\gamma c\left(\frac{l}{\bar{l}}\right)^{\eta}}$ and $e^{-\gamma(c-\zeta x)\left(\frac{l}{\bar{l}}\right)^{\eta}}$ for a given year, l.

To calculate the e-value of these quantities, we multiply each probability in Column 2 with its corresponding value in Columns 3 and 4 and then sum over all possible values. Table 34.3 shows that

$$< e^{-\gamma c\left(\frac{l}{\bar{l}}\right)^{\eta}} | \& > = 0.1152 \text{ and } < e^{-\gamma(c-\zeta x)\left(\frac{l}{\bar{l}}\right)^{\eta}} | \& > = 0.1161$$

Substituting gives

$$p = \frac{0.1161 - 0.1152}{1 - 0.1152} = 0.001$$

TABLE 34.3 White Pill Calculations for 30 Year Old Male

Age at Start of Year	Death Probability $\{l \mid \text{Alive at 30}, \&\}$	$e^{-\gamma c\left(\frac{l}{\bar{l}}\right)^{\eta}}$	$e^{-\gamma(c-\zeta x)\left(\frac{l}{\bar{l}}\right)^{\eta}}$
30	0	0.0000	0.0000
31	0.002	0.9983	0.9984
32	0.002	0.9934	0.9934
33	0.002	0.9852	0.9853
34	0.002	0.9738	0.9740
35	0.002	0.9594	0.9597
36	0.003	0.9421	0.9425
37	0.003	0.9220	0.9225
38	0.003	0.8994	0.9000
39	0.003	0.8744	0.8752
40	0.003	0.8473	0.8482
41	0.003	0.8184	0.8194
42	0.003	0.7878	0.7890
43	0.003	0.7558	0.7572
44	0.003	0.7228	0.7243
45	0.004	0.6889	0.6905
46	0.004	0.6544	0.6561
47	0.004	0.6196	0.6215
48	0.005	0.5847	0.5867
49	0.005	0.5499	0.5520
50	0.005	0.5155	0.5177

(Continued)

TABLE 34.3 (Continued)

Age at Start of Year	Death Probability {*l* | Alive at 30, &}	$e^{-\gamma c\left(\frac{l}{l}\right)^{\eta}}$	$e^{-\gamma(c-\zeta x)\left(\frac{l}{l}\right)^{\eta}}$
51	0.006	0.4816	0.4839
52	0.007	0.4485	0.4508
53	0.007	0.4163	0.4187
54	0.008	0.3851	0.3875
55	0.009	0.3551	0.3575
56	0.009	0.3263	0.3287
57	0.01	0.2989	0.3012
58	0.011	0.2729	0.2752
59	0.012	0.2483	0.2505
60	0.013	0.2252	0.2273
61	0.014	0.2035	0.2056
62	0.016	0.1834	0.1854
63	0.017	0.1646	0.1666
64	0.018	0.1473	0.1492
65	0.019	0.1314	0.1331
66	0.02	0.1168	0.1185
67	0.021	0.1035	0.1050
68	0.022	0.0914	0.0928
69	0.024	0.0805	0.0818
70	0.025	0.0706	0.0718
71	0.026	0.0617	0.0629
72	0.027	0.0538	0.0548
73	0.029	0.0467	0.0477
74	0.03	0.0405	0.0413
75	0.031	0.0349	0.0357
76	0.032	0.0300	0.0307
77	0.032	0.0257	0.0264
78	0.033	0.0220	0.0225
79	0.033	0.0187	0.0192
80	0.034	0.0159	0.0163
81	0.034	0.0135	0.0138
82	0.033	0.0113	0.0117
83	0.033	0.0095	0.0098
84	0.032	0.0080	0.0082
85	0.031	0.0067	0.0069
86	0.029	0.0055	0.0057
87	0.026	0.0046	0.0048
88	0.022	0.0038	0.0039
89	0.02	0.0031	0.0032
90	0.017	0.0026	0.0027
91	0.014	0.0021	0.0022
92	0.012	0.0017	0.0018
93	0.01	0.0014	0.0015

(Continued)

TABLE 34.3 (Continued)

Age at Start of Year	Death Probability $\{l \mid \text{Alive at 30}, \&\}$	$e^{-\gamma c\left(\frac{l}{l}\right)^{\eta}}$	$e^{-\gamma(c-\zeta x)\left(\frac{l}{l}\right)^{\eta}}$
94	0.008	0.0011	0.0012
95	0.006	0.0009	0.0010
96	0.005	0.0007	0.0008
97	0.004	0.0006	0.0006
98	0.002	0.0005	0.0005
99	0.001	0.0004	0.0004
100	0.001	0.0003	0.0003
101	0.001	0.0002	0.0002
102	0	0.0002	0.0002
103	0	0.0001	0.0002
104	0	0.0001	0.0001
105	0	0.0001	0.0001
	1	$< e^{-\gamma c\left(\frac{l}{l}\right)^{\eta}} \mid \& >$	$< e^{-\gamma(c-\zeta x)\left(\frac{l}{l}\right)^{\eta}} \mid \& >$
		0.1152	0.1161

PROBLEMS

Problems marked with an asterisk (*) are considered more challenging.

***1.** On your own, repeat the analysis of the Black Pill and White Pill problems presented in this chapter using the example of a 30-year old male, with constant annual consumption, $c = \$65{,}000$, $\eta = 2$, and doubling probability $= 0.8$. Use the remaining lifetime probability distribution in Table 34.1.

a. Plot the curve of Dollars vs. probability of death for both the White Pill and Black Pill cases.

b. Calculate the micromort value from the slope of the safety region.

***2.** Repeat problem 1 under the following scenarios. Determine the micromort value in each case.

a. $c = \$105{,}000$, $\eta = 2$, and doubling probability $= 0.8$.

b. $c = \$65{,}000$, $\eta = 4$, and doubling probability $= 0.8$.

c. $c = \$65{,}000$, $\eta = 2$, and doubling probability $= 0.95$.

d. What effect does c, η, and the doubling probability have on the micromort value?

CHAPTER

35

Discretizing Continuous Probability Distributions

CHAPTER CONCEPTS

After reading this chapter, you will be able to explain the following concepts:

- Discretizing continuous distributions
- The Equal Areas method
- The Moment matching method
- Precautions about discretization

35.1 INTRODUCTION

When we have a decision described by one or more continuous measures, it is often convenient to discretize the measures into distinctions with a finite number of degrees. This **discretization** simplifies the analysis and enables the calculations of the best decision alternative and the value of information using the traditional decision tree analysis. In this chapter, we discuss how to carry out this discretization procedure and examine the consequences of choosing to perform it in various ways.

Figure 35.1 shows an example of a probability distribution for a continuous measure (cost) and its discretization using a distinction with three degrees.

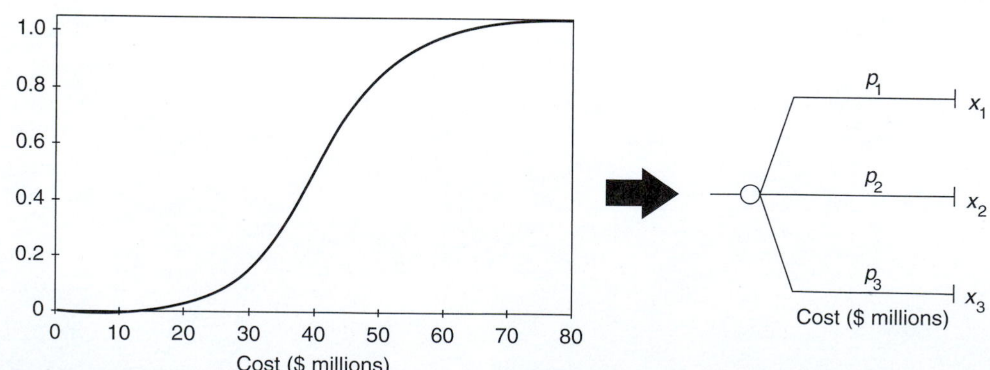

FIGURE 35.1 Discretizing Probability Distribution into Probability Tree with Three Degrees

As you might expect, there are a number of considerations to be made when discretizing a continuous distribution. First, we need to create a discretized probability tree. In the case of Figure 35.1, how do we specify the three-degree tree on the right side? We need the values of the measure, x_1, x_2, and x_3, and their corresponding probabilities p_1, p_2, and p_3.

Since

$$p_1 + p_2 + p_3 = 1$$

we only have five unknowns in this tree. In general, if we wish to discretize a probability distribution into n degrees, we need $(2n - 1)$ numbers to complete the discretization; n values of the measure and $(n - 1)$ probabilities (because the probabilities sum to 1).

The second consideration is the number of degrees we would like to have in the discretized tree. Choosing three degrees in Figure 35.1 was a mere illustration. As the number of degrees increases, the approximation improves, but the overall size of the discretized tree increases as well, as does the computational complexity. As we proceed, we must determine the most important features of the problem that we wish to preserve.

Since we will be maximizing the *e*-values of *u*-values at any decision node, preserving the *e*-values of these *u*-values as we move back through the tree will be extremely important. One common and simple discretization method involves choosing a procedure that will preserve the *e*-values of the measures as we move back through the tree. This means that for any measure the *e*-value of the approximation, or the **discretized distinction**, will be equal to the *e*-value of the continuous measure. This will be an excellent discretization procedure if the decision maker is risk-neutral.

However, while simple, we will find that such an approach may understate the variance of the deal and may lead to deviations between the exact certain equivalent and the certain equivalent resulting from the discretized tree for risk averse decision makers.

Another discretization method involves determining the values of the discrete distinction to preserve the moments of the continuous measure. We will discuss both methods in this chapter.

35.2 EQUAL AREAS METHOD

The **Equal Areas method** is a discretization method that preserves the *e*-values of the measures. As the name indicates, this method determines the discretized values by forming an "equal areas" analysis on the cumulative probability distribution. Refer back to our discussion in Chapter 24, which centered around calculating the *e*-value using an equal areas approach on the cumulative probability. The same type of analysis occurs here, except that we have several regions for which we conduct the equal areas analysis. The Equal Areas method uses the following steps.

1. Choose the number of degrees and required probabilities in the discretized tree. First, we decide on the number of degrees we desire in the discretized tree and the corresponding probabilities associated with each degree. Figure 35.2 shows an example of the choice of three degrees for the distinction (cost) on the cumulative probability curve. We want these degrees to be increasing in cost and to have probabilities 0.25, 0.50, and 0.25. We form three regions in the cumulative probability curve formed by the three horizontal lines at $y = 0.25$, $y = 0.75$, and $y = 1$, corresponding to the desired probabilities $p_1 = 0.25$, $p_2 = 0.5$, and $p_3 = 0.25$ in the discretized tree. The lines divide the graph into three areas of cumulative probabilities: the interval 0 to $0.25 = 0.25$, the interval 0.25 to $0.75 = 0.5$, and the interval 0.75 to $1 = 0.25$.

Note *The difference between these cumulative probabilities determines the chosen probabilities of* 0.25, 0.5, *and* 0.25, *respectively, for three degrees of the discretized tree in Figure* 35.2.

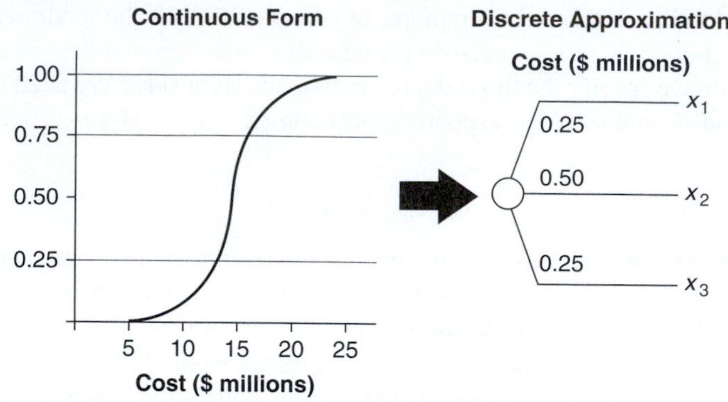

FIGURE 35.2 Continuous Distribution Represented by Three-Degree Distinctions with Probability Values 0.25, 0.5, and 0.25

This procedure can be conducted for any number of degrees and for any chosen probabilities. For example, if we had chosen four degrees with probabilities 0.2, 0.4, 0.3, and 0.1 for x_1, x_2, x_3, and x_4, with $x_1 < x_2 < x_3 < x_4$, then we would have drawn three horizontal lines at cumulative probabilities of 0.2, 0.6, and 0.9. On the other hand, if we had chosen four degrees with probabilities 0.1, 0.3, 0.4, and 0.2 for x_1, x_2, x_3, and x_4 with $x_1 < x_2 < x_3 < x_4$, then we would have drawn three horizontal lines at cumulative probabilities of 0.1, 0.4, and 0.8. For n degrees, we draw $n - 1$ horizontal lines for cumulative values at the vertical axis of the cumulative probability curve.

2. Choose the measure values corresponding to the chosen intervals. Next, we need to determine the values of the measure that will be assigned at the end of the tree in Figure 35.2. We now explain how to determine these values. For more details on the basis of these procedures, you can refer to Appendix A of this chapter.

To determine these measure values, we first highlight the rectangular regions obtained by the intersection of the horizontal lines with the cumulative probability curve. For example, Figure 35.3 shows the cumulative distribution with two horizontal lines at the cumulative probabilities of 0.25 and 0.75 with three shaded regions.

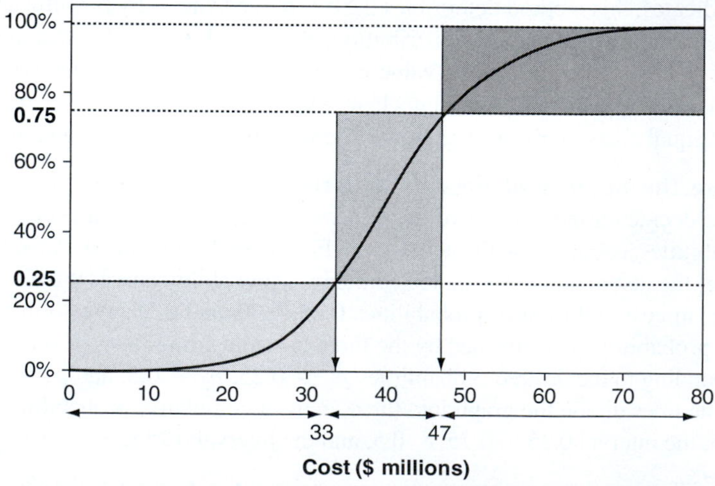

FIGURE 35.3 Selection of Areas Corresponding to Probabilities and Number of Degrees

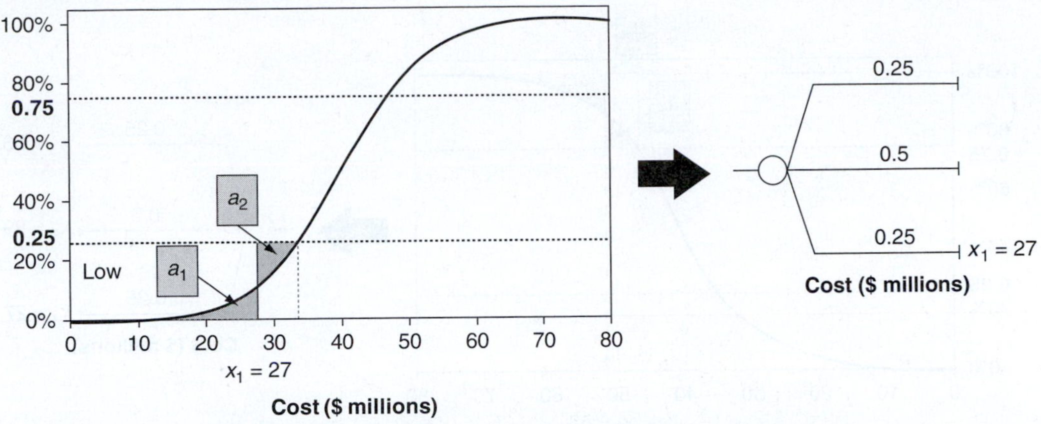

FIGURE 35.4 Value of Measure for First Region by Equal Areas

Figure 35.3 shows three regions: Region 1 is defined by $y = 0$ and $y = 0.25$, corresponds to the variable x_1 in the discretized tree, and represents values of x from 0 to 33 in the continuous measure. Region 2 is defined by $y = 0.25$ and $y = 0.75$, corresponds to the variable x_2 in the discretized tree, and has values of x from 33 to 47. Region 3 is defined by $y = 0.75$ and $y = 1$, corresponds to x_3 in the discretized tree, and has values of x above 47.

For Figure 35.3, we now choose a representative value of the measure for each degree in the discretized tree. Each measure value will be chosen from its corresponding region. We choose this value such that the shaded area in this region to the left of this point below the cumulative distribution curve is equal to the shaded area in this region to the right of this point and above the cumulative distribution. For example, in Figure 35.4, we choose the value of the measure such that the area, a_1, below the graph is equal to the area, a_2, above the graph and below the 25% line. This value of the measure, x_1, determines the value of the measure that corresponds to the 0% to the 25% cumulative probability in the discretized tree. This is why we called this method the Equal Areas method: The value of cost determined in this way is approximately equal to $x_1 = 27$.

Similarly, we determine the values of the measure for each of the other regions. Figures 35.5 and 35.6 show the values x_2 and x_3 for the probabilities of 0.5 and 0.25 in the discretized tree

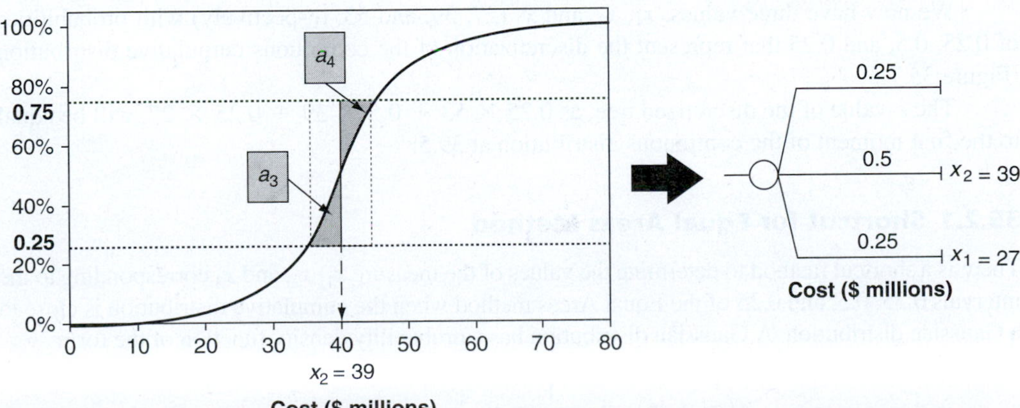

FIGURE 35.5 Value of Measure for 0.5 Probability of Second Degree

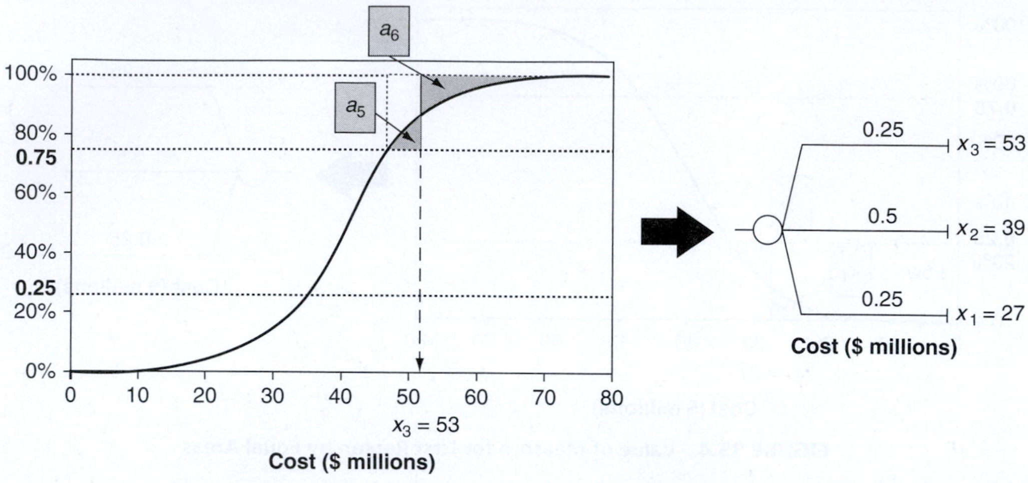

FIGURE 35.6 Value of Measure for 0.25 Probability of Third Degree

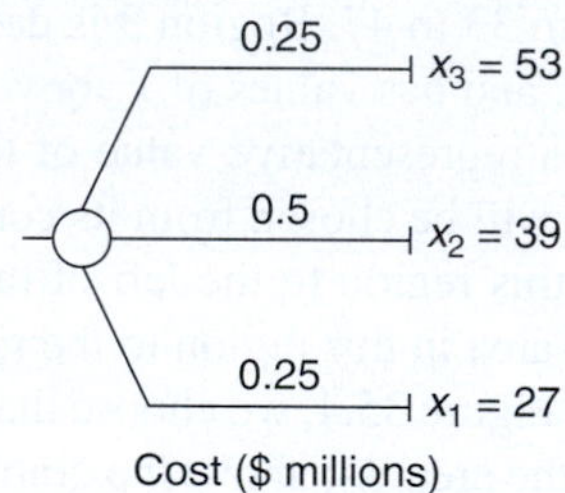

FIGURE 35.7 Discretized Tree using Equal Areas Method

obtained using the Equal Areas method. The value of x_2 is approximately 39, and the value of x_3 is approximately 53.

We now have three values, x_1, x_2, and x_3 (27, 39, and 53, respectively) with probabilities of 0.25, 0.5, and 0.25 that represent the discretization of the continuous cumulative distribution (Figure 35.7).

The *e*-value of the discretized tree, as $0.25 \times 53 + 0.5 \times 39 + 0.25 \times 27$, will be equal to the first moment of the continuous distribution at 39.5.

35.2.1 Shortcut for Equal Areas Method

There is a shortcut method to determine the values of the measure x_1, x_2, and x_3 corresponding to the intervals 0.25, 0.5, and 0.25 of the Equal Areas method when the cumulative distribution is close to a Gaussian distribution. A Gaussian distribution has a probability density function of the form

$$f(x|\mu, \sigma) = \frac{1}{\sigma\sqrt{2\pi}} e^{\frac{(x-\mu)^2}{2\sigma^2}}, -\infty < x < \infty$$

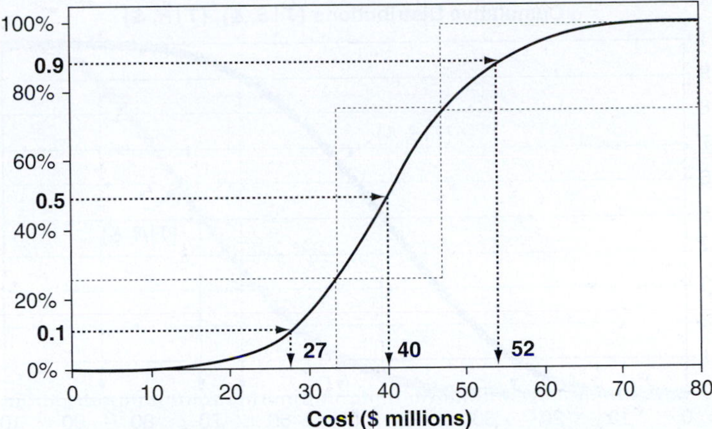

FIGURE 35.8 Shortcut for Equal Areas Method

where μ and σ are the mean and standard deviation (square root of the variance), respectively. This probability density is symmetric, so the skewness is 0.

The approximate discretization procedure simply selects the measure values that correspond to the cumulative probabilities of 0.1, 0.5, and 0.9. Figure 35.8 shows that these values are approximately 27, 40, and 52, respectively, and are very close to the values obtained from the equal areas approximation. We call those values the 10–50–90 values.

Note *This shortcut is convenient, because often you will have assessed only the 10–50–90 values, as they are very commonly assessed in practice. They occur with probabilities of* 0.25, 0.5, *and* 0.25, *respectively, in the discretized tree.*

Some people tend to call the 10–50–90 points the low–base–high assessments. We view this as a nomenclature error, since low and high might be confused with the lower and upper bounds of the measure (which are not the 0.1 and 0.9 fractiles).

35.3 CAUTION WITH DISCRETIZATION

Now, we will apply discretization to a practical problem and illustrate some of the issues that might arise. Refer back to the party problem first introduced in Chapter 9 and the continuous Beta detector of Chapter 20.

Figure 35.9 plots the cumulative distributions of the detector indications $\{T|S, \&\}$ and $\{T|R, \&\}$. The 10–50–90 points for each distribution are also highlighted in the figure.

Figure 35.10 shows the discretized tree for this continuous Beta detector, which is obtained by discretizing the indications $\{T|S, \&\}$ and $\{T|R, \&\}$ using the Equal Areas method.

The tree in Figure 35.10 is in assessed form. Now, suppose we wish to reverse the order of the distinctions in the tree to determine the posterior distributions of the weather given the detector indication.

Note *We can't reverse the tree of Figure 35.10 because the degrees are no longer the same.*

This discretized tree allows for only three possible indications if the weather is Sun (namely, $T = 25; 44; 65$) and three indications if the weather is Rain $(T = 51; 71; 87)$.

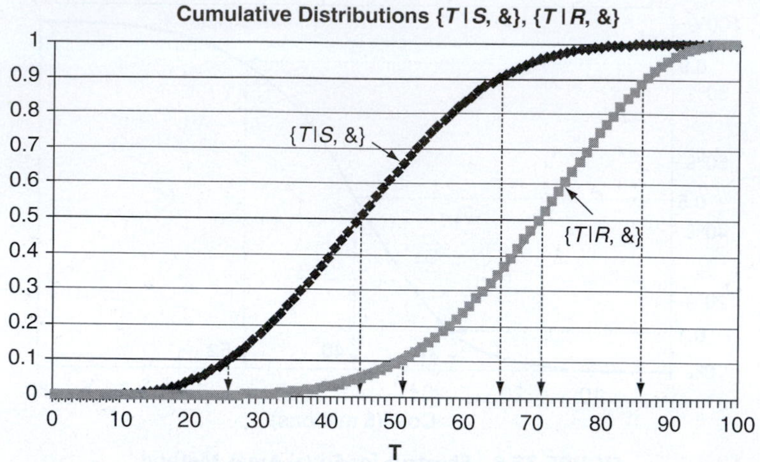

FIGURE 35.9 Cumulative Distributions for Continuous Beta Detector

This is not the true representation of this situation. In fact, even the clairvoyant we first met back in Chapter 2 can no longer answer which degree of this discretized tree will occur, since it is, in fact, a new representation. The clairvoyant will reveal the actual indication that will occur, and it can be any integer from 0 to 100.

What is a correct way of discretizing this tree to allow for tree reversals? The clairvoyant can answer whether or not the measure will lie in an interval. Therefore, the correct way of doing this would be to conduct the same analysis we did in Chapter 20 for Figure 20.23, which we repeat on the next page for convenience in Figure 35.11. We define intervals for each degree instead of particular discretized values representing each interval. Since the degrees are the same for each distinction, we can now reverse the order of the tree and determine the value of the detector.

Therefore, the Equal Areas discretization method would work well for conditional distributions if we wish to calculate the *certain equivalent* of a deal. However, it does not work well when we have *conditional distributions* that are relevant to a distinction of interest. In this latter case, we need to make sure that the conditional distributions are discretized into the same intervals and then flip the tree.

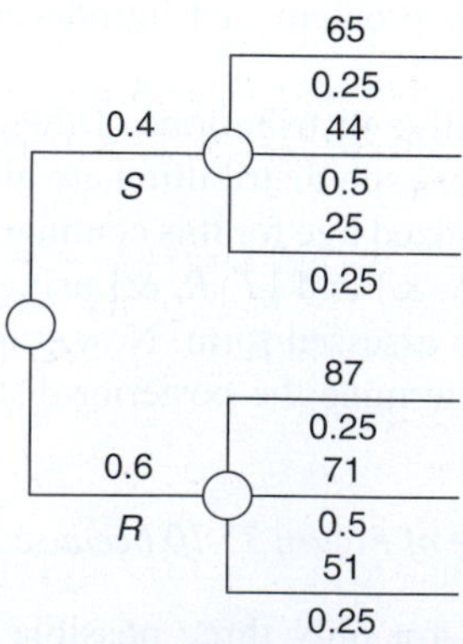

FIGURE 35.10 Discretized Tree for Continuous Detector

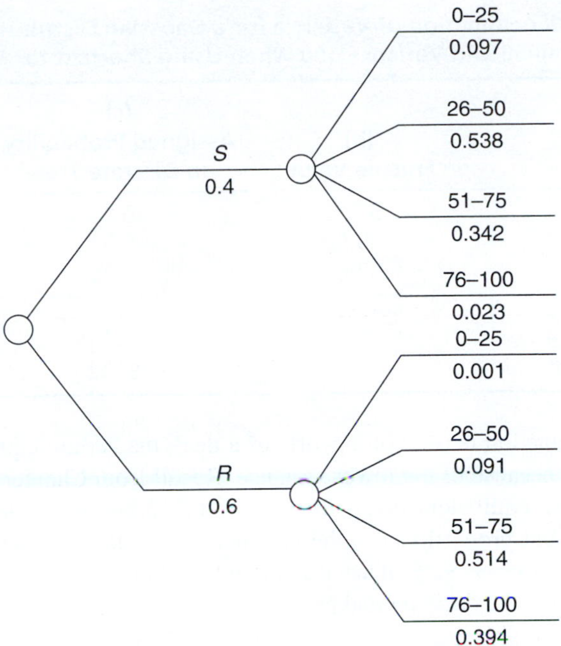

FIGURE 35.11 Probability Tree for Discretized Detector Analysis

35.4 ACCURACY OF 10–50–90 APPROXIMATE METHOD FOR EQUAL AREAS

The **Shortcut Approximation method** simply assigns the value of the measure corresponding to cumulative probabilities 0.1, 0.5, and 0.9 to a discrete tree with three degrees and probabilities equal to 0.25, 0.50, and 0.25. This section explores the accuracy of this approximation when representing the variance of the distribution.

EXAMPLE 35.1 Approximation of Gaussian Distribution Using 10–50–90 Discrete Approximation

For the purposes of this numeric illustration, consider a Gaussian distribution with the mean equal to 50 and standard deviation equal to 10. The variance of this distribution is 100.

Table 35.1 shows a numeric application. Column (a) shows the 10–50–90 probabilities. Column (b) shows the actual values of the 10–50–90 fractile values of a Gaussian distribution with mean 50 and standard deviation 10 $(\text{Variance} = 100)$. Column (c) shows the probabilities associated with the 10–50–90 fractile values in the discrete tree if we use the approximation method. Column (d) shows the fractile values squared. The approximation leads to a three-degree distinction whose mean is 50 (same as the continuous distribution) but whose variance is 82.12 (less than the variance of the continuous distribution, which was 100).

TABLE 35.1 Underestimation of Variance for a Gaussian Distribution with Mean 50 and Variance 100 when Using Shortcut for Equal Areas

(a) Fractile	(b) Fractile Value	(c) Assigned Probability in Discrete Tree	(d) Fractile Value Squared
10%	37.18	0.25	1382.69
50%	50.00	0.5	2500.00
90%	62.82	0.25	3945.79
e-value of Discrete Tree		50	
Second Moment of Discrete		2582.12	
Tree Variance of Discrete Tree		82.12	

If this approximation is used on payoffs of a deal, the certain equivalent may be considerably overstated because of the lower variance. Recall from Chapter 24 that the approximation for the certain equivalent given shows that the certain equivalent will be less than the mean by an amount proportional to the variance. Using the approximation would make the reduction in the variance 82% of what it should be and, therefore, make the deal artificially more attractive to the decision maker.

TABLE 35.2 Accuracy of Interval Means Using Shortcut for Equal Areas

(a) Fractile	(b) Fractile Value	(c) Assigned Probability in Tree	(d) Corresponding Interval	(e) Conditional Mean of Interval	(f) Difference
10%	37.18	0.25	Negative Infinity → 43.26	37.6	0.42
50%	50.00	0.5	43.26 → 56.74	50	0
90%	62.82	0.25	56.74 → Infinity	62.32	−0.5

We can also inspect the accuracy of the approximation for each interval of the 0.25–0.5–0.25 probability values using the approximation method. As shown in Table 35.2, Column (d), these intervals correspond to (i) negative infinity to 43.26 corresponding to the cumulative probability interval 0 to 0.25, (ii) 43.26 to 56.74 corresponding to the cumulative probability interval 0.25 to 0.75, and (iii) 56.74 to infinity corresponding to the cumulative probability interval 0.75 to 1. Column (e) shows the actual value of the conditional mean of x given that we know it lies within each interval for a Gaussian distribution with a mean of 50 and variance of 100. The interval means are 37.6, 50, and 62.32, respectively. Note that these conditional means are very close to the values read directly from the 10%, 50%, and 90% values of the cumulative probability. The differences in means are shown in Column (f).

35.4.1 Approximate 10–50–90 Method with Different Probability Values

In principle, we can still use the 10–50–90 fractiles for the discrete tree and get a better approximation of the variance, but we need to associate them with different probabilities, since they are only an approximation to the Equal Areas method.

TABLE 35.3 Accuracy of Variance Approximation Modified by Using Different Probabilities

(a) Fractile	(b) Fractile Value	(c) Assigned Probability in Discrete Tree	(d) Fractile Value Squared
10%	37.18	0.3	1382.69
50%	50.00	0.4	2500.00
90%	62.82	0.3	3945.79
e-value of Discrete Tree		50	
Second Moment of Discrete Tree		2598.54	
Variance of Discrete Tree		98.54	

Table 35.3 shows a numeric application. Column (a) shows the 10–50–90 fractiles. Column (b) shows the 10–50–90 fractile values of a Gaussian distribution with a mean of 50 and a standard deviation of 10 (Variance 100). Column (c) shows the set of different probabilities associated with them in the approximate tree. Column (d) shows the fractile values squared. The approximation leads to a three-degree distinction representation whose mean is 50 (same as continuous Gaussian) but whose variance is now 98.54, which is a much better approximation for the variance than that obtained using the standard 10–50–90 approach. Using this approximation will avoid the misrepresentation of variance, and, therefore, will avoid the overstatement of certain equivalent of deals that results from the 0.25–0.50–0.25 probabilities associated with the usual 10–50–90.

We can also inspect the accuracy of the modified approximation for each interval of the 0.3–0.4–0.3 probability values. These intervals correspond to the values of negative infinity to 44.76, 44.76 to 55.24, and 55.24 to infinity shown in Column (d) of Table 35.4. Column (e) shows the conditional mean of x given that it lies within each interval. These values are given by 38.7, 50.05, and 61.24, respectively. Note that these conditional means are also close to the values read directly from the 10%, 50%, and 90% values of the cumulative probability; the differences are shown in Column (f). Notice also that the conditional mean differences are larger, particularly for the first and third intervals, than for the case when we assign 0.25, 0.5, 0.25 probabilities.

By conducting similar analyses for other distributions, we can determine whether we would like to provide more accuracy for the variance estimate versus the values for each region.

TABLE 35.4 Accuracy of Interval Means Using Modified Shortcut Method

(a) Fractile	(b) Fractile Value	(c) Assigned Probability in Tree	(d) Corresponding Interval	(e) Conditional Mean of Interval	(f) Difference
10%	37.18	0.3	Negative Infinity → 44.76	38.7	1.52
50%	50.00	0.4	44.76 → 55.24	50	0
90%	62.82	0.3	55.24 → Infinity	61.24	−1.58

35.5 MOMENTS OF DISCRETE AND CONTINUOUS MEASURES

Another method to discretize a continuous probability distribution preserves the higher-order moments as well as the mean. This method calculates the probabilities and measure values of the discretized tree by solving a set of equations.

To illustrate, we first review the definition of moments from Chapter 24 and introduce some new concepts. The *m*th moment of a discrete measure as shown in Figure 35.12:

$$< x^m | \& > = \sum_{i=1}^{n} p_i x_i^m = p_1 x_1^m + p_2 x_2^m + \cdots + p_n x_n^m$$

where the summation symbol $\sum_{i=1}^{n}$ simply means we are summing n terms.

The moments also can be calculated for continuous measures using integrals. The first moment of a continuous distribution is obtained using the integral

$$< x | \& > = \int x f(x)\, dx$$

In a similar manner, we can define higher order moments for continuous measures as

$$< x^m | \& > = \int x^m f(x)\, dx$$

We can also define central moments for continuous measures as

$$<(x - < x | \& >)^m | \& > = \int (x - < x | \& >)^m f(x)\, dx$$

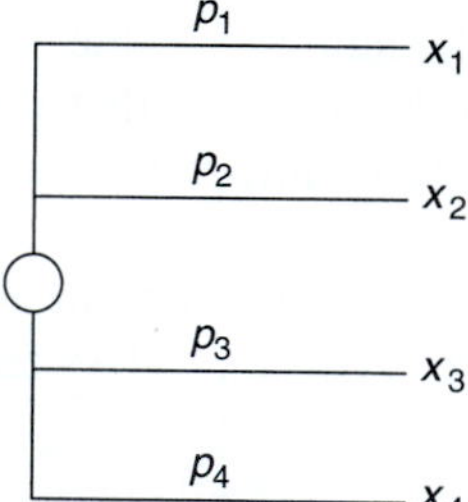

FIGURE 35.12 Probability Tree with Measure

35.6 MOMENT MATCHING METHOD

The **Moment Matching method** equates the moments of the discretized measure with the moments of the continuous measure. Suppose, for example, that we would like to discretize a continuous distribution into three degrees, p_1, p_2, p_3.

We know the probabilities must satisfy

$$p_1 + p_2 + p_3 = 1$$

If we know the first moment of the continuous distribution from $< x|\& > = \mu$, we can match that with the *e*-value of the discretized tree as

$$p_1x_1 + p_2x_2 + p_3x_3 = \mu$$

We now have two equations: $p_1 + p_2 + p_3 = 1$ and $p_1x_1 + p_2x_2 + p_3x_3 = \mu$ with six unknowns, $p_1, p_2, p_3, x_1, x_2,$ and x_3.

There are many ways to draw the discretized tree. One way is to choose a value (or values) for one of the unknown probabilities, or **measure values**, and solve to determine the unknown values. Alternatively, if we know higher moments of the measure, we can match those as well to determine the remaining values.

EXAMPLE 35.2 Moment Matching of a Gaussian Distribution

Consider a Gaussian distribution that has the probability density function

$$f(x|\mu,\sigma) = \frac{1}{\sigma\sqrt{2\pi}}e^{\frac{(x-\mu)^2}{2\sigma^2}}, -\infty < x < \infty$$

where μ and σ are the mean and standard deviation (square root of variance), respectively.

Suppose we have a Gaussian distribution where the first moment is equal to 0. The variance (second central moment) is equal to σ^2, and the skewness is equal to 0, since it is a symmetric distribution.

If we wish to find a two-degree tree representation (Figure 35.13) for this distribution, we have $(2n - 1) = 3$ unknowns. Therefore, we can to match the first three moments as

$$\{x_1|\&\}x_1 + \{x_2|\&\}x_2 = 0$$

$$\{x_1|\&\}x_1^2 + \{x_2|\&\}x_2^2 = \sigma^2$$

$$\{x_1|\&\}x_1^3 + \{x_2|\&\}x_2^3 = 0$$

Solving gives $\{x_1|\&\} = \{x_2|\&\} = 0.5$, $x_1 = \sigma$, and $x_2 = -\sigma$

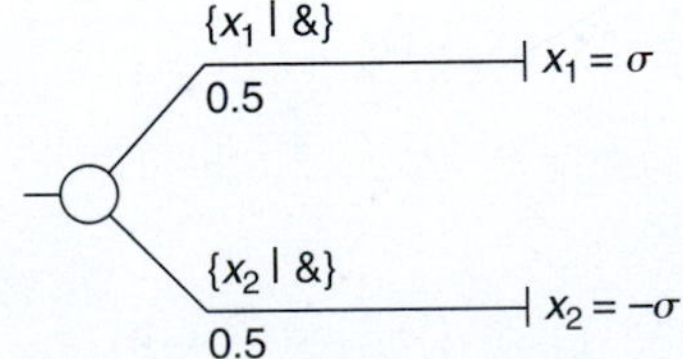

FIGURE 35.13 Two-Degree Representation of Gaussian Distribution with Mean 0 and Variance σ^2

Note *This approximation states that the chance of a result greater in magnitude than sigma is* 0. *For the original distribution it is* 0.317.

35.7 SUMMARY

We discretize a continuous measure to provide computational convenience and the use of decision trees to determine the best decision alternative.

Two methods to discretize continuous measures are the Equal Areas method and the Moment Matching method.

The Equal Areas method divides the cumulative distribution plot into intervals whose probabilities correspond to the probabilities chosen on the discretized tree. For each region, we chose the measure value such that the area in this region that is below the cumulative distribution curve to the left of the point is equal to the area in this region that is above the cumulative distribution curve to the right of this point. This method preserves the conditional mean over each region, and makes the *e*-value of the discretized tree equal to the mean of the continuous distribution.

There is an approximation for the Equal Areas method when the distribution is close to a Gaussian distribution. We read directly the 10–50–90 points from the cumulative distribution plot and assign to them probabilities of 0.25, 0.5, and 0.25 respectively. This shortcut will understate the variance of the distribution. We can sometimes compensate by assigning probabilities of 0.3, 0.4, and 0.3 to the intervals.

The Moment Matching method matches the moments of the continuous distribution and the discrete tree.

KEY TERMS

- Discretization
- Discretized distinction
- Equal Areas method
- Measure values
- Moment Matching method
- Shortcut Approximation method

APPENDIX A Rationale for Equal Areas Method

The idea behind the Equal Areas method of discretization is to preserve the conditional mean within each discretized interval and also to preserve the mean of the original continuous distribution.

Let us first start by showing why the mean of a cumulative distribution satisfies the Equal Areas approach (as we discussed in Chapter 24). There are many ways to show this, and we will provide a simple illustration of why.

We are given a measure with a probability density, $f(x)$, and cumulative probability function, $F(x)$. Let x_2 be the value where the cumulative probability curve is equal to 1 and x_3 be the value where the cumulative probability curve is equal to 0. The mean is given by

$$< x|\& > = \int_{x_3}^{x_2} xf(x)\,dx$$

Now, consider the three regions defined in Figure 35.14. Region A_1 is to the left of the vertical line $x = x_1$ and below the cumulative probability curve; Region A_2 is to the right of the vertical line $x = x_1$ and above the cumulative probability curve, and Region A_3 is to the left of the vertical line $x = x_1$ and above the cumulative probability curve.

First, we show how the sum of areas of Regions A_2 and A_3 relates to the mean. Using the rule of integration by parts, this area A_2 and A_3 above the cumulative probability curve is

$$\int_{x_3}^{x_2} (1 - F(x))\,dx = (x_2 - x_3) - \int_{x_3}^{x_2} F(x)\,dx = (x_2 - x_3) - \left\{ xF(x)\,\Big|_{x_3}^{x_2} - \int_{x_3}^{x_2} xf(x)\,dx \right\}$$

$$= -x_3 + \int_{x_3}^{x_2} xf(x)\,dx = -x_3 + < x|\& >$$

If the two areas A_1 and A_2 are equal, the sum of areas A_2 and A_3 must also be equal to the sum of areas A_1 and A_3. But the area of A_1 and A_3 is the area of a rectangle of breadth equal to $(x_1 - x_3)$ and height equal to 1, so this is equal to $(x_1 - x_3)$.

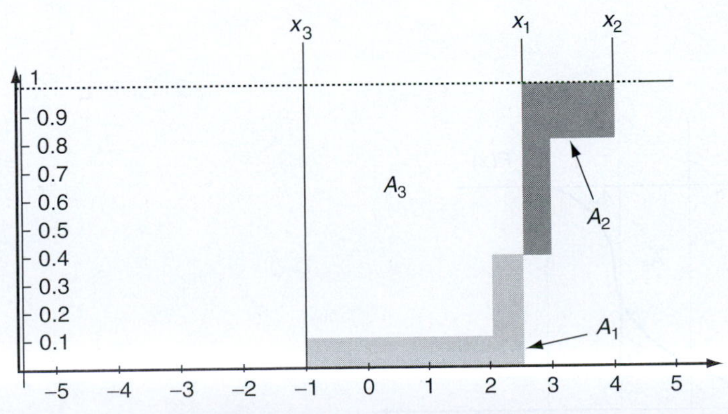

FIGURE 35.14 Mean of the Measure

Equating the two areas gives

$$-x_3 + <x|\&> = x_1 - x_3$$

This implies that $<x|\&> = x_1$

> *Therefore, if we can adjust the vertical line $x = x_1$ such that the areas A_1 and A_2 are equal, then x_1 is in fact the mean.*

The exact same idea applies to the conditional mean on an interval, although the math is a little more tedious.

The conditional mean of the measure, given that the measure takes on a value in the domain $[a, b]$, $a < b$ is given by

$$<x|x \in [a, b], \&> = \frac{\int_a^b xf(x)\,dx}{\int_a^b f(x)\,dx} = \frac{\int_a^b xf(x)\,dx}{F(b) - F(a)}$$

where $F(a)$ and $F(b)$ are the cumulative probability values of $x = a$ and b respectively.

> ***Note*** *When $a = -\infty$ and $b = \infty$, the formula reduces to the definition of the mean of a continuous measure*

$$<x|\&> = \frac{\int_{-\infty}^{\infty} xf(x)\,dx}{\int_{-\infty}^{\infty} f(x)\,dx} = \int_{-\infty}^{\infty} xf(x)\,dx$$

Consider the region defined in Figure 35.15 between two values of the measure x having values x_1 and x_2 with cumulative probabilities p_1 and p_2, respectively.

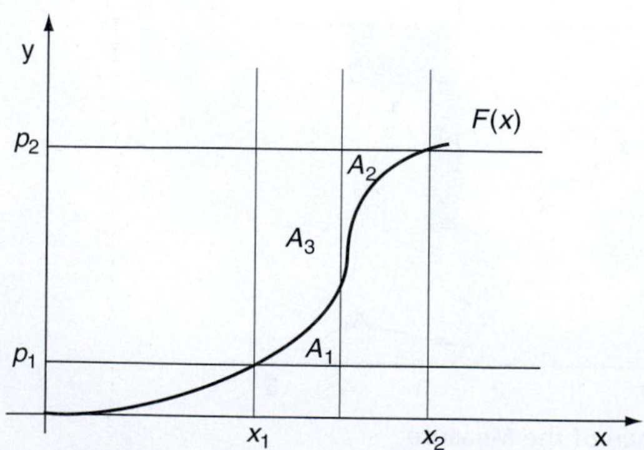

FIGURE 35.15 Conditional Mean at an Interval

First, let's calculate the area that is defined by (1) the line $y = p_2$, (2) the curve $y = F(x)$, and (3) the line $x = x_1$. This is the region defined by the areas A_2 and A_3 in Figure 35.14. This area must equal

$$\int_{x_1}^{x_2} (p_2 - F(x))\,dx = p_2(x_2 - x_1) - \int_{x_1}^{x_2} F(x)\,dx$$

Using the rule of integration by parts, we can express the last term as

$$\int_{x_1}^{x_2} F(x)\,dx = x_2F(x_2) - x_1F(x_1) - \int_{x_1}^{x_2} xf(x)\,dx$$

$$= x_2p_2 - x_1p_1 - (p_2 - p_1) < x|x \in [x_1, x_2], \& >$$

$$= x_2p_2 - x_1p_2 + x_1p_2 - x_1p_1 - (p_2 - p_1) < x|x \in [x_1, x_2], \& >$$

$$= (x_2 - x_1)p_2 + (x_1 - < x|x \in [x_1, x_2], \& >)(p_2 - p_1)$$

Substituting gives

$$\int_{x_1}^{x_2} (p_2 - F(x))\,dx = (< x|x \in [x_1, x_2], \& > - x_1)(p_2 - p_1)$$

Note that this area is equal to the area of the rectangle included between the conditional mean $< x|x \in [x_1, x_2], \& >$ and p_2, p_1. This is equivalent to the areas A_1 and A_3 in Figure 35.14.

Note also that the area A_3 is common to both the area above the curve $F(x)$ from x_1 to x_2, and the area of the rectangle obtained in the last integral. Consequently, at the conditional mean of the interval, we must have the area $A_1 = A_2$. At the conditional mean of the interval, $[x_1, x_2]$, the area to the left of the conditional mean that lies below the cumulative curve, $F(x)$ and above the boundary at $F(x_1)$ must be equal to the area that lies to the right of the conditional mean ad that lies below the cumulative curve, $F(x)$ and below the boundary, $F(x_2)$.

To determine this conditional mean graphically, we simply choose (by inspection) the value of the measure such that the area, A_1, below the graph in Figure 35.14 is equal to the area A_2 above the curve.

PROBLEMS

1. For the cumulative distribution provided in the figure, use the Equal Areas method to discretize it into the criteria given in problems 1 through 4. For each of the four cases, calculate the e-values of the discretized distribution.
   a. Three degrees with probabilities 0.25, 0.5, and 0.25.
   b. Three degrees with probabilities 0.3, 0.4, and 0.3.
   c. Four degrees with probabilities 0.25, 0.25, 0.25, and 0.25.
   d. Four degrees with probabilities 0.2, 0.3, 0.23, and 0.2.

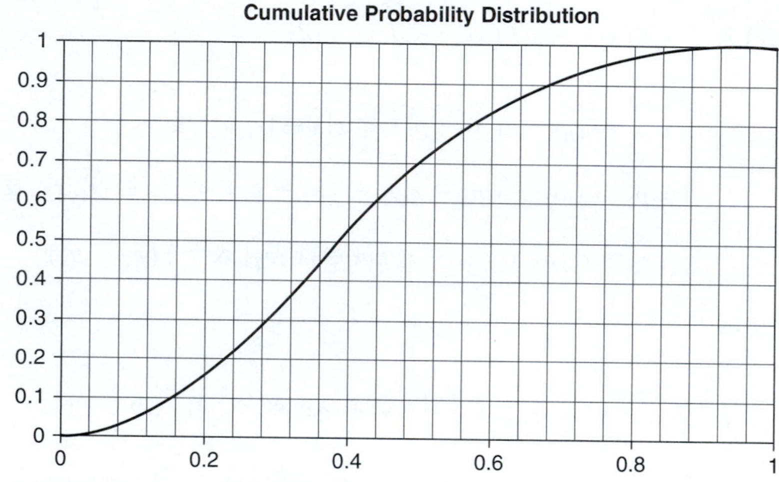

2. Using the moment matching method determine the two-degree approximation of the following distributions
   i. Beta (10, 10) on a scale $a = 0$ to $b = 100$
   ii. Beta (2, 10) on a scale $a = 0$ to $b = 100$
   iii. Beta (5, 10) on a scale $a = 0$ to $b = 100$

   Hint: For a Beta (r, k) distribution on the internal $[a, b]$;

$$\text{Mean} = (b - a)\frac{r}{r + k} + a$$

$$\text{Variance} = \frac{(b - a)^2}{(r + k)^2} \frac{rk}{r + k + 1}$$

$$\text{Skewness} = (b - a)^3 \frac{2(k - r)\sqrt{r + k + 1}}{(r + k + 2)\sqrt{rk}}$$

CHAPTER

36

Solving Decision Problems by Simulation

CHAPTER CONCEPTS

After reading this chapter, you will be able to explain the following concepts:

- Solving decision problems by simulation
- Simulating a decision with discrete distinctions
- Simulating a decision with continuous distinctions

36.1 INTRODUCTION

In the previous chapter, we illustrated how to simplify the computation of decision problems by using discretization. In this chapter, we present *simulation*, another method of computing decision problems. This method is particularly useful when the distinctions are continuous, but it can also be used when they are discrete. Our purpose in this chapter is not to provide a full description of the theory of simulation but to illustrate how to use it to compute decision problems.

36.2 USING SIMULATION FOR SOLVING PROBLEMS

Simulation means carrying out a process over and over again and deriving conclusions based on the results obtained. For example, instead of conducting the probability analysis for the chance of winning a craps game, as discussed in Chapter 7, we could play the game over and over again—perhaps hundreds (or even thousands) of times—and then see what fraction of those times we won.

If we played craps only a few times, this fraction might not be representative of our chances of winning, but the more we play, the closer it will get. In fact, this fractional number of successes will approach the probability of success at a rate that is proportional to the square root of the number of trials we perform. This means that, if we perform 10,000 trials, the deviation of those results over the results for 100 trials will be 1/10 as large. 100 times as many trials gives you about 1/10 the deviation.

While performing 10,000 trials would be a major task for a human, it is a fast and easy problem for any modern computer. Computing is also ideal for providing other outcomes, such as the **profitability** of a project. Any uncertain distinctions and corresponding probabilities are generated, repeated trials are performed, and the profits duly noted. The average profit will converge to the mean profit at the same rate as with the probability of any event. 10,000 trials will reduce the deviation in profit to about 1/10 the deviation of 100 trials. And that, in a nutshell, is simulation.

Solving decision problems by simulation usually involves the following steps.

1. Generating possible outcomes for each distinction with the same probabilities we assign to them in the decision tree.
2. Solving the problem using the generated outcomes to determine the best decision. This stage involves generating the *u*-values of the outcomes, and choosing the decision with the highest *u*-value.
3. Repeating the analysis with a different set of generated outcomes.
4. Calculating some measures based on the repeated analysis. Generally speaking, we calculate the average of the *u*-values generated. As the number of repeat trials increases, the average gets closer to the *e*-value of the *u*-values.

Simulating decisions does not require a computer. For example, a decision involving a payoff of a coin toss where you get \$100 for heads and \$0 for tails can be simulated by tossing the coin many times if you believe the outcomes of the coin you toss have the same probabilities as the outcomes in the decision. Then, by calculating the *u*-value for each toss and the average of the generated *u*-values, you can determine the *e*-value of *u*-values for that decision.

Given any other alternative, we can also simulate its *u*-values of the payoffs and compare the alternatives based on the highest average of the *u*-values we obtain. By doing this for a large number of times, we can identify the best decision alternative and whether or not it would be a good idea to go ahead with a decision alternative in the first place.

With modern computers, however, we can easily generate outcomes having the same probabilities that we have assigned to them in the decision tree, and therefore, we can simulate any decision we face.

The generation of the outcomes for the repeat analysis usually starts by uniformly generating a number between 0 and 1, such that the histogram of the generated numbers is uniform, and the successive numbers generated—for all practical purposes—are mutually irrelevant. Such uniform random number generators are widely available in software packages. This generated uniform random number, or simply **random number**, is then used to generate the outcomes of interest with their corresponding probabilities. We illustrate these ideas in more detail next.

36.3 SIMULATING DECISIONS HAVING A SINGLE DISCRETE DISTINCTION

Let us start by considering the following decision. A decision maker with an exponential *u*-curve is

$$U(x) = \frac{1 - e^{-\gamma x}}{1 - e^{-\gamma 100}}$$

where $\gamma = 0.01$ is the risk aversion coefficient and faces a choice between two deals shown in Figure 36.1.

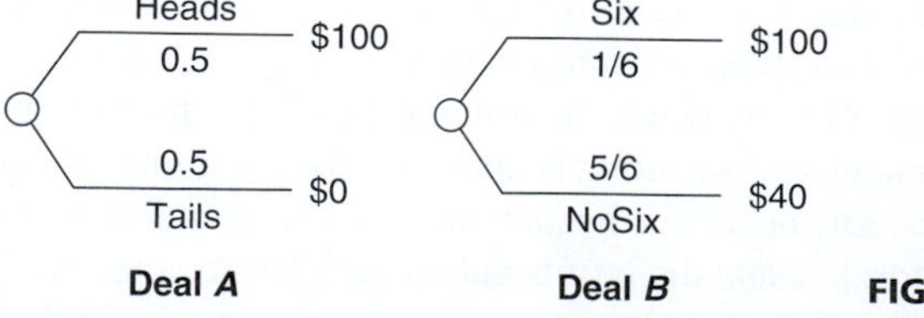

FIGURE 36.1

Deal A**:** Pays \$100 if the outcome of a coin toss is Heads, and pays 0 if the outcome is Tails. The decision maker believes that $\{\text{Heads} \mid \&\} = \{\text{Tails} \mid \&\} = 0.5$.

Deal B**:** Pays \$100 if the outcome of a roll of a die is Six, and pays \$40 otherwise. The decision maker believes that $\{\text{Six} \mid \&\} = 1/6$.

Which deal should this decision maker choose?

Of course, we can solve this problem using the tools we have developed so far by calculating the *u*-value of each deal as

$$\textbf{Deal } A\textbf{:}\ e\text{-value of } u\text{-values} = 0.5 * u(100) + 0.5 * u(0) = 0.5 * 1 + 0.5 * 0 = 0.5$$

$$\textbf{Deal } B\textbf{:}\ e\text{-value of } u\text{-values} = \frac{1}{6} u(100) + \frac{5}{6} u(40) = \frac{1}{6}.1 + \frac{5}{6}(0.52) = 0.601$$

Therefore, Deal B is preferred to Deal A. Moreover, the certain equivalent of Deal A is obtained by finding the value of x that has a *u*-value of 0.5, which is \$37.99, while the certain equivalent of Deal B is the value of x that has a *u*-value of 0.601, namely \$47.81.

> **Note** *Both of these deals have the same* e-value *of \$50, but Deal B is preferred due to the risk aversion of the decision maker.*

Now, we solve this problem using simulation. To simulate Deal A, we generate several outcomes of the coin toss (Heads and Tails) such that the ratio of these outcomes to the total number of repeated trials is the same probability that we assign to $\{\text{Heads} \mid \&\}$ and $\{\text{Tails} \mid \&\}$ for Deal A. For example, we would like to generate an equal number of Heads and Tails to solve the problem if we believe that $\{\text{Heads} \mid \&\} = \{\text{Tails} \mid \&\} = 0.5$. If we use a uniform random number generator in our analysis, we can simulate this situation as given here.

1. Use a uniform random number generator to generate a number between 0 and 1, call it RAND.
2. If RAND < 0.5, denote Outcome = Heads, ELSE Outcome = Tails.
3. If Outcome = Heads, Payoff = \$100, ELSE Payoff = \$0.
4. If Payoff = \$100, *u*-value = 1, ELSE *u*-value = \$0.
5. Go back to Step 1 and repeat.
6. Take the sum of *u*-values and divide by the number of trials. Note the average *u*-value generated.

As the number of trials increases, the average *u*-value becomes closer to the *e*-value of *u*-values of the deal. This is called the **Law of Large Numbers**.

Table 36.1 shows an example of sequences generated to solve this problem. Column (a) shows the number of trials. Column (b) generates a uniform random number between 0 and 1. Column (c) shows the outcome, Heads or Tails, as determined by whether the number generated is less than or greater than 0.5. Column (d) shows the payoff determined by each outcome. Column (e) shows the *u*-value of the payoff.

TABLE 36.1 Solving the Problem by Simulation

(a) Number	(b) Rand	(c) Outcome	(d) Payoff	(e) *u*-Value
1	0.8136	Tails	0	0
2	0.1532	Heads	100	1
3	0.6114	Tails	0	0
4	0.0068	Heads	100	1
5	0.9022	Tails	0	0
⋮	⋮	⋮	⋮	⋮
45	0.1436	Heads	100	1
46	0.9207	Tails	0	0
47	0.0138	Heads	100	1
48	0.1972	Heads	100	1
49	0.4111	Heads	100	1
50	0.8049	Tails	0	0
		Sum	2500	25
		Average	$ 50.00	0.5
		CE		$ 37.99

After repeating this analysis 50 times, we calculate the sum of the payoffs and sum of the *u*-values (2500 and 25 respectively). Then we divide by the number of trials, 50. The average payoff is $50 and the average *u*-value is 0.5. We then calculate the certain equivalent of the deal using the average *u*-value that has been generated i.e.,

$$\text{Average } u\text{-value} = \bar{u} = \frac{1}{50}\sum_{i=1}^{50} u_i$$

$$= 0.5$$

$$\text{Certain equivalent} = u^{-1}(\bar{u}) = u^{-1}(0.5)$$

$$= 37.99$$

Now, we simulate the second deal. Once again, we need to generate possibilities representative of the probabilities we assign to them.

1. Generate a random number from a uniform random number generator RAND.
2. If RAND $<1/6$, Outcome = "Six," ELSE Outcome = "NoSix."
3. If Outcome = "Six," Payoff = \$100, ELSE Payoff = \$40.
4. If Outcome = "Six," *u*-value = 1, ELSE *u*-value = 0.52.
5. Go back to Step 1 and repeat.
6. Take the sum of *u*-values and divide by the number of trials. Note the average *u*-value that has been generated.

The sequences generated and the calculations are shown in Table 36.2. After 50 trials, the ratio of sum of *u*-values divided by the number of trials is equal to 0.598 (compare this to the actual *e*-value of *u*-values calculated as 0.601).

Comparing the rightmost cell in the last row of each table shows that Deal *B* is a better deal than Deal *A*. This result matches our earlier calculations.

TABLE 36.2 Simulating the Deal with the Die

(a) Number	(b) Rand	(c) Outcome	(d) Payoff	(e) *u*-Value
1	0.2680	NoSix	40	0.521546
2	0.7497	NoSix	40	0.521546
3	0.0884	Six	100	1
4	0.3743	NoSix	40	0.521546
5	0.4673	NoSix	40	0.521546
⋮	⋮	⋮	⋮	⋮
45	0.4676	NoSix	40	0.521546
46	0.9028	NoSix	40	0.521546
47	0.6179	NoSix	40	0.521546
48	0.0644	Six	100	1
49	0.3533	NoSix	40	0.521546
50	0.7711	NoSix	40	0.521546
		Sum	2480	29.90493
		Average	$ 49.60	0.598099
		CE		$ 47.49

The previous example illustrated how to determine the best decision alternative using simulation. The basic idea is to simulate the *u*-value each time and then calculate the average *u*-values generated. Note how close the average *u*-values were to the *e*-value of the *u*-values after only 50 trials.

36.4 DECISIONS WITH MULTIPLE DISCRETE DISTINCTIONS

The same analysis can be applied to more complicated decisions, particularly those that involve multiple distinctions. To illustrate, let us refer back to the game of craps and use simulation to find the probability of winning the game.

The first step is to simulate the outcome of the first roll of two dice. To do this, we can simulate the outcome of each die separately. Starting from a Rand(), we can verify that the formula

$$\text{Integer}\ (6 * \text{Rand}(\ )) + 1$$

will give us integer outcomes between 1 and 6 with equal probabilities. The "Integer" operation just removes the decimal portion of the uniform random number generated.

We can now perform the same analysis to simulate the outcome of the second die. The sum determines the first roll of the two dice. Therefore, the sum of two dice is obtained from a Rand() by the formula

$$\text{Integer}\ (6 * \text{Rand}(\ )) + \text{Integer}\ (6 * \text{Rand}(\ )) + 2$$

An example of simulated outcomes using this method is

12, 8, 11, 7, 8, 9, 5, 3, 6, 9, 6, 5, 6, 9, 7, 8, 5, 4, 9, 4, 8, 7, 4, 7, 11, 7, 10, 4, 11, 10, 7, 4, 5, 11, 6, 12, 6, 7, 6, 5, 5, 8, 9, 11, 7, 7, 9, 6, 11, 6, 4, 8, 7, 8, 10, 4, 7, 2, 8, 5, 5, 7, 6, 9, 5, 9, 9, 4, 2, 5, 3, 9, 5, 7, 9, 8, 7, 7, 6, 5, 11, 5, 8, 7, 7, 3, 9, 8, 7, 9, 8, 5, 7, 8, 8, 8, 12, 9.

Note *The number 7 occurs more frequently than the other numbers.*

Table 36.3 shows a summary of the numbers generated in 5000 trials, and Figure 36.2 shows the histogram of generated numbers.

TABLE 36.3 Fraction of Times a Number is Generated by Simulation in 5000 Trials

Number Generated	Number of Times	Fraction of Times
2	135	0.027
3	301	0.060
4	385	0.077
5	576	0.115
6	668	0.134
7	833	0.167
8	713	0.143
9	585	0.117
10	402	0.080
11	261	0.052
12	141	0.028
Sum	5000	

Given the roll of two dice that we have simulated, we can determine whether or not the game of craps is over.

> *If the sum is 2, 3, or 12, then the game is over at the first roll (with Lose). If it is 7, or 11, then the game is over (with Win) at the first roll.*

In Table 36.3, the game was over (lose) at the first roll $135 + 301 + 141 = 577$ times, or a fraction of 0.1154 from the 5000 trials. The game was (win) at the first roll $833 + 261 = 1094$ times or a fraction 0.2188 of the 5000 trials. If the game is not over after the first roll, then you will have established a point. In Table 36.3 a point was established $5000 - 577 - 1094 = 3329$ times. The game will then be over the first time a 7 is thrown (lose) or the point is repeated (win). You can determine this by simulating several rolls of the dice and observing the outcomes generated.

Table 36.4 shows an example of a simulation of a full game of craps. The first roll results in a 10, so it is established as the point. The game ends by rolling a 7 and losing.

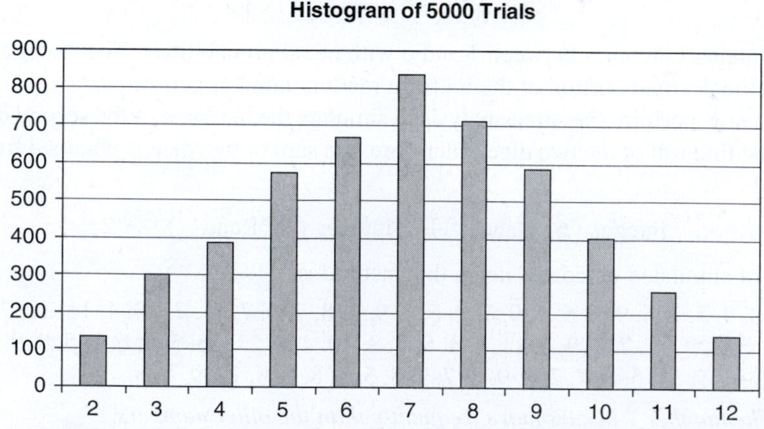

FIGURE 36.2 Histogram of Outcomes Generated in 5000 Repeat Trials

TABLE 36.4 Example of Simulation-Generated Craps Game

Sum of Dice	Game Over	Game Result
10	False	Still On
9	False	Still On
9	False	Still On
6	False	Still On
8	False	Still On
8	False	Still On
2	False	Still On
6	False	Still On
3	False	Still On
6	False	Still On
11	False	Still On
7	True	Lose

This is only one scenario. Of course, it does not have to occur this way every time. We might repeat this simulation again, for example, and end up with the results in Table 36.5.

Table 36.6 shows another example, where a 7 was generated on the first roll resulting in a win.

Table 36.7 shows another example where the point, 4, was established and repeated before a 7.

TABLE 36.5 Another Example of a Simulation-Generated Craps Game

Sum of Dice	Game Over	Game Result
10	False	Still On
6	False	Still On
9	False	Still On
7	True	Lose

TABLE 36.6 Craps Simulation—7 on First Roll Results in Win

Sum of Dice	Game Over	Game Result
7	True	Win

TABLE 36.7 Craps Simulation—Repeating Point Before 7 Results in Win

Sum of Dice	Game Over	Game Result
4	False	Still On
9	False	Still On
4	True	Win

By repeating this simulation several times, we can calculate the fraction of times we win versus the times we lose. By simulation, the fraction of wins obtained after two thousand times is 0.493464052. Notice how close this is to the probability of win calculated in Chapter 6.

Reflection

Repeat the analysis and determine the fraction of wins in three thousand times. You can also calculate other things. As an example, try calculating the mean time it takes to play the game of craps by calculating the average times the game was on during the simulation.

36.5 SIMULATING A MEASURE WITH A CONTINUOUS DISTRIBUTION

In many cases, the distinction may involve a continuous measure, such as the continuous detector we discussed in Chapter 20. We can also use simulation to generate outcomes of continuous measures with a uniform number generator, as we will illustrate here. This method works well when the distribution is continuous and strictly increasing to conduct such a simulation procedure.

1. Generate a number between 0 and 1 from a uniform random number generator, Rand.
2. Project the number on the cumulative distribution curve of the measure, and get the value of the measure whose cumulative distribution corresponds to this generated value.

For example, suppose you generate a value of Rand $= 0.75$ for the measure shown in Figure 36.3. Projecting back on the cumulative distribution gives a value of $x = 105$.

To get the u-value, we simply calculate $u(x)$ for the measure we have generated.

The e-value of u-values is equal to the average u-values generated when we repeat this process a large number of times.

Now, we will illustrate how to use simulation to determine the best decision alternative for a decision with continuous measures.

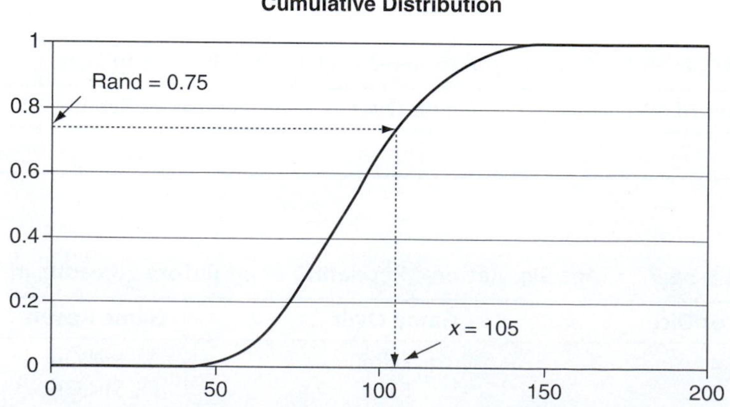

FIGURE 36.3 Simulating Measure with Continuous Distribution

EXAMPLE 36.1 Solving a Project Management Decision by Simulation

A project manager of an automobile manufacturing company is facing two design alternatives, as seen in Figure 36.4.

1. "Alternative Major," designing a new, innovative vehicle with major upgrades or
2. "Alternative Minor," performing minor upgrades to the previous year's vehicle.

In this example, the Major alternative has a triangular probability density function that is symmetric on the domain [0, \$200 M].

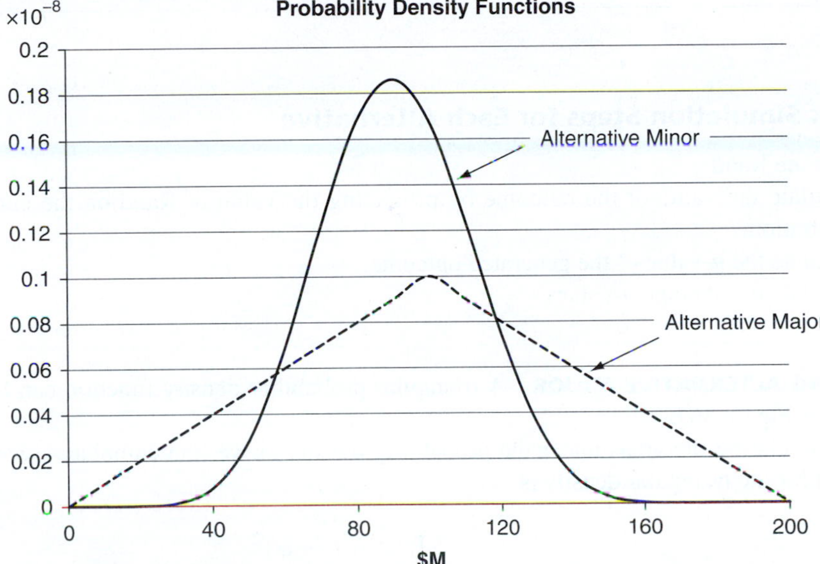

FIGURE 36.4 Cumulative Probability Distributions for Two Alternatives

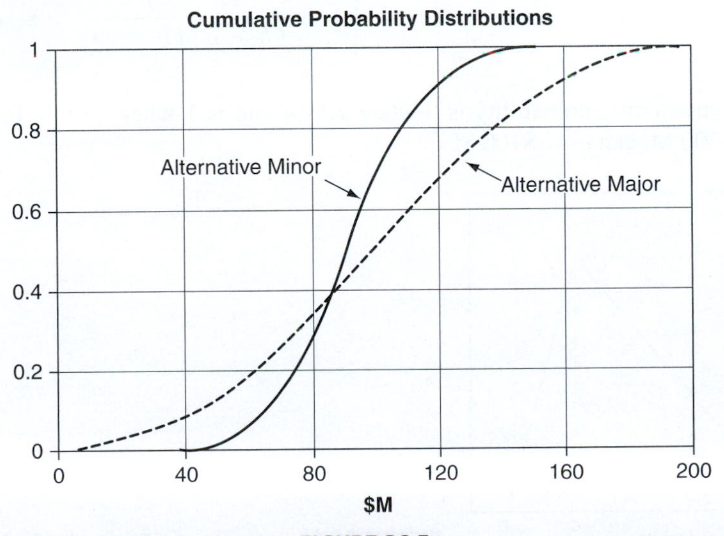

FIGURE 36.5

The Minor alternative has a probability distribution given by a scaled Beta (10,12) distribution on the interval [0, \$200 M].

Figures 36.4 shows and 36.5 show the probability density functions and the cumulative distributions of each design alternative.

The company has an exponential u-curve,

$$u(x) = \frac{1 - e^{-\gamma x}}{1 - e^{-\gamma 200}}, \; 0 \leq x \leq 200$$

with the risk-aversion coefficient of $\gamma = 0.03\,(\$M)^{-1}$. Determine the best decision alternative by simulation.

Solution: Simulation Steps for Each Alternative

1. Generate Rand.
2. Calculate the value of the outcome by projecting the value of Rand on the cumulative distribution.
3. Calculate the u-value of the generated outcome.
4. Calculate the average u-values.

SIMULATING ALTERNATIVE MAJOR A triangular probability density function can be represented as per Figure 36.6.

By calculating the areas under the probability density curve, the cumulative probability distribution for the triangular density is

$$\text{Cumulative probability distribution} = \begin{cases} \dfrac{(x-a)^2}{(b-a)(c-a)} & a \leq x \leq c \\ 1 - \dfrac{(b-x)^2}{(b-a)(b-c)} & c \leq x \leq b \end{cases}$$

Note that the cumulative probability is 0 when $x = a$ and is 1 when $x = b$. In our example, $a = 0$, $b =$ \$200 M, and $c =$ \$100 M.

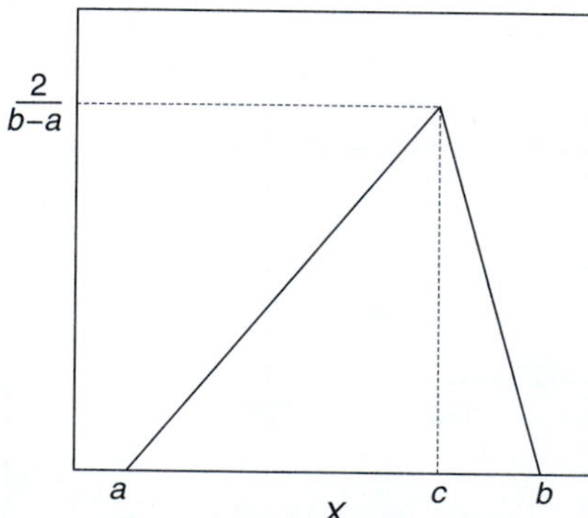

FIGURE 36.6 Example of a Triangular Probability Density

TABLE 36.8 Simulation Results for Alternative Major

(a) Number	(b) Rand	(c) Outcome ($M)	(e) *u*-Value
1	0.2663	72.98	0.8878
2	0.1097	46.84	0.7541
3	0.5259	78.87	0.9059
4	0.8582	78.87	0.9059
5	0.8844	78.87	0.9059
⋮	⋮	⋮	⋮
1996	0.3743	86.52	0.9252
1997	0.2661	72.95	0.8876
1998	0.3722	86.28	0.9247
1999	0.0872	41.76	0.7136
2000	0.4159	91.20	0.9350

To simulate an outcome from this distribution, we generate Rand and project it back to get the value of x. Since the domain is non-negative, the positive value of x is taken. Hence

$$x = \begin{cases} a + \sqrt{\text{Rand} \times (b-a)(c-a)}, & \text{Rand} \leq \dfrac{(c-a)}{(b-a)} \\[2ex] b - \sqrt{[1-\text{Rand}](b-a)(b-c)}, & \text{Rand} > \dfrac{(c-a)}{(b-a)} \end{cases}$$

Table 36.8 shows the simulation results. Column (a) represents the number of each simulation trial. Column (b) represents the Rand generated. Column (c) generates the value of x from the rand as per the formula. Column (d) calculates the u-value of the value of x generated in Column (c).

Sum of u-values generated in 2000 simulations $= 860$

Average u-values $= 0.43$

Certain equivalent $= \$18.87\,\text{M}$

SIMULATING ALTERNATIVE MINOR Once again, we start with a Rand and project back on the cumulative distribution (in this case, a scaled Beta distribution). Many programs have a built-in inverse Beta distribution. Table 36.9 shows the simulation results.

Sum of u-values generated in 2000 simulations $= 935.3$

Average u-value $= 935.3/2000 = 0.46$

Certain equivalent $= \$20.54\,\text{M}$

Conclusion: Minor is the preferred alternative.

TABLE 36.9 Simulation Results for Alternative Minor

(a) Number	(b) Rand	(c) Outcome ($M)	(e) *u*-Value
1	0.2488	76.28	0.8983
2	0.0136	47.12	0.7561
3	0.8586	113.71	0.9669
4	0.2274	74.85	0.8938
5	0.9733	131.39	0.9805
⋮	⋮	⋮	⋮
1996	0.0784	61.49	0.8415
1997	0.0104	45.37	0.7430
1998	0.9051	118.72	0.9715
1999	0.5125	91.30	0.9352
2000	0.5990	96.02	0.9438

36.6 SIMULATING MUTUALLY IRRELEVANT DISTINCTIONS

When more than one distinction is present, the simulation procedure will need to generate an *outcome* for each distinction. We now need to consider the relevance between the distinctions (arrows connecting them in the decision diagram) to produce possibilities representative of their probabilities in the decision tree. When no arrows are present (the distinctions are mutually irrelevant), the procedure is simply a matter of generating the outcome of each distinction separately using its cumulative distribution.

36.6.1 Choosing a Restaurant with Simulation

To illustrate, let us refer back to the peanut butter and jelly sandwich example in Chapter 28. Suppose we are uncertain about the sandwich we will get at two different restaurants. One restaurant might have the reputation of putting more bread and less of the other ingredients. The restaurants charge the same price for the sandwich.

We may represent our belief about the sandwich at each restaurant using a probability distribution for thickness of peanut butter, jelly, and bread. We can solve this problem by simulation to determine the best restaurant to visit.

In this section, we assume that the distributions of the various attributes of the sandwich are mutually irrelevant. To solve this problem we generate representative values for the peanut butter, jelly, and bread and then calculate the *u*-value of the sandwich at each restaurant.

RESTAURANT 1 Suppose that the distributions are scaled beta distributions Beta (r, k, Lowerbound, Upperbound) with the following values.

$$\text{Peanut Butter, p: Lowerbound} = 0.03, \text{ Upperbound} = 0.07, r = 2, k = 2$$

$$\text{Bread, b: Lowerbound} = 0.9, \text{ Upperbound} = 1.1, r = 2, k = 3$$

$$\text{Jelly, j: Lowerbound} = 0.08, \text{ Upperbound} = 0.15, r = 4, k = 5$$

TABLE 36.10 Simulation Results for Restaurant 1

(a) Number	(b) Rand 1	(c) Rand 2	(d) Rand 3	(e) *p*	(f) *b*	(g) *j*	(h) *f*	(i) *V*(*p*,*b*,*j*,*f*)	(j) *u*-Value
1	0.5958	0.1941	0.2253	0.053	0.9	0.1	0.5	9.93	0.99995
2	0.5040	0.4085	0.0705	0.050	1.0	0.1	0.5	9.93	0.99995
3	0.9143	0.9397	0.8438	0.063	1.0	0.1	0.5	8.84	0.99985
4	0.2591	0.5858	0.1495	0.043	1.0	0.1	0.4	9.66	0.99994
5	0.0325	0.0449	0.7527	0.034	0.9	0.1	0.3	7.09	0.99917
⋮	⋮	⋮	⋮	⋮	⋮	⋮	⋮	⋮	⋮
4995	0.0890	0.0882	0.3302	0.037	0.9	0.1	0.4	8.48	0.9998
4996	0.7261	0.4953	0.5989	0.056	1.0	0.1	0.5	9.65	0.9999
4997	0.1659	0.1403	0.2903	0.040	0.9	0.1	0.4	9.07	0.9999
4998	0.3665	0.7371	0.9390	0.046	1.0	0.1	0.4	8.42	0.9998
4999	0.5454	0.3595	0.1496	0.051	1.0	0.1	0.5	9.97	1.0000
5000	0.7329	0.0444	0.8295	0.056	0.9	0.1	0.5	9.24	0.9999
							Sum	45186.79	4999.096
							Average	$ 9.04	1.000
							CE		$ 8.62

Table 36.10 shows the simulation results. Column (a) shows the number of trials. Columns (b), (c), and (d) show three Rand numbers generated. Columns (e), (f), and (g) show the values of peanut butter, thickness of bread, and jelly generated using the Rand numbers. Column (h) shows the fraction of peanut butter to jelly. Column (i) calculates the value of the sandwich for the generated values. Column (j) calculates the u-value of the sandwich.

The average u-value is 0.9998 giving a certain equivalent of \$8.62.

RESTAURANT 2 Restaurant 2 gives more bread, less peanut butter, and less jelly. Suppose that the distributions are scaled beta distributions with the following values.

Peanut Butter, p: Lowerbound $= 0.02$, Upperbound $= 0.05$, $r = 2$, $k = 2$

Bread, b: Lowerbound $= 1.2$, Upperbound $= 1.5$, $r = 2$, $k = 3$

Jelly, j: Lowerbound $= 0.06$, Upperbound $= 0.1$, $r = 4$, $k = 5$

Table 36.11 shows the simulation results.

The average u-value is 0.998, and the certain equivalent of the sandwich in this restaurant is \$6.63. Therefore, it is better to go to Restaurant 1.

This example also shows that simulation can be used even when the problem has multiple attributes.

TABLE 36.11 Simulation Results for Restaurant 2

(a) Number	(b) Rand 1	(c) Rand 2	(d) Rand 3	(e) p	(f) b	(g) j	(h) f	(i) $V(p,b,j,f)$	(j) u-Value
1	0.1933	0.2025	0.9249	0.028	1.3	0.1	0.3	6.55	0.999
2	0.7102	0.9513	0.4834	0.039	1.4	0.1	0.5	7.41	0.999
3	0.8100	0.6121	0.1595	0.042	1.3	0.1	0.6	7.68	1.000
4	0.6003	0.1580	0.4906	0.037	1.3	0.1	0.5	8.26	1.000
5	0.1560	0.3931	0.3062	0.027	1.3	0.1	0.4	6.33	0.998
⋮	⋮	⋮	⋮	⋮	⋮	⋮	⋮	⋮	⋮
4995	0.2765	0.1849	0.4461	0.030	1.3	0.1	0.4	7.14	0.999
4996	0.0069	0.5366	0.8245	0.021	1.3	0.1	0.3	4.48	0.989
4997	0.2184	0.5140	0.5052	0.029	1.3	0.1	0.4	6.60	0.999
4998	0.5964	0.8768	0.2367	0.037	1.4	0.1	0.5	7.28	0.999
4999	0.2411	0.5892	0.3380	0.030	1.3	0.1	0.4	6.64	0.999
5000	0.0350	0.0193	0.3799	0.023	1.2	0.1	0.3	5.48	0.996
							Sum	36174.22	4993.394
							Average	$ 7.23	0.999
							CE		$ 6.63

Reflection

Repeat the analysis of the peanut butter and jelly sandwich using 10000 trials.

36.7 VALUE OF INFORMATION WITH SIMULATION

This section shows how to use simulation to calculate the value of information.

36.7.1 Party Problem: Deltaperson

Let us refer back to the Chapter 9 party problem, whose decision tree is shown as Figure 36.7 for convenience.

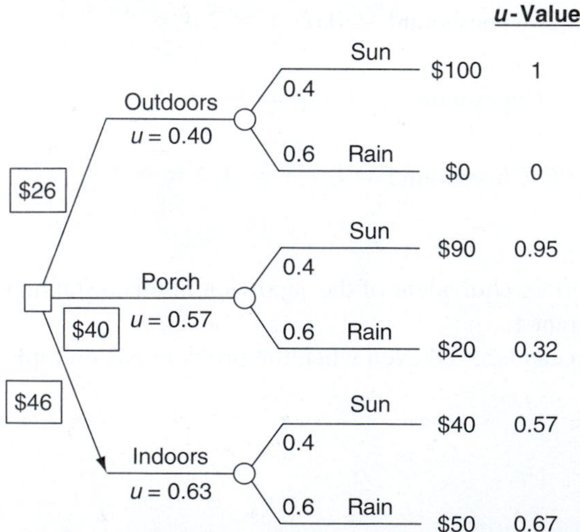

FIGURE 36.7 Decision Tree for Party Problem

To determine the party value and location with no information, we simulate the weather distinction Sun/Rain and calculate the *u*-values for each alternative. Using Rand, we can say

If Rand < 0.4, then Weather = Sun, otherwise, Weather = Rain

Table 36.12 shows the simulation results with no information. It is clear from the table that the highest *u*-value is 0.63 for the Indoors alternative, yielding a certain equivalent of $46.

To determine the value of clairvoyance on weather, we again use Rand to determine the clairvoyant report. If Rand generates Sun, we use a *u*-value of 1 for the Outdoors alternative. If it generates Rain, then we use a *u*-value of 0.67 for the Indoors alternative. The average *u*-value is the *u*-value of the party with free clairvoyance.

Table 36.13 shows that the *u*-value with free clairvoyance is 0.8 corresponding to a certain equivalent of $66.20. Since we know Kim to be a deltaperson, the value of clairvoyance is the value of the deal with free clairvoyance less its value with no clairvoyance, so $66.20 − $46 = $20.20.

TABLE 36.12 Simulation Results for Party Value with No Information

Trial	Rand	Weather	*u*-Value Outdoors	*u*-Value Porch	*u*-Value Indoors
1	0.346745	Sun	1	0.95	0.57
2	0.998081	Rain	0	0.32	0.67
3	0.6646	Rain	0	0.32	0.67
4	0.028792	Sun	1	0.95	0.57
5	0.584572	Rain	0	0.32	0.67
6	0.762655	Rain	0	0.32	0.67
7	0.13702	Sun	1	0.95	0.57
8	0.274785	Sun	1	0.95	0.57
9	0.64757	Rain	0	0.32	0.67
10	0.463797	Rain	0	0.32	0.67
⋮	⋮	⋮	⋮	⋮	⋮
990	0.590122	Rain	0	0.32	0.67
991	0.68379	Rain	0	0.32	0.67
992	0.723116	Rain	0	0.32	0.67
993	0.970438	Rain	0	0.32	0.67
994	0.849832	Rain	0	0.32	0.67
995	0.378497	Sun	1	0.95	0.57
996	0.827888	Rain	0	0.32	0.67
997	0.524129	Rain	0	0.32	0.67
998	0.325957	Sun	1	0.95	0.57
999	0.066519	Sun	1	0.95	0.57
1000	0.548125	Rain	0	0.32	0.67
		Average *u*	0.40	0.58	0.63

TABLE 36.13 Simulation Results for Party Value with Information

Trial	Rand	Weather	*u*-Value of Best Decision
1	0.975916	Rain	0.67
2	0.586798	Rain	0.67
3	0.84926	Rain	0.67
4	0.565929	Rain	0.67
5	0.085167	Sun	1.00
6	0.730241	Rain	0.67
7	0.292204	Sun	1.00
8	0.19473	Sun	1.00
9	0.823823	Rain	0.67
10	0.583113	Rain	0.67
⋮	⋮	⋮	⋮
995	0.549097	Rain	0.67
996	0.843327	Rain	0.67
997	0.923586	Rain	0.67
998	0.781193	Rain	0.67
999	0.127713	Sun	1.00
1000	0.772459	Rain	0.67
		Average *u*	0.80

36.7.2 Party Problem: Non-Deltaperson

We can also use simulation to calculate the value of clairvoyance when a person does not follow the delta property. Consider for example, a logarithmic decision maker with an initial wealth $72 as

$$u(x) = [\log(x + 72) - \log(72)]/[\log[100 + 72] - \log(72)]$$

The u-values of the party prospects for this u-curve are shown in Table 36.14.

Table 36.15 shows the simulation results.

The best decision is still to go indoors. Now, consider clairvoyance. Table 36.16 shows the analysis for $b = \$10$. The prospects are \$100 or \$50 less the value paid to the clairvoyant. Hence, u-value $= u(x - b)$. Note that $u(100 - 10) = u(90) = 0.9312$, and $u(50 - 10) = u(40) = 0.5074$.

TABLE 36.14 Party Prospects for *u*-Curve

$	*u*-Value
100	1.00
0	0.00
90	0.93
20	0.28
40	0.51
50	0.61

TABLE 36.15 Simulation Results for Party Problem for Non-Deltaperson and No Information

Trial	Rand	Weather	*u*-Value Outdoors	*u*-Value Porch	*u*-Value Indoors
1	0.420804	Rain	0	0.28	0.61
2	0.792401	Rain	0	0.28	0.61
3	0.144632	Sun	1	0.93	0.51
4	0.026849	Sun	1	0.93	0.51
5	0.234979	Sun	1	0.93	0.51
6	0.747008	Rain	0	0.28	0.61
7	0.456153	Rain	0	0.28	0.61
8	0.055378	Sun	1	0.93	0.51
9	0.904989	Rain	0	0.28	0.61
10	0.848608	Rain	0	0.28	0.61
⋮	⋮	⋮	⋮	⋮	⋮
995	0.243949	Sun	1	0.93	0.51
996	0.569529	Rain	0	0.28	0.61
997	0.721794	Rain	0	0.28	0.61
998	0.463854	Rain	0	0.28	0.61
999	0.445019	Rain	0	0.28	0.61
1000	0.732466	Rain	0	0.28	0.61
		Average *u*	0.38	0.53	0.57

Since the u-value of 0.6722 is higher than the u-value without clairvoyance of 0.57, the value of clairvoyance (VOC) must be higher than $b = 10$. We increase b by iteration until we get the same u-value as we obtained with no clairvoyance. Table 36.17 shows that this is achieved with $b = \$20$. If we have the results of the simulation stored, we can find the value of b without further simulation.

TABLE 36.16 Simulation Results for Party with Information Paying $10

b = $10

Trial	Rand	Weather	*u*-Value Outdoors	*u*-Value Porch	*u*-Value Indoors	Prospect with Clairvoyance	*u*-Value of Best Decision
1	0.625341	Rain	0	0.28	0.61	50	0.5074
2	0.711182	Rain	0	0.28	0.61	50	0.5074
3	0.298646	Sun	1	0.93	0.51	100	0.9312
4	0.43989	Rain	0	0.28	0.61	50	0.5074
5	0.528157	Rain	0	0.28	0.61	50	0.5074
⋮	⋮	⋮	⋮	⋮	⋮	⋮	⋮
995	0.56868	Rain	0	0.28	0.61	50	0.5074
996	0.413261	Rain	0	0.28	0.61	50	0.5074
997	0.029054	Sun	1	0.93	0.51	100	0.9312
998	0.791803	Rain	0	0.28	0.61	50	0.5074
999	0.030346	Sun	1	0.93	0.51	100	0.9312
1000	0.686303	Rain	0	0.28	0.61	50	0.5074
		Average *u*	0.39	0.53	0.57		0.6722

TABLE 36.17 Simulation Results for Party with Information Paying $20

				b = $20				
Trial	**Rand**	**Weather**	***u*-Value Outdoors**	***u*-Value Porch**	***u*-Value Indoors**	**Prospect with Clairvoyance**	***u*-Value of Best Decision**
1	0.420804	Rain	0	0.28	0.61	50	0.4000
2	0.792401	Rain	0	0.28	0.61	50	0.4000
3	0.144632	Sun	1	0.93	0.51	100	0.8581
4	0.026849	Sun	1	0.93	0.51	100	0.8581
5	0.234979	Sun	1	0.93	0.51	100	0.8581
⋮	⋮	⋮	⋮	⋮	⋮	⋮	⋮
995	0.84535	Rain	0	0.28	0.61	50	0.4000
996	0.593971	Rain	0	0.28	0.61	50	0.4000
997	0.151086	Sun	1	0.93	0.51	100	0.8581
998	0.640708	Rain	0	0.28	0.61	50	0.4000
999	0.868903	Rain	0	0.28	0.61	50	0.4000
1000	0.383085	Sun	1	0.93	0.51	100	0.8581
		Average *u*	0.39	0.53	0.57		0.5782

36.8 SIMULATING MULTIPLE DISTINCTIONS WITH RELEVANCE

Simulation can be useful in determining certain calculations that otherwise would require integrals or discretized numerical analysis. For example, let us refer back to the continuous detector of Chapter 20, which gives an indication, T, from 0 to 100. The conditional probability distribution for its indication given Sun is $\{T|S,\&\} = \text{Beta}(4.4, 5.6, 0, 100)$, and its indication given Rain is determined by the conditional probability $\{T|S,\&\} = \text{Beta}(7, 3, 0, 100)$.

In Chapter 20, we observed that the best decision would be Outdoors when $0 \le T < 36$, Porch when $36 \le T < 56$, and Indoors when $56 \le T \le 100$. Figure 36.8 shows the u-value of the best party location given the indication T.

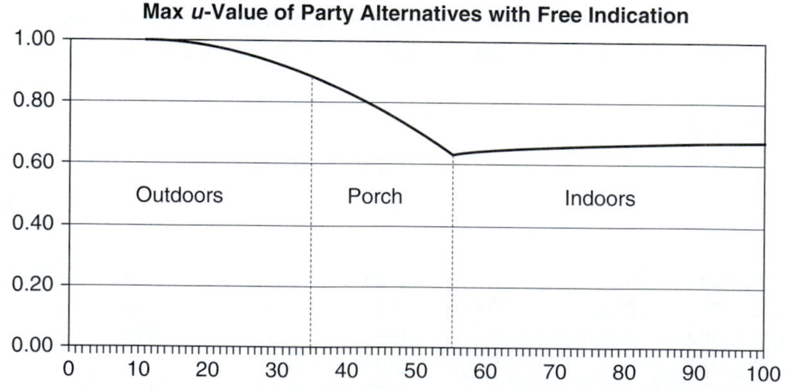

FIGURE 36.8 *u*-Value of Best Party for Different Continuous Detector Indications

TABLE 36.18 *u*-Value of Party with Free Continuous Detector

Trial	Rand_1	Weather	Rand_2	Detector Indication, *T*	Location	*u*-Value
1	0.775153	Rain	0.673229	77.60	I	0.662559
2	0.96065	Rain	0.232956	60.06	I	0.632973
3	0.730129	Rain	0.106517	51.59	P	0.694306
4	0.261943	Sun	0.235889	32.41	O	0.914826
5	0.812725	Rain	0.948592	90.12	I	0.669279
6	0.660354	Rain	0.678608	77.80	I	0.662559
7	0.80949	Rain	0.295603	63.15	I	0.639986
8	0.743298	Rain	0.809795	82.83	I	0.666364
9	0.185182	Sun	0.609276	48.07	P	0.73836
10	0.87876	Rain	0.729207	79.66	I	0.664309
⋮	⋮	⋮	⋮	⋮	⋮	⋮
990	0.599461	Rain	0.881598	86.09	I	0.668211
991	0.248173	Sun	0.722592	53.15	P	0.663395
992	0.213367	Sun	0.645011	49.60	P	0.724052
993	0.649735	Rain	0.768193	81.15	I	0.66575
994	0.049063	Sun	0.190017	30.10	O	0.931504
995	0.948464	Rain	0.177281	56.85	I	0.623032
996	0.064553	Sun	0.696514	51.92	P	0.694306
997	0.594887	Rain	0.30593	63.62	I	0.639986
998	0.131704	Sun	0.978367	74.51	I	0.659287
999	0.131894	Sun	0.577821	46.75	P	0.765633
1000	0.717819	Rain	0.94499	89.85	I	0.669072
			Average *T*	60.13	**Average *u***	0.718

We can now calculate the u-value of the deal with the free continuous detector using simulation. To do this, we first simulate the detector indication.

Generate Rand_1

If Rand_1 $<$ 0.4,

Weather $=$ Sun

otherwise

Weather $=$ Rain

Generate Rand_2

If Rand_1 $<$ 0.4,

$T =$ BetaInverse (Rand_2, 4.4, 5.6, 0, 100)

otherwise

$T =$ BetaInverse (Rand_2, 7, 3, 0, 100)

where BetaInverse in the inverse function of the Beta cumulative distribution.

Depending on the value of T, we then generate the u-value of the best party from Figure 36.8. Table 36.18 shows the simulation results. The average u-value generated is the e-value of the detector.

Compare the average u-value, 0.718, to the e-value of the u-values of the free detector of 0.714 calculated in Chapter 20 using discretization.

36.9 SUMMARY

Simulation can help compute decision problems especially with continuous distributions where integrals may be difficult to evaluate.

Starting with a uniform distribution, we can generate outcomes representing many useful distributions and decision situations.

KEY TERMS

- Law of large numbers
- Profitability
- Uniform random numbers
- Simulation

PROBLEMS

Problems marked with an asterisk (*) are considered more challenging.

*1. Repeat the simulation tables we presented in this chapter and verify the results.

*2. Refer back to the example *Let's Make a Deal* in Chapter 6. Using simulation, determine the fraction of times a person would get the car with and without switching doors. Verify that the best strategy is to switch after the TV host shows a door with a goat behind it.

*3. Simulate two continuous distinctions having first-order probabilistic dominance. For example, two exponential probability distributions, $\{X \leq x\} = 1 - e^{-\alpha x}$ and $\{Y \leq y\} = 1 - e^{-\beta y}$, for $x, y \geq 0$ exhibit first-order dominance for $\alpha \neq \beta$. Using simulation, calculate the *e*-value of *u*-values for each distribution by taking the average *u*-values for

a. a logarithmic *u*-curve with initial wealth \$100

b. an exponential *u*-curve with risk tolerance \$100

c. an exponential *u*-curve with risk tolerance \$ − 100

Verify that the *e*-value of *u*-values is always higher for the distribution that dominates the other distribution probabilistically in the first order.

*4. Chevalier de Mere gambled frequently to increase his wealth. He designed a deal where he bet that he would get at least one 6 in four roles of a die. Chevalier de Mere made money. Using simulation, determine the probability of getting at least one 6 in four rolls of a die.

*5. Chevalier de Mere designed another game. In the new game, he bet that he would get at least one 12 (6–6) in 24 rolls of two dice. Chevalier de Mere started losing money. Using simulation, determine the probability of winning this new game.

CHAPTER

37

The Decision Analysis Cycle

CHAPTER CONCEPTS

After reading this chapter, you will be able to explain the following concepts:

- Phases of the decision analysis cycle:
  - Formulation
  - Evaluation (deterministic and probabilistic)
  - Appraisal
  - Decision
- Value model
- Tornado diagrams and their limitations
- Open loop and closed loop sensitivity

37.1 INTRODUCTION

Decisions we face in practice may involve a large numbers of distinctions, possible frames, and decision stages. As compared to simpler decisions, these more complicated decisions usually require more complex computations as well as a more elaborate elicitation process. To deal with these decisions in an analytically tractable way, it is important to target the elements most important to making the decision. This is the purpose of the **decision analysis cycle**.

The decision analysis cycle consists of several phases for providing focus on what really matters in the decision. In previous chapters, we have discussed many elements of the decision analysis cycle. For example, as discussed extensively in Chapter 17, *framing* is one important element of providing focus. We shall now connect the topics we have covered in previous chapters to get a big picture view of the nature of the analysis of a decision. We shall also present some new tools to help with the analysis and provide a case study to be solved that involves many of the tools discussed so far.

37.2 THE DECISION ANALYSIS CYCLE

Figure 37.1 shows the four phases of the decision analysis cycle. The first phase is **Formulation**, where the decision is framed. The second phase is **Evaluation**, where the decision is analyzed both deterministically and probabilistically. The result is insight into the decision implications of the present formulation, and knowledge of the sensitivity of the results to problem inputs. The **Appraisal** phase uses this knowledge to choose whether there is now clarity of action, or instead to postpone action to refine the analysis. This refinement may require gathering new assessments, or pursuing information gathering because it is profitable given the cost and value

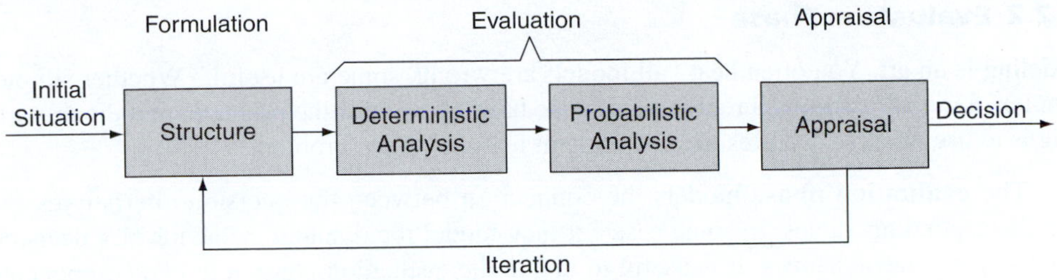

FIGURE 37.1 Decision Analysis Cycle

of information. Conducting this refinement will require returning to the formulation phase, as shown by the arrow in the figure, when the new findings are available, thus iterating the process. Of course, if there is no time for iteration before the decision must be made, then it is determined by the current analysis.

We shall now discuss each of these phases of the decision analysis cycle in detail.

37.2.1 Formulation Phase

The **formulation phase** answers the basic question, "What is the decision we face?"

As we discussed in Chapter 17, decision formulation is perhaps the most important phase in the entire analysis. The formulation phase begins with an unclear formulation of the problem and ends with a well-structured decision problem. To formulate the problem, we identify the appropriate scope of the decision using the **decision hierarchy**, and we identify the most critical alternatives. This step reduces the computational complexity and helps us focus on the most important decisions that we need to decide now as opposed to those decisions that we are taking as given and those that will be decided later.

As we discussed in Chapter 17, before we establish the scope of our decision, it is useful to expand and contract the frame to make sure we neither overlook any important decisions nor focus too narrowly on insignificant decisions.

In Chapter 39, we shall discuss another tool, the strategy table, which we can use to help identify important decision alternatives. It is also essential to think about possible sequential decisions during the formulation phase and not just a one-time decision formulation (for more about sequential decisions and creating decision options, refer back to Chapter 19).

The formulation phase identifies the uncertainties in the decision and creates useful distinctions about them. It also identifies the values that we care about. Recall the distinction between direct and indirect values that we discussed in Chapter 26.

Ultimately, the formulation phase provides an appropriate frame and a decision diagram that captures the important decisions, uncertainties, and values we have established. Within this phase, we are merely thinking at the distinction level in the diagram and not necessarily considering the degrees of distinction or any associated numbers. The success of the formulation phase is essential to the success of all related future analyses. Why is this useful? Because virtually everything is uncertain. If you try to capture every possible uncertainty you will be confused and exhausted by the effort. Sometimes people attempt to draw decision trees that continually expand as uncertainties are realized. This often results in a decision bush rather than a decision tree. It may be viewed that the decision analysis has failed because it is complicated. This is a rookie mistake. The proper procedure is to determine and include only those uncertainties that are material to the decision.

37.2.2 Evaluation Phase

Modeling is an art. You often hear "all models are wrong, some are useful." Whether simple or complex, a model is simply a representation we have chosen for the problem, and the important thing is to use one that captures the most salient features of the problem.

> The **evaluation phase** models the connection between the decision alternatives and the corresponding values. In some cases, it may model the dynamic behavior of a decision, such as the price dynamics. It is useful to divide the evaluation phase into two components: *deterministic analysis* and *probabilistic analysis*.

37.2.2.1 DETERMINISTIC ANALYSIS In the **deterministic analysis** phase, we derive the value model for the problem by specifying our value when the prospects are fixed. We also identify the important uncertainties that contribute most to the variation in value and the less important ones that can be modeled deterministically.

Value Model The **value model** (or the value function) provides a value for all the prospects of the problem. As we discussed in Chapter 26, it also makes deterministic trade-offs among the direct value attributes of the decision by moving along an *isopreference contour*.

Sometimes trade-offs rely on the physical principles of a system. For example, when building a bridge, we might want to reduce thickness of one of the pillars due to space limitations on one side of the bridge, while increasing the width of another to compensate for this reduction. This trade-off can be achieved by studying the forces affecting the stability of the bridge using basic static and possibly dynamic analysis. In this case, it is not a good idea to assess the trade-off directly without conducting an engineering study.

In other cases, the trade-offs depend entirely on the decision maker's preference. We have already discussed several examples of such value functions in Chapter 26, such as the value function for the peanut butter and jelly sandwich, the value function for life and death decisions, and, in Chapter 27, the value functions for cash flows.

The end of this value model task is a clear value function capturing the direct values, determining the deterministic trade-offs among them, and identifying a value measure that will be the basic measure for the value node in the decision diagram.

To illustrate a value model, consider the decision diagram of Figure 37.2 where a company has specified profit as its value measure. We may have two nodes connected to profit such as Revenue and Cost, where

$$\text{Profit} = \text{Revenue} - \text{Cost}$$

Revenue is further divided into Demand (quantity sold) and Price per unit. Similarly, Cost is divided into Distribution Cost, Overhead Cost, and Raw Material Cost.

The value function for profit is then

$$\text{Profit} = \text{Price} \times \text{Demand} - (\text{Distribution cost} + \text{Overhead cost} + \text{Raw material cost})$$

The deterministic phase requires specifying a range for each of the distinctions. The analysis then swings each uncertainty through its range to see its effect on the value measure for each alternative. The uncertainties that have a very small effect on the variation in value can be fixated at a nominal value (such as the median) for the purposes of making a decision. Those that have a substantial effect on value are selected for treatment in the probabilistic phase.

By the end of the deterministic phase of the analysis, we have constructed a value model and identified the uncertainties that contribute most to the change in variance of the value model.

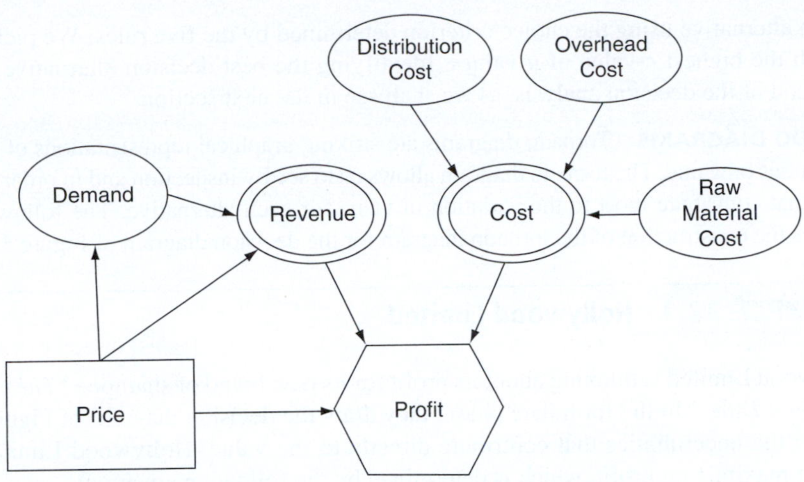

FIGURE 37.2 Decision Diagram showing Demand, Distribution Cost, Overhead Cost, and Raw Material Cost as Uncertainties that will Appear in the Tornado Diagram

The less important uncertainties are fixated. We have not yet assessed the probabilities on the probability tree. The next phase analyzes the problem taking the full joint probability into account.

PROBABILISTIC ANALYSIS Deterministic analysis characterized the uncertainties that contributed most to the change in values for each alternative as they vary from their lower bound to upper bound. By themselves, the lower and upper bounds are not precise descriptors of our uncertainty about a distinction of interest. Using a few fractile probability elicitations, as discussed in Chapter 15, we can better describe these uncertainties. With this more refined description, we are better able to see their actual effect on value using the notion of a "tornado diagram." A **tornado diagram** is a graph that describes the contribution of each uncertainty to value at specific fractiles.

Any distinctions that have been identified by the tornado diagram as contributing little to the variation in value are fixated and need no further elicitation. Those that are identified as major contributors to value will require further probability elicitations using the same method of probability encoding discussed in Chapter 15. Remember the biases of Chapter 16 that might be encountered during this elicitation phase. The analysis should strive to minimize these biases as much as possible following the procedure we discussed.

When the probability distributions of the important distinctions are assessed, we should also check for probabilistic dominance among the alternatives, as this can make assessing risk attitude by *u*-curves unnecessary. If there is no probabilistic dominance, we need to encode the risk preference at this point as well. This analysis follows the steps we discussed in Chapter 11—being aware of the saturation effect for an exponential decision maker if the prospects are much larger than the risk tolerance. A logarithmic or linear risk-tolerance *u*-curve also might be used and the parameters assessed, as we discussed in Chapters 22 and 23. When prospects have multiple attributes, such as health and wealth, the utility function is then assessed over the value model itself, as discussed in Chapter 28.

By the end of the probabilistic phase, we have all the information required to determine the best decision alternative. We have an analysis that focuses on what is important: the decisions as identified by the frame, the values as characterized by direct and indirect, and the material uncertainties. We are ready to compute the certain equivalent of each alternative and identify the best decision alternative using the choice criterion determined by the five rules: We pick the alternative with the highest *e*-value of *u*-values. Identifying the best decision alternative, however, is not the end of the decision analysis, as we shall see in the next section.

TORNADO DIAGRAMS Tornado diagrams are striking graphical representations of the effects of uncertainties on value. The tornado diagram allows us to see by inspection and in order, those uncertainties that contribute most to the variation in value for each alternative. The following example illustrates the construction of the tornado diagram for the decision diagram of Figure 37.2.

EXAMPLE 37.1 Hollywood Limited

Hollywood Limited is thinking about its profit for its new brand of shampoo "*The Clairvoyant Goes on a Date*." In the formulate phase, they draw the decision diagram of Figure 37.2 and identify the uncertainties that contribute directly to the value. Hollywood Limited is interested in maximizing profit, which is determined by the following equation:

$$\text{Profit} = \text{Price}(\text{Per unit}) \times \text{Demand} - (\text{Distribution cost} + \text{Overhead cost} + \text{Raw material cost})$$

The company needs to decide what price to charge for the new shampoo and considers two alternatives: \$10 or \$15. The first will result in a higher demand but also generates less revenue. Hollywood Limited is risk-neutral over this range of prospects.

To construct the tornado diagram, the first step is to elicit the 10–50–90 fractiles for each distinction for each alternative. This can be done using the methods we discussed earlier with the 20 questions exercise in Chapter 15. Recall that these values also correspond to the shortcut method for the equal areas approximation that we discussed in Chapter 35.

Table 37.1 shows the 10–50–90 values elicited for the uncertainties of the decision for the price alternative = \$10.

Now we evaluate the value—in this case, profit—at the different fractiles by varying each uncertainty across its 10–50–90 fractiles, while keeping the remaining uncertainties at their base (50%) value. By doing this, we are in effect assuming mutual irrelevance between the uncertainties. Otherwise varying the level of one uncertainty would change our belief about the rest.

To illustrate the profit calculation using this method, when the demand is at its 10% fractile of \$12 million (M), and when the remaining uncertainties are at their median values (Distribution cost = \$7 M, Overhead cost = \$90 M, Raw material cost = \$30 M), the value of Profit is

$$\text{Profit} = 10 \times 12 - 7 - 90 - 30 = -7(\$M)$$

The company will suffer a loss of \$7 million if the demand is its 10% fractile value and the remaining uncertainties are at their median values.

TABLE 37.1 10–50–90 Values when Price = \$10

	10%	50%	90%
Demand (\$M)	12	16	22
Distribution Cost (\$M)	4	7	12
Overhead Cost (\$M)	80	90	100
Raw Material Cost (\$M)	20	30	60

When the demand is at its 50% fractile of $16 million, the value of the profit obtained (when Distribution cost = $7 M, Overhead cost = $90 M, Raw material cost = $30 M) is

$$\text{Profit} = 10*16 - 7 - 90 - 30 = \$33 \text{ M}$$

Finally, when the demand is at its 90% fractile of $22 M, the profit obtained (Distribution cost = $7 M, Overhead cost = $90 M, Raw material cost = $30 M) is

$$\text{Profit} = 10*22 - 7 - 90 - 30 = \$93 \text{ M}$$

We can now plot the variation in profit for demand as it changes across its 10–50–90 values, while the remaining uncertainties are at their base (median) values. This is shown in Figure 37.3.

We now repeat the same calculation for profit and calculate the change in profit as the distribution cost varies across its 10–50–90 values (Table 37.2).

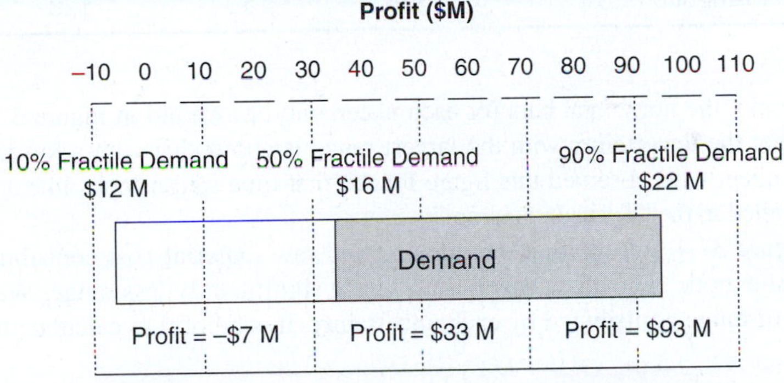

FIGURE 37.3 Plot of Profit for 10-50-90 Values of Demand

We now calculate the range in profit contributed by each uncertainty as it varies across its 10–50–90 values. We obtain this by calculating the difference between the maximum and minimum values of profit for each row. For example, the value of profit is −$7 M and $93 M, when demand is 10% and 90%, respectively. We keep the remaining uncertainties at their median values. The range in profit by varying demand is $93 − (−$7) = $100 M. Table 37.3 shows the range calculated for each direct value distinction.

We see that demand provides the highest range, then raw material cost, then overhead cost, and finally distribution cost. We now rearrange the order of the uncertainties according to their range as in Table 37.4.

TABLE 37.2 Calculating Profit by Varying Each Distinction

	Profit ($M)		
Varying Demand across its 10–50–90 values	($7)	$33	$93
Varying Distribution Cost across its 10–50–90 values	$36	$33	$28
Varying Overhead Cost across its 10–50–90 values	$43	$33	$23
Varying Raw Material Cost across its 10–50–90 values	$43	$33	$3

TABLE 37.3 Range in Profit by Changing Uncertainties Across 10–50–90 Values

	Profit ($M)			Range
Demand	($7)	$33	$93	$100
Distribution Cost	$36	$33	$28	$8
Overhead Cost	$43	$33	$23	$20
Raw Material Cost	$43	$33	$3	$40

TABLE 37.4 Ordering Distinctions Based on Range

	Profit ($M)		
Demand	($7)	$33	$93
Raw Material Cost	$43	$33	$3
Overhead Cost	$43	$33	$23
Distribution Cost	$36	$33	$28

Next, we plot the horizontal bars for each uncertainty, as we did in Figure 37.3. If we order the bars such that the uncertainty with the largest range is above that with a lower range, we get Figure 37.4. A client who observed this figure for the first time said it looks like a tornado. Since then, we have called it, the "Tornado diagram".

From Figure 37.4, we see that the demand and raw material cost contribute most to the range in the Value node. The other uncertainties have significantly less range. We can calculate the percentage of range contributed by each uncertainty. It is helpful to calculate this percentage

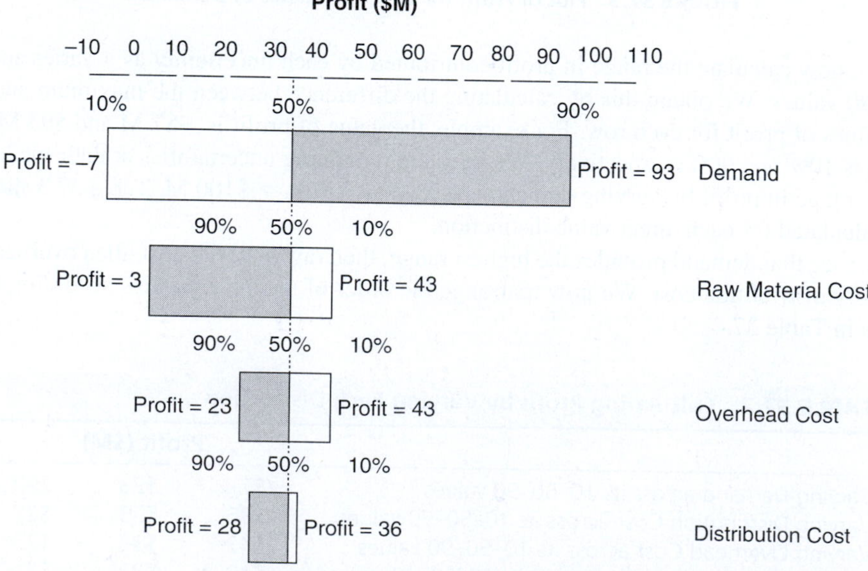

FIGURE 37.4 Tornado Diagram at Price = $10

TABLE 37.5 Calculating Percentage Range Squared

	Profit ($M)			Maximum Swing	(Maximum Swing)^2	%
Varying Demand	($7)	$33	$93	$100	10000.00	82.89
Varying Distribution Cost	$36	$33	$28	$8	64.00	0.53
Varying Overhead Cost	$43	$33	$23	$20	400.00	3.32
Varying Raw Material Cost	$43	$33	$3	$40	1600.00	13.26
					12064.00	100%

on range squared, since variance is proportional to this quantity. For example, if the uncertainties were Gaussian distributions, the difference between the 90% fractile and the 50% fractile is, in fact, 1.29σ, where σ is equal to the standard deviation. Moreover, the variance is what is used in the certain equivalent approximation (refer back to Chapter 24).

Note from Table 37.5 that demand contributes 82.89% of the swing, raw material contributes 13.26%, overhead cost contributes 3.32%, and distribution cost contributes 0.53% of the sum of maximum swing squared. A first order approximation would be to make distribution cost and overhead cost fixated at their base values and consider only demand and raw material cost as uncertainties. As we discussed, they are still uncertain, but their uncertainty will be ignored until further notice. The tornado diagram, therefore, simplifies the analysis of the problem by removing some of the less significant uncertainties.

If there are different uncertainties for each decision alternative or if there are the same uncertainties but with different probabilities, we will need to draw a tornado diagram for each decision alternative. For example, we can repeat this tornado diagram analysis for the other price alternative $15. We can also do such an analysis in higher dimensions to show the effects of related factors: two mutually relevant uncertainties or a decision variable-uncertainty combination. In the next section, we shall show how to treat such situations.

LIMITATIONS OF TORNADO DIAGRAMS We have discussed that tornado diagrams determine the nodes that contribute to the variance in value. We did this by varying the value of each node while keeping the remaining nodes fixated. If an indirect value node was changed while keeping the remaining nodes fixed then, as we discussed in Chapter 26, it will not change the deterministic order of the prospects or the value as determined by the value function.

In some cases, it also may be important to incorporate relevance between the uncertainties in the tornado diagram. For example, it might be that, when cost is at its 10% value, revenue is also at its 10% value. In situations like these, we can vary both nodes simultaneously.

The tornado diagram required only 10–50–90 assessments and did not require any relevance information. A more thorough analysis would update the uncertainties in the decision diagram for every variation in the level of any of the uncertainties. This, however, would require a complete elicitation of the joint distribution for each alternative instead of just the 10–50–90 values for each distinction.

We will discuss the effects of relevance between the uncertainties in the probabilistic phase of the analysis using the ideas of closed loop and open loop sensitivity analysis. We will also show how this provides several insights into the value of information on a given distinction.

37.2.3 Appraisal Phase

We now need to check whether the decision will change if we change any of the inputs entered into the model. For example, in Chapter 12, we discussed the idea of sensitivity analysis and illustrated how the best decision may change if we change the probability of sunshine and rain in the party problem. If the best decision does not change for a range of the probability that the decision maker believes contains all values he might assign, this will reduce the time and effort needed for more precise elicitation. We can perform a sensitivity analysis to the degree of risk aversion or time preference to determine how changing these preferences will affect the best decision. We can also see how gathering new information on uncertainties might change the decision and determine the value of the information.

There are many variants of this idea that can help reduce the complexity of the decision by reducing some of the uncertainties. For example, suppose we have identified an uncertainty and have observed that the best decision does not change, regardless of the degree of distinction. We refer to this distinction as an **immaterial distinction**. For simplicity, we might choose to make it fixated.

As we change the outcome of an uncertainty, several questions arise about the mechanism by which this change is carried out and the way it affects other uncertainties in the problem. For example, if the uncertainty is relevant to other distinctions, how should the other distinctions update as we vary the outcomes of an uncertainty? And if we are interested in the change in value due to the change in outcomes of the uncertainty, can we change the chosen alternative? As we discussed, tornado diagrams do not take this relevance effect into account. Our discussion of influences illuminates the issues.

The **appraisal phase** uses the idea of open loop and closed loop sensitivity analysis to provide insights into how the best decision changes when information is available about the outcomes of an uncertainty and when there is relevance between the uncertainties. Open loop sensitivity changes the outcomes while keeping the chosen decision alternative fixed. It observes how the value of a chosen alternative changes as we change the outcome. Closed loop sensitivity, on the other hand, changes the outcomes while allowing the decision maker to choose different decision alternatives using the knowledge of the change. It shows the effects of making a decision while knowing the outcomes of the uncertainty. We expect that the value obtained by closed loop sensitivity will be at least as high as that with an open loop. The difference between closed loop and open loop sensitivities illustrates the effect of receiving information and creates the value of that information.

OPEN LOOP SENSITIVITY Let us refer again to the Chapter 9 party problem. Suppose that Kim has already carried out the analysis and has decided to have the party indoors. We know her certain equivalent for this deal is \$45.83. Now, if Kim has made her indoor decision and must stick with it, then hearing that the weather will be sunny will give her the \$40 party appropriate to sunshine for an indoor party. If she learns that the weather will be rainy, she gets a \$50 value for an indoor party. If she cannot change her decision, this change in value corresponding to changing the degrees of the weather is referred to as open loop sensitivity.

CLOSED LOOP SENSITIVITY For closed loop sensitivity, Kim can change to the best alternative after hearing the weather. If she hears that the weather will be sunny, she will change from indoors to outdoors and receive the value of \$100; in the face of rain, she will continue to have the indoor party and receive \$50.

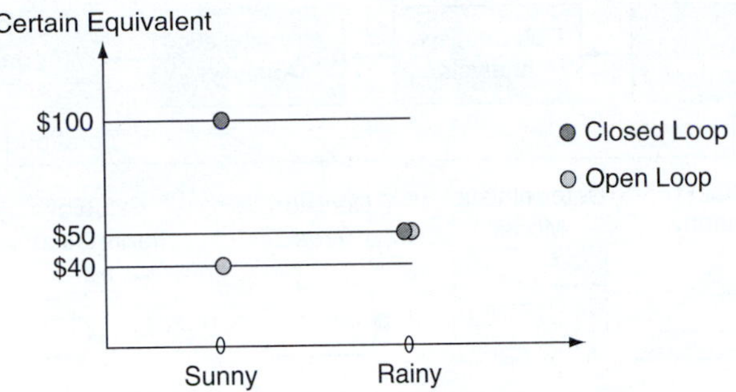

FIGURE 37.5 Open and Closed Loop Sensitivity

The difference between the open loop and closed loop sensitivity occurs only for the report of sunshine. If this change were zero, it might be worth considering the weather distinction as fixated. Figure 37.5 plots the values obtained versus different degrees of weather for the open loop and closed loop analyses.

> **Note: The value of Revelation**
>
> If weather was an immaterial distinction not affecting the decision alternative, the two curves would coincide, and the resulting value of information due to weather would be 0. We call the difference between the two curves—that cannot be negative—the **value of revelation**.

For a risk-neutral decision maker, the difference between the open loop and closed loop curves is multiplied by the probability of each outcome to produce the value of information. It is the *e*-value of the value of revelation.

For example, consider risk-neutral Jane in the party problem. She will be having the party on the Porch with a value of 90 if it is Sunny and 20 if it is Rainy and with an *e*-value of 48. If she knew the weather was to be Sunny, she would switch to Outdoors and enjoy an Outdoor party with value 100, which is an increase of 10. If the weather will be Rainy, she will switch to an Indoor party with value 50, which is an increase of 30 over the 20 value she would have experienced with a Porch party. Therefore, she will gain 10 with probability 0.4 and gain 30 with probability 0.6. Her *e*-value of this deal is 22, which is the value of clairvoyance (VOC) we had calculated previously.

> **Note**
>
> Even for a deltaperson, the VOC is *not* the certain equivalent of the deal represented by the probability distribution of the value of revelation. It *is* the difference between the certain equivalent of the closed loop deal (66.10) and the certain equivalent of the open loop deal (45.83), 20.27.

Appendix A provides a detailed explanation of open loop and closed loop sensitivity within the context of a bidding decision.

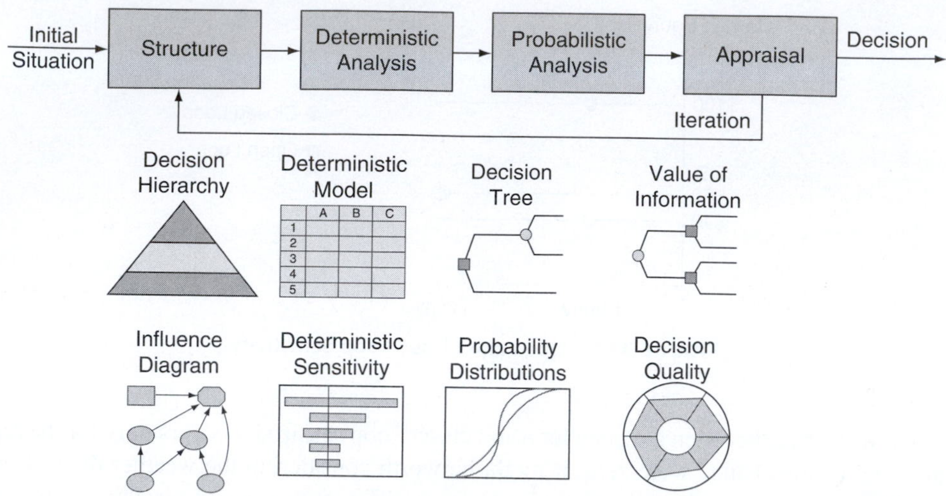

FIGURE 37.6 Expanded View for Phases of Decision Analysis Cycle

Each phase of the decision analysis cycle includes a set of tools that can help with the analysis. Figure 37.6 shows an expanded view of the cycle and some of the tools that can help with each phase.

37.3 THE MODEL SEQUENCE

The analyst can conduct decision analyses at various depths to meet different ends. We call this process the **model sequence**. The process is analogous to the one used in developing any system from initial design through testing, tuning, and finally, full-scale deployment. A chemical production facility begins with a pilot plant. The production of an aircraft begins with mathematical modeling of a new aircraft design, then proceeds with the building of an extensively tested prototype. As we show next, the decision analyst should proceed systematically through stages of refinement.

37.3.1 The Pilot Phase

The first step in the model sequence is the **pilot phase**. After agreeing on a frame, the analyst proceeds to develop and analyze a pilot decision basis that contains a simplified decision model, a tentative preference structure, and a rough characterization of major uncertainties. The purpose of the pilot phase is to understand and establish effective communication regarding the nature of the decisions and the associated major issues and uncertainties. While it may lead to preliminary decision recommendations, and, in some cases, produce immediate clarity of action obviating further analysis, the usual consequence of the pilot phase is guidance in conducting a full-scale representation of the decision.

37.3.2 Full-Scale Analysis

The purpose of **full-scale analysis** is to develop clarity of action based on a carefully constructed decision basis. Full-scale analysis requires a balanced, realistic decision model, certified preferences, and appropriate representation of important uncertainties. If done properly, the full-scale analysis should leave the decision maker convinced and committed to its results.

STAGES OF FULL-SCALE ANALYSIS Now, we will examine the three stages of full-scale analysis: the prototypical stage; the integrated stage; and the defensible stage.

1. **The Prototypical Stage** Full-scale analysis proceeds in stages with different purposes. The first is the **prototypical stage** that has all the required elements of the decision basis developed according to the guidance of the pilot phase. Exercising the prototype reveals the weaknesses and excesses in full-scale analysis that are worthy of correction. The definition of a prototype is an original type, form, or instance that serves as a model upon which later stages are based.

   For example, the first prototype in the development of an aircraft would sit on the runway looking much like the final production version. However, its interior would not look very similar to what passengers would see when it is in operation. It would be flown by a test pilot to check all aspects of performance to see what alterations and adjustments might be necessary. Similarly, exercising the prototype of the full-scale decision model would show the need to consider an additional alternative, introduce a sequential decision, refine a probability assessment, or adjust a value trade-off.

   Exercising the prototype has the goal of eliminating any refinements that are not material and economic and of adding only refinements that are. The goal is to achieve a consistent "wince level" regarding all aspects of the analysis. This means that if you had additional resources to devote to the analysis you would find it equally beneficial to use them on any of its elements.

2. **The Integrated Stage** The completion of the prototypical stage is an integrated decision model. The **integrated stage** provides the decision maker with a unified, balanced, and economical analysis that clearly indicates the correct decision.

   A doctoral student performed one of his first consulting jobs for the owner of a ferry service who wanted decision assistance in adding a new ship to his fleet. At the end of the project, the student presented the results of the integrated full-scale analysis to his client. The client thanked him, saying that he would follow the recommended course of action. The student, having been used to academic research where the final result is almost always a written report on the work, asked the client whether he should submit a report of the analysis. The client replied, "Why? I asked you to help me with this decision, and now I have made it. Thank you for your help." Such is the desired result of an integrated full-scale analysis.

3. **The Defensible Stage** Sometimes an additional stage of full-scale analysis is required, the **defensible stage**. The need for this stage arises when the decision maker, though certain of the recommended alternative, faces the task of demonstrating to supportive, doubtful, and possibly hostile stakeholders that the analysis provides an appropriate basis for decision. Stakeholders could be members of the board, others in the company who must implement the decision, or outside bodies whose cooperation is necessary for execution. For highly controversial decisions affecting the public, every element of the analysis will be scrutinized, and any discrepancy in the analysis exploited to delay or prevent the execution of the decision. Performing a defensible full-scale analysis requires many times the resources of an integrated full-scale analysis.

37.3.3 The Cyclone Enzyme Case Study

Case Study

Consider the following case, and then use the decision analysis cycle shown in Figure 37.7 to arrive at an appropriate decision. As part of your analysis, you must determine which concepts and tools are most appropriate for this case study and then use them effectively to bring clarity of action to the decision maker.

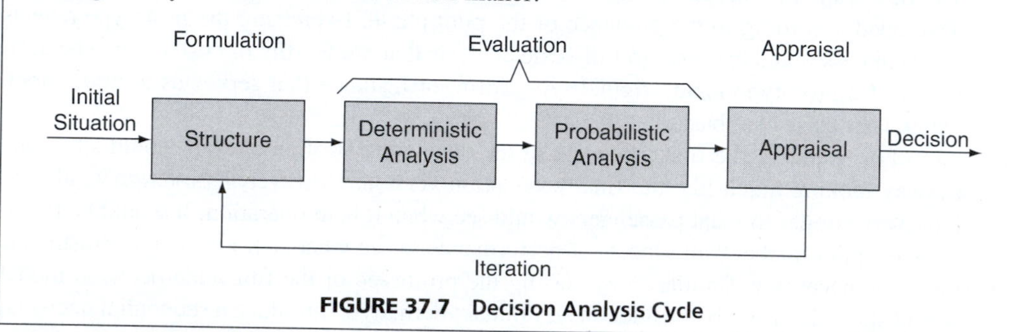

FIGURE 37.7 Decision Analysis Cycle

BACKGROUND Daylight Alchemy Corporation (DA Corp.) is a start-up biotechnology firm specializing in the development and production of synthetic enzymes. DA Corp. was originally a research laboratory and has recently started with production. However, they have very little experience in the marketing of synthetic enzymes.

Researchers at DA Corp. developed a new enzyme that allows Riesling grapes to be highly susceptible to the attack of the botrytis mold. When almost-ripe Riesling grapes are attacked by the botrytis mold, the grape skin becomes porous, allowing water to evaporate and sugar to remain. This situation is very attractive to winemakers, because the wine resulting from the harvesting and processing of such grapes has extraordinary complexity. Winemakers in the United States could profit enormously from using this enzyme to enhance the quality of the Riesling wines, since today such Rieslings are very rare and expensive.

The new enzyme is called Cyclone. While the final work is being completed in the lab, management is currently considering how best to proceed with production and marketing of Cyclone. To meet demand, they are considering expanding their current production facility.

DA Corp. has decided to launch Cyclone and knows which markets to target. Due to regulatory concerns, Cyclone will not be launched in any country besides the United States.

There are several decisions that need to be made but not during this analysis. For example, should DA Corp. produce Cyclone with the new NX-10 process? Should DA Corp. wait and "stack" the Cyclone enzyme with other enzymes in their development portfolio?

During the first pass through the decision analysis cycle, the management would like you to assume that Cyclone will be produced with the new NX-10 process, and DA Corp. will not "stack" the Cyclone enzyme.

STRATEGIC DECISIONS DA Corp. could produce and market the Cyclone enzyme itself. However, its current production capabilities are limited to 2.5 million units of enzyme per year and this may not be enough if demand is high. On the other hand, the 6-million unit expansion currently under consideration would not be needed if demand is low. The expansion would bring total capacity to 8.5 million units per year.

DA Corp. has considered strategic alliances with several larger chemical engineering firms, and has identified a potential candidate, EnzyTech. EnzyTech has enough capacity to meet any realistic demand and has made the following two offers.

- **Joint Venture**—EnzyTech will pay for the plant expansion in return for a 60% share of the net profits before tax (sales revenue − production cost). EnzyTech would perform all marketing. DA Corp. would produce Cyclone.
- **Royalty Partnership**—DA Corp. licenses the enzyme to EnzyTech. EnzyTech will produce and market the new product. DA Corp. would receive a royalty payment of 2% of sales revenue.

INFORMATION The management of DA Corp. wants to think about the sales of Cyclone in terms of a potential demand and capacity. For example, if the demand for Cyclone is 7 million units but only 2.5 million units of capacity are in place, then the sales are equal to 2.5 million units. DA Corp. is unsure what the demand for Cyclone will be in the first and subsequent years.

The management feels comfortable assessing an initial demand and a sales growth rate. For example, if the initial sales figure was 2 million units and the growth rate was 5%, DA Corp. would sell 2 million units the first year, so $2*(1 + 0.05) = 2.1$ million units the second and $2*(1 + 0.05)^2 = 2.1*(1 + 0.05) = 2.21$ the third year.

The initial demand will depend on who markets Cyclone, but the growth rate will not. In order to capture management's belief that initial demand will depend on who markets Cyclone, you have created the following distinctions: "Demand" if DA Corp. markets Cyclone, and "Multiple" if EnzyTech markets Cyclone. For example: If EnzyTech is going to market Cyclone, total demand would equal the demand if DA Corp. had marketed Cyclone times the multiple if EnzyTech markets it.

Cyclone will be ready for production at the start of the year 2020. This is the same time that an expansion of the current production facility would be completed. If the expansion is constructed, it must be paid for in full at the time of completion. The construction cost for the plant is uncertain.

In your analysis, management would like you to consider 10 years of Cyclone sales (final year of sales is 2029). They feel that after this time, a new and improved enzyme will be developed.

- Management does not want to consider any use for the expansion after the product cycle of Cyclone has ended. They do not want to consider any salvage value for the expansion either.
- DA Corp. is uncertain about the unit production costs and market price (unit revenue) of Cyclone.
- DA Corp. faces a tax rate of 38%. To calculate the taxes in each year, you multiply the net income (revenue − cost − expenses) by the tax rate. To determine the cash flow or profit in each year, you subtract the tax payment from the net income. Do not worry about depreciating the plant expansion. Just deduct the cost of the plant in the year it is completed. This will make the cash flow in the first year negative.
- Take out taxes. In this case, taxes will be negative in the year you finish the plant, so in effect, DA Corp. will be getting a tax credit the first year. DA Corp can apply this tax credit to subsequent years' taxes. (Do not discount the tax credit.) Only after this tax credit has been fully used does DA Corp. need to start paying taxes again.
- DA Corp. has a discount rate (time preference) of 8%. DA Corp. follows the delta property and has a risk tolerance of $30 million.
- Assume all cash flows occur at the start of the year. Please state your results in year 2020 dollars.

ASSESSMENTS You assessed the following 10–50–90 points, during a meeting with DA Corp.'s top management. The results are shown in Table 37.6.

TABLE 37.6 10–50–90 Fractiles of uncertainties

Distinction	Units	10%	50%	90%
Initial Demand				
DA Corp. Markets	Million Units/Year	1.0	2.5	5.5
EnzyTech Multiple	multiplier	1.3	1.6	1.8
Demand Growth Rate	%/year	−5%	7%	9%
Unit Revenue	$/Unit	9.75	10.75	11.50
Production cost	$/Unit	9.25	9.50	9.75
Plant Construction Cost	$ Million	13.0	14.5	17.0

CUMULATIVE DISTRIBUTIONS You can assess the full cumulative distribution on any of the above distinctions from DA Corp.'s experts. To simulate this process, we have included all the cumulative distributions in Figure 37.8. You should think of this as the cumulative distribution you would obtain *if* you were to go back to the expert and assess the full distribution.

> **Note** *When you go back to reassess an expert, the 10–50–90 points of the full distribution may not match your original low–base–high points.*

HINTS After the cumulative distributions, we have attached selected outputs from our value model for each alternative in Table 37.7. The values were obtained by setting all the distinctions to their base case value. Your base case calculations should match these values.

> **Note** *The number provided for royalty payments in the year 2002 was modeled as a payment of 98% of revenue to EnzyTech. This is the same as saying that DA Corp. gets 2%.*

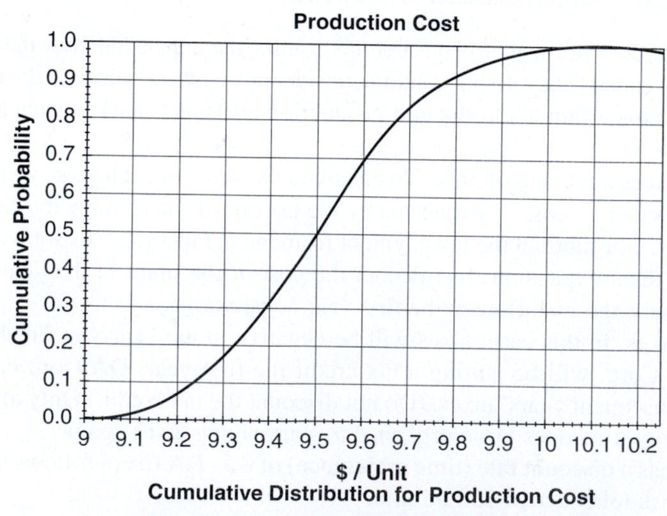

FIGURE 37.8 Cumulative Distributions

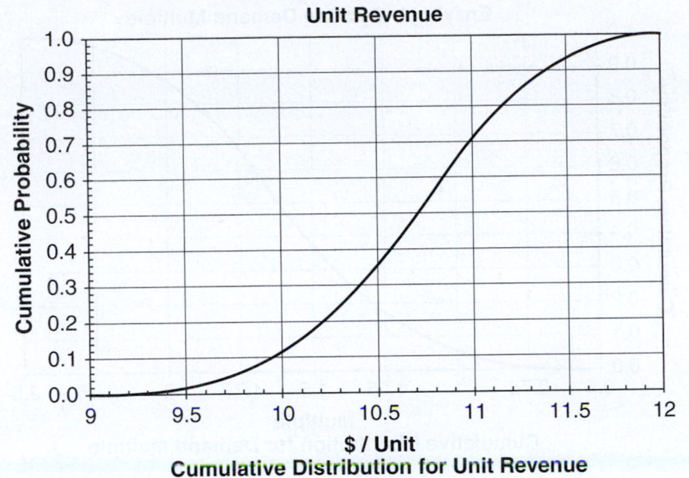

Cumulative Distribution for Unit Revenue

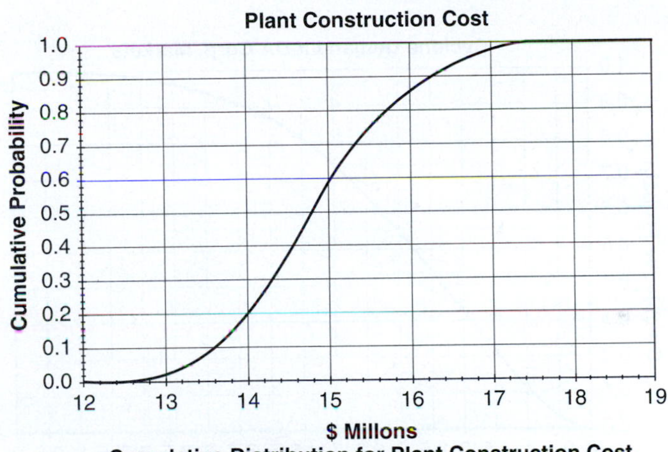

Cumulative Distribution for Plant Construction Cost

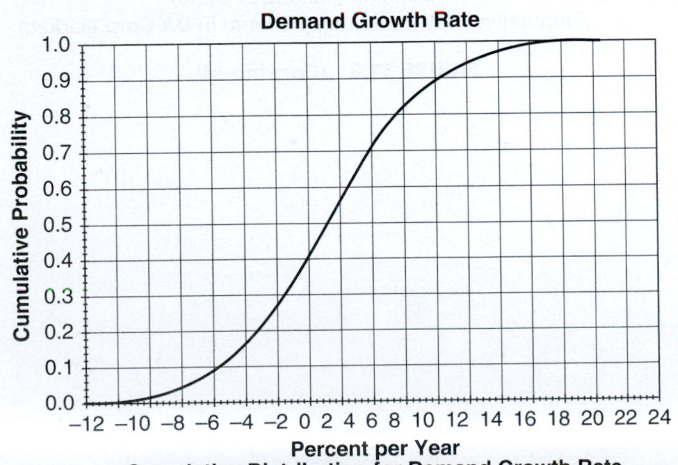

Cumulative Distribution for Demand Growth Rate

FIGURE 37.8 (Continued)

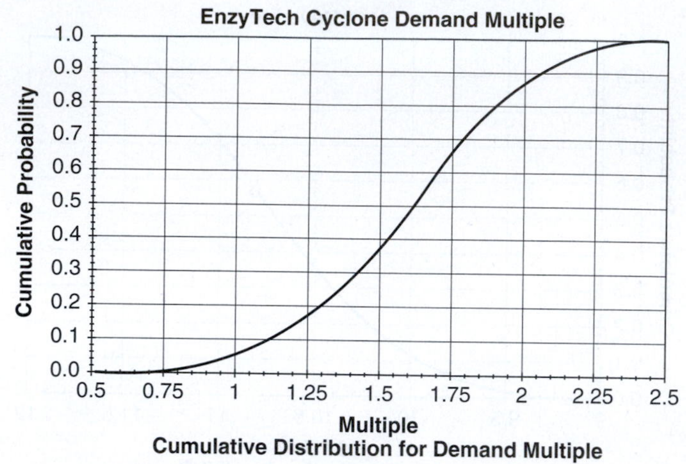

Cumulative Distribution for Demand Multiple

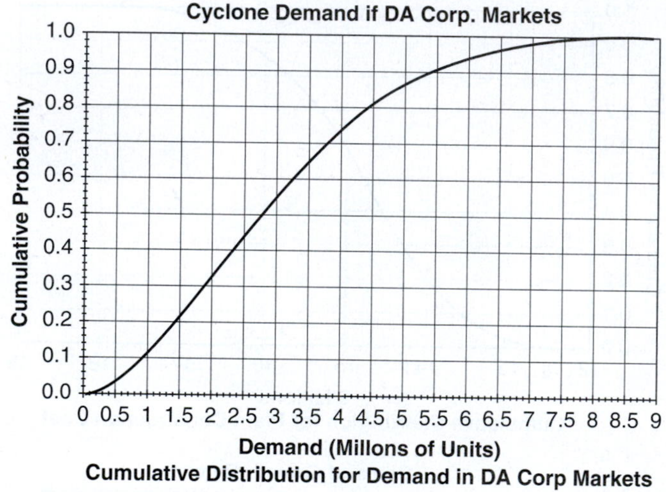

Cumulative Distribution for Demand in DA Corp Markets

FIGURE 37.8 (Continued)

TABLE 37.7 Section 4 - Calculations

No Partnership / No Expansion											
Year		**2020**	**2021**	**2022**	**2023**	**2024**	**2025**	**2026**	**2027**	**2028**	**2029**
Period		**0**	**1**	**2**	**3**	**4**	**5**	**6**	**7**	**8**	**9**
Sales											
Demand | DA Corp. Markets	MM units			2.86							
Demand | EnzyTech Markets	MM units			4.58							
Capacity	MM units										
Sales	MM units										
Revenue	MM $										
Production Cost	MM $										
Plant Cost	MM $										
JV Payment	MM $										
Royalty Payment	MM $										
Earnings Before Tax	MM $			3.13							
Taxes	MM $										
Tax Credit Available for Next Year	MM $										
Tax Payment	MM $			1.19							
Cash Flow	MM $										
Present Equivalent	MM 2020 $										

(Continued)

TABLE 37.7 Section 4 - Calculations (continued)

No Partnership / Expansion											
Year		**2020**	**2021**	**2022**	**2023**	**2024**	**2025**	**2026**	**2027**	**2028**	**2029**
Period		**0**	**1**	**2**	**3**	**4**	**5**	**6**	**7**	**8**	**9**
Sales											
Demand | DA Corp. Markets	MM units										
Demand | EnzyTech Markets	MM units										
Capacity	MM units										
Sales	MM units			2.86							
Revenue	MM $			30.77							
Production Cost	MM $										
Plant Cost	MM $										
JV Payment	MM $										
Royalty Payment	MM $										
Earnings Before Tax	MM $										
Taxes	MM $										
Tax Credit Available for Next Year	MM $										
Tax Payment	MM $										
Cash Flow	MM$										
Present Equivalent	MM 2020 $										

TABLE 37.7 Section 4 - Calculations (continued)

Joint Venture											
Year		**2020**	**2021**	**2022**	**2023**	**2024**	**2025**	**2026**	**2027**	**2028**	**2029**
Period		**0**	**1**	**2**	**3**	**4**	**5**	**6**	**7**	**8**	**9**
Sales											
Demand | DA Corp. Markets	MM units										
Demand | EnzyTech Markets	MM units										
Capacity	MM units										
Sales	MM units										
Revenue	MM $										
Production Cost	MM $										
Plant Cost	MM $										
JV Payment	MM $			3.43							
Royalty Payment	MM $										
Earnings Before Tax	MM $										
Taxes	MM $										
Tax Credit Available for Next Year	MM $										
Tax Payment	MM $										
Cash Flow	MM$			$1.42							
Present Equivalent	MM 2020 $										

(*Continued*)

TABLE 37.7 Section 4 - Calculations (continued)

Royalty											
Year		**2020**	**2021**	**2022**	**2023**	**2024**	**2025**	**2026**	**2027**	**2028**	**2029**
Period		**0**	**1**	**2**	**3**	**4**	**5**	**6**	**7**	**8**	**9**
Sales											
Demand | DA Corp. Markets	MM units										
Demand | EnzyTech Markets	MM units										
Capacity	MM units										
Sales	MM units										
Revenue	MM$										
Production Cost	MM$										
Plant Cost	MM$										
JV Payment	MM$										
Royalty Payment	MM$			45.02							
Earnings Before Tax	MM$										
Taxes	MM$										
Tax Credit Available for Next Year	MM$										
Tax Payment	MM$										
Cash Flow	MM$			$1.42							
Present Equivalent	MM 200$										
Net Present Value	MM$	$4.43									

37.4 SUMMARY

The essence of decision analysis is continuous focusing on the elements of the decision to determine clarity of action.

The decision analysis cycle is an iterative process for refining the analysis to produce this result.

The cycle incorporates the procedures previously discussed, such as framing, assessment of uncertainty, sensitivity analysis, valuation and risk attitude, value of information, and probabilistic dominance.

This chapter adds to the decision analysis cycle the use of tornado diagrams and the concepts of open-loop and closed-loop sensitivity analysis.

The discussion of stages of a decision analysis helps guide the level of detail required in different decision settings.

Ultimately, mastery of the decision analysis cycle allows the analyst to judge when to stop analyzing and decide

KEY TERMS

- Decision analysis cycle
- Formulation phase
- Evaluation phase
- Appraisal phase
- Value model
- Tornado diagrams
- Model sequence

APPENDIX A Open Loop and Closed Loop Sensitivity for the Bidding Decision

To further illustrate the idea of open loop and closed loop sensitivity in a practical setting, we now refer back to Chapter 31 and apply it to the bidding decision with the winner's curse. The decision diagram is repeated for convenience in Figure 37.9.

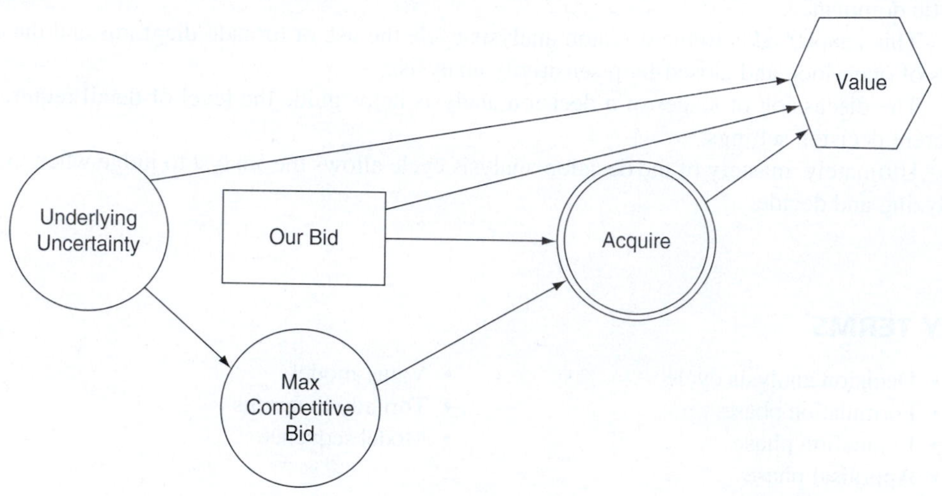

FIGURE 37.9 Decision Diagram for Bidding on Item with Underlying Uncertainty

Here there are two distinctions: maximum competitive bid (MCB) and "v" representing some underlying uncertainty about the value of the bidding item. As specified in the example, let v range from 0.5 to 0.9. Knowing the MCB updates our probability distribution of v, as shown in Table 37.8.

Now suppose you learn that the MCB is high, but you cannot change your bid (open loop condition). How does this change your value for the decision situation? Alternatively, suppose you learn it is high, and you can change your bid (closed loop condition). How does your value now change? We will discuss this and show how the difference between the two conditions provides insights into the value of information on the MCB.

The *e*-value of the bidding deal for any MCB can be calculated by taking the *e*-operation of v and the updated probability distribution. Table 37.9 shows these values.

OPEN LOOP SENSITIVITY IN THE BIDDING PROBLEM

The best bid for a risk-neutral decision maker was calculated in Chapter 31 as 0.525. Now suppose the decision maker has already decided to bid this amount. How does knowing the MCB change the value of the bidding opportunity?

As the MCB gets higher (but below the amount 0.525), it increases the value of both the item bid for and the bidding opportunity. If the MCB is higher than the 0.525 bid, the value of the bidding opportunity becomes 0 because the bidder has been outbid. Figure 37.10 shows this situation for a risk-neutral decision maker.

Note that the value is -0.025 when the MCB is 0. This is because the value of the item bid for is 0.5, and the risk-neutral bidder has already bid 0.525 for it.

TABLE 37.8 Rows Representing Conditional Probability Distributions {v|MCB,&}

MCB	0.5	0.6	0.7	0.8	0.9	{v|MCB}
0.000						
0.025	0.99999805	1.9502E-06	8.59235E-14	1.56531E-22	6.18E-33	
0.050	0.999930519	6.94805E-05	1.10067E-10	7.24221E-18	1.04E-26	
0.075	0.999410031	0.000589961	7.99406E-09	4.52051E-15	5.57E-23	
0.100	0.997180881	0.002818937	1.82415E-07	4.93775E-13	2.92E-20	
0.125	0.99013274	0.009865031	2.22907E-06	2.10376E-11	4.34E-18	
0.150	0.971854192	0.02812754	1.82672E-05	4.93558E-10	2.91E-16	
0.175	0.931411963	0.068476132	0.000111897	7.57082E-09	1.12E-14	
0.200	0.854789263	0.144670655	0.000539999	8.30655E-08	2.78E-13	
0.225	0.732923343	0.264976339	0.002099635	6.82801E-07	4.82E-12	
0.250	0.574669266	0.418673879	0.006652532	4.32293E-06	6.08E-11	
0.275	0.409065316	0.573448235	0.017464759	2.16888E-05	5.82E-10	
0.300	0.266953473	0.693892778	0.039064286	8.94586E-05	4.42E-09	
0.325	0.162540762	0.76030483	0.076839115	0.000315265	2.79E-08	
0.350	0.093607782	0.769137026	0.136276355	0.000978686	1.51E-07	
0.375	0.051225757	0.725076534	0.220970905	0.002726081	7.23E-07	
0.400	0.02655243	0.637465134	0.329108067	0.006871288	3.08E-06	
0.425	0.012950955	0.521005273	0.450304582	0.0157274	1.18E-05	
0.450	0.005911885	0.394903911	0.566340626	0.032802814	4.08E-05	
0.475	0.002520314	0.277756997	0.656876011	0.062718224	0.000128	
0.500	0.001003372	0.181734244	0.706136579	0.110753201	0.000373	
0.525	0.000372603	0.110753193	0.706136601	0.18173423	0.001003	
0.550	0.000128454	0.062718224	0.656876011	0.277756997	0.00252	
0.575	4.07642E-05	0.032802814	0.566340626	0.394903911	0.005912	
0.600	1.17898E-05	0.0157274	0.450304582	0.521005273	0.012951	
0.625	3.08094E-06	0.006871288	0.329108067	0.637465134	0.026552	
0.650	7.22798E-07	0.002726081	0.220970905	0.725076534	0.051226	
0.675	1.51203E-07	0.000978686	0.136276355	0.769137026	0.093608	
0.700	2.78603E-08	0.000315265	0.076839115	0.76030483	0.162541	
0.725	4.41887E-09	8.94586E-05	0.039064286	0.693892778	0.266953	
0.750	5.81974E-10	2.16888E-05	0.017464759	0.573448235	0.409065	
0.775	6.08183E-11	4.32293E-06	0.006652532	0.418673879	0.574669	
0.800	4.81786E-12	6.82801E-07	0.002099635	0.264976339	0.732923	
0.825	2.79979E-13	8.30655E-08	0.000539999	0.144670655	0.854789	
0.850	0	7.57081E-09	0.000111897	0.068476132	0.931412	
0.875	0	4.93583E-10	1.82672E-05	0.02812754	0.971854	
0.900	0	0	2.22907E-06	0.009865031	0.990133	
0.925	0	0	1.82414E-07	0.002818936	0.997181	
0.950	0	0	0	0.000589961	0.99941	
0.975	0	0	0	6.9535E-05	0.99993	
1.000	0	0	0	0	1	

TABLE 37.9 *e*-Value of Bidding Item Given MCB

MCB	<*v*|MCB, &>
0.000	0.5000
0.025	0.5000
0.050	0.5000
0.075	0.5001
0.100	0.5003
0.125	0.5010
0.150	0.5028
0.175	0.5069
0.200	0.5146
0.225	0.5269
0.250	0.5432
0.275	0.5608
0.300	0.5772
0.325	0.5915
0.350	0.6045
0.375	0.6175
0.400	0.6316
0.425	0.6469
0.450	0.6626
0.475	0.6780
0.500	0.6928
0.525	0.7072
0.550	0.7220
0.575	0.7374
0.600	0.7531
0.625	0.7684
0.650	0.7825
0.675	0.7955
0.700	0.8085
0.725	0.8228
0.750	0.8392
0.775	0.8568
0.800	0.8731
0.825	0.8854
0.850	0.8931
0.875	0.8972
0.900	0.8990
0.925	0.8997
0.950	0.8999
0.975	0.9000
1.000	0.9000

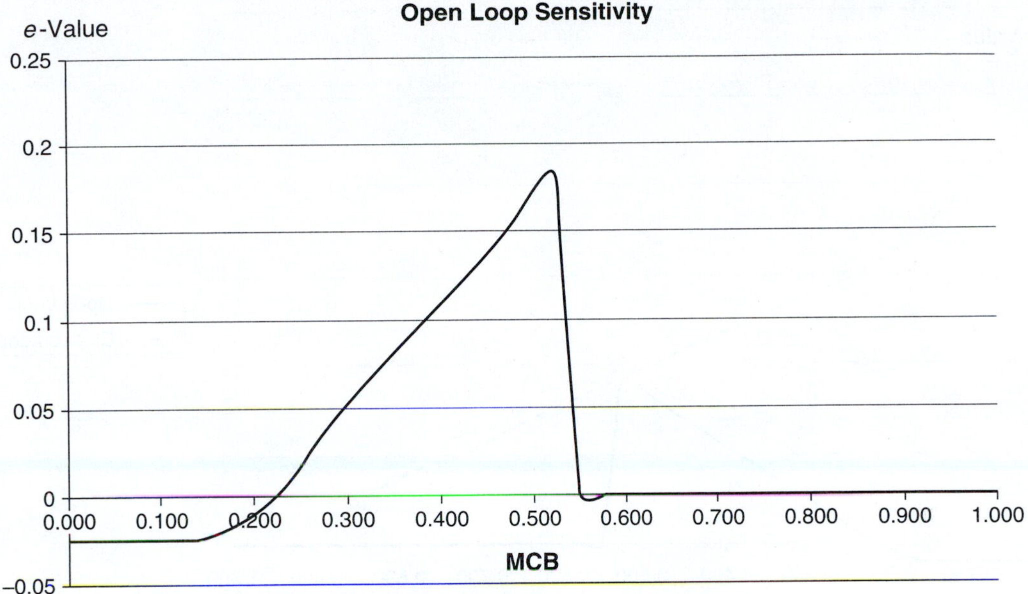

FIGURE 37.10 Open Loop Sensitivity Analysis

CLOSED LOOP SENSITIVITY IN THE BIDDING PROBLEM

If the bidder can change his bid, after knowing the MCB, he will bid just above it (incremental amount). As the MCB gets higher, the value of the bidding item is higher, but he also has to bid higher to acquire it. Figure 37.11 shows this situation for the risk-neutral decision maker.

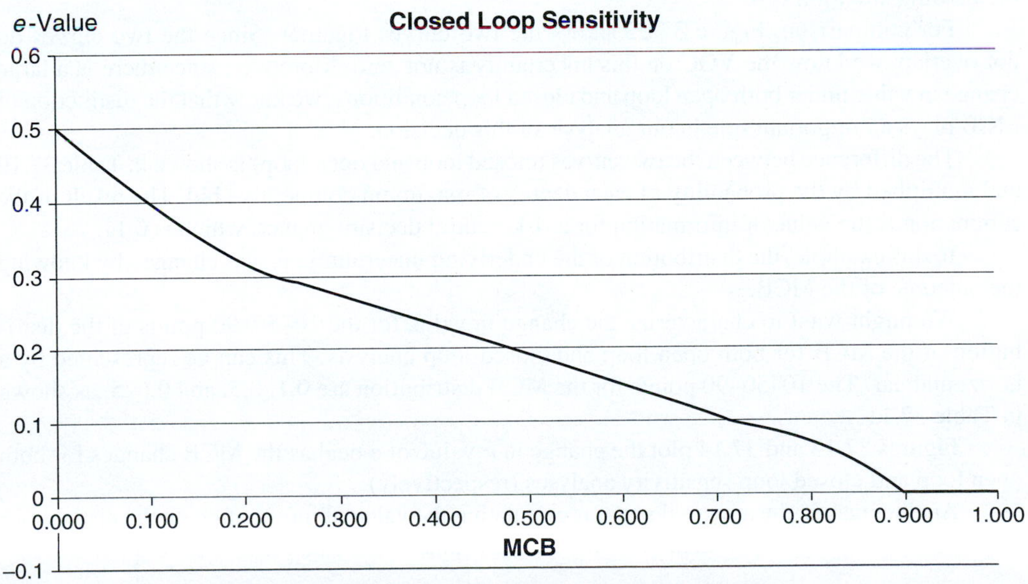

FIGURE 37.11 Closed Loop Sensitivity Analysis

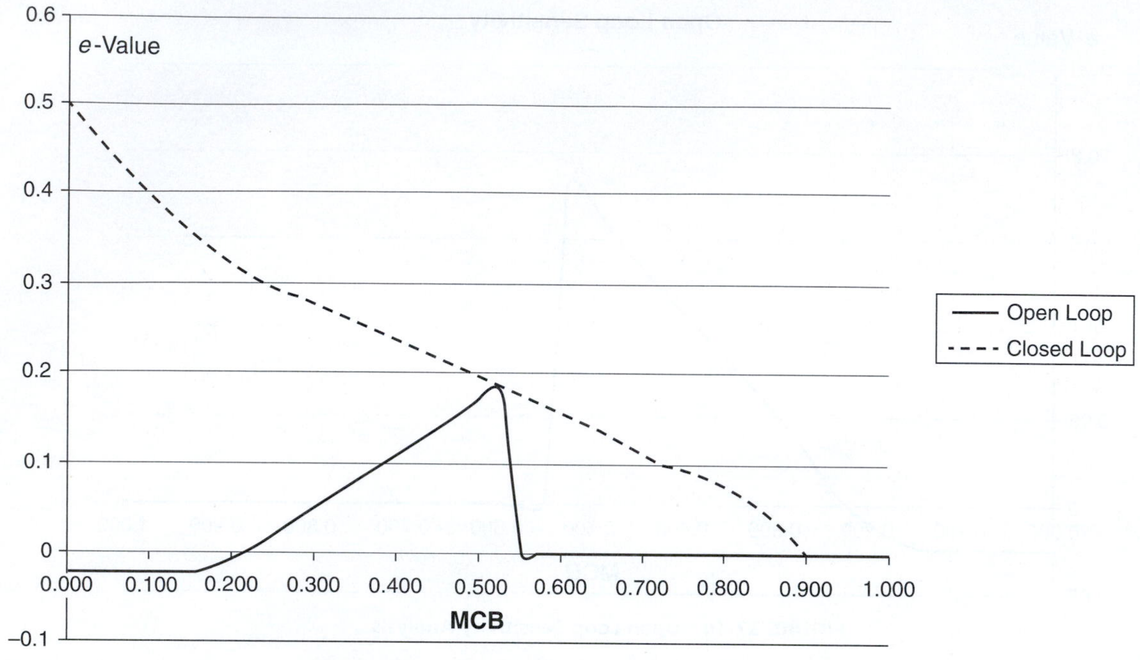

FIGURE 37.12 Comparing Open and Closed Loop Analysis

As we see, the resulting curve is decreasing with the MCB because the increase in value of the item does not exceed the increase in bid amount. This continues until a point where the MCB exceeds the value of the item, at which point it is not a good decision to bid, and the value of the bidding situation is 0.

For comparison, Figure 37.12 shows the two curves together. Since the two curves do not overlap, we know the VOC on this uncertainty is not zero. Moreover, since there is a large change in value under both open loop and closed loop conditions, we know that the distinction of MCB plays an important role in our analysis of this decision.

The difference between the two curves (closed loop and open loop) is shown in Table 37.10 and multiplied by the probability of each degree of maximum competitive bid. The result of the summation is the value of information for a risk-neutral decision maker, which is 0.11.

In this example, the distribution of the underlying uncertainty, *v*, was changed by knowing the outcome of the MCB.

We might want to characterize the change in value for the 10–50–90 points of the distribution of the MCB for both open loop and closed loop analysis. This can be represented by a horizontal bar. The 10–50–90 points for the MCB distribution are 0.1, 0.5, and 0.625, as shown in Table 37.11.

Figures 37.13 and 37.14 plot the change in *e*-value of a deal as the MCB changes for both open loop and closed loop sensitivity analyses (respectively).

Another task in the appraisal phase is calculating the value of information on the distinctions and seeing if we can get information for less than its value. If we can, then we should.

TABLE 37.10 Calculating Difference Between Open and Closed Loop Sensitivity Analysis

MCB	Open Loop	Closed Loop	Difference	{MCB|&}	Difference × {MCB|&}
0.000	−0.025	0.5	0.525	0	0
0.025	−0.024999805	0.475000195	0.5	1.64E-14	8.20009E-15
0.050	−0.024993052	0.450006948	0.475	2.36E-10	1.12244E-10
0.075	−0.024941002	0.425058998	0.45	4.45E-08	2.00393E-08
0.100	−0.02471807	0.40028193	0.425	1.38E-06	5.87588E-07
0.125	−0.024013051	0.375986949	0.4	1.56E-05	6.25721E-06
0.150	−0.022183592	0.352816408	0.375	9.28E-05	3.48152E-05
0.175	−0.018130005	0.331869995	0.35	0.000355	0.000124252
0.200	−0.01042491	0.31457509	0.325	0.001001	0.000325409
0.225	0.001917766	0.301917766	0.3	0.002299	0.000689761
0.250	0.018199191	0.293199191	0.275	0.004616	0.00126951
0.275	0.035844282	0.285844282	0.25	0.008451	0.002112865
0.300	0.052228974	0.277228974	0.225	0.01432	0.003221942
0.325	0.066492896	0.266492896	0.2	0.022461	0.004492253
0.350	0.07946264	0.25446264	0.175	0.03259	0.005703178
0.375	0.092519948	0.242519948	0.15	0.043939	0.006590829
0.400	0.106630745	0.231630745	0.125	0.055573	0.006946649
0.425	0.12188438	0.22188438	0.1	0.066636	0.006663556
0.450	0.137615665	0.212615665	0.075	0.076311	0.005723289
0.475	0.153017751	0.203017751	0.05	0.083659	0.004182963
0.500	0.167775742	0.192775742	0.025	0.087679	0.00219197
0.525	0.182224258	0.182224258	0	0.087679	0
0.550	0	0.171982249	0.171982249	0.083659	0.014387907
0.575	0	0.162384335	0.162384335	0.076311	0.012391632
0.600	0	0.15311562	0.15311562	0.066636	0.010202945
0.625	0	0.143369255	0.143369255	0.055573	0.007967487
0.650	0	0.132480052	0.132480052	0.043939	0.005821022
0.675	0	0.12053736	0.12053736	0.03259	0.003928263
0.700	0	0.108507104	0.108507104	0.022461	0.002437207
0.725	0	0.097771026	0.097771026	0.01432	0.001400056
0.750	0	0.089155718	0.089155718	0.008451	0.000753496
0.775	0	0.081800809	0.081800809	0.004616	0.000377625
0.800	0	0.073082234	0.073082234	0.002299	0.000168031
0.825	0	0.06042491	0.06042491	0.001001	6.0501E-05
0.850	0	0.043130005	0.043130005	0.000355	1.53114E-05
0.875	0	0.022183592	0.022183592	9.28E-05	2.05954E-06
0.900	0	0	0	1.56E-05	0
0.925	0	0	0	1.38E-06	0
0.950	0	0	0	4.45E-08	0
0.975	0	0	0	2.36E-10	0
1.000	0	0	0	1.64E-14	0
					0.11

TABLE 37.11 10-50-90 Fractiles for Maximum Competitive Bid Distribution

	MCB	Open Loop	Closed Loop
0.1	0.100	−0.025	0.400
0.5	0.500	0.168	0.193
0.9	0.625	0.000	0.143

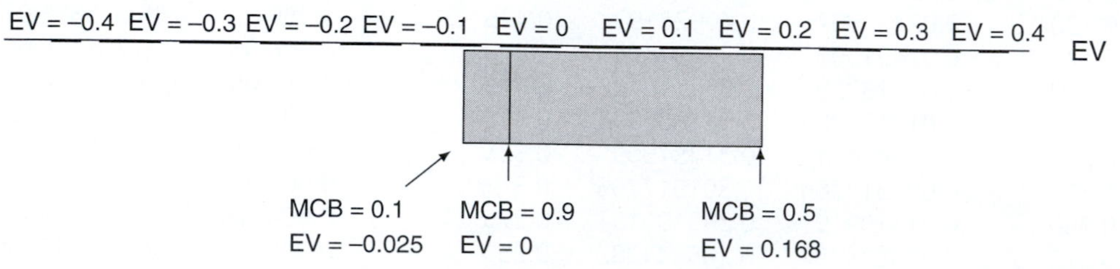

Open Loop Sensitivity

FIGURE 37.13 Open Loop Sensitivity Analysis for *E*-Value of Deal

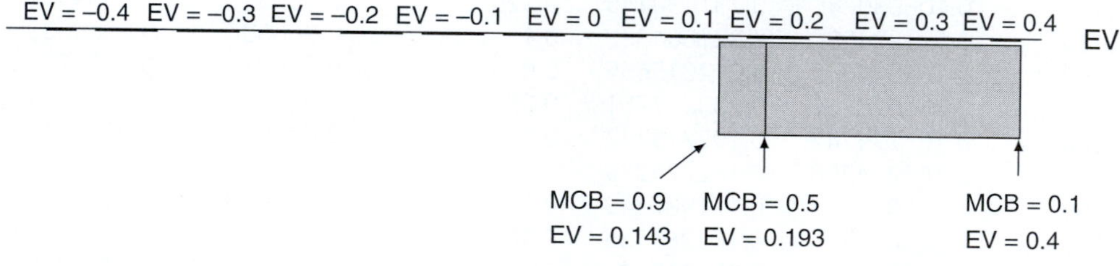

Closed Loop Sensitivity

FIGURE 37.14 Closed Loop Sensitivity Analysis for *E*-Value of Deal

CHAPTER

38

Topics in Organizational Decision Making

CHAPTER CONCEPTS

After reading this chapter, you will be able to explain the following concepts:

- Organizations should operate to maximize value
- Effects of budgets and incentive structures on decision making within organizations
- Problems with setting multiple independent requirements
- The need for a corporate risk tolerance
- Value gaps in organizations
- Common motivational biases in organizations

38.1 INTRODUCTION

In the previous chapters, we presented situations where a single decision maker was facing a decision. In contrast, decision making in organizations usually involves several decision makers. For example, if a large company is launching a new product, the marketing department would be interested in bringing the product to market faster than the competitor. The engineering department would want to make sure that the product meets certain quality standards before its release. When departmental decision makers have different interests, disputes may arise over the best course of action. The governing factor in organizational decision making should be the decision that creates the most value for the organization.

It is useful to think of the decision making in a company as if it were a single decision maker operating with a corporate value function. In business decisions, it will almost always be true that the organization prefers more value to less. What remains, then, is to identify the actual value function and risk tolerance that should be used for the organization. In doing this, it is important to recall the distinction between direct and indirect values that we discussed in Chapter 26.

Now, suppose that a group of individuals do act with the same value function. Should they operate with the same risk tolerance? What effects do the organization's incentive structures have on the quality of its decisions? What effects does operating with a unified value function and risk tolerance have on the decision-making capabilities of an organization? What are the benefits? Are there additional challenges in organizational decision making?

In this chapter, we discuss several topics in organizational decision making and the advantages of having a corporate value function and corporate risk tolerance. We also discuss the effects of incentive structures and motivational biases on organizational decision making.

38.2 OPERATING TO MAXIMIZE VALUE

Consider a manufacturing company designing a product. This product can be simple, such as the peanut butter and jelly sandwich discussed in Chapter 26, or a more complicated product, such as a photocopier, laptop, car, or airplane. For ease of presentation, let us start with a hypothetical peanut butter and jelly sandwich manufacturer. The same type of analysis applies to any other product with a specified value function.

In Chapter 26, the value function for the peanut butter and jelly sandwich reflected preferences of an individual decision maker purchasing the sandwich. In this case, it is associated with a production decision.

Suppose several different divisions are involved in manufacturing the peanut butter and jelly sandwich, and the company is interested in maximizing its profit. One division makes the bread, another division adds the peanut butter, and a third division adds the jelly. All divisions work for the same company, and their goal is to create *value.* What is the best setting for the thickness of peanut butter, bread, and jelly? These levels can be obtained by thinking of the levels of peanut butter, bread, and jelly as decisions and thinking about the profit (value) associated with each level.

The company might be uncertain about the demand for the sandwich for a given mixture of ingredients and price. The demand might also be relevant to the market size and the mixtures and ingredients in the competitors' sandwiches. Figure 38.1 shows a simple decision diagram for the company's decision. The decision node is associated with the choice of the ingredients of the sandwich as well as its selling price. The competitor performance is associated with how the demand for this company's sandwich fares in comparison to the competitor's offering. Cost will be determined by ingredients and the number of units produced.

The company will need to assess the conditional probability distribution of demand for different values of the two uncertainties: market size and competitor performance, as well as the different price and ingredient decisions. Sometimes, market surveys and promotions can be used to estimate demand under these various conditions. Sometimes value functions for individual preferences, as we discussed in Chapter 26, can help estimate what the demand would be for different ingredients and price for various individuals.

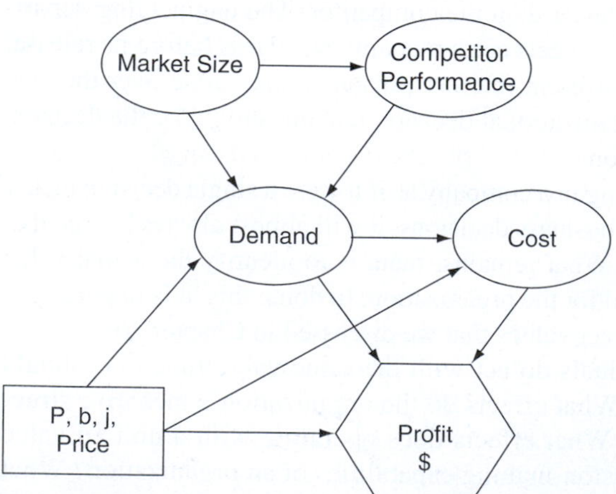

FIGURE 38.1 Simple Decision Diagram for Maximizing Company Profit

By maximizing the *e*-value of *u*-values of total profit, the company can determine the levels p^*, b^*, and j^* and the price that maximize its certain equivalent of profit. If the company is risk-neutral for this decision, it can maximize the *e*-value of profit.

The previous example, while simple, includes many features of organizational decision making. Operating this way has several benefits for the company. Suppose that the current settings of peanut butter, bread, and jelly currently manufactured by the company are p_0, b_0, and j_0. A division manager walks into the CEO's office one morning and tells him that there is a new machine that can produce any required level of peanut butter with a better accuracy than the previous one. The new machine has additional benefits in not wasting as much peanut butter as does the existing machine.

The CEO ponders the value of such a machine to the company and compares it to the machine's price. Given our previous analysis, his decision is straightforward. The CEO should calculate the certain equivalent of profit with the new machine, and if he satisfies the delta property, he should simply subtract that certain equivalent from the current certain equivalent produced by the existing machine.

This method of corporate decision making can also help resolve conflicts and set priorities among the divisions. For example, suppose that another division manager enters and mentions that he can bring the thickness level of bread to a desired level, b^*, through the purchase of another new machine. Should the company purchase this new machine? Should the company purchase both machines? Which division should get priority? Again, the answers to these questions are straightforward and can be determined by calculating the value of each machine and comparing those values to the cost of each machine.

The previous analysis also applies to much more complicated products. For example, consider a company manufacturing a laptop. Design decisions may include factors such as computer size, screen size, and extra features. By maximizing the *e*-value of *u*-values of profit for each design, the company can determine the best settings for these design decisions. For profit-maximizing firms, the demand for a product at a given price is an uncertainty that will play an important role determining the *e*-value of *u*-values.

As a further illustration, refer back Chapter 14 and the dual-support motorcycle design with the decision diagram repeated for convenience in Figure 38.2. The design decisions specify all the performance attributes entering the Market Share node—given all the other attributes that ultimately affect profit (like cost) that are a result of the design. The distinctions of Market Size and Competitive Price/Design would not be in the value function but are left uncertain, since they are indirect value nodes. Recall that the division of the design decision into the elements of engine size, frame design, and the provision of electric start was for graphical convenience: All of these elements affect each other.

When the certain equivalent profit of the decision diagram is solved for a proposed design, a sensitivity analysis would show how changing each of the performance attributes would affect the certain equivalent of profit. For example, how much is the provision of electric start worth given its effect on weight, cost, performance, and the reception by the market? If this feature is highly valued by the market, additional costs may be justified. All engineers must weigh potentially higher costs against the user value of superior product performance. That is why almost all electrical wiring is made of copper—even though silver is at least 10% more conductive and is the best electrical conductor of any known metal.

Suppose that the company is also interested in environmental issues concerning their product. The analysis is still the same. The company assigns a value function that takes into account profit and environmental impact for each prospect. Instead of maximizing the *u*-value

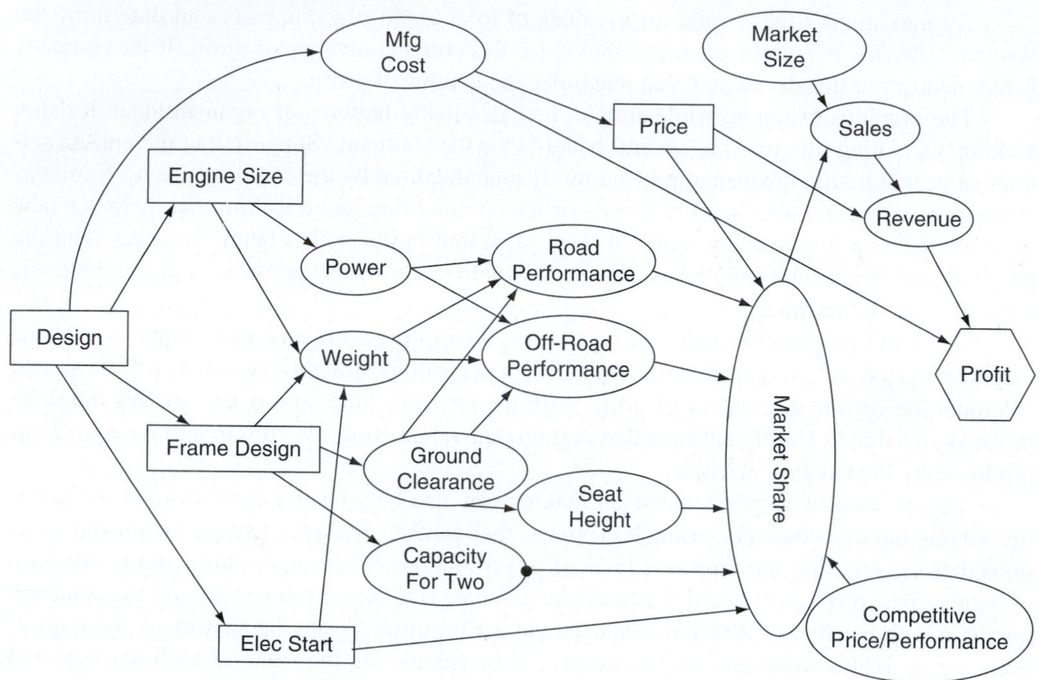

FIGURE 38.2 Decision Diagram for Motorcycle Design

of profit for a given design, the company maximizes the *e*-value of the *u*-value over the value function. The value function could include direct value attributes, such as money and carbon footprint. For this to be consistent within the organization, the company should have a corporate value function that specifies the trade-offs between profit and environmental impact. For example, the company may specify a trade-off between profit and a unit increase in its carbon foot print. Refer back to Chapter 26 for methods to specify the value function. The company should also specify the time preference with which it operates (Chapter 27).

38.3 ISSUES WHEN OPERATING WITH BUDGETS

We have illustrated a method of organizational decision making using a corporate value function in which each individual maximizes the *e*-value of profit. Quite often, however, a company will set a budget limit on expenditures for each division. Budgets have inherent problems. While they are easy to understand and implement, they can lead to a loss of value.

For example, a pharmaceutical company might miss the cure for a cancer because it did not allow spending an extra $1 million on research and development. If significant net value to the organization would be obtained by exceeding the budget, the opportunity may not be pursued because of the need to "stay within our budget." Managers often pride themselves in not exceeding a specified budget.

Suppose that, in the previous example, the division manager of the peanut butter was operating under a budget. He would not have even considered the purchase of a new machine if its cost was more than his specified budget amount, even though this expenditure could result in a much higher value for the company.

Another issue with budgets is that, if you end up with a surplus, the desire to preserve it or increase it for the next time period will lead to poorly justified expenditures. A well-known saying about a budget is "if you don't use it, you lose it."

Like any constraint, budgets can never add value; they can only limit it. They eliminate feasible alternatives that you might otherwise pursue. You should invest resources in any decision until the net value of an additional expenditure would be negative.

38.4 ISSUES WITH INCENTIVE STRUCTURES

The question of using incentives arises in many contexts. Every organization must decide the basis for compensating those who contribute to its success. Whenever many people share responsibility for the thriving of the enterprise, it is always a possibility that some will think they are more responsible for good outcomes and less responsible for bad outcomes than others. In the case of good outcomes, there may be a common perspective that "but for me," the good outcome would not have happened. If all contributions are necessary for success, how should managers allocate rewards?

Compensation schemes can range anywhere between two extremes. One extreme is the equal and predetermined division of rewards among all parties. In the era of the Nantucket whalers, for example, voyage profits were distributed among the owner, captain, officers, and crew in shares established at the outset. Except in the most unusual circumstances, exceptional contributions or deficiencies of the participants would not change the share.

At the other extreme is compensation based solely on individual performance, as when the income of sales people is derived solely from commissions on sales.

To see the consequences of compensation schemes, consider the case of a rowing team of several individuals competing for prize money. The team managers might begin by distributing net prize money equally among team members. However, some team member may notice that he is putting much more energy into his rowing than are others, and that his behavior is a major factor in winning the race. If the team manager agrees, a system might be installed in the oarlocks to show the total energy provided by each team member during the race. Prize money would then be distributed according to this measurement. Now that each person knows the basis for his compensation, each will strive to focus on energy input. However, other features contributing to success—like the direction of the boat, bailing of water, and training and encouragement of other team members—may be neglected, perhaps resulting in defeat in the next race.

This deficiency is in the nature of incentive systems. When an incentive structure is set, employees may focus on achieving the outcomes of a decision that maximizes their chances of meeting the incentive structure (and achieving their reward). Whatever is immediately rewarded becomes emphasized, while other factors important to long-run success, such as the development of future members of the organization, may be neglected. Incentive systems must be carefully designed to avoid unintended consequences.

Consider the example of a project manager in an automobile manufacturing organization who has to choose between the following two alternatives.

- Producing a new innovative vehicle
- Making improvements to the design of last year's vehicle (Figure 38.3)

The *e*-value for producing a new innovative vehicle is

$$0.4 \times \$100\text{ M} + 0.6 \times -(\$20\text{ M}) = \$28\text{ M}$$

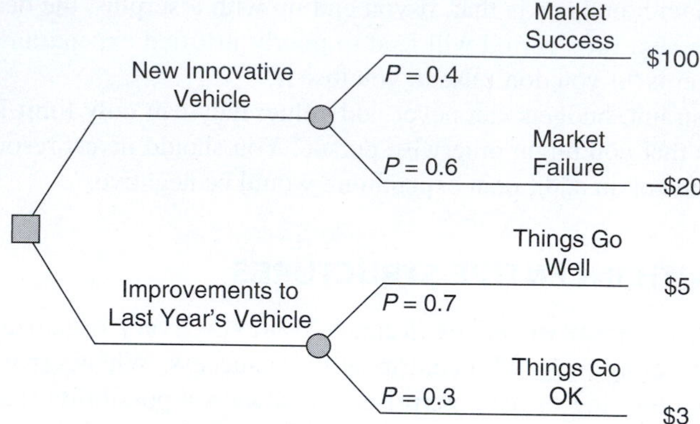

FIGURE 38.3 Choosing Optimal Decision Alternative based on Incentive Structure

On the other hand, the *e*-value for improving last year's vehicle design is

$$0.7 \times (\$5\text{ M}) + 0.3 \times (\$3\text{ M}) = \$4.4\text{ M}$$

If the company is risk neutral, then the best decision is to produce a new innovative vehicle. However, if the incentive structure is such that the project manager will be rewarded by the outcomes produced, this may lead him to choose the alternative that has the highest probability of achieving a positive profit. If the incentive is simply to produce profit in tough times, the project manager may choose to improve last year's design, as he will have a 100% chance of producing positive profit. The new innovative vehicle alternative, on the other hand, has a 60% chance of receiving negative profit, and therefore, he will have a 60% chance of not meeting his incentives.

If the incentive is to exceed $10 million however, the manager will choose the new innovative vehicle, as it gives him a higher chance of exceeding his target than does improving an existing design. Companies therefore need to be aware that their incentive structure may impede making good decisions, especially if immediate outcomes are the main focus.

38.5 A COMMON ISSUE: MULTIPLE SPECIFICATIONS VS. TRADEOFFS

In the previous section, we saw how setting a target based incentive can lead to decisions that may not be in the best interest of the company. It is common practice for engineers designing a system to be given performance specifications and a cost budget limit for the system.

For example, an engineer working for a photocopier manufacturer may be asked to design a new copy machine. The firm would specify (among other attributes) the desired number of copies per minute. Success would be exceeding this number, and failure would be achieving a lower number. Another specification might be obtaining a certain resolution for digital scanning. Again, success would be exceeding the specified resolution, and failure would be anything less.

This process simplifies total system design by dividing the total system into a set of subsystems that may be designed in isolation. Someone with an overview of the total system could see ways to improve total system performance by making trade-offs on resolution to achieve a certain copying speed. The problem is how to achieve this result when no such person exists. When setting multiple thresholds for each subsystem, the result might be a product

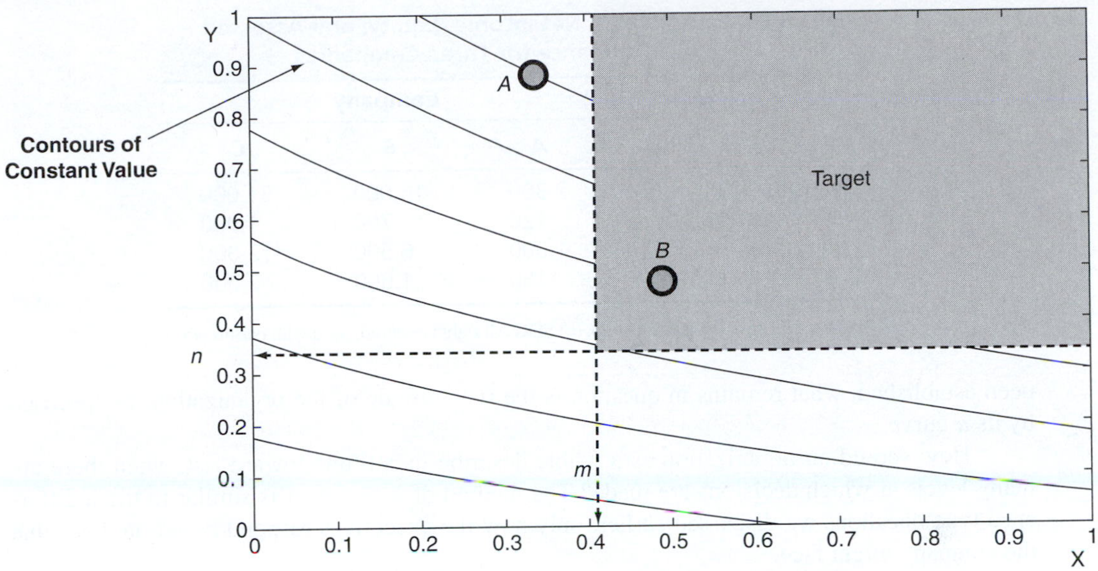

FIGURE 38.4 Multiple Targets and Poorer Decision Making (Rectangular Target Region)

that has less value for the company. To illustrate, suppose that the contours of constant value for the company are as shown in Figure 38.4. The contours correspond to the situation where more of every design attribute is preferred to less. The company sets fixed targets to exceed m units of Attribute X and n units of Attribute Y, as determined by the shaded rectangular region in the figure.

Consider Points A and B in the figure. Point A achieves higher value for the company, but it does not meet the threshold value of achieving m units of Attribute X. Point B, on the other hand, meets the design specifications on both attributes, but produces less profit for the company. An incentive structure that rewards engineers based on meeting independent design specifications would result in Point B and less value to the company.

If the design engineer is given the contours of constant value, he will see the value to the company of each number of copies per minute in the face of competitive offerings. High values of increasing this number would encourage the engineer to investigate a wide range of alternatives that might have been rejected as too costly. Low values of increase would direct attention to other performance attributes. Instituting value-guided design could be described by the maxim "Don't tell me what you want, just tell me the value of what I can provide."

38.6 NEED FOR A CORPORATE RISK TOLERANCE

Major corporations often tell their managers to be more "risk-taking." While all too common, this language leaves us to wonder exactly what managers should be doing with company money. Anyone making important financial decisions for a company should have clear guidance on the risk attitude to be used in making those decisions. Once the other elements of decision making, such as framing, have been properly specified, and once trade-offs and time preferences have

TABLE 38.1 Net Sales, Net Income, Equity, and Assessed Risk Tolerance for Three Companies

	Company		
	A	*B*	*C*
Net Sales ($M)	2,300	16,000	31,000
Net Income ($M)	120	700	1,900
Equity ($M)	1,000	6,500	12,000
Risk Tolerance ($M)	150	1,000	2,000

(Source: Strategic Decisions Group. (c) SDG

been established, what remains in question is the risk attitude of the organization, as specified by its *u*-curve.

How should an organization as a whole describe its attitude toward risk when there are many levels at which decisions are made? The method of assessment is similar to the indifference type questions we discussed earlier, only now the executives respond based on deals that the company might face.

For example, a team from Strategic Decisions Group once assessed the risk tolerance for Companies *A*, *B*, and *C*, all in the same industrial sector. First, the group analyzed the companies based on their accounting reports of net sales, net income, and equity. Then, they assessed their risk tolerance (Table 38.1).

The group then found that the ratios of the risk tolerance to sales, risk tolerance to net income, and risk tolerance to equity were close for all three companies (Table 38.2), with average values of

$$\text{Risk tolerance/Sales} = 0.064$$

$$\text{Risk tolerance/Net Income} = 1.24$$

$$\text{Risk tolerance/Equity} = 0.157$$

Soon after, the same team worked for another Company, *D*, which conducted most of its business in the same sector as did Companies *A*, *B*, and *C*. Company *D* was deciding whether or not to invest in a startup company in a different business sector. The time available for the analysis was very short, since the investment decision had to be made by a particular date. Unfortunately, the president and chairman of this company, who would be authorizing this investment, could not meet with the team until the very day the commitment was required. There would be no opportunity to discuss risk preference with them until the actual presentation of the decision analysis.

TABLE 38.2 Expressing Assessed Risk Tolerance as Ratio of Sales, Net Income, and Equity

	Company			
	A	*B*	*C*	Average
Risk Tolerance/Sales	0.0652	0.0625	0.0645	0.064
Risk Tolerance/Net Income	1.25	1.43	1.05	1.24
Risk Tolerance/Equity	0.15	0.154	0.167	0.157

(Source: Strategic Decisions Group. (c) SDG

TABLE 38.3 Company *D*. Inferring Risk Tolerance by Comparing Ratios

	Ratio (Risk Tolerance/ Measure)	Measure ABC 1979 ($M)	Risk Tolerance (measure × ratio) ($M)
Net Sales	0.064	3,307	212
Net Income	1.24	152	188
Equity	0.157	1,153	181
		Average ($M):	194
		ABC risk tolerance used in analysis ($M):	190

(Source: Strategic Decisions Group. (c) SDG All rights reserved; used with permission.)

As a result, the team decided to prepare for the presentation by referring to what they had learned about this industry from the work with Companies *A*, *B*, and *C*. For Company *D*, net sales were $3307 M, net income $152 M, and equity $1,153 M. Using the risk tolerance ratios derived for the other three companies, they obtained an average risk tolerance from the three methods equal to $194 M, as shown in Table 38.3.

On the day of the meeting, the team assessed the risk tolerance directly from the executives and found it to be $190 M, which was close to the value obtained by comparing the ratios of the other three companies. The analysis did not need revision, and the executives decided to fund the investment.

In practice, however, executives at the top of the organization often operate with a larger risk tolerance than do managers or engineers at the bottom of the hierarchy. The owners of a company would want everyone in the company to make decisions using the corporate risk preference curve; otherwise, there could be loss of value within the organization. Figure 38.5 shows

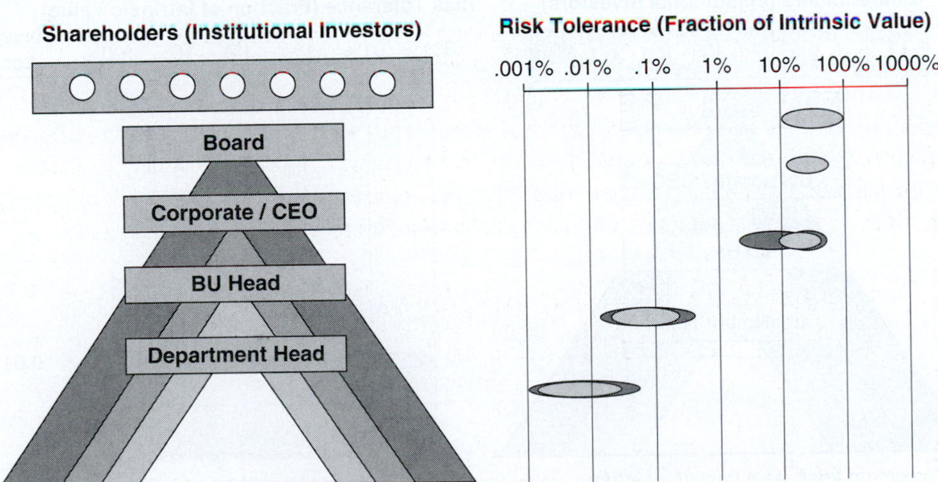

FIGURE 38.5 Risk Tolerance Assessment Across Hierarchy (Source: Strategic Decisions Group. (c) SDG All rights reserved; used with permission.)

an example of a risk tolerance assessment conducted in a variety of organizations by Strategic Decisions Group. Individuals at different levels of the organization were asked about the risk tolerance they would use in making decisions, as measured by equivalence rule assessments. The risk tolerance is presented as a fraction of the intrinsic value of the firm, which is the total value of the firm.

The results confirmed that people higher up in the hierarchy operate with larger risk tolerances than people at the bottom. This system has disadvantages. For example, it may result in new projects being rejected at the bottom of the organizational hierarchy that executives at the top could have found appealing. The result is a loss of value to the organization and what we call a **value gap**, which is the shaded region in Figure 38.6. This region represents loss of projects across the organizational hierarchy that board members would have approved.

Having different risk tolerances across the organization encourages arbitrage within different departments, where a division might value one project more highly than another. This problem is solved by having a unified corporate risk tolerance for all decisions.

If the company is in bidding competitions with other companies and for security reasons, it may not wish to have this known throughout the organization. Consequently, it may give lower levels of the company only the portion of the corporate *u*-curve that they will need. This may mean telling their employees to be risk-neutral for minor decisions, exponential with a specified risk tolerance for medium sized decisions, and reserving the actual corporate *u*-curve for board-level decisions.

A colleague who had served at the executive level of several organizations mentioned that after accepting a new management position, the first thing he did was to figure out how many dollars of company money was equivalent to one dollar of his own money. This allowed him to determine the approximate amount of time he should spend on deliberating any financial decision. In making corporate decisions, you do not want a manager to use his or her own personal risk preference. If a manager has a risk tolerance of R_M, and

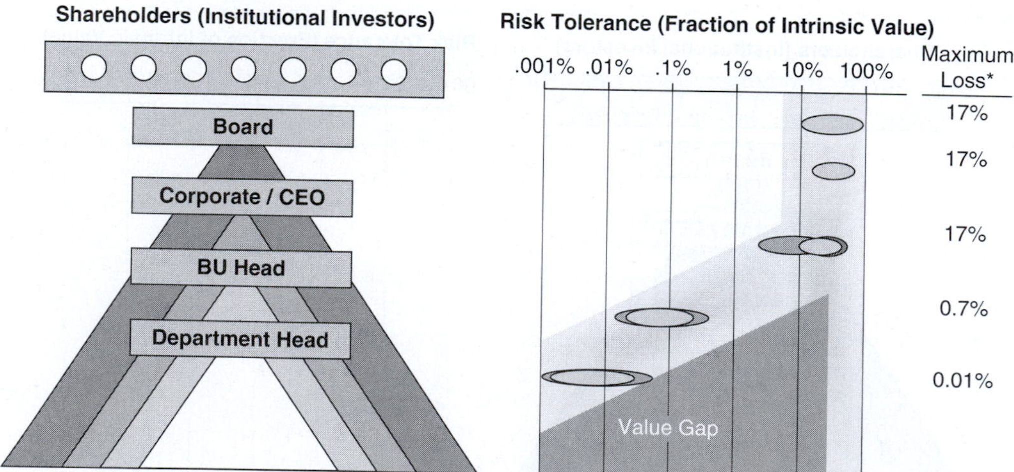

*Maximum Loss, as a fraction of intrinsic value for a 50–50 chance at doubling intrinsic value.

FIGURE 38.6 Value Gap from Varying Risk Tolerance Across Hierarchy (Source: Strategic Decisions Group. (c) SDG All rights reserved; used with permission.)

the corporate risk tolerance is R_C, he can use the idea of risk scaling that we described in Chapter 32. The ratio $\frac{R_M}{R_C}$ can convert any deal faced by the company into a deal for the manager. If the manager is willing to accept this deal based on his personal risk tolerance, R_M, then the company would accept the larger deal with its risk tolerance R_C.

38.7 COMMON MOTIVATIONAL BIASES IN ORGANIZATIONS

In Chapter 16, we discussed cognitive biases, often unconscious. Some biases also can be present consciously, and are due either to motivations of the person providing the judgment or from differing incentive structures in an organization. For example, consider the implications of the incentive set by a CEO hanging pictures of some of the sales force in frame with the words, "This person belongs to the 400% club." When asked what an employee needed to do to become a member of the 400% club, the CEO replied that these are sales people whose sales exceeded their forecast by 400%. Think of the implications this incentive could have on the forecasts obtained by sales managers.

Motivational biases can occur quite often in organizations and also can have consequences in assigning probabilities or influencing discussions. You can learn much about the decision analysis by critically examining another person's analysis and wondering whether such biases affected their thought processes. Consider the following example as an illustration.

EXAMPLE 38.1 The Executive's Problem

At the time of the 1970s oil embargo, the president of an American oil company that was a subsidiary of an international oil company spent the morning in his office with representatives of the sales and legal departments discussing what to do about supplying customers in this period of high oil prices. The contract with the customers specified that, in the case of unusual conditions, the company would be able to withhold 10% of the oil that it had promised to supply to the customers. This would allow the company to sell the withheld oil at the current high price rather than at the lower contracted price. However, the legal department warned that the customers might take legal action to prevent the withholding, and if they did, they might well prevail. The discussion seemed to be going around in circles.

At this point, the executive recalled that he had taken a two day course in decision analysis and that he might be able to apply it to this problem. He took out a sheet of paper and drew the decision tree shown in Figure 38.7 (his handwritten original has been redrawn for our purposes). The analysis showed that his company would be better off by \$728,000 (a gain of \$248,000 instead of a loss of \$480,000) if he invoked the "10% Not Supply" alternative, and he decided to do so. He was very pleased that he had been able to arrive so quickly at a course of action.

At the conclusion of the meeting, he faxed a copy of Figure 38.7 to those who had presented the decision analysis course and asked for their comments on his work.

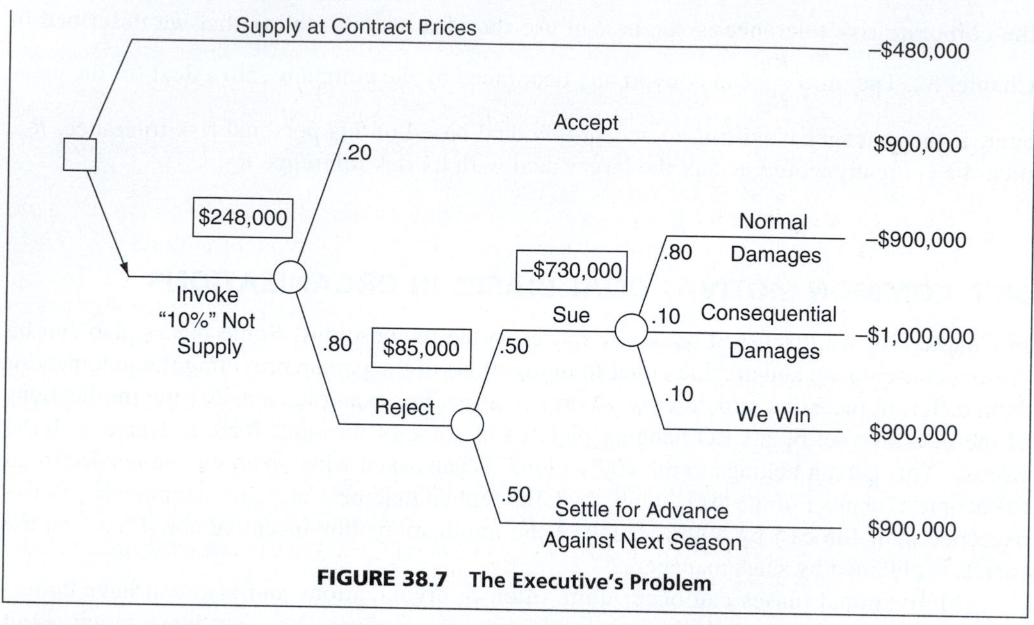

FIGURE 38.7 The Executive's Problem

Reflection

Given your extensive studies of decision analysis, suppose that you received this fax and were going to respond to the executive. Take some time to jot down all the comments you might make in constructively criticizing the analysis. Given this story, do you find any issues with the tree he faxed?

When previous students were faced with this same exercise, they had several questions and comments, all of which are listed below. Our responses follow, in italics.

- The analysis is conducted as if the company were risk-neutral. Is that reasonable? *Since this was a small resource decision compared to the wealth of the parent company, it is a "small potatoes" decision and risk neutrality may well be appropriate.*
- The analysis does not consider the effect on future customer relations of withholding the oil. *The appropriateness of such a frame depends on whether this is a market where loyalty is a factor or more like a transaction in a commodity market. The transaction frame could be appropriate, but the executive should realize that this is the frame used.*
- What is the meaning of the numbers at the end points? Why is supplying a $480,000 loss while not supplying is a $900,000 gain? Where is zero? Are these prospect value measures the actual changes in the cash flow of the company?
  Good questions.
- Why is there no cost to the company of the legal action? *This was a point mentioned by the executive in his fax transmittal. He felt that the lawyers as employees of the company would have to be paid whether they were engaged in litigation or not. However, there would be many other costs to the company if it faced a suit, including the time spent by executives in giving depositions.*

- Are the high chances of losing in a suit reasonable? *Here, we might ask about the motivational biases of the lawyers. As in-house attorneys, they would prefer not to face litigation, since there will always be a chance that they will not be able to defend the company successfully. They may be tempted to exaggerate the chances of losing. This can be highlighted in practice by considering the incentive structures that different people in the organization have when providing their responses.*
- If the customers have such high chances of winning a suit and receiving high damages, why is there only a 50–50 chance that they will sue? *Another good question: Might this be wishful thinking?*
- Once the company sees that the customers will sue, why does it not introduce the alternative of settling the suit? Why is there no downstream decision in this situation? *This is, perhaps, the most fundamental structural flaw in the chronology of the decision. Failure to recognize the option of settling at any point in a legal matter can substantially affect the results of the analysis.*

We can also inspect the executive's frame and observe that he has considered the effects of his decision on his company's profit in a purely economic way. Another frame could have considered the relationship he has with the customer company and the value he places on that.

38.8 SUMMARY

Operating with a corporate value function and corporate risk tolerance has several benefits in organizational decision making.

It is important to understand the exact value that a corporation wishes to maximize.

Budgets and incentive structures have several inherent problems that may affect the quality of the decision making within an organization.

Cognitive and motivational biases might affect the quality of decision making within an organization. It is important to be aware of them.

KEY TERMS

- Incentive structures
- Requirements
- Targets
- Corporate risk tolerance
- Value gap
- Motivational biases in organizations

PROBLEMS

Problems marked with an asterisk (*) are considered more challenging.

1. Draw the decision diagram for the executive's problem presented in Figure 38.7. Think of other issues or clarifications you might ask the decision maker in this situation.
2. For the tree in Figure 38.3, what is the risk tolerance that makes the manager indifferent between the two alternatives.

*3. Give examples of situations where incentive structures might lead to sub-optimal choices.

*4. What are the main issues with budgets as described in this chapter?

*5. What are the problems with setting multiple independent requirements in a firm?

*6. Explain the concept of a "Value Gap" in an organization.

CHAPTER

39

Coordinating the Decision Making of Large Groups

CHAPTER CONCEPTS

After reading this chapter, you will be able to explain the following concepts:

- Group think
- Analytical complexity vs organizational complexity
- The dialogue decision process (DDP)
- Issue raising
- The strategy table

39.1 INTRODUCTION

In Chapter 32, we discussed the decision made by a partnership, illustrated how the group can operate effectively with a group risk tolerance, and determined that they can share uncertain deals based on the risk tolerance of each individual member. In Chapter 38, we discussed some potential barriers to effective organizational decision making, such as imposed incentive structures, and we illustrated the benefits of operating with a single risk tolerance across the organization.

When a group (or a team) of decision makers commits to making a decision, the coordination of the decision-making process within the group becomes essential. While group members are all engaged in the same effort, some may have conflicting information and preferences, and they may possess different levels of resources. In this chapter, we highlight some common challenges for groups engaged in a common decision-making process and offer some recommendations for better coordination.

39.2 ISSUES CONTRIBUTING TO POOR GROUP DECISION MAKING

One common phenomenon that can lead to poor decision making within a group is known as **groupthink**. A famous example of groupthink is the "trip to Abilene," which is a parable based on a real experience[1] that is used to describe the issues surrounding how individuals

---

[1]Jerry B. Harvey, *The Abilene Paradox and Other Meditations on Management* (San Francisco: Jossey-Bass, 1988). The original publication of the Abilene Paradox appeared as: "The Abilene Paradox: The Management of Agreement," in *Organizational Dynamics* (Summer 1974).

reach agreement—or more accurately—*believe* they have reached agreement. The parable goes as follows.

> Four adults are sitting on a porch in 104° heat in the small town of Coleman, Texas, some 53 miles from Abilene. They are engaging in as little motion as possible, drinking lemonade, watching the fan spin lazily, and occasionally playing the game of dominoes. The characters are a married couple and the wife's parents. At some point, the wife's father suggests that they drive to Abilene to eat at a cafeteria there. The son-in-law thinks this is a crazy idea but doesn't see any need to upset the apple cart, so he goes along with it, as do the two women. They get in their un-airconditioned Buick and drive through a dust storm to Abilene. They eat a mediocre lunch at the cafeteria and return to Coleman exhausted, hot, and generally unhappy with the experience. Back on the porch, they chat and realize that *none* of them really wanted to go to Abilene—they were just going along because they thought the others were eager to go. When a group decides to take an action that nobody agrees with, but which no one is willing to question, we say that they're "taking a trip to Abilene."

Organizational decision making poses many additional challenges based on the diverse identities and perspectives of the people involved. For example, while discussing the decision, some members of the team may use their authority to exhibit influence over others or to steer the decision towards their own preferences. Some members also might be reluctant to reveal their reservations about a given project if higher management has already voiced its support. In addition, various departments within an organization may have very different preferences about the same decision. For example, the marketing group might favor an early introduction for a new product, while the engineering group might prefer to delay launch until the product has been thoroughly tested. The size and structure of an organization may itself prevent decision opportunities from being recognized.

Many of the issues we discussed in earlier chapters that contribute to poor decision making on an individual level might also exist within a team or an organization. For example, acknowledging the difference between a decision and its outcome may pose a major obstacle to the quality of decision making in an organization. It is very common to hear sentences like "We tried this last time and it didn't work, so this time, we'll try something else." Or you might hear the opposite: "This worked last time, so it's a safe bet. Let's just keep things the same." These types of arguments favor making decisions based on the outcomes of previous decisions—regardless of their quality at the time.

Another impediment to effective group decision making is the **sunk cost trap** that we discussed in Chapter 2. In organizations, this happens when decision makers continue to pursue a project simply because of the resources it has already consumed. In meetings, it is very common to hear things like, "We have spent so much money in this project already. There's no turning back," or "We have made a large investment in this project so we need to wait until it achieves a higher return." What they face now are choices that will determine the future and cannot affect the past. The sunk cost trap is commonly described as "throwing good money after bad." In business, as in life, you must summon the courage to admit that sometimes projects undertaken with great enthusiasm can ultimately fail.

Lack of an effective process for structuring and framing the decision problem in organizations can also impede quality decision making. Perhaps the most common way for a group to make a decision is to gather around a table to discuss the decision situation and eventually arrive at a conclusion by decree, vote, or agreement without using any kind of formal analysis. The absence of a **structured process** for decision making often results in selecting the wrong problem or not even starting with the correct problem, but then failing to recognize the changes in the decision situation that now render the problem obsolete.

Having a structured decision process, on the other hand, requires that decision makers be open to the participation of others. Some executives may not want their decision making process exposed to potential criticism. Consider, for example, political decision making where lack of clarity may be regarded as an advantage.

In other cases, executives may feel they have earned the right to "call the shots" and make these decisions. They do not want to share this privilege with others. Too often, *overconfidence biases* may impede having a quality decision process. Executives may feel that a record of success guarantees their eternal "magic touch" for making decisions. In some cases, there may be other ulterior motives that obstruct the quality of the decision making process, such as an epidemic of personal greed, ego, or fear of losing control. Some executives may be too operationally focused because of a previous operational position within the organization. The idea of designing a decision process may simply be out of their comfort zone.

39.3 CLASSIFYING DECISION PROBLEMS

When thinking about organizational decision problems, it is useful to classify them into two dimensions: (1) the **analytical complexity** required by the decision, and (2) the **organizational complexity** associated with those involved in the decision. This classification can help you determine the right tools needed for the decision.

When we describe *analytical complexity*, we refer to things such as the type of analysis needed to solve the problem, the mathematical tools needed, the number of variables involved, and the nature of the decision problems (static vs. dynamic; uncertain vs. deterministic). Based on the type of technical analysis that is appropriate, we can roughly classify problems as either low or high in analytic complexity.

Organizational complexity also ranges from low to high as we move from a small firm (or a single decision maker) to a large organization with multiple decision makers. The role of a **facilitator** is especially important in large settings with great organizational complexity. The facilitator is the person who runs the meetings, makes sure everyone's views are being expressed and considered, and aims for group agreement.

However, facilitation alone is not sufficient. Without a good decision-making process, the group might agree on the wrong thing. For ease of discussion, we now classify decision problems into four regions using the 2×2 matrix shown in Figure 39.1.

The decision regions described by the matrix of Figure 39.1 are given here.

39.3.1 Low Analytical Complexity–Low Organizational Complexity

These situations are encountered very often in organizations. For example, an engineer may be deciding on which photocopier to purchase for the department. If this choice is analyzed by each member in the department, it would be a time-consuming process. Instead, it will be much more efficient if the engineer independently conducts some research about existing copiers, solicits feedback, and makes a decision on behalf of the group rather than waiting for consensus among the group members.

In large organizations, there are thousands of similar types of decisions that need to be promptly decided. It is important to identify those decisions and act upon them immediately, without draining organizational resources.

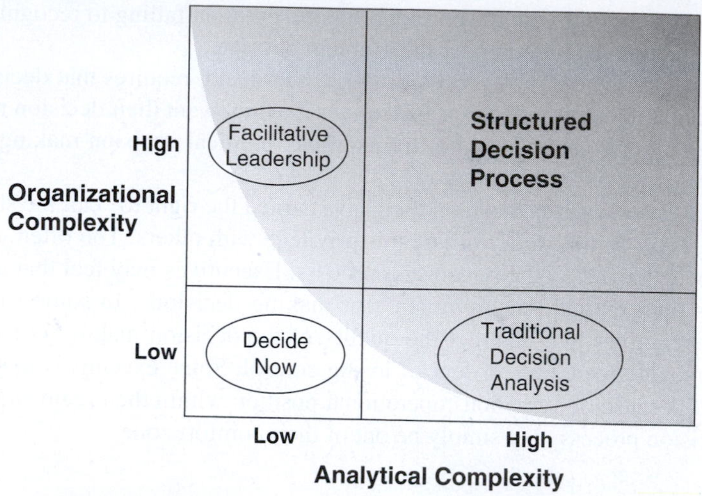

FIGURE 39.1 Classification Matrix for Organizational Decision Problems (Source: Strategic Decisions Group. (c) SDG All rights reserved; used with permission.)

39.3.2 Low Analytical Complexity–High Organizational Complexity

These types of situations also occur very frequently in organizations. For example, a company is holding a retreat to brainstorm possible growth scenarios. They might need a process for recording ideas and an environment where ideas are generated without fear or criticism. This is often referred to as a **facilitation activity**. Facilitative leadership requires involving the right people, acknowledging different perspectives, and coordinating discussions to ensure all team members have an opportunity to express their views.

39.3.3 High Analytical Complexity–Low Organizational Complexity

These types of decisions, too, are very common. For example, one store within a company chain requires an analysis to determine its optimal inventory level. Facilitation alone will not help. In this case, they will need an analyst with the appropriate skills to carry out the analysis and derive the important insights.

39.3.4 High Analytical Complexity–High Organizational Complexity

These types of decision situations arise in large organizations making analytically complicated decisions. For example, NASA may be reevaluating the design of its space shuttles or its entire supply chain of providers. Another example could be an automobile manufacturing company reevaluating its portfolio of vehicles currently in the market or allocating budgets to its current research and development (R&D) projects. Several departments may be involved in these situations, and the analysis may be extensive.

To work on decision situations in this region, we need a structured decision process that promotes a dialogue between the decision makers and the project team that is analyzing the alternatives. In these types of situations, one of the failure modes that may arise is ignoring the interactions between the project team and the decision makers until the very end of the project. This failure mode will reduce the commitment to follow through by the decision makers. Furthermore, the final recommendation from the project team may even come as a surprise to the decision team, and the project team may find out that the decision team has strong reasons not to act upon the recommendation. One method for coordinating the decision making within this region is known as the **dialogue decision process (DDP)** and will be the topic of the next section.

DIALOGUE DECISION PROCESS (DDP) The *dialogue decision process* (DDP) was designed at the decision analysis group at Stanford Research Institute (SRI) in the 1970s. This process has since been tested and applied to hundreds of projects spanning major industries such as oil and gas issues pharmaceutical companies, automobile manufacturing, and many others. The basic idea behind DDP is to make the decision maker the owner of the decision process, rather than to try to sell him a ready-made decision consulting service. Its main purpose is to involve the decision maker in the dialogue and build on his ideas.

The dialogue decision process starts with a kick-off meeting to identify the decision makers who will form the **decision board** as well as the decision analysts and the key experts in the fields involved in the decision who will constitute the **project team**. Together, these two bodies will provide the framework for the decision situation.

The Decision Board As we have discussed, decision making is more difficult when several decision makers are involved. Therefore, the decision board should be selected based on the *minimum* number of people whose approval is sufficient to approve a decision alternative. Another key element in the selection of the decision board is their availability and commitment to attend project meetings and answer questions as they arise. Otherwise, meetings of the whole decision team may be delayed by difficulty in scheduling. If members of the decision board have not been exposed to decision analysis and the dialogue decision process, the project may start with a tutorial and overview session. At this session, participants get a chance to see the overall process, agree on its scope, logistics, and communication process.

The Project Team This is the team that will be working with the decision board throughout the project. This team has the skills needed to perform the analysis and the facilitation. Members must have time for thinking carefully, gathering information, and preparing a clear picture for executive review.

Structure of DDP Once the composition of the two groups is settled, the next step is to discuss the flow of the project, the phases of the analysis, the timelines for the deliverables, and the scheduling of essential future meetings. Figure 39.2 shows an example of a typical process that shows the interactions between the decision board and the project team. Because of its shape, Figure 39.2 is often called the "snake diagram" in practice.

The snake diagram shows the flow of information between the decision board and the project team. The first role of the decision board is to agree to the analysis of the decision by providing a general description of the purpose of the analysis. This will lead the project team to assess the business situation faced by the company and to provide to the decision team a frame for their approval. As we have said, the frame will specify what is to be viewed as accepted, what will be decided later, and which stakeholders to be considered. The decision board will adjust the frame as necessary and authorize the project team to develop the **decision basis**: the alternatives, information, and preferences that will characterize the decision. The decision board will review the resulting basis with particular emphasis on alternatives and the specification of preferences.

The project team then performs the analysis to evaluate the alternatives by showing their certain present equivalents. As a result, they may suggest new alternatives or require additional information gathering before deciding. This step may require additional iteration between the

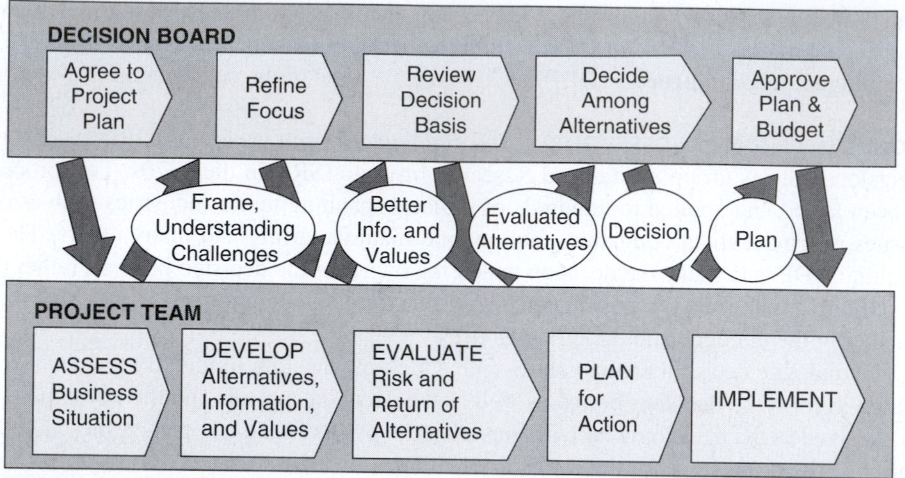

FIGURE 39.2 Dialogue Decision Process Snake Diagram (Source: Strategic Decisions Group. (c) SDG All rights reserved; used with permission.)

groups. Finally, the decision board must choose an alternative that the project team will plan to implement. Once the decision board approves their plan and authorizes the resource commitments, the implementation begins. If there is no implementation, there is no decision—only intention.

39.4 STRUCTURING DECISION PROBLEMS WITHIN ORGANIZATIONS

When applied in an organizational setting, the decision analysis cycle presented in Chapter 37 becomes more complicated. In particular, structuring and formulating the decision problem will require additional care. The following are some good operating practices and tools to help with the *formulate phase*.

39.4.1 Establish a Vision Statement

Before embarking on a decision analysis in an organization, one important step is to have a **vision statement**: a clear idea of what needs to be done or decided. This step is equally important for individual decision makers, but it is surprising that even after several meetings individuals within a team may have different frames and perspectives about the decision. The idea of the vision statement is to bring a meeting of the minds among team members about the exact nature of the decision.

A one page vision statement is usually a good tool to bring clarity at this point. At a minimum, it should answer the following three questions:

What are we doing?
Why are we doing it?
How will we know if we are successful?

These questions can be distributed to individual team members before the meeting. While these questions may seem straightforward and simple, different people within the team will answer them differently. Having clarity on the answers to these questions is essential for the future framing of the decision.

39.4.2 Raise the Issues

The next step in the formulate phase is called **issue raising**. This step is an opportunity to capture all aspects related to the decision situation in a brainstorming session that involves both the decision team and the project team.

Recall that the best decision alternative that emerges at the end of the analysis is usually one of the alternatives originally proposed, although sometimes a hybrid alternative formed from the proposed alternatives turns out to be optimal. Therefore, it is important to make sure that many significantly different alternatives have been identified before starting the analysis.

Typical issues raised during this session are uncertainties, worries, preferences of the decision makers, policies, facts, threats, opportunities, and the organization's strengths and weaknesses. This is the time to get thoughts potentially pertinent to the decision expressed discursively before filtering and refining them.

The presence of a facilitator during an issue raising session can help ensure that participants can introduce issues without fear of criticism, and that no one participant dominates the discussion. The process is enhanced by a *scribe* who records the issues and has the facility to project them to the team as they are being raised. This system allows instant correction of any misunderstandings, and documents the whole session. If no scribe is available, groups can often use Post-It® notes to write their issues and post them in front of everybody.

39.4.3 Sort the Issues

After the issues have been raised, the next step is to sort them in a way that facilitates their use in decision analysis. This requires categorizing the issues into four subgroups: *decisions, uncertainties, values* (to form the decision basis), and *other* issues, such as facts that are known such as government regulations or specifications (see Figure 39.3).

Decisions include choices that must be made as well as alternatives available for those choices. *Uncertainties* will include any distinctions participants would like to know because of their beneficent or maleficent effects on outcomes. Finally, the *values* of the decision makers need to be identified. Values should distinguish between direct and indirect value attributes and may include monetary prospects or total shareholder return, reputation of the company, safety of

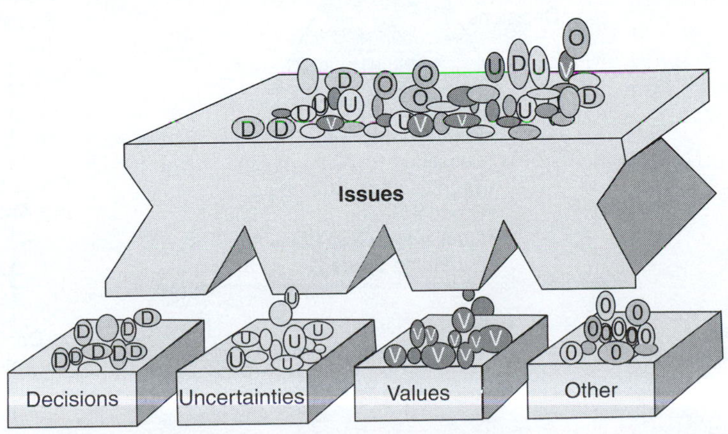

FIGURE 39.3 Sorting the Issues

employees, motivation, or developing a sense of community within the organization. The values category will include specifications of time and risk preference.

Categorization of issues plays an important role in framing the decision. Some teams naturally categorize most of the issues raised as decisions and values with very few uncertainties. Other teams may categorize most of the issues raised as uncertainties that are out of their control. The facilitator can make sure that the collection of issues is properly balanced. Keeping in mind who the decision maker is (the CEO, company, or department) can help categorize the issues.

After the issues have been categorized, the next step is to classify the decisions into policy decisions, strategic decisions, and tactical decisions (refer back to the decision hierarchy in Chapter 17). As we discussed, policy decisions concern the choices that already have been approved by the organization are now part of corporate policy and should be viewed as accepted.

While drawing the decision hierarchy, it is good practice to challenge the current frame. For example, a budget constraint may be viewed as accepted, but as we have discussed, it might also lead to sub-optimal alternatives for the company. If such a constraint is present, it should be validated.

As an example, a railroad company might have a policy decision that the company will remain in the railroad business even if this method of transportation might not be the most profitable. Policy decisions such as this should be challenged. The strategic decisions are then identified as those that need to be reviewed and adopted in this analysis and where the effort should be focused. Strategic decisions are the decisions that will be incorporated into the decision diagram. The tactical decisions are also identified as those that need to be reviewed and finalized later.

Figure 39.4 shows an example of a decision hierarchy that was constructed in an issue-raising session to help an automobile manufacturing company with the decision to launch a new product. As we discussed in Chapter 17, the decision hierarchy has three levels of decisions: (1) policy decisions that are viewed as accepted; (2) strategic decisions that are the focus of the current analysis; and (3) tactical decisions that will be made later.

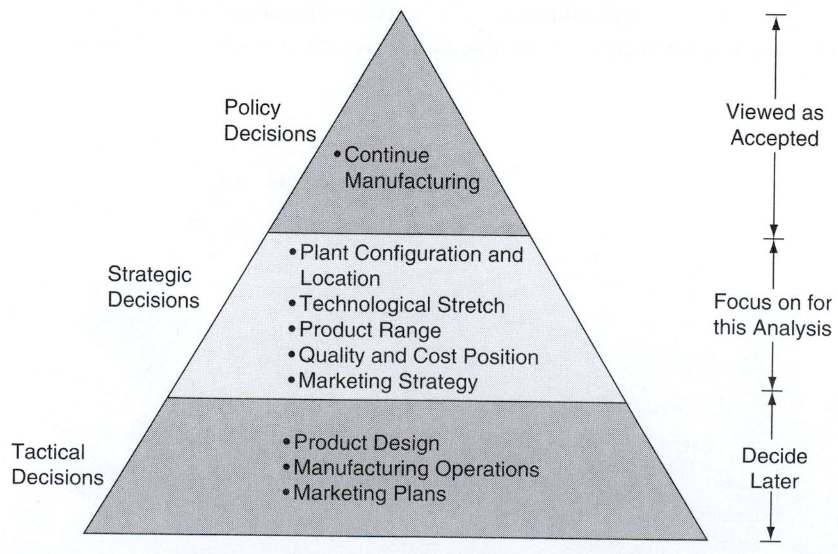

FIGURE 39.4 Issue-Raising Decision Hierarchy

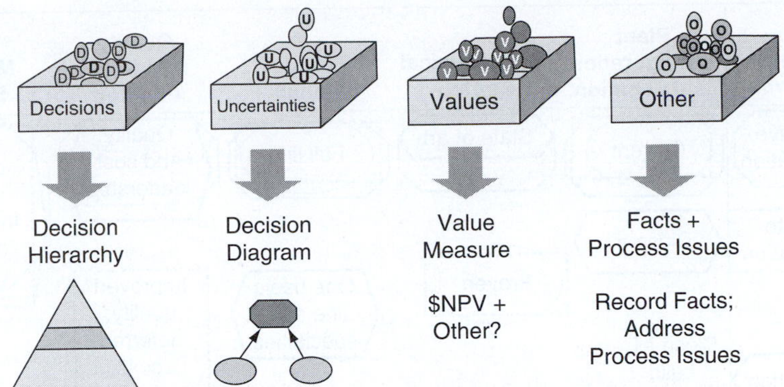

FIGURE 39.5 Formulating the Decision Situation

Uncertainties are used to create the distinctions needed in the decision diagram, and the values are used to determine the elements of the value node (Figure 39.5).

39.4.4 Construct a Strategy Table

After the decision hierarchy is agreed upon and the strategic decisions have been identified, the next step is to construct a **strategy table**. The purpose of this step is to identify key strategies that will be explored and analyzed. Strategy tables often result in *hybrid strategies* that add significant value to the corporation.

To construct a strategy table, each of the strategic decisions in the decision hierarchy is placed in the top cell of a separate column. For example, in Figure 39.6 we have the same decision hierarchy for the automobile manufacturing company. The strategic decisions are the plant location for manufacturing, the technology that will be used, the products that will be manufactured, the quality segment, and the marketing techniques. In each column, we list the set of possible alternatives that are available for each strategic decision.

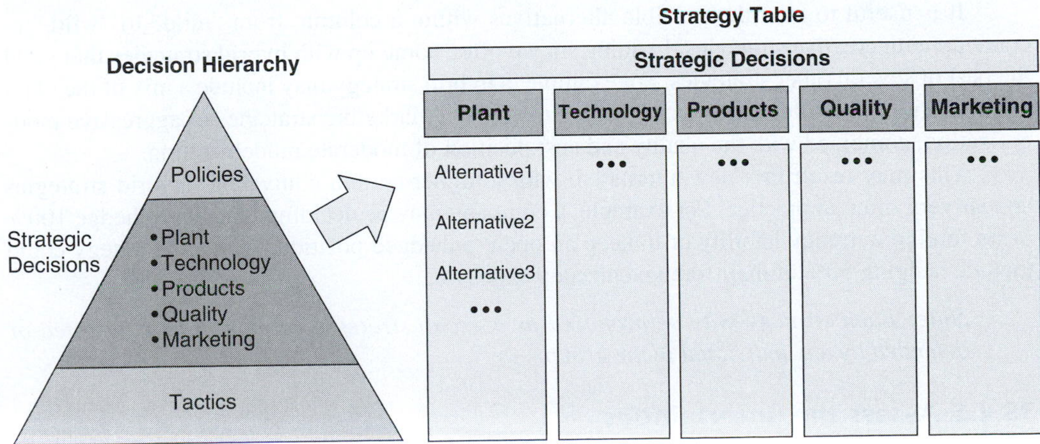

FIGURE 39.6 Strategy Table from Decision Hierarchy (Source: Strategic Decisions Group. (c) SDG All rights reserved; used with permission.)

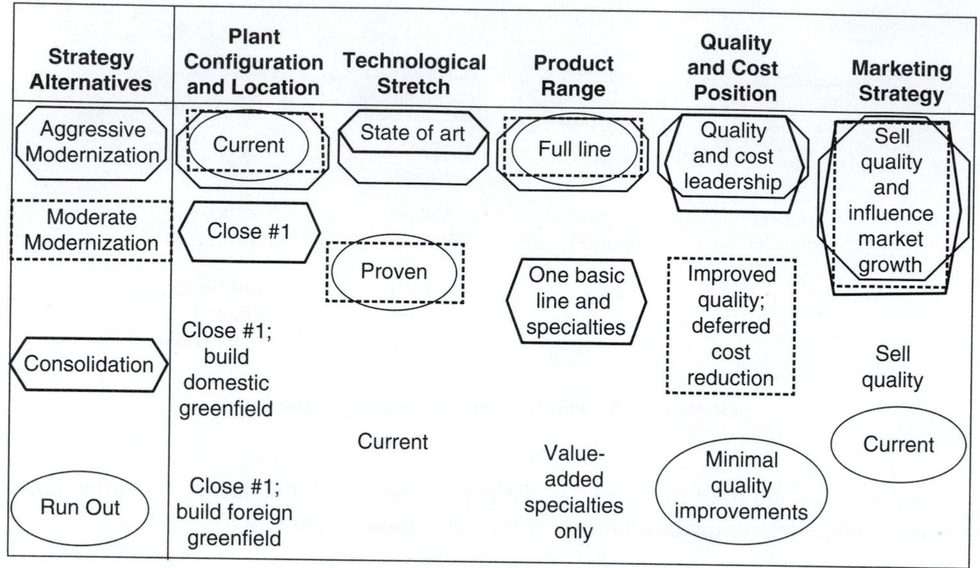

FIGURE 39.7 Identify Different Strategies

After the strategy table is filled out with possible alternatives for each decision, the next step is to choose strategies for analysis. Each strategy incorporates different combinations from the possible alternatives for each strategic decision. A strategy must be coherent. It is not just a matter of picking an element from each column—that would lead to a very large number of alternative strategies. Each strategy investigated must, of course, be a course of action that could actually be instituted, and it must cohere. A few coherent strategies will be selected for analysis, and each is typically given a name.

For example, Figure 39.7 shows an example of a strategy table with four strategies: aggressive modernization, moderate modernization, consolidation, and run out. Each strategy uses different alternatives for each of the strategic decisions being considered.

It is useful to scan the possible alternatives within a column, from "mild" to "wild," to come up with creative strategies. In doing so, we often come up with hybrid strategies that yield the best of several other strategies. For example, a hybrid strategy may include a mix of the plant configuration, technological stretch, product range, and marketing strategies of aggressive modernization, combined with the quality and cost position of moderate modernization.

This may result in a new alternative with a higher certain equivalent. Hybrid strategies appear very often in practice. For example, a company may be deciding whether to hedge 100% of its foreign currency liability or to keep an open, unhedged position. A hybrid strategy could include hedging 50% of their foreign currency exposure.

Note: *Each strategy is to be proposed as a group strategy, not one to be advocated or defended by any individual in the group.*

39.4.5 Assess the Uncertainties

We will now assess the probability distributions for the uncertainties that have been created. In doing so, we need to address and mitigate the effects of biases, such as the cognitive and

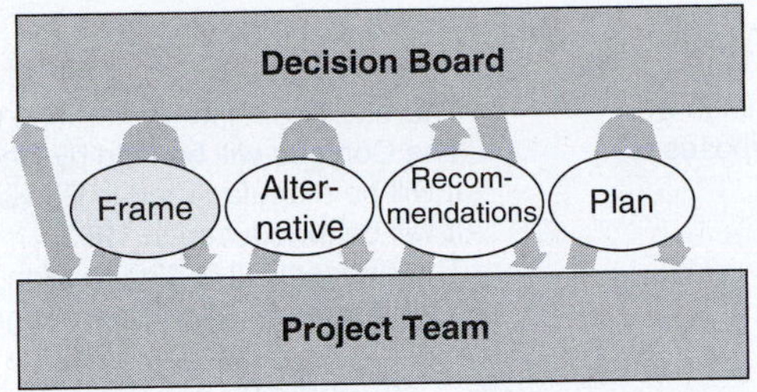

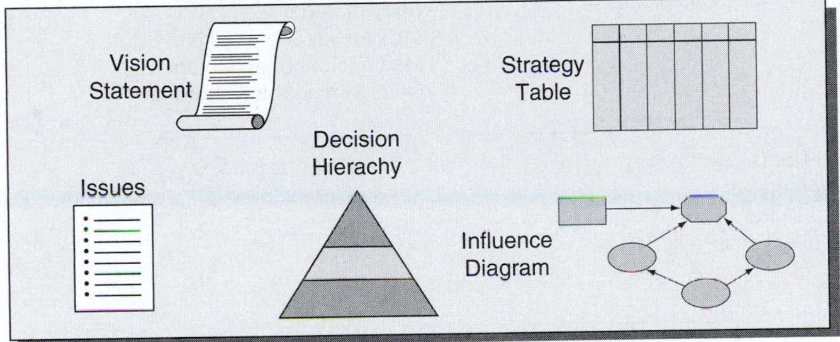

FIGURE 39.8 Summary of Deliverables after Formulate Phase

motivational biases we discussed earlier in Chapters 16 and 38. It is also good practice to provide a pedigree for each assessment (Chapter 15) and employ a certified probability encoder to conduct the assessments. Experts from within the company or consultants may be asked to provide probability assessments for distinctions in their domain.

Given the decisions, uncertainties, and values, we now have the complete decision diagram for the decision situation. Figure 39.8 shows a summary of the deliverables at the end of this coordinated decision process:

- Vision statement
- List of issues raised
- Decision hierarchy
- Strategy table
- Decision diagram

39.5 EXAMPLE: THE FIFTH GENERATION CORVETTE

In 1988, General Motors decided to design an all-new, fifth-generation Corvette. General Motors consulted Strategic Decisions Group where a project team was sent to help with the decision making process. The group structured the decision problem and started with an issue raising session. After the session was finished, the decisions were selected and a decision hierarchy was constructed. The decision hierarchy showing the policy decisions, strategic decisions, and tactical decisions is shown in Figure 39.9.

POLICY DECISIONS The company viewed as accepted that there would be a fifth-generation Corvette, that it would be sold by Chevrolet and no other manufacturer or retailer, that the budget

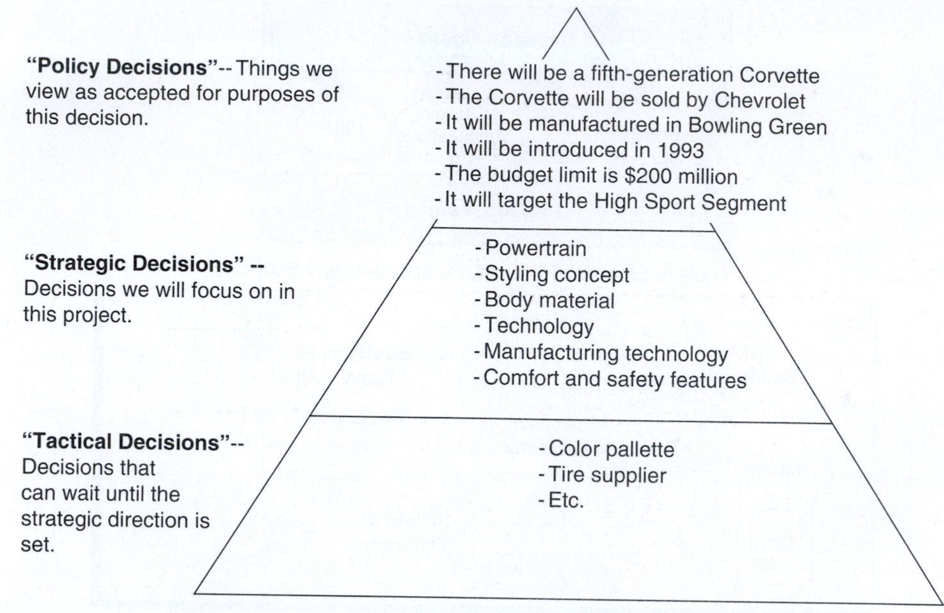
"Policy Decisions"-- Things we view as accepted for purposes of this decision.
- There will be a fifth-generation Corvette
- The Corvette will be sold by Chevrolet
- It will be manufactured in Bowling Green
- It will be introduced in 1993
- The budget limit is $200 million
- It will target the High Sport Segment
"Strategic Decisions" -- Decisions we will focus on in this project.
- Powertrain
- Styling concept
- Body material
- Technology
- Manufacturing technology
- Comfort and safety features
"Tactical Decisions"-- Decisions that can wait until the strategic direction is set.
- Color pallette
- Tire supplier
- Etc.

FIGURE 39.9 Decision Hierarchy for Fifth Generation Corvette

limit for the project would be $200 million (refer to our discussion of budgets), and that it would be tailored to the high sport segment.

STRATEGIC DECISIONS The design decisions that the company wanted to focus on were the powertrain, the styling concept, the material of the body, the manufacturing technology, and some of the comfort and safety features.

The issue-raising session also identified some tactical decisions for future discussions.

The strategy table involving three of the strategic decisions and their alternatives is shown in Figure 39.10. Several alternatives for performance, styling concept, and comfort/convenience were considered. Three main strategies were identified and named as follows.

Strategy 1: *New Improved Corvette 4*

350 HP; major face lift in styling; minor changes to comfort and convenience.

Strategy 2: *Kindler and Gentler Corvette*

350 HP; distinct but evolutionary in styling; major changes to comfort and convenience.

Strategy 3: *Corvette with the Right Stuff*

450 HP; radically new styling; lowering doors at least three inches.

The decision diagram involving the main uncertainties is shown in Figure 39.11.

The project team also developed a value model for various design decisions. They determined that much of the value came from having extra space and lowering the door sill. They decided to lower the sill and bring the transmission to the back of the vehicle. The tornado diagram (with some items concealed for confidentiality) is shown in Figure 39.12.

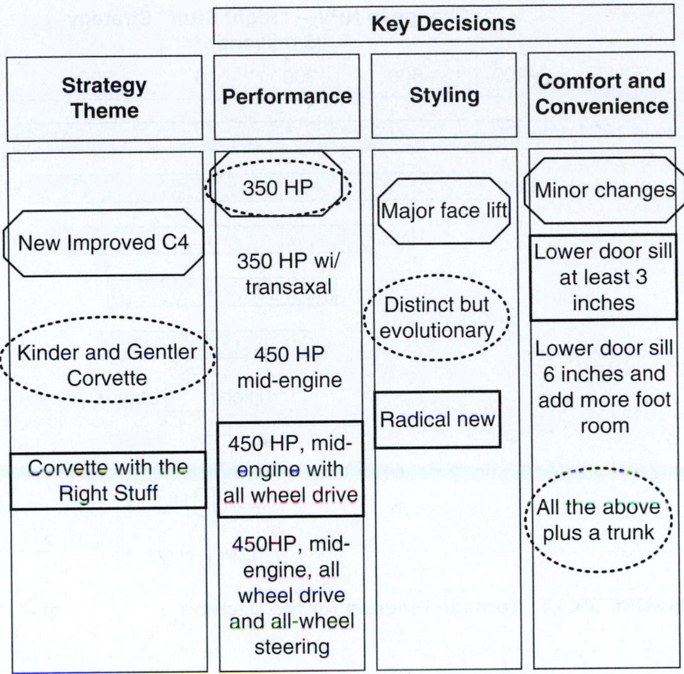

FIGURE 39.10 Corvette Decision Diagram

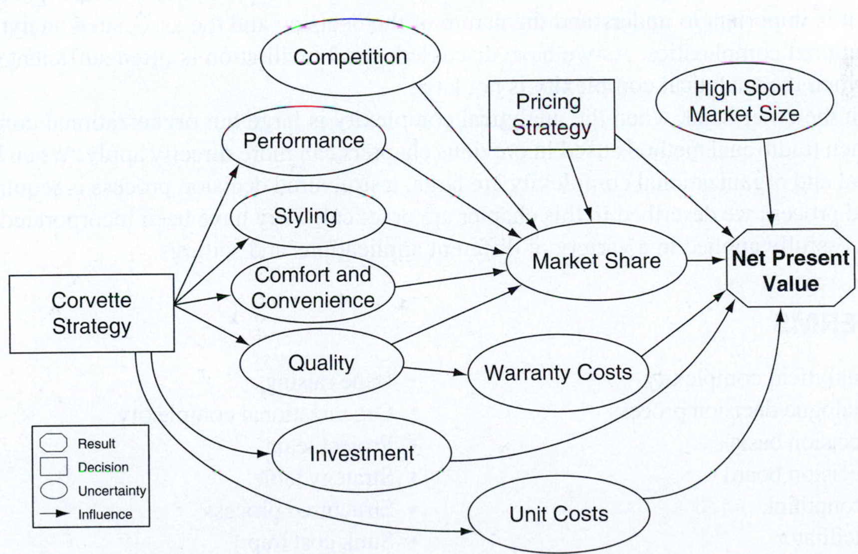

FIGURE 39.11 Decision Diagram for Corvette Design

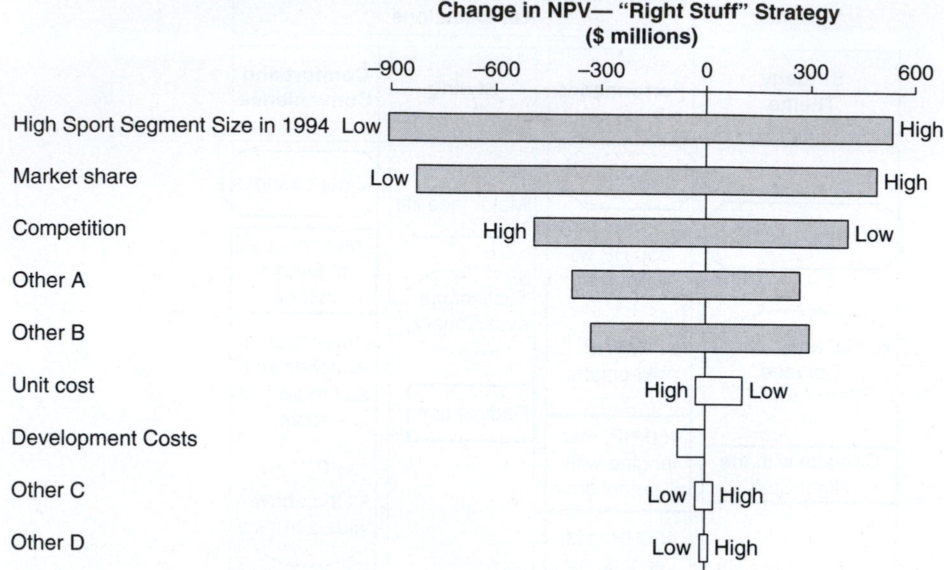

FIGURE 39.12 Tornado Diagram for the Decision

39.6 SUMMARY

Many of the decision-making tools we presented in earlier chapters can still be applied in a group setting. When coordinating the decision making of large groups, particularly in an organizational setting, it is important to understand the nature of the decision and the associated analytical and organizational complexities. As we have discussed, good facilitation is often sufficient to bring clarity when the analytical complexity is not large.

On the other hand, when the analytical complexity is large but organizational complexity is not, then traditional methods used in previous chapters can more directly apply. When both the analytical and organizational complexity are large, a structured decision process is required. The tools and process we described in this chapter are practical: They have been incorporated, tested, and successfully applied in a variety of different applications and settings.

KEY TERMS

- Analytical complexity
- Dialogue decision process
- Decision basis
- Decision board
- Groupthink
- Facilitator
- Facilitation activity
- Issue raising
- Organizational complexity
- Project team
- Strategy table
- Structured process
- Sunk cost trap
- Vision statement

CHAPTER

40

Decisions and Ethics

CHAPTER CONCEPTS

After reading this chapter, you will be able to explain the following concepts:

- The role of ethics in decision making
- Classifying actions:
  - Ethical
  - Legal
  - Prudential
- Action-based vs consequence-based ethics
- Positive vs. negative injunctions
- Telling the whole truth
- Building your ethical code

40.1 INTRODUCTION

Throughout this book, we have shown how to use decision analysis to think clearly about a decision and to choose the best course of action. However, we have not considered the ethical implications of our decisions. Decision analysis as presented to this point is amoral, that is, not admitting of moral distinctions or judgments. Just as someone can use a knife or fire to perform moral or immoral deeds, so too can decision analysis. A criminal could use decision analysis to choose the best way to steal or kill, balancing the benefits of achieving the goal with the associated risks. If ethical considerations are to enter into the analysis, the decision maker must introduce them. In this chapter, we present ethical terminology and concepts that support ethical considerations in decision making.

Ethics are your *personal* standards of right and wrong: Your code of *proper behavior*. The purpose of this chapter is not to provide a complete history and analysis of the field of ethics, but, rather, to present a background adequate to raise our ethical sensitivity when making decisions. Lack of awareness of ethical considerations is a common cause of both poor decision quality and a troubled life.

We recommend that you use this discussion of ethics as a basis for judging your own actions, rather than those of others: Focus on the self-examination that can produce sensitivity and consequent change. You will see that this chapter does not promote a specific set of ethics, but instead it provides distinctions for understanding and analyzing ethical situations so you can decide for yourself what role ethics should play in your decision making. For more details on this topic, refer to *Ethics for the Real World: Creating a Personal Code to Guide Decisions in Work and Life* (Howard and Korver, 2008).

ETHICS BY EXAMPLE In a classic story, Buddha hears about a man who had killed 999 people, and travels to the hill where he lives. When they meet, the murderer picks up his sword and says to the Buddha, "I don't know why you have come to me, but now I will kill you."

Buddha replies, "Before you kill me, please grant me two favors. First, show me your great strength by cutting off the limb of that tree." The murderer agrees, but emphasizes that these wishes will not change Buddha's fate. With a great blow from his sword, the murderer cleaves the massive limb from the tree. Buddha then asks, "With your great strength, please replace the limb." And the murder becomes enlightened and a monk.

The murderer was not awake. That is, he did not appreciate the implications of his actions. Buddha did not tell him that killing was wrong. Rather, he helped the murderer see the consequences of his actions by raising his sensitivities to issues of right and wrong.

If you are enlightened like the Buddha, you have no need of ethics: You will simply choose the right action. Until we achieve enlightenment, ethics can be our guide.

40.2 THE ROLE OF ETHICS IN DECISION MAKING

40.2.1 Ethics in Your Personal Decisions

When teaching ethics classes, we start by asking the students to describe ethically challenging situations they have experienced in their personal or professional lives. Almost all have experiences to offer. One student, for example, mentioned that he was at a post office and saw a man with a pile of stamps and a pile of letters. The man attached a stamp to each letter and then mailed them. After the man exited the post office, the student noticed that the man had left one stamp and one letter where he had been working. The student stuck the stamp on the letter and mailed it. Later, he had second thoughts about whether or not the man had actually intended to post this letter. Perhaps the man had decided not to mail his resignation from work. This is an example at one extreme of ethical concern.

Another student described a problem he faced as an employee of a software company that was planning for a major international product exposition. The company had developed a program that would solve standard problems twice as fast as its competitors and had designed a live demonstration in which machines would simultaneously run its software and its competitors' software to solve the same problem. The idea was that potential customers could see for themselves the advantages of the company's program. Shortly before leaving for the exposition, a problem arose that had nothing to do with the program, but it made the live demonstration impossible. However, they had saved files of earlier sessions that would make the machine in the demonstration look as if it were solving the problem, when it was actually just following the files. The company managers asked the student to run the demonstration without revealing that the machine was not running the code in real time.

Another student worked for a company that had promised to develop a product for a client. Early tests the student performed showed that the product was not performing nearly as well as specified in the contract. The managers asked the student not to mention these results to the client at an upcoming meeting. They explained to him that it didn't matter, because, of course, they were going to make sure that everything would be fine in the end.

Other ethical dilemmas described are more tragic. A student from a country in South America said that rebels had once kidnapped a worker from his family's farm. They asked for a large ransom to release the worker. The student sent his overseer to negotiate with the rebels and gave him half the ransom money, an amount the student thought was sufficient to assure the worker's release. The rebels killed the overseer.

Closer to our field was the case of a student who was supposed to perform research on an investment strategy under the direction of a boss who made it clear that the student's future in the company would depend on obtaining the results the boss had prescribed in advance.

In each of these cases, we might be tempted to judge what should have been done, telling the last student, "Refuse, and quit if necessary." But the decision was the student's responsibility. Students raising these experiences in class display the discomfort they have about previous actions. The class lets them ponder what they will do next time—either to deal with similar situations or to avoid being placed in such a position.

Thinking clearly about your ethical views is often difficult when you are in the throes of an ethical dilemma. Ethical situations are often emotionally charged and may involve a complex set of conflicting values. In addition, ethically sensitive situations sometimes require a fast response. For example, a student is at a party and someone passes her a joint: Either she inhales or she doesn't. There isn't time to contemplate the situation. The best way to incorporate ethics into your decision making is to develop an **ethical code** to guide actions in preparation for dealing with and avoiding ethical dilemmas.

Reflection

Before we proceed, take some time to think about an ethical dilemma you once faced or are currently facing. Why is it difficult? What are the important attributes?

40.3 ETHICAL DISTINCTIONS

40.3.1 Classifying Actions: Prudential, Legal, Ethical

In the course of our ethical discussions, we find it useful to classify actions according to whether they are *prudential, legal, or ethical*.

An action is **prudential** if that act is in the person's self-interest, where self-interest can include the effect on others, without regard to legal or ethical considerations. For example, the well-being of our children is, for most of us, in our self-interest. The act can be prudential in the short- or the long-term. Examples of prudential actions that are not ethically or legally sensitive would be changing the oil in your car's engine, buying a stereo, or getting an education. Examples of prudential actions that are ethically or legally sensitive would be continuing or ending a pregnancy depending on the sex of the fetus, and pocketing money found in a dropped wallet. Even a decision to return the money might be considered prudential if you think doing so, even anonymously, will encourage others to return your lost money in the future. "Honesty is the best policy" can be a purely prudential maxim.

For the enlightened person, there may be no distinction between prudential and ethical actions, and this would be a desirable state for everyone.

However, until we all attain that state, we will face ethical dilemmas. Consider the case of a parent stealing food to feed his or her child. We would call this act prudential and one that presents a conflict between the ethic of not stealing and the prudential consideration of caring for your children. Decision analysis as presented up to this point in the book has been primarily prudentially oriented.

An action is **legal** or illegal according to whether it is legally required or proscribed in your current location. The law is inherently coercive: The law always implies the use of physical force or the threat of physical force against people or their property. The law subjects you to physical harm or loss of property if you do what the law prohibits, such as commit assault or ingest certain substances. It also subjects you to physical harm or loss of property if you do not do what the law demands, such as file tax returns or report for military service.

Note

Each of these actions may or may not be prudential for the person in question. Murdering a rich relative in an attempt to inherit his money or using an illegal substance both may be seen as prudential by the individual involved.

An action is **ethical** if it is right, regardless of whether it is also legal. It is possible to have an ethic of doing only what is legal. One part of your ethical code should concern which of your ethics you will impose on others by force: Your desired *legal system.* An action may be ethical but not legal. For example, the distinction between legal and ethical actions was the basis for the Nuremberg trials. The people sheltering Anne Frank were performing an action that, in their view, was ethical but, in the view of their society, was certainly not legal. People in the United States sheltering illegal immigrants from Central America in their churches are in a similar, but less perilous, situation.

Conversely, an action may be legal but not ethical. For example, intentionally misdirecting a stranger may be reprehensible, but it is not a criminal offense.

The overall judgment you can make about an action is whether it is sagacious (wise). An otherwise sagacious action may be unwise because it is illegal—for example, the act of selling wine in the United States during prohibition. For some, an act of abortion would be prudential, unethical, legal, and unwise. For others, the same act could be prudential, ethical, illegal, and wise. To help clarify the distinction between ethical, legal, and prudential views, we refer to Figure 40.1.

There are seven possible regions defined by whether an act is or is not prudential, legal, or ethical.

Reflection

Consider an ethically sensitive decision in your life that is challenging to you. Classify alternative actions you might take in terms of their being ethical, legal, and prudential.

40.3.2 Eliminating Alternatives from the Decision Tree

Now that we have created ethical–legal–prudential distinctions, we can incorporate them into our decision making processes. For example, suppose that one of the alternatives is illegal, and you have decided never to commit an illegal act. This means you will only consider legal acts,

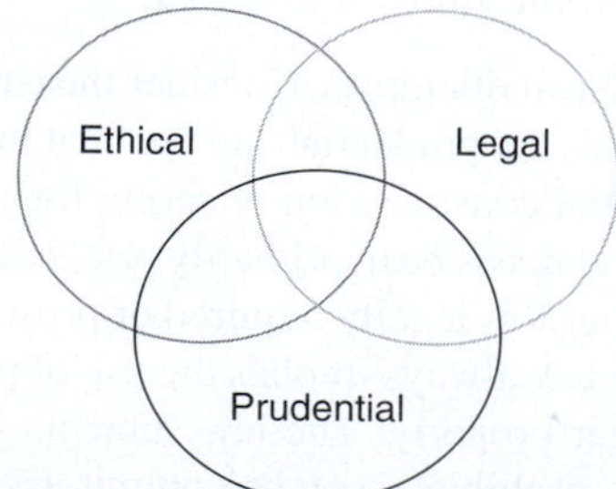

FIGURE 40.1 Classifying Situations as Ethical–Legal–Prudential

and so you can remove any illegal acts from your set of available alternatives. Other examples, such as those that arise in business, might involve receiving or giving bribes. Another example may be receiving information that is economic and material to your decision situation but also illegal, such as the exchange of insider trading information. You will eliminate all such alternatives if you have decided not to commit any illegal act.

Other situations are not so black-and-white. For example, you might have an alternative of merging with another company when you have ethical issues about how its owners run their business. While the merger may be legal and prudential, you may still decide to eliminate this alternative from your possible actions if it is against your ethical code.

40.3.3 Classifying Ethical Theories

Ethical dilemmas can be classified according to two main ethical theories: **Action-based ethics** or *ethical formalism,* and **consequence-based ethics** or *utilitarianism.*

ACTION-BASED ETHICS **Ethical formalism** is action-based. In this view, ethical responsibility attaches to the actions taken, regardless of their consequences. According to this theory, the crime of attempted murder would be equivalent to murder. The fact that fate intervened to thwart the perpetrator's intentions would not absolve him or her of responsibility for them.

Ethical formalism was devised by the philosopher Immanuel Kant. According to Kant, formalist ethics must be universal. He advocated following ethical rules that you would want everyone to follow. For example, the ethic "always tell the truth" is a proper ethic for you only if you would like everyone to follow it. This would mean wanting people to tell you what they believe and not necessarily what you would like to hear.

To make this point, we ask how many people in the class would like an optional new feature for their car. The special instrument will always tell you what you want to hear. If you are approaching a traffic police officer, the speedometer will always show that you are not exceeding the speed limit. If you are demonstrating your car's impressive performance to a friend, it will show you are going faster than you are. The fuel gauge will always indicate plenty of fuel for your purposes. No warning lights will ever annoy you. No one really wants these special instruments, yet these same people often justify lying to their friends because they do not want to hurt their feelings.

Excusing conditions, or exceptions, are allowable in ethical formalism so long as they can be consistently universalized. For example, an ethical formalist may believe that it is acceptable to kill in self-defense. The formalist may also believe that it is acceptable to lie in any situation when you are subject to force or the threat of force. As long as you can say that you would want everyone to be able to act according to these exceptions, they fit the test of acceptable excusing conditions.

Ethical formalists can use decision analysis to make their ethically sensitive decisions. Once they eliminate the ethically unacceptable alternatives, they face choices with only prudential and legal attributes.

CONSEQUENCE-BASED ETHICS The other major ethical theory, **utilitarianism**, is consequence-based. The responsibility attaches to the consequences—not to the person's actions. According to this theory, deaths that resulted from a crime are considered murder even if the perpetrator had no intent to kill and, in fact, had taken precautions to avoid harm to anyone. The crime of felony murder is consistent with this view. Utilitarianism derives from philosophers such as Jeremy Bentham and John Stuart Mill who believed that the calculus of world pleasure and happiness should be the justification for action. A consequence-based ethic is often characterized by the notion that the end justifies the means. Lying could be justified if you believed it would bring about a good end. Another utilitarian notion is "achieving the greatest good for the greatest number."

Utilitarians can use decision analysis to make their ethically sensitive decisions by applying the rules as usual and using their ethical preferences for prospects.

Decision analysis clarifies ethical discussions because it provides a formal structure for representing ethical choices, including the issues that arise because of uncertainty. The distinction between making a good decision and achieving a good outcome that is central to decision analysis is equally useful in ethical discussions. For ethical formalists the structure applies after all ethically unacceptable alternatives have been removed; for utilitarians, the structure applies without change.

40.3.4 Classifying Ethics: Positive vs. Negative Injunctions

Another important classification of ethical rules is whether they are negative or positive. Negative rules are prohibitions, such as "I will not." Following negative rules requires no energy. Most of us follow the rule that we will not murder. We followed it yesterday and the day before, and we expect to follow it today without exerting ourselves.

It is insightful to look into some of the teachings of different religions and classify them as either positive or negative. For example, in the Ten Commandments, we have several negative injunctions, such as thou shalt not murder, commit adultery, bear false witness, steal, or covet.

In contrast, positive rules are obligations, such as "I will." If your ethical code says, "I will feed the poor," then you have given yourself a full-time and global job that will seriously drain your energy and resources. When you spend forty dollars for your own dinner, have you acted consistent with your positive ethic? The challenge with positive ethics is knowing where to draw the line.

Including positive rules in your ethical code requires some circumspection. A milder form of expressing your sentiment might be to say, "I have a positive regard for feeding the poor." You will then be able to decide what actions are appropriate for you given your limited energy and resources.

In some countries, taking positive action in some situations is required by law. For example, if you see a person drowning, and you can throw him a life preserver, you must. In this case, the action has legal as well as ethical implications.

40.4 HARMING, STEALING, AND TRUTH TELLING

What are the main ethically sensitive actions? Topping the list for most would be killing or physically harming innocent people. Next would come depriving others of their property by stealing. Both actions were prevalent in the Nazi era, and for that reason, we study in class why such evil arose in a technically and culturally advanced country. Similar evils are not uncommon today, and we must be sensitive to our potential responsibility.

An example of an ethical issue involving potential harm concerns whether you would consider making or selling a product you believe is harmful to others. You may have a negative ethic of avoiding harmful activities yet still want to avoid imposing it on others by force through the legal system. For example, you may personally avoid cigarettes and believe that they are harmful yet not wish to make the manufacture and sale of cigarettes illegal.

As another example, you may have an ethic against the use of violence except in defending yourself or others.

These ethics affect your life—from the kind of employment you will pursue, to the projects to which you will contribute or endorse. One way that an organization can accommodate the ethics of its employees is to have a policy that permits employees to refuse to work on any projects they consider ethically offensive. We have known editors who refuse to work on reports concerning subjects they found ethically objectionable.

This leads us to the question of how close to an ethically objectionable activity you must be to hold yourself ethically responsible for it. In the days of the extermination camps, we doubt that we could ever have been guards. But what about being the locomotive engineer transporting victims? Or the baker from the neighboring town who brought pastries to the guards every morning?

When you lie you make statements you know to be false with the intention to mislead. Lying also includes telling so-called "**white lies**" that are assumed by some to be socially admissible, and even desirable.

Consider white lies and the rhubarb pie, a recently engaged student meeting his future in-laws for the first time. After sharing a meal, his future mother-in-law served her special dessert: a rhubarb pie. She somehow has the idea that he loves rhubarb pie, when, in fact, he detests it. However, wanting to make a good first impression, he told a "white lie." He said that he loved rhubarb pie and wanted a piece. After managing to eat it, he put the matter behind him.

For years now every time he visits his in-laws, his mother-in-law makes rhubarb pie. Sometimes, when rhubarb is not in season, she goes to special efforts to find the ingredients to make her son-in-law's favorite pie. The first few times this happened, the student told himself that it was similar to his first dinner and that it was important to make a good impression, and he obligingly choked down at least one piece of pie. However, the longer the deception continued, the more embarrassing it was to admit. As far as we know, the rhubarb pies continue to this day.

The real cost of this deception was not the occasional necessity to consume rhubarb pie, but the barriers it created in the relationship. The deception reduced his desire to have dinner and a deeper relationship with his in-laws. It is easy to imagine that reticence in your relationship with your in-laws could also have negative affects on your marriage.

From a prudential point of view, lying can often be more costly than telling the truth.

40.4.1 Deception

Deception is giving a false impression without strictly telling a lie. It is easy to deceive without lying by simply failing to correct inaccurate impressions or more actively by creating a false impression.

In certain games or pursuits, such as acting, the participants expect false statements. A great actor is one who can convincingly play a role the audience knows to be false. No truth-telling ethical issues arise from these activities. We do not expect an actor to announce to the audience, "I am not really Hamlet, Prince of Denmark."

You may wish to consider truth telling ethics beyond simply not lying or not deceiving. The ethic that we have found to cut through most ethical dilemmas involving expression almost immediately is the positive ethic of "tell the whole truth." The problem with telling the whole truth is that it is hard work as we must search within ourselves to see what is the truth before we can say it.

40.4.2 Telling the Whole Truth

Let us demonstrate what telling the whole truth might look like in the case of the rhubarb pie. The whole truth might sound something like this.

> "*Thank you for thinking of me when you made the rhubarb pie. It makes me feel like part of your family to have you do something special like this for me. This is difficult for me to say, because I want our relationship to get off to a good start, but I do not like rhubarb pie. I do want you to know that I appreciate your thoughtfulness and I look forward to being a part of your family.*"

Note here that the student had to realize why he was initially tempted to deceive: Namely, he wanted to make a good impression on his soon to be in-laws, and he thought the new relationship could not tolerate the truth. By facing up to this fear, the student has raised the relationship with his in-laws to a higher level and can now look forward to building a strong relationship based on trust and honesty.

To summarize, the major ethical issue that confronts students and colleagues in life and in business is that of truth telling. And the answer in my experience is always simply to tell the truth.

40.4.3 Truth Telling in Business

Let us see how to apply the "tell the truth" ethic in several business situations.

Situation 1: An employee asks, "Is it okay if I charge a client for an airplane ticket even though I am planning to drive? It would be no cheaper for him if I flew, and driving is more convenient for me."

The test is simple. Would you call the client and ask whether this was okay? If you have any hesitation in asking, then don't do it. It is usually very easy for us to know whether the other person would feel deceived if he or she found out. If there is even a suspicion that a feeling of deception could arise, then the action under consideration should not be taken.

Situation 2: A consultant says, "We are submitting a proposal for Phase *A* of a project, and we know that it will take $300,000 worth of work to accomplish it to professional standards, but the client doesn't want to spend more than $200,000. We're sure that if we begin with the $200,000 price, he will soon see the need for additional work and agree to pay our original price. Is it okay to estimate Phase *A* at $200,000 in our proposal?"

Here again we have a simple test. If you were the client, would you want a consultant telling you that he can do your work for $200,000 when in his best professional opinion it will require $300,000? Would you want a mechanic to tell you that fixing your car will cost $200 when in his best professional judgment it will cost $300? The obvious answer is no. Once more, the right course is to tell the whole truth. Call the client, tell him that in your best professional judgment the work will require ultimately $300,000 for completion and that you were tempted to say it would cost only $200,000 so that he would begin this work and later agree to the wisdom of your estimate. But tell him that you view this as the wrong course of action and that he should think of this as a $300,000 job.

The consultant may or may not get the job; that is not the issue. What is important is if the job is received, it will be received on a basis on which the consultant can be ethically and professionally proud. Furthermore, if this job is not received, the potential client will be left with the impression of an ethical and highly professional firm that he can trust in future business dealings.

Situation 3: A business client is calling competitors from your consulting company's premises and leaving the impression that he is your employee. The competitors are more forthcoming with information than they would be if they knew he was an employee of one of their competitors. This type of practice is clearly a violation of the whole truth standard. It has no place in an ethical organization.

In all of these cases, honesty is the answer.

40.4.4 Euphemisms

A final aspect of truth telling is the use of euphemisms. We can think of **euphemisms** as deceptive speech intended to avoid ethical evaluation of the topic under discussion. The Nazis used euphemisms liberally. Execution was "special treatment" and the retarded or insane were "useless eaters." Even today, we use "collateral damage" to refer to innocent people killed by mistake. "Friendly fire" is not a log burning in a fireplace but a euphemism for our armed forces killing our own troops. Politicians describe a discovered lie as "no longer operational." Companies that once used to fire people are now just "downsizing" or "rightsizing." One company even referred to this act as "returning resources to the economy." In personal life, the popularity of the term "white lie" tells us that small deceptions are understandable—if not praiseworthy. In essence, euphemisms are *ethical caution signs*. To ignore them is to dull your ethical sensitivity.

40.5 ETHICAL CODES

An **ethical code** is an expression of your internal compass. It points you in the right direction when you are lost or confused. The ethical distinctions from earlier provide a framework for analyzing various ethical situations and creating your own ethical code. When teaching ethics in class, we analyze many different ethical situations using these distinctions. Through discussion and reflection, students find ethics that work for them. From these, they develop personal ethical codes they can rely on when their roads become ethically challenging.

40.5.1 Basis for Your Ethical Code

Where does a personal ethical code come from? What are its origins? Some would say that being ethical is simply following an inner sense of right and wrong. But there is remarkable agreement on the essential characteristics of an ethical code. For example, if we examine the great religious teachings of Judeo-Christianity, Buddhism, and Islam, we find that, although there is disagreement on things such as religious practices, forms of worship, or dietary restrictions, all religions agree on not hurting people, not stealing their property, and not lying. In particular, not hurting people and not stealing are the basis for most legal systems—whether they be the English common law or the California vigilantes.

It may be helpful to look to others for inspiration and guidance. Parents, heroes, colleagues, and friends all help form your choice of ethics. Ultimately, however, decision analysis is a philosophy that recognizes that you are unique in your information and preferences, including ethical standards.

40.5.2 Components of an Ethical Code

The most useful ethical codes cover the ethical dilemmas you face most frequently. For most people, this means truth telling. Other components you might want to consider include reproductive issues (abortion, surrogate motherhood, and guardianship), suicide, what organizations you associate with, treatment of animals, and any special ethics associated with your profession.

40.5.3 Tests of an Ethical Code

Although it is tempting to create a praiseworthy, high-toned ethical code, you should strive to design codes for use—not just for admiration. When judging a code, make sure you can answer "yes" to the following four criteria:

- *Reciprocity:* Does each rule apply to you, whether you are initiating or receiving the action?
- *Universality:* Do you want each rule to apply to everyone?
- *Consistency:* Is the system of rules logically consistent?
- *Actualization:* Do the rules provide guidance for behavior?

When you have analyzed an ethically sensitive decision situation using your ethical code, then before acting, you may want to apply some internal, personal checks to your ethical code.

- Would you feel comfortable discussing your actions with those whose ethical judgment you respect?
- Would you feel good about looking at yourself in the mirror in the morning?
- Would you want your children to base their beliefs about what is right and wrong on your actions?
- Would you be comfortable having your actions reported by highly regarded news organizations?

If you fail any of these tests, you have an indication that that your ethical code requires revision.

40.6 ETHICAL SITUATIONS

40.6.1 Ethics in Professional Life

The central issue in the ethics of professional life is whether we use a different ethical code as professionals from the one we use with friends and relatives. Do we wear different labels with different ethical standards as we move from one activity to another? Are we each to have a single ethical code that we apply in every situation? Having a single ethical code simplifies life and prevents the difficulties that arise when you treat people in different ways because of different labels.

BUSINESS The most common ethical issues in business, as in personal life, involve truth telling.

The victim of deception may be a customer—perhaps someone limited in education or coping abilities. The victim may be an employee—someone told a position is permanent when it is not. The victim may be a business partner—one who is discarded when his knowledge has been absorbed. The victim may be an investor—maybe one only partially informed about factors affecting the investment. The victim may be the client of a consultant who intentionally underestimates the cost of the first phase of a project in the belief that the client will agree to subsequent phases at much higher rates after committing the organization to the project.

Business leaders, too, can be deceived. Some employees lie about their credentials, knowledge, or experience. A *Wall Street Journal* article portrayed a high-level corporate executive who had lied about his military career, graduate degree, and martial arts achievements, among other things. When challenged, he said, "I, in some sense, am guilty of exaggerating and embellishing for a purpose from a business standpoint."

Students may represent themselves as seeking permanent employment even when they plan to leave at the end of the summer; they rationalize this behavior by saying that if they told the truth, they would not be hired.

The challenge in truth telling is finding the whole truth, not just telling the truth. For example, for the student seeking a summer job the truth might be, "I am seriously considering returning to school in the fall but have not made a final decision. If you hire me and I do decide to return to school, I will be an excellent employee this summer and you will want me to return when I have finished my education. Furthermore, I will guarantee to train my replacement." Whether the applicant gets the job or not, he or she will be a person of integrity. Prudentially, imagine being a person who got the job by misrepresentation and then had to spend the summer telling further lies to maintain the fiction.

ADVERTISING One area that seems to test the whole truth principle is advertising. We have all seen cars advertised in television commercials that show people cruising down empty California coastal highways in the sunshine. Those of us who drive know that people spend most of their time in cars on crowded streets, highways, and parking lots. The kind of driving experience shown in the commercial is hardly representative of the experience of car ownership. Is it deceptive? We find the answer by examining the background of the people who view the advertisement. As long as the public knows that advertising presents a product in the best possible light and that the commercial is not representative of the actual use of the product, there is no deception.

This is probably true with car advertising. The viewers of the commercial are aware of the distortions in the truth that the commercial creates. If the viewers are so aware, there is no deception. If they are not, there is a violation of the whole truth principle.

ACADEMIA Academia has ethical problems, too, and they can extend beyond cheating on examinations. We have all heard of cases in which academics plagiarize or falsify experimental results. Here is a story that a doctoral student told of his dealings with his former professor. One day the student told the professor about a new result he had obtained that he wanted to include in his doctoral thesis. The professor said the result was unimportant and talked the student out of including it in his thesis. The student complied, completed his work, and received his degree. The professor then published a book that contained the student's result. The student, by then a professor himself, decided to publish his original work in a professional journal, but his former professor was on the editorial board of the journal and rejected the paper on the grounds that much of the work had already been published in his own book. Note that this is the student's side of the story. Regardless of the facts, however, the fact that a student could believe his professor could behave this way is a concern in itself. Professors, like doctors, psychiatrists, and religious leaders, have a great deal of power over those to whom they minister. They can use this power for good or evil.

Another story is that of a student who completed his doctoral program and became an assistant professor in a business school. The student felt an urgent need to publish research papers in top tier journals to obtain tenure. He welcomed his former doctoral advisor—a member of the editorial board of many journals—as a co-author on all of his papers, believing that this would expedite the review process and maximize the chance of acceptance.

At an annual conference, he heard unpublished research results and then wrote a paper with his former advisor without telling the advisor that he had plagiarized the ideas. He submitted the paper to a top tier journal. The paper was accepted in several days without even a review process, which would not have occurred had his former advisor not been a co-author. The ethical problem arose when the same research results were about to appear in another journal by different authors. The result was major embarassment not only to the student, but to his advisor, and to the academic journal for having a double standard for reviewing.

LEGAL CONFLICTS There are situations in which the legal system appears to be in conflict with the whole truth principle. For example, if an employee is strongly suspected of stealing but there is no evidence that would suffice for legal conviction, it may be more prudent for the company to fire him by using the excuse that his job has been phased out rather than to tell him the real reason. This is the case when the strong penalties that the employer faces for apparent arbitrary discharge under the legal system create pressure to dissemble rather than to tell the truth.

A similar situation arises in the case of providing references about former employees. If the former employee in fact had serious performance problems, the employer has nothing to gain and much potentially to lose by disclosing these problems in a reference. Therefore, the legal system once more creates a pressure either to lie or to refuse to give a reference. One answer is to create a legal system where the law does not punish people who tell the truth by making them pay to defend themselves in court even when they are ultimately found innocent.

ETHICS OF BEING AN ANALYST One test of whether an act is ethical is whether you would perform the act if it involved someone close to you rather than a stranger. For example, if your brother or sister asked you whether the results of a study you prepared were really so, you should be able to say, "Of course," rather than, "No, because there was an artificial constraint." Life is too short to live under false pretenses. Even if this test does not determine your behavior, failing it should cause you to reconsider what you are doing.

Ethically sensitive situations can also arise in the course of technical work. Many years ago, someone representing a large aircraft company approached our consulting group. He wanted a decision analysis of a new fighter airplane. He made it known that this would be a large contract and that—wink, wink—nod, nod—we all knew how it was going to turn out. We declined to participate.

Several years later, our group worked on a government study on how to achieve self-sufficiency in energy by means of synthetic fuel production. Even after taking account of what might be learned from early experimental work, it was clear that we could justify only a program of very modest size. Yet in the final private meeting, the government project director tried to persuade us to report that a large program would be best, because he was politically committed to a large program. Happily, here again we were able to resist this pressure and present the report that we supported.

40.7 SUMMARY

In reviewing these many areas of ethical concern, one conclusion is clear: Ethical difficulties are much easier to avoid than to resolve. You can avoid many ethical difficulties by following three practices: First, decline to be a part of organizations that have ethical codes and behaviors inconsistent with your own; second, avoid participating in ethically objectionable activities; and finally, treat all people as you would treat those closest to you.

In the situations for which you can't avoid the ethical dilemma, it is prudent to have a clearly articulated ethical code. Without an ethical code, you end up cutting off pieces of yourself to live with ethical compromises. The study of ethics therefore can be energizing. It allows you to live a more satisfying, complete life. It allows you to embrace your full self, knowing that your actions are consistent with the principles you know are right.

KEY TERMS

- Action-based ethics
- Consequence-based ethics
- Deception
- Ethical actions
- Ethical code
- Ethical formalism
- Ethics
- Euphemisms
- Legal actions
- Prudential actions
- Utilitarianism
- White lies

PROBLEMS

Problems marked with an asterisk (*) are considered more challenging.

1. Where would you classify yourself in terms of action-based versus consequence-based views?
2. Mention some areas where you have positive ethical obligations and others where you have negative ethical injunctions.

*3. Think of an ethical dilemma you are facing and classify it according to ethical–legal–prudential.

*4. Do you think it is acceptable to "oversell" yourself on your resume?

*5. You are seeking financial backing for your business, and you meet with a venture capitalist. Is it acceptable to make overly optimistic projections of profitability in order to increase his desire to invest?

*6. You are new to a company. Your boss has asked you to spend time doing some personal things for himself rather than for the company. How would you react?

*7. Do you expect your friends to lie for you?

INDEX

A

E

F

G

Q

R

S

T

U

W

Z

THE DECISION ANALYSIS CORE CONCEPTS MAP

The following map is a useful tool to help you understand the concepts we have described throughout the book. It is not a relevance diagram. It is a map that shows the main concepts. An arrow from one concept to another helps you identify what you need to know before understanding this concept. For example, the arrows from "Probability" to "*e*-value" and from "Measure" to "*e*-value" mean that it is important to first learn the concepts of probability and measures before learning the concept of *e*-value.

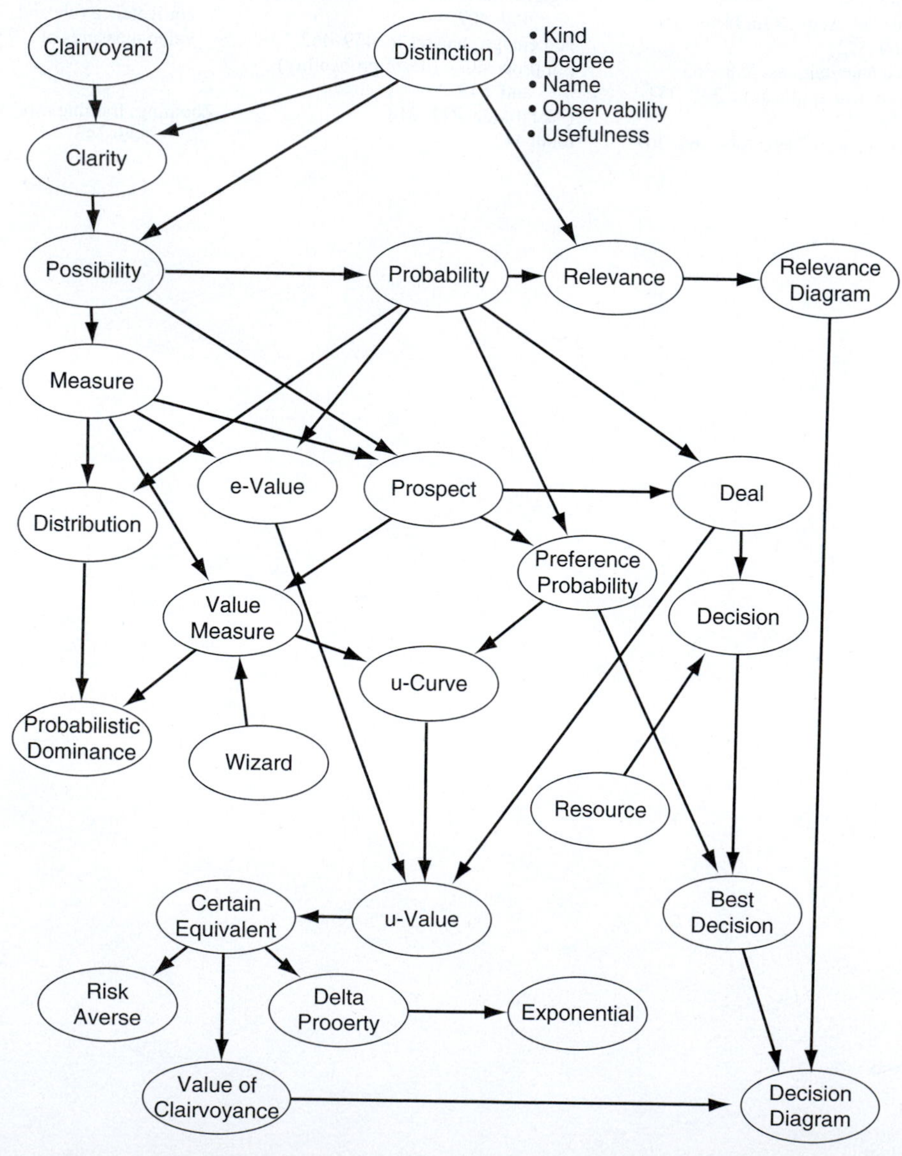